中 外 物 理 学 精 品 书 系

本 书 出 版 得 到 " 国 家 出 版 基 金 " 资 助

U0230629

国家出版基金项目
NATIONAL PUBLICATION FOUNDATION

中外物理学精品书系

引进系列 · 5

Modern Many-Particle Physics:
Atomic Gases, Nanostructures and Quantum Liquids
2nd Edition

现代多粒子物理
——原子气体、纳米结构和量子液体

第二版

（影印版）

〔意〕利帕里尼（E. Lipparini） 著

北京大学出版社
PEKING UNIVERSITY PRESS

著作权合同登记号　图字：01-2012-2820

图书在版编目(CIP)数据

现代多粒子物理：原子气体、纳米结构和量子液体：第 2 版：英文/（意）利帕里尼
(Lipparini，E.)著. —影印本. —北京：北京大学出版社，2012.12
(中外物理学精品书系·引进系列)
书名原文：Modern Many-Particle Physics：Atomic Gases，Nanostructures and
Quantum Liquids　2nd Edition
ISBN 978-7-301-21550-0

Ⅰ．①现… Ⅱ．①利… Ⅲ．①粒子物理学-英文 Ⅳ．①O572.2

中国版本图书馆 CIP 数据核字(2012)第 274115 号

书　　　名：Modern Many-Particle Physics：Atomic Gases，Nanostructures and Quantum Liquids　2nd
　　　　　　Edition(现代多粒子物理——原子气体、纳米结构和量子液体　第二版)(影印版)
著作责任者：〔意〕利帕里尼(E. Lipparini)　著
责 任 编 辑：刘　啸
标 准 书 号：ISBN 978-7-301-21550-0/O·0892
出 版 发 行：北京大学出版社
地　　　址：北京市海淀区成府路 205 号　100871
网　　　址：http://www.pup.cn
新 浪 微 博：@北京大学出版社
电 子 信 箱：zpup@pup.cn
电　　　话：邮购部 62752015　发行部 62750672　编辑部 62752038　出版部 62754962
印 　刷 　者：北京中科印刷有限公司
经 　销 　者：新华书店
　　　　　　730 毫米×980 毫米　16 开本　37.5 印张　693 千字
　　　　　　2012 年 12 月第 1 版　2012 年 12 月第 1 次印刷
定　　　价：101.00 元

序　言

　　物理学是研究物质、能量以及它们之间相互作用的科学。她不仅是化学、生命、材料、信息、能源和环境等相关学科的基础,同时还是许多新兴学科和交叉学科的前沿。在科技发展日新月异和国际竞争日趋激烈的今天,物理学不仅囿于基础科学和技术应用研究的范畴,而且在社会发展与人类进步的历史进程中发挥着越来越关键的作用。

　　我们欣喜地看到,改革开放三十多年来,随着中国政治、经济、教育、文化等领域各项事业的持续稳定发展,我国物理学取得了跨越式的进步,做出了很多为世界瞩目的研究成果。今日的中国物理正在经历一个历史上少有的黄金时代。

　　在我国物理学科快速发展的背景下,近年来物理学相关书籍也呈现百花齐放的良好态势,在知识传承、学术交流、人才培养等方面发挥着无可替代的作用。从另一方面看,尽管国内各出版社相继推出了一些质量很高的物理教材和图书,但系统总结物理学各门类知识和发展,深入浅出地介绍其与现代科学技术之间的渊源,并针对不同层次的读者提供有价值的教材和研究参考,仍是我国科学传播与出版界面临的一个极富挑战性的课题。

　　为有力推动我国物理学研究、加快相关学科的建设与发展,特别是展现近年来中国物理学者的研究水平和成果,北京大学出版社在国家出版基金的支持下推出了《中外物理学精品书系》,试图对以上难题进行大胆的尝试和探索。该书系编委会集结了数十位来自内地和香港顶尖高校及科研院所的知名专家学者。他们都是目前该领域十分活跃的专家,确保了整套丛书的权威性和前瞻性。

　　这套书系内容丰富,涵盖面广,可读性强,其中既有对我国传统物理学发展的梳理和总结,也有对正在蓬勃发展的物理学前沿的全面展示;既引进和介绍了世界物理学研究的发展动态,也面向国际主流领域传播中国物理的优秀专著。可以说,《中外物理学精品书系》力图完整呈现近现代世界和中国物理科学发展的全貌,是一部目前国内为数不多的兼具学术价值和阅读乐趣的经典物理丛书。

《中外物理学精品书系》另一个突出特点是,在把西方物理的精华要义"请进来"的同时,也将我国近现代物理的优秀成果"送出去"。物理学科在世界范围内的重要性不言而喻,引进和翻译世界物理的经典著作和前沿动态,可以满足当前国内物理教学和科研工作的迫切需求。另一方面,改革开放几十年来,我国的物理学研究取得了长足发展,一大批具有较高学术价值的著作相继问世。这套丛书首次将一些中国物理学者的优秀论著以英文版的形式直接推向国际相关研究的主流领域,使世界对中国物理学的过去和现状有更多的深入了解,不仅充分展示出中国物理学研究和积累的"硬实力",也向世界主动传播我国科技文化领域不断创新的"软实力",对全面提升中国科学、教育和文化领域的国际形象起到重要的促进作用。

值得一提的是,《中外物理学精品书系》还对中国近现代物理学科的经典著作进行了全面收录。20 世纪以来,中国物理界诞生了很多经典作品,但当时大都分散出版,如今很多代表性的作品已经淹没在浩瀚的图书海洋中,读者们对这些论著也都是"只闻其声,未见其真"。该书系的编者们在这方面下了很大工夫,对中国物理学科不同时期、不同分支的经典著作进行了系统的整理和收录。这项工作具有非常重要的学术意义和社会价值,不仅可以很好地保护和传承我国物理学的经典文献,充分发挥其应有的传世育人的作用,更能使广大物理学人和青年学子切身体会我国物理学研究的发展脉络和优良传统,真正领悟到老一辈科学家严谨求实、追求卓越、博大精深的治学之美。

温家宝总理在 2006 年中国科学技术大会上指出,"加强基础研究是提升国家创新能力、积累智力资本的重要途径,是我国跻身世界科技强国的必要条件"。中国的发展在于创新,而基础研究正是一切创新的根本和源泉。我相信,这套《中外物理学精品书系》的出版,不仅可以使所有热爱和研究物理学的人们从中获取思维的启迪、智力的挑战和阅读的乐趣,也将进一步推动其他相关基础科学更好更快地发展,为我国今后的科技创新和社会进步做出应有的贡献。

《中外物理学精品书系》编委会 主任
中国科学院院士,北京大学教授
王恩哥
2010 年 5 月于燕园

Atomic Gases,
Nanostructures and
Quantum Liquids

Modern
Many-Particle
Physics

2nd Edition

Enrico Lipparini

University of Trento, Italy

 **World Scientific**

NEW JERSEY · LONDON · SINGAPORE · BEIJING · SHANGHAI · HONG KONG · TAIPEI · CHENNAI

Preface

This book is the fruit of my lectures on "The Theory of Many-Body Systems," which I have been teaching for many years in the degree course on Physics at the University of Trento. As often happens, the outline of the book came from my students' notes; in particular, the notes of the students of the academic year 1999–2000, which were extremely useful to me. Chapter 6, on the Monte Carlo methods, is the work of Francesco Pederiva, a research assistant in our department. During the course Francesco, apart from illustrating the method, teaches the students all the computer programs (continually referred to in this book), by means of practical exercises in our computational laboratory. In particular, he teaches the Hartree–Fock, Brueckner–Hartree–Fock, Kohn–Sham and diffusion Monte Carlo programs for the static properties, the RPA, and time-dependent HF and the LSDA for Boson and Fermion finite systems. These programs are available to anyone who is interested in using them.

The book is directed toward students who have taken a conventional course on quantum mechanics and have some basic understanding of condensed matter phenomena. I have often gone into extensive mathematical details, trying to be as clear as possible, and I hope that the reader will be able to rederive many of the formulas presented without too much difficulty.

In the book, even though a lot of space is devoted to the description of the homogeneous systems, such as electron gas in different dimensions, quantum wells in an intense magnetic field, liquid helium and nuclear matter, the most relevant part is dedicated to the study of finite systems. Particular attention is paid to those systems realized recently in laboratories throughout the world: metal clusters, quantum dots and the condensates of cold and dilute atoms in magnetic traps. However, some space is also allotted to the more traditional finite systems, like the helium drops and the nuclei. I have tried to treat all these systems in the most unifying way possible, hoping to bring all the analogies to light. My intention was to narrow the gap between the usual undergraduate lecture course and the literature on these systems presented in scientific journals.

It is important to note that this book takes a "quantum chemist's" approach to many-body theories. It focuses on methods of getting good numerical approximations to energies and linear response based on approximations to first-principle Hamiltonians. There is another approach to many-body physics that focuses on symmetries and symmetry breaking, quantum field theory and renormalization groups, and aims to extract the emergent features of the many-body systems. This works with "effective" model theories, and does not attempt to do "*ab initio* computations." These two ways of dealing with many-body systems complement each other, and find common ground in the study of atomic gases, metal clusters, quantum dots and quantum Hall effect systems, which are the main application of the book.

I am indebted to many of my colleagues in the Physics Department of Trento for discussions and remarks. Specifically, I'm grateful to G. Bachelet, D.M. Brink, S. Giorgini, F. Iachello, W. Leidemann, R. Leonardi, F. Pederiva, G. Orlandini, S. Stringari, M. Traini, G. Viliani and A. Vitturi. Many aspects of the book were clarified during my stays in Barcelona, Paris and Palma de Mallorca, where I had the occasion to discuss many subjects with M. Barranco, A. Emperador, M. Pi, X. Campi, N. Van Giai, D. Vautherin, Ll. Serra and A. Puente, as well as during the frequent visits to our department by my friends A. Richter and K. Takayanagi.

Thanks are also due to Irene Diamond, for the English translation of the book.

This book has cost me a great investment in time, which recently has kept me from other research projects and, above all, from my family. It is dedicated to my wife, Giovanna, and to my children, Fiorenza, Filippo and Luigi. Filippo has been of enormous help in editing the figures.

Enrico Lipparini
January 2003

Preface to the Second Edition

In this edition the main changes are a new chapter on the spin–orbit coupling in semiconductor heterostructures and a considerable expansion of the chapters dealing with trapped atomic gases, density functional calculations, current response to an electromagnetic field, and the Brueckner–Hartree–Fock and Monte Carlo approaches.

The spin–orbit (SO) interaction in nanostructures has prompted intense activity in recent years since it is an essential mechanism for most spintronic devices. In fact, it links the spin and charge dynamics, opening up the possibility of spin control through an electric field. Indeed, recent experimental and theoretical investigations have shown that the SO coupling affects charge transport, far-infrared absorption, and electronic spin precession in a magnetic field, besides giving rise to the spin-Hall effect. All these topics are analyzed in the new Chapter 6 of this edition.

After the first experimental realization of Bose–Einstein condensation in dilute atomic gases, the field of ultracold gases has become a rapidly growing one. In the last few years a considerable amount of experimental and theoretical work has focused on ultracold Fermi gases. The description of the ground state and excited state properties of these systems has been added in many new sections of the book.

The illustration of density functional calculations in quantum wires and molecules has been subjected to much more detailed examination than before. Particular attention has been given to the description of noncollinear local spin density approximation calculations in nanostructures in the presence of SO interaction.

The sections illustrating current response to an electric field have been expanded to give a detailed description of the conductivity problem, with particular emphasis on Landauer conductance, magnetoconductivity and spin-Hall conductivity. A section on the problem of Hall conductivity in graphene has been added.

The Monte Carlo chapter has been revised and expanded to include numerical applications to trapped Fermi gases and many-nucleon systems. A similar revision and expansion has been carried out for the chapter dealing with the Brueckner–Hartree–Fock theory.

Apart from the above main additions and expansions, the remainder of the book has undergone slight revisions and corrections.

Enrico Lipparini
June 2007

Contents

9.9   The Adiabatic Time-Dependent LSDA (TDLSDA) 438
       9.9.1   The TDLSDA Longitudinal Response Function 440
       9.9.2   The TDLSDA Transverse Response Function   445
9.10   RPA and TDLSDA Commutators and Symmetry
       Restoration 448
9.11   The Linear Response Based on the Green Functions; RPAE 451
9.12   The Screened Response Function and the Dielectric Constant 454
9.13   Examples of Application of the TDLSDA Theory 456
       9.13.1   Quantum Wells Under a Very High External Magnetic Field 456
       9.13.2   Quantum Dots Under a Magnetic Field 468

Chapter 10   Dynamic Correlations and the Response Function   491
10.1   Introduction 491
10.2   Interaction Energy and Correlation Energy 491
10.3   The RPA Correlation Energy 494
       10.3.1   The RPA Correlation Energy for the Cold and Dilute Gas
                of Bosons and Fermions 495
10.4   Theories Beyond the *RPA* 497
10.5   The STLS Theory 500
10.6   Comparison of Different Theories for the Electron Gas in 2D 502
10.7   Quasiparticle Properties 504
10.8   Nonlocal Effects 506
10.9   Mean Energy of Many-Particle Excitations 515
10.10 The Polarization Potential Model 516
10.11 The Gross–Kohn Model 519
10.12 The Method of Lorentz Transforms 523

Chapter 11   The Hydrodynamic and Elastic Models   527
11.1   The Hydrodynamic Model for Bosons 527
       11.1.1   Backflow 530
       11.1.2   Compression and Surface Modes of Spherical Drops 530
       11.1.3   Compression and Surface Modes of a Bose Gas in a
                Magnetic Trap 534
       11.1.4   Compression and Surface Modes of a Superfluid Trapped
                Fermi Gas 535
       11.1.5   The Moment of Inertia and the Scissor Mode of a Bose Gas
                in a Magnetic Trap 536
       11.1.6   Vortices in the Bose Gas in a Magnetic Trap 541
11.2   The Fluidodynamic and Hydrodynamic Model for Fermions 544
       11.2.1   Dipolar Modes in Metal Clusters 552
       11.2.2   Spin Oscillations in Trapped Fermi Gases 554
       11.2.3   The Scalar Quadrupole Mode in Confined Systems 554
       11.2.4   The Scissor Mode in Fermi Systems 556

Chapter 1

The Independent-Particle Model

1.1 Introduction

The main purpose of many-body theory in nonrelativistic quantum mechanics is the study of the properties of the solutions to Schrödinger's equation with the Hamiltonian

$$H = \sum_{i=1}^{N} \left(\frac{p_i^2}{2m} + v_{\text{ext}}(\mathbf{r}_i) \right) + \sum_{i<j=1}^{N} v(\mathbf{r}_{ij}), \tag{1.1}$$

which describes a system composed of N identical particles, which interact with an external field through a one-body potential $v_{\text{ext}}(\mathbf{r}_i)$, and among themselves through a two-body potential $v(\mathbf{r}_{ij})$.

The simplest model case is when the Hamiltonian contains only one-body terms, and is referred to as the independent-particle model:

$$H_0 = \sum_{i=1}^{N} \left(\frac{p_i^2}{2m} + v_{\text{ext}}(\mathbf{r}_i) \right). \tag{1.2}$$

In this approximation the eigenfunctions of H_0 may be written as the product of single-particle wave functions, each of which satisfies the equation ($\hbar = 1$)

$$\left(-\frac{\nabla^2}{2m} + v_{\text{ext}}(\mathbf{r}) \right) \varphi_k(\mathbf{r}, \sigma) = \varepsilon_k \varphi_k(\mathbf{r}, \sigma), \tag{1.3}$$

where k indicates the set of quantum numbers that characterize the single-particle state, and $\mathbf{r}$ and σ are the position and spin variables, respectively. A further variable is introduced in nuclear physics — the isospin τ. In what follows we will indicate as x the variable set $\mathbf{r}, \sigma, \tau$. For example, for electrons in a central external field,

$$\varphi_k(x) = \varphi_{n,\ell,m}(\mathbf{r}) \chi_{m_s}(\sigma). \tag{1.4}$$

Here $\chi_{m_s}(\sigma)$ is the spinor which satisfies the equations

$$s^2\chi_{m_s}(\sigma) = \frac{3}{4}\chi_{m_s}(\sigma), \quad s_z\chi_{m_s}(\sigma) = m_s\chi_{m_s}(\sigma), \tag{1.5}$$

with $s^2 = s_x^2 + s_y^2 + s_z^2$, $s_{x,y,z} = \frac{1}{2}\sigma_{x,y,z}$, and $\sigma_{x,y,z}$ are the Pauli matrices $\sigma_x = \begin{pmatrix} 0 & 1 \\ 1 & 0 \end{pmatrix}$, $\sigma_y = \begin{pmatrix} 0 & -i \\ i & 0 \end{pmatrix}$ and $\sigma_z = \begin{pmatrix} 1 & 0 \\ 0 & -1 \end{pmatrix}$; the spinor has the components

$$\chi_{\frac{1}{2}}(1) = \chi_{-\frac{1}{2}}(-1) = 1, \qquad \chi_{\frac{1}{2}}(-1) = \chi_{-\frac{1}{2}}(1) = 0. \tag{1.6}$$

The wave functions φ_k constitute a complete set which fulfills the conditions

$$\int \varphi_k^*(x)\varphi_\ell(x)dx \equiv \sum_\sigma \int \varphi_k^*(\mathbf{r},\sigma)\varphi_\ell(\mathbf{r},\sigma)d\mathbf{r} = \delta_{k,\ell}, \tag{1.7}$$

where we have assumed that the single-particle energies ϵ_k are discrete (in the case of continuum values, Kronecker's symbol is to be replaced by Dirac's delta function).

An eigenfunction of H_0 with eigenvalue E is therefore written as

$$\psi(x_1,\ldots,x_N) = \prod_{i=1}^{N}\varphi_{k_i}(x_i), \tag{1.8}$$

with

$$E = \sum_{i=1}^{N}\varepsilon_{k_i}. \tag{1.9}$$

However, the solution expressed as a product of single-particle wave functions should satisfy the symmetry or antisymmetry requisite under particle exchange, depending on whether we have a system of Bosons or Fermions, respectively.

1.2   Bosons

Symmetric wave functions under the exchange of the x coordinates of any two particles are obtained by considering all possible permutations P of N objects, and taking the following linear combination:

$$\psi_B(x_1,\ldots,x_N) = \sqrt{\frac{n_1!n_2!\cdots}{N!}}\sum_P P\prod_{i=1}^{N}\varphi_{k_i}(x_i). \tag{1.10}$$

The coefficients n_i which appear in the normalization constant of Eq. (1.10) represent the number of times the wave function φ_{k_i} appears in the product $\Pi_{i=1}^{N}\varphi_{k_i}$ and thus the average occupation number of state φ_{k_i}. Therefore, such a number may be larger than 1, thanks to the Boson nature of the particles.

The ground state of a system of N Bosons is obtained by occupying N times the lowest-energy state φ_{k_1}:

$$\psi_B(x_1, \ldots, x_N) = \varphi_{k_1}(x_1) \cdots \varphi_{k_1}(x_N). \tag{1.11}$$

This situation, which is characterized by a macroscopic occupation of only one single-particle state, is known as Bose–Einstein condensation and has been realized experimentally by cooling — by means of laser techniques and evaporation of hotter atoms — a diluted alkali atom gas in a magnetic trap (Anderson *et al.*, 1995; Davis *et al.*, 1995; Bradley *et al.*, 1995). However, a realistic description of this phenomenon demands the solution of the many-body problem with the two-body interaction included in the Hamiltonian (1.1). This description will be presented in the forthcoming chapters.

1.3   Fermions

In the case of a system of N Fermions, starting from the product functions of Eq. (1.8), we obtain a fully antisymmetric wave function under the exchange of the coordinates of two particles with the following combination:

$$\psi_F(x_1, \ldots, x_N) = \frac{1}{\sqrt{N!}} \sum_P (-1)^P P \prod_{i=1}^N \varphi_{k_i}(x_i). \tag{1.12}$$

In Eq. (1.12), the factor $(-1)^P$ is equal to 1 if the permutation is an even one. In the case of Fermions, the occupation number of a single-particle state φ_{k_i} is either 0 or 1, otherwise the function (1.12) vanishes identically (Pauli exclusion principle). This can be seen very easily by putting the wave function (1.12) in determinant form (Slater determinant):

$$\psi_F(x_1, \ldots, x_N) = \frac{1}{\sqrt{N!}} \det \begin{vmatrix} \varphi_{k_1}(x_1) & \varphi_{k_1}(x_2) & \cdots & \varphi_{k_1}(x_N) \\ \varphi_{k_2}(x_1) & \varphi_{k_2}(x_2) & \cdots & \varphi_{k_2}(x_N) \\ \vdots & \vdots & \ddots & \vdots \\ \varphi_{k_N}(x_1) & \varphi_{k_N}(x_2) & \cdots & \varphi_{k_N}(x_N) \end{vmatrix}. \tag{1.13}$$

The determinant is zero if two rows are equal (i.e. if two sets of quantum numbers are equal) and changes its sign under the exchange of two columns (i.e. the coordinates of two particles).

In the independent-particle model, the ground state for a system of N Fermions is obtained by occupying the N lowest-energy single-particle states characterized by the sets of quantum numbers $k_1, k_2, \ldots, k_N$. The highest-energy occupied state defines the Fermi level, and its energy defines the Fermi energy.

For example, if we consider a single-particle Hamiltonian for a three-dimensional harmonic oscillator,

$$H_0 = \sum_{i=1}^{N} \left(\frac{p_i^2}{2m} + \frac{m\omega_0^2 r_i^2}{2} \right), \tag{1.14}$$

the single-particle spectrum is given by $(\hbar = 1)$

$$\epsilon_{nl} = \left(N + \frac{3}{2} \right) \omega_0, \quad \text{with } N = 2(n-1) + l,$$

where $N = 0, 1, 2, \ldots$, $l = N, N-2, \ldots, 0$ or 1. Up to $l = 3$, in order of increasing energy, the orbitals and their degeneracy $[2(2l+1)]$ which for fixed values of n and l determines the number of particles that can be accommodated in the orbital without violating the Pauli principle (we have $2l+1$ different values of m, and for each m value two possible values of m_s), are reported in Table 1.1.

Thus, if we have eight particles, then the Fermi level coincides with the $1p$ orbital, and the Fermi energy is $5/2\omega_0$. The ground state of the independent-particle model is the Slater determinant of the 8×8 matrix built up by means of the single-particle wave functions characterized by the sets of quantum numbers (n, l, m, m_s):

$$(1,0,0,1/2),\ (1,0,0,-1/2),\ (1,1,1,1/2),\ (1,1,1,-1/2),$$
$$(1,1,0,1/2),\ (1,1,0,-1/2),\ (1,1,-1,1/2),\ (1,1,-1,-1/2).$$

Finally, the ground state energy is $E_0^F = [2(3/2) + 6(5/2)]\omega_0 = 18\omega_0$.

The same system, but now made up of Bosons, has the ground state wave function $(1/a = \sqrt{m\omega_0})$

$$\psi(r_1, \ldots, r_8) = \varphi_{1s}(r_1)\varphi_{1s}(r_2) \cdots \varphi_{1s}(r_8)$$

$$= (\pi^{-3/4}a^{3/2})^8 \exp\left(-1/2a^2 \sum_{i=1}^{8} r_i^2 \right), \tag{1.15}$$

to which corresponds the energy $E_0^B = 8(3/2)\omega_0 = 12\omega_0$.

Table 1.1. Harmonic oscillator orbitals and their degeneracy up to $N = 3$ in order of increasing energy.

N	E	n	l	Orbital	Degeneracy
3	$(9/2)\omega_0$	1 / 2	3 / 1	$1f$ / $2p$	$14 + 6$
2	$(7/2)\omega_0$	1 / 2	2 / 0	$1d$ / $2s$	$10 + 2$
1	$(5/2)\omega_0$	1	1	$1p$	6
0	$(3/2)\omega_0$	1	0	$1s$	2

The excited states of the Fermion system are obtained from the ground state in the following way: one-particle–one-hole ($1p$–$1h$) states are obtained by occupying $N-1$ single-particle states below the Fermi level and 1 state above the Fermi level. In the following we will indicate such states as $|i^{-1}m\rangle$, where φ_i is the unoccupied state below the Fermi level (hole), and φ_m is the occupied state above the Fermi level (particle). The wave function of the state $|i^{-1}m\rangle$ is the Slater determinant obtained by substituting the row characterized by the set of quantum numbers i with the row characterized by the set of quantum numbers m. Its energy is given by $E_0 + \epsilon_m - \epsilon_i$, where E_0 is the ground state energy and ϵ_m, ϵ_i are the energies of the single-particle states φ_m and φ_i.

Two-particle–two-hole states ($2p$–$2h$) are obtained by occupying $N-2$ single-particle states below the Fermi level and two single-particle states above the Fermi level, and are indicated by $|i^{-1}j^{-1}mn\rangle$, which means the Slater determinant obtained from the ground state by substituting the two rows i and j by the rows m and n. The energy of state $|i^{-1}j^{-1}mn\rangle$ is given by $E_0 + \epsilon_m + \epsilon_n - \epsilon_i - \epsilon_j$. In the same way, it is possible to build up three-particle–three-hole states, and so on. In the following, we will always indicate hole states by the letters i, j, k, l, and particle states by the letters m, n, p, q. Greek letters α, β, γ indicate both particle and hole states.

In the case of the Boson system in order to build $1p$–$1h$ states, or $2p$–$2h$ ones, we proceed in the same way, but starting from the ground state of the Boson condensate of (1.11), where all particles occupy the same lowest-energy state. For example, to set up a $1p$–$1h$ state it is enough to replace a single-particle wave function φ_{k_1} by a wave function corresponding to a higher-energy state, and then to symmetrize the product wave function so obtained.

Note that, since the above N-particle states are a complete set of eigenstates of H_0, a sum of Slater determinants (symmetric products) forms the general element of the many-Fermion (Boson) Hilbert space when the effects of the interaction are included in the theory. The general Hamiltonian (1.1) normally conserves the number of particles N, the total angular momentum L, the total spin S, and their components along the z direction L_z and S_z, respectively. The N-particle correlated states with defined (L, S, L_z, S_z) can be constructed starting from the N-independent particle model states described above, and then the diagonalization problem for H can be solved, yielding the energies and the corresponding eigenstates. As an example of this procedure [configuration interaction (CI) technique], let us consider a laterally confined (in the x, y plane) two-dimensional electron gas in a strong magnetic field which fully spin-polarizes the electrons (see Chapter 5). In this case $S = S_z = N/2$ and the relevant angular momentum is L_z. The many-particle wave functions Ψ_{L_z} eigenfunctions of the relevant Hamiltonian [see Eq. (5.1)] are linear combinations of Slater determinants of total spin $S = N/2$ and a given algular momentum $L_z = \sum_i \ell_i$:

$$\Psi_{L_z} = \sum_\alpha c_\alpha \Phi_\alpha^{L_z}, \tag{1.16}$$

where ℓ_i are the angular momentum quantum numbers of the single-particle wave functions used to construct each Slater determinant. To have a good basis, it is necessary that $\langle \Phi_\alpha | \Phi_\beta \rangle = \delta_{\alpha\beta}$, which is fulfilled if the single-particle wave functions are orthonormal. This is the case of the Fock–Darwin states of Eq. (5.7). The diagonalization problem,

$$\sum_\alpha (\langle \Phi_\beta | H | \Phi_\alpha \rangle - E\delta_{\beta\alpha})c_\alpha = 0, \tag{1.17}$$

is then solved and one obtains the energies E and the corresponding eigenvectors $\{c_\alpha\}$. By increasing the size of the basis, the eigenvalues E approach asymptotically the energies of the eigenstates of the Hamiltonian. The most important parameter that determines the quality and the feasibility of the calculation is the number of configurations necessary to obtain a stable result.

Calculations of this type have been done by many authors. For example, see Maksym and Chakraborty (1990), MacDonald and Johnson (1993), Wójs and Hawrylak (1997), Reimann, Koskinen and Manninen (2000), Tavernier *et al.*, (2003), Yannouleas and Landman (2003), Emperador *et al.*, (2005).

1.4 Matrix Elements of One-Body Operators

By the term "one-body operator" we mean an operator of the kind

$$F_1 = \sum_{i=1}^N f(x_i), \tag{1.18}$$

where f is a function of the variables x. In what follows we will evaluate the matrix elements of such operators between N-particle states in the independent-particle model. In general, we have

$$\langle \phi | F_1 | \psi \rangle = \int \phi^*(x_1, \ldots, x_N) \sum_{i=1}^N f(x_i)\psi(x_1, \ldots, x_N)dx_1 \cdots dx_N$$

$$= N \int \phi^*(x_1, \ldots, x_N)f(x_1)\psi(x_1, \ldots, x_N)dx_1 \cdots dx_N, \tag{1.19}$$

where ϕ and ψ are two N-particle states and in the last passage we have exploited the symmetry or antisymmetry property of the two wave functions. By writing down explicitly, for both Fermions and Bosons, the wave functions ϕ and ψ as products of single-particle wave functions [see Eqs. (1.10)–(1.13)], it is easily seen that the matrix element (1.19) vanishes if ϕ and ψ differ by more than one pair of single-particle states. The only two possibilities are then: (1) $\phi = \psi$, i.e. the same single-particle states are occupied in the two wave functions; (2) the same $N - 1$ single-particle states are occupied in ϕ and ψ, while the Nth state of ϕ is φ_i and that of ψ is φ_m $(m \neq i)$. By taking ϕ to be the ground state $(|\phi\rangle = |0\rangle)$ and specializing to Fermions, in the first case we obtain

$$\langle 0|F_1|0\rangle = \sum_{s=1}^{N} \int \varphi_{i_s}^*(x) f(x) \varphi_{i_s}(x) dx, \tag{1.20}$$

and in the second case

$$\langle 0|F_1|i^{-1}m\rangle = \int \varphi_i^*(x) f(x) \varphi_m(x) dx. \tag{1.21}$$

Equations (1.20) and (1.21) are easily obtained by employing in (1.19) the expansion of the determinant (1.13) according to the elements of the first column.

In the Boson case, where the ground state is a single-particle state φ_{i_1} which is N-fold-occupied, we obtain

$$\langle 0|F_1|0\rangle = N \int \varphi_{i_1}^*(x) f(x) \varphi_{i_1}(x) dx, \tag{1.22}$$

and

$$\langle 0|F_1|i_1^{-1}m\rangle = \sqrt{N} \int \varphi_{i_1}^*(x) f(x) \varphi_m(x) dx, \tag{1.23}$$

where we have made use of

$$|i_1^{-1}m\rangle = \frac{1}{\sqrt{N}} \sum_P P \varphi_m(x_1) \varphi_{i_1}(x_2) \varphi_{i_1}(x_3) \cdots \varphi_{i_1}(x_N). \tag{1.24}$$

As an example of the mean value of a one-body operator, let us compute the mean square radius of a system of N Fermions in a three-dimensional harmonic potential. The mean square radius is defined as $\langle r^2 \rangle = \langle 0|\frac{1}{N}\sum_{i=1}^{N} r_i^2|0\rangle$, and by (1.20) we have

$$\langle r^2 \rangle = \frac{1}{N} \sum_{n,l,m,m_s} \langle n,l|r^2|n,l\rangle, \tag{1.25}$$

where $\langle n,l|r^2|n,l\rangle = \int R_{n,l}^2 r^2 dr$. To evaluate the matrix element $\langle n,l|r^2|n,l\rangle$, it is convenient to use the virial theorem:

$$\langle n,l|\frac{p^2}{2m}|n,l\rangle = \langle n,l|\frac{m\omega_0^2 r^2}{2}|n,l\rangle = \frac{1}{2}\epsilon_{n,l} = \frac{1}{2}\left[2(n-1)+l+\frac{3}{2}\right]\omega_0. \tag{1.26}$$

Using Eqs. (1.25) and (1.26), we finally obtain

$$\langle r^2 \rangle = \frac{2}{Nm\omega_0} \sum_{n,l}(2l+1)\left[2(n-1)+l+\frac{3}{2}\right]. \tag{1.27}$$

In the Boson case, since in the ground state only the $1s$ state is occupied, which has degeneracy N [instead of $2(2l+1)$], we have

$$\langle r^2 \rangle = \frac{3}{2m\omega_0},\tag{1.28}$$

independent of the number of particles.

Note that the virial theorem can be considered as a consequence of the identity

$$0 = \langle 0|[H, F]|0\rangle,\tag{1.29}$$

with the operator F given by

$$F = \sum_{i=1}^{N} \mathbf{r}_i \cdot \nabla_{\mathbf{r}_i} \ .$$

For the single particle Hamiltonian (1.2) one gets

$$2\langle 0|T|0\rangle = \langle 0|\sum_{i=1}^{N} \mathbf{r} \cdot \nabla v_{\text{ext}}(\mathbf{r}_i)|0\rangle.\tag{1.30}$$

1.5   Matrix Elements of Two-Body Operators

By the term "two-body operator" we mean an operator of the following type:

$$F_2 = \sum_{i<j=1}^{N} f(x_i, x_j),\tag{1.31}$$

with $f(x_i, x_j) = f(x_j, x_i)$.

A particular case of two-body operator is the two-body interaction

$$F_2 = \sum_{i<j=1}^{N} v(r_{i,j}),$$

with $r_{i,j} = |\mathbf{r}_i - \mathbf{r}_j|$. We are interested in evaluating the matrix elements

$$\langle \phi|F_2|\psi\rangle = \int \phi^*(x_1, \ldots, x_N) \sum_{i<j=1}^{N} f(x_i, x_j)\psi(x_1, \ldots, x_N)dx_1 \cdots dx_N$$

$$= \frac{N(N-1)}{2} \int \phi^*(x_1, \ldots, x_N)f(x_1, x_2)\psi(x_1, \ldots, x_N)dx_1 \cdots dx_N.$$
$$\tag{1.32}$$

If we take state ϕ as the ground state $|0\rangle$, it is easily realized that in this case there exist only three nonvanishing matrix elements, and more precisely $\langle 0|F_2|0\rangle$, $\langle 0|F_2|i^{-1}, m\rangle$ and $\langle 0|F_2|i^{-1}, k^{-1}, m, n\rangle$. As in the case of a one-body operator, in this case the calculation is different for Boson and Fermions.

Bosons

When $\phi = \psi = |0\rangle$, from Eqs. (1.11) and (1.32), and taking into account the normalization of the single-particle wave functions, we obtain immediately

$$\langle 0|F_2|0\rangle = \frac{N(N-1)}{2} \int \varphi_{i_1}^*(x_1)\varphi_{i_1}^*(x_2)f(x_1,x_2)\varphi_{i_1}(x_1)\varphi_{i_1}(x_2)dx_1dx_2. \quad (1.33)$$

When $\phi = |0\rangle$ and $\psi = |i_1^{-1}m\rangle$, using Eq. (1.24) we have

$$\langle 0|F_2|i_1^{-1}m\rangle = \sqrt{N}(N-1)\int \varphi_{i_1}^*(x_1)\varphi_{i_1}^*(x_2)f(x_1,x_2)\varphi_m(x_1)\varphi_{i_1}(x_2)dx_1dx_2. \quad (1.34)$$

Finally, if $\phi = |0\rangle$ and $\psi = |i_1^{-1}i_1^{-1}mn\rangle$, then using

$$|i_1^{-1}i_1^{-1}mn\rangle = \frac{1}{\sqrt{N(N-1)}}\sum_P P\varphi_m(x_1)\varphi_n(x_2)\varphi_{i_1}(x_3)\cdots\varphi_{i_1}(x_N) \quad (1.35)$$

we easily obtain

$$\langle 0|F_2|i_1^{-1}i_1^{-1}mn\rangle$$
$$= \sqrt{N(N-1)}\int \varphi_{i_1}^*(x_1)\varphi_{i_1}^*(x_2)f(x_1,x_2)\varphi_m(x_1)\varphi_n(x_2)dx_1dx_2.$$
$$(1.36)$$

Fermions

In the Fermion case, we use the representation of the wave functions as Slater determinants, and expanding with respect to the two first columns,

$$\Phi(x_1,\ldots,x_N) = \frac{1}{\sqrt{N!}} \begin{vmatrix} \varphi_{i_1}(x_1) & \varphi_{i_1}(x_2) & \varphi_{i_1}(x_3) & \cdots & \varphi_{i_1}(x_N) \\ \varphi_{i_2}(x_1) & \varphi_{i_2}(x_2) & \varphi_{i_2}(x_3) & \cdots & \varphi_{i_2}(x_N) \\ \varphi_{i_3}(x_1) & \varphi_{i_3}(x_2) & \varphi_{i_3}(x_3) & \cdots & \varphi_{i_3}(x_N) \\ \vdots & \vdots & \vdots & \ddots & \vdots \\ \varphi_{i_N}(x_1) & \varphi_{i_N}(x_2) & \varphi_{i_N}(x_3) & \cdots & \varphi_{i_N}(x_N) \end{vmatrix}$$

$$= \frac{1}{\sqrt{N!}}\sum_{i_s<i_t}^{N}[\varphi_{i_s}(x_1)\varphi_{i_t}(x_2) - \varphi_{i_s}(x_2)\varphi_{i_t}(x_1)]$$
$$\times M_{\{i_k\}\neq i_s,i_t}^{N-2}(x_3,\ldots,x_N),$$

where $M_{\{i_k\}\neq i_s,i_t}^{N-2}(x_3,\ldots,x_N)$ means the minor of order $N-2$ which depends on the space variables $x_3,\ldots,x_N$, from which the first and second columns, as well as the i_s-th and i_t-th rows, have been removed. Hence it depends on a set of quantum numbers $\{i_k\}$ which are all different from i_s and i_t. In order to avoid cumbersome notation, we will write $\mathcal{M}(x_3, \ldots, x_N)$ instead of $M_{\{i_k\} \neq i_s, i_t}^{N-2}(x_3, \ldots, x_N)$. Let us now compute the matrix element:

$$\langle \phi | F_2 | \psi \rangle = \frac{N(N-1)}{2N!} \sum_{i_s < i_t}^{N} \sum_{i_{\tilde{s}} < i_{\tilde{t}}}^{N} \int \phi^*(x_1, x_2, i_s, i_t) \mathcal{M}^*(x_3, \ldots, x_N) f(x_1, x_2)$$

$$\times \psi(x_1, x_2, i_{\tilde{s}}, i_{\tilde{t}}) \tilde{\mathcal{M}}(x_3, \ldots, x_N) dx_1 \cdots dx_N, \qquad (1.37)$$

where $\phi(x_1, x_2, i_s, i_t) = \varphi_{i_s}(x_1)\varphi_{i_t}(x_2) - \varphi_{i_s}(x_2)\varphi_{i_t}(x_1)$ and similarly for ψ. Starting from Eq. (1.37), it is easily seen that the only nonvanishing matrix elements are, as mentioned, $\langle 0|F_2|0\rangle$, $\langle 0|F_2|i^{-1}, m\rangle$, $\langle 0|F_2|i^{-1}, k^{-1}, m, n\rangle$. One finally obtains

$$\langle 0|F_2|0\rangle = \frac{1}{2} \sum_{i_s, i_t}^{N} \int \varphi_{i_s}^*(x_1)\varphi_{i_t}^*(x_2) f(x_1, x_2)$$

$$\times [\varphi_{i_s}(x_1)\varphi_{i_t}(x_2) - \varphi_{i_s}(x_2)\varphi_{i_t}(x_1)] dx_1 dx_2, \qquad (1.38)$$

$$\langle 0|F_2|i^{-1}, m\rangle = \sum_{i_t}^{N} \int \varphi_i^*(x_1)\varphi_{i_t}^*(x_2) f(x_1, x_2)$$

$$\times [\varphi_m(x_1)\varphi_{i_t}(x_2) - \varphi_m(x_2)\varphi_{i_t}(x_1)] dx_1 dx_2, \qquad (1.39)$$

$$\langle 0|F_2|i^{-1}, j^{-1}, m, n\rangle = \int \varphi_i^*(x_1)\varphi_j^*(x_2) f(x_1, x_2)$$

$$\times [\varphi_m(x_1)\varphi_n(x_2) - \varphi_m(x_2)\varphi_n(x_1)] dx_1 dx_2. \qquad (1.40)$$

Note the presence of the exchange term $x_1 \to x_2$ and $x_2 \to x_1$, which is proper for Fermions since it originates from the antisymmetrization of the wave function.

1.6 Density Matrices

Generally speaking, the one-body and two-body properties of a many-body system are embodied in the one-body and two-body density matrices, respectively. The one-body density matrix is defined as

$$\rho^{(1)}(x_1, x_1') = N \int \phi_0^*(x_1, x_2, \ldots, x_N)\phi_0(x_1', x_2, \ldots, x_N) dx_2 \cdots dx_N, \qquad (1.41)$$

and the two-body one as

$$\rho^{(2)}(x_1, x_2, x_1', x_2') = N(N-1) \int \phi_0^*(x_1, x_2, x_3, \ldots, x_N)$$

$$\times \phi_0(x_1', x_2', x_3, \ldots, x_N) dx_3 \cdots dx_N. \qquad (1.42)$$

The following properties hold:

$$\int \rho^{(2)}(x, x_2, x', x_2)dx_2 = (N-1)\rho^{(1)}(x, x'), \tag{1.43}$$

$$\int \rho^{(1)}(x, x)dx = N. \tag{1.44}$$

The diagonal part of the one-body density matrix (when $x = x'$) is related to the particle density $\rho(\mathbf{r})$,

$$\rho(\mathbf{r}) = \sum_{\sigma} \rho^{(1)}(\mathbf{r}, \sigma, \mathbf{r}, \sigma), \tag{1.45}$$

while the nondiagonal part is related to the momentum distribution $n(\mathbf{p})$:

$$n(\mathbf{p}) = \frac{1}{(2\pi)^3} \sum_{\sigma} \int \rho^{(1)}(\mathbf{r}, \sigma, \mathbf{r}', \sigma)e^{i\mathbf{p}\cdot(\mathbf{r}-\mathbf{r}')}d\mathbf{r}d\mathbf{r}'. \tag{1.46}$$

The densities $\rho^{(1)}$ and $\rho^{(2)}$ can be used in order to calculate the mean values of one-body and two-body operators in the ground state:

$$\langle 0| \sum_i f(x_i)|0\rangle = \int \rho^{(1)}(x, x)f(x)dx, \tag{1.47}$$

$$\langle 0| \sum_{i<j} f(x_i, x_j)|0\rangle = \frac{1}{2}\int \rho^{(2)}(x_1, x_2, x_1, x_2)f(x_1, x_2)dx_1 dx_2. \tag{1.48}$$

In the case where the ground state is that of the independent-particle model, we obtain

$$\rho_B^{(1)}(x_1, x_1') = N\varphi_{i_1}^*(x_1)\varphi_{i_1}(x_1'), \tag{1.49}$$

$$\rho_B^{(2)}(x_1, x_2, x_1', x_2') = N(N-1)[\varphi_{i_1}^*(x_1)\varphi_{i_1}^*(x_2)\varphi_{i_1}(x_1')\varphi_{i_1}(x_2')] \tag{1.50}$$

for Bosons (φ_{i_1} is the lowest-energy single-particle state), and

$$\rho_F^{(1)}(x_1, x_1') = \sum_{i=1}^N \varphi_i^*(x_1)\varphi_i(x_1'), \tag{1.51}$$

$$\rho_F^{(2)}(x_1, x_2, x_1', x_2') = \rho_F^{(1)}(x_1, x_1')\rho_F^{(1)}(x_2, x_2') - \rho_F^{(1)}(x_1, x_2')\rho_F^{(1)}(x_2, x_1') \tag{1.52}$$

for Fermions. Note that the following property holds for Slater determinants:

$$\int \rho_{\text{SD}}^{(1)}(x, x')\rho_{\text{SD}}^{(1)}(x', x'')dx' = \rho_{\text{SD}}^{(1)}(x, x''). \tag{1.53}$$

As an example of the calculation of density matrices, let us compute $\rho^{(1)}$ for a system of Fermions in a spherical one-body potential, so that the single-particle wave functions can be written in the form $\varphi_i = R_{n,l}Y_{l,m}\chi_{m_s}$. Let us carry out the calculation for a system with N such that the $2(2l + 1)$ degenerate states of each level n, l are completely occupied (closed-shell system). By using the properties

$$\sum_m Y_{lm}(\vartheta, \varphi) Y_{lm}(\vartheta', \varphi') = \frac{2l + 1}{4\pi} P_l(u),$$

(1.54)

where $P_l(u)$ are the Legendre polynomials and $u = \cos(\theta - \theta')$, and

$$\sum_{m_s} \chi^*_{m_s}(\sigma) \chi_{m_s}(\sigma') = \delta_{\sigma, \sigma'},$$

(1.55)

we have

$$\rho^{(1)}(\mathbf{r}, \sigma, \mathbf{r}', \sigma') = \frac{1}{4\pi} \sum_{n,l} (2l + 1) R_{n,l}(r) R_{n,l}(r') P_l(u) \delta_{\sigma, \sigma'} .$$

(1.56)

Therefore, the density is written as

$$\rho(\mathbf{r}) = \sum_\sigma \rho^{(1)}(\mathbf{r}, \sigma, \mathbf{r}, \sigma) = 2 \sum_{n,l} \frac{2l + 1}{4\pi} |R_{n,l}(r)|^2 ,$$

(1.57)

and it is spherical.

1.7   The Ideal Bose Gas Confined in a Harmonic Potential

The noninteracting Bose gas confined in a magnetic trap is described by (1.3) taking

$$v_{\text{ext}}(\mathbf{r}) = \frac{1}{2} m (\omega_x^2 x^2 + \omega_y^2 y^2 + \omega_z^2 z^2) .$$

(1.58)

This potential exactly reproduces all possible experimental situations. The single-particle energies for this potential are given by

$$\epsilon_{n_x n_y n_z} = \left(n_x + \frac{1}{2} \right) \omega_x + \left(n_y + \frac{1}{2} \right) \omega_y + \left(n_z + \frac{1}{2} \right) \omega_z ,$$

(1.59)

where n_x, n_y, n_z are nonnegative integers and the single-particle wave function is given by

$$\varphi_{n_x n_y n_z}(x, y, z) = \phi_{n_x}(x) \phi_{n_y}(y) \phi_{n_z}(z) ,$$

(1.60)

with

$$\phi_{n_x}(x) = (m\omega_x/\pi)^{1/4} (1/\sqrt{2^{n_x} n_x!}) e^{\frac{-m\omega_x}{2} x^2} H_{n_x}(x\sqrt{m\omega_x}) ,$$

where H_{n_x} are the Hermite polynomials, and similarly for the functions depending on y and z.

In the ground state, at zero temperature, all the Bosons occupy the state $(n_x, n_y, n_z) = (0, 0, 0)$ at the energy $\epsilon(0, 0, 0) \equiv \epsilon_0$ and with the wave function

$$\phi_0 = \left(\frac{m\bar{\omega}}{\pi}\right)^{3/4} e^{-m/2(\omega_x x^2 + \omega_y y^2 + \omega_z z^2)},$$

where we have defined $\bar{\omega} = (\omega_x \omega_y \omega_z)^{1/3}$, and one has the Bose–Einstein condensate at zero temperature. One also talks about condensate at finite temperature, when there is a macroscopic occupation of state ϵ_0. As temperature is increased, a critical temperature is reached at which the lowest-energy state is no longer macroscopically occupied.

In the following we calculate, in the limit of large number N of Bosons, the critical temperature T_c at which phase transition takes place, as well as the temperature dependence of the depopulation of the condensate state.

The Boson distribution function at temperature T is given by Bose statistics:

$$n_{n_x, n_y, n_z} = \frac{1}{e^{[\beta(\epsilon_{n_x, n_y, n_z} - \mu)]} - 1}, \tag{1.61}$$

where $\beta = 1/KT$ (K is Boltzmann's constant) and μ is the chemical potential, which is determined by the condition that the summation of the number of occupied states coincides with the total number of Bosons:

$$N = \sum_{n_x, n_y, n_z} n_{n_x, n_y, n_z}. \tag{1.62}$$

Once N is fixed, let us find the condition that must be fulfilled by the chemical potential of the confined gas in order that there is a macroscopic occupation of state ϵ_0. The occupation n_0 of this state is given by

$$n_0 = \frac{1}{e^{-\beta\mu} e^{\beta(\frac{\omega_x}{2} + \frac{\omega_y}{2} + \frac{\omega_z}{2})} - 1}, \tag{1.63}$$

which shows that there is condensation when

$$\mu = \mu_c = \left(\frac{\omega_x}{2} + \frac{\omega_y}{2} + \frac{\omega_z}{2}\right) + \alpha, \tag{1.64}$$

with $\alpha \to 0$ when $N \to \infty$. Due to confinement, this condition differs from the one for homogeneous ideal gas, in which case condensation takes place in the limit $\mu \to 0$.

Now we can calculate the number of Bosons outside the condensate $N - N_0$, by using for the chemical potential its limiting value for which Bose–Einstein condensation takes place, in (1.61) and (1.62):

$$N - N_0 = \sum_{n_x, n_y, n_z \neq 0,0,0} \frac{1}{e^{[\beta(\omega_x n_x + \omega_y n_y + \omega_z n_z)]} - 1}. \tag{1.65}$$

In order to compute expression (1.65) we replace the summation $\sum_{n_x, n_y, n_z}$ by the integral $\int dn_x dn_y dn_z$. This approximation turns out to be a good one if $KT \gg$

$\omega_x, \omega_y, \omega_z$ and the main contribution to the summation comes from states whose energy is much higher than $\omega_x, \omega_y, \omega_z$. In this case, we obtain

$$N - N_0 \simeq \int_0^\infty \frac{dn_x dn_y dn_z}{e^{[\beta(\omega_x n_x + \omega_y n_y + \omega_z n_z)]} - 1}, \tag{1.66}$$

and thus

$$N - N_0 \simeq \left(\frac{KT}{\bar{\omega}}\right)^3 \zeta(3), \tag{1.67}$$

where $\zeta(n)$ is the Riemann function. The temperature dependence of the condensate depopulation, T^3, is different from the homogeneous gas case, where the law is $T^{3/2}$. The critical temperature for the transition to the condensate is obtained from Eq. (1.67) by putting $N_0 = 0$. We find that

$$KT_c \simeq 0.94 \bar{\omega} N^{1/3}. \tag{1.68}$$

For temperatures higher than T_c the chemical potential is smaller than μ_c and depends on N, and the population of the lowest-energy state is no longer of the order of N. Note that the thermodynamic limit for confined systems is obtained by taking $N \to \infty$ and $\bar{\omega} \to 0$ while keeping the product $N\bar{\omega}^3$ constant. With this definition, T_c is well defined in the thermodynamic limit.

If we insert the expression of T_c in the equation for $N - N_0$, we obtain the T dependence of the condensate fraction for $T \leq T_c$:

$$\frac{N_0}{N} = 1 - \left(\frac{T}{T_c}\right)^3. \tag{1.69}$$

For the magnetic traps that are realized nowadays, for which $\bar{\omega}/K$ is of the order of a few nanokelvin, and for N values in the range of 10^3 to 10^7, the critical temperature turns out to be of the order of tens or a few hundred nanokelvin. Such low temperatures are realized by means of laser cooling and evaporation techniques.

The effects of interaction on the critical temperature and on the behavior of the condensate depopulation have been discussed in detail by Dalfovo *et al.*, (1999).

1.8 The Fermi Gas

The noninteracting Fermi gas is obtained from Eq. (1.3) by putting $v_{\text{ext}} = 0$ and imposing periodicity conditions on the plane-wave-like solutions,

$$\varphi_i(x) = \frac{1}{\sqrt{V}} e^{i\mathbf{k}\cdot\mathbf{r}} \chi_{m_s}(\sigma), \quad i \equiv k_x, k_y, k_z, m_s, \tag{1.70}$$

at the boundaries of a cube having volume V and side L:

$$e^{ik_x(x+L)} = e^{ik_x x}, \quad e^{ik_y(y+L)} = e^{ik_y y}, \quad e^{ik_z(z+L)} = e^{ik_z z}. \tag{1.71}$$

Therefore, the particle momentum $\mathbf{k}$ can only take on the values

$$k_\alpha = \frac{2\pi}{L} n_\alpha \quad (\alpha = x, y, z), \tag{1.72}$$

where n_x, n_y, n_z are positive and negative integers, including zero. In general, if D is the dimensionality of the problem, with $D = 3, 2, 1$, then we will have $V = L^D$ and $\mathbf{k} = (2\pi/L)(n_x, n_y)$ for $D = 2$ and $k_x = (2\pi/L) n_x$ for $D = 1$.

The single-particle energies are given by

$$\epsilon_\mathbf{k} = \frac{k^2}{2m}. \tag{1.73}$$

In the $V \to \infty$ limit, the allowed values $\mathbf{n}$ in momentum space become closely spaced, and since their density is given by $d\mathbf{n} = L^D/(2\pi)^D d\mathbf{k}$, we can write

$$\sum_{n_x, n_y, n_z, m_s} \longrightarrow \frac{L^D}{(2\pi)^D} \sum_{m_s} \int d\mathbf{k}. \tag{1.74}$$

Therefore, the ground state is set up by filling a sphere (3D), or a circle (2D), or a segment (1D), with radius k_F, within which the single-particle states are all occupied, and outside which they are all unoccupied. The Fermi energy, i.e. the energy on the Fermi surface, boundary or extremum, is given by $\epsilon_F = k_F^2/2m$. The value of k_F is connected to the density $\rho = N/L^D$. In order to find such a relationship, let us calculate the number N of particles using Eq. (1.74):

$$N = \sum_{n_x, n_y, n_z, m_s}^{N} = \frac{L^D}{(2\pi)^D} \sum_{m_s} \int n_\mathbf{k} d\mathbf{k}, \tag{1.75}$$

where $n_\mathbf{k}$ is the ground state distribution:

$$n_\mathbf{k} = \begin{cases} 1 & \text{if } k \le k_F \\ 0 & \text{if } k > k_F \end{cases}. \tag{1.76}$$

For the densities $\rho = N/L^D$, we obtain

$$\rho^{3D} = \frac{k_F^3}{3\pi^2}, \quad \rho^{2D} = \frac{k_F^2}{2\pi}, \quad \rho^{1D} = \frac{2k_F}{\pi}. \tag{1.77}$$

The energy per particle of the Fermi gas is given by

$$\frac{E}{N} = \frac{1}{N} \sum_i \epsilon_i = \frac{2}{N} \frac{L^D}{(2\pi)^D} \int n_\mathbf{k} \frac{k^2}{2m} d\mathbf{k}. \tag{1.78}$$

From Eq. (1.78), we obtain

$$\frac{E}{N} = \frac{D}{D+2} \frac{k_F^2}{2m}. \tag{1.79}$$

If we add a particle to the system, we have the ground state energy of a system with $N+1$ particles if the added particle is put in the available state of lowest energy, i.e. on the Fermi surface. The system's chemical potential μ, defined by

$$\mu = E(N+1) - E(N) = \left(\frac{\partial E}{\partial N}\right)_V,\qquad(1.80)$$

is therefore given by $\mu = \epsilon_F$.

The energy of the system may be written as a function of volume $V = L^D$, from which it is possible to write down the pressure P, defined as

$$P = -\frac{\partial E}{\partial V} = \rho^2 \frac{\partial E/N}{\partial \rho}.\qquad(1.81)$$

Moreover, the compressibility K is given by

$$\frac{1}{K} = -V\frac{\partial P}{\partial V} = \rho^2\left(\rho\frac{\partial^2 E/N}{\partial\rho^2} + 2\frac{\partial E/N}{\partial\rho}\right),\qquad(1.82)$$

and is connected to the sound velocity v by

$$v^2 = \frac{1}{Km\rho}.\qquad(1.83)$$

For the Fermi gas in three dimensions, for example, one obtains $\frac{1}{K} = \frac{1}{9m\pi^2}k_F^5$.

The one-body density matrix is given by

$$\rho^{(1)}(\mathbf{r},\sigma,\mathbf{r}',\sigma') = \sum_i \varphi_i^*(\mathbf{r},\sigma)\varphi_i(\mathbf{r}',\sigma') = \delta_{\sigma,\sigma'}\frac{1}{(2\pi)^D}\int n_\mathbf{k}e^{-i\mathbf{k}\cdot(\mathbf{r}-\mathbf{r}')}d\mathbf{k}.\quad(1.84)$$

The calculation of the integral leads to

$$\rho^{(1)}(\mathbf{r},\sigma,\mathbf{r}',\sigma') = \delta_{\sigma,\sigma'}\frac{k_F^3}{2\pi^2}\frac{C(k_F|\mathbf{r}-\mathbf{r}'|)}{k_F|\mathbf{r}-\mathbf{r}'|}\qquad\text{(in 3D)},\qquad(1.85)$$

$$\rho^{(1)}(\mathbf{r},\sigma,\mathbf{r}',\sigma') = \delta_{\sigma,\sigma'}\frac{k_F^2}{2\pi}\frac{J_1(k_F|\mathbf{r}-\mathbf{r}'|)}{k_F|\mathbf{r}-\mathbf{r}'|}\qquad\text{(in 2D)},\qquad(1.86)$$

$$\rho^{(1)}(x,\sigma,x',\sigma') = \delta_{\sigma,\sigma'}\frac{k_F}{\pi}\frac{\sin(k_F(x-x'))}{k_F(x-x')}\qquad\text{(in 1D)},\qquad(1.87)$$

where $C(x) = \frac{\sin x - x\cos x}{x^2}$ and J_1 is the Bessel function. Putting $\mathbf{r} = \mathbf{r}'$ and $\sigma = \sigma'$ in the previous equations and summing over σ, one reobtains the diagonal density of (1.77).

By using Eq. (1.52), let us now calculate the diagonal part of the two-body density matrix $\rho_F^{(2)}(x_1,x_2,x_1,x_2)$ in the three-dimensional case taken as an example.

We obtain

$$\rho_F^{(2)}(\mathbf{r}_1, \sigma_1, \mathbf{r}_2, \sigma_2, \mathbf{r}_1, \sigma_1, \mathbf{r}_2, \sigma_2)' = \frac{k_F^6}{36\pi^4}\left[1 - 9\delta_{\sigma_1,\sigma_2}\left(\frac{C(k_F|\mathbf{r}_1 - \mathbf{r}_2|)}{k_F|\mathbf{r}_1 - \mathbf{r}_2|}\right)^2\right]$$

$$\equiv \frac{k_F^6}{36\pi^4} g_{\sigma_1,\sigma_2}(\mathbf{r}_1 - \mathbf{r}_2)\,, \tag{1.88}$$

where we have introduced the pair correlation function

$$g_{\sigma_1,\sigma_2}(\mathbf{r}) = 1 - 9\delta_{\sigma_1,\sigma_2}\left(\frac{C(k_F|\mathbf{r}_1 - \mathbf{r}_2|)}{k_F|\mathbf{r}_1 - \mathbf{r}_2|}\right)^2. \tag{1.89}$$

The function g is proportional to the probability of finding two particles at distance $r = |\mathbf{r}_1 - \mathbf{r}_2|$ apart, with spin variables σ_1 and σ_2.

If we are interested in the probability that two particles are at distance r apart, irrespective of the spin, we should calculate

$$\sum_{\sigma_1,\sigma_2} g_{\sigma_1,\sigma_2} = 4 - 18\left(\frac{C(k_F r)}{k_F r}\right)^2. \tag{1.90}$$

If, on the other hand, we are interested in the probability that two particles with the same spin variables are at distance r apart, we should calculate

$$\sum_{\sigma_1=\sigma_2} g_{\sigma_1,\sigma_2} = 2 - 18\left(\frac{C(k_F r)}{k_F r}\right)^2. \tag{1.91}$$

Finally, for the probability that two particles with opposite spin are at distance r apart, we have

$$\sum_{\sigma_1\neq\sigma_2} g_{\sigma_1,\sigma_2} = 2\,. \tag{1.92}$$

The above three probabilities are plotted as a function of r in Figs. 1.1–1.3. The fact that the probability in (1.91) tends to zero for $r \to 0$ is a consequence of the antisymmetrization of the wave function and of the presence of an exchange term in $\rho_F^{(2)}$. The Pauli principle prevents the two particles with the same spin from being at the same place ($r = 0$). When the two particles have opposite spins like in (1.92), they become distinguishable and the Pauli principle does not operate, so that the probability is independent of r like in the classical case. The situation described by Eq. (1.90) is intermediate between the two previous ones, and results from the copresence in $\rho_F^{(2)}$ of both symmetric and antisymmetric components under the exchange of $\mathbf{r}_1 \to \mathbf{r}_2$.

As an example of application of the Fermi gas model, let us calculate the expectation value of the Hamiltonian

$$H = \sum_{i=1}^{N} \frac{p_i^2}{2m} + \sum_{i<j}^{N}\left(t_0 + \frac{t_3}{6}\rho^\sigma\right)\delta(\mathbf{r}_i - \mathbf{r}_j) \tag{1.93}$$

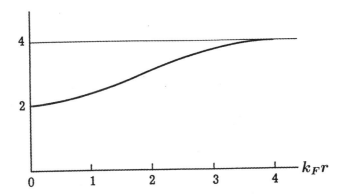

Fig. 1.1   Pair correlation function $\sum_{\sigma_1,\sigma_2} g_{\sigma_1,\sigma_2}$ of Eq. (1.90).

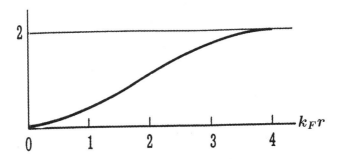

Fig. 1.2   Pair correlation function $\sum_{\sigma_1=\sigma_2} g_{\sigma_1,\sigma_2}$ of Eq. (1.91).

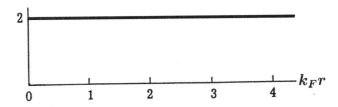

Fig. 1.3   Pair correlation function $\sum_{\sigma_1\neq\sigma_2} g_{\sigma_1,\sigma_2}$ of Eq. (1.92).

on the Fermi gas ground state as a function of the density ρ. Hamiltonian (1.93) gives a first modelization of liquid ^{3}He at zero temperature (see Chapter 4). From Eqs. (1.38), (1.88) and (1.89) one gets

$$\langle 0|\sum_{i<j}^N v_{i,j}|0\rangle = \frac{1}{2}\sum_{\sigma_1,\sigma_2}\int \frac{\rho^2}{4}\left[1-9\delta_{\sigma_1,\sigma_2}\left(\frac{C(k_F r_{12})}{k_F r_{12}}\right)^2\right]\left(t_0+\frac{t_3}{6}\rho^\sigma\right)$$

$$\times\delta(\mathbf{r}_1-\mathbf{r}_2)d\mathbf{r}_1 d\mathbf{r}_2 = \frac{1}{4}\rho^2\left(t_0+\frac{t_3}{6}\rho^\sigma\right)V, \qquad (1.94)$$

where $V = L^D$ is the volume of the system and we have used $C(x)/x \to 1/3$ when $x \to 0$. The energy per particle is then given by

$$\frac{E}{N} = \frac{1}{2m}\frac{3}{5}(3\pi^2)^{2/3}\rho^{2/3} + \frac{1}{4}\rho\left(t_0 + \frac{t_3}{6}\rho^\sigma\right), \tag{1.95}$$

from which, by using Eqs. (1.80)–(1.82), one gets

$$\mu = \frac{1}{2m}(3\pi^2)^{2/3}\rho^{2/3} + \frac{1}{4}\left(2t_0\rho + \frac{\sigma+2}{6}t_3\rho^{\sigma+1}\right), \tag{1.96}$$

$$P = \frac{1}{2m}\frac{2}{5}(3\pi^2)^{2/3}\rho^{5/3} + \frac{1}{4}t_0\rho^2 + \frac{\sigma+1}{24}t_3\rho^{\sigma+2}, \tag{1.97}$$

$$\frac{1}{K} = \frac{1}{2m}\frac{2}{3}(3\pi^2)^{2/3}\rho^{5/3} + \frac{1}{2}t_0\rho^2 + \frac{(\sigma+1)(\sigma+2)}{24}t_3\rho^{\sigma+2}, \tag{1.98}$$

and for the sound velocity

$$mv^2 = \frac{\partial P}{\partial\rho} == \frac{1}{2m}\frac{2}{3}(3\pi^2)^{2/3}\rho^{2/3} + \frac{1}{2}t_0\rho + \frac{(\sigma+1)(\sigma+2)}{24}t_3\rho^{\sigma+1}. \tag{1.99}$$

The saturation density ρ_0 is the density for which at zero temperature the energy per particle has a minimum ($\frac{\partial E/N}{\partial\rho} = 0$) and the pressure is zero. By inserting $\rho = \rho_0$ in the first three equations above and comparing with the experimental data, one can extract the values of the parameters t_0, t_3 and σ and then give predictions for the sound velocity and for the density dependence of the energy and compressibility (for the details see Chapter 4).

1.8.1 *Excited States*

The excited states of the Fermi gas, for example the $1p$–$1h$ state $|i^{-1}, m\rangle$, obtained by operating on the ground state with the density operator

$$\hat{\rho} = \sum_{i=1}^{N} e^{i\mathbf{q}\cdot\mathbf{r}_i}, \tag{1.100}$$

are set up by annihilating an occupied hole state with momentum $|\mathbf{p}| < k_F$ in the ground state, and creating a particle state with momentum $|\mathbf{p} + \mathbf{q}| > k_F$. This can be easily seen from Eq. (1.21), which yields a nonzero result only under these conditions. Moreover, since the matrix element $\int \varphi_m^*(\mathbf{r}, \sigma)e^{i\mathbf{q}\cdot\mathbf{r}}\varphi_i(\mathbf{r}, \sigma)d\mathbf{r}$ between plane-wave states is equal to 1, the dynamic form factor defined as ($\omega_{n0} = E_n - E_0$)

$$S(\mathbf{q}, \omega) = \sum_{n\neq 0} |\langle n|\hat{\rho}|0\rangle|^2 \delta(\omega - \omega_{n0}) \tag{1.101}$$

in the Fermi gas may be written as

$$S(\mathbf{q}, \omega) = 2 \sum_{p \langle k_F, |\mathbf{p}+\mathbf{q}| \rangle k_F} \delta(\omega - \omega_{\mathbf{pq}}), \qquad (1.102)$$

where

$$\omega_{\mathbf{pq}} = \frac{(\mathbf{p}+\mathbf{q})^2}{2m} - \frac{p^2}{2m} = \frac{q^2}{2m} + \frac{\mathbf{p} \cdot \mathbf{q}}{m}, \qquad (1.103)$$

and the factor 2 takes into account the summation on the spin index. In this way, the calculation of $S(\mathbf{q}, \omega)$ is reduced to a simple integration over the region in momentum space such that $p < k_F$ and $|\mathbf{p} + \mathbf{q}| > k_F$, from which $1p$–$1h$ pairs with total momentum $\mathbf{q}$ may be excited. This region, in the $D = 3$ case that we took as an example, corresponds to the shaded area in Fig. 1.4, which is bounded by the Fermi sphere and by the sphere obtained by shifting S_F by the quantity $-\mathbf{q}$. Even without carrying out any calculation, it is clear that the excitation spectrum for the $1p$–$1h$ states with momentum $\mathbf{q}$ will consist in a continuum between the limits

$$0 \leq \omega_{\mathbf{pq}} \leq \frac{q^2}{2m} + \frac{qk_F}{m}, \quad \text{for } q \leq 2k_F,$$

$$\frac{q^2}{2m} - \frac{qk_F}{m} \leq \omega_{\mathbf{pq}} \leq \frac{q^2}{2m} + \frac{qk_F}{m}, \quad \text{for } q \geq 2k_F, \qquad (1.104)$$

in the $D = 3$ and $D = 2$ cases, while in the 1D case the limits are

$$-\frac{q^2}{2m} + \frac{qk_F}{m} \leq \omega_{\mathbf{pq}} \leq \frac{q^2}{2m} + \frac{qk_F}{m}, \quad \text{for } q \leq 2k_F,$$

$$\frac{q^2}{2m} - \frac{qk_F}{m} \leq \omega_{\mathbf{pq}} \leq \frac{q^2}{2m} + \frac{qk_F}{m}, \quad \text{for } q \geq 2k_F. \qquad (1.105)$$

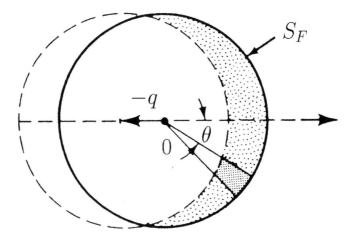

Fig. 1.4 Region in momentum space from which the $1p$–$1h$ states with momentum q may be excited.

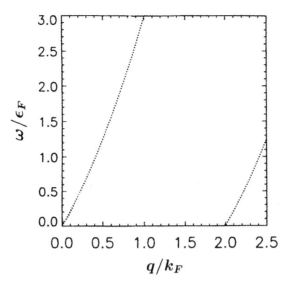

Fig. 1.5 Excitation spectrum of the $1p$–$1h$ states as a function of q, in the 2D and 3D cases.

The above-allowed energies correspond to the areas within the dotted lines in Figs. 1.5 and 1.6. Outside these zones the dynamic form factor vanishes. This behavior is characteristic of noninteracting systems and does not hold in the presence of interaction. Notice the relevance of dimensionality in determining the single-particle spectrum. The relevant feature of the one-dimensional spectrum, as compared to two and three dimensions, is that the $1p$–$1h$ excitations are forbidden in a large part of the low-energy phase space, i.e. the one that lies below the dashed line ABC in Fig. 1.6.

The explicit calculation of $S(\mathbf{q}, \omega)$ is very simple in the case where q is much smaller than k_F. In fact, in this limit the regions in momentum space from which the $1p$–$1h$ states can be excited are very thin. Taking as an example the three-dimensional case, the number of states in the elementary volume having polar angles between θ and $\theta + d\theta$ can be written as

$$dn = \frac{V}{(2\pi)^3} 2\pi k_F^2 q \cos\theta \sin\theta d\theta, \qquad (1.106)$$

and the excitation energy for the pair is approximately given by

$$\omega_{\mathbf{pq}} = q v_F \cos\theta, \qquad (1.107)$$

where $v_F = k_F/m$ is the Fermi velocity. Replacing the summation in (1.102) by an integration on θ, and using the property

$$\delta(cx - x_0) = \delta\left(x - \frac{x_0}{c}\right)/|c|,$$

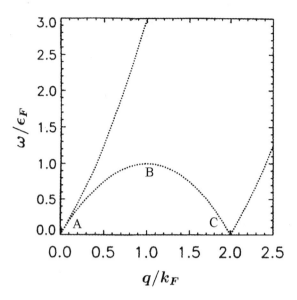

Fig. 1.6   Excitation spectrum of the $1p - 1h$ states as a function of q, in the 1D case.

one finds that

$$S_{3D}(\mathbf{q}, \omega) = V \frac{m^2 \omega}{2\pi^2 q}, \quad \text{for } \omega < qv_F,$$

$$S_{3D}(\mathbf{q}, \omega) = 0, \quad \text{for } \omega > qv_F. \tag{1.108}$$

Note that for all q values $S(\mathbf{q}, \omega)$ goes to zero when $\omega \to 0$. This behavior stems from the Pauli principle, which limits the number of low-energy excitations.

Finally, let us calculate the static form factor

$$S(\mathbf{q}) = \frac{1}{N} \int d\omega S(\mathbf{q}, \omega) = \frac{2}{N} \sum_{p\langle k_F, |\mathbf{p}+\mathbf{q}|\rangle k_F} 1 \tag{1.109}$$

which is proportional to the area of the shaded region in Fig. 1.4. A simple calculation yields

$$\text{(3D):} \qquad S(\mathbf{q}) = \begin{cases} \dfrac{3q}{4k_F} - \dfrac{q^3}{16k_F^3} & \text{if } q < 2k_F \\ 1 & \text{if } q \geq 2k_F \end{cases}, \tag{1.110}$$

$$\text{(2D):} \quad S(\mathbf{q}) = \begin{cases} \dfrac{2}{\pi} \sin^{-1}\left(\dfrac{q}{2k_F}\right) + \dfrac{q}{\pi k_F}\sqrt{1 - \left(\dfrac{q}{2k_F}\right)^2} & \text{if } q < 2k_F \\ 1 & \text{if } q \geq 2k_F \end{cases}, \tag{1.111}$$

(1D):

$$S(\mathbf{q}) = \begin{cases} \dfrac{q}{2k_F} & \text{if } q < 2k_F \\ 1 & \text{if } q \geq 2k_F \end{cases} . \tag{1.112}$$

1.8.2 Polarized Fermi Gas

The Fermi gas may be (partially) spin-polarized, the polarization being defined by

$$\xi = \frac{\rho_+ - \rho_-}{\rho}, \tag{1.113}$$

where $\rho_\pm$ are the densities with up and down spin, respectively. Using Eq. (1.113) together with

$$\rho = (\rho_+ + \rho_-), \tag{1.114}$$

one may write

$$\rho_\pm = \frac{\rho}{2}(1 \pm \xi). \tag{1.115}$$

Here, $\xi = 0$ means that the system is unpolarized, while $\xi = 1$ means that the Fermi gas is completely polarized.

Using

$$N_+ = \sum_{n_x,n_y,n_z,m_s=1/2}^{N_+} = \frac{L^D}{(2\pi)^D} \int n_{\mathbf{k}}^+ d\mathbf{k}, \tag{1.116}$$

where $n_{\mathbf{k}}^+$ is the distribution $n_{\mathbf{k}}^+ = 1$ for $k < k_F^+$ and $n_{\mathbf{k}}^+ = 0$ for $k > k_F^+$, and analogously for N_-, we obtain

$$k_F^\pm(\text{3D}) = [3\pi^2 \rho(1 \pm \xi)]^{1/3},$$

$$k_F^\pm(\text{2D}) = [2\pi \rho(1 \pm \xi)]^{1/2}, \tag{1.117}$$

$$k_F^\pm(\text{1D}) = \frac{\pi \rho}{2}(1 \pm \xi).$$

If we now use the generalization of Eq. (1.78), it is easy to obtain the energy of the polarized Fermi gas:

$$\frac{E^{\text{3D}}}{N} = \frac{3}{20m}(3\pi^2 \rho)^{2/3}[(1 + \xi)^{5/3} + (1 - \xi)^{5/3}],$$

$$\frac{E^{\text{2D}}}{N} = \frac{\pi \rho}{2m}(1 + \xi^2), \tag{1.118}$$

$$\frac{E^{\text{1D}}}{N} = \frac{\pi^2 \rho^2}{48m}[(1 + \xi)^3 + (1 - \xi)^3].$$

It is interesting to note that the above formulas hold also in the case of isospin polarization that occurs in asymmetric nuclear matter. In this case N_+ and N_- represent the number of neutrons N and protons Z, respectively, and $k_F^{\pm}$ are the Fermi momenta of neutrons and protons. The polarization ξ is given by $\frac{N-Z}{A}$, where A is the mass number, and formulas (1.118), expanded at the lowest order in ξ, yield the kinetic energy contribution to the symmetry energy of the nuclear matter (see for example the book by Bohr and Mottelson).

As an application of the obtained formulas, let us calculate the spin susceptibility of the Fermi gas. For this purpose, let us put the Fermi gas in a magnetic field $\mathbf{B}$ in the direction of the z axis. The total energy per particle of the system is given by

$$\frac{E}{N} = \frac{E_T(\rho, \xi)}{N} + \mu_0 B \xi, \tag{1.119}$$

where $E_T(\rho, \xi)/N$ are the kinetic energies per particle of Eq. (1.118), μ_0 is Bohr's magneton $e\hbar/2mc$, and the Fermions interact with the magnetic field via the interaction Hamiltonian

$$\mu_0 \mathbf{B} \cdot \sigma.$$

The polarization ξ is obtained by requiring that the variation of the energy (1.119) with respect to ξ be zero. The magnetic susceptibility χ_σ is given by

$$\chi_\sigma = \frac{\langle \sum_i \sigma_z^i \rangle}{B} = \frac{\xi \rho L^D}{B}. \tag{1.120}$$

For example, in the case of a two-dimensional gas we obtain $\xi = -m\mu_0 B/\pi\rho$ and

$$\frac{\chi_\sigma}{N} = -\frac{m\mu_0}{\pi\rho} = -\frac{\mu_0}{\epsilon_F}. \tag{1.121}$$

1.8.3 *The Fermi Gas in Two Dimensions with Rashba Interaction*

Consider the two-dimensional Rashba Hamiltonian (Rashba, 2004)

$$H_0^R = \sum_{i=1}^{N} \left[\frac{p_x^2 + p_y^2}{2m} + \lambda_R(p_y\sigma_x - p_x\sigma_y) \right]_i = \sum_{i=1}^{N} h_i, \tag{1.122}$$

where λ_R is a strength parameter. Since $[h, \mathbf{p}] = 0$, the solution of the single-particle Schrödinger's equation $h\varphi = \epsilon\varphi$ is translationally invariant in the (x, y) plane and has the form

$$\varphi = \frac{1}{\sqrt{S}} e^{i\mathbf{k}\cdot\mathbf{r}} \begin{pmatrix} \alpha \\ \beta \end{pmatrix}, \tag{1.123}$$

where $\begin{pmatrix} \alpha \\ \beta \end{pmatrix}$ is a spinor to be determined. One gets

$$\frac{k^2}{2m} \begin{pmatrix} \alpha \\ \beta \end{pmatrix} + \lambda_R \left[k_y \begin{pmatrix} \beta \\ \alpha \end{pmatrix} + i k_x \begin{pmatrix} \beta \\ -\alpha \end{pmatrix} \right] = \epsilon \begin{pmatrix} \alpha \\ \beta \end{pmatrix}, \tag{1.124}$$

from which the following solutions are recovered:

$$\epsilon_{k\pm} = \frac{k^2}{2m} \pm \lambda_R |k|,$$

$$\varphi_\pm = \frac{1}{\sqrt{2S}} e^{i\mathbf{k}\cdot\mathbf{r}} \begin{pmatrix} 1 \\ \mp i e^{i\psi} \end{pmatrix}, \tag{1.125}$$

where $\psi = \mathrm{arctg}\frac{k_y}{k_x}$. Note that the wave function (1.125) is an eigenstate of a component of the spin which is perpendicular to the direction of $\mathbf{k}$. The two branches of excitation $\epsilon_{k\pm}$ are plotted in Fig. 1.7 as a function of the momentum k. For $\mu > 0$, both the branches are filled up to the Fermi momenta k_+^F and k_-^F. Starting from

$$E_+ = \frac{S}{(2\pi)^2} \int \epsilon_{k_+} n_{\mathbf{k}}^+ d\mathbf{k},$$

$$E_- = \frac{S}{(2\pi)^2} \int \epsilon_{k_-} n_{\mathbf{k}}^- d\mathbf{k}, \tag{1.126}$$

it is immediate to get for a density ρ of particles

$$\frac{E}{N} = \frac{E_+ + E_-}{N} = \frac{\pi\rho}{4m} \left[(1-\xi)^2 + (1+\xi)^2 \right]$$

$$+ \frac{\lambda_R}{3} \sqrt{2\pi\rho} \left[(1-\xi)^{3/2} - (1+\xi)^{3/2} \right], \tag{1.127}$$

where ξ is related to the Fermi momenta by $k_\pm^F = \sqrt{2\pi\rho(1 \mp \xi)}$. From the minimum condition $\frac{\partial E/N}{\partial \xi} = 0$ one then gets $\xi = \frac{2m}{\sqrt{2\pi\rho}}\lambda_R$, $k_-^F - k_+^F = 2m\lambda_R$ and

$$\frac{E}{N} = \frac{\pi\rho}{2m} \left(1 - \frac{2m^2\lambda_R^2}{\pi\rho} \right),$$

$$\mu = \frac{\partial E/S}{\partial \rho} = \frac{\pi\rho}{m} - m\lambda_R^2. \tag{1.128}$$

Note that $\mu > 0$ if $\rho \geq \frac{m\lambda_R^2}{\pi}$.

1.9 Finite Temperature and Quasiparticles

By definition the ground state of the system is the zero-temperature state. If the system temperature is increased, the system is automatically excited. In the limit of low temperatures, the number of allowed excited states for the Fermion system is strongly reduced by the Pauli exclusion principle. At (low) temperature T, only

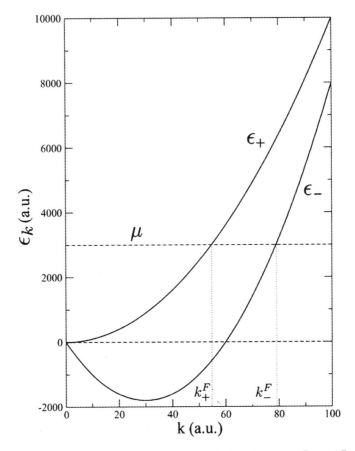

Fig. 1.7 Schematic of the energy spectrum of the Rashba Hamiltonian. k_+^F and k_-^F are the Fermi momenta for the upper (ϵ_+) and lower (ϵ_-) branches of excitation, respectively. μ is the chemical potential.

those single-particle states whose energy differs from the Fermi energy ϵ_F by KT (where K is Boltzmann's constant) are affected by a temperature change.

It is convenient to describe the structure of these excited states in terms of the changes of their occupation numbers as compared to those in the ground state. If we refer to the unoccupied hole states with $\epsilon < \epsilon_F$ and occupied particle states with $\epsilon > \epsilon_F$, as the quasiparticle states, it is possible to describe the statistical properties of the degenerate Fermi gas in terms of these elementary excitations. In fact, the creation of an excited state through the excitation of a given number of particles across the Fermi surface S_F is equivalent to the creation of the same number of particles outside S_F and of holes inside S_F. The excitation energy of the Fermi gas at equilibrium at temperature T is characterized by the quantity

$$\delta n_{\mathbf{k}} = n_{\mathbf{k}}(T, \mu) - n_{\mathbf{k}}(0, \mu), \qquad (1.129)$$

which measures how much the quasiparticle distribution function has been changed with respect to its value in the ground state, which is given by (1.76). The excitation of a particle with momentum $k' > k_F$ corresponds to $\delta n_{\mathbf{k}} = \delta_{\mathbf{k},\mathbf{k}'}$, and the excitation of a hole with momentum $k' < k_F$ corresponds to $\delta n_{\mathbf{k}} = -\delta_{\mathbf{k},\mathbf{k}'}$. Note that, since in an isolated system the number of particles is conserved, the number of excited particles must equal that of excited holes, which entails

$$\sum_{\mathbf{k}} \delta n_{\mathbf{k}} = 0. \tag{1.130}$$

For a noninteracting system, the excitation energy is given by

$$E - E_0 = \sum_{\mathbf{k}} \frac{k^2}{2m} \delta n_{\mathbf{k}}, \tag{1.131}$$

where E_0 is the ground state energy. In Eq. (1.129), the quasiparticle distribution function at temperature T is given by

$$n_{\mathbf{k}} = \frac{1}{1 + \exp[\beta(\epsilon_{\mathbf{k}} - \mu)]}, \tag{1.132}$$

where $\beta = 1/KT$ and μ is the chemical potential, which for a noninteracting system and in the limit of low temperature is equal to the Fermi energy. A simple calculation [see Huang (1963)] yields for the excitation energy

$$E - E_0 = \frac{\pi^2}{4\epsilon_F} N(KT)^2, \tag{1.133}$$

and for the specific heat

$$C_v = \left(\frac{\partial E}{\partial T}\right)_N = \frac{\pi^2}{2\epsilon_F} NK^2 T. \tag{1.134}$$

The quasiparticles introduced above are the low-energy elementary excitations for the degenerate Fermi systems. In the limit of low energy the quantum states of the degenerate Fermi liquids, which in nature are realized by liquid ^{3}He (above 4 m $^\circ$K), by the conduction electrons of non-superconducting metals and in nuclear matter, may be described in terms of such elementary excitations. The theory of these states in terms of quasiparticles was established in 1956 by Landau. A detailed description of this theory is found in the book by Pines and Nozières; in the following we report only its main steps.

The basic hypothesis of the Landau theory is that the dynamic and thermodynamic properties of a Fermi liquid can be traced back to those of a gas of elementary excitations near the Fermi surface, i.e. the quasiparticles. The theory studies the effect of the deviations $\delta n(\mathbf{k}, \mathbf{r})$ of the local distribution function $n(\mathbf{k}, \mathbf{r})$, which provides the quasiparticle distribution in a unitary volume centered at $\mathbf{r}$, with respect to the equilibrium distribution $n_{\mathbf{k}}^0$, which is assumed to have the same form as that of a noninteracting Fermion gas. Therefore, the concept of Fermi surface is still valid, and in order that the idea of the quasiparticle makes sense, the deviations $\delta n(\mathbf{k}, \mathbf{r})$ must be appreciably large only for values $k \simeq k_F$. Stated otherwise, the energy ω and the momentum q associated with the quasiparticles should be much smaller than the Fermi energy and momentum.

The deviations $\delta n(\mathbf{k}, \mathbf{r})$ produce changes in the quasiparticle gas energy which, to second order in $\delta n(\mathbf{k}, \mathbf{r})$, are written in the form

$$E(n(\mathbf{k}, \mathbf{r})) - E_0 = \int \delta E(\mathbf{r}) dr,$$

$$\delta E(\mathbf{r}) = \sum_{\mathbf{k}} \epsilon_{\mathbf{k}} \delta n(\mathbf{k}, \mathbf{r}) + \frac{1}{2} \sum_{\mathbf{k}, \mathbf{k}'} f_{\mathbf{k}, \mathbf{k}'} \delta n(\mathbf{k}, \mathbf{r}) \delta n(\mathbf{k}', \mathbf{r}),$$

(1.135)

where ϵ and f are the first and second functional derivatives of the energy with respect to the quasiparticle distribution. They have the respective meanings of the energy associated with a single quasiparticle excitation and interaction energy between quasiparticles. By means of $\epsilon_{\mathbf{k}}$ it is possible to define the group velocity of quasiparticles $\mathbf{v_k} = \nabla_{\mathbf{k}} \epsilon_{\mathbf{k}}$, and hence their effective mass is introduced via $\mathbf{v_k} = \mathbf{k}/m^*$.

The basic equation for the dynamics of the quasiparticles is the transport equation of the type similar to the Boltzmann equation for the deviations of the distribution function $n(\mathbf{k}, \mathbf{r}, t)$, which gives the probability of finding a quasiparticle located at $\mathbf{r}$ with momentum $\mathbf{k}$ at time t:

$$\frac{\partial}{\partial t} \delta n(\mathbf{k}, \mathbf{r}, t) + \mathbf{v_k} \cdot \nabla_{\mathbf{r}} \delta n(\mathbf{k}, \mathbf{r}, t)$$

$$- \nabla_{\mathbf{k}} n_{\mathbf{k}}^0 \cdot \sum_{\mathbf{k}'} f_{\mathbf{k}, \mathbf{k}'} \nabla_{\mathbf{r}} \delta n(\mathbf{k}', \mathbf{r}, t) = I(\delta n)_{\text{coll}}.$$

(1.136)

With respect to the usual Boltzmann equation, the novel features of the Landau equation are the effective quasiparticle mass which appears in $\mathbf{v_k} = \mathbf{k}/m^*$ and, most of all, the external force term $\nabla_{\mathbf{k}} n_{\mathbf{k}}^0 \cdot F_{\text{ext}}$, which in the case of Eq. (1.136) has the form

$$F_{\text{ext}} = \sum_{\mathbf{k}'} f_{\mathbf{k}, \mathbf{k}'} \nabla_{\mathbf{r}} \delta n(\mathbf{k}', \mathbf{r}, t),$$

and is not a force external to the system but rather is produced by the other quasiparticles. This term must be computed in a self-consistent way since it depends on δn. The external force in the Landau equation is characterized by the function $f_{\mathbf{k}, \mathbf{k}'}$. The force acting on the quasiparticle with momentum $\mathbf{k}$ is due to all other quasiparticles with momentum $\mathbf{k}'$. It is assumed that $f_{\mathbf{k}, \mathbf{k}'}$ is continuous as $\mathbf{k}$ or $\mathbf{k}'$ crosses the Fermi surface. Since the whole theory is valid in the neighborhood of the Fermi surface, what is needed in practice is only the values of f on the Fermi surface, so that $f_{\mathbf{k}, \mathbf{k}'}$ depends only on the directions of $\mathbf{k}$ and $\mathbf{k}'$ and on the spins of the two quasiparticles. It is convenient to introduce spin-symmetric $(f_{\mathbf{k}, \mathbf{k}'}^s)$ and spin-antisymmetric ($f^a_{\mathbf{k},\mathbf{k'}}$) components of the quasiparticle interaction by means of the equations

$$f^{\uparrow\uparrow}_{\mathbf{k},\mathbf{k'}} = f^s_{\mathbf{k},\mathbf{k'}} + f^a_{\mathbf{k},\mathbf{k'}}, \quad f^{\uparrow\downarrow}_{\mathbf{k},\mathbf{k'}} = f^s_{\mathbf{k},\mathbf{k'}} - f^a_{\mathbf{k},\mathbf{k'}}. \qquad (1.137)$$

Moreover, if the system is isotropic the f depend only on the angle θ between the directions $\mathbf{k}$ and $\mathbf{k'}$, and may be expanded in series of Legendre polynomials:

$$f^{s,a}_{\mathbf{k},\mathbf{k'}} = \sum_{l=0}^{\infty} f^{s,a}_l P_l(\cos\theta). \qquad (1.138)$$

The coefficients $f^{s,a}_l$ fix the interaction. Finally, it is opportune to put these coefficients in terms of dimensionless coefficients:

$$\nu(0)f^{s,a}_l = F^{s,a}_l, \qquad (1.139)$$

where

$$\nu(0) = \sum_{\mathbf{k}} \delta(\epsilon_{\mathbf{k}} - \epsilon_F) = \frac{Vm^*k_F}{\pi^2} \qquad (1.140)$$

is the quasiparticle density of states at the Fermi surface. In Eq. (1.136), the term $I(\delta n)_{\text{coll}}$ accounts for the changes $\delta n(\mathbf{k}, \mathbf{r}, t)$ due to quasiparticle collisions. These collisions are similar to those between molecules in the normal kinetic theory of gases. Their importance is evaluated qualitatively by the collision time τ, i.e. the time elapsed between two collisions. This depends on the nature of the particle interaction and on temperature T. At low temperature, which is the case we are interested in, collisions are quenched by the Pauli principle; there are few of them and τ is large. It can be shown that its behavior is approximately T^{-2}. Therefore, collisions play a role only in phenomena characterized by low frequency ω such that $\omega\tau \ll 1$ (viscosity, thermal conduction, etc.). This regime is referred to as the hydrodynamic or classical regime. In the opposite limit, $\omega\tau \gg 1$, the frequency is so high that the system can undergo many oscillation periods without any collision. In this case the collision term can be neglected and the system is said to be in the elastic regime. In this regime, the interaction plays a crucial role. To study these two regimes, it is possible to change the temperature and, consequently, τ.

The Landau theory has been very successful especially because it can explain a large number of physical phenomena with a very small number of parameters in the interaction $F^{s,a}_l$. For example, for a systematic description of the dynamic properties of ^{3}He, only four parameters are needed (F^s_0 and F^s_1 for the spin-symmetric quantities, and F^a_0 and F^a_1 for the spin-antisymmetric ones). These parameters can be connected to quantities known from experiments, as we will discuss briefly in the following.

First we should remember that by integrating the distribution function $n(\mathbf{k}, \mathbf{r}, t)$ in momentum space we obtain the quasiparticle spatial density:

$$\rho(\mathbf{r}, t) = \sum_{\mathbf{k}} n(\mathbf{k}, \mathbf{r}, t) = \frac{V}{(2\pi)^3} \int n(\mathbf{k}, \mathbf{r}, t) d\mathbf{k},$$

while by integrating it in coordinate space we obtain the momentum density $n(\mathbf{k}, t) = \int n(\mathbf{k}, \mathbf{r}, t) d\mathbf{r}$. By integrating the Landau equations in momentum space we have

$$\frac{\partial}{\partial t} \rho(\mathbf{r}, t) + \nabla_{\mathbf{r}} \left[\sum_{\mathbf{k}} \left(\delta n(\mathbf{k}, \mathbf{r}, t) - \frac{\partial n_{\mathbf{k}}^0}{\partial \epsilon_{\mathbf{k}}} \sum_{\mathbf{k}'} f_{\mathbf{k}, \mathbf{k}'} \delta n(\mathbf{k}', \mathbf{r}, t) \right) \cdot \mathbf{v}_{\mathbf{k}} \right]$$

$$= \sum_{\mathbf{k}} I(\delta n(\mathbf{k}, \mathbf{r}, t))_{\text{coll}}, \qquad (1.141)$$

where we have used $\nabla_{\mathbf{k}} n_{\mathbf{k}}^0 = \mathbf{v}_{\mathbf{k}} \frac{\partial n_{\mathbf{k}}^0}{\partial \epsilon_{\mathbf{k}}}$. Equation (1.141) yields the continuity equation

$$\frac{\partial}{\partial t} \rho(\mathbf{r}, t) + \nabla_{\mathbf{r}} \mathbf{J}(\mathbf{r}, t) = 0, \qquad (1.142)$$

with

$$\mathbf{J}(\mathbf{r}, t) = \sum_{\mathbf{k}} \left[\delta n(\mathbf{k}, \mathbf{r}, t) - \frac{\partial n_{\mathbf{k}}^0}{\partial \epsilon_{\mathbf{k}}} \sum_{\mathbf{k}'} f_{\mathbf{k}, \mathbf{k}'} \delta n(\mathbf{k}', \mathbf{r}, t) \right] \mathbf{v}_{\mathbf{k}}$$

$$= \sum_{\mathbf{k}} \delta n(\mathbf{k}, \mathbf{r}, t) \left(\mathbf{v}_{\mathbf{k}} - \sum_{\mathbf{k}'} f_{\mathbf{k}, \mathbf{k}'} \frac{\partial n_{\mathbf{k}'}^0}{\partial \epsilon_{\mathbf{k}'}} \mathbf{v}_{\mathbf{k}'} \right), \qquad (1.143)$$

where we have used the result $\sum_{\mathbf{k}} I(\delta n(\mathbf{k}, \mathbf{r}, t))_{\text{coll}} = 0$, because the continuity equation expresses the quasiparticle number conservation and collisions can change neither the number of quasiparticles nor the current.

The current expression (1.143) shows that $\mathbf{J}(\mathbf{r}, t) \neq \sum_{\mathbf{k}} \delta n(\mathbf{k}, \mathbf{r}, t) \mathbf{v}_{\mathbf{k}}$, i.e. the current is not determined by quasiparticle velocity, but there exist nonlocal dynamic contributions, due to the interaction. On the other hand, if the system is translationally invariant, the current cannot have any contribution from the interaction, but only from the kinetic energy, because it is a constant of motion which commutes with the interaction. This means that one must have

$$\mathbf{J}(\mathbf{r}, t) = \sum_{\mathbf{k}} \delta n(\mathbf{k}, \mathbf{r}, t) \frac{\mathbf{k}}{m}, \qquad (1.144)$$

where m is the bare quasiparticle mass. Comparison of Eqs. (1.143) and (1.144) implies that

$$\frac{\mathbf{k}}{m} = \mathbf{v}_{\mathbf{k}} - \sum_{\mathbf{k}'} f_{\mathbf{k}, \mathbf{k}'} \frac{\partial n_{\mathbf{k}'}^0}{\partial \epsilon_{\mathbf{k}'}} \mathbf{v}_{\mathbf{k}'}. \qquad (1.145)$$

Eq. (1.145) is a condition which is imposed on $f_{\mathbf{k},\mathbf{k}'}$ by translational invariance. For an isotropic system both the current and the velocity are parallel to $\mathbf{k}$; therefore, Eq. (1.145) involves only the spin-symmetric component with $l = 1$ in the expansion (1.138). Using the relationship

$$\frac{\partial n_{\mathbf{k}}^0}{\partial \epsilon_{\mathbf{k}}} = -\delta(\epsilon_k - \epsilon_F) , \qquad (1.146)$$

which stems from the fact that the equilibrium quasiparticle distribution function is a step function, and from the definition of effective mass $v_{k_F} = k_F/m^*$, and (1.139) and (1.140), we can change Eq. (1.145) into the relation

$$\frac{m^*}{m} = 1 + \frac{F_1^s}{3} . \qquad (1.147)$$

Since the effective mass can be evaluated by specific heat experiments, by using relation (1.134), which still holds for the gas of quasiparticles with $\epsilon_F = k_F^2/2m^*$, Eq. (1.147) enables the determination of the Landau parameter F_1^s.

In order to establish a connection between F_0^s and a measurable quantity, let us assume that the hydrodynamic regime holds, and let us look for Euler-like equations for the density variations. To this end, let us multiply the Landau equation by k_j/m, where k_j is one of the momentum components, and let us perform a summation on $\mathbf{k}$. We obtain

$$\frac{\partial}{\partial t} J_j(\mathbf{r}, t) + \nabla_{r_i} \sum_{\mathbf{k}} \delta n(\mathbf{k}, \mathbf{r}, t) v_i \frac{k_j}{m} - \nabla_{r_i} \sum_{\mathbf{k}\mathbf{k}'} \frac{\partial n_{\mathbf{k}}^0}{\partial \epsilon_{\mathbf{k}}} f_{\mathbf{k},\mathbf{k}'} \delta n(\mathbf{k}', \mathbf{r}, t) v_i \frac{k_j}{m} = 0, \quad (1.148)$$

where we have used the result $\sum_{\mathbf{k}} k_j I(\delta n(\mathbf{k}, \mathbf{r}, t))_{\text{coll}} = 0$, which arises from the current being conserved during collisions. However, the fact that the collision term does not change the quasiparticle density and their current, does not mean that it is not relevant in the hydrodynamic regimes where $\omega\tau \ll 1$. In the hydrodynamic regime the collisions are important and tend to re-establish the equilibrium distribution n^0. They tend to keep the normal displacement of the Fermi surface $u_{\mathbf{k}}$ at point $\mathbf{k}$, which, for isotropic systems, is defined by

$$\delta n(\mathbf{k}, \mathbf{r}, t) = \delta(\epsilon_p - \epsilon_F) v_F u_{\mathbf{k}}(\mathbf{r}, t) , \qquad (1.149)$$

at small values, setting to zero all its components associated with deformations with $l > 1$. In this limit $u \simeq a + b\cos\theta$ and it corresponds to an oscillation of the density superimposed on a uniform translation of the fluid. This is what is expected for the propagation of acoustic waves. We are in a regime of local equilibrium.

If we consider isotropic systems for which $\delta n(\mathbf{k}, \mathbf{r}, t)$ is spherical, only terms with $i = j$ survive in (1.148), and using the relationship (1.146) as well as Eqs. (1.139) and (1.140), we easily obtain the equation

$$\frac{\partial}{\partial t} \mathbf{J}(\mathbf{r}, t) + \frac{2}{3} \frac{\epsilon_F}{m} (1 + F_0^s) \nabla_{\mathbf{r}} \rho(\mathbf{r}, t) = 0 , \qquad (1.150)$$

which holds only in the hydrodynamic regime. In fact, in the conservation equation (1.148), the $l = 2$ term would survive. However, we put it to zero because we have assumed that we are in the hydrodynamic regime. Equation (1.150) resembles the classical one, and together with the continuity equation (1.142) leads to the sound equation

$$\frac{\partial^2}{\partial t^2} \rho(\mathbf{r}, t) - v_1^2 \nabla^2 \rho(\mathbf{r}, t) = 0 \,, \tag{1.151}$$

with v_1 being the sound velocity in the quantum liquid, given by

$$v_1^2 = \frac{k_F^2}{3mm^*}(1 + F_0^s) \,. \tag{1.152}$$

Equation (1.151) describes the propagation of ordinary sound in quantum liquid, and the corresponding propagation velocity (1.152) is connected to the system compressibility by (1.83).

In the case of ^{3}He, the measurements of specific heat and ordinary sound velocity at vapor pressure $(P = 0)$ provide the following values of the Landau parameters: $F_0^s = 10$ and $F_1^s = 6$. From these values it is immediately clear that ^{3}He is very different from a noninteracting Fermi gas. For example, the effective mass turns out to be three times as large as the bare mass.

The Landau parameters F_0^a and F_1^a are connected to the magnetic properties of quasiparticles. For the study of such properties, it is necessary to study the effect of the deviations $\delta n^a(\mathbf{k}, \mathbf{r}) = \delta n_\uparrow(\mathbf{k}, \mathbf{r}) - \delta n_\downarrow(\mathbf{k}, \mathbf{r})$ of the local spin-antisymmetric distribution function $n^a(\mathbf{k}, \mathbf{r}) = n_\uparrow(\mathbf{k}, \mathbf{r}) - n_\downarrow(\mathbf{k}, \mathbf{r})$, with respect to the equilibrium spin-antisymmetric distribution function $n_{\mathbf{k}\uparrow}^0 - n_{\mathbf{k}\downarrow}^0$, which is assumed to have the same form as the distribution of a noninteracting and spin-polarized Fermion gas. In this case, the integral of the distribution $n^a(\mathbf{k}, \mathbf{r}, t)$ in momentum space gives the spatial magnetization of the quasiparticles, $m(\mathbf{r}, t)$. The equations for $\delta n^a(\mathbf{k}, \mathbf{r})$ have the same form as those for $\delta n^s(\mathbf{k}, \mathbf{r}) = \delta n_\uparrow(\mathbf{k}, \mathbf{r}) + \delta n_\downarrow(\mathbf{k}, \mathbf{r})$ which we have studied above, but are characterized by the Landau parameters F_0^a and F_1^a.

The system magnetic susceptibility per particle can be expressed through F_0^a as

$$\frac{\chi_\sigma}{N} = -\mu_0 \frac{3m^*}{k_F^2} \frac{1}{1 + F_0^a} \,, \tag{1.153}$$

which should be compared to the corresponding value for the noninteracting Fermi gas $\chi_\sigma^{\mathrm{FG}}/N = -\mu_0(3m/k_F^2)$. The effect of interaction comes in through m^* and F_0^a. In the case of ^{3}He, F_0^a has a *very* negative value $\simeq -0.7$. Recalling that $m^* = 3m$, we find that the magnetic susceptibility is approximately 20 times larger than in the Fermi gas. The system is quasi-ferromagnetic, in the sense that it is very easily magnetized. As for the F_1^a parameter, its value can be derived from specific heat measurements in the nonlinear temperature regime. In fact, the magnetic excited states have low energy and affect the specific heat. From such analysis we find that $F_1^a = -0.55$.

The Landau theory is easily generalized to the two-dimensional case. In this case the Landau equations keep the same form as Eqs. (1.136), and Eqs. (1.138) and (1.140) should be replaced by

$$f^{s,a}_{\mathbf{k},\mathbf{k}';2D} = \sum_{l=0}^{\infty} f^{s,a}_l \cos l(\theta - \theta') \tag{1.154}$$

and

$$\nu^{2D}(0) = \sum_{\mathbf{k}} \delta(\epsilon_{\mathbf{k}} - \epsilon_F) = \frac{Vm^*}{\pi}. \tag{1.155}$$

It is not simple to connect the Landau parameters to measurable quantities in two dimensions, and in general the parameters themselves are derived from the physical interaction. As happened in the 3D case, translational invariance imposes a condition on the quasiparticle current in 2D which results in a relationship between the effective mass and the Landau parameter F_1^s:

$$\frac{m^*}{m} = 1 + \frac{F_1^s}{2}. \tag{1.156}$$

It is also possible to derive the sound equations, which keep the form (1.151) unchanged, but where the ordinary sound velocity is given by

$$v^2_{1,2D} = \frac{k_F^2}{2mm^*}(1 + F_0^s). \tag{1.157}$$

Finally, it is possible to link F_0^a to the system magnetic susceptibility per particle as follows

$$\frac{\chi_{\sigma,2D}}{N} = -\mu_0 \frac{2m^*}{k_F^2} \frac{1}{1 + F_0^a}. \tag{1.158}$$

As a last remark, recall that there also exist Bose quantum liquids (e.g. ^{4}He), and that recently in various laboratories Bose–Einstein condensates of alkali atoms have been realized. The elementary excitations of Bose systems are completely different from those of Fermi systems, and have a phonon nature. The different nature of elementary excitations has the consequence that the specific heat of these systems, contrary to (1.134), is proportional to T^3 as $T \to 0$. In the following chapters, we will treat the phonon nature of the elementary excitations of Bose systems in detail.

References

Anderson, M.H., J.R. Ensher, M.R. Matthews, C.E. Wieman and E.A. Cornell, *Science* **269**, 198 (1995).

Bohr, A. and B. Mottelson, *Nuclear Structure* (Benjamin, New York, 1975).

Bradley, C.C., C.A. Sackett, J.J. Tollet and R.G. Hulet, *Phys. Rev. Lett.* **75**, 1687 (1995).

Dalfovo, F., S. Giorgini, L.P. Pitaevskii and S. Stringari, *Rev. Mod. Phys.* **71**, 463 (1999).

Davis, K.B., M.O. Mewes, M.R. Andrews, N.J. van Druten, D.S. Durfee, D.M. Kurn and W. Ketterle, *Phys. Rev. Lett.* **75**, 3969 (1995).

Emperador, A., E. Lipparini and F. Pederiva, *Phys. Rev.* **B72**, 033306 (2005).

Huang, K., *Statistical Mechanics*, second edition (John Wiley and Sons, New York, 1987).

Landau, L.D., *Sov. Phys. JEPT* **3**, 920 (1956); **5**, 101 (1957).

MacDonald, A.H. and M.D. Johnson, *Phys. Rev. Lett.* **70**, 3107 (1993).

Maksym, P.A. and T. Chakraborty, *Phys. Rev. Lett.* **65**, 108 (1990).

Pines, D. and P. Nozières, *The Theory of Quantum Liquids* (Benjamin, New York, 1966).

Rashba, E.I., *Phys. Rev.* **B70**, 201309(R) (2004).

Reimann, S.M., M. Koskinen and M. Manninen, *Phys. Rev.* **B62**, 8108 (2000).

Tavernier, M.B., E. Anisimovas, F.M. Peeters, B. Szafran, J. Adamowski and S. Bednarek, *Phys. Rev.* **B68**, 205305 (2003).

Wójs, A. and P. Hawrylak, *Phys. Rev.* **B56**, 13227 (1997).

Yannouleas, C. and U. Landman, *Phys. Rev.* **B68**, 035326 (2003).

Chapter 2

The Hartree–Fock Theory

2.1 Introduction

When the interaction term of Eq. (1.1) is included in the theory, the many-Fermion (Boson) wave function is a combination of Slater determinants (symmetric products) that must be determined by means of variational or other, more sophisticated techniques. However, even in the presence of the interaction, the ground state of the system can be forced to be a single Slater determinant (symmetric product). This is the idea at the basis of the Hartree–Fock (HF) theory, which tries to describe some physical properties of the system with a wave function which, as in the independent-particle model (IPM), is the product of single-particle functions. The single Slater determinant (symmetric product) is chosen from among all the possible ones in order to minimize the energy of the system, calculated as the mean value of the Hamiltonian (1.1) (including the interaction) on the product of single-particle functions. Clearly, the only correlations included in the method are the statistical ones, deriving for example from the Pauli exclusion principle. Dynamic correlations, which are due to the interactions and are responsible for the divergence of the wave function from a product of single-particle functions and of the two-body density from the predictions of the IPM of Eqs. (1.49)–(1.52), are not included in the theory.

One important consequence of this fact is that the HF theory is applicable only to those systems in which the interaction is well behaved and, for example, it does not diverge too fast as the interparticle distance tends to zero. This is clearly seen by calculating, for example, the average value of a two-body potential in the Fermi gas model:

$$\left\langle \text{SD} \Big| \sum_{i<j}^{N} v(\mathbf{r}_{ij}) \Big| \text{SD} \right\rangle = \frac{1}{8}\rho N \sum_{\sigma_1 \sigma_2} \int g_{\sigma_1 \sigma_2}(\mathbf{r}) v(\mathbf{r}) d\mathbf{r}, \qquad (2.1)$$

where $r_{ij} = |\mathbf{r}_i - \mathbf{r}_j|$, ρ is the constant density of the gas (i.e. the diagonal part of the one-body density), while $\sum_{\sigma_1 \sigma_2} g_{\sigma_1 \sigma_2}$ is given by Eq. (1.90) in the 3D case. It is observed that whether the integral in the interaction term will diverge or not will depend on the shape of the potential $v(\mathbf{r})$ or on the dimensionality of the system under study. This is, for example, the case of the Lennard–Jones potential, which describes the interaction between two atoms and contains terms of the form $1/r^{12}$, or of the nuclear interaction with a hard core, or of the Coulomb potential for one-dimensional systems.

Therefore, we see that the HF method can be used only with potentials having a soft behavior, which do not cause the divergence of the integrals appearing in the average value of the Hamiltonian on the product functions of Eqs. (1.10) and (1.12).

One evolution of this method is the theory of Brueckner–Hartree–Fock, which introduces an effective potential g^{eff}, defined by

$$v|0\rangle = g^{\text{eff}}|\text{HF}\rangle, \qquad (2.2)$$

where $|0\rangle$ is the true ground state on which operates the true potential (which may be divergent), while g^{eff} is nondivergent and operates on a noncorrelated (mean field) wave function. This theory will be considered in Chapter 3.

2.2   The Hartree–Fock Method for Fermions

Let us calculate the energy of the system as the mean value of the Hamiltonian with two-body interaction of equation (1.1) on a generic Slater determinant, by using the formulas of the previous Chapter 1:

$$E = \langle \text{SD}|H|\text{SD}\rangle = \sum_{i\sigma} \int d\mathbf{r}\,\varphi_i^*(\mathbf{r},\sigma)\left(\frac{-\nabla^2}{2m} + v_{\text{ext}}(\mathbf{r})\right)\varphi_i(\mathbf{r},\sigma)$$

$$+ \frac{1}{2}\sum_{ij\sigma\sigma'}\int d\mathbf{r}d\mathbf{r}'\,\varphi_i^*(\mathbf{r},\sigma)\varphi_j^*(\mathbf{r}',\sigma')v(\mathbf{r}-\mathbf{r}')[\varphi_i(\mathbf{r},\sigma)\varphi_j(\mathbf{r}',\sigma')$$

$$- \varphi_i(\mathbf{r}',\sigma')\varphi_j(\mathbf{r},\sigma)], \qquad (2.3)$$

where φ_i are the single-particle wave functions which appear in the Slater determinant, which we will determine by using the HF theory.

Let us minimize the energy (2.3) with respect to arbitrary variations of the φ_i functions. Since the φ_i must be normalized, the quantity to be minimized is

$$E - \sum_{i\sigma}\varepsilon_i\int|\varphi_i|^2 d\mathbf{r}, \qquad (2.4)$$

where the ε_i are N Lagrange multipliers. If we require that Eq. (2.4) be stationary with respect to variations of φ_i^*,

$$\delta\left(E - \sum_{i,\sigma}\varepsilon_i\int|\varphi_i(\mathbf{r},\sigma)|^2 d\mathbf{r}\right) = 0 \qquad (2.5)$$

we obtain the HF equations :

$$\left(\frac{-\nabla^2}{2m} + v_{\text{ext}}(\mathbf{r})\right)\varphi_i(\mathbf{r},\sigma) + \sum_{j\sigma'}\int d\mathbf{r}'\varphi_j^*(\mathbf{r}',\sigma')v(\mathbf{r}-\mathbf{r}')$$

$$\times\left[\varphi_i(\mathbf{r},\sigma)\varphi_j(\mathbf{r}',\sigma') - \varphi_i(\mathbf{r}',\sigma')\varphi_j(\mathbf{r},\sigma)\right] = \varepsilon_i\varphi_i(\mathbf{r},\sigma). \qquad (2.6)$$

These equations can be put in a compact way by using the following relationships for the one-body diagonal and nondiagonal densities respectively:

$$\rho(\mathbf{r}') = \sum_{j\sigma'}|\varphi_j(\mathbf{r}',\sigma')|^2$$

and

$$[\rho^{(1)}(\mathbf{r},\sigma,\mathbf{r}',\sigma')]^* = \sum_j \varphi_j(\mathbf{r},\sigma)\varphi_j^*(\mathbf{r}',\sigma').$$

In this way we obtain

$$\left(\frac{-\nabla^2}{2m} + v_{\text{ext}}(\mathbf{r})\right)\varphi_i(\mathbf{r},\sigma) + U(\mathbf{r})\varphi_i(\mathbf{r},\sigma)$$

$$- \sum_{\sigma\|\sigma'}\int d\mathbf{r}'(\rho^{(1)}(\mathbf{r},\sigma,\mathbf{r}',\sigma'))^*v(\mathbf{r}-\mathbf{r}')\varphi_i(\mathbf{r}',\sigma') = \varepsilon_i\varphi_i(\mathbf{r},\sigma), \qquad (2.7)$$

where

$$U(\mathbf{r}) = \int d\mathbf{r}'\rho(\mathbf{r}')v(\mathbf{r}-\mathbf{r}'), \qquad (2.8)$$

and $\sigma\|\sigma'$ means that the spins must be parallel, since otherwise the exchange term vanishes.

Note that the HF potential

$$v^{\text{HF}}\varphi_i(\mathbf{r},\sigma) = v_{\text{ext}}(\mathbf{r})\varphi_i(\mathbf{r},\sigma) + U(\mathbf{r})\varphi_i(\mathbf{r},\sigma)$$

$$- \sum_{\sigma\|\sigma'}\int d\mathbf{r}'[\rho^{(1)}(\mathbf{r},\sigma,\mathbf{r}',\sigma')]^*v(\mathbf{r}-\mathbf{r}')\varphi_i(\mathbf{r}',\sigma') \qquad (2.9)$$

is a self-consistent field which depends on the φ_i functions which are occupied by the N particles, and that the exchange term [i.e. the last term in (2.9)] is nonlocal. In the Hartree theory only the local (direct) term $U(\mathbf{r})$ was considered.

The HF equations form a system of N coupled integral-differential equations. The ε_i have the physical meaning of single-particle energies.

The HF equations are solved by means of the *self-consistent iterative method*, i.e. starting from N known single-particle functions $\varphi_i^{(0)}$, which are used to determine

$v_{(0)}^{\mathrm{HF}}$. Equations (2.7) are then solved and N new wave functions are determined, by means of which a new potential, $v_{(1)}^{\mathrm{HF}}$, is built up. Equations (2.7) are subsequently solved again, and so on, until solutions are obtained which coincide with those used to set up the potential of the previous iterations (self-consistency). The Slater determinant obtained with the solutions of the last iteration is the HF ground state.

Properties of the Solutions

The solutions to the HF equations for the wave functions φ_i and the single-particle energies ε_i have the following properties:

- When self-consistency is attained, Eqs. (2.7) provide a complete set of solutions φ_i; only the N states corresponding to the lowest values of the ε_i are used to set up the Slater determinant that gives the HF ground state. However, the remaining states provide a natural basis for describing the $1p$–$1h$, $2p$–$2h$, ..., excited states.
- Wave functions φ_i corresponding to different energies are mutually orthogonal.
- The particles are non-self-interacting. The potential acting on the particle described by the state φ_i has no contribution from such a state, but only from the states corresponding to all other particles. This is due to the mutual cancellation of the direct and exchange terms in the HF potential (2.9) when $j = i$. This result follows from the anti-antisymmetrization of the wave function. In the Hartree theory, in which the exchange term is neglected, one has the unphysical interaction of a particle with itself.
- The mean value of the system Hamiltonian in the HF state is different from the sum of the single-particle energies ε_i corresponding to the occupied states in the HF ground state:

$$\langle \mathrm{HF}|H|\mathrm{HF}\rangle \neq \sum_i \varepsilon_i. \tag{2.10}$$

In fact, using Eqs. (2.3) and (2.6) we obtain

$$\langle \mathrm{HF}|H|\mathrm{HF}\rangle = \frac{1}{2}\left(\sum_i \varepsilon_i + \sum_i t_i + \sum_i v_i\right), \tag{2.11}$$

where

$$t_i = \sum_\sigma \int \varphi_i^*(\mathbf{r},\sigma)\left(\frac{-\nabla^2}{2m}\right)\varphi_i(\mathbf{r},\sigma)d\mathbf{r},$$

$$v_i = \sum_\sigma \int \varphi_i^*(\mathbf{r},\sigma)v_{\mathrm{ext}}(\mathbf{r})\varphi_i(\mathbf{r},\sigma)d\mathbf{r}. \tag{2.12}$$

Residual Interaction

Apart from a constant term, the Hartree–Fock equations define the HF Hamiltonian $H^{\mathrm{HF}} = \sum_i H_i^{\mathrm{HF}}$:

$$H_i^{\mathrm{HF}} \varphi_i = (\varepsilon_i + \mathrm{const.}) \varphi_i. \tag{2.13}$$

It is easily verified that the HF ground state is an eigenstate of H^{HF} and its eigenvalue is the Hartree–Fock energy: $E^{\mathrm{HF}} = \sum_i (\varepsilon_i + \mathrm{const.})$. But it is always possible to choose the value of the constant such that the HF energy coincides with the mean value of the two-body Hamiltonian H in the HF state:

$$E^{\mathrm{HF}} = \langle \mathrm{HF}|H|\mathrm{HF}\rangle = \langle \mathrm{HF}|H^{\mathrm{HF}}|\mathrm{HF}\rangle. \tag{2.14}$$

This means that if we write

$$\mathrm{H} = H^{\mathrm{HF}} + V_{\mathrm{res}}, \tag{2.15}$$

then V_{res} is a residual two-body interaction whose mean value in the HF state is zero:

$$\langle \mathrm{HF}|V_{\mathrm{res}}|\mathrm{HF}\rangle = 0. \tag{2.16}$$

Moreover, it is possible to show that the matrix elements of the residual interaction between the HF ground state and the $1p$–$1h$ excited states vanish as well:

$$\langle \mathrm{HF}|V_{\mathrm{res}}|i^{-1}, m\rangle = 0. \tag{2.17}$$

On the other hand, the matrix elements of the residual interaction between the HF ground state and the $2p$–$2h$ excited states do not vanish:

$$\langle \mathrm{HF}|V_{\mathrm{res}}|i^{-1}j^{-1}, mn\rangle \neq 0. \tag{2.18}$$

Equation (2.17) can be rewritten as:

$$\langle \mathrm{HF}|H|i^{-1}, m\rangle = 0, \tag{2.19}$$

and is a particular case of the more general condition

$$\langle \mathrm{HF}|[H, F]|\mathrm{HF}\rangle = 0, \tag{2.20}$$

where F is any one-body operator. This condition means that $|\mathrm{HF}\rangle$ is the Slater determinant which minimizes the system energy. In fact, if Eq. (2.20) did not hold, by using the transformation $e^{\alpha[H,F]}$ on the HF state it would be possible to lower the HF energy, which is impossible by definition of the HF state.

2.2.1   *Examples of Physical Systems Treated by the Hartree–Fock Method*

Metal Clusters

The HF method was employed initially in the field of atomic and molecular physics, and provided very good results for atoms with a small number of electrons and simple molecules like diatomic ones. In recent years the method has also been used for calculations on metal clusters and quantum dots. In the following we will describe the latter two systems, describing their main characteristics and showing the results of HF calculations.

Metal clusters, which we will consider here, are clusters of alkali metals which are obtained experimentally by evaporating the metal inside a container where a noble gas is also introduced, which acts as a catalyser (Knight *et al.*, 1985; de Heer *et al.*, 1987). If a microhole (of the order of a few μm) is made in the container, the gas (which initially has given values of pressure and temperature) undergoes a supersonic expansion which produces a lowering of temperature and the formation of metallic aggregates called clusters.

The resulting beam consists of different aggregates containing different numbers of atoms; it is of interest to measure the relative abundance of clusters of a given type as a function of the number N of constituting atoms. This kind of experiments (see Fig. 2.1) provides a spectrum of abundance, showing a series of peaks corresponding to some values of N, which are a sort of magic numbers analogous to those observed, for example, in the ionization energies of the atoms and nuclear separation.

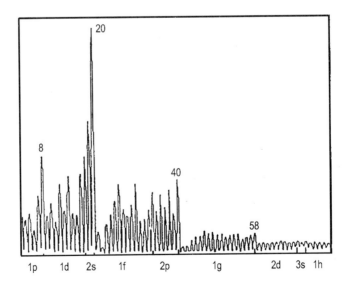

Fig. 2.1   Abundance spectrum for Na clusters.

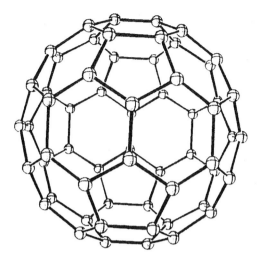

Fig. 2.2 Geometrical structure of fullerene C_{60}.

One of the best-known results obtained by this technique is the discovery of the fullerene (Kroto *et al.*, 1985; Kratschmer *et al.*, 1990), i.e. an allotropic state of carbon consisting of 60 atoms placed on the vertices of the icosahedral structure with 12 pentagonal and 20 hexagonal faces shown, in Fig. 2.2. This structure is the one used to make soccer balls, and this is the reason for the name "soccer-like carbon cluster" commonly used in the literature to indicate the fullerene.

The simplest model for these systems is the so-called jellium model (Ekardt, 1984; Yannouleas *et al.*, 1989; Bertsch *et al.*, 1991; Van Giai and Lipparini, 1992; Guet and Johnson, 1992; Lipparini *et al.*, 1994; Broglia *et al.*, 2002), obtained by considering the aggregate as a set of electrons (the valence electrons of the cluster atoms) which feel the electrostatic field produced by a homogeneous distribution of positive ions inside a tridimensional sphere of radius R (metal clusters) or on the surface of a sphere (fullerene). The electrons interact by the repulsive Coulomb force. This model is expected to work rather well, especially for alkali metal aggregates, where the single valence electron is nearly free. In this case (which is the only one that we consider in the following) the number of valence electrons coincides with the number N of atoms in the aggregate and with the number of ions as well. The ions are considered as rigid, in the sense that they do not interact with possible external fields.

Therefore, in the case of metal clusters the ion density can be schematized as follows:

$$\rho_I(\mathbf{r}) = \rho_b \Theta(R - \mathbf{r}), \tag{2.21}$$

where ρ_b is the bulk density, i.e. the constant density of ions (which coincides with the electron density) in the homogeneous system, and Θ is the step function. Using atomic units [where $m = e^2 = \hbar = 1$, the energy is in Hartree units H

$(1H = 27.2\,\text{eV})$, and the length is in units of the Bohr radius a_0 $(1a_0 = 0.529\,\text{Å}^\circ)]$, the Hamiltonian of the N electrons is given by

$$H = \sum_{i=1}^{N} \left(\frac{\mathbf{p}_i^2}{2} - \int \frac{\rho_I(\mathbf{r}')}{|\mathbf{r}_i - \mathbf{r}'|} d\mathbf{r}' \right) + \sum_{i<j}^{N} \frac{1}{|\mathbf{r}_i - \mathbf{r}_j|}, \qquad (2.22)$$

where the confinement potential of the ions

$$V_+(\mathbf{r}) = - \int \frac{\rho_I(\mathbf{r}')}{|\mathbf{r} - \mathbf{r}'|} d\mathbf{r}' \qquad (2.23)$$

for the distribution (2.21) is given by

$$V_+(\mathbf{r}) = 2\pi\rho_b \begin{cases} \dfrac{1}{3}r^2 - R^2 & \text{if } r \le R \\[2mm] -\dfrac{2R^3}{3r} & \text{if } r > R \end{cases}, \qquad (2.24)$$

and the radius of the sphere where the ions are uniformly distributed is fixed by the relation

$$\frac{4}{3}\pi\rho_b R^3 = N_I, \qquad (2.25)$$

being N_I being the number of ions in the cluster.

The HF equations for the Hamiltonian (2.22) become

$$\left[-\frac{\nabla^2}{2} + \int \frac{\rho(\mathbf{r}') - \rho_I(\mathbf{r}')}{|\mathbf{r} - \mathbf{r}'|} d\mathbf{r}' \right] \varphi_i(\mathbf{r}, \sigma)$$
$$- \sum_{\sigma \| \sigma'} \int d\mathbf{r}' (\rho^{(1)}(\mathbf{r}, \sigma, \mathbf{r}', \sigma'))^* \frac{1}{|\mathbf{r} - \mathbf{r}'|} \varphi_i(\mathbf{r}', \sigma') = \varepsilon_i \varphi_i(\mathbf{r}, \sigma), \qquad (2.26)$$

where

$$\rho(\mathbf{r}) = \sum_{i,\sigma} |\varphi_i(\mathbf{r}, \sigma)|^2$$

and

$$[\rho^{(1)}(\mathbf{r}, \sigma, \mathbf{r}', \sigma')]^* = \sum_j \varphi_j(\mathbf{r}, \sigma)\varphi_j^*(\mathbf{r}', \sigma').$$

Note that for the homogeneous system, to which the cluster tends when $N \to \infty$, one has $\rho_e = \rho_I$, and in the HF equation only the kinetic energy and exchange terms survive. It is then clear that the bonding is due to the exchange term, which is a quantum effect. The same holds for clusters (i.e. the nonhomogeneous system), because the exchange term is stronger than the sum of the ionic potential with the direct term deriving from electron-electron interaction.

In Fig. 2.3 we compare the single-particle spectra of sodium clusters formed by 20, 40, 58 and 92 atoms, as obtained by the Coulomb Hartree-Fock method

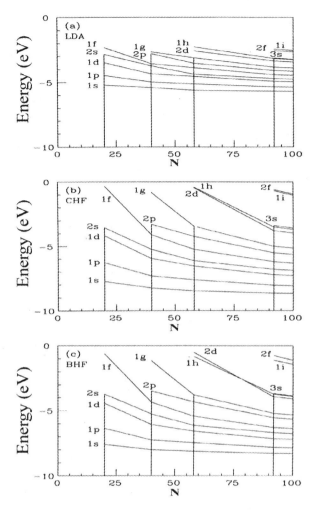

Fig. 2.3 Single-particle energies of Na clusters with 20,40,58 and 92 atoms in the spherical jellium model. Part (a) is within LDA, part (b) within Coulomb HF, and part (c) within BHF. The height of the vertical lines shows the filled levels in each cluster. The (n, l) of the levels are indicated.

(CHF), to those derived by more refined theories [Brueckner–Hartree–Fock (BHF); see Chapter 3] and by the density functional theory (LDA) (see Chapter 4), which take into account the dynamic interactions in an approximate way. In all kinds of calculations the confinement potential of Eq. (2.24) with $R = r_s N^{1/3}$ and Wigner–Seitz radius $r_s = 4$ (in atomic units) is used (r_s is by definition the radius in atomic units of the sphere which encloses one electron). It can be noted that the LDA spectrum is strongly compressed with respect to the other two, and that the gap between the highest occupied level and the lowest unoccupied one is much larger in HF and BHF than is in LDA. The strong bonding of lowest energy levels in HF and BHF derives from the fact that the Coulomb exchange term in the HF scheme depends strongly on the state. Inclusion of the correlation term in BHF causes a stronger compression of the spectrum than it does in HF, where the dynamic correlation terms are completely neglected. The difficulty that the LDA theory encounters in predicting that $N = 40$ is a magic number for sodium clusters (in agreement with the experimental abundance spectra), is absent in the HF scheme, where the $2s$–$1f$ and $2p$–$1g$ gaps are much larger than in LDA. On the other hand, as will be discussed in the following, the LDA theory is not expected to be a good theory for the single-particle spectrum, but only for the total energies and for the density of the ground state. In HF and BHF the highest energy occupied states have energy close to the ionization potentials determined experimentally, in agreement with the Koopmans theorem, which proves their coincidence in the $N \to \infty$ case. For example, in Na_{20} (sodium cluster with 20 atoms) the experimental value is 3.75 eV, which should be compared with the HF and BHF predictions of 3.54 and 3.72 eV, respectively. In Na_{40} the corresponding values are 3.60 eV, 3.28 eV and 3.47 eV, respectively. The LDA predictions are 4.03 eV in Na_{20} and 3.75 eV in Na_{40}.

In Table 2.1 we report the total energies per particle (in eV) for some sodium clusters and for the homogeneous system ($N = \infty$), as predicted by th HF calculation, together with the LDA and BHF results (Lipparini *et al.*, 1994). In the same table are also reported the Monte Carlo (see Chapter 7) results for the homogeneous system (Ceperley, 1978; Ceperley and Alder, 1980), and for Na_{20} (Ballone *et al.*, 1992). The Monte Carlo results should be considered as those of the best-possible calculation. From the table we can see the importance of the dynamic correlations in determining the total binding energy. It can also be noted that the HF and BHF results have an oscillatory behavior, and that BHF tends to the bulk value more slowly than the LDA theory.

In Fig. 2.4 we report the electronic densities of Na_{40}, computed by the three different methods. The three densities are similar but show some marked difference in the central part, reflecting the smaller amplitude of the s state wave function at the center in HF and BHF with respect to LDA. It can be shown that the inclusion in the HF density of the effects due to long range correlations (RPA; see Chapter 10) increases the density profile at the center, in agreement with the LDA approach.

As was discussed at the beginning of this subsection, the HF and LDA calculations, in the jellium approximation of Eqs. (2.21) and (2.23) for the electron–ion

Table 2.1. Total energies per particle (eV) for some sodium clusters and for the homogenous system ($N = \infty$) as predicted by different methods.

N	HF	BHF	LDA	MC
20	−1.237	−1.846	−2.019	−1.860
40	−1.200	−1.840	−2.048	
58	−1.241	−1.926	−2.093	
92	−1.237	−1.931	−2.119	
∞	−1.236	−2.035	−2.154	−2.154

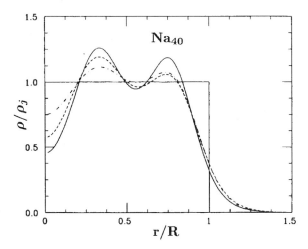

Fig. 2.4   Electronic density profiles in Na$_{40}$ in the three approaches. The dotted–dashed, dashed and full lines refer to the LDA, HF and BHF results, respectively. The right-angled line shows the jellium density ρ_j.

interaction potential, became of common use during the 1990s. In spite of its simplicity, the jellium model provides a good description of the optical properties of some metal clusters, in particular sodium and potassium ones (see Chapter 9). However, for other clusters like lithium ones, this approach cannot reproduce experimental data. This fact urged some authors (Blundell and Guet, 1993; Serra *et al.*, 1993; Yabana and Bertsch, 1995; Lipparini *et al.*, 1994; Alasia *et al.*, 1995) to modify the electron–ion potential of the jellium model so as to take into account the effects of ionic structure. This was accomplished by means of the use of pseudo-potentials, which during the atomic calculations allow an atom and its core to be replaced by a pseudoatom with only the valence electrons. In this way, Eq. (2.23) is replaced by the expression $V_+(\mathbf{r}) = \int \rho_I(\mathbf{r}')v_{ps}(|\mathbf{r} - \mathbf{r}'|)d\mathbf{r}'$, where the ionic density is still given by Eq. (2.21), i.e. the atoms are still uniformly distributed on a sphere, but instead of describing the electron–ion interaction by $-|\mathbf{r} - \mathbf{r}'|^{-1}$ one uses the pseudopotential $v_{ps}(|\mathbf{r} - \mathbf{r}'|)$. In the literature this model is commonly named pseudojellium. In order to show the differences between the jellium and pseudojellium results, we report in Table 2.2 the LDA values of the static polarizability of some metal clusters, normalized to the classical polarizability $\alpha_{\mathrm{cl}} = R^3 = r_s^3 N_I$, in the different approximations. The LDA polarizability is obtained by adding to the Kohn–Sham equations (see Chapter 4) a static electric dipole constraint (see also Chapter 9).

The values $r_s = 3.25, 3.93$ and 4.86 a.u. have been used for Li, Na and K, respectively. The experimental values were taken from Knight *et al.* (1985). PPJM indicates the approximation with pseudopotential of Bachelet *et al.* (1982), PHJM indicates the approximation with the pseudo-Hamiltonian of Bachelet *et al.* (1989). As can be seen from the table, the pseudopotential and pseudo-Hamiltonian models,

Table 2.2. LDA values of the static polarizability of some metal clusters, normalized to the classical polarizability, in the different approximations (see text).

$\alpha/\alpha_{\rm cl}$	PPJM	PHJM	JM	Expt.
Na_{20}	1.92	1.90	1.37	1.76 ± 0.10
Na_{40}	1.75	1.73	1.33	1.71 ± 0.10
K_{20}	1.92	1.91	1.28	1.73 ± 0.20
Li_{20}	2.14	2.07	1.47	
Li_{139}^{+}	1.48	1.45	1.20	

when compared to the jellium model (JM), reduce a great deal the discrepancy between experimental and LDA results.

Quantum Dots

Recently, nanotechnologies (1 nanometer $= 10^{-9}$ meters) based on semiconductor heterostructures (for example, $GaAs/Ga_xAl_{1-x}As$) allowed the realization of quantum wires and dots: these are specific objects where to study the physics of low dimensionality electrons. Quantum wires and dots are realized by the lateral confinement of the two-dimensional electron gas (2DEG) at the boundary of the heterostructure. The procedure to construct these systems can be explained schematically, as follows. The heterostructure is the union of two materials having different electronic properties. In general, the bottom of the conduction band E_C and the top of the valence band E_V lie at different energies in the two materials. Therefore, there is a discontinuity in both E_C and E_V at the contact surface. Electric equilibrium between the two materials is reached when the Fermi levels E_F of the two materials coincide. This equilibrium can be reached by a charge (electron) transfer, which produces a potential gap in the contact area, which in turn levels the E_F's, and at the same time bends the conduction and valence bands of the semiconductor. In some cases, between the curvature and the discontinuity it is possible that in the conduction band a quantum hole is produced, in which the electric charges are trapped being at an energy lower than E_F. The trapped electrons are obliged to move in a very limited region and, in practice, in the contact plane between the two semiconductors. In this way, we have a two-dimensional electron gas. Using lithographic techniques, it is then possible to make structures (wires or dots) having a size of the order of, or less than, a few hundred nanometers. For a recent review in quantum dots, see Reimann and Manniner (2002). At present, these techniques allow the construction of quantum dots, or artificial atoms, with a number N of electrons ranging from zero to a few hundred.

This enabled detailed study of the correlated ground state of the quantum dot; the many-particle ground state strongly depends on the number N of electrons

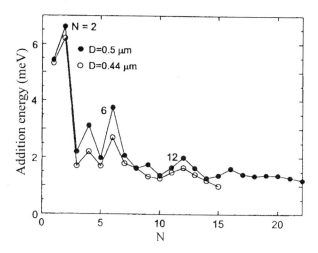

Fig. 2.5 Addition energy (Tarucha *et al.*, 1996) as a function of the number of electrons for two dots of different sizes.

and obeys the Hund rules, well known in atomic physics. These characteristics are particularly evident in the experiments (see Fig. 2.5) on addition energy (Tarucha *et al.*, 1996) necessary for putting an additional electron on the quantum dot, $\Delta_N = \mu(N+1) - \mu(N)$, where $\mu(N) = E(N) - E(N-1)$ is the chemical potential of the system.

The theoretical predictions of the $E(N)$ energy are based on the study of the following Hamiltonian:

$$H = H_0 + H_{\text{int}} = \sum_{i=1}^{N} \left(\frac{p_i^2}{2m} + v_{\text{ext}}(\mathbf{r}_i) \right) + \frac{e^2}{\epsilon} \sum_{i<j}^{N} \frac{1}{|\mathbf{r}_i - \mathbf{r}_j|}, \qquad (2.27)$$

which describes the motion of N electrons confined in the $z = 0$ plane $[\mathbf{r} \equiv (r, \vartheta)]$. In Eq. (2.27), $m = m^* m_e$ is the effective mass of the electron inside the semiconductor, ϵ is the dielectric constant of the semiconductor, and $v_{\text{ext}}(\mathbf{r})$ is the confinement potential, which in the following will be of parabolic shape:

$$v_{\text{ext}}(\mathbf{r}) = m\omega_0^2 r^2 / 2.$$

Moreover, in the following we will employ effective atomic units or dot units (d.u.), defined by $\hbar = e^2/\varepsilon = m = 1$. In this system of units the length unit is Bohr radius a_0 times ε/m^*, that we will indicate by a_0^*, and the energy unit H^* is the Hartree times m^*/ε^2. For GaAs quantum dots which we consider in this section, we have $\varepsilon = 12.4$, $m^* = 0.067$, $a_0^* = 97.94$ Å and $H^* \simeq 11.86$ meV.

First of all, let us check that the independent-particle model does not provide results for Δ_N which agree with experimental data. We solve the eigenvalue problem

$$H_0 \varphi_i(\mathbf{r}, \sigma) = \varepsilon_i \varphi_i(\mathbf{r}, \sigma). \qquad (2.28)$$

Since $[H, L_z] = [H_0, L_z] = 0$, we can choose common eigenfunctions for H_0 and for the z component of the angular momentum $L_z = \sum_{i=1}^{N} l_{z_i}$, where l_{z_i} is the single-particle angular momentum. Remembering that in polar coordinates $l_z = -i\frac{\partial}{\partial\vartheta}$, we find that the eigenfunctions of l_z are $e^{-il\vartheta}$ with eigenvalues $l = 0, \pm1, \pm2, \ldots$ and that the eigenfunctions of H_0 may be written as

$$\varphi_{n,l,m_s}(\mathbf{r}, \sigma) = R_{nl}(\mathbf{r})e^{-il\vartheta}\chi_{m_s}(\sigma), \qquad (2.29)$$

where R_{nl} is the radial part which depends on the n and l quantum numbers.

Subsequent solution of the Schrödinger equation for R_{nl} with a confinement potential given by the two-dimensional harmonic oscillator provides the single-particle energies:

$$\varepsilon_{nl} = (2n + 1 + |l|)\omega_0 = (N + 1)\omega_0, \qquad (2.30)$$

with $N = 0, 1, 2, \ldots$. The filling pattern of the orbitals is schematized in Table 2.3:

From the table we see that the independent-particle model predicts the existence of the magic numbers: 2, 6, 12, 20 etc. (note that these are different from those of the tridimensional case, and that dimensionality plays an important role). Using $E(N) = \sum_{i=1}^{N} \varepsilon_i$, which holds in the independent-particle model, it is easily found that in such a model the addition energy has peaks of height ω_0 at the magic numbers, while it vanishes for the other values of N. The agreement with experimental data of Fig. 2.5 is quite poor and all the features observed in the figure are due to interaction.

The HF equations for the Hamiltonian (2.27) are written as

$$\left(\frac{-\nabla^2}{2} + \frac{\omega_0^2 r^2}{2} + \int \frac{\rho(\mathbf{r}')}{|\mathbf{r} - \mathbf{r}'|}d\mathbf{r}'\right)\varphi_i$$
$$- \sum_{\sigma\|\sigma'} \int d\mathbf{r}' (\rho^{(1)}(\mathbf{r}, \sigma, \mathbf{r}', \sigma'))^* \frac{1}{|\mathbf{r} - \mathbf{r}'|}\varphi_i(\mathbf{r}', \sigma') = \varepsilon_i\varphi_i. \qquad (2.31)$$

In what follows we discuss the results obtained on the basis of their solutions.

In Table 2.4 we report the HF energies (in H^* units and with $\omega_0 = 0.28$) for the GaAs dots with $N = 2, 3, 4, 6, 8, 10, 13$, and compare them to the LSDA and Monte Carlo (MC) results (Pederiva *et al.*, 2000).

Table 2.3. Filling pattern of the orbitals for the two-dimensional harmonic oscillator potential.

$N = 2n + |l|$	n, l	degeneracy
0	$(0,0)$	2
1	$(0, \pm1)$	4
2	$(1, 0), (0, \pm2)$	6
3	$(1, \pm1), (0, \pm3)$	8
.	$\cdots$	.

Table 2.4. Comparison of ground state energies (in H*) for the dots with $2 \leq N \leq 13$ computed by the Hartree–Fock, LSDA and Monte Carlo method.

N	HF	LSDA	MC
2	1.1420	1.0468	1.02167(7)
3	2.4048	2.2631	2.2339(3)
4	3.9033	3.6864	3.7135(5)
6	8.0359	7.6349	7.5996(8)
8	13.1887	12.7276	12.6903(7)
10	19.4243	18.7636	18.7244(5)
13	30.4648	29.5363	29.4942(7)

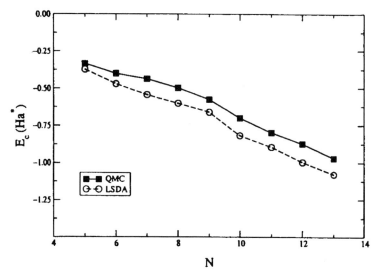

Fig. 2.6 Correlation energies E_c for a series of parabolic dots, computed by diffusion Monte Carlo (full squares) and the LSDA (open circles).

In Fig. 2.6, we plot the correlation energy computed as the difference between the MC total energy and the HF energy, as a function of the number of electrons N in the dot. The dashed line indicates the LSDA results, and gives an idea of how good the local density approximation (see Chapter 4) is.

From the table we see that HF overestimates by some percent the MC energy.

In Fig. 2.7, we compare the electronic densities for $N = 20$ in HF, BHF, LDA and diffusion Monte Carlo (DMC). The agreement between the LDA and MC curves is very good, and covers the whole r range, including the boundary where the density gradients are large. It is also seen that the BHF theory definitely improves the HF results when compared to MC.

Finally, in Fig. 2.8 we plot the addition energies of the various methods. Structures and peaks are observed at the electron numbers $2, 4, 6, 9$ and 12, in good

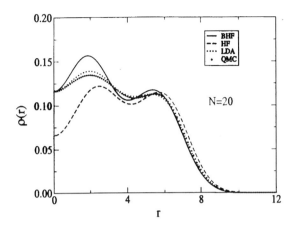

Fig. 2.7 Electronic density profiles for a dot with 20 electrons in different theories.

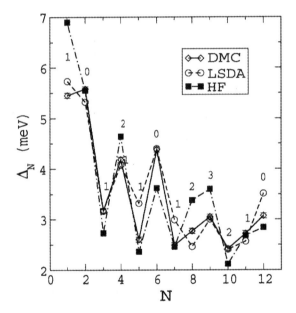

Fig. 2.8 Addition energies Δ_N as a function of the number of electrons in the dot. In the plot are reported the spin polarizations $P = N_\uparrow - N_\downarrow$, which are the same for all curves.

agreement with the experimental results of Fig. 2.5. The spin polarizations $P = N \uparrow - N \downarrow$ are also reported,

$$P = \left\langle 0 \Big| \sum_i \sigma_i^z \Big| 0 \right\rangle, \qquad (2.32)$$

which turn out to be the same for all curves, in agreement with Hund's rule, according to which the total spin takes on the maximum value consistent with the fact that the electrons are all in the same shell and with Pauli's principle. For HF this is a consequence of the fact that the exchange term is attractive and different from zero only for parallel spins.

2.2.2 Examples of Infinite Systems Treated by the Hartree–Fock Method

Electron Gas

The interacting-electron gas model can be used with good results to describe in three dimensions the valence electrons of an alkali metal. The electrons are considered as quasifree electrons which move in the electrostatic field produced by the uniform distribution of the positive charge metallic ions, and which interact through the Coulomb force.

In two dimensions, an analogous model is used to describe the electrons which are confined on the boundary surface of a semiconductor heterostructure (quantum well).

In the case of an infinite system the HF equations become particularly simple since it is possible to exploit the translational invariance of the system, which requires $\rho_I = \rho_e = \text{const}$. The constant-density solutions to the HF equations exist and are plane waves

$$\varphi_i(\mathbf{r}, \sigma) = \frac{1}{\sqrt{L^D}} e^{i\mathbf{k}\cdot\mathbf{r}} \chi_{m_s}(\sigma),$$

where D is the system dimensionality.

The electron gas is described by Hamiltonian (2.22), and the HF equations (2.26) for the homogeneous system become

$$-\frac{\nabla^2}{2}\varphi_i(\mathbf{r}, \sigma) - \frac{1}{(2\pi)^D} \int d\mathbf{r}' \int d\mathbf{k}' n_{\mathbf{k}'} \frac{e^{i\mathbf{k}'\cdot(\mathbf{r}-\mathbf{r}')}}{|\mathbf{r}-\mathbf{r}'|}\varphi_i(\mathbf{r}', \sigma) = \varepsilon_i\varphi_i(\mathbf{r}, \sigma), \qquad (2.33)$$

where for the one-body nondiagonal density we have used the result of Eq. (1.84).

By inserting the plane wave solutions in equation (2.33) we get the following result for the single-particle energies:

$$\varepsilon_{\mathbf{k}} = \frac{k^2}{2} - \frac{1}{(2\pi)^D} \int d\mathbf{k}' n_{\mathbf{k}'} v(|\mathbf{k}-\mathbf{k}'|), \qquad (2.34)$$

where $v(|\mathbf{k}-\mathbf{k}'|) = 4\pi/|\mathbf{k}-\mathbf{k}'|^2$ in 3D, and $v(|\mathbf{k}-\mathbf{k}'|) = 2\pi/|\mathbf{k}-\mathbf{k}'|$ in 2D.

The second term in equation (2.34) is the self-energy term $\Sigma_x(k)$ due to the exchange interaction, which introduces a k dependence of ε, which in turn modifies the quadratic law of the kinetic term. For example, in 3D we obtain

$$\Sigma_x(k) = -\frac{k_F}{\pi}\left(1 + \frac{1-y^2}{2y}\ln\left|\frac{1+y}{1-y}\right|\right), \qquad (2.35)$$

with $y = k/k_F$.

The self-energy (2.35) produces an effective mass defined by $(\varepsilon_k^0 = k^2/2)$

$$\frac{1}{m^*} = \frac{1}{k}\frac{\partial \varepsilon_k}{\partial k} = 1 + \frac{\partial \Sigma_x(k)}{\partial \varepsilon_k^0}, \qquad (2.36)$$

which in the present case turns out to be given by

$$\frac{1}{m^*} = \frac{1}{2\pi k_F}\frac{1}{y^2}\left(\frac{1+y^2}{2y}\ln\left|\frac{1+y}{1-y}\right| - 2\right). \qquad (2.37)$$

The effective mass in Eq. (2.37) diverges at the Fermi energy, where $y \to 1$. The cause of this logarithmic divergence is the long range nature of the Coulomb interaction, together with the HF approximation. It is a nonphysical divergence, and is removed in theories which treat the Coulomb interaction at higher order than HF, such as the RPA which will be discussed later on. In this regard, it should be noted that for homogeneous systems (and only for them) the HF theory coincides with the first-order perturbation theory (in the interaction). As for the Coulomb energy, the second order perturbation correction for the ground state energy diverges, and the lowest-order nontrivial and consistent approximation for the electron gas energy is given by the RPA theory; the latter sums a series of diagrams which are individually divergent but whose sum is finite.

The electron gas HF energy per particle is obtained by summing to the kinetic energy per particle of the Fermi gas the exchange energy given by

$$\frac{E_x}{N} = \frac{1}{\rho}\int\frac{d\mathbf{k}}{(\pi)^D}n_\mathbf{k}\Sigma_x(k). \qquad (2.38)$$

From Eq. (2.38), we see that the self-energy $\Sigma_x(k)$ can be obtained from E_x by taking the functional derivative of E_x with respect to $n_\mathbf{k}$.

The electron gas exchange energy per particle in different dimensions is given by

$$\frac{E_x}{N} = -\frac{2}{\pi}\frac{D}{D^2-1}\frac{1}{\alpha r_s}, \qquad (2.39)$$

where the Wigner–Seitz radius r_s is connected to the density by

$$r_s = \begin{cases} (3/4\pi\rho)^{1/3} & \text{for } D = 3 \\ (1/\pi\rho)^{1/2} & \text{for } D = 2 \\ 1/2\rho & \text{for } D = 1 \end{cases} \qquad (2.40)$$

and

$$\alpha = \begin{cases} (4/9\pi)^{1/3} & \text{for } D = 3 \\ 1/\sqrt{2} & \text{for } D = 2 \end{cases}. \qquad (2.41)$$

Note that in the 1D case the exchange integral is not defined, and one should introduce the width a and change the potential $1/\sqrt{(x-x')^2}$ with

$$1/\sqrt{(x-x')^2 + (y-y')^2} \simeq 1/\sqrt{(x-x')^2 + a^2}.$$

These cases are called quasi-1D (Q1D) cases, and have applications in the quantum wires. For the potential $\sqrt{(x-x')^2 + a^2}$, one obtains

$$\frac{E_x^{Q1D}}{N} = -\frac{k_F}{2\pi} \int_{-\infty}^{+\infty} \frac{j_0^2(k_F x)}{\sqrt{x^2 + a^2}} dx, \qquad (2.42)$$

where $j_0(y) = \sin(y)/y$. A more sophisticated approximation for the Q1D electron gas is obtained by the substitution

$$1 \Big/ \sqrt{(x-x')^2 + (y-y')^2} \rightarrow \int_{-\infty}^{+\infty} 1 \Big/ \sqrt{(x-x')^2 + (y-y')^2} |\varphi(y)|^2 |\varphi(y')|^2 dy dy'$$

of the 2D potential with the direct matrix element in the y direction, between single-particle states which are the lowest-energy states of some confinement potential.

The electron gas exchange energy per particle can also be put in an analytic form for the case of a partially polarized system with polarization ξ. Using the formalism of Subsection 1.8.2, we obtain:

$$\frac{E_x}{N}(r_s, \xi) = -\frac{2\sqrt{2}}{3\pi r_s}[(1+\xi)^{3/2} + (1-\xi)^{3/2}] \quad \text{(in 2D)} \qquad (2.43)$$

and

$$\frac{E_x}{N}(r_s, \xi) = -\frac{3}{8\pi r_s}\left(\frac{9\pi}{4}\right)^{1/3}[(1+\xi)^{4/3} + (1-\xi)^{4/3}] \quad \text{(in 3D)} \qquad (2.44)$$

In the 3D case, the HF energy per particle for the nonpolarized system can be written as (in Hartree units)

$$\frac{E_{3D}^{HF}}{N} = \frac{1.105}{r_s^2} - \frac{0.458}{r_s}, \qquad (2.45)$$

where the two contributions are the kinetic and exchange ones, respectively. One may wonder about the limits of validity of such an expression. To this end, let us consider the system Hamiltonian and write it in terms of the variable $\xi = \mathbf{r}/r_s$:

$$H = \sum_{i=1}^{N}\left(-\frac{1}{2}\frac{\partial^2}{\partial \mathbf{r}_i^2}\right) + \sum_{i<j}^{N}\frac{1}{|\mathbf{r}_i - \mathbf{r}_j|} = \frac{1}{r_s^2}\left[\sum_{i=1}^{N}\left(\frac{-\partial^2}{2\partial \xi_i^2}\right) + r_s\sum_{i<j}^{N}\frac{1}{|\xi_i - \xi_j|}\right]. \qquad (2.46)$$

For large density ($r_s \rightarrow 0$), and from Eq. (2.46), we see that the kinetic term dominates the potential one. Moreover, the exchange term gives the first-order correction in the potential, and from Eq. (2.45) it is intuitive that in this limit the ground state energy is a power series of the variable r_s. Although it is in general very unreliable to extrapolate from only two terms, this is just what happens in this case: for small r_s the energy actually is a series in r_s. This result was obtained by Gell-Mann and Brueckner (1957), and is known as the result of the RPA theory:

$$\frac{E_{3D}^{RPA}}{N} = \frac{1.105}{r_s^2} - \frac{0.458}{r_s} - 0.047 + 0.0311 \ln(r_s) + \cdots. \tag{2.47}$$

If we define the correlation energy as the difference between the ground state energy as provided by a given theory, and the HF energy, $E_c/N = E/N - E_{HF}/N$, we see that the RPA correlation energy is given by

$$E_c^{RPA}(3D) = -0.047 + 0.0311 \ln(r_s) + O(r_s).$$

For small density ($r_s \to \infty$), the kinetic term is negligible with respect to the potential one, and the system tends to find a configuration such as to minimize the repulsive Coulomb interaction. Wigner first showed that the most effective way to minimize this energy is to localize the electrons on the sites of a lattice. This phenomenon, which is called Wigner crystallization, shows that the lowest-energy solution may lose the initial (translational) symmetry of the Hamiltonian (see Subsection 4.11). For a cubic lattice Wigner found that in the $r_s \to \infty$ limit the potential energy of the lattice is given by $-0.9/r_s$. This is the total Coulomb energy, i.e. including exchange and correlation. If we subtract from this the exchange energy $E_x = -0.46/r_s$, we find a correlation energy given by $E_c^{WIG}(3D) = -0.44/r_s$.

Metals have values of r_s in the range 2–6. It is then clear that the two limiting cases cannot be applied to physical systems. Therefore, Wigner produced a formula for the correlation energy that extrapolates between the high-density and low-density limits,

$$\frac{E_c(3D)}{N} = -\frac{0.44}{r_s + 7.8}, \tag{2.48}$$

which was largely employed in past years. Nowadays, interpolations obtained from Monte Carlo results (see Subsection 4.5) are normally used.

For the two-dimensional electron gas the energies which correspond to Eqs. (2.45) and (2.47) are (see for example Isihara, 1989)

$$\frac{E_{2D}^{HF}}{N} = \frac{0.5}{r_s^2} - \frac{0.6}{r_s}, \tag{2.49}$$

$$\frac{E_{2D}^{RPA}}{N} = \frac{0.5}{r_s^2} - \frac{0.6}{r_s} - 0.192 - 0.086 r_s \ln(r_s) + \cdots. \tag{2.50}$$

From these we see that the kinetic energy of the system in two dimensions is about 1/2 of the 3D case at the same r_s, while the exchange energy in 2D is slightly larger than in 3D, and the correlation energy in RPA is much larger. The RPA correlations are stronger in 2D than in 3D.

2.3 The Hartree–Fock Method for Bosons

In the case of a system of N Bosons with Hamiltonian

$$H = \sum_{i=1}^{N}\left(\frac{p^2}{2m} + v_{\text{ext}}\right)_i + \sum_{i<j}^{N} v(\mathbf{r}_i, \mathbf{r}_j),$$

the HF equations are obtained starting from the energy, obtained as the mean value of H on the following symmetrized product of single-particle wave functions:

$$\phi(\mathbf{r}_1\mathbf{r}_2,\ldots,\mathbf{r}_N) = \varphi(\mathbf{r}_1)\varphi(\mathbf{r}_2)\ldots\varphi(\mathbf{r}_N), \tag{2.51}$$

where the φ are all equal. Using the formulas of chapter 1, we find that

$$\langle\phi|H|\phi\rangle = N\int \varphi^*(\mathbf{r})\left(-\frac{\nabla^2}{2m}\right)\varphi(\mathbf{r})d\mathbf{r} + N\int|\varphi(\mathbf{r})|^2 v_{\text{ext}} d\mathbf{r}$$
$$+ \frac{N(N-1)}{2}\int\int d\mathbf{r}d\mathbf{r}'\varphi^*(\mathbf{r})\varphi^*(\mathbf{r}')v(\mathbf{r}-\mathbf{r}')\varphi(\mathbf{r})\varphi(\mathbf{r}'). \tag{2.52}$$

Note that the structure is very similar to the one for the Fermion case, but it is much simpler since the single-particle wave functions are all equal in the ground state; in the last term the exchange part, produced by the antisymmetrization of the Fermion wave function, is obviously missing.

If we apply the variational principle with the constraint that the single-particle wave function is normalized,

$$\delta(E - \mu N\int|\varphi(\mathbf{r})|^2 d\mathbf{r}) = 0, \tag{2.53}$$

where μ is the only Lagrange multiplier that appears in the Boson case, and its physical meaning is that of a chemical potential, we find the HF equation:

$$\left\{-\frac{\nabla^2}{2m} + v_{\text{ext}} + \frac{N-1}{N}\int d\mathbf{r}'\rho(\mathbf{r}')v(\mathbf{r}-\mathbf{r}')\right\}\varphi(\mathbf{r}) = \mu\varphi(\mathbf{r}). \tag{2.54}$$

This equation looks like a nonlinear Schrödinger equation, where the non-linearity is derived from the mean field term proportional to the density $\rho(\mathbf{r}) = N|\varphi(\mathbf{r})|^2$. The solution to Eq. (2.54) is obtained by iteration (with the self-consistent method). The HF energy is given by

$$E^H = \frac{1}{2m}\int|\nabla\sqrt{\rho(\mathbf{r})}|^2 d\mathbf{r} + \int v_{\text{ext}}\rho(\mathbf{r})d\mathbf{r} + \frac{N-1}{2N}\int d\mathbf{r}d\mathbf{r}'\rho(\mathbf{r})\rho(\mathbf{r}')v(\mathbf{r}-\mathbf{r}'), \tag{2.55}$$

where the density is the self-consistent one. The first term of equation (2.55) corresponds to the quantum kinetic energy due to the uncertainty principle. It is usually named "quantum pressure" and vanishes for homogeneous systems.

By direct integration of equations (2.54) one finds that

$$\mu = \int d\mathbf{r}\varphi^*(\mathbf{r}) \left\{ -\frac{\nabla^2}{2m} + v_{\text{ext}} + \frac{N-1}{N} \int d\mathbf{r}'\rho(\mathbf{r}')v(\mathbf{r}-\mathbf{r}') \right\} \varphi(\mathbf{r}), \qquad (2.56)$$

that is,

$$\mu = \frac{1}{N}(E_{\text{kin}} + 2E_{\text{int}} + E_{\text{ext}}), \qquad (2.57)$$

where E_{kin} is the kinetic energy, E_{int} is the interaction energy and E_{ext} is the energy of the external field.

2.4   The Gross–Pitaevskii Equations

The solutions to the HF equations for Bosons, in general, are a worse starting point for more refined theories than in the Fermion case, since they contain no correlation of any kind (not even the statistical ones due to the Pauli principle which are present in the Fermion solutions). There exists, however, a case in which they provide an extremely good solution to the problem, i.e. the case of the diluted and cold atomic gas which has recently been realized with the Bose–Einstein condensation (BEC) of alkali atoms in magnetic traps. For such systems, only low-energy collisions between pairs of atoms are important. These collisions are characterized by only one parameter, the s-wave scattering length, irrespective of the details of the two-body potential. This allows the two-body interaction in the Hamiltonian to be replaced by an effective interaction given by:

$$V(\mathbf{r}-\mathbf{r}') = g\delta(\mathbf{r}-\mathbf{r}'), \qquad (2.58)$$

where the coupling constant g is related to the scattering length a by

$$g = \frac{4\pi\hbar^2 a}{m}. \qquad (2.59)$$

This replacement works well in the limit where the (diluted gas) condition holds: $\rho|a|^3 \ll 1$.

Using the potential (2.58) in the HF equations (2.54), and taking

$$(N-1)/N = 1,$$

we obtain the Gross–Pitaevskii equations (Gross, 1961; Pitaevskii, 1961) (see also Subsection 9.6) for the ground state,

$$\left\{ -\frac{\nabla^2}{2m} + v_{\text{ext}} + g\rho(\mathbf{r}) \right\} \varphi(\mathbf{r}) = \mu\varphi(\mathbf{r}), \qquad (2.60)$$

which are known to provide the good theory for diluted systems (see Dalfovo *et al.*, 1999). For example, the corrections to the mean field energy due to dynamic correlations behave as $(\rho|a|^3)^{1/2}$ and are typically of the order of 1% (see Chapter 10).

The numerical solutions to the Gross–Pitaevskii equations, obtained using Eq. (2.60), in the case of a harmonic oscillator external potential with parameter ω_{ho}, and for different values of the parameter $N|a|/a_{ho}$ $[a_{ho} = (\hbar/m\omega_{ho})^{1/2}]$, are plotted in Fig. 2.9 for attractive interaction among the atoms $(a < 0)$ and in Fig. 2.10 for repulsive interaction $(a > 0)$. The effects of interaction are manifested in the departure from the Gaussian shape predicted by the model without interaction. Comparison of these solutions with the low-temperature experimental profiles, as well as with Monte Carlo calculations, gives excellent agreement.

If the forces are attractive $(a < 0)$, the system density tends to increase at the trap center in order to lower the interaction energy (see Fig. 2.9). This tendency is balanced by the kinetic term which can stabilize the system. In any case, if the central density increases too much the kinetic term cannot avoid gas collapse. For a given kind of atoms, and of external potential (i.e. the trap), one expects collapse when the number of particles in the condensate overcomes a given critical value N_{cr} of the order of $a_{ho}/|a|$. Clearly, in the homogeneous system where quantum pressure due to the kinetic term is absent, the condensate is always unstable. N_{cr}

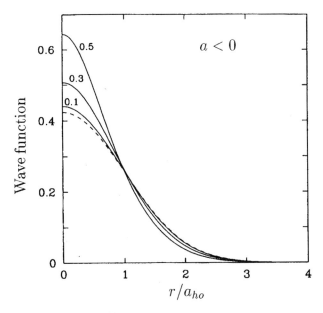

Fig. 2.9 Wave function of the condensate, at $T = 0$, obtained by numerical solution of the Gross–Pitaevskii equations (2.60) in a spherical harmonic oscillator trap with negative scattering length. The three curves correspond to $N|a|/a_{ho} = 0.1, 0.3$ and 0.5. The dashed line is the ideal gas prediction. The curves are normalized to 1.

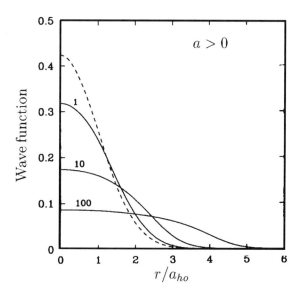

Fig. 2.10   The same as Fig. 2.9, but for repulsive interaction ($a > 0$) and $N|a|/a_{ho} = 1, 10$ and 100.

can be calculated by the Gross–Pitaevskii equations and turns out to be given by $N_{cr} = 0.575 a_{ho}/|a|$, in good agreement with experimental data.

2.5   Hartree–Fock in Second Quantization Language

Given a complete set of single-particle states φ_i, in second quantization formalism (see the appendix at the end of the chapter) the Slater determinant built with the $\varphi_1 \ldots \varphi_N$ states is written as

$$|\text{SD}\rangle = c_1^+ \cdots c_N^+ |\rangle, \tag{2.61}$$

where $|\rangle$ is the particle vacuum state and c_δ^+ are the creation operators in the state φ_δ. The requirement that $|\text{SD}\rangle$ be antisymmetric produces the following anticommutation relations for the c and c^+ operators:

$$\{c_\delta, c_\gamma^+\} = \delta_{\lambda,\gamma}, \quad \{c_\delta, c_\gamma\} = 0, \quad \{c_\delta^+, c_\gamma^+\} = 0, \tag{2.62}$$

where c_δ, the Hermitian conjugate of c_δ^+, is the annihilation operator of the particle in state δ. The operator $n_\delta = c_\delta^+ c_\delta$ has the property $n_\delta^2 = n_\delta$, and gives the occupation number of state δ, which can be either 0 or 1.

In the case of Boson systems, the function $|\Phi\rangle$, which is the product of single-particle functions φ_i, is written as

$$|\Phi\rangle = a_1^+ \cdots a_N^+ |\rangle, \tag{2.63}$$

where $|\rangle$ is again the particle vacuum state, and a_δ^+ and a_δ are the creation and annihilation operators of a Boson in state φ_δ, analogous to c_δ^+ and c_δ. The requirement that $|\Phi\rangle$ be symmetric leads to commutation relations for the a and a^+ operators, instead of the anticommutation relations of the Fermion case:

$$[a_\delta, a_\gamma^+] = \delta_{\lambda,\gamma}, \quad [a_\delta, a_\gamma] = 0, \quad [a_\delta^+, a_\gamma^+] = 0. \tag{2.64}$$

In second quantization language the basic Hamiltonian for a many-Fermion system [see Eq. (1.1)] is written as

$$H = \sum_{\delta\gamma} h_{\delta\gamma} c_\delta^+ c_\gamma + \sum_{\delta\gamma\alpha\beta} \frac{1}{2} V_{\delta\gamma\alpha\beta} c_\gamma^+ c_\beta^+ c_\alpha c_\delta, \tag{2.65}$$

where

$$h_{\delta\gamma} = \sum_\sigma \int d\mathbf{r}\, \varphi_\delta^*(\mathbf{r},\sigma) \left(\frac{-\nabla^2}{2m} + v_{\text{ext}}(\mathbf{r}) \right) \varphi_\gamma(\mathbf{r},\sigma) \tag{2.66}$$

and

$$V_{\delta\gamma\alpha\beta} = \sum_{\sigma\sigma'} \int d\mathbf{r} d\mathbf{r}'\, \varphi_\gamma^*(\mathbf{r},\sigma) \varphi_\beta^*(\mathbf{r}',\sigma') v(\mathbf{r}-\mathbf{r}') \varphi_\delta(\mathbf{r},\sigma) \varphi_\alpha(\mathbf{r}',\sigma'). \tag{2.67}$$

The HF ground state is determined by the equation

$$\langle \text{HF}|[H, c_\delta^+ c_\gamma]|\text{HF}\rangle = 0, \tag{2.68}$$

where δ and γ represent both hole and particle states. This equation is completely equivalent to the HF equations (2.6), and may be derived (see for example Rowe, 1970) starting from the variational condition $\delta\langle \text{SD}|H|\text{SD}\rangle = 0$, by using the Thouless theorem (Thouless, 1972). This theorem states that any Slater determinant $|\text{SD}'\rangle$ which is not orthogonal to $|\text{SD}\rangle$ can be written as

$$|\text{SD}'\rangle = e^{\sum_{mi} b_{mi} c_m^+ c_i}|\text{SD}\rangle, \tag{2.69}$$

where i and m are hole and particle states for $|\text{SD}\rangle$ but not for $|\text{SD}'\rangle$. Therefore, an arbitrary and infinitesimal variation $\delta|\text{SD}\rangle$ can be written as

$$\delta|\text{SD}\rangle = \sum_{mi} \delta b_{mi} c_m^+ c_i |\text{SD}\rangle.$$

For a homogeneous system, the natural basis set is provided by plane waves, so that the set of quantum numbers that defines the state δ is given by $(\mathbf{p}, \lambda)$, where λ is the spin component ($\uparrow$ or $\downarrow$), and the Hamiltonian takes the form of

$$\begin{aligned} H = &\sum_{\mathbf{p},\lambda} \epsilon_p c_{\mathbf{p},\lambda}^+ c_{\mathbf{p},\lambda} + \sum_{\mathbf{p},\mathbf{q},\lambda} v_{\text{ext}}(\mathbf{q}) c_{\mathbf{p}+\mathbf{q},\lambda}^+ c_{\mathbf{p},\lambda} \\ &+ \frac{1}{2} \sum_{\mathbf{p}_1,\mathbf{p}_2,\mathbf{q},\lambda,\lambda'} V(\mathbf{q}) c_{\mathbf{p}_1+\mathbf{q},\lambda}^+ c_{\mathbf{p}_2-\mathbf{q},\lambda'}^+ c_{\mathbf{p}_2,\lambda'} c_{\mathbf{p}_1,\lambda}, \end{aligned} \tag{2.70}$$

where $V(q)$ is the Fourier transform of the interaction and $\epsilon_p = p^2/2m$.

Equations (2.65)–(2.70) are easily generalized to the Boson case.
For the electron gas (jellium model) the Hamiltonian becomes

$$H = \sum_{\mathbf{p},\lambda} \epsilon_p c^+_{\mathbf{p},\lambda} c_{\mathbf{p},\lambda} + \frac{1}{2} \sum_{\mathbf{p}_1,\mathbf{p}_2,\mathbf{q}\neq 0,\lambda,\lambda'} V(\mathbf{q}) c^+_{\mathbf{p}_1+\mathbf{q},\lambda} c^+_{\mathbf{p}_2-\mathbf{q},\lambda'} c_{\mathbf{p}_2,\lambda'} c_{\mathbf{p}_1,\lambda}, \quad (2.71)$$

where $\mathbf{q} = 0$ is missing in the electron–electron interaction because this part of the potential is canceled by the positive charge background field. The HF energy for the electron gas is obtained by taking the mean value of Hamiltonian (2.71) in the ground state of the Fermi gas, in which all the plane wave states with momentum smaller than or equal to the Fermi momentum are occupied. The occupation number operator, defined by

$$c^+_{\mathbf{p},\lambda} c_{\mathbf{p},\lambda} |\text{HF}\rangle = n_{\mathbf{p},\lambda} |\text{HF}\rangle, \quad (2.72)$$

takes the form of (1.76). When the potential energy term in (2.71) acts on the ground state, it destroys a pair of particles, $(\mathbf{p}_1, \lambda)$ $(\mathbf{p}_2, \lambda')$, inside the Fermi sphere. One obtains a nonvanishing result only if these particles are simultaneously created by the operators $c^+_{\mathbf{p}_1+\mathbf{q},\lambda} c^+_{\mathbf{p}_2-\mathbf{q},\lambda'}$. This happens if $\mathbf{q} = 0$ [but this case must be excluded, because in our case $\mathbf{q}$ is always different from zero (jellium model)] or if $\mathbf{p}_2 - \mathbf{q} = \mathbf{p}_1$ and $\lambda' = \lambda$. Therefore, the HF energy is

$$E^{\text{HF}} = \sum_{\mathbf{p},\lambda} \epsilon_p n_{\mathbf{p},\lambda} + \frac{1}{2} \sum_{\mathbf{p}_1,\mathbf{q}\neq 0,\lambda} V(\mathbf{q}) \langle \text{HF} | c^+_{\mathbf{p}_1+\mathbf{q},\lambda} c^+_{\mathbf{p}_1,\lambda} c_{\mathbf{p}_1+\mathbf{q},\lambda} c_{\mathbf{p}_1,\lambda} | \text{HF} \rangle. \quad (2.73)$$

Using the anticommutation relations (2.62), one obtains the following expression in terms of the distribution function $n_{\mathbf{p},\lambda}$:

$$E^{\text{HF}} = \sum_{\mathbf{p},\lambda} \epsilon_p n_{\mathbf{p},\lambda} - \frac{1}{2} \sum_{\mathbf{p}_1,\mathbf{p}_2,\lambda} V(\mathbf{p}_2 - \mathbf{p}_1) n_{\mathbf{p}_2,\lambda} n_{\mathbf{p}_1,\lambda}. \quad (2.74)$$

The first term in Eq. (2.74) is the kinetic energy of the noninteracting Fermi gas; the second term is the exchange energy which in 3D, for example, may be rewritten as

$$E_x = - \sum_{p_1 \leq k_F, p_2 \leq k_F} \frac{4\pi e^2}{|p_1 - p_2|^2}, \quad (2.75)$$

and once integration is performed it reproduces the result of Eq. (2.39).

2.6   Hartree–Fock at Finite Temperature

At finite temperature the equilibrium state of a system of particles is a statistical mixture defined by the density matrix

$$D = \sum_n p_n |n\rangle \langle n|, \quad (2.76)$$

where p_n represents the occupation probability of state $|n\rangle$ at the considered temperature.

Using the density matrix, it is possible to write the expectation value of a generic observable O as

$$\langle O \rangle = \sum_n p_n \langle n|O|n \rangle = \sum_{ijn} \langle n||j \rangle \langle j|O|i \rangle \langle i||n \rangle p_n$$

$$= \sum_{ij} \langle j|O|i \rangle \langle i|D|j \rangle = \mathrm{Tr}(DO). \tag{2.77}$$

The observable that we will consider in the following is the thermodynamic grand potential

$$G = E - TS - \mu \langle N \rangle, \tag{2.78}$$

where μ is the chemical potential, $\langle N \rangle$ is the average number of particles in the system,

$$E = \langle H \rangle = \mathrm{Tr}(D\,H) = \sum_n p_n E_n \tag{2.79}$$

is the system mean energy, and

$$S = -K\,\mathrm{Tr}(D \ln D) = -K \sum_n p_n \ln p_n \tag{2.80}$$

is its entropy (K is Boltzmann's constant). The value of μ determines the average number of particles at temperature T.

Using the above equations we can write

$$G = \mathrm{Tr}(D[H - \mu N - \beta^{-1} \ln D]), \tag{2.81}$$

where $\beta = 1/KT$. For given values of β and μ, D is determined by the functional equation

$$\frac{\delta G}{\delta D} = 0. \tag{2.82}$$

This equation has the formal solution

$$D = Z^{-1} e^{-\beta(H - \mu N)}, \tag{2.83}$$

with

$$Z = \mathrm{Tr}(e^{-\beta(H - \mu N)}). \tag{2.84}$$

Owing to the structure of H, the operator D obtained in this way is very complex, and it is practically impossible to carry out calculation starting from this expression.

In the HF approximation, the following form is imposed *a priori* on D, in the case of Fermions [see for example Des Cloiseaux (1967)]:

$$D_{\mathrm{HF}} = \prod_{\nu} [f_{\nu}(c_{\nu}^{\dagger} c_{\nu}) + (1 - f_{\nu})(c_{\nu} c_{\nu}^{\dagger})], \qquad (2.85)$$

where c_{ν} and $c_{\nu}^{\dagger}$ are the annihilation and creation operators of the single-particle state $|\nu\rangle$, and f_{ν} its occupation probability. Using expression (2.85), the entropy is written as

$$S = -K \sum_{\nu} [f_{\nu} \ln f_{\nu} + (1 - f_{\nu}) \ln(1 - f_{\nu})], \qquad (2.86)$$

which is the usual expression for the entropy of free Fermions. Now the problem is that of determining the basis $|\nu\rangle$, the energies ϵ_{ν} (which are the parameters of the theory) and the occupation numbers f_{ν}, by minimizing the HF grand potential, defined as

$$G_{\mathrm{HF}} = E_{\mathrm{HF}} - T S_{\mathrm{HF}} - \mu \langle N \rangle, \qquad (2.87)$$

with $\langle N \rangle = \sum_{\nu} f_{\nu}$, and where E_{HF} is the HF energy given by

$$E = \frac{1}{2m} \int \tau(\mathbf{r}) d\mathbf{r} + \int \rho(\mathbf{r}) v_{\mathrm{ext}}(\mathbf{r}) d\mathbf{r}$$
$$+ \frac{1}{2} \int [\rho(\mathbf{r})\rho(\mathbf{r}') - |\rho^{(1)}(\mathbf{r}, \mathbf{r}')|^2] v(\mathbf{r} - \mathbf{r}') d\mathbf{r} d\mathbf{r}', \qquad (2.88)$$

where the densities are defined in terms of the occupation numbers f_{ν} and of the single-particle wave functions ϕ_{ν} as follows:

$$\rho(\mathbf{r}) = \sum_{\nu} f_{\nu} |\phi_{\nu}(\mathbf{r})|^2,$$
$$\tau(\mathbf{r}) = \sum_{\nu} f_{\nu} |\nabla \phi_{\nu}(\mathbf{r})|^2, \qquad (2.89)$$
$$\rho^{(1)}(\mathbf{r}, \mathbf{r}') = \sum_{\nu} f_{\nu} \phi_{\nu}^{*}(\mathbf{r}) \phi_{\nu}(\mathbf{r}').$$

By minimizing $\hat{G} = G - \sum_i \lambda_i \langle \phi_{\nu} | \phi_{\nu} \rangle$ with respect to the wave functions, we obtain the HF equations at finite temperature ($\hbar = 1$):

$$\left(-\frac{\nabla^2}{2m} + v_{\mathrm{ext}} + \int \rho(\mathbf{r}') v(\mathbf{r} - \mathbf{r}') d\mathbf{r}' \right) \phi_{\nu}(\mathbf{r}, \sigma)$$
$$- \sum_{\sigma \| \sigma'} \int d\mathbf{r}' (\rho^{(1)}(\mathbf{r}, \sigma, \mathbf{r}', \sigma'))^{*} \phi_{\nu}(\mathbf{r}', \sigma') v(\mathbf{r} - \mathbf{r}')$$
$$= \frac{\lambda_{\nu}}{f_{\nu}} \phi_{\nu}(\mathbf{r}, \sigma) \equiv \varepsilon_{\nu} \phi_{\nu}(\mathbf{r}, \sigma). \qquad (2.90)$$

By minimizing $\hat{G}$ with respect to the f_ν and using the HF equations, we obtain

$$\varepsilon_\nu + KT \ln \frac{f_\nu}{1 - f_\nu} - \mu = 0, \tag{2.91}$$

and thus

$$f_\nu = \frac{1}{1 + \exp[(\varepsilon_\nu - \mu)/KT]}. \tag{2.92}$$

The chemical potential is subsequently determined by the condition

$$\langle N \rangle = \sum_\nu \frac{1}{1 + \exp[(\varepsilon_\nu - \mu)/KT]}. \tag{2.93}$$

In the case of Bosons, in the HF approximation, the following form is imposed *a priori* to D:

$$D_{\mathrm{HF}} = \prod_\nu [f_\nu(a_\nu^\dagger a_\nu) - (1 + f_\nu)(a_\nu a_\nu^\dagger)], \tag{2.94}$$

which yields the expression

$$S = -k \sum_\nu [f_\nu \ln f_\nu - (1 + f_\nu) \ln(1 + f_\nu)], \tag{2.95}$$

for the entropy. In the thermodynamic limit ($f_\nu/N = 0$ for all $\nu \neq 0$), one gets for the HF energy of the Bosons (Huse and Siggia, 1982)

$$E = N_0 \langle 0|H_0|0 \rangle + \sum_{\nu \neq 0} f_\nu \langle \nu|H_0|\nu \rangle + \frac{1}{2} N_0^2 \langle 00|V|00 \rangle$$
$$+ 2N_0 \sum_{\nu \neq 0} f_\nu \langle 0\nu|V|0\nu \rangle + \sum_{\nu \neq 0} \sum_{\mu \neq 0} f_\nu f_\mu \langle \nu\mu|V|\nu\mu \rangle, \tag{2.96}$$

where $f_0 = N_0$ is the occupation number of the lowest single-particle state φ_0 and

$$\langle \nu|H_0|\mu \rangle = \int d\mathbf{r}\, \varphi_\nu^*(\mathbf{r}) \left(\frac{-\nabla^2}{2m} + v_{\mathrm{ext}}(\mathbf{r}) \right) \varphi_\mu(\mathbf{r}),$$
$$\langle \nu\mu|V|\alpha\lambda \rangle = \int d\mathbf{r}\, d\mathbf{r}'\, \varphi_\nu^*(\mathbf{r}) \varphi_\mu^*(\mathbf{r}') v(\mathbf{r} - \mathbf{r}') \varphi_\alpha(\mathbf{r}) \varphi_\lambda(\mathbf{r}'), \tag{2.97}$$

and in the following we will approximate the interparticle potential $v(\mathbf{r} - \mathbf{r}')$ as a δ function with weight $g = 4\pi a/m$, where a is the s wave scattering length. By minimizing $\hat{G} = G - \sum_i \lambda_i \langle \phi_\nu|\phi_\nu \rangle$ with respect to φ_0 and φ_ν with $\nu \neq 0$, we obtain the HF equations for trapped Bosons at finite temperature (Goldman *et al.*, 1981; Huse and Siggia, 1982):

$$\left(-\frac{\nabla^2}{2m} + v_{\mathrm{ext}} + 2\rho_T(\mathbf{r})g + \rho_0(\mathbf{r})g \right) \phi_0(\mathbf{r}) = \frac{\lambda_0}{N_0} \phi_0(\mathbf{r}) \equiv \varepsilon_0 \phi_0(\mathbf{r}),$$
$$\left(-\frac{\nabla^2}{2m} + v_{\mathrm{ext}} + 2[\rho_T(\mathbf{r}) + \rho_0(\mathbf{r})]g \right) \phi_\nu(\mathbf{r}) = \frac{\lambda_\nu}{f_\nu} \phi_\nu(\mathbf{r}) \equiv \varepsilon_\nu \phi_\nu(\mathbf{r}) \quad \nu \neq 0,$$

$$\tag{2.98}$$

where $\rho_T = \sum_{\nu \neq 0} |\phi_\nu(\mathbf{r})|^2$ is the density of uncondensed particles (the thermal component) and $\rho_0 = N_0 |\phi_0(\mathbf{r})|^2$ is the condensate density. Note that in the form (2.98), the HF equations do not guarantee that $\langle \nu | 0 \rangle = 0$ for $\nu \neq 0$. For macroscopic systems this has negligible consequences for all but possibly a very few $\nu \neq 0$ single-particle states and, from a practical point of view, one can safely ignore the problem. Formally, however, one can guarantee the condition $\langle \nu | 0 \rangle = 0$ following the prescriptions of Huse and Siggia (1982).

By minimizing $\hat{G}$ with respect to the f_ν and using the HF equations, we finally obtain

$$\varepsilon_\nu + KT \ln \frac{f_\nu}{1 + f_\nu} - \mu = 0, \tag{2.99}$$

and thus

$$f_\nu = \frac{1}{\exp[(\varepsilon_\nu - \mu)/KT] - 1}. \tag{2.100}$$

The chemical potential is subsequently determined by the condition

$$\langle N \rangle = \sum_\nu \frac{1}{\exp[(\varepsilon_\nu - \mu)/KT] - 1}, \tag{2.101}$$

with $N_0 = (\exp[(\varepsilon_0 - \mu)/KT] - 1)^{-1}$.

The HF equations and the normalization conditions (2.93), (2.101) are to be solved in a self-consistent way by using iterative methods. The fluctuation of the particle number can be expressed in terms of the thermal occupation numbers. For example, in the case of Fermions one gets

$$(\Delta N)^2 = \langle N^2 \rangle - \langle N \rangle^2 = \sum_\nu f_\nu (1 - f_\nu). \tag{2.102}$$

The quantity $\Delta N/N$ should be small. Otherwise, the use of the grand-canonical ensemble may be inadequate to describe a finite system [see for example Gross (1997)].

As we have already discussed at the end of Chapter 1, in the homogeneous case the normal spectrum of a system of Bosons has a phonon-like behavior. For such a system the HF description at finite temperature is then expected to be incorrect. However, for confined systems, like the one recently built in magnetic traps, the HF description at finite temperature for Bosons works quite well, apart from the description of a very few low-energy states. This point is discussed in detail in the paper by Giorgini, Pitaevskii and Stringari (1997) and will be the object of further discussion in Chapter 9.

2.7 Hartree–Fock–Bogoliubov and BCS

Normal metals have a single-particle excitation spectrum which is analogous to that of a noninteracting electron gas, i.e. a continuum starting from zero. The degeneracy associated with this spectrum produces a linear temperature dependence for the specific heat near absolute zero, and to large electric and thermal conductivities for metals. In the superconducting phase, which is observed below a critical temperature T_c, the single-particle excitation spectrum is completely different from that of normal metals. In order to create a single-particle excitation from the ground state in superconductors, one needs a minimum energy of 2Δ, known as the energy gap. This mere fact is at the origin of all superconductivity-related phenomena, such as the different (nonlinear) T dependence of the specific heat of superconductors when $T \leq T_c$. This difference between the excitation spectra is reflected in a difference between the wave functions of the two phases. In fact, it can be shown that while in the normal phase the probability that two single-particle states $(\mathbf{k}, \lambda)$ and $(\mathbf{k}', \lambda')$ are simultaneously occupied is a smooth function of $\mathbf{k}$ and $\mathbf{k}'$, in the superconducting phase such a probability increases by a finite amount for given pairs of states. This singular behavior of the pair correlation function actualizes London's idea that superconductivity is due to electron condensation in momentum space. When one considers the residual interaction, which is usually neglected in describing normal states, these pair correlations which lead to superconductivity emerge clearly. The Hartree–Fock–Bogoliubov (HFB) theory is the one that introduces such correlations in the wave functions. The HFB theory generalizes the HF theory since it minimizes the system energy not in the space of Slater determinants, but in the space of the following states:

$$|\Psi\rangle = \prod_{\nu > 0} (u_\nu + v_\nu c_\nu^+ c_{\bar{\nu}}^+)|\rangle, \tag{2.103}$$

where $\bar{\nu}$ defines the time-reversed state, $u_\nu = u_{\bar{\nu}}$ and $v_\nu = -v_{\bar{\nu}}$; u_ν and v_ν are real and $\nu > 0$ indicates that the pair ν, $\bar{\nu}$ appears only once in (2.103). If u_ν and v_ν satisfy the condition

$$u_\nu^2 + v_\nu^2 = 1 \quad \forall \nu, \tag{2.104}$$

then the state $|\Psi\rangle$ is normalized to 1. Moreover, the state $|\Psi\rangle$ has the property

$$\alpha_\mu |\Psi\rangle = 0 \quad \forall \mu, \tag{2.105}$$

with

$$\alpha_\mu = u_\mu c_\mu - v_\mu c_\mu^+. \tag{2.106}$$

Therefore, this represents the annihilation operator of a quasiparticle which occupies state φ_μ. The operators α and α^+ obey anticommutation relations identical to those in (2.62). The usual Slater determinants are obtained from $|\Psi\rangle$ by taking $u_\nu = 0$ and

$v_\nu = 1$ if ν lies below the Fermi level, and $u_\nu = 1$ and $v_\nu = 0$ if ν is above the Fermi level. In general $|\Psi\rangle$ contains pair correlations. Moreover, it can be shown that the state $|\Psi\rangle$ does not have a definite number of particles, i.e. it violates the particle number conservation symmetry. Clearly, in order that the theory be realistic, the particle number fluctuations should be small. In order to apply the variational principle, the system Hamiltonian is written in terms of the quasiparticle operators (2.106) and their Hermitian conjugates, by utilizing the inverse transformations, and the following quantity is minimized:

$$\langle \Psi | H - \mu N | \Psi \rangle, \tag{2.107}$$

where $|\Psi\rangle$ is given by Eq. (2.103) and

$$N = \sum_{\nu > 0} (c_\nu^+ c_\nu + c_{\bar\nu}^+ c_{\bar\nu}) \tag{2.108}$$

is the particle number operator, while the parameter μ is fixed at the end of the calculation to reproduce the correct mean value of N. The minimization procedure provides the set of HFB wave functions φ_ν and the weights u_ν and v_ν.

The BCS theory is a simplified approach to the HFB theory, in which the pairs, correlated by (2.103), are made up of the plane wave states $(\mathbf{k}, \uparrow)$ and $(-\mathbf{k}, \downarrow)$, having energy close to the Fermi energy, and the effective Hamiltonian is the pairing Hamiltonian

$$H = \sum_{\mathbf{k},\lambda} \frac{k^2}{2m} c_{\mathbf{k},\lambda}^+ c_{\mathbf{k},\lambda} + \sum_{k,l} V_{kl} c_{\mathbf{k}\uparrow}^+ c_{-\mathbf{k}\downarrow}^+ c_{-l\downarrow} c_{l\uparrow}, \tag{2.109}$$

with $V_{kl} = \langle k, -k | V | l, -l \rangle$. In order that the phenomenon of superconductivity takes place, it is necessary that the matrix element V_{kl} is mainly negative near the Fermi surface. This attraction is due to the presence of the crystal lattice and to the electron–phonon interaction which screens the repulsive electron–electron interaction, until the sign of the effective interaction is changed.

At mean field level, the Hamiltonian (2.109) is written as

$$H = \sum_{\mathbf{k},\lambda} \frac{k^2}{2m} c_{\mathbf{k},\lambda}^+ c_{\mathbf{k},\lambda} + \sum_{k,l} V_{kl} X_{\mathbf{k}}^+ c_{-l\downarrow} c_{l\uparrow} + \sum_{k,l} V_{kl} X_l c_{-\mathbf{k}\uparrow}^+ c_{\mathbf{k}\downarrow}^+, \tag{2.110}$$

where

$$X_{\mathbf{k}}^+ = -\langle c_{-\mathbf{k}\downarrow}^+ c_{\mathbf{k}\uparrow}^+ \rangle = -\mathrm{Tr}(D c_{-\mathbf{k}\downarrow}^+ c_{\mathbf{k}\uparrow}^+), \tag{2.111}$$

and the density matrix is given by Eq. (2.83). The grand-canonical Hamiltonian $K = H - \mu N$ can be written in the form

$$K = \sum_{\mathbf{k}} \begin{pmatrix} c_{\mathbf{k}\uparrow}^+ & c_{-\mathbf{k}\downarrow} \end{pmatrix} \begin{pmatrix} \epsilon_k & \Delta_{\mathbf{k}} \\ \Delta_{\mathbf{k}}^+ & -\epsilon_k \end{pmatrix} \begin{pmatrix} c_{\mathbf{k}\uparrow} \\ c_{-\mathbf{k}\downarrow}^+ \end{pmatrix}, \tag{2.112}$$

where $\epsilon_k = k^2/2m - \mu$,

$$\Delta_{\mathbf{k}} = \sum_1 V_{kl} X_1, \tag{2.113}$$

and energies are measured starting from the Fermi energy. The Hamiltonian (2.112) is a Hamiltonian for the quasiparticles meant as the ensemble of particles created outside the Fermi sphere and the holes created inside the Fermi sphere. This Hamiltonian can be put into diagonal form by applying a Bogoliubov unitary transformation:

$$c_{\mathbf{k}\uparrow} = u_{\mathbf{k}}\alpha_{\mathbf{k}-} + v_{\mathbf{k}}\alpha_{\mathbf{k}+}^+, \quad c_{-\mathbf{k}\downarrow}^+ = u_{\mathbf{k}}\alpha_{\mathbf{k}+}^+ - v_{\mathbf{k}}\alpha_{\mathbf{k}-}, \tag{2.114}$$

where the quasiparticle operator $\alpha_{\mathbf{k}-}$ lowers the system momentum by $\mathbf{k}$, and lowers the spin by unity [it annihilates a particle with quantum numbers $(\mathbf{k}\uparrow)$ and creates one with quantum numbers $(-\mathbf{k}\downarrow)$], while $\alpha_{\mathbf{k}+}$ increases the system momentum by $\mathbf{k}$ and increases the spin by unity. If we require the transformation to diagonalize the Hamiltonian, i.e. it should result in

$$K = \sum_{\mathbf{k}\lambda} E_{\mathbf{k}\lambda}\alpha_{\mathbf{k}\lambda}^+\alpha_{\mathbf{k}\lambda}, \tag{2.115}$$

we obtain

$$E_{\mathbf{k}\mp} = \pm\omega_{\mathbf{k}} = \pm\sqrt{\epsilon_k^2 + \Delta_{\mathbf{k}}^2}, \tag{2.116}$$

and

$$u_{\mathbf{k}}^2 = \frac{1}{2}\left(1 + \frac{\epsilon_k}{\sqrt{\epsilon_k^2 + \Delta_{\mathbf{k}}^2}}\right), \quad v_{\mathbf{k}}^2 = \frac{1}{2}\left(1 - \frac{\epsilon_k}{\sqrt{\epsilon_k^2 + \Delta_{\mathbf{k}}^2}}\right). \tag{2.117}$$

From the transformations (2.114), we obtain for the system density

$$\rho = \frac{N}{V} = \frac{1}{V}\left\langle \sum_{\mathbf{k}} c_{\mathbf{k}}^+ c_{\mathbf{k}} \right\rangle = \frac{1}{V}\sum_{\mathbf{k}}\left(\frac{u_{\mathbf{k}}^2}{e^{\beta\omega_{\mathbf{k}}}+1} + \frac{v_{\mathbf{k}}^2}{e^{-\beta\omega_{\mathbf{k}}}+1}\right)$$
$$= \frac{1}{V}\sum_{\mathbf{k}}\frac{1}{2}\left(1 - \frac{\epsilon_k}{\omega_{\mathbf{k}}}\tanh\left(\frac{\beta\omega_{\mathbf{k}}}{2}\right)\right), \tag{2.118}$$

and using Eq. (2.113) as well, the self-consistent equation for the gap

$$\Delta_1 = -\frac{1}{2}\sum_{\mathbf{k}} V_{kl}\frac{\Delta_{\mathbf{k}}}{\omega_{\mathbf{k}}}\tanh\left(\frac{\beta\omega_{\mathbf{k}}}{2}\right). \tag{2.119}$$

An explicit solution to Eq. (2.119) is obtained by approximating V_{kl} with the potential:

$$V_{kl} = \begin{cases} -g/V & \text{if } \left|\mu - \frac{k^2}{2m}\right| \le \Delta\epsilon \text{ and } \left|\mu - \frac{l^2}{2m}\right| \le \Delta\epsilon \\ 0 & \text{otherwise} \end{cases}, \tag{2.120}$$

where g is a constant, so that V_{kl} is attractive in a layer of thickness $2\Delta\epsilon$ centered on the Fermi surface. In this case, since the right hand side of (2.119) does not depend on l, the gap itself must be a constant depending solely on temperature:

$$
\Delta_l = \begin{cases} \Delta(T) & \text{if } \left|\mu - \dfrac{l^2}{2m}\right| \le \Delta\epsilon \\ 0 & \text{otherwise} \end{cases}
\tag{2.121}
$$

By changing the summation into integration and approximating $k^2 dk = k_F^2 dk$, where k_F is the Fermi momentum, we obtain

$$
1 = gN(0) \int_0^{\Delta\epsilon} d\epsilon_{\mathbf{k}} \frac{\tanh(\beta\omega_{\mathbf{k}}/2)}{\omega_{\mathbf{k}}},
\tag{2.122}
$$

where $N(0) = mk_F/(2\pi^2)$ is the quasiparticle density of the states at the Fermi surface per unit volume. At the critical temperature the gap is zero and we can write

$$
1 = gN(0) \int_0^{\Delta\epsilon} d\epsilon_{\mathbf{k}} \frac{\tanh(\beta_c\epsilon_{\mathbf{k}}/2)}{\epsilon_{\mathbf{k}}} = gN(0) \ln(2\gamma\beta_c\Delta\epsilon/\pi),
\tag{2.123}
$$

with $\beta_c = (KT_c)^{-1}$ and $\ln\gamma = 0.5772$. Therefore, from Eq. (2.123) the following result is obtained,

$$
KT_c = 2\frac{\gamma}{\pi}\Delta\epsilon e^{-1/N(0)g},
\tag{2.124}
$$

for the critical temperature of the BCS transition. From this equation we see that the critical temperature is an analytic function of the interaction intensity, so that a perturbative treatment starting from the normal phase cannot produce this result until an infinite number of graphs of a given class is summed.

Equation (2.122) can also be used to find the gap at zero temperature, where the hyperbolic tangent is equal to 1. We obtain

$$
\Delta(T = 0) = 2\Delta\epsilon e^{-1/N(0)g},
\tag{2.125}
$$

and therefore the important result

$$
\frac{\Delta(T = 0)}{KT_c} = 1.764,
\tag{2.126}
$$

which is valid in the limit of weak coupling $N(0)g \le 1/4$. This ratio is in reasonable agreement with the experimental results on superconductors with weak coupling, while it is too small in the case of lead and mercury where damping effects, which we have neglected, are important.

For a detailed discussion on the HFB and BCS theory, one should see the books by Schrieffer (1964) and Fetter and Walecka (1971). For applications of the BCS theory to nuclear physics, see the book by Rowe (1970).

In the case of a uniform diluted Fermi gas interacting with a delta force fixed by the coupling constant $4\pi\hbar^2 a/m$, where $a < 0$ is the s wave scattering length relative to the low-energy collision between particles of opposite spin, $V_{kl} = -g/V = -4\pi\hbar^2|a|/mV$ and

$$\Delta(T = 0) = 2\Delta\epsilon e^{-\frac{\pi}{2k_F|a|}} = \frac{\pi}{\gamma}KT_c. \tag{2.127}$$

The above results provide the $T = 0$ value of the gap and critical temperature to exponential accuracy. In fact, the value of the pre-exponential factor in (2.127) is unknown. In order to evaluate it, one needs to renormalize the coupling constant g, similar to what is done in the calculation of the RPA correlation energy for the cold and dilute gas of Bosons and Fermions (see Subsection 10.3.1). The proper calculation was done by Gorkov and Melik-Barkhudarov (1961), yielding the result

$$2\Delta\epsilon = \frac{1}{2}\left(\frac{2}{e}\right)^{\frac{7}{3}} \epsilon_F. \tag{2.128}$$

Results (2.127) and (2.128) show that, due to the exponential factor, the value of KT_c is much smaller than the Fermi energy ϵ_F. The interacting Fermi gas becomes superfluid at a temperature much smaller than the one characterizing a Bose condensate. According to the BCS result (2.127), large negative scattering lengths, which can be realized working close to a Feshbach resonance, should enhance the value of T_c. However, the validity of BCS is uncertain in this case (Randeria, 1995; Combescot, 1999; Holland *et al.*, 2001; Chiofalo *et al.*, 2002; Ohashi and Griffin, 2002). For trapped atomic Fermi gases, the BCS theory can be extended in the local density approximation only if the size R of the system is much larger than the characteristing interaction length $\xi = \hbar/\sqrt{2mg\rho}$ (healing length). Since in a superfluid Fermi gas, ξ can be defined in terms of the inverse of the momentum width $\delta k \simeq \Delta/v_F$, by using $\epsilon_F = 1/2m\omega_{ho}R^2$, one finds that $\xi/R = \hbar\omega_{ho}/\Delta(T = 0)$ and the BCS theory is valid only if the energy gap is larger than the oscillator energy. If the gap becomes comparable with ω_{ho}, then the BCS theory requires a further development (Bruun *et al.*, 1999).

2.8 Appendix: Second Quantization

Ordinarily, to describe the state of a system of N particles, we have used a wave function $\psi(x_1, \ldots, x_N)$, depending on the variables (of position, spin etc.) of the particles in the configuration, spin etc. space. This representation is only one of the possible representations, and normally not the more practical. In the "occupation number" representation a state of the system, described in the ordinary representation, for example for Fermions, by the Slater determinant of Eq. (1.13), is specified if we give the single-particle states φ_α, φ_β,... and their order. Hence in the "occupation number" representation we write

$$\frac{1}{\sqrt{N!}}\det\begin{vmatrix} \varphi_\alpha(x_1) & \varphi_\alpha(x_2) & \cdots & \varphi_\alpha(x_N)\\ \varphi_\beta(x_1) & \varphi_\beta(x_2) & \cdots & \varphi_\beta(x_N)\\ \varphi_\gamma(x_1) & \varphi_\gamma(x_2) & \cdots & \varphi_\gamma(x_N)\\ \vdots & \vdots & \ddots & \vdots\\ \cdots & \cdots & \cdots & \cdots \end{vmatrix} \equiv |\alpha\,\beta\,\gamma\cdots\rangle. \tag{2.129}$$

With the goal of manipulating the wave functions, evaluating matrix elements and so on, it is necessary to build an algebra. This is conveniently done in the second quantization language. First of all we define the creation operator c_ν^+ which creates a particle in the state ν. Hence one has

$$|\alpha\,\beta\,\gamma\cdots\rangle = c_\alpha^+ c_\beta^+ c_\gamma^+ \cdots |\,\rangle, \tag{2.130}$$

where $|\,\rangle$ is the particle vacuum state. The requirement that $|SD\rangle$ be antisymmetric produces

$$c_\alpha^+ c_\beta^+ c_\gamma^+ \cdots |\,\rangle = -c_\beta^+ c_\alpha^+ c_\gamma^+ \cdots |\,\rangle,$$

and since two Fermions cannot occupy the same state one gets

$$c_\alpha^+ c_\alpha^+ c_\gamma^+ \cdots |\,\rangle = 0.$$

These two requirements can be simultaneously expressed by the anticommutation relation for the c^+ and c operators:

$$\{c_\alpha^+, c_\beta^+\} = \{c_\alpha, c_\beta\} = 0. \tag{2.131}$$

If we build the adjoint of Eq. (2.130),

$$\langle\alpha\,\beta\,\gamma\cdots| = \langle|\cdots c_\gamma c_\beta c_\alpha,$$

then

$$\langle\alpha\,\beta\,\gamma\cdots|\alpha\,\beta\,\gamma\cdots\rangle = \langle|\cdots c_\gamma c_\beta c_\alpha c_\alpha^+ c_\beta^+ c_\gamma^+ \cdots|\,\rangle.$$

It follows that

$$|\,\rangle = c_\alpha c_\alpha^+ |\,\rangle = c_\beta c_\alpha c_\alpha^+ c_\beta^+ |\,\rangle \cdots$$

and then that the Hermitian conjugate c_ν of c_ν^+ is the destruction operator of the particle in state ν. We also have

$$c_\beta c_\alpha^+ c_\beta^+ c_\gamma^+ \cdots|\,\rangle = -c_\beta c_\beta^+ c_\alpha^+ c_\gamma^+ \cdots|\,\rangle = -c_\alpha^+ c_\gamma^+ \cdots|\,\rangle = -c_\alpha^+ c_\beta c_\beta^+ c_\gamma^+ \cdots|\,\rangle$$

or

$$\{c_\beta, c_\alpha^+\}|\beta\,\gamma\cdots\rangle = 0 \quad \alpha \neq \beta.$$

In this equation the anticommutator operates on the most favorable wave function, i.e. on that with the state β occupied and the state α unoccupied. Operating on any other wave function, the two elements of the anticommutatotor must vanish separately. Hence

$$\{c_\beta, c_\alpha^+\} = 0 \quad \alpha \neq \beta.$$

Since one has

$$c_\alpha c_\alpha^+ |\alpha\ \beta\ \gamma \cdots\rangle = 0, \quad c_\alpha c_\alpha^+ |\beta\ \gamma\ \delta \cdots\rangle = |\beta\ \gamma\ \delta \cdots\rangle,$$
$$c_\alpha^+ c_\alpha |\alpha\ \beta\ \gamma \cdots\rangle = |\alpha\ \beta\ \gamma \cdots\rangle \quad c_\alpha^+ c_\alpha |\beta\ \gamma\ \delta \cdots\rangle = 0,$$

(2.132)

the general anticommutation relation is

$$\{c_\alpha^+, c_\beta\} = \delta_{\alpha,\beta}.$$

(2.133)

It can be seen from Eq. (2.131) that $c_\nu^2 = 0$; a state can accommodate only a single particle. Furthermore, from the above equations one finds that the operator $n_\nu = c_\nu^+ c_\nu$ has the property $n_\nu^2 = n_\nu$, and gives the occupation number of the state ν, which can be either 0 or 1.

We now consider a system of Bosons; we denote the creation operator for a particle in the state ν by a_ν^+, the destruction operator by a_ν, and proceed to build up a set of states in a fashion directly analogous to the Fermion case. The difference between the two systems is manifested in the symmetry property of the many-body wave function. That for Bosons must be symmetric on the interchange of any two particles; the operators a_ν must therefore satisfy the commutation relations

$$[a_\alpha^+, a_\beta^+] = [a_\alpha, a_\beta] = 0 \quad [a_\alpha, a_\beta^+] = \delta_{\alpha,\beta}.$$

(2.134)

In the case of Bosons, the occupation operator $n_\nu = a_\nu^+ a_\nu$ gives the number of particles in the state ν.

Let us come back to Fermions and to the problem of writing operators in second quantization; generalization to Bosons is straightforward.

From the operator n_ν defined above, we see immediately that the operator measuring the total number of particles is given by

$$N = \sum_\nu n_\nu = \sum_\nu c_\nu^+ c_\nu.$$

(2.135)

Let us now consider a general one-body operator, $F_1 = \sum_{i=1}^N f(x_i)$. The expectation value of F_1 on the state $|\alpha\ \beta\ \gamma \cdots\rangle$ is given by (see Subsection 1.4)

$$\langle \alpha\ \beta\ \gamma \cdots |F_1|\alpha\ \beta\ \gamma \cdots\rangle = f_{\alpha\alpha} + f_{\beta\beta} + f_{\gamma\gamma} + \cdots,$$

where $f_{\alpha\alpha} = \int \varphi_\alpha^* f(x)\varphi_\alpha\ dx$. The matrix elements of F_1 between states that differ only for a single particle state are given by

$$\langle \alpha\ \beta\ \gamma \cdots |F_1|\alpha\ \beta'\ \gamma \cdots\rangle = f_{\beta\beta'}, \quad \beta \neq \beta',$$

and the ones between states that differ for more than a single particle state vanish:

$$\langle \alpha \, \beta \, \gamma \cdots | F_1 | \alpha' \, \beta' \, \gamma \cdots \rangle = 0, \quad \alpha, \beta \neq \alpha', \beta'.$$

It is then clear that all these expressions are reproduced by

$$F_1 = \sum_{\nu\nu'} f_{\nu\nu'} c_\nu^+ c_{\nu'}. \tag{2.136}$$

Let us then consider a two body-operator, $F_2 = 1/2 \sum_{i \neq j=1}^{N} f(x_i, x_j)$, and define the matrix elements:

$$F_{\alpha\beta\alpha'\beta'} = \int \varphi_\beta^*(x)\varphi_\alpha(x) f(x, x') \varphi_{\beta'}^*(x')\varphi_{\alpha'}(x') \, dx \, dx'. \tag{2.137}$$

In the second quantization scheme, F_2 is expressed as

$$F_2 = \frac{1}{2} \sum_{\mu\nu\mu'\nu'} F_{\mu\nu\mu'\nu'} c_\nu^+ c_{\nu'}^+ c_{\mu'} c_\mu, \tag{2.138}$$

and one can verify that this expression gives the correct matrix elements of F_2 in all the cases. For example, let us consider the matrix element $\langle \alpha \, \beta \, \gamma \, \delta \cdots | F_2 | \alpha' \, \beta' \, \gamma \, \delta \cdots \rangle$, where $\alpha, \beta \neq \alpha', \beta'$:

$$\langle \alpha \, \beta \, \gamma \, \delta \cdots | F_2 | \alpha' \, \beta' \, \gamma \, \delta \cdots \rangle$$
$$= \frac{1}{2} \sum_{\mu\nu\mu'\nu'} F_{\mu\nu\mu'\nu'} \langle | \cdots c_\delta c_\gamma c_\beta c_\alpha c_\nu^+ c_{\nu'}^+ c_{\mu'} c_\mu c_{\alpha'}^+ c_{\beta'}^+ c_\gamma^+ c_\delta^+ \cdots | \rangle$$
$$= \frac{1}{2} \left(F_{\alpha'\alpha\beta'\beta} - F_{\beta'\alpha\alpha'\beta} - F_{\alpha'\beta\beta'\alpha} + F_{\beta'\beta\alpha'\alpha} \right)$$
$$= \int \varphi_\alpha^*(x)\varphi_\beta^*(x') f(x, x') \left(\varphi_{\alpha'}^*(x)\varphi_{\beta'}(x') - \varphi_{\alpha'}(x')\varphi_{\beta'}^*(x) \right) dx \, dx', \tag{2.139}$$

which is the right result (see Subsection 1.5), and to get it we have used the anti-commutation rules of Eqs. (2.133).

In the case of a homogeneous system of N Fermions, each state is characterized by a wave vector $\mathbf{p}$ and a spin index λ. The creation and destruction operators thus become $c_{\mathbf{p}\lambda}^+$ and $c_{\mathbf{p}\lambda}$. They satisfy the anticommutation relations

$$\{c_{\mathbf{p}\lambda}^+, c_{\mathbf{p}'\lambda'}^+\} = \{c_{\mathbf{p}\lambda}, c_{\mathbf{p}'\lambda'}\} = 0, \quad \{c_{\mathbf{p}\lambda}^+, c_{\mathbf{p}'\lambda'}\} = \delta_{\mathbf{p}\mathbf{p}'}\delta_{\lambda\lambda'}. \tag{2.140}$$

The normalized wave function for a planar wave of momentum $\mathbf{p}$ is

$$\frac{1}{\sqrt{V}} e^{i\mathbf{p}\cdot\mathbf{r}}.$$

The probability amplitude for the destruction of a Fermion of momentum $\mathbf{p}$ and spin λ taking place at the point $\mathbf{r}$ is thus represented by the operator

$$\frac{1}{\sqrt{V}} e^{i\mathbf{p}\cdot\mathbf{r}} c_{\mathbf{p}\lambda}$$

(with a corresponding expression for the case of a Boson). This leads us to introduce the destruction operator at the point $\mathbf{r}$, $\hat{\Phi}_\lambda(\mathbf{r})$, defined by

$$\hat{\Phi}_\lambda(\mathbf{r}) = \frac{1}{\sqrt{V}} \sum_\mathbf{p} c_{\mathbf{p}\lambda} e^{i\mathbf{p}\cdot\mathbf{r}},$$

$$c_{\mathbf{p}\lambda} = \frac{1}{\sqrt{V}} \int d\mathbf{r} \hat{\Phi}_\lambda(\mathbf{r}) e^{-i\mathbf{p}\cdot\mathbf{r}}. \tag{2.141}$$

The operator $\hat{\Phi}_\lambda(\mathbf{r})$ describes the destruction at $\mathbf{r}$ of a particle of spin λ of any momentum. Similarly, the creation operator $\hat{\Phi}_\lambda^+(\mathbf{r})$ is defined by

$$\hat{\Phi}_\lambda^+(\mathbf{r}) = \frac{1}{\sqrt{V}} \sum_\mathbf{p} c_{\mathbf{p}\lambda}^+ e^{-i\mathbf{p}\cdot\mathbf{r}}. \tag{2.142}$$

The two operators $\hat{\Phi}_\lambda(\mathbf{r})$ and $\hat{\Phi}_\lambda^+(\mathbf{r})$ satisfy the anticommutation relations

$$\{\hat{\Phi}_\lambda(\mathbf{r}), \hat{\Phi}_{\lambda'}(\mathbf{r}')\} = \{\hat{\Phi}_\lambda^+(\mathbf{r}), \hat{\Phi}_{\lambda'}^+(\mathbf{r}')\} = 0,$$

$$\{\hat{\Phi}_\lambda(\mathbf{r}), \hat{\Phi}_{\lambda'}^+(\mathbf{r}')\} = \delta(\mathbf{r} - \mathbf{r}')\delta_{\lambda,\lambda'}, \tag{2.143}$$

and analogous commutation relations for Bosons. With the operators $\hat{\Phi}_\lambda(\mathbf{r})$ and $\hat{\Phi}_\lambda^+(\mathbf{r})$ one can express the physical quantities of the system. For example, the operator

$$\rho((\mathbf{r}, \lambda) = \hat{\Phi}_\lambda^+(\mathbf{r})\hat{\Phi}_\lambda(\mathbf{r}) \tag{2.144}$$

represents the density of particles at point $\mathbf{r}$. The total number of particles of the system is given by

$$N = \sum_\lambda \int d\mathbf{r} \; \hat{\Phi}_\lambda^+(\mathbf{r})\hat{\Phi}_\lambda(\mathbf{r}) = \sum_{\mathbf{p},\lambda} c_{\mathbf{p}\lambda}^+ c_{\mathbf{p}\lambda}. \tag{2.145}$$

By using Eqs. (2.136), (2.138) and (2.141) one can write the Hamiltonian of N interacting Fermions (and Bosons) as

$$H = -\sum_\lambda \int d\mathbf{r} \hat{\Phi}_\lambda^+(\mathbf{r}) \frac{\nabla^2}{2m} \hat{\Phi}_\lambda(\mathbf{r})$$

$$+ \frac{1}{2} \sum_{\lambda,\lambda'} \int d\mathbf{r} d\mathbf{r}' \hat{\Phi}_\lambda^+(\mathbf{r}) \hat{\Phi}_{\lambda'}^+(\mathbf{r}') v(\mathbf{r} - \mathbf{r}') \hat{\Phi}_\lambda(\mathbf{r}) \hat{\Phi}_{\lambda'}(\mathbf{r}')$$

$$= \sum_{\mathbf{p},\lambda} \frac{p^2}{2m} c_{\mathbf{p}\lambda}^+ c_{\mathbf{p}\lambda} + \frac{1}{2} \sum_{\mathbf{p}\mathbf{k}\mathbf{q},\lambda,\lambda'} V(\mathbf{q}) c_{\mathbf{p}+\mathbf{q}\lambda}^+ c_{\mathbf{k}-\mathbf{q}\lambda'}^+ c_{\mathbf{k}\lambda'} c_{\mathbf{p}\lambda}, \tag{2.146}$$

where $V(\mathbf{q}) = \int d\mathbf{r} \; e^{-i\mathbf{q}\cdot\mathbf{r}} V(\mathbf{r})$.

In this representation one gets for the Fourier component of the density $\rho_q = \int d\mathbf{r}\, e^{-i\mathbf{q}\cdot\mathbf{r}}\rho(\mathbf{r})$

$$\rho_q = \sum_{\mathbf{p},\lambda} c^+_{\mathbf{p}\lambda} c_{\mathbf{p}+\mathbf{q}\lambda}, \qquad (2.147)$$

and of the current density $\mathbf{j}_q$

$$\mathbf{j}_q = \sum_{\mathbf{p},\lambda} \frac{\mathbf{p}}{m} c^+_{\mathbf{p}\lambda} c_{\mathbf{p}+\mathbf{q}\lambda}. \qquad (2.148)$$

References

Alasia, F. *et al.*, *Phys. Rev.* **B52**, 8488 (1995).

Bachelet, G.B., D.M. Ceperley and M.G.B. Chiocchetti, *Phys. Rev. Lett.* **62**, 2088 (1989).

Bachelet, G.B., D.R. Hamann, and M. Schluter, *Phys. Rev.* **B26**, 4199 (1982).

Ballone, P., C.J. Umrigar and P. Delaly, *Phys. Rev.* **B45**, 6293 (1992).

Bertsch, G.F., A. Bulgac, D. Tomanek and Y. Wang, *Phys. Rev. Lett.* **67**, 2690 (1991).

Blundell, S.E. and C. Guet, *Z. Phys.* **D28**, 81 (1993).

Broglia, R.A., G. Colo, G. Onida and H.E. Roman, *Solid State Physics of Finite Systems: Metal Clusters, Fullerenes, Atomic Wires*, CUSL, Milan, 2002.

Bruun, G. *et al.*, *Eur. Phys. J.* **D7**, 433 (1999).

Ceperley, D., *Phys. Rev.* **B18**, 3126 (1978); D.M. Ceperley and B.J. Alder, *Phys. Rev. Lett.* **45**, 566 (1980).

Chiofalo, M.L. *et al.*, *Phys. Rev. Lett.* **88**, 090402 (2002).

Combescot, R., *Phys. Rev. Lett.* **83**, 3766 (1999). .

Dalfovo, F., S. Giorgini, L. P. Pitaevskii and S. Stringari, *Rev. Mod. Phys.* **71**, 463 (1999).

Des Cloiseaux, J., in *Many-Body Physics, Les Houches 1967* (Gordon and Breach, New York).

Ekardt, W., *Phys. Rev. Lett.* **52**, 1925 (1984); *Phys. Rev.* **B32**, 1961 (1985).

Fetter, A.L. and J.D. Walecka, *Quantum Theory of Many-Body Systems*, (Mc Graw Hill, 1971).

Gell-Mann, M. and K.A. Brueckner, *Phys. Rev.* **106**, 364 (1957).

Giorgini, S., L.P. Pitaevskii and S. Stringari, *J. Low Temp. Phys.* **109**, 309 (1997).

Goldman, V.V., I.F. Silvera and A.J. Legget, *Phys. Rev.* **B24**, 2870 (1981).

Gorkov, L.P. and T.K. Melik-Barkhudarov, *Sov. Phys. JETP* **13**, 1018 (1961).

Gross, E.P., *Nuovo Cimento* **20**, 454 (1961).

Gross, D.H.E., *Phys. Rep.* **279**, 119 (1997).

Guet, C. and W.R. Johnson, *Phys. Rev.* **B45**, 11283 (1992).

Holland, M.J. *et al.*, *Phys. Rev. Lett.* **87**, 120406 (2001).

Huse, D.A. and E.D. Siggia, *J. Low Temp. Phys.* **46**, 137 (1982).

Isihara, A., *Solid State Physics* **42**, 271 (1989).

Knight, W.D., K. Clemenger, W.A. de Heer and W.A. Saunders, *Phys. Rev.* **B31**, 445 (1985); W.A. de Heer, W.D. Knight, M.Y. Chou and M.L. Cohen, *Solid State Physics* **40**, 93 (1987).

Knight, W.D., K. Clemenger, W.A. de Heer, and W.A. Saunders, *Phys. Rev.* **B31**, 445 (1985).

Kratschmer, W., L.D. Lamb, K. Fostiropoulos and D.R. Huffman, *Nature* **347**, 354 (1990).

Kroto, H.W., J.R. Heath, S.C. O'Brien, R.F. Curl and R.E. Smalley, *Nature* **318**, 162 (1985).

Lipparini, E., Ll. Serra and K. Takayanagi, *Phys. Rev.* **B49**, 16733 (1994).

Ohashi, Y. and A. Griffin, *Phys. Rev. Lett.* **89**, 130402 (2002).

Pederiva, F., C.J. Umrigar and E. Lipparini, *Phys. Rev.* **B62**, 8120 (2000).

Pitaevskii, L.P., *Sov. Phys. JEPT* **13**, 451 (1961).

Randeria, M., in *Bose–Einstein Condensation*, eds. A. Griffin, D.W. Snoke and S. Stringari, (Cambridge University Press, 1995).

Reimann, S.M. and M. Manninen, *Rev. Mod. Phys.* **74**, 1283 (2002).

Rowe, D.J., *Nuclear Collective Motion* (Methuen, London, 1970).

Schrieffer, J.R., *Theory of Superconductivity* (Benjamin, Reading, Massachusetts, 1964).

Serra, Ll., G.B. Bachelet, Nguyen Van Giai, and E. Lipparini, *Phys. Rev.* **B48**, 14708 (1993).

Tarucha, S., D.G. Austing, T. Honda, R.J. van der Hage, and L.P. Kouwenhoven, *Phys. Rev. Lett.* **77**, 3613 (1996).

Thouless, D.J., *The Quantum Mechanics of Many-Body Systems* (Academic, 1972).

Van Giai, N. and E. Lipparini, *Z. Phys.* **D27**, 193 (1993).

Wigner, E.P., *Phys. Rev.* **46**, 1002 (1934).

Yabana, K. and G.F. Bertsch, *Z. Phys.* **D32**, 329 (1995).

Yannouleas, C., R.A. Broglia, M. Brack, and P.F. Bortignon, *Phys. Rev. Lett.* **63**, 255 (1989).

Chapter 3

The Brueckner–Hartree–Fock Theory

3.1 Introduction

In several many-body problems, the strongly repulsive nature of the short-range interactions does not allow direct use of either mean field theories or ordinary perturbation theory. As a consequence, during the 1950's and 1960's the g matrix theory was developed by Brueckner. In this theory, the effective interaction of two particles in the presence of all the others is defined in a self-consistent way. This theory is particularly simple in the homogeneous system, where the single-particle states are plane waves, and we are only faced with the problem of determining g. The problem is much more complex in finite systems because the states need to be determined self-consistently together with g (double self-consistency problem).

A possible way of simplifying the problem in finite systems is to use the results of the homogeneous problem, by using the local density approximation (LDA). In the LDA, one first determines the g matrix in the homogeneous system, and then uses it as an effective interaction in HF self-consistent calculations in finite systems.

Calculations of this kind have been carried out in nuclear physics, on ^{3}He and on the electron gas. In the first two systems the g matrix theory does not work as well as in the electron gas case. This is due to the fact that the g matrix takes into account only two-body interactions, and three-body and four-body interactions are important for nucleons and ^{3}He, while they are negligible in the electron case.

3.2 The Lippman–Schwinger Equation

Let us consider the elastic scattering of two particles having the same mass, m. Let us separate the system Hamiltonian into a one-body term, H_0, plus the two-body interaction V: $H = H_0 + V$. Let us assume that the scattering takes place at energy E (i.e. the sum of the energies of the two particles in the initial state). Let ϕ be the wave function of H_0 with energy E, i.e. $H_0\phi = E\phi$, where $E = \epsilon_1 + \epsilon_2$. Let ψ be the wave function corresponding to the same energy E, which describes the interacting system: $H\psi = E\psi$.

From the two equation:

$$(H_0 - E)\phi = 0 \tag{3.1}$$

and

$$(H_0 - E)\psi = -V\psi, \tag{3.2}$$

by subtraction one gets

$$\psi = \phi - \frac{V}{H_0 - E}\psi, \tag{3.3}$$

which is an exact relationship between the correlated wave function ψ and the non-correlated wave function ϕ, in the form of an integral equation. Then we introduce the operator g such that

$$g\phi = V\psi. \tag{3.4}$$

From Eqs. (3.3) and (3.4) we immediately obtain

$$g\phi = V\phi - \frac{Vg}{H_0 - E}\phi,$$

which, in operator form, can be rewritten as

$$g = V - \frac{V}{H_0 - E}g, \tag{3.5}$$

and is the Lippman–Schwinger equation. Note that, formally, we can write

$$g = \frac{V}{1 + \frac{V}{H_0 - E}},$$

and from this equation it follows that g is finite even if V is divergent at short distance.

Taking Eq. (3.5) between the initial state, where the particles have momenta $\mathbf{k}_1, \mathbf{k}_2$, and the final one characterized by momenta $\mathbf{k}_1', \mathbf{k}_2'$, we obtain the following integral equation:

$$\langle \mathbf{k}_1, \mathbf{k}_2 | g | \mathbf{k}_1', \mathbf{k}_2' \rangle = \langle \mathbf{k}_1, \mathbf{k}_2 | V | \mathbf{k}_1', \mathbf{k}_2' \rangle$$

$$+ \sum_{\mathbf{t}_1, \mathbf{t}_2} \langle \mathbf{k}_1, \mathbf{k}_2 | V | \mathbf{t}_1, \mathbf{t}_2 \rangle \frac{1}{E - \epsilon_{t_1} - \epsilon_{t_2}} \langle \mathbf{t}_1, \mathbf{t}_2 | g | \mathbf{k}_1', \mathbf{k}_2' \rangle, \tag{3.6}$$

where we have introduced a complete set of two-particle intermediate states $|\mathbf{t}_1, \mathbf{t}_2\rangle$ which are eigenstates of H_0 (plane waves) with eigenvalue $\epsilon_{t_1} + \epsilon_{t_2} = \frac{t_1^2}{2m} + \frac{t_2^2}{2m}$, and the initial energy E is given by $E = \frac{k_1^2}{2m} + \frac{k_2^2}{2m}$.

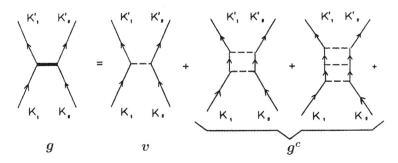

Fig. 3.1 Graphical representation of Eq. (3.6). The dashed lines represent the bare interaction.

The solution of (3.6) represents the summation of the ladder diagrams of Fig. 3.1, where the dashed line represents the interaction V. The first-order diagram [i.e. the first term in Eq. (3.6)] represents the Born approximation to the elastic scattering of two particles. All other diagrams are obtained by iteration. As in the RPA (see Chapter 10), the individual diagrams may diverge, but the sum converges.

Taking into account that in the scattering the sum of momenta is conserved, the integral equation (3.6) can be rewritten so as to stress the dependence of g on three momenta, i.e. the initial momenta $\mathbf{k}_1$ and $\mathbf{k}_2$, and the transfer momentum $\mathbf{q} = \mathbf{k}_1' - \mathbf{k}_1 = \mathbf{k}_2 - \mathbf{k}_2'$. In this way we obtain

$$g(\mathbf{k}_1, \mathbf{k}_2, \mathbf{q}) = V(\mathbf{q}) + \int \frac{d\mathbf{k}}{(2\pi)^D} \frac{V(|\mathbf{q} - \mathbf{k}|)g(\mathbf{k}_1, \mathbf{k}_2, \mathbf{k})}{\frac{k_1^2}{2m} + \frac{k_2^2}{2m} - \frac{(\mathbf{k}_1 + \mathbf{k})^2}{2m} - \frac{(\mathbf{k}_2 - \mathbf{k})^2}{2m}}. \tag{3.7}$$

This integral equation needs to be solved numerically for each $\mathbf{q}$, once $\mathbf{k}_1$ and $\mathbf{k}_2$ are fixed.

3.3   The Bethe–Goldstone Equation

The theory reported in the previous section solves the problem of elastic scattering of two particles in vacuum. The Brueckner–Hartree–Fock (BHF) theory, which is usually employed for Fermions, replaces the vacuum with a medium of particles of the same type. Due to the presence of the medium, the interaction between the two particles is modified. The way this modification is accounted for requires that interactions other than two-body ones be negligible, and approximates the many-body medium by a noninteracting Fermi gas characterized by momenta which are all smaller than the Fermi momentum.

The role of the Fermi sea is that of limiting the summation on the intermediate states in (3.6) and (3.7), so as not to include occupied states below the Fermi surface. This idea is implemented in the Bethe–Goldstone equation for the effective interaction $g(\mathbf{k}_1, \mathbf{k}_2, \mathbf{q})$ between any two particles of momenta $\mathbf{k}_1$ and $\mathbf{k}_2$:

$$g(\mathbf{k}_1, \mathbf{k}_2, \mathbf{q}) = V(\mathbf{q}) + \int \frac{d\mathbf{k}}{(2\pi)^D} V(|\mathbf{q} - \mathbf{k}|)$$

$$\times \frac{(1 - n_{\mathbf{k}_1 + \mathbf{k}})(1 - n_{\mathbf{k}_2 - \mathbf{k}})}{\frac{k_1^2}{2m} + \frac{k_2^2}{2m} - \frac{(\mathbf{k}_1 + \mathbf{k})^2}{2m} - \frac{(\mathbf{k}_2 - \mathbf{k})^2}{2m}} g(\mathbf{k}_1, \mathbf{k}_2, \mathbf{k}), \qquad (3.8)$$

where $n_{\mathbf{k}}$ is the usual Fermi distribution function at zero temperature [see Eq. (1.76)]. The $n_{\mathbf{k}}$ introduce a dependence on the medium density in the effective interaction through the Fermi momentum.

In the Boson case there is no limitation on the intermediate states because, as known, the Pauli principle plays no role in this case. Therefore, Eq. (3.7) still holds. Moreover, for Bosons, in the ground state it is possible to take $\mathbf{k}_1 = \mathbf{k}_2 = 0$. Therefore, we obtain the integral equation

$$g(\mathbf{q}) = V(\mathbf{q}) - m \int \frac{d\mathbf{k}}{(2\pi)^D} V(|\mathbf{q} - \mathbf{k}|) \frac{g(\mathbf{k})}{k^2}. \qquad (3.9)$$

The Bethe–Goldstone integral Eq. (3.8) for Fermions can be solved by standard techniques (see for example Suwa, 2002) and yields the matrix $g(\mathbf{k}_1, \mathbf{k}_2, \mathbf{q})$. From this quantity it is possible to calculate the interaction energy in the ground state, or directly the correlation energy, by taking the expectation value of the second term on the right hand side of (3.8) in a Slater determinant of plane waves.

For example, in the case where the interaction depends solely on the interparticle distance $[V = V(r)]$, the correlation energy is given by (Nagano, Singwi and Ohnishi, 1984)

$$E_c = 4E_d + 2E_x, \qquad (3.10)$$

where E_d and E_x are the direct and exchange contributions respectively (the factors 4 and 2 result from summation on spin variables), and are given by

$$E_d = \frac{1}{2} \sum_{\mathbf{k}_1, \mathbf{k}_2, \mathbf{q}} V(\mathbf{q}) H(\mathbf{k}_1, \mathbf{k}_2, \mathbf{q}) g(\mathbf{k}_1, \mathbf{k}_2, \mathbf{q}) \qquad (3.11)$$

and

$$E_x = -\frac{1}{2} \sum_{\mathbf{k}_1, \mathbf{k}_2, \mathbf{q}} V(\mathbf{k}_1 - \mathbf{k}_2 + \mathbf{q}) H(\mathbf{k}_1, \mathbf{k}_2, \mathbf{q}) g(\mathbf{k}_1, \mathbf{k}_2, \mathbf{q}), \qquad (3.12)$$

where

$$H(\mathbf{k}_1, \mathbf{k}_2, \mathbf{q}) = \frac{n_{\mathbf{k}_1}(1 - n_{\mathbf{k}_1 + \mathbf{q}}) n_{\mathbf{k}_2}(1 - n_{\mathbf{k}_2 - \mathbf{q}})}{\frac{k_1^2}{2m} + \frac{k_2^2}{2m} - \frac{(\mathbf{k}_1 + \mathbf{q})^2}{2m} - \frac{(\mathbf{k}_2 - \mathbf{q})^2}{2m}}. \qquad (3.13)$$

3.4   Examples of Application of the BHF Theory

3.4.1   *The One-Dimensional Fermion System*

As an example of the application of the BHF theory, in this subsection we consider the case of a one-dimensional system of Fermions interacting through a potential whose Fourier transform is a constant given by $V(\mathbf{q}) = 2C$ (Nagano and Singwi, 1983). In coordinate space the interaction is given by $V(x) = 2C\delta(x)$, where x is the distance between Fermions on the x axis.

Though this system is very simple and of little relevance as regards interesting physical systems such as quasi-one-dimensional conductors and quantum wires, it is very instructive to study because the exact solution for the ground state energy is known, and can be solved analytically for many models, including the BHF one.

In this case the Bethe–Goldstone equation becomes

$$g(k_1, k_2) = 2C + 2C \int \frac{dk}{(2\pi)} \frac{(1 - n_{k_1+k})(1 - n_{k_2-k})}{\frac{k_1^2}{2m} + \frac{k_2^2}{2m} - \frac{(k_1+k)^2}{2m} - \frac{(k_2-k)^2}{2m}} g(k_1, k_2), \quad (3.14)$$

whose solution is

$$g(k_1, k_2) = \frac{2C}{1 - 2C A(k_1, k_2)}, \quad (3.15)$$

where

$$A(k_1, k_2) = \int \frac{dk}{(2\pi)} \frac{(1 - n_{k_1+k})(1 - n_{k_2-k})}{\frac{k_1^2}{2m} + \frac{k_2^2}{2m} - \frac{(k_1+k)^2}{2m} - \frac{(k_2-k)^2}{2m}}. \quad (3.16)$$

The integral in Eq. (3.16) is analytic and turns out to be equal to

$$A(k_1, k_2) = \left\{ \begin{array}{ll} \dfrac{m}{\pi} \dfrac{1}{k_1 - k_2} \ln \left| \dfrac{k_2 + k_F}{k_1 + k_F} \right| & \text{if } 0 \le k_1 + k_2 \le 2k_F \\[3mm] \dfrac{m}{\pi} \dfrac{1}{k_2 - k_1} \ln \left| \dfrac{k_2 - k_F}{k_1 - k_F} \right| & \text{if } -2k_F \le k_1 + k_2 < 0 \end{array} \right\}, \quad (3.17)$$

where the Fermi momentum is connected to the linear density by $N/L = \rho = 2k_F/\pi$ (see Subsection 1.8).

The ground state energy per unit length E/L is the kinetic energy plus the interaction energy. Using $E_{\text{kin}}/L = 2\epsilon_0^F k_F/(3\pi)$ [see Eq. (1.79)], where ϵ_0^F is the Fermi energy, and the expectation value of (3.15) for the interaction energy, we find that

$$\frac{E}{L} = \frac{2}{3} \frac{k_0^F \epsilon_0^F}{\pi} + \int_{-k_F}^{k_F} \frac{dk_1}{2\pi} \int_{-k_F}^{k_F} \frac{dk_2}{2\pi} g(k_1, k_2). \quad (3.18)$$

A simple calculation yields the energy per unitary length expressed in dimensionless units $\epsilon = E/(L\epsilon_0^F k_F)$:

$$\epsilon = \frac{2}{3\pi} + \frac{C_p}{2\pi^2} \int_{-1}^{1} dx \left(\int_{-x}^{1} dy \frac{1}{1 + C_p h(x,y)} + \int_{-1}^{-x} dy \frac{1}{1 + C_p h(-x,-y)} \right), \quad (3.19)$$

where

$$h(x,y) = -\frac{1}{\pi} \frac{1}{x-y} \ln \left(\frac{y+1}{x+1} \right),$$

and we have introduced the dimensionless parameter

$$C_p = \frac{C k_F}{\epsilon_F} = \frac{4mC}{\pi \rho}.$$

For $C_p \ll 1$ ($\rho \to \infty$), Eq. (3.19) reduces to

$$\epsilon = \frac{2}{3\pi} + 8\frac{C_p}{\pi^2}, \quad (3.20)$$

which is the result one obtains in HF.

For $C_p = \infty$ ($\rho \to 0$), we have

$$\epsilon = \frac{2}{3\pi} + \frac{1}{2\pi} \times 5.120 = 1.03. \quad (3.21)$$

The exact result is $\epsilon = 8/3\pi$. Therefore, even in the extreme case of an infinite coupling constant, the BHF result does not differ from the exact one by more than 20%. For values of $C_p \leq 6$, the BHF calculation yields values which depart from the exact ones by less than 3%.

In Fig. 3.2 we show the plot of the energy ϵ as a function of the parameter C_p for the different available many-body theories and for the exact calculation. We note that the RPA and HF theories work well only for small C_p values, i.e. for small $\mathbf{r}_s$.

The STLS approximation, (which will be described later), though better than the previous ones, is not quantitatively good for $C_p > 2$. The BHF approximation works really well with the interaction used.

3.4.2 *Ultracold Highly Polarized Fermi Gases*

In this subsection we consider the problem of a single spin-down $\downarrow$ atom in the presence of a Fermi sea of ultracold spin-up $\uparrow$ atoms with density $\rho_\uparrow = k_F^3/6\pi^2$ in the vicinity of a Feshbach resonance. The $\uparrow - - \downarrow$ interaction is characterized by a zero-range force $c\delta(r)$, which in momentum space is given by the expression

$$c = \frac{4\hbar^2 \pi a/m}{1 - 4\hbar^2 \pi a/m \int \frac{d\mathbf{p}}{(2\pi)^3} \frac{m}{p^2}}, \quad (3.22)$$

which allows one to remove the ultraviolet divergences typical of calculations with a tridimensional delta force, by a suitable renormalization of the bare coupling

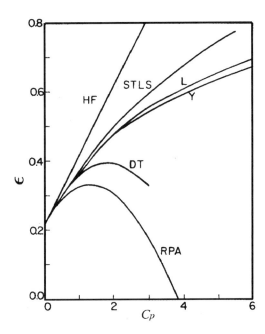

Fig. 3.2 Energy density ϵ as a function of the dimensionless parameter C_P in different theories: HF, STLS, $L \equiv$ BHF, RPA and Y, which indicates the exact result based on the Yang integral equation (Yang, 1967). DT is the dynamic theory developed by Nagano and Singwi (1983).

constant c (see also Subsection 10.3.1). The s wave scattering length a can be tuned via a Feshbach resonance from the BEC ($1/k_F a \gg 1$) to the BCS ($1/k_F a \ll -1$) limits, and the Fermi gas can be assumed to be ideal due to the suppression of higher angular momentum scattering at low temperature.

This problem is much simpler than the case of two equal spin populations in the BEC–BCS crossover (see Subsection 7.15.1) and is the simplest realization of the moving impurity problem. Moreover, it has strong similarity to other condensed matter problems, such as the Kondo problem. Highly polarized systems of this kind have been experimentally observed by many groups (see, for example, Partridge *et al.*, 2006; Zwierlein *et al.*, 2006).

To calculate the interaction energy of the $\downarrow$ atom with the Fermi sea of ultracold $\uparrow$ atoms, we shall use the Bethe-Goldstone equation

$$g(\mathbf{k}_1, \mathbf{k}_2) = c + c \int \frac{d\mathbf{k}}{(2\pi)^D} \frac{1 - n_{\mathbf{k}_1 + \mathbf{k}}}{\frac{k_1^2}{2m} + \frac{k_2^2}{2m} - \frac{(\mathbf{k}_1 + \mathbf{k})^2}{2m} - \frac{(\mathbf{k}_2 - \mathbf{k})^2}{2m}} g(\mathbf{k}_1, \mathbf{k}_2), \quad (3.23)$$

where $\mathbf{k}_2$ is the momentum of the $\downarrow$ atom and $\mathbf{k}_1$ stands for the momentum of the generic $\uparrow$ atom of the Fermi sea. Without any loss of generality we can take $\mathbf{k}_2 = 0$, and write the solution to the above equation as

$$g(\mathbf{k}_1) = \frac{c}{1 - cA(\mathbf{k}_1)}, \quad (3.24)$$

where

$$A(\mathbf{k}_1) = -\frac{m}{(2\pi)^3} \int_{|\mathbf{p}| \geq k_F} d\mathbf{p} \frac{1}{\mathbf{p} \cdot (\mathbf{p} - \mathbf{k}_1)}. \tag{3.25}$$

The interaction energy is then given by

$$\begin{aligned}
\epsilon_{\text{int}} &= \frac{1}{2} \frac{1}{(2\pi)^3} \int d\mathbf{k}_1 g(\mathbf{k}_1) \\
&= \frac{\hbar^2 a}{m(2\pi)^2} \int \frac{d\mathbf{k}_1}{1 - 4\hbar^2 \pi a/m \left[A(\mathbf{k}_1) + \int \frac{d\mathbf{p}}{(2\pi)^3} \frac{m}{p^2} \right]},
\end{aligned} \tag{3.26}$$

where we have used Eq. (3.22). Note that, in Eq. (3.26), the term $\int \frac{d\mathbf{p}}{(2\pi)^3} \frac{m}{p^2}$ removes the ultraviolet divergence arising in the calculation of $A(\mathbf{k}_1)$. One finally gets the result

$$\frac{\epsilon_{\text{int}}}{\frac{3}{5}\epsilon_F} = \frac{10 a k_F}{3\pi} \int_0^1 dx \frac{x^2}{1 - \frac{a k_F}{\pi} \left(1 - \frac{1 - x^2}{2x} \ln \frac{1-x}{1+x} \right)}. \tag{3.27}$$

In Fig. 3.3 we report the numerical result for $\frac{\epsilon_{\text{int}}}{\frac{3}{5}\epsilon_F}$ of Eq. (3.27) as a function of $1/ak_F$ together with the Monte Carlo calculation of Pilati and Giorgini (2007) for comparison. For simplicity, we have restricted ourselves to the case where there is no bound state between the ↓ and ↑ atoms. Such a bound state exists in the BEC limit $1/k_f a \to \infty$ and its effect in the intermediate regime can be taken into account in the above formalism. As one can see from the figure, the agreement of the Bethe–Goldstone calculation with the Monte Carlo one is good. It becomes

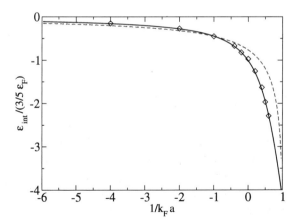

Fig. 3.3 Interaction energy (in units of free energy) of a ↓ atom in a Fermi sea of ↑ atoms as a function of $1/k_F a$. Diamonds: Monte Carlo results by Pilati and Giorgini (2007). Dashed line: The approximate Bethe–Goldstone calculation of Eqs. (3.23)–(3.27). Full line: The full Bethe–Goldstone calculation by Combescot *et al.* (2007) (see text). Courtesy of A. Recati.

almost perfect if one introduces in the g matrix the interaction energy dependence, i.e. if one calculates the g matrix with an initial energy E containing the interaction energy, $E = \epsilon_{\text{int}} + (k_1^2 + k_2^2)/2m$, instead of using the free particle initial energy, $(k_1^2 + k_2^2)/2m$, as done in Eq. (3.23). In this case $g = g(\mathbf{k}_1, \epsilon_{\text{int}})$, $A = A(\mathbf{k}_1, \epsilon_{\text{int}})$ and Eq. (3.26) for ϵ_{int} becomes an implicit equation to be solved self-consistently. This calculation has been carried out by Combescot *et al.* (2007) and their numerical result is reported in Fig. 3.3.

3.5   Numerical Results of BHF Calculation in Different Systems

Numerical solutions to the Bethe–Goldstone equation were obtained for nuclear matter, ^{3}He, and the two- and three-dimensional electron gas.

In the case of nuclear matter, the first step is to take a "realistic" nucleon–nucleon interaction, for example of the type of the soft core Reid potential (Reid, 1968). This is a phenomenological interaction which fits the properties of deuterium and the scattering data in the nonrelativistic energy range. Subsequently, one solves numerically the integral equation for the g matrix using this interaction.

The next step is to approximate the g matrix, which in general is nonlocal in coordinate space, by a sum of local functions which depend on the Fermi momentum k_F. More precisely, the effective interaction is subdivided into a central part $V_c(r)$, a spin–orbit part $V_{LS}(r)\mathbf{L} \cdot \mathbf{S}$ and a tensor part $V_T(r)S_{12}$. Then, in each spin–isospin channel (S, T) (S and T are the spin and the isospin of the pair of nucleons, respectively), one looks for a parametrization of each part in the form

$$V(r) = \sum_i C_i(k_F) e^{-\frac{r^2}{\lambda_i^2}}, \tag{3.28}$$

where the coefficients $C_i(k_F)$ and the ranges λ_i are fitted to obtain, for each value of the density:

- a good approximation for the diagonal matrix elements of the g matrix;
- general agreement with the nondiagonal matrix elements of the g matrix.

This program has been carried out, for example by Sprung and Banerjee (1971), who looked for different k_F dependencies of the g matrix.

The energy per nucleon, E/A, obtained by the g matrix as computed directly from the Reid potential either without any parametrization or with some of its parametrizations ($G_0 - G_3$), is plotted in Fig. 3.4 as a function of k_F.

The saturation point (i.e. the minimum of the curves) is found to correspond to a density value $k_F = 1.43\,\text{fm}^{-1}$, which is too high with respect to the experimental one, $1.37\,\text{fm}^{-1}$, and to an energy value which is too small ($E/A = -11\,\text{MeV}$, to be compared to the experimental value of $-17.04\,\text{MeV}$). This is due to the fact that the g matrix takes into account only the two-body contributions to the ladder

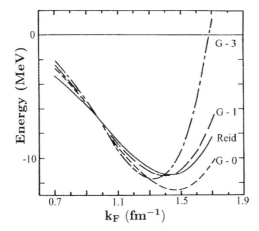

Fig. 3.4 Energy per nucleon as a function of k_F obtained in BHF using the Reid potential and some of its parametrizations $G_0 - G_3$.

diagrams. The three-body and four-body contributions to E/A are important, and when added to the previous ones make the calculated energy values agree well with the experimental value. Since it is very difficult to calculate accurately the three-body and four-body contributions to the effective interaction, what one does is simply to renormalize in an empirical way the effective interaction by slightly changing some of the Gaussian ranges in (3.28). In this way, by introducing two further parameters, it is possible to have saturation at $k_F = 1.35$ fm^{-1} and $E/A = -16.5$ MeV. Such renormalized interaction is then used for HF calculations in finite nuclei using the local density approximation, in which k_F is replaced by its relation with density.

The HF calculations in finite nuclei, using the g matrix parametrizations and the local density approximation, were developed by Negele (1970) and by Campi and Sprung (1972). These calculations yield very satisfying results. In particular, the binding energies and the densities of closed shell nuclei are very well reproduced. More recent g matrix calculations, as well as further references, may be found in Song *et al.* (1998).

In the case of ^{3}He and of the electron gas, the problem of the g matrix is simpler because the bare interaction is much simpler than in nuclear matter. Nonetheless, for ^{3}He the numerical solution yields unsatisfying results for the binding energy. This is due to the fact that ^{3}He is a highly correlated system, in which many-body interactions play a fundamental role. For example, they produce an effective mass that is more than three times greater than the bare mass of ^{3}He atoms. The effective interaction as computed using the Bethe–Goldstone equation and a two-body potential is completely unable to reproduce such a value. Therefore, in the following, we will concern ourselves with the electron gas, for which the situation is much more satisfying.

Table 3.1. Correlation energies per particle (eV) of the 2DEG in different theories.

r_s	ϵ_c^{MC}	ϵ_c^{L}	ϵ_c^{RPA}
1	−2.99	−2.60	−5.40
5	−1.34	−1.23	−3.10
10	−0.83	−0.76	−2.24
20	−0.48	−0.44	−1.61

In the case of the Coulomb potential, once Eq. (3.8) for g is solved, one computes the correlation energy directly from Eqs. (3.10)–(3.13). This was done by Nagano, Singwi and Ohnishi (1984), Takayanagi and Lipparini (1996) and Suwa, Takayanagi and Lipparini (2003) in 2D, and by Lowy and Brown (1975), Bedell and Brown (1978) and Lipparini, Serra and Takayanagi (1994) in 3D.

The results for the 2D electron gas are reported in Table 3.1, together with the Monte Carlo results of Tanatar–Ceperley (Tanatar and Ceperley, 1989), as well as the RPA results, for the sake of comparison. Energies are in eV. From the table we see that there is very good agreement between Monte Carlo and the ladder theory for large values of r_s. At small values of r_s long-range correlations, which are taken into account by the RPA theory, are important. As we discussed previously, and as can be seen from the table, RPA is the good theory only for $r_s \to 0$. A similar level of agreement between the ladder approximation and Monte Carlo calculations is obtained in three dimensions.

In order to stress further the role of dynamic correlations, in Fig. 3.5 we plot the pair correlation function $g_{\uparrow\downarrow}(r)$ for antiparallel spins in 3D, computed in the ladder approximation and compared to the RPA (see Chapter 10), for different density values. In this case statistical exchange correlations are absent, and the HF calculation yields a constant $g_{\uparrow\downarrow}(r)$ value, equal to 1.

From this figure we can see that at short distance the correlation function is very different from the classical value, and that the RPA correlations become unphysical (i.e. negative) at short distance, faster and faster as the r_s value is increased.

3.6 The g Matrix for the 2D Electron Gas

In this subsection we report in detail the solution of the Bethe–Goldstone equation for the g matrix, for the case of the 2D electron gas, and derive some approximations for g which allow us to use it in a simple and practical way for both static and dynamic calculations. These approximations were used also in nuclear physics and for the 3D electron gas.

For the 2D and 3D electron gas, where the HF theory is well defined, it is better to solve the Bethe–Goldstone integral equation for the quantity

$$g^c = g - V, \qquad (3.29)$$

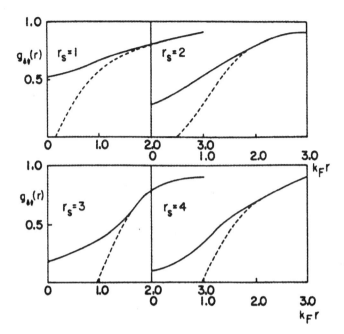

Fig. 3.5 Pair correlation function $g(r)$ for antiparallel spins. This function represents the correlations in the absence of exchange. The full line is the result of the ladder approximation. The dashed line is the RPA result.

rather than for the g matrix, where V is the Coulomb potential [i.e. the first term in Eq. (3.8)], and g^c is the correlation due to multiple-scattering processes [i.e. the second term in Eq. (3.8)]. g^c obeys the integral equation

$$g^c = V\frac{Q}{E - H_0}V + V\frac{Q}{E - H_0}g^c, \tag{3.30}$$

where, for simplicity, we have indicated by Q the Pauli operator which limits the intermediate states to the excited states with momentum greater than the Fermi momentum, and changes the Lippman–Schwinger equation (3.6) into the Bethe–Goldstone (3.8).

The solution of the integral Eq. (3.30), evaluated on a plane wave Slater determinant, gives directly the correlation energy of the electron gas.

3.6.1 *Decomposition in Partial Waves*

As we saw previously, the solution of the Bethe–Goldstone equation depends on three momenta: those of the initial states $\mathbf{k}_1$ and $\mathbf{k}_2$ and the transfer momentum $\mathbf{q}$. These three momenta may be replaced by $\mathbf{p}_1 = \frac{\mathbf{k}_1 - \mathbf{k}_2}{2}$, $\mathbf{p}_2 = \frac{\mathbf{k}_1' - \mathbf{k}_2'}{2}$, $\mathbf{P} = \mathbf{k}_1 + \mathbf{k}_2$, i.e. the relative momenta of the initial and final states and the total momentum. This means that it is possible to put the matrix elements between two-particle states of Fig. 3.1 into the form

$$\langle \mathbf{k}_1 \mathbf{k}_2 | g^c | \mathbf{k}_1' \mathbf{k}_2' \rangle = (2\pi)^2 \delta(\mathbf{k}_1 + \mathbf{k}_2 - \mathbf{k}_1' - \mathbf{k}_2') \langle \mathbf{p}_1 | g_P^c | \mathbf{p}_2 \rangle, \qquad (3.31)$$

where $\mathbf{p}_1 = (p_1, \alpha_1)$ and $\mathbf{p}_2 = (p_2, \alpha_2)$.

The angles α_1 and α_2 are measured starting from the fixed direction of $\mathbf{P}$. Note that g_P^c can be interpreted as an effective interaction operating on the relative wave function with fixed $\mathbf{P}$. In what follows we describe the techniques suitable for solving Eq. (3.30) and for obtaining the matrix elements $\langle \mathbf{p}_1 | g_P^c | \mathbf{p}_2 \rangle$ in (relative) momentum space, for a fixed value of $\mathbf{P}$.

The Coulomb interaction in (relative) coordinate space is given by

$$\langle \mathbf{r}_1 | V | \mathbf{r}_2 \rangle = \delta(\mathbf{r}_1 - \mathbf{r}_2) \frac{1}{|\mathbf{r}_1 - \mathbf{r}_2|} \exp(-\mu |\mathbf{r}_1 - \mathbf{r}_2|), \qquad (3.32)$$

where $\mathbf{r}_1$ and $\mathbf{r}_2$ are the relative coordinates of the initial and final states, respectively, and we have introduced the damping factor $\exp(-\mu |\mathbf{r}_1 - \mathbf{r}_2|)$ using the cutoff parameter $\mu (> 0)$, which will be discussed later. Using the following expansion for the plane wave, which is valid in 2D,

$$e^{i\mathbf{p}\cdot\mathbf{r}} = \sum_{n=-\infty}^{\infty} i^n J_n(pr) e^{in(\alpha-\phi)}, \qquad (3.33)$$

where $\mathbf{p} = (p, \alpha)$ and $\mathbf{r} = (r, \phi)$, and J_n is the Bessel function, we may put the matrix element in momentum space in the form

$$\langle \mathbf{p}_1 | V | \mathbf{p}_2 \rangle = \int d\mathbf{r}_1 d\mathbf{r}_2 e^{-i\mathbf{p}_1 \cdot \mathbf{r}_1} \langle \mathbf{r}_1 | V | \mathbf{r}_2 \rangle e^{i\mathbf{p}_2 \cdot \mathbf{r}_2}$$

$$= \sum_{n_1, n_2 = -\infty}^{\infty} (-i)^{n_1} \frac{e^{in_1\alpha_1}}{\sqrt{2\pi}} \langle p_1 n_1 | V | p_2 n_2 \rangle i^{n_2} \frac{e^{-in_2\alpha_2}}{\sqrt{2\pi}}, \qquad (3.34)$$

where the partial wave components are calculated by

$$\langle p_1 n_1 | V | p_2 n_2 \rangle = \delta_{n_1 n_2} (2\pi)^2 \int r dr \, J_{n_1}(p_1 r) J_{n_1}(p_2 r) \frac{e^{-\mu r}}{r}$$

$$= \delta_{n_1 n_2} \frac{4\pi}{\sqrt{p_1 p_2}} Q_{|n_1| - \frac{1}{2}} \left(\frac{p_1^2 + p_2^2 + \mu^2}{2 p_1 p_2} \right), \qquad (3.35)$$

and Q_n is the Legendre function of the second kind. Note that the above expression becomes very large for $p_1 \sim p_2$. In fact, it would diverge for $p_1 = p_2$ in the absence of the cutoff μ.

Let us now consider the Pauli operator

$$\langle \mathbf{p}_1 | Q_P | \mathbf{p}_2 \rangle = (2\pi)^2 \delta(\mathbf{p}_1 - \mathbf{p}_2) \theta \left(\left| \frac{\mathbf{P}}{2} + \mathbf{p}_1 \right| - k_F \right) \theta \left(\left| \frac{\mathbf{P}}{2} - \mathbf{p}_1 \right| - k_F \right), \qquad (3.36)$$

which depends explicitly on the center-of-mass momentum $\mathbf{P}$ and on the Fermi momentum k_F, and makes the g matrix dependent on these quantities as well. The partial wave decomposition of the Pauli operator is given by

$$\langle \mathbf{p}_1 | Q_P | \mathbf{p}_2 \rangle = (2\pi)^2 \frac{\delta(p_1 - p_2)}{p_1}$$

$$\times \sum_{n_1, n_2} (-i)^{n_1} \frac{e^{in_1\alpha_1}}{\sqrt{2\pi}} \langle n_1 | Q_P(p_1) | n_2 \rangle i^{n_2} \frac{e^{-in_2\alpha_2}}{\sqrt{2\pi}}, \qquad (3.37)$$

where

$$\langle n_1 | Q_P(p) | n_2 \rangle = \begin{cases} \delta_{n_1,n_2} & \cdots \quad k_F \leq \left| \dfrac{P}{2} - p \right| \\ \dfrac{1 + (-1)^N}{2} \dfrac{2\gamma}{\pi} \dfrac{\sin N\gamma}{N\gamma} & \cdots \quad \left| \dfrac{P}{2} - p \right| \leq k_F \leq \sqrt{\dfrac{P^2}{4} + p^2} \\ 0 & \cdots \text{ otherwise.} \end{cases}$$

$$(3.38)$$

In the above expression we have defined

$$\gamma = \sin^{-1} \left(\frac{\frac{P^2}{4} + p^2 - k_F^2}{Pp} \right), \qquad N = n_1 - n_2, \qquad (3.39)$$

and we have assumed that

$$\frac{\sin N\gamma}{N\gamma} = 1 \text{ for } N = 0. \qquad (3.40)$$

Note that the Pauli operator $\langle \mathbf{p}_1 | Q_P | \mathbf{p}_2 \rangle$ depends separately on the angles α_1 and α_2, and thus is not invariant under rotations. Therefore, the above expression for the exact Pauli operator mixes the components with different angular momentum n, and makes the numerical calculation rather complex.

The problem of the coupling of different partial waves may be avoided by defining a Pauli operator which is averaged over angles, $\bar{Q}_P$, by averaging over angle α_1 in the previous expression (Bethe and Goldstone, 1957; Lowy and Brown, 1975). In this way, Eq. (3.38) is replaced by

$$\langle n_1 | \bar{Q}_P(p) | n_2 \rangle = \delta_{n_1,n_2} \begin{cases} 1 & \cdots \quad k_F \leq \left| \dfrac{P}{2} - p \right| \\ 2\gamma/\pi & \cdots \quad \left| \dfrac{P}{2} - p \right| \leq k_F \leq \sqrt{\dfrac{P^2}{4} + p^2} \\ 0 & \cdots \text{ otherwise,} \end{cases} \qquad (3.41)$$

which is diagonal with respect to the angular momentum n. The numerical calculations that we report here (Suwa *et al.*, 2003) were carried out both with the exact

Pauli operator of (3.38) and with the angle-averaged Pauli operator of (3.41). As we will discuss in what follows, the difference between the two calculations is small.

The matrix g_P^c can then be expanded in the same way as Eq. (3.34) for V, as follows:

$$\langle \mathbf{p}_1 | g_P^c | \mathbf{p}_2 \rangle = \sum_{n_1, n_2 = -\infty}^{\infty} (-i)^{n_1} \frac{e^{in_1\alpha_1}}{\sqrt{2\pi}} \langle p_1 n_1 | g_P^c | p_2 n_2 \rangle i^{n_2} \frac{e^{-in_2\alpha_2}}{\sqrt{2\pi}}. \qquad (3.42)$$

Using the above partial wave expansions, Eq. (3.30) reduces to the following one-dimensional integral equation:

$$\langle p_1 n_1 | g_P^c | p_2 n_2 \rangle = \sum_{n,n'} \int_0^\infty \frac{p\,dp}{(2\pi)^2} \langle p_1 n_1 | V | p n \rangle \frac{\langle n | Q_P(p) | n' \rangle}{p_1^2 - p^2 + i\eta} \langle p n' | V | p_2 n_2 \rangle$$

$$+ \sum_{n,n'} \int_0^\infty \frac{p\,dp}{(2\pi)^2} \langle p_1 n_1 | V | p n \rangle \frac{\langle n | Q_P(p) | n' \rangle}{p_1^2 - p^2 + i\eta} \langle p n' | g_P^c | p_2 n_2 \rangle,$$

$$(3.43)$$

where we have put $E = p_1^2$ for the (vector) initial state with relative momentum p_1 (note that the reduced mass is one half). In other words, we choose the initial energy E in such a way that the vector initial state is on its energy shell. As a consequence, if the two particles in the initial states are below the Fermi energy, then the energy denominator of (3.43) cannot vanish, and the matrix elements of the g matrix are real and symmetric. Since we wish to determine an effective interaction to describe the ground state and the low-energy excited states, we may assume that the above condition is fulfilled in cases we are interested in. In any case, in order to study the overall structure of g, it is worth calculating it also in the cases where one or both particles in the ground states are outside the Fermi surface. In these cases, the initial energy is large enough to allow real scattering processes to take place, and the g matrix becomes complex. In these cases too, we retain only the real part of the g matrix, and g is still real and symmetric in momentum space. It is easy to show that

$$\langle p_1 n_1 | g_P^c | p_2 n_2 \rangle = \langle p_1 - n_1 | g_P^c | p_2 - n_2 \rangle.$$

The coupled integral Eq. (3.43) may be solved, for example, by the method of matrix inversion to yield $\langle p_1 n_1 | g_P^c | p_2 n_2 \rangle$, and thus the matrix element $\langle \mathbf{p}_1 | g_P^c | \mathbf{p}_2 \rangle$ of Eq. (3.42). In any case, it is clear that the dependence of the g^c matrix on $\mathbf{P}$ in momentum space leads to a g^c matrix in coordinate space which depends on $\mathbf{R}_1 - \mathbf{R}_2$, i.e. the difference of the coordinates of the center of mass of the two particles between the initial and final states. Obviously, an effective interaction of this kind is not convenient for practical purposes. In order to get an effective interaction which depends solely on the relative coordinates of the two particles, in (3.42) the center-of-mass momentum P is replaced by the square root of its mean square value $\bar{P} = \sqrt{\langle \mathbf{P}^2 \rangle}$ for all two-particle states below the Fermi level. At the same time Eq. (3.42) is averaged over the directions of $\mathbf{P}$ in order to make the resulting expression invariant under rotations. After these manipulations it is possible to define the matrix element of g^c as a function of $\mathbf{p}_1$ and $\mathbf{p}_2$:

$$\langle \mathbf{p}_1 | \, g^c \, | \mathbf{p}_2 \rangle = \frac{1}{2\pi} \sum_{n=-\infty}^{\infty} e^{in(\alpha_1 - \alpha_2)} \langle p_1 n | \, g_{\bar{P}}^c \, | p_2 n \rangle$$

$$= \frac{1}{2\pi} \sum_{n=0}^{\infty} \epsilon_n \langle p_1 n | g^c | p_2 n \rangle \cos n(\alpha_1 - \alpha_2), \qquad (3.44)$$

where we have defined

$$\epsilon_n = \begin{cases} 1 & \text{for } n = 1 \\ 2 & \text{for } n \neq 1 \end{cases}, \qquad (3.45)$$

and have used the fact that the matrix elements

$$\langle p_1 n | g^c | p_2 n \rangle = \langle p_1 - n | g^c | p_2 - n \rangle$$

are real. The above expression for $\langle \mathbf{p}_1 | g^c | \mathbf{p}_2 \rangle$ is real and depends only on the combination $\alpha_1 - \alpha_2$, thus showing explicitly that it is invariant under the simultaneous rotation of $\mathbf{p}_1$ and $\mathbf{p}_2$. This means that in coordinate space the g^c matrix depends only on the initial and final relative coordinates and is invariant under rotations. By taking the Fourier transform of (3.44), one arrives at the effective interaction

$$\langle \mathbf{r}_1 | g^c | \mathbf{r}_2 \rangle = \frac{1}{2\pi} \sum_{n=0}^{\infty} \epsilon_n \, \langle r_1 n | g^c | r_2 n \rangle \cos n(\phi_1 - \phi_2), \qquad (3.46)$$

where $\mathbf{r}_i = (r_i, \phi_i)$ and

$$\langle r_1 n | g^c | r_2 n \rangle = \frac{1}{(2\pi)^2} \int p_1 dp_1 p_2 dp_2 \, J_n(p_1 r_1) J_n(p_2 r_2) \langle p_1 n | g^c | p_2 n \rangle. \qquad (3.47)$$

The interaction (3.46) has a finite range and is not diagonal in the initial and final relative coordinates (nonlocal interaction), rendering this rather useless for practical applications. Therefore, in the following, we show how to turn it into a simple and useful interaction.

3.6.2 *The Separable Approximation*

In order to put the matrix g^c ($\langle \mathbf{r}_1 | g^c | \mathbf{r}_2 \rangle$) into the form of an effective interaction that can be used for practical purposes, we need two successive steps. The first one corresponds to the so-called separable approximation, which will be explained and justified in what follows; the second step is the g^c expansion, which will be described in the next subsection.

In the separable approximation it is assumed that the matrix elements of g^c of Eq. (3.44) in momentum space may be approximated by the product of a local factor $v(q)$ and a nonlocal factor $c(p)$, as follows:

$$\langle \mathbf{p}_1 | g^c | \mathbf{p}_2 \rangle = c\left(\frac{1}{2} |\mathbf{p}_1 + \mathbf{p}_2| \right) v(|\mathbf{p}_1 - \mathbf{p}_2|) = c(p)v(q), \qquad (3.48)$$

where we have introduced $p = |\mathbf{p}| = |\mathbf{p}_1 + \mathbf{p}_2|/2$ and $q = |\mathbf{q}| = |\mathbf{p}_1 - \mathbf{p}_2|$. This expression assumes that the p and q dependencies of g^c are separable. The form of g^c in (3.48) is a different way of representing the g matrix, which, contrary to the one described in Subsection (3.5) in the case of nuclear matter, takes into account the nonlocal parts of g^c.

Taking the Fourier transform of the above separable form, we obtain the g^c matrix in coordinate space:

$$\langle \mathbf{r}_1 | g^c | \mathbf{r}_2 \rangle = \tilde{c}(|\mathbf{r}_1 - \mathbf{r}_2|)\tilde{v}\left(\frac{1}{2} |\mathbf{r}_1 + \mathbf{r}_2| \right) = \tilde{c}(s)\tilde{v}(r), \qquad (3.49)$$

where $s = |\mathbf{s}| = |\mathbf{r}_1 - \mathbf{r}_2|$ and $r = |\mathbf{r}| = |\mathbf{r}_1 + \mathbf{r}_2|/2$.

This is still an intricate nonlocal interaction depending on the s and r coordinates separately. We note that in the case where $\tilde{c}(|\mathbf{r}_1 - \mathbf{r}_2|) \propto \delta(\mathbf{r}_1 - \mathbf{r}_2)$, the above interaction becomes the usual two-body interaction which is diagonal in the relative coordinates of the two interacting particles.

In the separable approximation, the local part $v(q)$ and the nonlocal part $c(p)$ of Eq. (3.48) are computed as follows. First of all, we eliminate the ambiguity in the normalization of $v(q)$ and $c(p)$ as defined in (3.48), by putting $c(0) = 1$. Next, if we let $\mathbf{p}_1 = \mathbf{q}/2$ and $\mathbf{p}_2 = -\mathbf{q}/2$ in (3.44) and (3.48), we have

$$\left\langle \frac{\mathbf{q}}{2} \middle| g^c \middle| -\frac{\mathbf{q}}{2} \right\rangle = \frac{1}{2\pi} \sum_{n=0}^{\infty} \epsilon_n (-1)^n \left\langle \frac{q}{2} n_1 \middle| g^c \middle| \frac{q}{2} n_2 \right\rangle$$

$$= c(0)v(q) = v(q). \qquad (3.50)$$

In the same way we have

$$\langle \mathbf{p} | g^c | \mathbf{p} \rangle = \frac{1}{2\pi} \sum_{n=0}^{\infty} \epsilon_n \langle pn_1 | g^c | pn_2 \rangle = c(p)v(0). \qquad (3.51)$$

From these equations we derive the following expressions for $v(q)$ and $c(p)$:

$$v(q) = \frac{1}{2\pi} \sum_{n=0}^{\infty} \epsilon_n (-1)^n \left\langle \frac{q}{2} n \middle| g^c \middle| \frac{q}{2} n \right\rangle, \qquad (3.52)$$

$$c(p) = \frac{1}{v(0)} \frac{1}{2\pi} \sum_{n=0}^{\infty} \epsilon_n \langle pn | g^c | pn \rangle. \qquad (3.53)$$

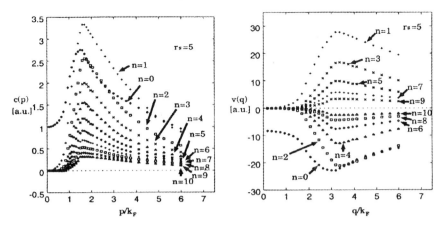

Fig. 3.6 Partial waves of $c(p)$ and $v(q)$ based on Eqs. (3.53) and (3.52) at $r_s = 5$ in atomic units, where $p = |\mathbf{p}_1 + \mathbf{p}_2|/2$ and $q = |\mathbf{p}_1 - \mathbf{p}_2|$.

It is clear that $v(q)$ is evaluated using the exchange process for forward scattering and thus it has the phase factor $(-1)^n$, which is absent in $c(p)$, since $c(p)$ is computed by the direct process.

In Fig. 3.6, for $r_s = 5$ we display how each partial wave contributes to the sum in (3.52) and (3.53), for values of q and p up to several times the Fermi momentum to show the overall structure of the interaction. It can be seen that all the partial wave contributions have a rapid increase at about $p \sim k_F$ for $c(p)$, and $q/2 \sim k_F$ for $v(q)$. This behavior can be explained in the following way. Assume that we have two particles below the Fermi energy; then the condition $p \sim k_F$ requires that the relative initial and final momenta of the two particles fulfill $p_2 \sim p_1 \sim k_F$, which in turn means that both particles are near the Fermi surface. Then these particles may be easily excited from the Fermi sea with a large amplitude because only a small transfer momentum is required, and this leads to a rapid increase in each partial wave component $\langle pn| \, g^c \, |pn\rangle$ for $p \sim k_F$. It is evident that this argument explains the increase of each partial wave component of $v(q)$ for $q/2 \sim k_F$.

The resulting interactions, $c(p)$ and $v(q)$, are shown in Fig. 3.7. It is seen that all partial waves contribute coherently in $c(p)$, so that convergence is slow and exhibits a very pronounced peak for $p \sim k_F$. For $v(q)$, Eq. (3.52) is the sum of an alternate series and converges quickly; the resulting peak at $q/2 \sim k_F$ is not very high. Note, further, that if one looks at the region where $q, p \ll k_F$ and from which the effective interaction will be derived later, one sees that only a few partial waves are needed in order to obtain a converging result.

The separable approximation can be justified only *a posteriori*, by comparing the matrix elements using the two expressions for the g^c matrix, i.e. the original

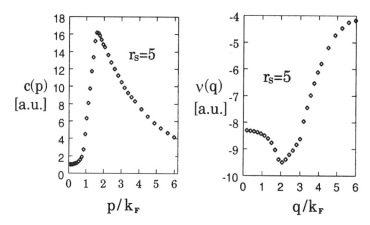

Fig. 3.7 Local $v(q)$ and nonlocal $c(p)$ components, in atomic units, of the correction to the g matrix, g^c, written in the separable form of Eq. (3.48), as functions of the momenta $q = |\mathbf{p}_1 - \mathbf{p}_2|$ and $p = |\mathbf{p}_1 + \mathbf{p}_2|/2$.

one (3.44) and the separable form (3.48). Here we will report some comparisons. In Fig. 3.8 we compare $\langle \mathbf{p}_1|g^c|\mathbf{p}_2 \rangle$ of Eq. (3.44) and $c(\frac{1}{2}|\mathbf{p}_1 + \mathbf{p}_2|)v(|\mathbf{p}_1 - \mathbf{p}_2|)$ for the cases where the two vectors $\mathbf{p}_1$ and $\mathbf{p}_2$ obey $p_1 = p_2 = p$, and for the angles $\alpha = \alpha_1 - \alpha_2$ equal to $\pi/4$, $\pi/2$ and $3\pi/4$. Note that for $\alpha = 0, \pi$ the matrix elements of g^c are used to compute $c(p)$ and $v(q)$, and so can be deduced exactly from the separable form. From the figure it is seen that the separable approximation works extremely well for $p \leq k_F$, where the two particles are below the Fermi surface in the initial state and no real scattering can take place. For $p \geq k_F$ there is no way of putting both particles within the Fermi circle. In this case the initial energy is high enough to allow real scattering processes and to produce a phase shift in the asymptotic form of the relative wave function. This change in the physical situation produces a slope change in the matrix element $\langle \mathbf{p}_1|\, g^c \,|\mathbf{p}_2 \rangle$ of Eq. (3.44) to $p \sim k_F$ (threshold effect), which cannot be reproduced by the separable approximation, as seen in the figure. Moreover, in the cases where $p_1 \neq p_2$, one finds that the separable approximation works well for all cases where $p_1, p_2 \leq k_F$. This means that the matrix element $\langle \mathbf{p}_1|\, g^c \,|\mathbf{p}_2 \rangle$ is very well approximated by the separable potential $c(p)v(q)$ in the range $p \leq k_F$ and $q \leq 2k_F$, i.e. for all particles below the Fermi surface.

Finally, we note that the matrix element g^c is negative, as can be seen from Figs. 3.6 and 3.7. This fact can be understood as follows: the multiple scattering processes account for the effects of short range interactions which cause distortion of the uncorrelated many-body function, which are such as to minimize the repulsive potential of other electrons, and so to lower the total system energy. The g matrix turns these effects into an effective interaction for the uncorrelated states, and so is necessarily an attractive interaction.

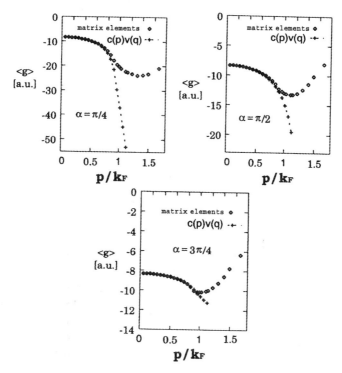

Fig. 3.8 Comparison between the matrix elements $\langle \mathbf{p}_1|g^c|\mathbf{p}_2\rangle$ and the separable potential $c(p)v(q) = c(\frac{1}{2}|\mathbf{p}_1 + \mathbf{p}_2|)v(|\mathbf{p}_1 - \mathbf{p}_2|)$, as a function of $p_1 = p_2 = p$ for $\alpha = \alpha_1 - \alpha_2$ equal to $\pi/4, \pi/2$ and $3\pi/4$ to $r_s = 5$.

3.6.3   *The g Matrix Expansion*

From the separable potential $c(p)v(q)$ in momentum space it is in principle possible to compute the potential $\tilde{c}(s)\tilde{v}(r)$ of Eq. (3.49) in coordinate space. However, this potential would be a complex nonlocal interaction with a finite range for both r and s, and would be of little use for practical applications. Therefore, a final approximation is made (i.e. the g matrix expansion) which stems from the inspection of the matrix elements of the separable potential $\tilde{c}(s)\tilde{v}(r)$, in order to get a simpler expression of the interaction. The matrix elements of g^c between any two states $|\psi\rangle$ and $|\varphi\rangle$ in the separable approximation can be written as

$$\langle\psi|g^c|\varphi\rangle = \int d\mathbf{r}_1 d\mathbf{r}_2 \psi^*(\mathbf{r}_1)\langle\mathbf{r}_1|g^c|\mathbf{r}_2\rangle\varphi(\mathbf{r}_2)$$

$$= \int d\mathbf{r}_1 d\mathbf{r}_2 \psi^* \left(\mathbf{r} + \frac{\mathbf{s}}{2}\right) \tilde{c}(s)\tilde{v}(r)\varphi \left(\mathbf{r} - \frac{\mathbf{s}}{2}\right)$$

$$= \int d\mathbf{r}\psi^*(\mathbf{r})g^c(\mathbf{r}, \nabla)\varphi(\mathbf{r}). \qquad (3.54)$$

In the previous expression we have defined

$$g^c(\mathbf{r}, \nabla) = \int d\mathbf{s} \exp\left(\frac{\mathbf{s}}{2} \cdot \overleftarrow{\nabla}\right) \tilde{c}(s)\tilde{v}(r) \exp\left(-\frac{\mathbf{s}}{2} \cdot \overrightarrow{\nabla}\right), \qquad (3.55)$$

where $\nabla = i\mathbf{p} = \partial/\partial\mathbf{r}$ operates on the relative coordinate of the two interacting particles and $\overleftarrow{\nabla}$ operates on the left, whereas $\overrightarrow{\nabla}$ operates on the right.

In the above expression for $g^c(\mathbf{r}, \nabla)$ we can assume that the maximum value of ∇ is of the order of k_F for particles in the ground state and in the low-energy excited states. Therefore, we can evaluate Eq. (3.55) by expanding the exponential and retaining the first two terms (g matrix expansion), provided that $k_F r_c \le 1$, where r_c is the range of $\tilde{c}(s)$. From Fig. 3.6, this range is estimated to be of the order of $r_c \sim 1/k_F$ due to the behavior of $c(p)$ in the small momentum region, where the separable approximation holds. Therefore, the above condition for expanding Eq. (3.55) is only marginally fulfilled. We will discuss this point again in the next subsection, where we will compare the correlation energies as computed with and without the expansion. We will find that the correlation energy is underestimated by the g matrix expansion, and this is the price to be paid in order to obtain a simple expression for the effective interaction.

By expanding the exponential functions of (3.55) and retaining only terms up to second order in ∇, we have

$$g^c(\mathbf{r}, \nabla) = \int d\mathbf{s}\left(1 + \frac{\mathbf{s}^2}{16}\overleftarrow{\nabla}^2 + \frac{\mathbf{s}}{2}\cdot\overleftarrow{\nabla}\right)\tilde{c}(s)\tilde{v}(r)\left(1 + \frac{\mathbf{s}^2}{16}\overrightarrow{\nabla}^2 - \frac{\mathbf{s}}{2}\cdot\overrightarrow{\nabla}\right)$$

$$= c_1\tilde{v}(r) + \frac{c_3}{16}\{\overleftarrow{\nabla}^2\tilde{v}(r) + \tilde{v}(r)\overrightarrow{\nabla}^2 - 2\overleftarrow{\nabla}\tilde{v}(r)\cdot\overrightarrow{\nabla}\}, \qquad (3.56)$$

where we have defined the momenta $c(s)$ as

$$c_i = 2\pi\int_0^\infty ds\, s^i \tilde{c}(s), \quad i = 1, 3, \ldots. \qquad (3.57)$$

Here $\tilde{v}(r)$ is further expanded around the zero-range interaction as follows. The Fourier transform of $\tilde{v}(r)$ can be developed in power series of q^2 as

$$v(q) = \int d\mathbf{r}\, e^{-i\mathbf{q}\cdot\mathbf{r}}\tilde{v}(r) = 2\pi\int r\, dr\, \tilde{v}(r)J_0(qr)$$

$$= v_1 - \frac{v_3}{4}q^2 + \frac{v_5}{64}q^4 + \cdots, \qquad (3.58)$$

where the momenta v_i of $\tilde{v}(r)$ are defined as

$$v_i = 2\pi\int_0^\infty dr\, r^i \tilde{v}(r), \quad i = 1, 3, \ldots. \qquad (3.59)$$

We then Fourier-transform (3.58) and retain terms up to second order in ∇, which yields

$$\tilde{v}(r) = \int \frac{d\mathbf{q}}{(2\pi)^2} e^{i\mathbf{q}\cdot\mathbf{r}} v(q) \cong v_1\delta(\mathbf{r}) + \frac{v_3}{4}\{\nabla^2\delta(\mathbf{r})\}$$

$$= v_1\delta(\mathbf{r}) + \frac{v_3}{4}\{\overleftarrow{\nabla}^2\delta(\mathbf{r}) + \delta(\mathbf{r})\overrightarrow{\nabla}^2 + 2\overleftarrow{\nabla}\delta(\mathbf{r})\cdot\overrightarrow{\nabla}\}. \tag{3.60}$$

It can be shown that the maximum value of q in (3.58), which is relevant for particles below the Fermi surface, is about $2k_F$. It can be further noted from Fig. 3.6 that the range of $\tilde{v}(r)$ is of the order of $r_v \sim 1/2k_F$, which shows that the maximum value of the exponent $\mathbf{q}\cdot\mathbf{r}$ in (3.60) is about $2k_F r_v \sim 1$. This justifies the above expansion at the same level of accuracy as the expansion of Eq. (3.55).

By substituting (3.60) into (3.56), we arrive at the following final expression for the effective interaction:

$$g^c(\mathbf{r}, \nabla) = u\delta(\mathbf{r}) + v\{\overleftarrow{\nabla}^2\delta(\mathbf{r}) + \delta(\mathbf{r})\overrightarrow{\nabla}^2\} + 2w\overleftarrow{\nabla}\delta(\mathbf{r})\cdot\overrightarrow{\nabla}, \tag{3.61}$$

where the coefficients u, v and w are defined as

$$u = v_1, \tag{3.62}$$

$$v = \frac{1}{16}v_1c_3 + \frac{1}{4}v_3, \tag{3.63}$$

$$w = -\frac{1}{16}v_1c_3 + \frac{1}{4}v_3. \tag{3.64}$$

The effective interaction of (3.61) is a zero-range interaction depending on the momentum, of the same type as the Skyrme ones largely used in nuclear physics (Vautherin and Brink, 1972; see Chapter 4). The c_i and v_i momenta can be computed numerically from the potential in momentum space as

$$c_1 = c(0) = 1,$$

$$c_3 = -2\left.\frac{d^2}{dp^2}c(p)\right|_{p=0},$$

$$v_1 = v(0),$$

$$v_3 = -2\left.\frac{d^2}{dq^2}v(q)\right|_{q=0}.$$

These expressions, together with (3.52) and (3.53), show that in order to obtain the effective interaction of (3.61), it is enough to know the diagonal matrix elements $\langle pn|g^c|pn\rangle$ of the g^c matrix in the low-energy limit ($p \sim 0$). A closer inspection of (3.44) shows that only the $\langle pn|g^c|pn\rangle$ matrix elements with $n = 0, 1$ contribute to the above coefficients. This is a direct consequence of the fact that the first two terms on the right hand side of (3.61) represent the s wave interaction ($n = 0$), and the third term the p wave one ($n = 1$). In other words, the g matrix expansion only simulates the s and p wave matrix elements of the separable potential, and neglects all other components with higher angular momentum. However, it should be noted that the s and p components are enough to describe the low-energy phenomena.

3.6.4   *Numerical Results and Discussion*

First of all, we discuss the role of the Pauli operator in numerical calculations. If we take the angle-averaged operator of (3.41), it is enough to retain only terms with $n = 0, 1$ in the integral equation (3.43) in order to calculate the effective interaction of (3.61). On the other hand, if we use the exact expression (3.38) for the Pauli operator, all partial wave components are coupled and we will need to consider many partial waves to obtain a convergent result. The largest n value required to have a convergent result in the case of the 2D electron gas is typically about 10, depending on the value of r_s.

Let us next discuss the cutoff parameter μ in the definition of the Coulomb interaction in (3.32). The value of μ should be so small that all numerical results are stable against variations of μ itself. In order to show that this is indeed possible, we report in Fig. 3.9 the numerical results of u, v and w for r_s ($= 1, 3$ and 5), as

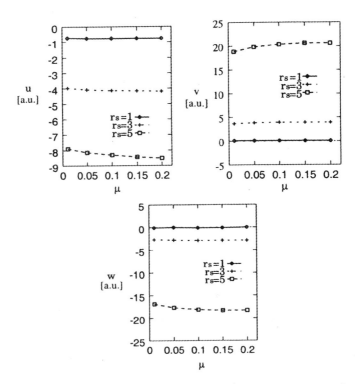

Fig. 3.9   Dependence of the u, v and w coefficients on the cutoff parameter μ (in units of k_F) for $r_s = 1, 3, 5$.

Table 3.2. Numerical values (in atomic units), of the parameters u, v and w of $g^c(\mathbf{r}, \mathbf{v})$ for several values of r_s.

r_s	u	v	w
1	-0.800	0.124	-0.059
3	-4.363	5.411	-3.745
5	-8.828	28.602	-23.831
10	-21.253	258.71	-269.89
20	-47.890	2262.7	-2795.6

a function of the cutoff parameter μ. From this figure it is seen that u, v and w approach their convergent values smoothly as μ decreases and, thus, that the use of a damping factor is legitimate. The numerical values of u, v and w in atomic units are reported in Table 3.2 for several values of r_s.

In the g matrix theory the correlation energy per electron is expressed as follows in terms of the effective interaction of (3.61):

$$\varepsilon_c^{\text{DME}} = \frac{1}{\rho}\langle 0|g^c|0\rangle = \rho\left(\frac{u}{4} - \frac{v}{8}k_F^2 + \frac{3w}{8}k_F^2\right), \tag{3.65}$$

where $|0\rangle$ indicates the ground state of the noninteracting system.

The correlation energies of (3.65) computed either with the exact Pauli operator or with the angle-averaged one turn out to be practically the same, and the difference between the two results increases slightly with increasing r_s. This shows that the mixing of states with different angular momenta becomes relevant only in the low-density region.

If we use the separable potential of (3.48) without any expansion, we get the following expression instead of (3.65) (see appendix):

$$\varepsilon_c = \frac{1}{2\rho}\int^{k_F}\frac{d\mathbf{k}_1}{(2\pi)^2}\frac{d\mathbf{k}_2}{(2\pi)^2}\{4c(|\mathbf{k}_1 - \mathbf{k}_2|/2)v(0) - 2c(0)v(|\mathbf{k}_1 - \mathbf{k}_2|)\}. \tag{3.66}$$

The correlation energies of Eqs. (3.66) ($\langle c(p)v(q)\rangle$) and (3.65) (ϵ_c^{DME}) are reported in Table 3.3 in eV/electron, together with the Monte Carlo results.

Table 3.3. Correlation energies of Eqs. (3.65) and (3.66) in unit of eV/electron. ϵ_c^{MC} gives the Monte Carlo results.

r_s	ϵ_c^{MC}	$\langle c(p)v(q)\rangle$	ϵ_c^{DME}
1	-2.99	-2.60	-2.34
3	-1.83	-1.65	-1.49
5	-1.34	-1.23	-1.11
10	-0.83	-0.76	-0.69
20	-0.48	-0.44	-0.40

It can be noted that the QMC correlation energies can be well reproduced by the separable potential, especially in the high r_s range. In the region of small r_s, where the long range RPA correlations are important, there is an expected discrepancy between the g matrix and Monte Carlo results. It is also possible to show that the correlation energy of (3.65) underestimates that of (3.66) by about 10%, and this is the price to be paid when passing from the separable potential to the Skyrme-like interaction.

As is clear from the derivation of the effective interaction of (3.61), this error originates from (i) neglecting the contributions of partial waves with $n \geq 2$, and (ii) inaccurate treatment of the s wave and p wave contributions.

3.7   The g Matrix for Confined Electron Systems

In this subsection we develop the BHF theory in finite size systems. We will mainly refer to systems of confined electrons; however, the same approach can be applied to atomic nuclei and helium drops.

Let us consider a system of N electrons with Hamiltonian

$$H = \sum_{i}^{N} h_0(i) + \sum_{i<j}^{N} v_{ij}, \tag{3.67}$$

where $h_0 = t_i + v_{\text{ext}}(r_i)$, t_i is the single-particle kinetic energy which can include the effect of the magnetic field (see Chapter 5), v_{ext} is the confining potential and $v_{ij} = e^2/\epsilon|\mathbf{r}_i-\mathbf{r}_j|$, with ϵ the dielectric constant. The ground state $|\Psi\rangle$ is the solution of the many-body Schrödinger equation

$$(H - E)|\Psi\rangle = 0, \tag{3.68}$$

where E is the ground state energy. Let us now consider an independent-particle model (the HF model in the following) where the eigenstates $|\Phi_n\rangle$ are Slater determinant solutions to the equation

$$(H_{\text{HF}} - W_n)|\Phi_n\rangle = 0, \tag{3.69}$$

where $H_{\text{HF}} = \mathcal{C} + \sum_i^N (h_0(i) + U_i)$, with U_i the HF potential and $\mathcal{C}$ a constant. The HF ground state determinant and energy are given by $|\Phi_0\rangle \equiv |\text{HF}\rangle$ and $W_0 \equiv E_{\text{HF}}$, respectively.

Remember that the residual interaction V_{res} fulfills

$$H = H_{\text{HF}} + V_{\text{res}}, \tag{3.70}$$

$$V_{\text{res}} = \sum_{i<j}^{N} v_{ij} - \sum_{i}^{N} U(i) - \mathcal{C}, \tag{3.71}$$

and that the HF theory yields the general matrix elements

$$\langle HF|V_{res}|HF\rangle = 0, \tag{3.72}$$

$$\langle HF|V_{res}|i^{-1}m\rangle = 0, \tag{3.73}$$

$$\langle HF|V_{res}|i^{-1}j^{-1}mn\rangle \neq 0, \tag{3.74}$$

where we have used the standard notation for particle–hole (ph) excitations, i.e. indexes i, j (m, n) refer to orbitals below (above) the Fermi energy and $|i^{-1}m\rangle$ is the Slater determinant obtained promoting one electron from orbital i to orbital m in $|HF\rangle$. Note that only 2p–2h excitations yield nonvanishing transition matrix elements since the two-body nature of V_{res} ensures that matrix elements between determinants differing in more than two orbitals will again vanish. Another immediate consequence from Eqs. (3.72)–(3.74) is that $E_{HF} = \langle HF|H|HF\rangle$.

We can write

$$|\Psi\rangle = |HF\rangle + \sum_{n\neq0} a_n|\Phi_n\rangle. \tag{3.75}$$

From Eqs. (3.68), (3.70) and (3.75) one easily finds that

$$(H_{HF} - E)\left(|HF\rangle + \sum_{n\neq0} a_n|\Phi_n\rangle\right) + V_{res}|\Psi\rangle = 0 . \tag{3.76}$$

Multiplying Eq. (3.76) by $\langle HF|$ on the left, one gets

$$E = E_{HF} + \langle HF|V_{res}|\Psi\rangle . \tag{3.77}$$

If multiplying by $\langle \Phi_n|$ one finds that $a_n = \frac{\langle \Phi_n|V_{res}|\Psi\rangle}{E-W_n}$ and hence the following implicit equation is obtained:

$$|\Psi\rangle = |HF\rangle + \sum_{n\neq0} \frac{\langle \Phi_n|V_{res}|\Psi\rangle}{E - W_n}|\Phi_n\rangle. \tag{3.78}$$

This equation can be solved by iteration taking as starting energy the HF one:

$$|\Psi\rangle = |HF\rangle + \sum_{n\neq0} \frac{\langle \Phi_n|V_{res}|HF\rangle}{E_{HF} - W_n}|\Phi_n\rangle + \cdots, \tag{3.79}$$

yielding for the energy

$$E = E_{HF} + \sum_{n\neq0} \frac{|\langle \Phi_n|V_{res}|HF\rangle|^2}{E_{HF} - W_n} + \cdots. \tag{3.80}$$

At the first order in V_{res} this equation gives a result for the energy which coincides with the one of first order perturbation theory; summing all the orders we get the correlation energy in the ladder approximation. This is clearer, defining the G matrix by the relation

$$G|\text{HF}\rangle = V_{\text{res}}|\Psi\rangle. \tag{3.81}$$

We then get the Bethe–Goldstone implicit equation for G:

$$G = V_{\text{res}} + \sum_{n \neq 0} V_{\text{res}} \frac{|\Phi_n\rangle\langle\Phi_n|}{E - W_n} G, \tag{3.82}$$

and from Eq. (3.77)

$$\begin{aligned} E &= E_{\text{HF}} + \langle\text{HF}|G|\text{HF}\rangle \\ &= E_{\text{HF}} + \sum_{n \neq 0} \frac{\langle\text{HF}|V_{\text{res}}|\Phi_n\rangle\langle\Phi_n|G|\text{HF}\rangle}{E - W_n}. \end{aligned} \tag{3.83}$$

Only 2p–2h determinants yield a nonvanishing contribution to the sum of Eq. (3.83), which can thus be reduced to a sum of two-body matrix elements. Assuming that $E = E_{\text{HF}}$ on the right hand side, as in the ladder approximation, one has

$$\begin{aligned} E = E_{\text{HF}} &+ \frac{1}{2} \sum_{ijmn} \frac{\langle ij|v|mn\rangle}{\epsilon_i + \epsilon_j - \epsilon_m - \epsilon_n} \\ &\times \left(\langle mn|g|ij\rangle - \langle mn|g|ji\rangle \right), \end{aligned} \tag{3.84}$$

where ϵ_α are the HF single-particle energies and we have associated the G matrix with an effective two-body interaction g.

In order to have a practical computational scheme, it remains for us now to specify the two-body matrix elements of g in Eq. (3.84). This is accomplished within the BHF independent-pair model [see, for example, Preston and Badhuri (1982)], where the off-diagonal matrix elements are found from

$$\begin{aligned} \langle mn|g|ij\rangle &= \langle mn|v|ij\rangle \\ &+ \sum_{pq} \frac{\langle mn|v|pq\rangle\langle pq|g|ij\rangle}{\epsilon_i + \epsilon_j - \epsilon_p - \epsilon_q}. \end{aligned} \tag{3.85}$$

The ground state energy of Eq. (3.84) with the matrix elements obtained from Eq. (3.85) is the BHF energy which sums all the ladder diagrams corresponding to the iterated solutions to Eq. (3.80).

In the following we report the numerical calculation by Emperador, Lipparini and Serra (2006), in which one is restricted to circular symmetry cases, where the HF orbitals can be factorized as

$$\langle \mathbf{r}\sigma|i\rangle \equiv R_{n_i m_i}(r) \frac{e^{im_i\theta}}{\sqrt{2\pi}} \chi_{\mu_i}(\sigma), \tag{3.86}$$

where $n_i = 0, 1, \ldots$, $m_i = 0, \pm 1, \ldots$ and $\mu_i = \pm 1/2$ are the principal, angular momentum and spin quantum numbers, respectively. In this situation the angular and spin parts of the two-body matrix elements yield selection rules on the corresponding quantum numbers, and the matrix elements reduce to

$$\langle ab|v|cd\rangle = \delta_{m_a+m_b,m_c+m_d}\,\delta_{\mu_a\mu_c}\,\delta_{\mu_b\mu_d}$$
$$\times I_r(R_{n_a m_a}, R_{n_b m_b}, R_{n_c m_c}, R_{n_d m_d}), \qquad (3.87)$$

where I_r is a radial integral that one computes numerically. Note that through Eq. (3.85) the same angular momentum selection rules apply to $\langle ab|g|cd\rangle$ and that both matrix elements are real.

The two-body matrix elements of g required for the evaluation of the total energy, Eq. (3.84), are found by solving Eq. (3.85) as a linear system for the unknowns $\langle mn|g|ij\rangle$. For each pair ij we have an independent linear system, and the above-mentioned selection rules are very important as they allow a big reduction in the number of effectively coupled equations. Since the space of particle states must be truncated, the convergence of the calculation with the number of empty HF states has to be controlled. Another check of the numerical accuracy must be done regarding the number of radial points used in the evaluation of the integrals I_r of Eq. (3.87).

Table 3.4 compares the energies of $B = 0$ ground states of N-electron dots in BHF with the results of HF, LSDA (using the Tanatar–Ceperley parametrization for the correlation energy), diffusion QMC (Pederiva, Umrigar and Lipparini, 2000) and configuration interaction (CI) calculations. The external confinement is taken to be of the parabolic type, $v_{\text{ext}}(r) = m\omega_0^2 r^2/2$, with m the electron effective mass. We refer all energies to the confinement energy $\hbar\omega_0$ and characterize the interaction strength by the repulsion-to-confinement ratio R, defined as

$$R \equiv \frac{e^2/\epsilon\ell_0}{\hbar\omega_0}, \qquad (3.88)$$

Table 3.4. Ground state energies for the dots with $2 \leq N \leq 13$ computed by HF, BHF, LSDA, QMC, and CI methods. The energies are in units of the confinement energy $\hbar\omega_0$. A fixed value $R = 1.89$ of the interaction-to-confinement ratio has been used.

N	E_{HF}	E_{BHF}	E_{LSDA}	E_{QMC}	E_{CI}
2	4.078	3.832	3.739	3.650	3.646
3	8.589	8.289	8.082	7.979	7.957
4	13.94	13.63	13.16	13.26	13.06
5	20.96	20.38	19.91	19.76	19.53
6	28.70	27.72	27.27	27.14	26.82
7	37.46	36.61	35.96	35.86	–
8	46.93	46.11	45.46	45.32	–
9	57.89	56.71	55.79	55.64	–
10	69.29	67.75	67.00	66.86	–
11	81.54	79.86	78.96	78.86	–
12	94.82	94.36	91.71	91.64	–
13	108.64	106.36	105.50	105.32	–

with ℓ_0 indicating the oscillator length $(\hbar\omega_0 = \hbar^2/m\ell_0^2)$. The results in Table 3.4 correspond to $R = 1.89$. Taking, for instance, typical GaAs values $\epsilon=12.4$ and $m = 0.067m_e$ the chosen R value would correspond to a confinement energy of $\hbar\omega_0 = 3.32$ meV, which reproduces the experimental value of Tarucha *et al.* (1996). The number of electrons is varied from $N = 2$ to 13.

At $B = 0$ the BHF energies obviously improve the HF ones, although they are still appreciably higher than the QMC and LSDA values. This is due to the fact that in BHF long range collective correlations are missed. The importance of short range correlations is expected to increase as the system is more tightly confined, for a fixed strength of the Coulomb repulsion and, thus, a better performance of BHF is expected when increasing the confinement strength. Indeed, this is shown to be the case in Fig. 3.10, where for $N = 2$ and 6 at $B = 0$ we have varied the value of R. Note that although the correlation energy is globally reduced when the ratio decreases, BHF accounts for a larger part of it, reproducing the exact value in the limit of small R. It can also be seen from Fig. 3.10 that at a given value of R, BHF correlations for the $N = 2$ dot are a somewhat larger piece of the total correlation energy than for $N = 6$. We attribute this difference to a more important role of the collective effects leading to long range correlations not included in BHF for the six-electron dot.

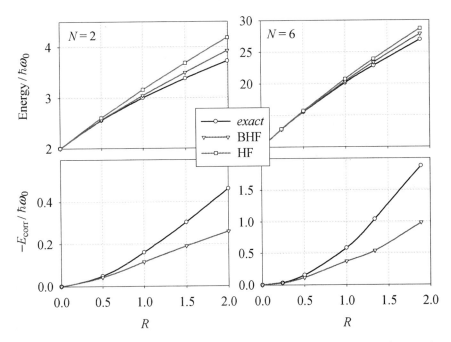

Fig. 3.10 Upper panels: Total energies in different methods for two and six-electron dots as a function of the interaction-to-confinement ratio (see text). Lower panels: Correlation energies within each model for the same two dots.

We focus next on the influence of a magnetic field and how the BHF energies are affected by it. In the symmetric gauge, a magnetic field in the z direction (perpendicular to the dot plane) induces a modification of the effective confinement from ω_0 to $\Omega = \sqrt{\omega_0^2 + \omega_c^2/4}$, with ω_c the cyclotron frequency. For increasing magnetic fields one then expects short range correlations to be enhanced due to the stronger confinement and, therefore, an improved performance of the BHF method. Figure 3.11 displays the evolution of the total energy with the ratio ω_c/ω_0 for a fixed value of $R = 1.5$ and for a dot with two electrons. Note that the correlation energy is about an order of magnitude higher for the singlet than for the triplet. As expected, for increasing values of ω_c/ω_0 BHF is accounting for a higher part of the correlation energy, although the increase is rather moderate. For the singlet state BHF correlations range from 73% to 81% of the total correlation energy when ω_c goes from 0 to $5\omega_0$. The evolution is even flatter for the triplet, where BHF correlations remain at $\approx 75\%$ for all values of ω_c. Note also that due to the sizeable energy correction for the singlet, the singlet–triplet transition point is remarkably improved in BHF with respect to HF.

Figure 3.12 shows the evolution with magnetic field of the results for a six-electron dot with a fixed interaction-to-confinement ratio of $R = 1.89$. As compared

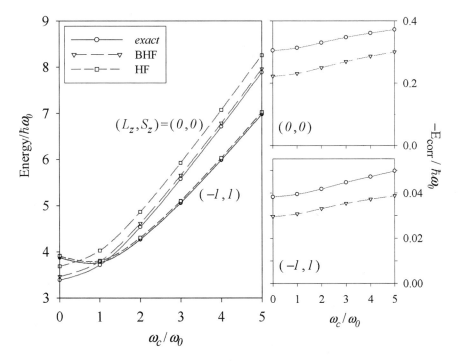

Fig. 3.11 Left (right) panels show total (correlation) energies for the models and states indicated by the corresponding labels. The results correspond to the $N = 2$ dot in a magnetic field, shown as a function of the cyclotron frequency (in units of ω_0). The interaction-to-confinement ratio R (see text) is chosen as $R = 1.5$.

to the $N = 2$ case, this dot shows a much richer phase diagram, with large variations in ground state angular momentum and spin when increasing the magnetic field. Most remarkable is the comparison of BHF and LSDA energies: while LSDA is clearly superior to BHF at low fields, the situation is reversed when entering the fully polarized phase corresponding to the maximum density droplet (MDD). In the MDD region the LSDA energy is actually higher than the HF one and only BHF is able to provide an energy correction in the right direction with respect to HF; total BHF energies being approximately halfway of CI and HF.

All the BHF results shown above have been obtained using a large enough space of empty HF states, always checking that for the given accuracy convergence in Eq. (3.84) has been achieved. In practice, one includes the lowest N_p particle states and repeats the calculation increasing this number. Figure 3.13 shows the evolution of the BHF energy with N_p, on a scale proportional to $1/N_p$, for three selected cases: the two upper panels correspond to six and nine electrons with the MDD configuration in a strong magnetic field and a moderate confinement, while the lower panel shows the case of four electrons in a very strong confinement and a zero field. The results have been fitted with a polynomial including powers up to $1/N_p^3$. In the chosen scale, the $N_p \rightarrow \infty$ limit is given by the intersection of the polynomial fit with the left vertical axis. It can be seen from Fig. 3.13 that the convergence is somewhat faster for the moderate confinement cases. However, in both examples the evolution with space dimension is quite smooth, indicating that the correlation energy builds up by gathering contributions from many states. The $N = 9$ results of Fig. 3.13 provide additional support to our preceding conclusion from Fig. 3.12 that BHF performs much better than LSDA in the MDD region.

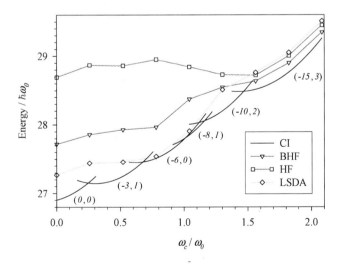

Fig. 3.12 Evolution of the total energy of a six-electron dot with the magnetic field. The different phases are indicated by the angular momentum labels (L_z, S_z). A fixed value of the parameter $R = 1.89$ has been used (see text).

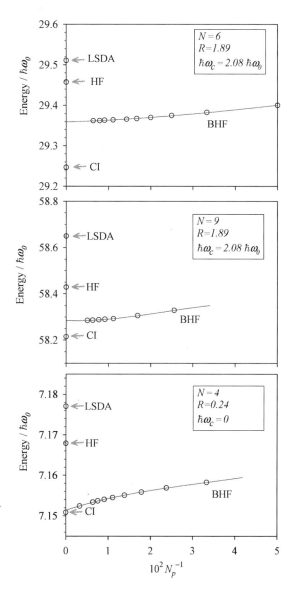

Fig. 3.13 Evolution of the total BHF energy with the number N_p of empty HF states included in the solution of Eq. (3.85). Each panel shows the results for a different quantum dot. The solid line is a cubic fit, in powers of $1/N_p$, allowing extrapolation to the $N_p \to \infty$ limit. The LSDA, HF and CI energies for each case are also shown for comparison.

As already mentioned, for strong confinement BHF reproduces the exact energies. Indeed, the $N = 4$ results of Fig. 3.13 are very illustrative in this respect since the extrapolated BHF energy essentially coincides with the CI result. This strongly confined system mimics a self-assembled InAs dot with $\hbar\omega_0 = 50$ meV. Following

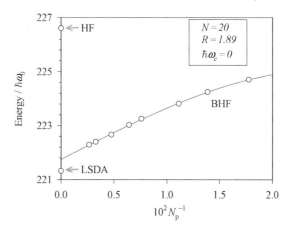

Fig. 3.14 The same as Fig. 3.13, for a dot with 20 electrons and the additional parameters given in the inset.

Brocke *et al.* (2003), we take for this material $m = 0.024\,m_e$ and $\epsilon = 15.15$, giving, for $N = 4$, a small Wigner–Seitz radius of $r_s \sim 0.12$.

To emphasize the possibility of calculating the energies of larger systems in BHF theory, we end this subsection by showing in Fig. 3.14 the results for an $N = 20$ dot. For this number of electrons exact methods like CI or QMC become extremely demanding and we have not attempted to make comparison with them. The evolution of the BHF energy with the number of particle states resembles that of Fig. 3.13 although, as one could expect, larger values of N_p need to be considered for a similar degree of convergence. The proximity of the extrapolated-BHF and LSDA energies for $N = 20$ is a bit surprising since, as shown in Fig. 3.12, for six electrons in the same confinement the difference is larger. BHF correlations are thus a more important contribution for $N = 20$ than for $N = 6$. This can be understood as a different degree of *magicity* for these two dots. Indeed, a highly magic system is characterized by a distribution of single-particle orbitals whose energies group in bunches corresponding to quasidegenerate shells, with large energy gaps between the shells. One expects a quenching of independent-pair motions in a highly magic system, with respect to collective motions, and therefore a worse performance of the BHF theory for them. Note also that since the 20-electron dot has a higher density than the 6-electron dot, for the same confinement, a better performance of BHF theory in the former agrees with our preceding results regarding the high density limit.

3.7.1 *Effective Interaction in Confined Electron Systems*

Another way to study the dynamic and static properties of finite systems such as electron systems confined in an external field is the one which uses the effective interaction of (3.61) together with the local density approximation. For these systems the effective Hamiltonian to be used in BHF-like calculations is given by (in effective atomic units)

$$H^{\text{eff}} = \sum_{i=1}^{N} \left(-\frac{\vec{\nabla}^2}{2} + v_{\text{ext}} \right) + \sum_{i<j} \left(\frac{1}{|\mathbf{r}_i - \mathbf{r}_j|} + g^c(\mathbf{r}_{ij}) \right), \tag{3.89}$$

where $g^c(\mathbf{r}_{ij})$ is given by Eq. (3.61) and the dependence on the density of the u, v and w coefficients is to be treated in the local density approximation.

The BHF energy functional for a parabolic external potential is given by the expression

$$E^{\text{BHF}} = \frac{1}{2} \int \tau d\mathbf{r} + \frac{1}{2}\omega_0^2 \int r^2 \rho d\mathbf{r} + \frac{1}{4} \int u(\rho)\rho^2 d\mathbf{r} - \int t_1(\rho)(\rho\tau - j^2) d\mathbf{r}$$

$$- \int t_2(\rho)(\nabla\rho)^2 d\mathbf{r} + \frac{1}{2} \int \frac{\rho(\mathbf{r})\rho(\mathbf{r}') - |\rho^{(1)}(\mathbf{r},\mathbf{r}')|^2}{|\mathbf{r} - \mathbf{r}'|} d\mathbf{r} d\mathbf{r}', \tag{3.90}$$

where

$$t_1 = \frac{1}{4}(v - 3w), \quad t_2 = \frac{3}{16}\left(v + w + 2/3\rho\frac{\partial v}{\partial \rho} \right),$$

and

$$\tau(\mathbf{r}) = \sum_i |\nabla\varphi_i(\mathbf{r})|^2, \quad \mathbf{j}(\mathbf{r}) = -\frac{1}{2i}\sum_i (\varphi_i(\mathbf{r})\nabla\varphi_i^*(\mathbf{r}) - \text{h.c.}),$$

$$\rho^{(1)}(\mathbf{r},\mathbf{r}') = \sum_i \varphi_i^*(\mathbf{r})\varphi_i(\mathbf{r}'), \quad \rho(\mathbf{r}) = \sum_i |\varphi_i(\mathbf{r})|^2$$

are the kinetic energy, current, one-body nondiagonal and diagonal densities, respectively. The $\varphi_i(\mathbf{r})$ are the single-particle wave functions, and the last terms in Eq. (3.90) are the electron–electron (both direct and exchange), Coulomb contributions. Besides the pure HF functional, the energy functional (3.90) has the correlation term

$$E_c(\rho, \tau) = \int \left(\frac{1}{4}u\rho^2 - t_2(\nabla\rho)^2 - t_1(\rho\tau - j^2) \right) d\mathbf{r}. \tag{3.91}$$

This term is very easily handled in both static and dynamic calculations. Clearly, the term in $(\nabla\rho)^2$ vanishes for homogeneous systems, for which ρ is a constant. On the other hand, the current term is zero for time-reversal invariant systems.

The single-particle wave functions which appear in (3.90) are to be determined in a self-consistent way by solving the BHF equations:

$$\left\{ -\frac{1}{2}\vec{\nabla}[1 - 2t_1(\rho)\rho(\mathbf{r})] \cdot \vec{\nabla} + \frac{1}{2}\omega_0^2 r^2 + V_c(\rho, \tau) + \int \frac{\rho(\mathbf{r}')}{|\mathbf{r} - \mathbf{r}'|} d\mathbf{r}' \right\} \varphi_i(\mathbf{r})$$

$$+ \sum_{j\|i}^{N} \int \frac{\varphi_j^*(\mathbf{r}')\varphi_i(\mathbf{r}')}{|\mathbf{r} - \mathbf{r}'|} d\mathbf{r}' \varphi_j(\mathbf{r}) = \varepsilon_i\varphi_i(\mathbf{r}), \tag{3.92}$$

where $j\|i$ means that the spins must be parallel, and

$$V_c(\rho,\tau) = \frac{1}{2}u\rho + \frac{1}{4}u'\rho^2 + \frac{1}{4}(3w - v)\tau + \frac{1}{4}(3w' - v')\rho\tau$$

$$+ \frac{1}{16}(5v' + 3w' + 2v''\rho)\left(\frac{d\rho}{dr}\right)^2 + \frac{3}{8}(v + w + 2/3v'\rho)\nabla^2\rho \quad (3.93)$$

is the correlation potential, and the $'$ and $''$ mean first and second derivatives of the u v and w functions, with respect to the density.

Equations (3.92) constitute a set of coupled integral-differential equations; they have been solved for quantum dots (T. Suwa *et al.*, 2003) and the resulting density in the case of a dot with 20 electrons was shown in Fig. 2.7. The resulting energies are generally very close to the ones obtained in the previous section and are reported in Table 3.4.

Calculations analogous to the 2D ones reported in this section were carried out in the 3D case by Lipparini *et al.* (1994) in metal clusters, and the results concerning the single-particle energies, the ionization energies, the total energies and the densities of some sodium clusters were shown in Subsection 2.2.1.

In general, it may be concluded that the BHF theory is a very good microscopic theory, in the sense that it improves the HF theory by including the ladder correlations, which are important ones, especially when the system is more tightly confined.

3.8   The BBP Method

We conclude this chapter by illustrating the Bethe, Brandow and Petschech (BBP) method for the calculation of the g matrix of the Bethe–Goldstone theory. This method has been applied in the past to nuclear matter (Bethe, Brandow and Petschech, 1963) and finite nuclei (see, for example, Preston and Badhuri, 1982) and is matter of investigation in other systems.

In the BBP method one defines a one-body problem:

$$h_{\text{BBP}}|\alpha\rangle = e_\alpha|\alpha\rangle, \quad (3.94)$$

where $h_{\text{BBP}} = t + u$, with t the usual kinetic energy and u a one-body potential. Originally, the method was developed for "self-bound" systems, like atomic nuclei, where there is no external field v_{ext}. In confined systems with this contribution, one has simply to make the substitution of the one-body operator $t \rightarrow t + v_{\text{ext}}$. Therefore

$$h_{\text{BBP}} = t + v_{\text{ext}} + u. \quad (3.95)$$

The BBP theory gives us the expression of the u potential but only in the self-consistent basis that we are looking for, which is not known in advance. In fact, we know its matrix elements in this basis [note: levels are denoted by i, j, k, l (occupied), m, n, p, q (unoccupied) and Greek letters (general)]:

$$\langle i|u|j \rangle = \frac{1}{2} \sum_k \Big\{ \langle ik|g(e_i + e_k)|jk \rangle - \langle ik|g(e_i + e_k)|kj \rangle$$

$$+ \langle ik|g(e_j + e_k)|jk \rangle - \langle ik|g(e_j + e_k)|kj \rangle \Big\},$$

$$\langle i|u|m \rangle = \sum_k \Big\{ \langle ik|g(e_i + e_k)|mk \rangle - \langle ik|g(e_i + e_k)|km \rangle \Big\}. \tag{3.96}$$

Recall that the HF prescription for the one-body potential is also given by similar matrix elements. It is, however, simpler since a single expression is valid for the matrix elements between occupied and/or unoccupied orbitals, and one has the bare interaction $v = 1/r_{12}$ instead of the g matrix. Namely,

$$\langle \alpha|u_{\mathrm{HF}}|\beta \rangle = \sum_k \left(\langle \alpha k|v|\beta k \rangle - \langle \alpha k|v|k\beta \rangle \right). \tag{3.97}$$

In Eq. (3.96) we have the energy-dependent g matrix. It is always evaluated at the energy of a hole–hole pair. This is more easily seen by rewriting Eq. (3.96) as

$$\langle i|u|j \rangle = \frac{1}{2} \sum_k \Big\{ \langle jk|g(e_i + e_k)|ik \rangle^* - \langle kj|g(e_i + e_k)|ik \rangle^*$$

$$+ \langle ik|g(e_j + e_k)|jk \rangle - \langle ki|g(e_j + e_k)|jk \rangle \Big\}$$

$$\langle i|u|m \rangle = \sum_k \Big\{ \langle mk|g(e_i + e_k)|ik \rangle^* - \langle km|g(e_i + e_k)|ik \rangle^* \Big\}. \tag{3.98}$$

Notice that the energy argument in the above matrix elements is always the sum of the hole energies for states on the right of the matrix element. That is, for $\langle \alpha\beta|g(\omega)|ij \rangle$, we take $\omega = e_i + e_j$. This is the *on-shell* prescription for the determination of the g matrix.

The on-shell matrix elements are given by the Bethe–Goldstone equation

$$\langle \alpha\beta|g(e_i + e_j)|ij \rangle = \langle \alpha\beta|v|ij \rangle + \sum_{ab} \frac{\langle \alpha\beta|v|ab \rangle \langle ab|g(e_i + e_j)|ij \rangle}{e_i + e_j - e_a - e_b}. \tag{3.99}$$

In Eq. (3.96) the hole–hole matrix elements are manifestly Hermitian, $\langle j|u|i \rangle = \langle i|u|j \rangle^*$, provided we have

$$\langle \alpha\beta|g(e_\gamma + e_\delta)|\gamma\delta \rangle = \langle \gamma\delta|g(e_\gamma + e_\delta)|\alpha\beta \rangle^*. \tag{3.100}$$

The particle–hole matrix elements of u have been defined in Eq. (3.96) for the hole on the left and the particle on the right. They will be Hermitian if take as the definition the fact that the reversed element is given by $\langle m|u|i \rangle = \langle i|u|m \rangle^*$. That is,

$$\langle m|u|i \rangle = \sum_k \left(\langle mk|g(e_i + e_k)|ik \rangle - \langle km|g(e_i + e_k)|ik \rangle \right). \tag{3.101}$$

Notice that g is also on-shell for the particle–hole matrix elements.

It remains for us now to specify the particle–particle elements, which are not fixed by the BBP scheme. One option commonly used in nuclear physics is to put them to zero: $\langle m|u|n \rangle = 0$. This is called the *discontinuous* choice. Another option is the on-shell prescription

$$\langle m|u|n \rangle = \frac{1}{2} \sum_k \Big(\langle mk|g(e_m + e_k)|nk \rangle - \langle mk|g(e_m + e_k)|kn \rangle$$
$$+ \langle mk|g(e_n + e_k)|nk \rangle - \langle mk|g(e_n + e_k)|kn \rangle \Big), \qquad (3.102)$$

which is just extending the above formula for hole–hole elements. Yet another possibility could be to use the HF prescription with the bare interaction v,

$$\langle m|u|n \rangle = \sum_k \Big(\langle mk|v|nk \rangle - \langle mk|v|kn \rangle \Big). \qquad (3.103)$$

We can call these three schemes the discontinuous, the on-shell and the particle–HF (pHF) prescriptions, respectively. In nuclear physics the pHF prescription is not considered. This is because for hard-core interactions (thus also for helium systems) the bare interaction yields divergences and it does not make sense to consider the unscreened matrix elements. In Coulomb systems the situation is different.

We also need to specify the new equation for g acting on a particle–hole pair [note that in Eq. (3.99) we had hole–hole pairs]. The obvious extension is

$$\langle \alpha\beta|g(e_m + e_i)|mi \rangle = \langle \alpha\beta|v|mi \rangle + \sum_{ab} \frac{\langle \alpha\beta|v|ab \rangle \langle ab|g(e_m + e_i)|mi \rangle}{e_m + e_i - e_a - e_b}. \qquad (3.104)$$

Technically, the evaluation of the matrix elements $\langle \alpha\beta|g(e_m + e_i)|mi \rangle$ requires the solution of an additional linear system for each mi pair. This is costly, since we have many additional pairs. In the pHF and discontinuous choices these elements are not needed and it suffices to solve one linear system for each hole–hole pair ij.

The total energy is given by

$$E = \sum_k \langle k|t + v_{\text{ext}}|k \rangle + \frac{1}{2} \sum_{kl} \langle kl|g(e_k + e_l)|kl \rangle - \langle kl|g(e_k + e_l)|lk \rangle. \qquad (3.105)$$

Note that the above expression is exactly equal to the usual HF energy if we replace $g(\omega)$ by v. This similarity motivates the name "effective two-body interaction" for g. The u expression also recovers the HF one by the replacement $g \to v$ in the on-shell and pHF choices. But not for the discontinuous choice, since one forces the particle–particle matrix elements to vanish. In general, the BBP scheme yields a modified orbital set and total energy as compared with HF.

3.8.1 *Appendix*

Starting from Eq. (3.48) it is possible to calculate the ground state energy E^{BHF} of the electron gas. To this end, it is convenient to use the second quantization formalism, in which the two-body interaction is written as (see Subsection 2.5)

$$\frac{1}{2}\sum_{\mathbf{k}_1,\mathbf{k}_2,\mathbf{q},\sigma,\sigma'} V(\mathbf{q})c^+_{\mathbf{k}_1+\mathbf{q},\sigma}c^+_{\mathbf{k}_2-\mathbf{q},\sigma'}c_{\mathbf{k}_2,\sigma'}c_{\mathbf{k}_1,\sigma},$$

where the operators c^+_λ and c_λ are creation and annihilation operators which obey the anticommutation relations

$$\{c_\lambda,c^+_{\lambda'}\}=\delta_{\lambda,\lambda'},\qquad \{c_\lambda,c_{\lambda'}\}=0,\qquad \{c^+_\lambda,c^+_{\lambda'}\}=0.$$

Using the transformations $\mathbf{p}_1=1/2(\mathbf{k}_1-\mathbf{k}_2)$ and $\mathbf{p}_2=1/2(\mathbf{k}'_1-\mathbf{k}'_2)$, where $\mathbf{k}'_1=\mathbf{k}_1+\mathbf{q}$ and $\mathbf{k}'_2=\mathbf{k}_2-\mathbf{q}$, we find that $\mathbf{p}_2-\mathbf{p}_1=\mathbf{q}$ and $1/2(\mathbf{p}_1+\mathbf{p}_2)=1/2(\mathbf{k}_1-\mathbf{k}_2+\mathbf{q})$, and finally

$$g^c=\frac{1}{2}\sum_{\mathbf{k}_1,\mathbf{k}_2,\mathbf{q},\sigma,\sigma'} c(|\mathbf{k}_1-\mathbf{k}_2+\mathbf{q}|/2)v(|\mathbf{q}|)c^+_{\mathbf{k}_1+\mathbf{q},\sigma}c^+_{\mathbf{k}_2-\mathbf{q},\sigma'}c_{\mathbf{k}_2,\sigma'}c_{\mathbf{k}_1,\sigma}.\qquad(3.106)$$

From this expression, one finds for the BHF energy

$$E^{\mathrm{BHF}}=E^{\mathrm{HF}}+\frac{1}{2}\sum_{\mathbf{k}_1,\mathbf{k}_2,\sigma,\sigma'}v(0)c(|\mathbf{k}_1-\mathbf{k}_2|/2)\langle\mathrm{HF}|c^+_{\mathbf{k}_1,\sigma}c^+_{\mathbf{k}_2,\sigma'}c_{\mathbf{k}_2,\sigma'}c_{\mathbf{k}_1,\sigma}|\mathrm{HF}\rangle$$

$$+\frac{1}{2}\sum_{\mathbf{k}_1,\mathbf{q},\sigma}c(0)v(|\mathbf{q}|)\langle\mathrm{HF}|c^+_{\mathbf{k}_1+\mathbf{q},\sigma}c^+_{\mathbf{k}_1,\sigma}c_{\mathbf{k}_1+\mathbf{q},\sigma}c_{\mathbf{k}_1,\sigma}|\mathrm{HF}\rangle,\qquad(3.107)$$

where E^{HF} is the HF energy,

$$E^{\mathrm{HF}}=\sum_{\mathbf{k},\sigma}\epsilon^{\mathrm{HF}}_{\mathbf{k},\sigma}n_{\mathbf{k},\sigma},\qquad(3.108)$$

written in terms of the distribution functions $n_{\mathbf{k},\sigma}$ ($=1,\ k\le k_F,\ =0\ k>k_F$) and of the single-particle HF energies, and the other terms give the correlation energy.

From Eq. (3.107), and using the commutation relations for operators c and c^+, and $c^+_{\mathbf{k},\sigma}c_{\mathbf{k},\sigma}|\mathrm{HF}\rangle=n_{\mathbf{k},\sigma}|\mathrm{HF}\rangle$, we obtain the following expression for E^{BHF}:

$$E^{\mathrm{BHF}}=E^{\mathrm{HF}}+\frac{1}{2}\sum_{\mathbf{k}_1,\mathbf{k}_2,\sigma,\sigma'}v(0)c(|\mathbf{k}_1-\mathbf{k}_2|/2)n_{\mathbf{k}_1,\sigma}n_{\mathbf{k}_2,\sigma'}$$

$$-\frac{1}{2}\sum_{\mathbf{k}_1,\mathbf{k}_2,\sigma,\sigma'}c(0)v(|\mathbf{k}_1-\mathbf{k}_2|)n_{\mathbf{k}_1,\sigma}n_{\mathbf{k}_2,\sigma'}\delta_{\sigma,\sigma'}.\qquad(3.109)$$

Next, the single-particle energies and the self-energy in BHF are obtained by differentiation with respect to $n_{\mathbf{k},t}$:

$$\epsilon_{\mathbf{k},t}^{\mathrm{BHF}} = \epsilon_{\mathbf{k},t}^{\mathrm{HF}} + \sum_{\mathbf{k}_2,\sigma'}[v(0)c(|\mathbf{k}-\mathbf{k}_2|/2)n_{\mathbf{k}_2,\sigma'} - c(0)v(|\mathbf{k}-\mathbf{k}_2|)n_{\mathbf{k}_2,\sigma'}\delta_{t,\sigma'}]$$

$$+\frac{1}{2}\sum_{\mathbf{k}_1,\mathbf{k}_2,\sigma,\sigma'}\left[\frac{\partial}{\partial\rho}(v(0)c(|\mathbf{k}_1-\mathbf{k}_2|/2)-c(0)v(|\mathbf{k}_1-\mathbf{k}_2|)\delta_{\sigma,\sigma'})\right]$$

$$\times n_{\mathbf{k}_1,\sigma}n_{\mathbf{k}_2,\sigma'}. \tag{3.110}$$

In the derivation of Eq. (3.110) we took into account the dependence of the correlation energy on density, and used

$$\frac{\partial}{\partial n_{\mathbf{k},t}} = \frac{\partial}{\partial\rho}\frac{\partial\rho}{\partial n_{\mathbf{k},t}},$$

with $\partial\rho/\partial n_{\mathbf{k},t} = 1$.

The effective mass in BHF is therefore given by

$$\frac{1}{m^*} = \frac{1}{k}\frac{\partial\epsilon_k^{\mathrm{BHF}}}{\partial k} = 1 + \frac{m}{k}\frac{\partial}{\partial k}\Sigma_x(k) + \frac{m}{k}\frac{\partial}{\partial k}\Sigma_c^{\mathrm{BHF}}(k), \tag{3.111}$$

where $\Sigma_x(k)$ is the exchange self-energy and $\Sigma_c^{\mathrm{BHF}}(k)$ is the BHF correlation contribution. Although in the literature there exists no accurate numerical calculation of $\Sigma_c^{\mathrm{BHF}}(k)$, an approximate estimate of such a quantity shows that the correlation contribution to the effective mass has the opposite sign with respect to the exchange one, and tends to cancel it.

References

Bethe, H.A., B.H. Brandow and A.G. Petschek, *Phys. Rev.* **129**, 225 (1963).
Brocke, T., M.T. Bottsmann, M. Tews, B. Wunsch, D. Pfannkuche, Ch. Heyn, W. Hansen, D. Heitmann and C. Schüller, *Phys. Rev. Lett.* **91**, 257401 (2003).
Brueckner, K.A. and C.A. Levinson, *Phys. Rev.* **97**, 1344 (1955); K.A. Brueckner, J.L. Gammel and H. Weitzner, *Phys. Rev.* **110**, 431 (1958).
Campi, X. and D.W.L. Sprung, *Nucl. Phys.* **A194**, 401 (1972).
Combescot, R., A. Recati, C. Lobo and F. Chevy, *Phys. Rev. Lett.* (2007), in publication.
Emperador, A., E. Lipparini and Ll. Serra, *Phys. Rev.* **B73**, 235341 (2006).
Lipparini, E., Ll. Serra and K. Takayanagi, *Phys. Rev.* **B49**, 16733 (1994); K. Takayanagi and E. Lipparini, *Phys. Rev.* **B54**, 8122 (1996).
Lowy, D.N. and G.E. Brown, *Phys. Rev.* **B12**, 2138 (1975); K. Bedell and G.E. Brown, *Phys. Rev.* **B17**, 4512 (1978).
Nagano, S. and K. Singwi, *Phys. Rev.* **B27**, 6732 (1983).
Nagano, S., K. Singwi and S. Ohnishi, *Phys. Rev.* **B29**, 1209 (1984).
Negele, J., *Phys. Rev.* **C1**, 1260 (1970).
Partridge, G.B. *et al.*, *Science* **311**, 503 (2006).
Pederiva, F., C.J. Umrigar and E. Lipparini, *Phys. Rev.* **B62**, 8120 (2000).

Pilati, S. and S. Giorgini (2007), to be published.

Preston, M.A. and R.K. Badhuri, *Structure of the Nucleus* (Addison-Wesley, USA, 1982), Chap. 8.

Reid, R.V., Ann. Phys. **50**, 411 (1968).

Song, H.Q., M. Baldo, G. Giansiracusa and U. Lombardo, *Phys. Rev. Lett.* **81**, 1584 (1998).

Sprung, D.W.L. and P.K. Banerjee, *Nucl. Phys.* **A168**, 273 (1971).

Suwa, T., doctoral thesis (Sophia University, Tokyo, 2002.)

Suwa, T., K. Takayanagi and E. Lipparini, *Phys. Rev.* **B69**, 115105 2004; *J. Phys. Soc. Jap.* **73**, 2781 (2004).

Suwa, T., K. Takayanagi, E. Lipparini and F. Pederiva (2003), unpublished.

Tanatar, B. and D.M. Ceperley, *Phys. Rev.* **B39**, 5005 (1989).

Tarucha, S., D.G. Austing, T. Honda, R.J. van der Hage and L.P. Kouwenhoven, *Phys. Rev. Lett.* **77**, 3613 (1996).

Vautherin, D. and D. Brink, *Phys. Rev.* **C5**, 626 (1972).

Yang, C.N., *Phys. Rev. Lett.* **19**, 1312 (1967).

Zwierlein, M.W. *et al.*, *Science* **311**, 492 (2006).

Chapter 4

The Density Functional Theory

4.1 Introduction

Hohenberg and Kohn, in a famous paper of 1964, were the first to prove in a rigorous way that the properties of the ground state of an N-particle system can be expressed as functionals of its density $\rho(\mathbf{r})$, i.e. they are determined by the knowledge of the density. The total energy E can be expressed in terms of such a functional and $E(\rho)$ obeys a variational principle. The Thomas–Fermi model for the atom is a special case of this formalism, and the same is true for the Gross–Pitaevskii theory for the ground state of a dilute Boson gas.

The density functional theory has been applied most of all to systems of electrons like atoms, molecules, homogeneous solids, surfaces and interfaces, quantum wells and quantum dots, etc. As we will discuss in the following, this theory has been used for the description of liquid helium and atomic nuclei as well.

The self-consistent single-particle equations that were derived from the variational principle by Kohn and Sham (1965) describe the properties of the ground state (energy and density) in a way that, formally, takes into account all many-body effects. Their solution, in the local density approximation, is as simple as the solution of the Hartree equations (without exchange terms) and leads to results better than the Hartree–Fock ones, though the latter are much more complicated, and in some cases even almost impossible, to obtain. The reason for this is that the Kohn and Sham solutions treat accurately both exchange and correlation effects, while the HF ones treat exchange exactly but completely neglect correlations.

4.2 The Density Functional Formalism

In what follows we report Levy's (1979) proof of the theorems on which is based the formalism of the density functional theory. Let us consider N interacting particles that move in an external field $v_{\text{ext}}(\mathbf{r})$ whose Hamiltonian will be written as

$$H = T + V + \sum_{i=1}^{N} v_{\text{ext}}(\mathbf{r}_i), \tag{4.1}$$

where T is the kinetic energy and V the interparticle interaction.

Let us consider all possible densities $\rho(\mathbf{r})$, i.e. those that can be obtained from any N-particle wave function $\Psi(\mathbf{r_1}, \mathbf{r_2}, \ldots, \mathbf{r_N})$ through

$$\rho(\mathbf{r}) = \langle \Psi | \sum_{i=1}^{N} \delta(\mathbf{r} - \mathbf{r}_i) | \Psi \rangle, \tag{4.2}$$

and let us define the functional

$$F(\rho) = \min_{\Psi \to \rho} \langle \Psi | T + V | \Psi \rangle, \tag{4.3}$$

where the minimum is with respect to all $|\Psi\rangle$ which produce the density ρ. $F(\rho)$ is universal in the sense that it does not refer to any specific system or to the external potential. Indicating by E_{GS}, Ψ_{GS} and ρ_{GS} the energy, wave function and density of the ground state, respectively, the two basic theorems of the density functional theory are

$$E(\rho) = \int d\mathbf{r} v_{\text{ext}}(\mathbf{r})\rho(\mathbf{r}) + F(\rho) \geq E_{\text{GS}}, \tag{4.4}$$

for any density (4.2), and

$$\int d\mathbf{r} v_{\text{ext}}(\mathbf{r})\rho_{\text{GS}}(\mathbf{r}) + F(\rho_{\text{GS}}) = E_{\text{GS}}. \tag{4.5}$$

In order to prove the variational principle (4.4), let us indicate by Ψ_{min}^{ρ} a wave function that minimizes (4.3), so that

$$F(\rho) = \langle \Psi_{\text{min}}^{\rho} | T + V | \Psi_{\text{min}}^{\rho} \rangle. \tag{4.6}$$

By writing

$$V_{\text{ext}} = \sum_{i=1}^{N} v_{\text{ext}}(\mathbf{r}_i),$$

we have

$$\int d\mathbf{r} v_{\text{ext}}(\mathbf{r})\rho(\mathbf{r}) + F(\rho) = \langle \Psi_{\text{min}}^{\rho} | T + V_{\text{ext}} + V | \Psi_{\text{min}}^{\rho} \rangle \geq E_{\text{GS}}, \tag{4.7}$$

as follows from the minimum properties of the ground state. This proves the inequality (4.4). Using again the minimum property, we then find that

$$E_{\text{GS}} = \langle \Psi_{\text{GS}} | T + V_{\text{ext}} + V | \Psi_{\text{GS}} \rangle \leq \langle \Psi_{\text{min}}^{\rho_{\text{GS}}} | T + V_{\text{ext}} + V | \Psi_{\text{min}}^{\rho_{\text{GS}}} \rangle. \tag{4.8}$$

By subtracting V_{ext}, we obtain

$$\langle \Psi_{\text{GS}} | T + V | \Psi_{\text{GS}} \rangle \leq \langle \Psi^{\rho_{\text{GS}}}_{\min} | T + V | \Psi^{\rho_{\text{GS}}}_{\min} \rangle. \tag{4.9}$$

On the other hand, the definition of $\Psi^{\rho_{\text{GS}}}_{\min}$ leads to an opposite inequality between the two sides of (4.9). This is possible only if

$$\langle \Psi_{\text{GS}} | T + V | \Psi_{\text{GS}} \rangle = \langle \Psi^{\rho_{\text{GS}}}_{\min} | T + V | \Psi^{\rho_{\text{GS}}}_{\min} \rangle. \tag{4.10}$$

Therefore we have

$$\begin{aligned}
E_{\text{GS}} &= \int d\mathbf{r} v_{\text{ext}}(\mathbf{r}) \rho_{\text{GS}}(\mathbf{r}) + \langle \Psi_{\text{GS}} | T + V | \Psi_{\text{GS}} \rangle \\
&= \int d\mathbf{r} v_{\text{ext}}(\mathbf{r}) \rho_{\text{GS}}(\mathbf{r}) + \langle \Psi^{\rho_{\text{GS}}}_{\min} | T + V | \Psi^{\rho_{\text{GS}}}_{\min} \rangle \\
&= \int d\mathbf{r} v_{\text{ext}}(\mathbf{r}) \rho_{\text{GS}}(\mathbf{r}) + F(\rho_{\text{GS}}), \tag{4.11}
\end{aligned}$$

which completes the proof of the basic theorems. From Eq. (4.10) follows the important result that if the ground state is not degenerate, then $\Psi^{\rho_{\text{GS}}}_{\min} = \Psi_{\text{GS}}$. If, on the contrary, it is degenerate, then $\Psi^{\rho_{\text{GS}}}_{\min}$ equals one of the ground state wave functions, and the others can be obtained as well. Therefore, the ground state density determines the wave function (or the wave functions) through which it is possible to compute all the properties of the ground state. Thus, these properties are functions of the density. This theorem is the formal justification for dealing with the density instead of the wave functions.

These theorems provide a general framework for calculating the ground state properties. One needs to find an approximation for $F(\rho)$ and to minimize $E(\rho)$ in Eq. (4.4) for the potential of interest V_{ext}. This leads to the corresponding approximation for E_{GS} and ρ_{GS}. Then, if we have an approximation for the functional $X(\rho)$ describing a given property X of the ground state, the same procedure leads to approximations for X itself.

4.3   Examples of Application of the Density Functional Theory

4.3.1   *The Thomas–Fermi Theory for the Atom*

In this approximation, the atom electrons are treated as independent particles, so that the system energy is given by the average value of the Hamiltonian (expressed in atomic units)

$$H = \sum_{i=1}^{N} \left(-\frac{\vec{\nabla}_i^2}{2} + v_{\text{ext}}(\mathbf{r}_i) \right) + \sum_{i<j} \frac{1}{|\mathbf{r}_i - \mathbf{r}_j|} \tag{4.12}$$

[where $v_{\text{ext}}(\mathbf{r}_i) = -Z/r_i$ is the Coulomb potential of the nucleus], taken on a Slater determinant:

$$E = \int \frac{\tau}{2} d\mathbf{r} + \int v_{\text{ext}} \rho(\mathbf{r}) d\mathbf{r} + \frac{1}{2} \int d\mathbf{r} d\mathbf{r}' \frac{\rho(\mathbf{r})\rho(\mathbf{r}')}{|\mathbf{r} - \mathbf{r}'|}$$
$$- \frac{1}{2} \sum_{\sigma,\sigma'} \int d\mathbf{r} d\mathbf{r}' \frac{|\rho^{(1)}(\mathbf{r}, \sigma, \mathbf{r}', \sigma')|^2}{|\mathbf{r} - \mathbf{r}'|}. \tag{4.13}$$

Further approximations consist in neglecting the exchange term [i.e. the last one in Eq. (4.13)] and employing the local density approximation for the kinetic energy density τ:

$$\tau = \tau(\rho) = \frac{3}{5}(3\pi^2)^{2/3}\rho(\mathbf{r})^{5/3}, \tag{4.14}$$

which means that we assume that τ is given by the kinetic energy of a non-interacting electron gas with density ρ [see Eqs. (1.77) and (1.79)]. This is a good approximation if $\rho(\mathbf{r})$ varies slowly enough in space, so that an electron located at $\mathbf{r}$ practically feels a homogeneous medium of density $\rho(\mathbf{r})$. Therefore, the Thomas–Fermi density functional is

$$E(\rho) = \frac{3}{10}(3\pi^2)^{2/3} \int \rho(\mathbf{r})^{5/3} d\mathbf{r} + \int v_{\text{ext}} \rho(\mathbf{r}) d\mathbf{r} + \frac{1}{2} \int d\mathbf{r} d\mathbf{r}' \frac{\rho(\mathbf{r})\rho(\mathbf{r}')}{|\mathbf{r} - \mathbf{r}'|}. \tag{4.15}$$

The equation that controls the system behavior is obtained by minimizing the energy functional under the constraint that the number of electrons $N = \int \rho d\mathbf{r}$ is constant:

$$\delta(E - \mu \int \rho d\mathbf{r}) = 0. \tag{4.16}$$

In this way the Thomas–Fermi equations are obtained

$$\frac{5}{3}\alpha\rho^{2/3} + v_{\text{ext}} + \int d\mathbf{r}' \frac{\rho(\mathbf{r}')}{|\mathbf{r} - \mathbf{r}'|} - \mu = 0, \tag{4.17}$$

where $\alpha = \frac{3}{10}(3\pi^2)^{2/3}$. Once solved, these equations yield ρ as a function of μ; the chemical potential is subsequently found by using the relationship that fixes the number of particles $N = \int \rho d\mathbf{r}$.

The Thomas–Fermi method describes only roughly the density and electrostatic potential of the atomic electrons. The main shortcoming of the model is that the density diverges at the nucleus and does not decay exponentially at infinity, but rather as r^{-6}. When applied to molecules it does not produce binding, and cannot predict the shell structure of atoms.

4.3.2 The Gross–Pitaevskii Theory for the Ground State of a Dilute Gas of Bosons

In Subsection 2.3, we saw that the mean field theories for Bosons in an external field lead to the density functional

$$E(\rho) = \frac{1}{2m} \int (\vec{\nabla}\sqrt{\rho})^2 d\mathbf{r} + \int v_{\text{ext}} \rho d\mathbf{r} + \frac{N-1}{2N} \int d\mathbf{r} d\mathbf{r}' \rho(\mathbf{r})\rho(\mathbf{r}')v(|\mathbf{r}-\mathbf{r}'|),$$
(4.18)

and that in the case of dilute systems, where the condition $\rho|a|^3 \ll 1$ is fulfilled, it is possible to replace the two-body interaction by an effective interaction given by $V(\mathbf{r}-\mathbf{r}') = g\delta(\mathbf{r}-\mathbf{r}')$, where the coupling constant g is connected to the scattering length a by $g = 4\pi\hbar^2 a/m$. Thus we obtain the Gross–Pitaevskii functional,

$$E(\rho) = \frac{1}{2m} \int (\vec{\nabla}\sqrt{\rho})^2 d\mathbf{r} + \int v_{\text{ext}} \rho d\mathbf{r} + \frac{g}{2} \int \rho^2 d\mathbf{r},$$
(4.19)

for the Bose–Einstein condensate, which is known to be the good theory, in the sense that the corrections to the mean field energy, due to dynamic correlations, behave as $(\rho|a|^3)^{1/2}$ (see Chapter 10) and are typically of the order of 1%.

The above functional is probably the best realization of the density functional theory.

In the case of atoms subject to repulsive interaction and confined in a parabolic external field, in the limit $Na/a_{\text{ho}} \gg 1$ $[a_{\text{ho}} = (\hbar/m\omega_{\text{ho}})^{1/2}]$, which turns out to be accomplishable in the experiments, the functional (4.19) leads to particularly simple solutions. In fact, in this limit the atoms of the condensate are pushed outward and the central density is very flat (see Fig. 2.10), and the kinetic term (which is proportional to the gradient of the density) is negligible. In this approximation, which is known as the Thomas–Fermi approximation, minimization of the functional with the constraint of a constant number of electrons yields the equation

$$v_{\text{ext}} + g\rho - \mu = 0,$$
(4.20)

whose solution is

$$\rho = \begin{cases} g^{-1}[\mu - v_{\text{ext}}] & \mu > v_{\text{ext}} \\ 0 & \text{otherwise} \end{cases}.$$
(4.21)

The constraint on the particle number gives the value of μ:

$$\mu = \frac{\hbar\omega_{\text{ho}}}{2}\left(\frac{15Na}{a_{\text{ho}}}\right)^{2/5}.$$
(4.22)

Moreover, since $\mu = \partial E/\partial N$, the energy per particle is given by

$$\frac{E}{N} = \frac{5}{7}\mu.$$
(4.23)

In this case, the Thomas–Fermi solutions turn out to be a very good approximation for the density and for the energy per particle of the condensate, when the number of atoms is of the order of 10^5–10^6.

The Thomas–Fermi approximation is employed also in the case of anisotropic traps where the external potential is described by

$$v_{\text{ext}}(\mathbf{r}) = \frac{1}{2}m(\omega_\perp^2(x^2 + y^2) + \omega_z^2 z^2). \tag{4.24}$$

This choice of the confinement potential reproduces very well most of the experimental conditions.

The Thomas–Fermi solution yields values of the average values of one-body operators which are quite different from those obtained from the independent-particle model, i.e. by neglecting the interaction. For example, for the potential (4.24), a simple calculation leads to the following results for the mean square radii and anisotropy of the velocity distributions:

$$\frac{\langle p_z^2 \rangle}{\langle p_x^2 \rangle} = \frac{\langle x^2 \rangle}{\langle z^2 \rangle} = \frac{\omega_z^2}{\omega_\perp^2} \tag{4.25}$$

in the case of the Thomas–Fermi approximation, and

$$\frac{\langle p_z^2 \rangle}{\langle p_x^2 \rangle} = \frac{\langle x^2 \rangle}{\langle z^2 \rangle} = \frac{\omega_z}{\omega_\perp} \tag{4.26}$$

in the case of the independent-particle model.

In the Thomas–Fermi approximation, one can evaluate analytically also the momentum distribution of the condensate, by taking the Fourier transform of the Thomas–Fermi wave function $\varphi = \sqrt{g^{-1}[\mu - v_{\text{ext}}]}$ (Baym and Pethick, 1996). One finds the result

$$n(\mathbf{k}) = N\frac{15}{16\hbar^3}R^3 \left(\frac{J_2(\hat{k})}{\hat{k}^2} \right)^2, \tag{4.27}$$

where $R = (R_x R_y R_z)^{1/3}$ and R_k are the radii of the density elipsoid, J_2 is the Bessel function of order 2 and $\hat{k} = \sqrt{k_x^2 R_x^2 + k_y^2 R_y^2 + k_z^2 R_z^2}/\hbar$ is the dimensionless momentum variable. Equation (4.27) gives a width of the momentum distribution much narrower then the width predicted by the noninteracting gas, which in the limit of large N approaches a delta function.

4.3.3 The Thomas–Fermi Approximation for the Fermi Gas Confined in a Harmonic Potential

In the last few years important experimental effort has been devoted to the realization of degenerate atomic Fermi gas in traps. Since the Pauli exclusion principle gives a zero s wave scattering length for collisions among atoms occupying the same hyperfine state, special effort has been devoted to the trapping of different spin components. The main goal of these studies was to reach the conditions of very low temperature where one can investigate the crossover from a Bose–Einstein condensate (BEC) to a Bardeen–Cooper–Schrieffer (BCS) superfluid by magnetically tuning the two-body scattering amplitude. For positive values of the s wave scattering length a, atoms with different spins are observed to pair into bound molecules which, at low enough temperature, form a Bose condensate (Jochim *et al.*, 2003; Greiner, Regal and Jin, 2003; Zwierlein *et al.*, 2003). The molecular BEC state is adiabatically converted in an ultracold Fermi gas with $a < 0$ and $k_F|a| \ll 1$ (Bartenstein *et al.*, 2004; Bourdel *et al.*, 2004), where standard BCS theory is expected to apply. In the crossover region the value of $|a|$ can be orders of magnitude larger than the inverse Fermi wave vector $1/k_F$ and one enters a new strongly correlated regime known as the unitary limit (Bartenstein *et al.*, 2004; Bourdel *et al.*, 2004; O'Hara *et al.*, 2002).

In this subsection we study the Fermi gas in the harmonic potential in the limit of large N where the motion of the atomic gas can be described in the Thomas–Fermi approximation. This approximation is compatible with the Pauli principle and provides a basic model for trapped atomic gases. As already discussed, it misses shell effects which are important for small samples at extremely low temperature.

The Thomas–Fermi equations of a given spin component of the trapped *ideal* Fermi gas read

$$\frac{1}{2m}(6\pi^2)^{2/3}\rho^{2/3} + v_{\text{ext}} - \mu = 0, \tag{4.28}$$

where v_{ext} is the trapping potential that will be chosen to be of the harmonic type and μ is the chemical potential fixed by the normalization condition $N = \int \rho d\mathbf{r}$. If more spin states are occupied, then Eq. (4.28) characterizes the Thomas–Fermi equation of each component separately. It differs from one component to another if the number of atoms in each spin state or the corresponding trapping potentials are different. The solution to Eq. (4.28) is

$$\rho(\mathbf{r}) = \frac{1}{6\pi^2}\left[\frac{2m}{\hbar^2}(\mu - v_{\text{ext}}(\mathbf{r}))\right]^{3/2}\Theta(\mu - v_{\text{ext}}(\mathbf{r}))$$

$$\equiv \frac{1}{6\pi^2}k_F^3(\mathbf{r})\Theta(\mu - v_{\text{ext}}(\mathbf{r})), \tag{4.29}$$

where Θ is the step function, and we have defined the local momentum by

$$k_F(\mathbf{r}) = \left[\frac{2m}{\hbar^2}(\mu - v_{\text{ext}}(\mathbf{r}))\right]^{1/2}. \tag{4.30}$$

The Fermi energy, defined by the zero temperature value of the chemical potential $\mu(T = 0) = \epsilon_F = KT_F$ (T_F is the Fermi temperature), is given by the relation

$$N = \int \rho d\mathbf{r} = \frac{1}{6}\left(\frac{\epsilon_F}{\hbar\omega_{\text{ho}}}\right)^3, \tag{4.31}$$

where $\omega_{\text{ho}} = (\omega_x \omega_y \omega_z)^{1/3}$ is the geometric average of the three trapping frequencies. Note that the Fermi energy resulting from Eq. (4.31) has the same dependence on the number of atoms and on the oscillator frequency as the critical temperature for Bose–Einstein condensation of trapped Bosons [see Eq. (1.68)].

For an axial symmetric trap $v_{\text{ext}}(\mathbf{r}) = 1/2m(\omega_\perp^2 r_\perp^2 + \omega_z z^2)$, the Fermi energy defined by Eq. (4.31) can be used to define the radial and axial widths $R_\perp$ and Z, respectively, of the density distribution through

$$\epsilon_F = \frac{1}{2}m\omega_\perp^2 R_\perp^2 = \frac{1}{2}m\omega_z^2 Z^2, \tag{4.32}$$

and one gets for the density

$$\rho(\mathbf{r}) = \frac{8}{\pi^2}\frac{N}{R_\perp^2 Z}\left(1 - \frac{r_\perp^2}{R_\perp^2} - \frac{z^2}{Z^2}\right)^{3/2}. \tag{4.33}$$

The density (4.29) can be recovered as the local limit $(\mathbf{r}' \to \mathbf{r})$ of the one-body nondiagonal density matrix [see Eq. (1.85)]:

$$\rho^{(1)}(\mathbf{r},\mathbf{r}') = \frac{1}{2\pi^2}k_F^3(\mathbf{q})\frac{C(k_F(\mathbf{q})s)}{k_F(\mathbf{q})s}, \tag{4.34}$$

with

$$\mathbf{q} = \frac{1}{2}(\mathbf{r}+\mathbf{r}'), \quad \mathbf{s} = \mathbf{r}-\mathbf{r}'.$$

Expression (4.34) allows the calculation of all the quantities of interest, such as the spectral density matrix

$$g^\epsilon(\mathbf{r},\mathbf{r}') = \langle\mathbf{r}|\delta(\epsilon - H)|\mathbf{r}'\rangle = \sum_n \varphi_n^*(\mathbf{r})\varphi_n(\mathbf{r}')\delta(\epsilon - \epsilon_n), \tag{4.35}$$

which, since $\rho^{(1)}(\mathbf{r},\mathbf{r}',\epsilon) = \sum_n \varphi_n^*(\mathbf{r})\varphi_n(\mathbf{r}')\Theta(\epsilon - \epsilon_n)$, is given by

$$g^\epsilon(\mathbf{r},\mathbf{r}') = \frac{\partial}{\partial\epsilon}\rho^{(1)}(\mathbf{r},\mathbf{r}',\epsilon). \tag{4.36}$$

Consequently the density of states $g(\epsilon)$ (the trace of the spectral density) in the Thomas–Fermi approximation is given by

$$g(\epsilon) = \frac{1}{4\pi^2}\left(\frac{2m}{\hbar^2}\right)^{3/2}\int d\mathbf{r}(\epsilon - v_{\text{ext}})^{1/2}\Theta(\epsilon - v_{\text{ext}}). \tag{4.37}$$

For the harmonic oscillator potential, the integral (4.37) can be evaluated analytically and one gets

$$g(\epsilon) = \frac{\epsilon^2}{2(\hbar\omega_{\text{ho}})^3}, \tag{4.38}$$

from which by integrating in ϵ, one gets again for the total number of atoms $N = \int_0^{\epsilon_F} d\epsilon\, g(\epsilon)$ the result (4.31). For the ground state energy one gets

$$E = \sum \epsilon_n = \int_0^{\epsilon_F} d\epsilon\, \epsilon\, g(\epsilon) = \frac{3}{4} N \epsilon_F. \tag{4.39}$$

Finally, starting from the Thomas–Fermi one-body nondiagonal density matrix, one can calculate the momentum distribution for the axially symmetric trapping. A simple calculation yields the result

$$n(\mathbf{k}) = \frac{8}{\pi^2} \frac{N}{k_F^3} \left(1 - \frac{k^2}{k_F^2}\right)^{3/2}, \tag{4.40}$$

where k_F is defined by $\epsilon_F = k_F^2/2m$. Result (4.40) is the analog of the momentum distribution $3N/(4\pi k_F^3)\Theta\left(1 - \frac{k^2}{k_F^2}\right)$ which characterizes the uniform Fermi gas.

It is interesting to compare Eqs. (4.29) and (4.40) with the analogous results (4.21) and (4.27) which hold for a trapped Boson gas. The shape of the profiles is similar in coordinate space and the radius of the atomic cloud increases with N in both cases, even if the value of the chemical potential for Bosons is fixed by the interaction while for Fermions it is determined by the Fermi motion. On the other hand, in momentum space the two distributions differ in a deep way. Whereas the momentum distribution of the Fermi gas is isotropic, that the Bose gas is anisotropic. Furthermore the width of a trapped Fermi gas scales like $m\omega_{\mathrm{ho}}R$ and increases with N while the momentum width of a trapped Bose–Einstein condensate scales like $\hbar/R$ and decreases with increasing N. The different behavior reflects the difference of the Heisenberg uncertainty inequality in the two cases: close to unity for the Bose–Einstein condensate and much larger than unity for the Fermi gas.

The above results provide a good description of a cold spin-polarized Fermi gas for which the interactions are strongly inhibited by the Pauli exclusion principle. When the system is formed by a mixture of atomic species in different spin states, two-body interactions may become important. As we have already discussed, they can bring the system into the BCS phase, but even in the normal phase they can influence the modes of excitation of the system and the collisional processes which are responsible for thermalization and the damping of collective oscillations. The dynamic behavior of a trapped Fermi gas will be discussed in the following chapters. As far as ground state properties are concerned, normally, interaction effects play a much less crucial role than in the case of Bosons. This can be easily seen by starting from the Hamiltonian

$$H = \sum_i \left(\frac{p^2}{2m} + V_{\uparrow,\mathrm{ho}}(\mathbf{r})\right)_i + \sum_j \left(\frac{p^2}{2m} + V_{\downarrow,\mathrm{ho}}(\mathbf{r})\right)_j + g \sum_{i,j} \delta(\mathbf{r}_i - \mathbf{r}_j), \tag{4.41}$$

valid for a two-component system occupying spin-up (characterized by the index i) and spin-down (characterized by the index j) states. In Eq. (4.41), $g = 4\pi\hbar^2 a/m$ is the interaction strength fixed by the s wave scattering length a associated with the scattering of atoms with opposite spin, and the trapping potentials (of the harmonic type) may be different for different species due to the difference in the corresponding magnetic moments.

By evaluating the expectation value E_{int} of the interaction term in the ground state of the noninteracting Hamiltonian and comparing it with the corresponding value E_{ho} of the oscillator potential, one finds the result

$$\frac{E_{\text{int}}}{E_{\text{ho}}} = 0.56N^{1/6}\frac{a}{a_{\text{ho}}} = 0.29k_F a, \qquad (4.42)$$

where $a_{\text{ho}} = \sqrt{\hbar/m\omega_{\text{ho}}}$, $k_F = 1.91N^{1/6}/a_{\text{ho}}$, N is the number of atoms of each species and we have assumed that $\rho^\uparrow(\mathbf{r}) = \rho^\downarrow(\mathbf{r}) = \rho(\mathbf{r})$, with $\rho(\mathbf{r})$ given by the unperturbed density of Eq. (4.33). Since $k_F a$ for dilute Fermi gas is very small, the effects of the interaction on the ground state properties are very small under normal conditions. The situation changes if the scattering length becomes very large, as happens when one is working close to a Feshback resonance. The equation of state of a Fermi gas in the BEC–BCS crossover where the interaction strength is varied over a wide range will be discussed in Chapter 7.

A detailed treatment of the results of the Thomas–Fermi theory when applied to the trapped Bose and Fermi gases can be found in the book by Pitaevskii and Stringari (2003).

4.4 The Kohn–Sham Equations

In 1965 Kohn and Sham proposed a method to obtain an exact, single-particle-like description of a many-body system. The method is based on the following separation of the energy functional:

$$E(\rho) = T_0(\rho) + \int d\mathbf{r}\rho(\mathbf{r})\left[v_{\text{ext}}(\mathbf{r}) + \frac{1}{2}U(\mathbf{r})\right] + E_{\text{xc}}(\rho), \qquad (4.43)$$

where $T_0(\rho)$ is the kinetic energy of the noninteracting system with density ρ, $U(\mathbf{r}) = \int d\mathbf{r}'\rho(\mathbf{r}')v(|\mathbf{r}-\mathbf{r}'|)$ is the Hartree potential (which in the case of electrons is the classical Coulomb potential) and $E_{\text{xc}}(\rho)$ can be thought of as the definition of the exchange-correlation energy. Though $T_0(\rho)$ is different from the real kinetic energy, it has a comparable magnitude and is treated exactly in this method. The exact treatment of $T_0(\rho)$ eliminates some of the shortcomings of the Thomas–Fermi approach to the Fermion system, such as the lack of shell effects or the absence of bonding in molecules and solids. All terms of (4.43), except E_{xc}, can be treated exactly. Therefore, the unavoidable approximations of the method concern only the exchange-correlation energy.

Application of the variational principle to (4.43), with the constraint of particle number conservation, leads to

$$\frac{\partial T_0(\rho)}{\partial \rho} + v_{\text{ext}}(\mathbf{r}) + U(\mathbf{r}) + \frac{\partial E_{\text{xc}}(\rho)}{\partial \rho} = \mu, \qquad (4.44)$$

where μ is the Lagrange multiplier related to the conservation of N. By comparing this equation with the corresponding one for a system of particles moving in an effective potential $V_{\text{eff}}(\mathbf{r})$, but without any two-body interaction, i.e.

$$\frac{\partial T_0(\rho)}{\partial \rho} + V_{\text{eff}}(\mathbf{r}) = \mu, \qquad (4.45)$$

we see that the two problems are identical provided that

$$V_{\text{eff}}(\mathbf{r}) = v_{\text{ext}}(\mathbf{r}) + U(\mathbf{r}) + \frac{\partial E_{\text{xc}}(\rho)}{\partial \rho}. \qquad (4.46)$$

Therefore, the solution to Eq. (4.45) can be found simply by solving the single-particle equation for the noninteracting particles:

$$\left(-\frac{\vec{\nabla}^2}{2m} + V_{\text{eff}}(\mathbf{r}) \right) \varphi_i(\mathbf{r}) = \epsilon_i \varphi_i(\mathbf{r}), \qquad (4.47)$$

which yields

$$\rho(\mathbf{r}) = \sum_{i=1}^{N} |\varphi_i(\mathbf{r})|^2. \qquad (4.48)$$

The constraint (4.46) is fulfilled by a self-consistent procedure. Finally, one obtains the density and energy of the ground state, and all quantities related to these two. In particular, to compute the energy one uses

$$\sum_i \epsilon_i = \sum_i \langle \varphi_i | -\frac{\vec{\nabla}^2}{2m} + V_{\text{eff}}(\mathbf{r}) | \varphi_i \rangle = T_0(\rho) + \int d\mathbf{r}\rho(\mathbf{r})V_{\text{eff}}(\mathbf{r}) \qquad (4.49)$$

and

$$E = \sum_i \epsilon_i - \frac{1}{2} \int d\mathbf{r}\rho(\mathbf{r})U(\mathbf{r}) + E_{\text{xc}}(\rho) - \int d\mathbf{r}\rho(\mathbf{r})\frac{\partial E_{\text{xc}}(\rho)}{\partial \rho}. \qquad (4.50)$$

From the numerical point of view, the solution of the Kohn–Sham Eqs. (4.46) and (4.47) is much simpler than that of the HF equations, since, contrary to the HF potential, the effective potential is local.

Note that, contrary to the HF theory, in the density functional theory the single-particle wave functions and energies have no physical meaning, but are rather just mathematical tools through which the physical quantities, ρ and the energy, are calculated. In particular, the occupied single-particle energies do not represent the physical energies required to excite the particles to the continuum.

The physical excitation energies $\hat{\epsilon}_i$ are the solutions to the Dyson equation:

$$-\frac{\vec{\nabla}^2}{2m}\hat{\varphi}_i(\mathbf{r}) + \int d\mathbf{r}'\Sigma(\mathbf{r},\mathbf{r}';\hat{\epsilon}_i)\hat{\varphi}_i(\mathbf{r}') = \hat{\epsilon}_i\hat{\varphi}_i(\mathbf{r}), \qquad (4.51)$$

where $\Sigma(\mathbf{r},\mathbf{r}';\hat{\epsilon})$ is the nonlocal self-energy operator. This operator is clearly different from the local operator $V_{\text{eff}}(\mathbf{r})$ of the Kohn–Sham equations, and contrary to the real eigenvalues ϵ_i, the $\hat{\epsilon}_i$ of Eq. (4.51) are in general complex, which reflects the finite lifetime of the ionized states. There are examples showing that the real parts of ϵ_i and $\hat{\epsilon}_i$ differ as well.

However, there exists a very important special case, i.e. that in which N is made to tend to infinity. In this case, it is possible to show that the highest energy occupied KS eigenvalue ϵ_N equals the true chemical potential $\mu = \hat{\epsilon}_N$. For the density functional theory, this is the analog of the Koopmans, theorem (Koopmans, 1933).

4.5   The Local Density Approximation for the Exchange-Correlation Energy

For a system with slowly varying density, the local density approximation can be made:

$$E_{\text{xc}}(\rho) = \int \epsilon_{\text{xc}}(\rho(\mathbf{r}))\rho(\mathbf{r})d\mathbf{r}, \qquad (4.52)$$

where $\epsilon_{\text{xc}}(\rho(\mathbf{r}))$ is the exchange-correlation energy per particle of the uniform system having density ρ. We have

$$\epsilon_{\text{xc}}(\rho(\mathbf{r})) = \epsilon_{\text{x}}(\rho(\mathbf{r})) + \epsilon_{\text{c}}(\rho(\mathbf{r})). \qquad (4.53)$$

For electron systems, the exchange energy per particle as a function of density is given by (2.39). For example, in 2D we have $\epsilon_{\text{x}} = -4/3\sqrt{2\rho/\pi}$. The best correlation energy per particle is surely the one given by the Monte Carlo calculations of Ceperley and coworkers. The calculated numerical values at different r_s are fitted by Padè approximants which, in the two-dimensional case taken as an example, are of the type

$$\epsilon_{\text{c}}(r_s,\xi) = a_0\frac{1+a_1 x}{1+a_1 x + a_2 x^2 + a_3 x^3}, \qquad (4.54)$$

with $x = (r_s)^{1/2}$ and $r_s = (\pi\rho)^{-1/2}$. This Padè form behaves as $\sim a+br_s$ for $r_s \to 0$, which is the correct high density expansion, and has the asymptotic form

$$\epsilon_{\text{c}} \sim \frac{a}{r_s} + \frac{b}{r_s^{3/2}} + \frac{c}{r_s^2} \qquad (4.55)$$

Table 4.1. Coefficients for the correlation energies of Eq. (4.54) for the normal ($\xi = 0$) and polarized ($\xi = 1$) fluids.

	$\xi = 0$	$\xi = 1$
a_0	-0.3568	-0.0515
a_1	1.1300	340.5813
a_2	0.9052	75.2293
a_3	0.4165	37.0170

for $r_s \to \infty$. The best fit a_i coefficients for the correlation energies for the normal ($\xi = 0$) and polarized ($\xi = 1$) fluids are reported in Table 4.1 in Rydberg units (1 atomic unit = 2 Rydberg).

Most of the calculations with the Kohn–Sham equations were carried out under the local density approximation, which produces surprisingly good results even in the cases where the density does not change slowly. The quality of the LDA can be appreciated from the results of Table 2.1 and Fig. 2.7, where the results of this approach for the energies and densities are compared to those of Monte Carlo, HF and BHF, for metal clusters and quantum dots. The LDA also produces very good results for the binding energy of atoms and diatomic molecules, and for the cohesion energies of metals [for an exhaustive discussion see Lundqvist and March (1983)].

4.6 The Local Spin Density Approximation (LSDA)

The formalism of the density functional shows that, in principle, it is possible to determine the total energy by using a functional which depends only on density, and not on spin density. However, the task of determining a good approximation for the exchange-correlation energy is much simpler if the functional is expressed in terms of the spin densities. For electron systems, this is the simplest way of fulfilling the demand that high-spin states tend to be energetically favored (Hund's rule). Contrary to the LDA, the LSDA approach obeys Hund's rule and gives results in good agreement with Monte Carlo. This is shown in Table 2.4 and Figs. 2.6–2.8 and 4.1, where the energies and densities of this approach are compared to those of Monte Carlo for metal clusters and quantum dots.

The LSDA can be expressed as

$$E_{xc}^{LSDA} = \int \epsilon_{xc}(\rho(\mathbf{r}), \xi(\mathbf{r}))\rho(\mathbf{r})d\mathbf{r}, \qquad (4.56)$$

where $\epsilon_{xc}(\rho(\mathbf{r}), \xi(\mathbf{r}))$ is the exchange-correlation energy per particle of a homogeneous, spin-polarized system with spin-up and spin-down densities $\rho_+(\mathbf{r})$ and $\rho_-(\mathbf{r})$, respectively, and $\xi = (\rho_+ - \rho_-)/\rho$ is the spin polarization. The parametrization most widely used in the literature for the exchange-correlation energy is that of Tanatar

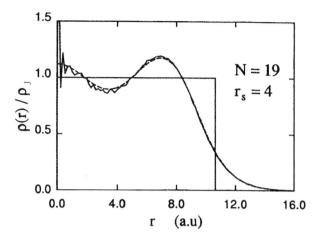

Fig. 4.1 Electronic density profiles of the sodium cluster with 19 electrons in the LSDA (dashed line), and in variational Monte Carlo (full line) (Ballone *et al.*, 1992). The right-angled line shows the jellium density ρ_j.

and Ceperley (1988) of Eq. (4.54) in 2D, and that of Ceperley and Alder (1980) in 3D for the nonpolarized ($\xi = 0$) and ferromagnetic ($\xi = 1$) cases. For intermediate polarizations one uses (von Barth and Hedin, 1972)

$$\epsilon_{xc}(\rho, \xi) = \epsilon_{xc}(\rho, 0) + f(\xi)[\epsilon_{xc}(\rho, 1) - \epsilon_{xc}(\rho, 0)]. \tag{4.57}$$

The function $f(\xi)$ is the same one that, used in (4.57) for the exchange energy alone, reproduces the exact calculation of Eqs. (2.43) and (2.44), i.e. we have

$$f(\xi) = \frac{(1 + \xi)^{3/2} + (1 - \xi)^{3/2} - 2}{2^{3/2} - 2} \quad \text{(in 2D)} \tag{4.58}$$

and

$$f(\xi) = \frac{(1 + \xi)^{4/3} + (1 - \xi)^{4/3} - 2}{2^{4/3} - 2} \quad \text{(in 3D)}. \tag{4.59}$$

Like in the case of unpolarized systems, from Eq. (4.43), with $E_{xc} = E_{xc}^{LSDA}$, we can derive a set of self-consistent equations, by treating the single-particle kinetic energy functional T_0 exactly. The following set of coupled equations for spin-up and spin-down particles is obtained:

$$\left\{ -\frac{\vec{\nabla}^2}{2m} + v_{ext}(\mathbf{r}) + \int d\mathbf{r}' \rho(\mathbf{r}') v(|\mathbf{r} - \mathbf{r}'|) + v_{xc}(\mathbf{r}) + \sigma W_{xc}(\mathbf{r}) \right\} \varphi_{i,\sigma}(\mathbf{r})$$

$$= \epsilon_{i,\sigma} \varphi_{i,\sigma}(\mathbf{r}), \tag{4.60}$$

where $\sigma = \pm 1$, and

$$v_{xc}(\mathbf{r}) = \frac{\partial E_{xc}(\rho, m)}{\partial \rho(\mathbf{r})} = \frac{\partial \rho \epsilon_{xc}(\rho, \xi)}{\partial \rho(\mathbf{r})}, \quad W_{xc}(\mathbf{r}) = \frac{\partial E_{xc}(\rho, m)}{\partial m(\mathbf{r})} = \frac{\partial \epsilon_{xc}(\rho, \xi)}{\partial \xi(\mathbf{r})}, \tag{4.61}$$

and

$$\rho(\mathbf{r}) = \sum_{\sigma} \rho_\sigma(\mathbf{r}), \quad m(\mathbf{r}) = \rho_+ - \rho_-, \quad \rho_\sigma(\mathbf{r}) = \sum_{i=1} |\varphi_{i,\sigma}(\mathbf{r})|^2. \qquad (4.62)$$

The LSDA theory was applied very successfully to atoms, molecules and metal clusters with nonzero spin and thus nonzero magnetization. An exhaustive description of the results obtained in these cases can be found in Lundqvist and March (1983), and Jones and Gunnarsson (1989).

The theory was also employed to calculate the magnetic susceptibility χ_σ of paramagnetic metals. To do this one has to solve Eqs. (4.60) in the presence of a weak magnetic field B which introduces a further term, $\mu_0 B\sigma$, in (4.60), where $\mu_0 = e\hbar/2mc$ is Bohr's magneton, and then calculate the total magnetization m [see also Eq. (1.120)]:

$$m = \frac{\chi_\sigma B}{L^D}. \qquad (4.63)$$

A method which is equivalent to the solution of the LSDA equations in a magnetic field is that of calculating the linear response function for the spin density operator $\hat{\rho} = \sum_i e^{i\mathbf{q}\cdot\mathbf{r}_i} \sigma_i^z$, by using the time-dependent form of the LSDA theory (see Subsection 9.9), and then computing the static response in the limit of zero transfer momentum q. In this way we find ($\chi_\sigma = \mu_0 \chi^{m,m}$ of Subsection 9.9.1) in atomic units

$$\chi_\sigma = \frac{\mu_0 \chi_0}{1 - \frac{F_0^a}{L^D} \chi_0}, \qquad (4.64)$$

where $F_0^a = \frac{\partial^2 E_{\text{xc}}(\rho,m)}{\partial m^2}|_{m=0}$ and χ_0 is the static response of the free electron gas in the $q \to 0$ limit:

$$\frac{\chi_0}{L^D} = \begin{cases} -mk_F/\pi^2 & \text{for D} = 3 \\ -m/\pi & \text{for D} = 2 \\ -2m/\pi k_F & \text{for D} = 1 \end{cases} \qquad (4.65)$$

From Eqs. (4.64) and (4.65), by using the expressions for the exchange energies of the partially polarized electron gas of Chapter 2, we find for the spin susceptibility of the electron gas

$$\frac{\mu_0 \chi_0}{\chi_\sigma} = \begin{cases} 1 - (2/3\pi^2)^{2/3} r_s + 3(4/9\pi)^{2/3} r_s^2 (\partial^2 \epsilon_c/\partial\xi^2)_{\xi=0} & \text{for D} = 3 \\ 1 - (\sqrt{2}/\pi) r_s + r_s^2 (\partial^2 \epsilon_c/\partial\xi^2)_{\xi=0} & \text{for D} = 2 \\ 1 + (16 r_s^2/\pi^2)(\partial^2 \epsilon_{\text{xc}}/\partial\xi^2)_{\xi=0} & \text{for D} = 1 \end{cases} \qquad (4.66)$$

In Table 4.2 we compare the results of (4.66) with the correlation energy obtained from Monte Carlo calculations, and the ξ dependence of (4.57) with the results of the variational calculation of Vosko *et al.*, (1975) as well as with the experimental results for alkali metals which, in nature, are the closest realization of the three-dimensional interacting electron gas.

Table 4.2. Spin susceptibility (in unit of the free one) of alkali metals. Experimental results are compared with variational and linear response theory results.

Metal	Variational	Response	Exp.
Na	1.62	1.59	1.65
K	1.79	1.78	1.70
Rb	1.78	1.87	1.72
Cs	2.20	1.99	1.76–2.24

As can be seen from this table, there is a rather good agreement between theory and experiment.

4.7 Inclusion of Current Terms in the DFT (CDFT)

The LSDA formalism was extended by Vignale and Rasolt (1987, 1988), who included in the exchange-correlation energy, terms depending on the strength of the magnetic field applied to the system. This theory, which is known as the current density functional theory (CDFT), provides a method for describing a system of interacting electrons in a gauge field.

The most important case of application of the CDFT concerns quantum dots in a magnetic field. In this case one has N electrons moving in the $z = 0$ plane, where they are confined by a circularly symmetric potential $v_{\text{ext}}(r)$, with $r = \sqrt{x^2 + y^2}$. A constant magnetic field B is applied normal to the plane, which in the symmetric gauge is described by the vector potential $\mathbf{A} = \frac{B}{2}(-y, x, 0)$. By introducing the cyclotron frequency $\omega_c = eB/mc$, we can write the CDFT grand potential as (see also Subsection 2.6) $G = E - TS - \mu N$ by adding to the total energy (we use effective atomic units with Boltzmann's constant $K = 1$):

$$E = \frac{1}{2} \int d\mathbf{r}\, \tau(\mathbf{r}) + \frac{\omega_c}{2} \int d\mathbf{r}\, r j_p(\mathbf{r}) + \frac{1}{8}\omega_c^2 \int d\mathbf{r}\, r^2 \rho(\mathbf{r})$$

$$+ g^* \mu_0 B \sum_i f_i s_{z_i} + \int d\mathbf{r}\, v_{\text{ext}}(\mathbf{r})\rho(\mathbf{r})$$

$$+ \frac{1}{2} \int \int d\mathbf{r}\, d\mathbf{r}' \frac{\rho(\mathbf{r})\rho(\mathbf{r}')}{|\mathbf{r} - \mathbf{r}'|} + \int d\mathbf{r}\, \mathcal{E}_{\text{xc}}(\rho(\mathbf{r}), \xi(\mathbf{r}), \mathcal{V}(\mathbf{r})), \qquad (4.67)$$

the temperature times the total entropy S, given by

$$S = -\sum_i [f_i \ln f_i + (1 - f_i) \ln(1 - f_i)], \qquad (4.68)$$

and the electron chemical potential μ times the number of electrons $N = \sum_i f_i$. The introduction of temperature in the CDFT formalism is necessary in the case of a nonvanishing magnetic field, even when one describes the system at zero temperature. This is due to the fact that at zero temperature, for some values of B (see Chapter 5) the energy levels near to the Fermi surface are very close in energy and it is impossible to achieve good convergence of the Kohn–Sham equations other than by first obtaining convergence at finite temperature, and by subsequent gradual lowering of the temperature itself, until convergence of the zero-temperature solution is attained.

The second and third terms of (4.67) originate from the modified kinetic energy term of the Hamiltonian of the electrons with charge $-e$ in the magnetic field $[\mathbf{p}^2/2 \to (\mathbf{p}+e/c\mathbf{A})^2/2]$. The fourth term represents the Zeeman interaction energy between the magnetic moment of the electrons (whose effective gyromagnetic factor is g^*) and the external magnetic field. In Eqs. (4.67) and (4.68), s_{z_i} is the z components of the spin ($\pm 1/2$), and f_i is the occupation number of the ith single-particle level. The density $\rho(\mathbf{r})$, the kinetic energy density $\tau(\mathbf{r})$ and the paramagnetic current density $\mathbf{j}_p(\mathbf{r})$ are all defined in terms of the single-particle occupation numbers f_i and wave functions ϕ_i, as follows:

$$\rho(\mathbf{r}) = \sum_i f_i \, |\phi_i(\mathbf{r})|^2, \tag{4.69}$$

$$\tau(\mathbf{r}) = \sum_i f_i \, |\nabla\phi_i(\mathbf{r})|^2, \tag{4.70}$$

$$\mathbf{j}_p(\mathbf{r}) = j_p(r)\hat{e}_\theta = -\frac{1}{r}\sum_i f_i l_i \, |\phi_i(\mathbf{r})|^2 \, \hat{e}_\theta, \tag{4.71}$$

where $\hat{e}_\theta$ is the azimuthal unit vector.

The exchange-correlation energy in (4.67) is written in the local density approximation (LDA):

$$E_{\mathrm{xc}} = \int d\mathbf{r}\rho(r)\epsilon_{\mathrm{xc}}[\rho(r), \xi(r), \mathcal{V}(r)], \tag{4.72}$$

and is a functional of the density ρ, of the local spin polarization ξ and of the local vorticity $\mathcal{V}(r)$:

$$\mathcal{V}(r) = \mathcal{V}(r)\hat{e}_z = -\frac{c}{er}\frac{\partial}{\partial r}(r\frac{j_p}{\rho})\hat{e}_z. \tag{4.73}$$

Following Ferconi and Vignale's (1994) indications, we take

$$\epsilon_{\mathrm{xc}}[\rho, \xi, \mathcal{V}] = \frac{1}{1+\nu^4}\epsilon_{\mathrm{xc}}^{\mathrm{LWM}}[\rho, \nu] + \frac{\nu^4}{1+\nu^4}\epsilon_{\mathrm{xc}}^{\mathrm{TC}}[\rho, \xi], \tag{4.74}$$

where $\nu = 2\pi\mathcal{L}^2\rho$ is the local filling factor (see Chapter 5) and $\mathcal{L}$ is the magnetic length $\mathcal{L}^2 = \hbar c/eB = \hbar/m\omega_c$. This expression is a Padè interpolation between Levesque, Weis and MacDonald's (1984) results for the exchange-correlation energy at a high magnetic field, $\epsilon_{\text{xc}}^{\text{LWM}}[\rho, \nu]$, and those of Tanatar and Ceperley at a zero magnetic field, $\epsilon_{\text{xc}}^{\text{TC}}[\rho, \xi]$ [see Eqs. (4.57) and (4.58)].

In the LDA, the vorticity $\mathcal{V}$ is taken to be proportional to the applied magnetic field $\mathcal{V} = -e\mathbf{B}/mc$, which allows the filling factor to be related to the vorticity as follows:

$$\nu = 2\pi\rho\frac{\hbar}{m}\frac{1}{\mathcal{V}}. \tag{4.75}$$

This completely defines the CDFT functional.

In order to obtain the single-particle wave functions $\phi_i(\mathbf{r})$ and the occupation numbers f_i, one needs to minimize G under the constraint that the single-particle wave functions are orthonormal. By minimizing with respect to the single-particle wave functions one gets the Kohn–Sham (KS) equations:

$$\left[-\frac{1}{2}\left(\frac{d^2}{dr^2} + \frac{1}{r}\frac{d}{dr} - \frac{l^2}{r^2}\right) - \frac{\omega_c}{2}l + \frac{1}{8}\omega_c^2 r^2 + v_{\text{ext}}(r) \right.$$

$$\left. + \int d\mathbf{r}' \frac{\rho(\mathbf{r}')}{|\mathbf{r} - \mathbf{r}'|} - \frac{e}{c}l\frac{A_{\text{xc}}(r)}{r} + V_{\text{xc}\sigma}(r) + \frac{1}{2}g^*\mu_0 B\sigma \right] u_{nl\sigma} = \epsilon_{nl\sigma}u_{nl\sigma}, \tag{4.76}$$

where we have used the circular symmetry of the wave functions

$$\phi_i(\mathbf{r}) = e^{-\imath l\theta}u_{nl\sigma}(r), \tag{4.77}$$

with $\sigma = \pm 1$, $n = 0, 1, 2, \ldots$ and $l = 0, \pm 1, \pm 2, \pm 3, \ldots$. These wave functions are eigenstates of l_z [see also Eq. (2.29)] with eigenvalue $-l$. At $B \neq 0$, the single-particle level i is nondegenerate, and in the above expression we used the notation $i \equiv \{n, l, \sigma\}$.

In Eq. (4.76) one has

$$V_{\text{xc}\sigma}(r) = \left.\frac{\delta E_{\text{xc}}(\rho, \xi, \mathcal{V})}{\delta\rho_\sigma}\right|_{\rho_{-\sigma}, \mathcal{V}} - \frac{e}{c}\mathbf{A}_{\text{xc}}(r) \cdot \frac{\mathbf{j}_p}{\rho}, \tag{4.78}$$

where $\mathbf{A}_{\text{xc}}$ is the exchange-correlation vector potential:

$$\frac{e}{c}\mathbf{A}_{\text{xc}}(r) = \frac{e}{c}A_{\text{xc}}\hat{e}_\theta = \frac{c}{e\rho}\frac{d}{dr}\left(\rho\left.\frac{\delta\mathcal{E}_{\text{xc}}}{\delta\mathcal{V}}\right|_{n,\xi}\right)\hat{e}_\theta. \tag{4.79}$$

By minimizing G with respect to f_i, and using the KS equations, one obtains the occupation numbers:

$$f_i = \frac{1}{1 + e^{(\epsilon_i - \mu)/T}}. \tag{4.80}$$

Next, the chemical potential is determined by normalization:

$$N = \sum_i f_i = \sum_i (1 + e^{(\epsilon_i - \mu)/T})^{-1}.$$

As can be seen by comparing the present subsection with Subsection 2.6, the finite temperature treatment is formally the same for HF and for the density functional theories.

The KS differential equations (4.76) and the normalization condition were solved in a self-consistent way by several authors (Ferconi and Vignale, 1994; Lipparini *et al.*, 1997; Pi *et al.*, 1998; Steffens *et al.*, 1998). The numerical results will be reported in the next chapter, which treats quantum dots under a magnetic field in the different theories.

4.8   The Ensemble Density Functional Theory (EDFT)

The density functional theory has been generalized to treat the case of degenerate Kohn–Sham ground states, producing the ensemble DFT (EDFT) (Dreizler and Gross, 1995). In fact, there are cases in which the true density of the system cannot be represented by a single Slater determinant of single-particle wave functions; for example, in the case where the KS orbitals are degenerate at the Fermi energy (which coincides with the highest single-particle energy of the occupied orbitals), so that there exists ambiguity as to how the degenerate orbitals should be occupied. As mentioned in the previous subsection, this case is realized for KS orbitals of quantum dots in the presence of a strong magnetic field. A way of tackling the problem is that of carrying out finite temperature calculations, and then lowering the temperature gradually; an alternative method is to use the EDFT, which we describe in the following.

The EDFT is the theory that describes the true density of the system as an ensemble of Slater determinants of KS orbitals. While it is possible to show that the representation is rigorous, it is not possible to show how the degenerate KS orbitals at the Fermi energy should be occupied, i.e. there exists no practical calculation procedure for the EDFT. Only recently has a practical generalization of the EDFT been proposed and applied to the fractional quantum Hall effect. In what follows we briefly describe such a method (Heinonen *et al.*, 1995).

In the EDFT, any physical density $\rho(\mathbf{r})$ can be represented by

$$\rho(\mathbf{r}) = \sum_{mn} f_{mn} |\phi_{mn}(\mathbf{r})|^2,$$

where f_{mn} are occupation numbers which obey $0 \leq f_{mn} \leq 1$, and the orbitals follow the generalized KS equation

$$H^{\mathrm{KS}} \phi_{mn}(\mathbf{r}) = \epsilon_{mn} \phi_{mn}(\mathbf{r}), \qquad (4.81)$$

where H^{KS} is the effective KS Hamiltonian which depends on the various densities. The problem lies in the determination of the KS orbitals and of their occupation numbers in the presence of degeneracy. In the scheme proposed by Heinonen *et al.*, we start from a given set of input orbitals and occupation numbers, and iterate the system N_{eq} times using the KS scheme. N_{eq} is large enough to ensure that the density is close to the final density after N_{eq} iterations (in practice, 20–40 iterations). If it were possible to represent the system density by a single Slater determinant of KS orbitals, the calculation would be finished. However, in the system there are many degenerate or quasidegenerate orbitals at the Fermi energy. After each iteration step, the KS scheme chooses to occupy the N_{eq}-orbitals with the lowest eigenvalues, which corresponds to building up with these orbitals a Slater determinant different from all the others. However, there are small density fluctuations between successive iteration steps, which cause a different occupation of these degenerate (or quasidegenerate) orbitals after each step. This corresponds to building up different Slater determinants after each step, and the occupation numbers f_{mn} of these orbitals are either 0 or 1 after each step, nearly at random. This means that the calculation will never converge. In any case, the averaged occupations, i.e. the occupation numbers as averaged over many iteration steps, are well defined and tend toward a well-defined value. As a consequence, one uses these average occupation numbers to build up an ensemble by storing average occupations $\langle f_{mn} \rangle$ after the first N_{eq} iteration steps:

$$\langle f_{mn} \rangle = \frac{1}{N_{it} - N_{eq}} \sum_{i=N_{eq}+1}^{N_{it}} f_{mn,i}, \qquad (4.82)$$

where $f_{mn,i}$ is the occupation number (either 0 or 1) of orbital ϕ_{mn} after the ith step; this is used to build up the density.

Basically, this algorithm takes a different degenerate (or quasidegenerate) Slater determinant after each iteration step, and all of these determinants are equally weighted in the ensemble.

Clearly, this scheme reduces to the KS one for systems whose density can be represented by a single Slater determinant of KS orbitals. Moreover, it can be verified numerically that the finite temperature version of the method reproduces the finite temperature CDFT distributions down to extremely low temperatures (see Subsection 5.6).

4.9   The DFT for Strongly Correlated Systems: Nuclei and Helium

In the case of strongly correlated systems like nuclei and liquid ^{3}He, dynamic correlations are responsible for nonlocal effects which result in effective masses m^* which are very different from the bare mass m, and depend on the system density. For example, for liquid ^{3}He one finds that m^* is strongly density-dependent and equal to about three times the bare mass at saturation density (zero-pressure value).

Nonlocal effects of this kind are predicted in the Brueckner–Hartree–Fock theory, which is the only applicable "mean field" theory because the nucleon–nucleon and atom–atom interactions cannot be treated in HF and one needs to find effective interactions (see Chapter 3).

As can be seen from Eq. (3.91), if we expand the effective interaction as was done in Subsection 3.6, the energy not only depends on the density ρ, but also on the kinetic energy density τ and on terms in $\nabla\rho$. Moreover, expression (3.91) generates, in the BHF Eqs. (3.92), an effective mass,

$$m^* = \frac{m}{1 + 2am\rho}, \tag{4.83}$$

with $a = -t_1$. These considerations suggested, for nuclei with the same number of neutrons (N) and protons (Z) and closed shells (Vautherin and Brink, 1972) and for liquid helium (Stringari, 1984; Dalfovo, 1989), the use of density functionals of the type

$$E[\rho, \tau] = \int \left\{ \frac{\tau}{2m^*} + \frac{1}{2}b\rho^2 + d(\nabla\rho)^2 + \frac{1}{2}c\rho^{2+\gamma} \right\} d\mathbf{r}, \tag{4.84}$$

where the last term has been added to take into account more-than-two-body correlations which are neglected in the BHF theory, but are important for nuclei and helium.

In the case of nucleons, the effective mass is parametrized exactly as in (4.83), while for ^{3}He one uses

$$m^* = \frac{m}{(1 - \frac{\rho}{\rho_c})^2}, \tag{4.85}$$

where ρ_c is a parameter. In the case of liquid ^{4}He, in the functional (4.84) it is necessary to take $m^* = m$ and $\tau = (\nabla\rho)^2/4\rho$, which is the correct expression for the kinetic energy density for Boson systems (see also Subsection 2.3).

In the case of nuclei the functional (4.84) is next generalized to take into account the spin–orbit interaction, Coulomb interaction between protons, and the isospin degrees of freedom when $N \neq Z$ (see Subsection 4.10).

The parameters that appear in the effective mass, as well as b, d, c and γ, are fitted to reproduce the values of some known experimental quantities and are reported in Table 4.3 for the various systems (SIII is the effective interaction of the nuclei).

Table 4.3. Parameters of the density functional of Eq. (4.84) for the various systems. SIII is the functional of nuclei.

Syst.	a	ρ_c	b	c	d	γ
SIII	44.38	–	−846.56	1750	62.97	1
^{3}He	–	0.0406	−683.0	1.405057×10^6	2222	2.1
^{4}He	–	–	−888.81	1.04554×10^7	2383	2.8

The parameter values in the table are in MeV fm^5 for a and d, in MeV fm^3 for b and in MeV fm^6 for c in the case of the SIII for the nuclei; for ^{3}He and ^{4}He, ρ_c is in Å^{-3}, b in °K Å^3, c in °K Å$^{3(1+\gamma)}$ and d in °K Å^5.

From Eq. (4.84), for homogeneous Fermi systems we obtain the following form for the energy per particle:

$$\frac{E}{N} = \frac{3}{5}\frac{(3\pi^2)^{2/3}}{2m^*(\rho)}\rho^{2/3} + \frac{1}{2}b\rho + \frac{1}{2}c\rho^{1+\gamma}, \qquad (4.86)$$

where $\rho = vk_F^3/6\pi^2$ with $v = 2$ in the case of ^{3}He and $v = 4$ in the case of nucleons, while for ^{4}He the first term vanishes identically and the density is a constant. From the above expression we derive the pressure

$$\frac{P}{\rho} = \rho\frac{\partial E/N}{\partial\rho}, \qquad (4.87)$$

and the compressibility

$$\frac{1}{K\rho} = \frac{\partial P}{\partial\rho} = \frac{\partial}{\partial\rho}\rho^2\frac{\partial E/N}{\partial\rho}. \qquad (4.88)$$

The saturation density ρ_0 is determined by putting $P = 0$ in (4.87). By inserting $\rho = \rho_0$ in Eqs. (4.86) and (4.88), we obtain the binding energy and the compressibility at $P = 0$, which are to be compared with the experimental data in order to extract the values of some of the parameters in Table 4.3. The other values are obtained by fitting, in the case of helium the surface tensions, and in the case of nucleons some properties of magic nuclei. The obtained numerical values for the saturation densities, binding energy per particle, compressibilities, and effective masses of the homogeneous systems, and the surface tensions σ of the semi-infinite systems, are reported in Table 4.4. For the homogeneous and semi-infinite nuclear matter, the values reported in the table refer to systems with $N = Z$.

The values of the parameters in the table are in fm^{-3} for ρ_0, in MeV for E/N and $(K\rho_0)^{-1}$, and in MeV fm^{-2} for σ in the case of the SIII parametrization for nuclei; on ^{3}He and ^{4}He, ρ_0 is in Å^{-3}, E/N and $(K\rho_0)^{-1}$ in °K, and σ in °K Å^{-2}.

The equations of state for the ^{3}He and ^{4}He systems, predicted from Eq. (4.87), are reported in Figs. 4.2 and 4.3 respectively, together with the experimental data of Wheatley (1975) and Watson *et al.*, (1969).

Table 4.4. Numerical values for saturation densities, binding energies per particle, compressibilities, effective masses and surface tensions obtained with the density functional of Eq. (4.84).

Syst.	ρ_0	E/N	$(K\rho_0)^{-1}$	m^*/m	σ
SIII	0.145	-15.85	355.4/9	0.763	1.07
^{3}He	1.6347×10^{-2}	-2.49	12.1	2.8	0.113
^{4}He	2.1836×10^{-2}	-7.15	27.2	1	0.274

The Density Functional Theory

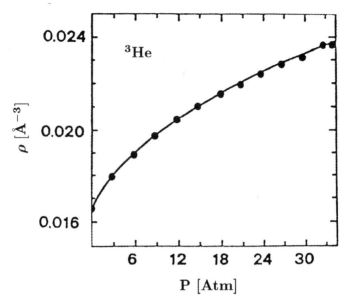

Fig. 4.2 Equation of state for liquid ^{3}He. The full line is from Eq. (4.87) and the experimental points are taken from Wheatley (1975).

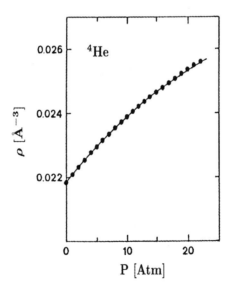

Fig. 4.3 Equation of state of ^{4}He. The full line is deduced from Eq. (4.87) and the experimental points are taken from Watson *et al.* (1969).

As can be seen from these figures, there is good agreement between theory and experiment over the whole pressure range. The same is true for compressibility.

The surface tension may be calculated starting from the functional (4.84), using the Wilet method; the latter, in the case of Fermion systems, uses the extended Thomas–Fermi approximation for the kinetic energy density:

$$\tau = \frac{3}{5}(3\pi^2)^{2/3}\rho^{5/3} + \beta\frac{(\nabla\rho)^2}{\rho} + \delta\nabla^2\rho, \tag{4.89}$$

where $\beta = 1/18$ and $\delta = 1/3$. In the homogeneous limit, τ reduces to the Thomas–Fermi term (4.14). Expression (4.89) is deduced by a semiclassical expansion of the energy of a Fermion system (Ring and Schuck, 1980). The surface tension of the semi-infinite system in the $z \leq 0$ half space with a free surface is written as

$$\sigma = \int_{-\infty}^{+\infty} [\epsilon(\rho, \rho') - \mu\rho]dz, \tag{4.90}$$

where z is the coordinate perpendicular to the system, ρ' is the density derivative with respect to z (in the other directions, the density is a constant and equals the bulk one), and

$$\epsilon(\rho, \rho') = h(\rho) + \frac{1}{2m^*(\rho)}\beta\frac{(\rho')^2}{\rho} + d_1(\rho')^2, \tag{4.91}$$

with

$$h(\rho) = \frac{1}{2m^*(\rho)}\frac{3}{5}(3\pi^2)^{2/3}\rho^{5/3} + \frac{1}{2}b\rho^2 + \frac{1}{2}c\rho^{2+\gamma} \tag{4.92}$$

and

$$d_1 = d - \frac{\delta}{2}\left(\frac{\partial}{\partial\rho}\frac{1}{m^*(\rho)}\right). \tag{4.93}$$

The equilibrium density, which is used to calculate the surface tension through Eq. (4.90), is found using the Euler equation

$$\frac{\partial}{\partial\rho(z)}\int_{-\infty}^{+\infty}[\epsilon(\rho, \rho') - \mu\rho]dz = 0. \tag{4.94}$$

Clearly, the Wilet method can be used in the case of ^{4}He as well, where one replaces Eq. (4.89), with the expression $\tau = (\rho')^2/4\rho$, i.e. the right one for the kinetic energy density of Bosons. The surface tension of metal surfaces has been studied by Lang and Kohn (1970) within the jellium model and by techniques similar to the previous ones, using the density functional theory.

The Kohn–Sham equations which derive from the functional in (4.84) are written as

$$\left\{-\vec{\nabla}\frac{1}{2m^*}\cdot\vec{\nabla} + U(\rho, \tau)\right\}\varphi_i(\mathbf{r}) = \varepsilon_i\varphi_i(\mathbf{r}), \tag{4.95}$$

where the self-consistent field U is given by

$$U(\rho, \tau) = \frac{1}{2}\tau \frac{\partial}{\partial \rho} \frac{1}{m^*(\rho)} + b\rho - 2d\nabla^2 \rho + \frac{1}{2}c(2+\gamma)\rho^{1+\gamma}. \qquad (4.96)$$

Assuming spherical symmetry, the single-particle states which enter the definition of the various densities of Eqs. (4.95) and (4.96) can be factorized in the form $\varphi_i(\mathbf{r}) = R_{n,l}(r)Y_{l,m}(\theta, \varphi)\chi_{m_s}(\sigma)$ and Eq. (4.95) turns into a nonlinear system of differential equations for the radial part of the wave functions, to be solved by iterative self-consistent methods. The calculation is carried out for nuclei and helium clusters, in which the $2(2l+1)$ degenerate states of the (n,l) level are completely occupied (closed shell nuclei and clusters) and for which the spherical symmetry of the density and of the average potential is guaranteed. In the case of ^{4}He, the Kohn–Sham equations become a differential equation for the density:

$$-\frac{1}{4m}\nabla^2 \rho + \frac{1}{8m}\frac{(\nabla \rho)^2}{\rho} + b\rho^2 - 2d\rho\nabla^2\rho + \frac{1}{2}c(2+\gamma)\rho^{2+\gamma} = \mu\rho. \qquad (4.97)$$

Equation (4.97) is solved numerically by requiring that the density vanish at large r values, and that its first derivative be zero at the center ($r = 0$). The chemical potential μ is determined by the normalization condition of the density.

Tables 4.5 and 4.6 report the results, relative to different N values, for the binding energy per particle, the chemical potential, the surface thickness t, the density at the center ρ_0, and the unit radius defined as

$$r_0(N) = \sqrt{\frac{5}{3}\langle r^2(N)\rangle}N^{-1/3}, \qquad (4.98)$$

where $\langle r^2 \rangle$ is the mean square radius for ^{4}He and ^{3}He clusters, respectively. In the case of ^{3}He, μ^- and μ^+ are single-particle energies of the highest occupied level and of the lowest nonoccupied level, respectively, and coincide with the chemical potential in the limit $N \to \infty$. The surface thickness t is defined as the distance

Table 4.5. Numerical results for ^{4}He clusters with N atoms obtained by density functional calculations.

N	E/N [°K]	$\mu(N)$ [°K]	$r_0(N)$ [Å]	$t(N)$ [Å]	$\rho_0(N)$ [10^{-2} Å^{-3}]
8	−0.40	−1.06	4.19	6.7	0.76
20	−1.27	−2.52	3.11	8.8	1.61
40	−2.18	−3.57	2.74	9.0	2.06
70	−2.93	−4.24	2.57	9.2	2.22
112	−3.51	−4.70	2.47	9.3	2.28
168	−3.96	−5.04	2.41	9.3	2.30
240	−4.31	−5.26	2.36	9.3	2.30
330	−4.60	−5.46	2.33	9.3	2.30
728	−5.19	−5.87	2.28	8.8	2.28
∞	−7.15	−7.15	2.22	7.0	2.19

Table 4.6.   Numerical results for ^{3}He clusters with N atoms obtained by density functional calculations.

N	E/N [°K]	$\mu^-(N)$ [°K]	$\mu^+(N)$ [°K]	$r_0(N)$ [Å]	$t(N)$ [Å]	$\rho_0(N)$ [10^{-2}Å^{-3}]
20	+0.08	−0.48	−0.15	3.99	8.6	0.85
40	−0.19	−1.04	−0.68	3.13	8.8	1.42
70	−0.50	−1.34	−0.94	2.90	8.8	1.61
112	−0.75	−1.53	−1.13	2.78	9.5	1.69
168	−0.95	−1.60	−1.28	2.69	9.6	1.73
240	−1.11	−1.65	−1.47	2.64	9.5	1.75
330	−1.25	−1.67	−1.63	2.61	9.4	1.75
∞	−2.49	−2.49	−2.49	2.44	8.3	1.65

Table 4.7.   Numerical results for ^{4}He and ^{3}He clusters with N atoms obtained by Monte Carlo methods.

	^{4}He		^{3}He	
N	E/N [°K]	$r_0(N)$ [Å]	E/N [°K]	$r_0(N)$ [Å]
8	−0.60	−	−	−
20	−1.57	2.77	0.21	−
40	−2.39v	2.54	−0.044	3.29
70	−3.03	2.48 − 2.50	−0.28	3.01
112	−3.50	2.43	−0.46	2.89
168	−	−	−0.62	2.78
240	−4.19	2.37	−0.74	2.71
728	−4.94	2.32	−	−

between the points at which the density decreases from 90% to 10% of the central density.

For comparison, we report in Table 4.7 the results of calculations carried out using the variational Monte Carlo method, for the binding energy and the unit radius, and in Figs. 4.4 and 4.5 the densities of two clusters, as obtained with both methods.

Let us first discuss the ^{4}He results. From Table 4.5 and Fig. 4.4 we see that:

- The ^{4}He clusters are bound for all values of N.
- For large N values the surface thickness, the central density and the unit radius tend to the values which correspond to the semi-infinite matter (i.e. the values in the three last columns of the row labeled ∞). The convergence of the binding energy toward the bulk value is quite slow, and can be represented by a mass formula typical of drop models for quantum liquids:

$$E(N) = a_v N + a_s N^{2/3} + \left(a_c - \frac{2}{9}a_s^2 \rho_0 K\right) N^{\frac{1}{3}} + a_0, \qquad (4.99)$$

with $a_v = -7.15 \,°$K, $a_s = 16.95 \,°$K, $a_c = 10.45 \,°$K and $a_0 = 29 \,°$K.

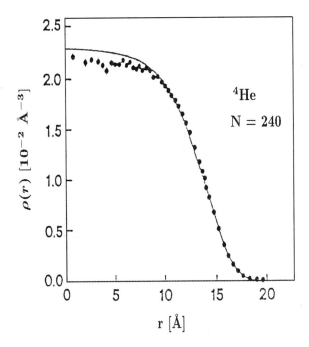

Fig. 4.4 Density profile for a cluster with 240 ^{4}He atoms. The dots refer to the MC calculation of Pandharipande *et al.* (1986).

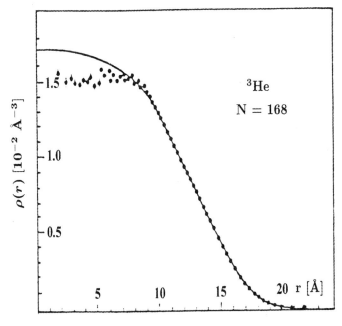

Fig. 4.5 Density profile for a cluster with 168 ^{3}He atoms. The dots refer to the MC calculation of Pandharipande *et al.* (1986).

- The results of the density functional method are in good agreement with those of Monte Carlo calculations (Pandharipande *et al.*, 1986).

  As from ^{3}He clusters, from the results of Table 4.6 and Fig. 4.5, it emerges that:

- The ^{3}He clusters are bound only for N values larger than $N_{\min} \simeq 30$. This result is consistent with the value $N_{\min} \simeq 40$ obtained by Monte Carlo calculations of Table 4.7. Clusters with $N < N_{\min}$ can exist as metastable states clusters, as long as the chemical potential (which is proportional to the derivative of the energy per particle with respect to N) is negative. The lower limit for metastability is $N \simeq 16$.
- The ^{3}He clusters have a shell structure. The average field has a form that is similar to the potential of a spherical harmonic oscillator (only for very large clusters is the surface thickness small with respect to the radius, and the potential approximately a square well). The single-particle levels, i.e. the solutions to Eqs. (4.95) and (4.96), turn out to be grouped following the order of the degenerate levels of the harmonic oscillator, and the most bound clusters correspond to the occupation of the respective major shells, i.e. $N = 40, 70, 112, 168, 240, \ldots$ (see also the table in Subsection 1.3). The gap per particle $(\mu^- - \mu^+)$ for the closed shell clusters is of the order of 0.2–0.4 °K, and decreases as $N^{-1/3}$ for large N values. The gap value depends strongly on the value of the effective mass. The lower the mass, the greater the kinetic contribution to (4.95).
- Contrary to the case of metal clusters (see Subsection 2.2.1) and of atomic nuclei, the shell structure of ^{3}He clusters does not entail density oscillations due to the filling of single-particle levels. These oscillations are energetically unfavored due to the relatively high value of the surface tension and to the low system compressibility.
- The convergence of the binding energy toward the homogeneous system value can be expressed by the mass formula (4.99) with $a_v = -2.49°$K, $a_s = 8.42°$K, $a_c = 5.39°$K and $a_0 = -19.8°$K.
- In the case of ^{3}He as well, the method of the density functional yields results in good agreement with the Monte Carlo calculations (Pandharipande *et al.*, 1986).

In Table 4.8 we report the binding energy and the charge radius of some closed shell nuclei for the SIII parametrization (Beiner *et al.*, 1975), together with the experimental values. The charge radius is defined as

$$r_c(N) = \sqrt{\langle r_c^2(N) \rangle}, \qquad \langle r_c^2(N) \rangle = \frac{1}{N} \int \rho_c(\mathbf{r}) r^2 d\mathbf{r}, \qquad (4.100)$$

where $\rho_c(\mathbf{r})$ is the charge density of the nucleus.

The table shows that the experimental results for the binding energies and the charge radii are very well reproduced by the SIII parametrization, for both light and heavy nuclei. In general, the nuclei have a strong shell structure which is strongly affected by the spin–orbit and isospin terms in the mean field, which for

Table 4.8. Comparison of binding energies and charge radii of some closed shell nuclei, obtained by density functional calculations, with the experimental values.

Nucleus	E^{exp} [MeV]	E^{SIII} [MeV]	r_c^{exp} [fm]	r_c^{SIII} [fm]
^{16}O	127.6	128.2	2.73	2.69
^{40}Ca	342.1	341.9	3.49	3.48
^{48}Ca	416.0	418.2	3.48	3.53
^{90}Zr	783.9	785.2	4.27	4.32
^{140}Ce	1172.7	1172.4	4.88	—
^{208}Pb	1636.5	1636.6	5.50	5.57

simplicity were omitted in (4.96). These terms are discussed in detail in the papers by Vautherin and Brink (1972) and Beiner *et al.* (1975), and in the following subsection. As in the case of ^{3}He, the value of the effective mass strongly affects the single-particle spectrum, and there exist parametrizations other than the one discussed here, i.e. SIII, which reproduce the binding energies and the charge radii as precisely as SIII itself, but which yield rather different single-particle spectra. However, recall that in theories based on the density functional, the single-particle spectrum is not expected to have physical meaning. The nuclear densities exhibit oscillations due to the filling of single-particle levels; this effect is shown in Fig. 4.6.

Finally, the SIII parametrization produces a mass formula (4.99) with $a_v = -15.849$ MeV, $a_s = 18.8$ MeV, $a_c = 10$ MeV and $a_0 = -5$ MeV.

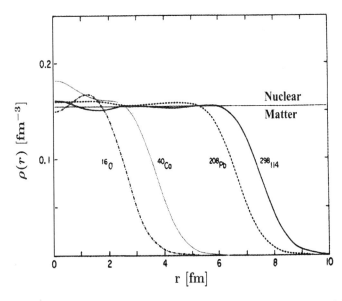

Fig. 4.6 Nuclear density profiles computed by the density functional derived from the SIII effective interaction.

4.10 The DFT for Mixed Systems

The density functional theory has also been used to study the properties of two-component systems. Examples of two-component systems where the DFT has been employed successfully are protons and neutrons in nuclei that we mentioned previously, electron–hole drops in semiconductors, and the ^{3}He–^{4}He mixtures. The DFT for many-component systems can be found in the book edited by Lundqvist and March (1983) and in Baym and Pethick (1978); here we limit ourselves to illustrating the problem taking as specific examples the ^{3}He–^{4}He mixtures (Dalfovo and Stringari, 1985) and nuclei.

Let us start with the ^{3}He–^{4}He mixtures. Let us indicate with

$$\rho = \rho_4 + \rho_3, \quad x = \frac{\rho_3}{\rho} \tag{4.101}$$

the total system density and the concentration of ^{3}He atoms, respectively. The density functional is set up by requiring that for $x = 0$ and $x = 1$ it coincide with the respective functionals of pure ^{4}He and ^{3}He [Eq. (4.84), with the parameters of Table 4.3]. The addition of a small number of mixed terms accounts for the effective interaction typical of the mixture. The functional is written as

$$E = \int d\mathbf{r} \left[\frac{\tau_4}{2m_4} + \frac{1}{2}b_4\rho_4^2 + d_4(\nabla\rho_4)^2 + \frac{1}{2}c_4\rho_4^2\rho^{\gamma_4} + \frac{\tau_3}{2m_3^*} + \frac{1}{2}b_3\rho_3^2 \right.$$
$$+ d_3(\nabla\rho_3)^2 + \frac{1}{2}c_3'\rho_3^2\rho^{\gamma_3} + \frac{1}{2}c_3''\rho_3^{2+\gamma_3} + b_{34}\rho_3\rho_4 + c_{34}\rho_3\rho_4\rho^{\gamma_{34}}$$
$$\left. + d_{34}(\nabla\rho_3)(\nabla\rho_4) \right], \tag{4.102}$$

where

$$m_3^* = \frac{m_3}{\left(1 - \frac{\rho_3}{\rho_{3c}} - \frac{\rho_4}{\rho_{4c}}\right)^2}. \tag{4.103}$$

The parameter ρ_{4c} accounts for mass renormalization of ^{3}He atoms due to the excitations induced in the surrounding ^{4}He, and may be determined by the value of the effective mass of a ^{3}He atom in solution, in the $x \to 0$ limit. The parametrization (4.103) turns out to be consistent with the experimental data relative to dilute solutions and yields the correct value of the effective mass of pure ^{3}He. For homogeneous solutions the energy per particle takes the form

$$\frac{E}{N} = \rho^{-1}\left[\frac{1}{2}b_4\rho_4^2 + \frac{1}{2}c_4\rho_4^2\rho^{\gamma_4} + \frac{3}{5}\frac{(3\pi^2)^{2/3}}{2m_3^*}\rho_3^{5/3} + \frac{1}{2}b_3\rho_3^2 \right.$$
$$\left. + \frac{1}{2}c_3'\rho_3^2\rho^{\gamma_3} + \frac{1}{2}c_3''\rho_3^{2+\gamma_3} + b_{34}\rho_3\rho_4 + c_{34}\rho_3\rho_4\rho^{\gamma_{34}} \right]. \tag{4.104}$$

From this we see that for $\rho = \rho_4$ one obtains again the expression of E/N for pure ^{4}He, while for $\rho = \rho_3$ one obtains again the expression of E/N for pure ^{3}He,

Table 4.9. Parameters for mixed systems of ^{3}He and ^{4}He to be used in functional (4.102).

ρ_{4c} [Å^{-3}]	b_{34}[°K Å^3]	c_{34}[°K Å$^{3(1+\gamma_{34})}$]	c_3''[°K Å$^{3(1+\gamma_3)}$]	γ_{34}
0.062	-777.29	4.564748×10^6	-5.0683×10^4	2.5

provided that $c_3 = c_3' + c_3''$. Five interaction parameters are left to be determined in the functional (4.102). The parameters b_{34}, c_{34} and c_3'' are fitted to reproduce the experimental values at $P = 0$ of the maximum solubility of ^{3}He in ^{4}He at zero temperature, of the chemical potential of ^{3}He in the $x \to 0$ limit, and of the ratio of the specific volumes of ^{3}He and ^{4}He in the same limit. For γ_{34} we take a value intermediate between $\gamma_3 = 2.1$ and $\gamma_4 = 2.8$. Finally, for the parameter d_{34} we take $d_{34} = d_3 + d_4$. As concerns the choice of these values, we note that the precise value of the parameters γ_{34} and d_{34} has little influence on the results derived from the functional (4.102). The values of the above parameters are reported in Table 4.9.

The physical quantities that distinguish the behavior of homogeneous solutions at equilibrium are all deduced from Eq. (4.104). For example, for pressure we have

$$\frac{P}{\rho} = \rho \left(\frac{\partial E/N}{\partial \rho} \right)_x, \tag{4.105}$$

from which we extract the equilibrium density $\rho(P, x)$. The volume increase induced by ^{3}He is described by the volume excess parameter defined by

$$\alpha_0(P) = -\lim_{x \to 0} \frac{1}{\rho} \left(\frac{\partial \rho}{\partial x} \right)_P. \tag{4.106}$$

The enthalpy per particle is given by

$$\frac{H}{N} = \frac{E}{N} + \frac{P}{N}, \tag{4.107}$$

from which we obtain the chemical potentials of ^{3}He and ^{4}He,

$$\mu_3(x, P) = \frac{H}{N} + (1 - x) \left(\frac{\partial}{\partial x} \frac{H}{N} \right)_P, \tag{4.108}$$

$$\mu_4(x, P) = \frac{H}{N} + x \left(\frac{\partial}{\partial (1 - x)} \frac{H}{N} \right)_P, \tag{4.109}$$

and the osmotic pressure

$$\Pi(x, P) = \rho(0, P)[\mu_4(0, P) - \mu_4(x, P)], \tag{4.110}$$

which is the pressure difference between the two liquids when a container holding the ^{3}He–^{4}He mixture is connected to a pure ^{4}He reservoir through a capillary.

In Figs. 4.7–4.9 we compare the results for the excess volume, the chemical potentials and the osmotic pressure as obtained starting from functional (4.104) with

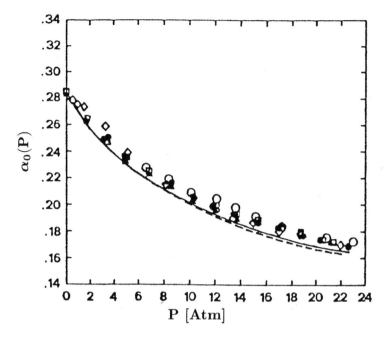

Fig. 4.7 Volume excess parameter as a function of pressure. The full line is the prediction of the theories obtained from the functional (4.104). The experimental data are from Watson *et al.* (1969).

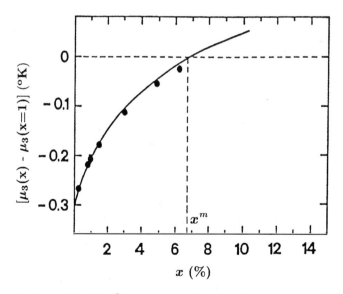

Fig. 4.8 Chemical potential of the ^{3}He atoms as a function of concentration. The full line is the prediction of the theories obtained from the functional (4.104). The experimental data are from Seligmann *et al.* (1969). The concentration x_m indicates the maximum solubility.

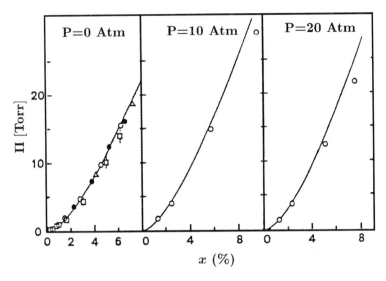

Fig. 4.9 Osmotic pressure as a function of x for three different pressures. The full line is the prediction of the theories obtained from the functional (4.104). The experimental data are from Ghozlan and Varoquaux (1979) and Ebner and Edwards (1970).

the corresponding experimental results (Ebner and Edwards, 1970; Watson *et al.*, 1969; Seligmann *et al.*, 1969; Ghozlan and Varoquaux, 1979). As can be seen, the agreement is quite good as it was for pure ^{4}He and ^{3}He systems. However, it must be stressed that functionals of the types, (4.84) and (4.102) lack the asymptotic $1/r^6$ tail of the He–He interaction. Even more important, zero-range functionals miss the fact that the fluid has a characteristic microscopic length, the radius of the interaction repulsive core. This renders them useless for describing situations where the liquid is probed on a microscopic scale, for example near the core of a quantized vortex, or close to an impurity. Improved functionals that include these effects were first proposed by Dupont-Roc *et al.* (1990) for ^{4}He, and by Garcia-Recio *et al.* (1992), Weisgerber and Reinhard (1992) and Barranco *et al.* (1993) for ^{3}He. The density gradient term was dropped and the $b\rho(r)^2$ term was replaced with a nonlocal term,

$$\int d\mathbf{r}' \rho(\mathbf{r})\rho(\mathbf{r}')V_{\text{eff}}(|\mathbf{r}-\mathbf{r}'|), \qquad (4.111)$$

where the effective two-body interaction V_{eff} is a Lennard–Jones potential, conveniently screened at short distances. The c term is generalized by replacing the density with a coarse-grained density $\hat{\rho}(\mathbf{r})$, as in the study of classical liquids (Tarazona, 1983), defined as

$$\hat{\rho}(\mathbf{r}) = \int d\mathbf{r}' \rho(\mathbf{r}')w(|\mathbf{r}-\mathbf{r}'|), \qquad (4.112)$$

where the weight function $w(r)$ has been taken as a constant function inside a sphere of radius h' and zero outside, normalized to unity. Each of these new forms introduces new parameters which should be fixed by adjusting selected properties in the bulk. A density functional for ^{4}He systems was later constructed (Dalfovo *et al.*, 1995) to describe the static and dynamical response of the homogeneous liquid. This functional improves by construction the description of properties such as the momentum dependence of the static response function and the phonon–roton dispersion relation. Besides, it is remarkable that these functionals predict a surface tension of the liquid at zero T which is in very good agreement with experiment. Very recently, a fully variational density functional method has been proposed (Ancilotto *et al.*, 2005) that can accurately describe the solid phase of ^{4}He and the freezing transition of liquid ^{4}He at zero temperature. This opens the possibility of using unbiased density functional methods to study highly nonhomogeneous systems, like ^{4}He drops doped with very strongly attractive impurities or substrates. The process of fixing a density functional for liquid ^{3}He is much more involved, because spin properties have also to be considered. Some attempts have been made by Stringari (1986), Weisgerber and Reinhard (1988) and Barranco, Hernandez and Navarro (1996). Finally, in the case of systems formed by a mixture of the two isotopes, a finite range energy functional was proposed by Barranco *et al.* (1997). All these finite range density functionals have been extensively used in the last few years to study the static and dynamic properties of pure and doped helium droplets, isolated or on absorbing substrates (for a review see Barranco *et al.*, 2006). Their predictions agree very well with the results of recent diffusion Monte Carlo calculations (Boronat, 2002).

In the case of nuclei with $N \neq Z$, the functional (4.84) is generalized as follows:

$$E = \int d\mathbf{r} \left[\frac{\tau}{2} + a(\rho\tau - j^2) + \frac{1}{2}b\rho^2 + d(\nabla\rho)^2 + \frac{1}{2}c\rho^{2+\gamma} \right.$$

$$\left. + a_1(\rho_1\tau_1 - j_1^2) + \frac{1}{2}b_1\rho_1^2 + d_1(\nabla\rho_1)^2 + \frac{1}{2}c_1\rho^\gamma\rho_1^2 \right] + E_{\text{so}} + E_{\text{C}},$$

$$(4.113)$$

where

$$\rho = \rho_n + \rho_p = \sum_{i\sigma,q} \varphi_i^*(\mathbf{r},\sigma,q)\varphi_i(\mathbf{r},\sigma,q),$$

$$\rho_1 = \rho_n - \rho_p = \sum_{i\sigma,q} q\varphi_i^*(\mathbf{r},\sigma,q)\varphi_i(\mathbf{r},\sigma,q),$$

$$(4.114)$$

with ρ_n and ρ_p the neutron and proton densities, respectively ($q = \pm 1$, where $+1$ indicates the neutron and -1 the proton), and the same for the kinetic energy densities τ and τ_1, and for the current densities j and j_1. E_{so} and E_{C} are the contributions of the spin–orbit and Coulomb interactions to the energy, respectively. The parameters a_1, b_1, c_1, d_1 for the SIII parametrization are reported in Table 4.10.

Table 4.10. Parameters of the functional (4.113) for $N \neq Z$ nuclei.

$a_1 [\mathrm{MeV fm^5}]$	$b_1 [\mathrm{MeV fm^3}]$	$c_1 [\mathrm{MeV fm^6}]$	$d_1 [\mathrm{MeV fm^5}]$
-30.6	536.2	-3500.0	-17.0

From the functional (4.113) one obtains the following set of coupled equations for the single-particle wave functions of neutrons and protons:

$$\left[-\frac{\vec{\nabla}}{2m^*} \cdot + a\mathbf{j} \cdot i\nabla\vec{\nabla} - a_1 \nabla\rho_1 \cdot \nabla + qa_1\mathbf{j_1} \cdot i\nabla + U(\mathbf{r}) + qW(\mathbf{r}) \right.$$

$$\left. + V_{\mathrm{so}} + V_{\mathrm{C}} \right] \varphi_{i,q}(\mathbf{r})c = \varepsilon_{i,q}\varphi_{i,q}(\mathbf{r}), \tag{4.115}$$

with the self-consistent U and W fields given by

$$U = b\rho + \frac{1}{2}c(2+\gamma)\rho^{1+\gamma} + \frac{1}{2}c_1\gamma\rho^{\gamma-1}\rho_1^2 + a\tau - 2d\nabla^2\rho \tag{4.116}$$

and

$$W = b_1\rho_1 + c_1\rho^\gamma\rho_1 + a_1\tau_1 - 2d_1\nabla^2\rho_1, \tag{4.117}$$

the effective mass from Eq. (4.83) and where V_{so} and V_{C} are the spin–orbit and Coulomb potentials, respectively. Equations (4.115)–(4.117) are solved by iteration and their solutions have been widely used to study the properties of finite nuclei.

It is interesting to note the analogy existing between the LSDA Eqs. (4.60) for the systems of spin-polarized electrons, and Eqs. (4.115) for nuclei with $N \neq Z$ (isospin-polarized). In particular, the exchange-correlation potential W_{xc}, which is a functional of the system magnetization m, is the analog of the symmetry potential W of Eq. (4.117), which is a functional of the isovector density ρ_1. As we will see later, these potentials play a fundamental role in the propagation of spin and isospin waves in electron and nucleon systems, respectively.

4.11   Symmetries and Mean Field Theories

The Hamiltonian H of the N-body system has the symmetry S if H and S commute:

$$[H, S] = 0. \tag{4.118}$$

From Eq. (4.118) it follows that if $|n\rangle$ is a nondegenerate eigenstate of H, it has the symmetry S. Spontaneous symmetry breaking can occur in the exact case only if there is degeneracy. For example, if the ground state of H is degenerate and S is a symmetry of the problem, by linearly combining the degenerate states it is possible to set up an eigenstate of H which is not an eigenstate of S.

In mean field theories, such as the HF and the Kohn–Sham theories, it is by no means granted that the single-particle Hamiltonian H^{SP}, which determines the single-particle states through the solution of the HF or Kohn–Sham equations, has the same symmetries as H. Indeed, it may be that

$$[H^{\mathrm{SP}}, S] \neq 0. \tag{4.119}$$

In general, this means that also the mean field ground state does not have the symmetry S, and so there may occur a symmetry breaking. Examples of symmetry breaking in mean field theories are the breakdown of translational invariance in the electron gas (Wigner crystallization), the breakdown of rotational invariance in deformed nuclei, and the violation of particle number conservation in HF–Bogoliubov calculations (superconductivity).

Recently, some symmetry breaking effects have been pointed out in HF and LSDA calculations on quantum dots. In particular, in the absence of an external magnetic field, some quantum dots have zero total spin, but exhibit a locally non-vanishing magnetization, a so-called spin density wave (SDW) (Koskinen *et al.*, 1997; Yannouleas and Landman, 1999; Puente and Serra, 1999). This can be observed in Fig. 4.10, where the polarization $\xi(x, y)$ is plotted for the case of the $S_z = 0$ ground state of a 34-electron dot. As seen in the figure, though $\int \xi d\mathbf{r} = 0$, ξ exhibits a pronounced radial oscillation, which just recalls the phenomenon of spin density waves in the bulk. Another symmetry breaking effect evidenced by CDFT calculations on quantum dots is the breakdown of time reversal (Steffens *et al.*, 1998). In some cases this breakdown gives rise to spontaneously induced orbital

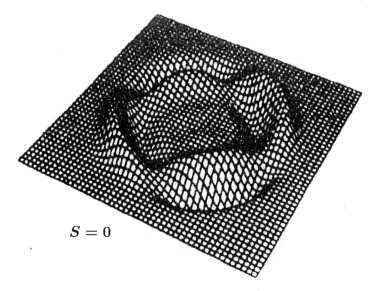

Fig. 4.10 Polarization $\xi(x, y)$ of the ground state with $S = 0$ of the 34-electron dot, computed in the LSDA without requiring circular symmetry.

currents and to nonzero magnetization also for ground states which have zero total angular momentum. Moreover, at low densities breakdown of the circular symmetry has been found in quantum dots, which is analogous to the ones observed in nuclear physics for deformed nuclei. This symmetry breaking has been confirmed by calculations based on the exact diagonalization of the Hamiltonian, which exhibits rotational-band-like spectra.

Note also that symmetry breaking of (4.119) affects the dynamic properties of mean field theories. If we want to study the system excitations by employing the eigenstates of H^{sp}, it is possible to obtain nonzero excitation energies for operators S which are symmetries of the system. This leads to spurious states in the excitation spectrum of the system itself. Use of the RPA theory (see Subsection. 9.7) for excited states can eliminate such spurious states and restore the symmetry of the ground state. In fact, the RPA separates out the collective excitation associated with each broken symmetry as a zero energy or spurious RPA mode. This mode plays a central role in the restoration of the symmetry broken at the mean field level.

This has been explicitly shown by Serra, Nazmitdinov and Puente (2003) in the case of small quantum dots described by Hamiltonian (2.27). The HF densities for the $N = 2,6,12$ quantum dots and different values of the adimensional parameter $R_W = \frac{e^2/\epsilon\ell_0}{\hbar\omega_0}$, giving a measure of the relative importance of electron–electron interaction to confinement potential strength $[\ell_0 = (\hbar/m\omega_0)^{1/2}]$, are plotted in Figs. 4.11 and 4.12.

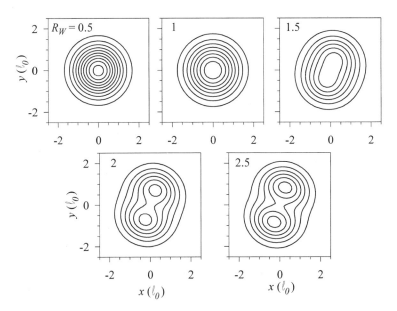

Fig. 4.11 HF densities for the $N = 2$ quantum dot with a varying R_W parameter. From the outermost contour line inward each line corresponds, respectively, to a density of $0.05,010,0.15,\ldots$, etc. in units of ℓ_0^{-2}. The cutoff in the basis has been chosen as $E_c \simeq 10.6E_0$, with $E_0 = \hbar\omega_0$.

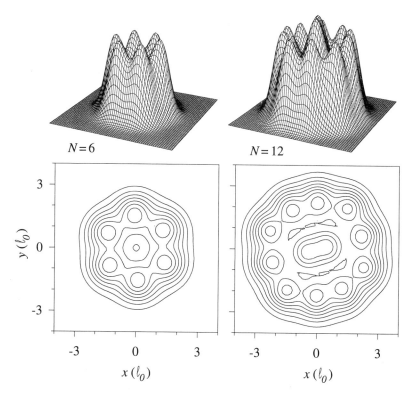

Fig. 4.12 The same as Fig. 4.11, for the $N = 6$ and 12 quantum dots. For clarity the upper plots display a 3D view of the corresponding densities. The contour lines are defined as in Fig. 4.11 and we have used $R_W = 1.89$. The basis cutoffs for $N = 6$ and 12 are $E_c = 16E_0$ and $18E_0$, respectively.

In Fig. 4.11, one notes that between $R_W = 1$ and 1.5 the system changes from a circularly symmetric solution to a spontaneously broken one, where the two electrons localize in opposite positions in the mean field frame. This intrinsic localization becomes more and more conspicuous as the R_W parameter is increased: an evident manifestation that if electron repulsion is strong enough the favored solution consists of particles located as far as possible from each other. A clear symmetry breaking is also seen in Fig. 4.12, using $R_W = 1.89$, which is a value adjusted to reproduce the experiments by Tarucha *et al.* (1996). In a second step the equation of motion

$$[H, O_\lambda^\dagger] = \omega_\lambda O_\lambda^\dagger \tag{4.120}$$

is solved in the RPA for the vibron operators $O_\lambda^\dagger$ to find the RPA eigenstates and excitation energies (see Subsections 9.7 and 9.10). After that the RPA ground state is determined from the condition that is the vacuum for all vibrons:

$$O_\lambda|0\rangle = 0. \tag{4.121}$$

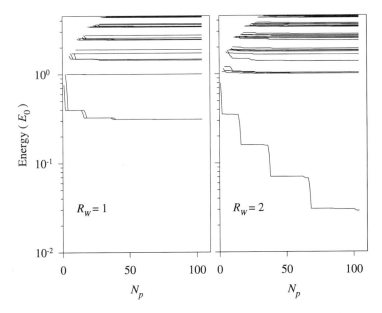

Fig. 4.13 Evolution of the RPA excitation spectra with the number of particle states included. The results correspond to $N = 2$ with $R_W = 1$ (*left*) and 2 (*right*). The same scale has been used in both panels for a better comparison. The basis size and cutoff of Fig. 4.11 have been used.

Figure 4.13 displays the eigenvalue ω_λ for two cases of the $N = 2$ quantum dot as a function of the number N_p of particle states included in the model space where the RPA equations are solved. The first case ($R_W = 1$) is circularly symmetric at the mean field level while the second one ($R_W = 2$) has rotational broken symmetry (see Fig. 4.11). The evolution of the eigenvalue set with N_p shows a remarkable difference in these two cases. While all the eigenvalues of the circular dot stabilize for a high enough value of N_p, the symmetry-broken solution exhibits one state whose energy keeps decreasing as the number of particle states is increased. It clearly corresponds to the appearance of an RPA spurious mode connected with the broken rotational symmetry, which in the limit $N_p \to \infty$ should lie at zero energy. Figure 4.14 shows the corresponding spectra for the $N = 6$ and 12 quantum dots. Both HF solutions correspond to broken symmetry cases (see Fig. 4.12) and, therefore, their RPA spectra should display spurious solutions going to zero energy as $N_p \to \infty$. This is clearly the situation for $N = 6$, with a well-separated low energy mode. For $N = 12$ the situation is a little bit less clear, due to the impossibility of extending the calculation to higher N_p values.

The RPA density is plotted and compared with the exact one in Fig. 4.15 for the two-electron dot. For all R_W parameter values the RPA density is circularly symmetric, fulfilling the symmetry restoration discussed above. Therefore in Fig. 4.15 we focus exclusively on the radial dependence. While the dot edge is well reproduced, a conspicuous feature is the underestimation of the central density by the

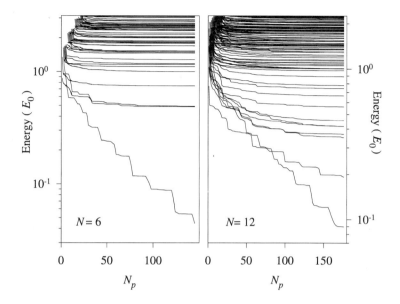

Fig. 4.14 The same as Fig. 4.13, for the $N = 6$ and 12 quantum dots.

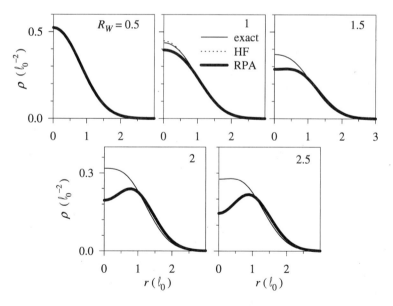

Fig. 4.15 Comparison of RPA and exact radial densities for the $N = 2$ dot with different R_W parameter values. In the $R_W = 0.5$ and 1 cases the HF density, which is circular, is also displayed.

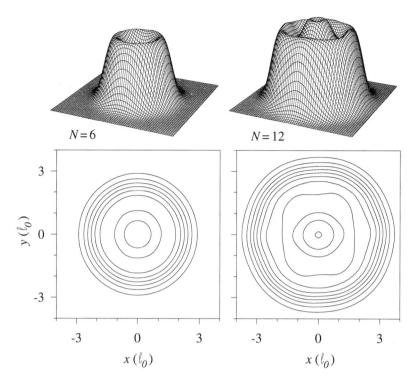

$N=6$ $N=12$

Fig. 4.16   RPA symmetry restoration of the HF densities displayed in Fig. 4.12.

RPA, specially at large R_W values. This is due to the fact (Reinhard, 1992) that the RPA correctly descrbes the low q components (large r's) of the form factor $F(q) = \int dr \exp(iqr)\rho(r)$, but it fails for the large q contributions (low r's). The overcorrection of both central density and correlation energy seems thus to be peculiarities of the RPA. Analogous results for the 6- and 12-electron dots are presented in Fig. 4.16. An excellent restoration of the circular symmetry is obtained for the $N=6$ dot (the corresponding HF density is shown in Fig. 4.12). For $N=12$ the RPA density, though more circular than the HF one, still has some residual deformation. This can surely be attributed to incompleteness of the RPA space considered in the numerical calculation. In fact, for this dot the spurious mode separation (Fig. 4.14), the ground state energy and the density (Fig. 4.16) are all indicating that convergence with the number of particle states is rather slow and, therefore, very difficult to be achieved in numerical calculations. It should be noted that the restoration of broken symmetries can also be attained via projection techniques. Examples demonstrating their use for the case of quantum dots have been presented by Yannouleas and Landman (2002).

Finally, we stress that the spontaneous breakdown of circular symmetry as found in HF calculation of quantum dots can be physical, like the one obtained in nuclear physics for deformed nuclei. This symmetry breaking corresponds to the formation of Wigner moleculae, for which symmetries are preserved in the laboratory frame and one must consider an intrinsic (rotating) frame to "see" the underlying deformation. In exact calculations (configuration interaction calculations) this situation occurs when the excitation spectrum exhibits a clean rotational band structure well decoupled by vibrational states.

A careful analysis of these phenomena in two-electron quantum dots has been carried out by Puente, Serra and Nazmitdinov (2004), who have performed an exact calculation of the excitation spectrum and compared it to the result of the HF and RPA calculations. CI calculations of the rotovibrational spectrum have also been performed in other systems of confined electrons. For a recent review see, for example, Reimann *et al.* (2002).

References

Ancilotto, F., M. Barranco, F. Caupin, R. Mayol and M. Pi, *Phys. Rev.* **B72**, 214522 (2005).

Ballone, P., C.J. Umrigar and P. Delaly, *Phys. Rev.* **B45**, 6293 (1992).

Barranco, M., R. Guardiola, E.S. Hernández, R. Mayol, J. Navarro and M. Pi, *J. Low Temp. Phys.* **142**, 1 (2006).

Barranco, M., E.S. Hernández and J. Navarro, *Phys. Rev.* **B54**, 7394 (1996).

Barranco, M., D.M. Jezek, E.S. Hernández, J. Navarro and Ll. Serra, *Z. Phys.* **D28**, 257 (1993).

Barranco, M., M. Pi, S. Gatica, E.S. Hernández and J. Navarro, *Phys. Rev.* **B52**, 8997 (1997).

Bartenstein, M. *et al.*, *Phys. Rev. Lett.* **92**, 120401 (2004).

Baym, G. and C. Pethick, *Phys. Rev. Lett.* **76**, 6 (1996).

Baym, G. and C.J. Pethick, in *The Physics of Liquid and Solid Helium*, Part ii, eds. K.H. Bennemann and J.B. Ketterson (Wiley, 1978), pp. 1–175.

Beiner, M., H. Flocard, Nguyen Van Giai and Ph. Quentin, *Nucl. Phys.* **A238**, 29 (1975).

Boronat, J., in *Microscopic Approaches to Quantum Liquids in Confined Geometries*, eds. E. Krotscheck and J. Navarro (World Scientific, 2002), p. 21.

Bourdel, T. *et al.*, *Phys. Rev. Lett.* **93**, 050401 (2004).

Ceperley, D., *Phys. Rev.* **B18**, 3126 (1978); D.M. Ceperley and B.J. Alder, *Phys. Rev. Lett.* **45**, 566 (1980); B. Tanatar and D.M. Ceperley, *Phys. Rev.* **B39**, 5005 (1989).

Dalfovo, F., Tesi di Dottorato dell'Università di Trento, 1989.

Dalfovo, F., A. Lastri, L. Pricaupenko, S. Stringari and J. Treiner, *Phys. Rev.* **B52**, 1193 (1995).

Dalfovo, F. and S. Stringari, *Phys. Lett.* **A112**, 171 (1985).

Dreizler, R.M. and E.K.U. Gross, *Density Functional Theory* (Springer-Verlag, Berlin, 1990).

Dupont-Roc, J., M. Himbert, N. Pavloff and J. Treiner, *J. Low Temp. Phys.* **81**, 31 (1990).

Ebner, C. and D.O. Edwards, *Phys. Rep.* **2**, 77 (1970).

Ferconi, M. and G. Vignale, *Phys. Rev.* **B50**, 14722 (1994).

Garcia-Recio, C., J. Navarro, Nguyen Van Giai and L.L. Salcedo, *Ann. Phys. (NY)* **214**, 293 (1992).

Ghozlan, A. and E. Varoquaux, *Ann. Phys. Fr.* **4**, 239 (1979).

Greiner, M., C.A. Regal and D.S. Jin, *Nature (London)* **426**, 537 (2003).

Heinonen, O., M.I. Lubin and M.D. Johnson, *Phys. Rev. Lett.* **75**, 4110 (1995).

Hohenberg, P. and W. Kohn, *Phys. Rev.* **B136**, 864 (1964).

Jochim, S. *et al.*, *Science* **302**, 2101 (2003).

Jones, R.O. and O. Gunnarsson, *Rev. Mod. Phys.* **61**, 689 (1989).

Kohn, W. and L.J. Sham, *Phys. Rev.* **A140**, 1133 (1965).

Koopmans, T.C., *Physica* **1**, 104 (1933).

Koskinen, M., M. Manninen and S.M. Reimann, *Phys. Rev. Lett.* **79**, 1389 (1999).

Lang, N.D. and W. Kohn, *Phys. Rev.* **B1**, 4555 (1970).

Levesque, D., J.J. Weiss and A.H. MacDonald, *Phys. Rev.* **B30**, 1056 (1984).

Levy, M., *Proc. Natl. Acad. Sci. (USA)* **76**, 6062 (1979).

Lipparini, E., N. Barberan, M. Barranco, M. Pi and Ll. Serra, *Phys. Rev.* **B56**, 12375 (1997).

Lundqvist, S. and N.H. March, *Theory of the Inhomogeneous Electron Gas* (Plenum, New York, 1983).

O'Hara, K.M. *et al.*, *Science* **298**, 2179 (2002).

Pandharipande, V.R., S.C. Pieper and R.B. Wiringa, *Phys. Rev.* **B34**, 4571 (1986).

Pi, M., M. Barranco, A. Emperador, E. Lipparini and Ll. Serra, Phys. Rev. B **57**, 14783 (1998).

Pitaevskii, L.P. and S. Stringari, *Bose–Einstein Condensation* (Clarendon, Oxford, 2003).

Puente, A. and Ll. Serra, *Phys. Rev. Lett.* **83**, 3266 (1999).

Reimann, S.M. and M. Manninen, *Rev. Mod. Phys.* **74**, 1283 (2002).

Reinhard, P.G., *Phys. Lett.* **A169**, 281 (1992).

Ring, P. and P. Schuck, *The Nuclear Many-Body Problem* (Springer-Verlag, New York, 1980).

Seligmann, P., D.O. Edwards, R.E. Sarwinsky and J.T. Tough, *Phys. Rev.* **181**, 415 (1969).

Serra, Ll., R.G. Nazmitdinov and A. Puente, *Phys. Rev.* **B68**, 035341 (2003).

Steffens, O., U. Rössler and M. Suhrke, *Europhys. Lett.* **42**, 529 (1998); O. Steffens, M. Suhrke and U. Rössler, *Europhys. Lett.* **44**, 222 (1998).

Stringari, S., *Phys. Lett.* **A106**, 267 (1984).

Stringari, S., *Europhys. Lett.* **2**, 639 (1986).

Tarazona, P., *Phys. Rev.* **A31**, 2672 (1983).

Tarucha, S., D.G. Austing, T. Honda, R.J. van der Hage and L.P. Kouwenhoven, *Phys. Rev. Lett.* **77**, 3613, 1996.

Vautherin, D. and D.M. Brink, *Phys. Rev.* **C5**, 626 (1972).

Vignale, G. and M. Rasolt, *Phys. Rev. Lett* **59**, 2360 (1987); *Phys. Rev.* **B37**, 10 685 (1988).

von Barth, V. and L. Hedin, *J. Phys.* **C5**, 1629 (1972).

Vosko, S.H., J.P. Perdew and A.H. MacDonald, *Phys. Rev. Lett.* **35**, 1725 (1975).

Watson, A.E., J.D. Reppy and R. Richardson, *Phys. Rev.* **188**, 384 (1969).

Weisgerber, S. and P.G. Reinhard, *Z. Phys.* **D23**, 275 (1992).

Weisgerber, S., P.G. Reinhard and C. Toepfer, *Spin-Polarized Quantum System.* ed. S. Stringari (World Scientific, Singapore, 1988), p. 121.

Wheatley, J.C., *Rev. Mod. Phys.* **47**, 467 (1975).

Yannouleas, C. and U. Landman, *Phys. Rev. Lett.* **82**, 5325 (1999).

Yannouleas, C. and U. Landman, *J. Phys.: Condens. Matter* **14**, L591 (2002); *Phys. Rev.* **B66**, 115315 (2002).

Zwierlein, M.W. *et al.*, *Phys. Rev. Lett.* **91**, 250401 (2003).

Chapter 5

The Confined 2D Electron Gas in a Magnetic Field

5.1 Introduction

In this chapter we will apply the theories discussed in the previous chapters to the case of quantum dots, wires, rings and molecules in an external magnetic field. This field, which is applied perpendicularly to the nanostructure itself, can completely spin-polarize the electrons of the system. Its presence strongly affects the motion of the electrons in the system and gives rise to substantial fluctuations in the static and dynamic properties of the nanostructure. These fluctuations have been related to the integer and fractional quantum Hall effect that is exhibited by the resistivity of a two-dimensional surface in the presence of a perpendicular magnetic field. The detailed microscopic study of such correspondence is one of the main purposes of this chapter.

5.2 Quantum Dots in a Magnetic Field

Let us consider the electrons belonging to a two-dimensional quantum dot, confined by a harmonic potential in the plane $z = 0$. The electrons are subject to a magnetic field in the z direction: $\mathbf{B} = B\hat{k}$ described by a vector potential $\mathbf{A} = \frac{B}{2}(-y, x, 0)$. Suppose that the electrons can be described by the Pauli Hamiltonian:

$$H = \sum_i \left\{ \frac{1}{2m} \left[\mathbf{p} + \frac{e}{c}\mathbf{A}(\mathbf{r}) \right]^2 + \frac{1}{2}m\omega_0^2 r^2 + \frac{e\hbar}{2m_e c}\sigma \cdot (\nabla \times \mathbf{A}) \right\}_i$$

$$+ \frac{e^2}{\varepsilon} \sum_{i<j}^N \frac{1}{|\mathbf{r}_i - \mathbf{r}_j|}, \tag{5.1}$$

where $m = m^* m_e$ is the effective mass of the electrons, $-e$ their charge, ε the dielectric constant of the semiconductor, and σ Pauli's vector matrix. Note that $\mathbf{A} = (\mathbf{B} \times \mathbf{r})/2$, so that $\nabla \cdot \mathbf{A} = 0$. In Eq. (5.1) there is a one-body part, which in the following we will indicate as H_0, and a two-body part given by the Coulomb interaction. For the moment, let us focus our attention on H_0, and let us study its eigenvalues and eigenfunctions. If we define

$$\Omega^2 \equiv \omega_0^2 + \frac{1}{4}\omega_c^2, \qquad (5.2)$$

where $\omega_c = eB/mc$ is the cyclotron frequency, and recalling the definition of angular momentum relative to the z axis $l_z = -i\hbar\partial/\partial\theta$, the one-body part of (5.1) can be rewritten as

$$H_0 = \sum_i \left\{ \frac{\mathbf{p}^2}{2m} + \frac{1}{2}\omega_c l_z + \frac{1}{2}m^*\Omega^2 r^2 + \frac{e\hbar}{2m_e c}B\sigma_z \right\}_i. \qquad (5.3)$$

Since

$$[H_0, L_z] = 0, \quad [H_0, S_z] = 0, \qquad (5.4)$$

where $L_z = \sum_i l_z^i$ and $S_z = \frac{1}{2}\sum_i \sigma_z^i$ are the z components of the angular momentum and of the total spin, it is possible to write the single-particle wave functions as $\phi(\mathbf{r}, \sigma) = e^{-il\theta}u(r)\chi_\sigma$. These functions correspond to a quantum number of the angular momentum along z equal to $-l$, and to a z spin component equal to $\pm 1/2$; by introducing the effective gyromagnetic factor g^* (for the bare electron $g^* = g = 2$), Bohr's magneton $\mu_0 = \hbar e/2m_e c$, and $\sigma \equiv \pm 1$, $u(r)$ is the solution to the radial Schrödinger equation:

$$\left[-\frac{\hbar^2}{2m}\left(\frac{d^2}{dr^2} + \frac{1}{r}\frac{d}{dr} - \frac{l^2}{r^2} \right) + \frac{m}{2}\Omega^2 r^2 - \frac{\hbar\omega_c}{2}l + \frac{g^*}{2}\mu_0 B\sigma \right] u_{nl\sigma} = \epsilon_{nl\sigma}u_{nl\sigma}. \qquad (5.5)$$

The physically acceptable solutions to this equation were determined by Fock (1928) and Darwin (1930). One finds that

$$\epsilon_{nl\sigma} = \hbar\Omega(2n + |l| + 1) - \frac{1}{2}\hbar\omega_c l + \frac{1}{2}g^*\mu_0 B\sigma, \qquad (5.6)$$

with $n = 0, 1, 2, \ldots$ and $l = 0, \pm 1, \pm 2, \ldots$. The normalized wave functions are written in terms of generalized Laguerre polynomials:

$$\phi(\mathbf{r})_{n,l,\sigma} = \frac{e^{-il\theta}}{\sqrt{2\pi}}\frac{1}{a}\sqrt{\frac{n!}{2^{|l|}(n+|l|)!}}\left(\frac{r}{a} \right)^{|l|}e^{-\left(\frac{r}{2a} \right)^2}L_n^{|l|}\left(\frac{r^2}{2a^2} \right)\chi_\sigma, \qquad (5.7)$$

where $a \equiv \sqrt{\hbar/2m\Omega}$. In the following, we study the two relevant limiting cases $\omega_0 \gg \omega_c$ e $\omega_c \gg \omega_0$.

5.2.1 The $\omega_0 \gg \omega_c$ Case

If we set ω_c equal to zero, Eq. (5.6) becomes

$$\epsilon_{nl} = \hbar\omega_0(N+1), \qquad (5.8)$$

where we have defined $N \equiv 2n + |l|$, with $N = 0, 1, 2, \ldots$. These are the levels of a two-dimensional harmonic oscillator. Each level is $2(N + 1)$-fold degenerate (by taking due account of the spin). As has already been discussed in Subsection 2.2.1, dots with a number of electrons $N_e = 2, 6, 12, 20, 30, 42, 56, \ldots$ are closed shell dots. These values are part of the sequence $(p+1)(p+2)$ with $p = 0, 1, 2, \ldots$, which equals the number of fully occupied shells. The shell structure is such that for a given N, n varies from zero to the integer value of $N/2$, and there are values of $|l|$ up to N, with the same parity as N. For example, when $N = 6$ we have $n = 0, 1, 2, 3$, $l = 0, \pm 2, \pm 4, \pm 6$, while with $N = 5$ we have $n = 0, 1, 2$, $l = \pm 1, \pm 3, \pm 5$. This means that not all transitions of a given multipolarity L can take place between adjacent shells. Dipole transitions with $L = 1$, which involve jumps with odd ΔN, can always occur, but for quadrupole transitions ($L = 2$) to take place jumps with even ΔN are required.

Equation (5.8) shows that for a fixed n, the single-particle levels are distributed according to straight lines as a function of $|l|$. If one used a different confinement potential, like the one produced by a uniform distribution of positive charges on a disk of fixed radius, there would be some distortion, but the levels would still be distributed in a very regular way.

Furthermore, retaining terms linear in B we find that

$$\epsilon_{nl} = \hbar\omega_0(N + 1) - \frac{\hbar\omega_c}{2}l + \frac{1}{2}g^*\mu_0 B\sigma. \tag{5.9}$$

The l and σ degeneracies are lifted by the B-linear terms, which push upward the levels with $l < 0$ and $\sigma < 0$ (provided $g^* < 0$ like in GaAs), and push downward the levels with $l > 0$ and $\sigma > 0$, thus introducing a pattern of intersections in the single-particle spectrum as a function of B, which can result in a shell structure for the dots which is quite different from the $B = 0$ one (see Fig. 5.1).

5.2.2   The $\omega_c \gg \omega_0$ Case

If we put ω_0 equal to zero, Eq. (5.6) becomes

$$\epsilon_{nl\sigma} = \hbar\omega_c\left(M + \frac{1}{2}\right) + \frac{1}{2}g^*\mu_0 B\sigma, \tag{5.10}$$

where we have introduced the label of the Landau levels $M \equiv n + \frac{1}{2}(|l| - l)$. We can see that the electronic states with the same spin are distributed in infinitely l-degenerate levels, characterized by the values of $M = 0, 1, 2, \ldots$. If $g^* < 0$ like in GaAs, then the levels $(M, \uparrow)$ are lower in energy than the levels $(M, \downarrow)$. For each of the two spin values, we see that the first Landau level consists of states with $n = 0$ and nonnegative l values, $l = 0, 1, 2, \ldots$. The second Landau level $M = 1$ is built up by states with $n = 1$, and nonnegative l values plus the state $n = 0, l = -1$. The third Landau level $M = 2$ is built up by $n = 2$, nonnegative l values plus the states $n = 1, l = -1$ and $n = 0, l = -2$, and so on.

At the lowest order in ω_0/ω_c one gets

$$\epsilon_{nl\sigma} = \hbar\omega_c\left(M + \frac{1}{2}\right) + \hbar\frac{\omega_o^2}{\omega_c}(2n + |l| + 1) + \frac{1}{2}g^*\mu_0 B\sigma. \qquad (5.11)$$

The l degeneracy is broken by a positive linear term. Once B, g^* and the electron number N are fixed, the filling of the Landau levels occurs according to the energy balance of the three terms of (5.11).

In Fig. 5.1 we show the single-particle energy levels of (5.6) in ω_0 units, as a function of ω_c/ω_0, and by neglecting the Zeeman term. The evolution of the harmonic oscillator spectrum toward the Landau spectrum as ω_c increases is clearly observed, together with the level degeneracy. The full line shows the Fermi energy of a 30-electron dot in this approximation.

It is important to stress again that the harmonic oscillator level structure and the Landau one are very different as regards the respective contents of single-particle angular momentum.

We mentioned above that in the case of the harmonic oscillator, transitions with $\Delta N = 1$ are impossible for any multipolarity of the excitation operator. On the other hand, since the Landau levels have practically no limitation as to the l value of their single-particle components, any multipole transition between Landau

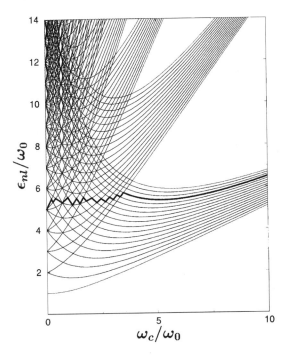

Fig. 5.1 Single-particle energy levels of Eq. (5.6) in ω_0 units, as a function ω_c/ω_0. The Zeeman term has been neglected for simplicity.

levels with $\Delta M = 1$ is possible. The implications of this fact are both simple and important for the interpretation of the experimental spectra:

If the electron–electron interaction is neglected, at low B the energies of the dipole ($L = 1$), quadrupole ($L = 2$) and octupole ($L = 3$) ... modes are $\hbar\omega_0(\Delta N = 1)$, $2\hbar\omega_0(\Delta N = 2)$ and $3\hbar\omega_0(\Delta N = 3)$ ..., respectively. At high B, all of the above modes are at energy $\hbar\omega_c(\Delta M = 1)$. The residual interaction and finite dimension effects (N is finite) change quantitatively this scenario, which, however, is still valid from a qualitative point of view (Schüller *et al.*, 1998; Barranco *et al.*, 1999).

5.2.3 *The Maximum Density Droplet (MDD) State*

Let us assume that the value of B is so high that the gap between Landau levels is such that only the first Landau level ($M = 0, \uparrow$) may be occupied, both in the ground state and for a large number of N-electron excited states.

Putting $M = 0$, and thus $n = \frac{1}{2}(l - |l|)$ in (5.11), we obtain

$$\epsilon_l = \frac{1}{2}\hbar\omega_c + \hbar\frac{\omega_o^2}{\omega_c}(l+1) + \frac{1}{2}g^*\mu_0 B. \tag{5.12}$$

From Eq. (5.12) we derive the energy of the eigenstates of H_0:

$$E = \sum_{l}^{\text{occ}} \epsilon_l = \text{const} + \gamma L, \tag{5.13}$$

where $\gamma = \hbar\frac{\omega_o^2}{\omega_c}$ and L is the eigenvalue (with opposite sign) of the total angular momentum operator

$$L = -\langle\Psi|\sum_{i=1}^{N} l_{iz}|\Psi\rangle = \sum_{l}^{\text{occ}} l. \tag{5.14}$$

Since the energy is linear in angular momentum, it follows that the ground state is the one with the lowest value of L. Let this value be L_0; this will result from the summation of lowest possible values of the single-particle angular momenta:

$$L_0 = \sum_{l=0}^{N-1} l = \frac{N(N-1)}{2}. \tag{5.15}$$

All states with $L > L_0$ are excited states. Apart from the $L_0 + 1$ state (dipole excited state), all other excited states that can be built up starting from the single-particle states belonging to the first Landau level are degenerate. In the case of three electrons, some of these states are reported, as an example, in Table 5.1.

As seen from the table, only the $L = 3$ and $L = 4$ states are not degenerate.

The Coulomb interaction modifies this scheme. However, if B is so large that it is possible to disregard the Coulomb mixing with other Landau levels, and consider

Table 5.1. Lower energy single particle states of the first Landau level and their total angular momentum L for the three-electron system.

l_1	l_2	l_3	L
0	1	2	$3 = L_0$
0	1	3	4
0	1	4	5
0	2	3	5

only the first Landau level for the description of the system at such a B value, the $|L_0\rangle$ and $|L_0 + 1\rangle$ states, which are nondegenerate eigenstates of H_0 and L_z, are eigenstates of the true Hamiltonian H (i.e. including the Coulomb interaction) as well, since H commutes with L_z. Therefore, it appears that values of B and N exist, for which the above conditions are realized, and so the true ground state of the system is the Slater determinant:

$$\Psi_0 = \frac{1}{\sqrt{N!}} \begin{vmatrix} \phi_0(\mathbf{r}_1) & \cdots & \phi_{N-1}(\mathbf{r}_1) \\ \vdots & \ddots & \vdots \\ \phi_0(\mathbf{r}_N) & \cdots & \phi_{N-1}(\mathbf{r}_N) \end{vmatrix}, \tag{5.16}$$

where, for $\omega_c >> \omega_0$,

$$\phi_l = \frac{1}{\sqrt{2\pi\mathcal{L}^2}} \frac{1}{\sqrt{l!}} \left(\frac{re^{-i\vartheta}}{\sqrt{2}\mathcal{L}}\right)^l e^{-\frac{r^2}{4\mathcal{L}^2}}, \tag{5.17}$$

and $\mathcal{L} = \sqrt{\frac{\hbar c}{eB}}$ is the magnetic length. In the literature the state in (5.16) is called the maximum density droplet state (MDD state). It is a completely spin-polarized state ($\langle \Psi_0 | S_z | \Psi_0 \rangle = \frac{N}{2}$) and its one-body diagonal density is given by

$$\rho = \sum_{l=0}^{N-1} |\phi_l|^2 = \sum_{l=0}^{N-1} \frac{1}{2\pi\mathcal{L}^2} \frac{1}{l!} \left(\frac{r}{\sqrt{2}\mathcal{L}}\right)^{2l} e^{-\frac{r^2}{2\mathcal{L}^2}}, \tag{5.18}$$

while the one-body nondiagonal density is given by

$$\rho^{(1)}(\mathbf{r}_1, \mathbf{r}_2) = \sum_{l=0}^{N-1} \phi_l^*(\mathbf{r}_1)\phi_l(\mathbf{r}_2) = \sum_{l=0}^{N-1} \frac{1}{2\pi\mathcal{L}^2} \frac{1}{l!} \left(\frac{r_1 r_2}{2\mathcal{L}^2}\right)^l e^{-\frac{r_1^2+r_2^2}{4\mathcal{L}^2}} e^{il(\vartheta_1-\vartheta_2)}. \tag{5.19}$$

For large N values the one-body diagonal density is characterized by the constant value

$$\rho = \frac{1}{2\pi\mathcal{L}^2} = \frac{eB}{2\pi\hbar c} \tag{5.20}$$

within the circle of radius $R = \mathcal{L}\sqrt{2N}$, and by a boundary thickness of the order of the magnetic length.

The diagonal two-body density of the MDD state for $N \to \infty$ is given by

$$\rho^{(2)}(|\mathbf{r_1} - \mathbf{r_2}|) = \rho^2 - |\rho^{(1)}(\mathbf{r_1}, \mathbf{r_2})|^2 = \left(\frac{1}{2\pi\mathcal{L}^2}\right)^2 \left(1 - e^{-\frac{|\mathbf{r_1} - \mathbf{r_2}|^2}{2\mathcal{L}^2}}\right), \quad (5.21)$$

and it allows the evaluation of the average value of two-body operators in the ground state of the system. For example, the Coulomb exchange energy per particle

$$\frac{E_x}{N} = \frac{1}{2}\frac{1}{2\pi\mathcal{L}^2}\int d^2r \frac{e^2}{r}[g(r) - 1] = -\frac{1}{2}\frac{e^2}{\mathcal{L}}\sqrt{\frac{\pi}{2}}, \quad (5.22)$$

where $g(|\mathbf{r_1} - \mathbf{r_2}|) = 1 - e^{-\frac{|\mathbf{r_1} - \mathbf{r_2}|^2}{2\mathcal{L}^2}}$ is the pair correlation function for the MDD state.

In the limit of infinite N, expression (5.20) allows us to define the filling factor of Landau levels. In fact, once the density is fixed, it is necessary that the magnetic field has the value given by Eq (5.20) (which we indicate as B_0) in order that the system is in the MDD state, where all the single-particle states of the first Landau level are occupied. Therefore, as an indicator of the filling of Landau levels we introduce the filling factor ν, defined as

$$\nu = \rho\frac{2\pi\hbar c}{eB}, \quad (5.23)$$

whose value is 1 when all the electrons are in the first Landau level ($B = B_0$). By progressively decreasing the magnetic field, some electrons will be placed in the second Landau level until, when $\nu = 2$, in the independent-particle model the second level is completely filled, and at $\nu = 3$ the third level will be filled, and so on. The corresponding, closed shell states are very stable and are called incompressible, integer filling states.

5.3   The Fractional Regime

If the external magnetic field is higher than B_0, the filling factor is smaller than 1: in this case we enter the so-called fractional regime, for which the ground state is no longer the MDD state, but is rather a strongly correlated state whose wave function, in some cases, was guessed by Laughlin (1983).

In order to understand the form of the ground state for the fractional regime suggested by Laughlin, it is convenient to write the Slater determinant (5.16) describing the MDD state in the following way (Vandermonde way):

$$\Psi(q_1, q_2 \cdots q_N) = \text{const} \begin{vmatrix} q_1^0 & \cdots & q_1^{N-1} \\ \vdots & \ddots & \vdots \\ q_N^0 & \cdots & q_N^{N-1} \end{vmatrix} e^{-\sum_{i=1}^{N} \frac{|q_i|^2}{4}}$$

$$= \text{const} \prod_{i<j=1}^{N} (q_i - q_j)e^{-\sum_{i=1}^{N} \frac{|q_i|^2}{4}}, \quad (5.24)$$

where $q_i = (x_i - iy_i)/\mathcal{L}$ (lengths are in units of the magnetic length $\mathcal{L}$); the normalization constant is given by

$$\text{Const} = (N!)^{-1/2}(2\pi\mathcal{L}^2)^{-N/2}2^{-N(N-1)/4}\prod_{j=0}^{N-1}(j!)^{-1/2}.$$

Laughlin suggested the following generalization of the wave function (5.24), valid for $B \geq B_0$, and where m is an odd integer in order that Fermi statistics hold:

$$\Psi_m(q_1, q_2 \cdots q_N) = \prod_{i<j=1}^{N}(q_i - q_j)^m e^{-\sum_{i=1}^{N}\frac{|q_i|^2}{4}}. \qquad (5.25)$$

This ansatz for the wave function implies that all the electrons are in the first Landau level, and that all spins are oriented along the magnetic field.

The Laughlin states have the following properties:

- They are eigenstates of the angular momentum with eigenvalue $m\frac{N(N-1)}{2}$.
- They are closely connected with the one-component classical plasma. In fact, writing

$$|\Psi_m|^2 = e^{-\beta\Phi}, \qquad (5.26)$$

with $\beta = 1/m$ playing the role of $\beta = 1/KT$, we find that

$$\Phi = -\sum_{i<j=1}^{N}2m^2\ln|q_i - q_j| + \frac{1}{2}m\sum_{i=1}^{N}|q_i|^2. \qquad (5.27)$$

We see that Φ is the potential energy of the two-dimensional plasma of N particles with charge $\sqrt{2}m$. For $\mathcal{L} = 1$, the first term is the repulsive interaction while the second is the interaction with a background of positive charges, uniformly distributed in the system with density $1/\sqrt{2\pi}$. For some density values this plasma is expected to have a uniform charge distribution. Actually, Monte Carlo simulations (Caillol *et al.*, 1982) showed that such a system is a fluid if $\Gamma = \sqrt{\pi}\rho e^2\beta = \sqrt{2}m$ is smaller than 140; for higher values it forms a hexagonal crystal. It is expected that the Laughlin state also has a constant density and may be connected with a quantum-liquid-like state. $|\Psi_m|^2$ corresponds to a system whose constant density is given by

$$\rho = \frac{1}{2\pi}\frac{1}{m}, \qquad (5.28)$$

which in turn corresponds to a fractional filling factor: $\nu = 1/m$.

- In the Laughlin states the energy per particle is lower than in any other state that has been proposed, such as HF and the charge density wave, and is very close to the fixed phase Monte Carlo calculations.

Starting from the Laughlin state (5.25), one can create a quasihole state at q_0, piercing the system with an infinitely thin solenoid and passing through it a flux quantum $\phi_0 = hc/e$ adiabatically. The state with such a quasihole is represented by

$$\Psi_m^{+1} = \prod_{i=1}^{N} (q_i - q_0) \Psi_m(q_1, q_2 \cdots q_N), \qquad (5.29)$$

as can be seen by writing

$$|\Psi_m^{+1}|^2 = e^{-\Phi^{+1}/m},$$

$$\Phi^{+1} = -\sum_{i<j=1}^{N} 2m^2 \ln|q_i - q_j| + \frac{1}{2} m \sum_{i=1}^{N} |q_i|^2 - 2m \sum_{i=1}^{N} \ln|q_i - q_0|. \quad (5.30)$$

Φ^{+1} represents the same plasma as before except for the quasihole at q_0, with its charge being $|e|/m$. The wave function (5.29) can also be obtained by applying to Ψ_m the creation operator (in units of magnetic length)

$$O^+(q_0) = \prod_{i=1}^{N} (c_i^+ - q_0),$$

$$c_i = q_i^* + 2\partial_{q_i}, \quad c_i^+ = q_i + 2\partial_{q_i^*}. \qquad (5.31)$$

The conjugate operator $O^-(q_0)$, which creates a quasiparticle, is then given by

$$O^-(q_0) = \prod_{i=1}^{N} (c_i - q_0^*), \qquad (5.32)$$

and the state with a quasiparticle at q_0 in the $1/m$ fractional state is represented by the wave function

$$\Psi_m^{-1} = \prod_{i=1}^{N} (2\partial_{q_i} - q_0^*) \Psi_m(q_1, q_2 \cdots q_N). \qquad (5.33)$$

The differentiation is performed only on the polynomial part of Ψ_m. Note that the adjoint of q is $2\partial_q$.

The above quasiholes and quasiparticles carry charges. However, they can be localized like the electrons in the original $1/m$ state, and carry no current in the direction of the applied electric field. One can then think that they form new fractional states when their densities reach certain velues. For a discussion on this interesting point, see the review paper by Isihara (1989).

5.4 The Hall Effect

The Hall effect manifests itself in two-dimensional electron systems under a uniform magnetic field perpendicular to the electron mobility plane (for example, the xy

plane). By applying an electric field in the x direction (in this way inducing a current density j_x), because of the Lorentz force a difference of potential V_y is created between the two surface boundaries parallel to the x axis. It is then possible to define the Hall transverse resistivity:

$$\rho_{yx} \equiv \frac{E_y}{j_x}. \tag{5.34}$$

From classical calculations, this quantity is proportional to the magnetic field and inversely proportional to the electron density:

$$\rho_{yx} = \frac{F_y}{e}\frac{1}{e\rho v_x} = \frac{ev_x B}{ce}\frac{1}{e\rho v_x} = \frac{B}{\rho ec}. \tag{5.35}$$

In experiments (von Klitzing, Dorda and Pepper, 1980; Ebert *et al.*, 1982), such linear dependence of the resistivity on B is observed at low B values. However, at low temperature, by increasing the value of B, one observes plateaux in the B dependence of the resistivity, when the magnetic field value is the inverse of an integer times the electron density:

$$B = \frac{2\pi\hbar c}{e}\rho\frac{1}{\nu}, \quad \nu = 1, 2, 3, \dots. \tag{5.36}$$

Moreover, the longitudinal resistivity $\rho_{xx} = E_x/j_x$ exhibits a minimum for these values of the magnetic field.

These facts can be explained by the theory developed in the previous subsections, by considering the quantum motion of the electrons. When the Landau levels are filled, the many-electron wave function produces very stable incompressible states and robust against changes in B, and the Hall resistivity stays constant, thereby causing the plateau. Moreover, the longitudinal resistivity becomes small because there exists a large energy gap for exciting the electrons, due to the closed shells. These closed shells, i.e. the above incompressible states, are formed when the magnetic field value is the inverse of an integer number times the density.

Plateaux of the Hall resistivity and minima of the longitudinal resistivity have also been observed (Tsui, Stormer and Gossard, 1982; Willet *et al.*, 1987) at low temperature and high field, when B is the inverse of a fractional number times the density:

$$B = \frac{2\pi\hbar c}{e}\rho\frac{1}{1/m}, \quad m = 1, 2, \dots. \tag{5.37}$$

This fractional quantum Hall effect is observed when the first Landau level is partially filled, and is due to the electron correlations induced by the mutual interaction, contrary to the case of the integer quantum Hall which is explained within the independent-particle model. The fractional Hall effect is described by the Laughlin correlated states, and by the ones which can be obtained creating quasiparticle and quasihole states inside them.

5.5   Elliptical Quantum Dots

In this subsection we will consider the properties of quantum dots in the case of anisotropic confinement. There exist many experimental (Dahl *et al.*, 1991; Sasaki *et al.*, 1998; Austing *et al.*, 1999) and theoretical (Peeters, 1990; Madhav and Chakraborty, 1994) studies of the electronic properties of quantum dots of this kind, as a function of anisotropy and an applied external magnetic field. In the case of a zero field, there are some theoretical works (Reimann *et al.*, 1998; Puente and Serra, 1999) in the LSDA for quantum dots with complex shapes. In what follows we will study the independent-particle model for elliptical dots under an external magnetic field, for which it is possible to deduce analytical results. At the end of this subsection we will briefly discuss some LSDA results for anisotropic systems.

If we define

$$\Omega_x^2 \equiv \omega_x^2 + \frac{1}{4}\omega_c^2, \quad \Omega_y^2 \equiv \omega_y^2 + \frac{1}{4}\omega_c^2, \tag{5.38}$$

the one-body Hamiltonian for elliptical quantum dots under a magnetic field turns out to be given by

$$H_{xy} = \sum_i h_{xy}^i = \sum_i \left\{ \frac{p_x^2 + p_y^2}{2m} + \frac{1}{2}\omega_c(xp_y - yp_x) \right.$$

$$\left. + \frac{1}{2}m(\Omega_x^2 x^2 + \Omega_y^2 y^2) + \frac{1}{2}g^*\mu_0 B\sigma_z \right\}_i. \tag{5.39}$$

The Schrödinger equation

$$h_{xy}\Psi_{n_x n_y}(x,y) = \epsilon_{n_x n_y}\Psi_{n_x n_y}(x,y) \tag{5.40}$$

may be rewritten in the form (Dippel *et al.*, 1994)

$$h_3\Phi_{n_x n_y}(x,y) = \epsilon_{n_x n_y}\Phi_{n_x n_y}(x,y), \tag{5.41}$$

where

$$h_3 = U^{-1}h_{xy}U, \quad \Psi_{n_x n_y} = (U\Phi_{n_x n_y})(x,y), \tag{5.42}$$

and the unitary operator U is given by

$$U = e^{i\alpha xy}e^{i\beta p_x p_y}, \tag{5.43}$$

with

$$\alpha = -\frac{m}{2}\left(\frac{\omega_x^2 - \omega_y^2}{\omega_c}\right) + \text{sgn}[\omega_x^2 - \omega_y^2]\frac{m}{2\omega_c}\sqrt{(\omega_x^2 + \omega_y^2 + \omega_c^2)^2 - 4\omega_x^2\omega_y^2},$$

$$\beta = \frac{1}{m}\text{sgn}[\omega_x^2 - \omega_y^2]\frac{\omega_c}{\sqrt{(\omega_x^2 + \omega_y^2 + \omega_c^2)^2 - 4\omega_x^2\omega_y^2}}, \tag{5.44}$$

where $\mathrm{sgn}[\omega_x^2 - \omega_y^2]$ is $+1$ if $\omega_x^2 \geq \omega_y^2$, and -1 if $\omega_x^2 \leq \omega_y^2$. Therefore, the transformed Hamiltonian can be written in terms of squared coordinate and momentum operators alone:

$$h_3 = \frac{p_x^2}{2M_1} + \frac{p_y^2}{2M_2} + \frac{1}{2}M_1\Omega_1^2 x^2 + \frac{1}{2}M_2\Omega_2^2 y^2 + \frac{1}{2}g^*\mu_0 B\sigma_z, \tag{5.45}$$

with

$$M_{1,2} = -\frac{2m\sqrt{(\omega_x^2 + \omega_y^2 + \omega_c^2)^2 - 4\omega_x^2\omega_y^2}}{\mathrm{sgn}[\omega_x^2 - \omega_y^2](\omega_x^2 - \omega_y^2 \pm \omega_c^2) + \sqrt{(\omega_x^2 + \omega_y^2 + \omega_c^2)^2 - 4\omega_x^2\omega_y^2}}, \tag{5.46}$$

$$\Omega_{1,2} = \frac{1}{\sqrt{2}}\left[\omega_x^2 + \omega_y^2 + \omega_c^2 \pm \mathrm{sgn}[\omega_x^2 - \omega_y^2]\sqrt{(\omega_x^2 + \omega_y^2 + \omega_c^2)^2 - 4\omega_x^2\omega_y^2}\right]^{1/2}.$$

These results can be easily obtained using the Baker–Hausdorff formula:

$$e^{i\lambda G}Ae^{-i\lambda G} = A + i\lambda[G, A] + \frac{i^2\lambda^2}{2!}[G, [G, A]]$$

$$+ \cdots + \frac{i^n\lambda^n}{n!}[G, [G, \cdots [G, A]\cdots]] + \cdots, \tag{5.47}$$

and by requiring that in the transformed Hamiltonian nonquadratic terms vanish.

The eigenvalues of h_3, which are eigenvalues of h_{xy} as well, are given by (we have dropped the spin part for simplicity)

$$\epsilon_{n_x n_y} = \left(n_x + \frac{1}{2}\right)\Omega_1 + \left(n_y + \frac{1}{2}\right)\Omega_2, \quad n_x, n_y = 0, 1, 2, \ldots. \tag{5.48}$$

The eigenfunctions of h_3 are products $\Phi_{n_x n_y} = \varphi_{n_x}(x)\varphi_{n_y}(y)$ of one-dimensional harmonic oscillator eigenfunctions φ_i, with frequencies Ω_i and masses M_i. The eigenfunctions of h_{xy} are very complex and may be obtained by transformations (5.43) and (5.44). The same transformation allows an easy evaluation of interesting quantities such as the mean square radii:

$$\langle x^2 \rangle = \frac{1}{N}\sum_{n_x, n_y, \sigma} \langle n_x n_y | x^2 | n_x n_y \rangle_\Psi,$$

$$\langle y^2 \rangle = \frac{1}{N}\sum_{n_x, n_y, \sigma} \langle n_x n_y | y^2 | n_x n_y \rangle_\Psi,$$

without explicit knowledge of the wave function. In fact, we have

$$\langle n_x n_y | x^2 | n_x n_y \rangle_\Psi = \langle n_x n_y | U^{-1} x^2 U | n_x n_y \rangle_\Phi,$$
$$\langle n_x n_y | y^2 | n_x n_y \rangle_\Psi = \langle n_x n_y | U^{-1} y^2 U | n_x n_y \rangle_\Phi, \tag{5.49}$$

and using Eq. (5.47),

$$U^{-1}x^2U = x^2 - 2\beta x p_y + \beta^2 p_y^2, \qquad U^{-1}y^2U = y^2 - 2\beta y p_x + \beta^2 p_x^2, \tag{5.50}$$

with $\langle n_x n_y | x p_y | n_x n_y \rangle_\Phi = \langle n_x n_y | y p_x | n_x n_y \rangle_\Phi = 0$ and the virial theorem

$$\langle n_x n_y | \frac{p_x^2}{M_1} | n_x n_y \rangle_\Phi = \langle n_x n_y | M_1 \Omega_1^2 x^2 | n_x n_y \rangle_\Phi = \left(n_x + \frac{1}{2} \right) \Omega_1,$$

$$\langle n_x n_y | \frac{p_y^2}{M_2} | n_x n_y \rangle_\Phi = \langle n_x n_y | M_2 \Omega_2^2 y^2 | n_x n_y \rangle_\Phi = \left(n_y + \frac{1}{2} \right) \Omega_2,$$

(5.51)

we obtain

$$\langle x^2 \rangle = \frac{1}{N} \sum_{n_x, n_y, \sigma} \left(\frac{1}{M_1 \Omega_1} \left(n_x + \frac{1}{2} \right) + \beta^2 M_2 \left(n_y + \frac{1}{2} \right) \Omega_2 \right),$$

$$\langle y^2 \rangle = \frac{1}{N} \sum_{n_x, n_y, \sigma} \left(\frac{1}{M_2 \Omega_2} \left(n_y + \frac{1}{2} \right) + \beta^2 M_1 \left(n_x + \frac{1}{2} \right) \Omega_1 \right).$$

(5.52)

The energies (5.48) have the following limiting behavior: at a zero magnetic field the system behaves like a pair of harmonic oscillators in the x, y directions. For a high magnetic field ($\omega_c \gg \omega_x, \omega_y$; $\omega_y > \omega_x$), we obtain

$$\Omega_1 \to 0 \quad \text{and} \quad \epsilon_{n_x, n_y} = (n_y + 1/2)\omega_c,$$

i.e. Landau levels are formed as in the case of isotropic parabolic confinement. When $\omega_x = \omega_y$ and the confinement is isotropic and parabolic, $n_x = n + |l|/2 - l/2$ and $n_y = n + |l|/2 + l/2$, where n and l are the principal and azimuthal quantum numbers, respectively. Also in the case $\omega_x \simeq \omega_y$, the energy levels are very similar to the isotropic ones, apart from the fact that the $(2n + |l| + 1)$-fold degeneracy at $B = 0$ is broken by deformation.

In Fig. 5.2(a) we show (Madhav and Chakraborty, 1994) the magnetic field dependence of the energy levels (5.48) for a dot with $\omega_x = 1.0\,\text{meV}$ and $\omega_y = 1.1\,\text{meV}$.

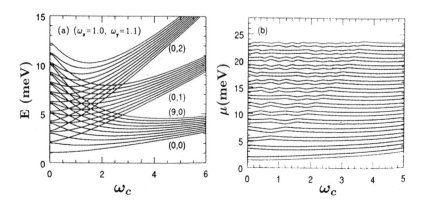

Fig. 5.2 (a) Energy levels of an anisotropic dot as a function of ω_c (meV) for $\omega_x = 1.0\,\text{meV}$ and $\omega_y = 1.1\,\text{meV}$. The lines are drawn in increasing order of (n_x, n_y), as indicated. (b) Chemical potential in the CI approximation for the energy levels of (a).

For this choice, the deviation from circular symmetry is very small and the energy levels are very similar to those of the circular dot, apart from the origin where a shift of the degenerate level due to deformation shows up. In Fig. 5.2(b) we show the chemical potential calculated in the constant interaction (CI) model, where the total energy is written as

$$E(N) = \sum_i^N \epsilon_i + \frac{1}{2}N^2 V,$$ (5.53)

where V is the electron–electron interaction energy, and ϵ_i is the energy of the ith electron. Therefore, in this model the chemical potential is

$$\mu = E(N) - E(N-1) = \epsilon_N + \left(N - \frac{1}{2}\right)V.$$ (5.54)

The result of the calculation reported in the figure (for N values in the range of 1 to $\simeq 30$) also includes the Zeeman energy suitable for GaAs. We have used $V = 0.6\,\text{meV}$, taken from the work of Ashoori *et al.* (1993).

The energies and chemical potentials for $\omega_x = 1.0\,\text{meV}$ and $\omega_y = 5$ and $10\,\text{meV}$ are plotted in Figs. 5.3 and 5.4. We can clearly see that as ω_y increases, the level crossing takes place at higher energies, and the oscillations of the chemical potentials are quenched at lower energies. For example, when $\omega_y = 5$, the oscillations are quenched for $N = 1 - 12$, and when $\omega_y = 10$, for $N = 1 - 22$. Moreover, the amplitude of the oscillations strongly decreases with increasing anisotropy. On the other hand, the magnetic field value beyond which the chemical potential oscillations disappear increases by increasing ω_y. As the threshold magnetic field value does so, the oscillations also shift toward higher magnetic fields, similar to what is observed experimentally (Ashoori, 1993).

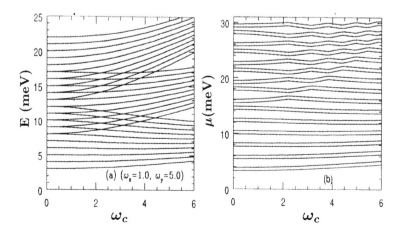

Fig. 5.3   The same as Fig. 5.2, but for $\omega_y = 5.0\,\text{meV}$.

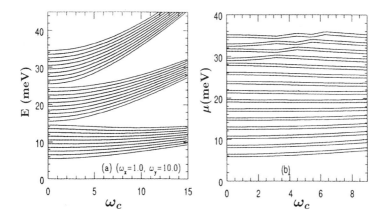

Fig. 5.4 The same as Fig. 5.2, but for $\omega_y = 10.0$ meV.

LSDA calculations in elliptical quantum dots have been carried out by Austing *et al.* (1999) and Serra, Puente and Lipparini (2002). These calculations show that the deformation has the effect of quenching the shell structure, which is very evident for circular dots in the addition energy spectra of Figs. 2.5 and 2.8, exhibiting very prominent peaks at the magic numbers $2, 6, 12, 20, \ldots$. In elliptical dots the resulting spectra consist of a series of low, similar peaks, without clear shell effects. Moreover, the deformation favors the appearance of spin density wave (SDW) states (see Subsection 4.11).

Finally, we note that the frequencies (5.46) yield the exact excitation energies of the electric dipole modes excited by the center-of-mass operators $D_x = \sum_{i=1}^{N} x_i$ and $D_y = \sum_{i=1}^{N} y_i$. In fact, due to the translational invariance of the interaction, the two-body interaction gives no contribution to the excitation energy of dipolar modes, which is affected solely by the external potential (see also Subsection 8.9.1). These excitations are produced by transitions with $\Delta n_x = 0$ and $\Delta n_y = 1$, or with $\Delta n_x = 1$ and $\Delta n_y = 0$, and therefore occur at energies Ω_1 and Ω_2.

5.5.1 *Analogies with the Bose–Einstein Condensate in a Rotating Trap*

Let us consider a dilute gas of Bosons interacting through repulsive forces, at zero temperature and placed in a harmonic trap with frequencies $\omega_x, \omega_y \ll \omega_z$. If the confinement potential rotates around the z axis with angular velocity Ω, the Hamiltonian of this N-particle, "quasi-two-dimensional" system, in the rotating reference frame, is written as

$$H = \sum_i \left\{ \frac{p_x^2 + p_y^2}{2m} - \Omega l^z + \frac{1}{2}m(\omega_x^2 x^2 + \omega_y^2 y^2) \right\}_i + \frac{2\pi\hbar^2}{m} g \sum_{i<j=1}^{N} \delta(\mathbf{r}_i - \mathbf{r}_j), \quad (5.55)$$

where the position vector in the delta-like interaction is in the x–y plane.

The two-dimensional Hamiltonian (5.55), in the $\Omega = 0$ case, has been studied by several groups (Haugset and Haugerud, 1998; Pitaevskii and Rosch, 1997; Bhaduri *et al.*, 1999; Petrov, Holzmann and Shlyapnikov, 2000), and in the three-dimensional case and $\Omega \neq 0$ by A. Recati, F. Zambelli and S. Stringari (2001).

The dimensionless coupling constant g in the two-dimensional, δ-like interaction may be related to the scattering length a for the s wave three-dimensional scattering. This is done starting from the three-dimensional δ-like potential (see Subsection 4.3.2) and calculating the expectation value of the δ function in the z direction, in the ground state wave function of the one-dimensional harmonic oscillator. In this way we obtain the effective two-dimensional interaction of (5.55), with coupling constant

$$g = \sqrt{\frac{2}{\pi}} \frac{a}{b_z}, \tag{5.56}$$

where $b_z = \sqrt{\hbar/m\omega_z}$ is the confinement length in the z direction. This quasi-two-dimensional description of the trap in three dimensions is valid only for dilute systems, whose scattering length fulfills the condition $a \ll b_z \ll b_x, b_y$, where $b_x = \sqrt{\hbar/m\omega_x}$ and $b_y = \sqrt{\hbar/m\omega_y}$. Therefore, from Eq. (5.56) it follows that in (5.55) we have $g \ll 1$. From simple estimates it turns out that Hamiltonian (5.55) is experimentally realizable for a gas of rubidium atoms with $N \simeq 10^4$ and $g \simeq 0.05$.

By comparing the one-body part of the Hamiltonian (5.55) with Eq. (5.39), we note that the formulas previously derived for the electron gas in an elliptical trap under an external magnetic field can be generalized to the case of a Boson system in a rotating trap. The analogs of Eqs. (5.44) and (5.46) are ($\omega_x > \omega_y$)

$$\alpha = \frac{m}{4}\left(\frac{\omega_x^2 - \omega_y^2}{\Omega}\right) - \frac{m}{4\Omega}\sqrt{(\omega_x^2 - \omega_y^2)^2 + 8\Omega^2(\omega_x^2 + \omega_y^2)},$$

$$\beta = -\frac{2}{m}\frac{\Omega}{\sqrt{(\omega_x^2 - \omega_y^2)^2 + 8\Omega^2(\omega_x^2 + \omega_y^2)}}, \tag{5.57}$$

and

$$M_{1,2} = \frac{2m\sqrt{(\omega_x^2 - \omega_y^2)^2 + 8\Omega^2(\omega_x^2 + \omega_y^2)}}{(\omega_x^2 - \omega_y^2 \pm 2\Omega^2) + \sqrt{(\omega_x^2 - \omega_y^2)^2 + 8\Omega^2(\omega_x^2 + \omega_y^2)}}, \tag{5.58}$$

$$\Omega_{1,2} = \frac{1}{\sqrt{2}}\left[\omega_x^2 + \omega_y^2 + 2\Omega^2 \pm \sqrt{(\omega_x^2 - \omega_y^2)^2 + 8\Omega^2(\omega_x^2 + \omega_y^2)}\right]^{1/2}.$$

From Eqs. (5.57) and (5.58), it follows directly that the Schrödinger equation for the noninteracting Bose gas has no stationary solutions in the rotating system in the regions $\Omega > \omega_y$ and $\Omega < \omega_x$, where $\Omega_{1,2}$ become imaginary. This is the main difference with respect to the electron gas under a magnetic field, for which the frequencies (5.46) are always real. In the case of the dilute Boson gas, the δ-like interaction has important effects and may change the situation by generating stationary solutions different from the ones predicted by the noninteracting model (A. Recati, F. Zambelli and S. Stringari, 2001). However, it does not affect the oscillation frequencies of the dipolar modes which are given by Ω_1 and Ω_2 of Eq. (5.58).

It is also very interesting to note that, in the case $\Omega = \omega_x = \omega_y$, the one-body part of the Hamiltonian (5.55) yields a single-particle spectrum:

$$\epsilon_{nl} = \hbar 2\Omega \left(M + \frac{1}{2} \right), \tag{5.59}$$

where we have introduced the label $M \equiv n + \frac{1}{2}(|l| + l)$. Therefore, we see that the Boson states in the rotating system are distributed among infinitely l-degenerate levels, which are the analogs of the Landau levels for electrons, characterized by the values of $M = 0, 1, 2, \ldots$. The $M = 0$ level contains states with $n = 0$, and nonpositive values of l, i.e. $l = 0, -1, -2, \ldots$. Therefore, in this situation there is a very close analogy with the quantum Hall effect. Such an analogy leads to the identification, also in the case of Bosons in a rotating trap, of ground states of the Laughlin type. For a discussion on this point, see for example Peredes *et al.* (2001), Bertsch and Papenbrock (1999) and Manninen *et al.* (2001).

5.6   The DFT for Quantum Dots in a Magnetic Field

In this subsection we present the results for two GaAs quantum dots ($g^* = -0.44$, $\epsilon = 12.4$, $m^* = 0.067$) with $N = 25$ and $N = 210$ electrons, under a constant external magnetic field (Lipparini *et al.*, 1997; Pi *et al.*, 1998); the results were obtained using the CDFT described in Subsection 4.7. The reason for this choice is that these dots have been studied experimentally (Demel *et al.*, 1990), and are good examples of small and large dots. We will also briefly discuss the results obtained by HF and EDFT calculations (see Subsection 4.8).

The CDFT calculations were carried out at low temperature, $T \le 0.1$ °K. This makes the calculation possible in the case of configurations with noninteger filling factor ν, and produces no appreciable thermal effect. For integer ν, the calculation was carried out at $T = 0$.

Two different models of the $N = 25$ dot were employed. In the first one, the external confinement potential was the one produced by a jellium disk with a radius R of 100 nanometers, on which one can find uniformly distributed $N^+ = 28$ positive charges. This potential is analytic:

$$v_{\text{ext}}(r) = \frac{4N^{+'}}{\pi R^2} \times \begin{cases} RE(r/R) & r < R \\ r\{\mathbf{E}(R/r) - [1 - (R/r)^2]\mathbf{K}(R/r)\} & r > R. \end{cases} \tag{5.60}$$

In this equation, **K** and **E** are the complete elliptic integrals of the first and second kinds respectively. In the second model, a harmonic oscillator potential was employed with $\omega_0 = 2.78$ meV. For the $N = 210$ dot only the disk was used, with radius $R = 160$ nm and $N^+ = N$.

The density and the single-particle energies of the $N = 210$ at $B = 0$ are plotted in Fig. 5.5. The single-particle energies, as a function of l, are distributed on parabolic curves, each curve being characterized by a different quantum number n. The figure shows the degeneracy of each single-particle level, corresponding to the possible choices $\pm|l|$, $\sigma = \pm 1$. Figure 5.6 shows the same results for the 25-electron dot and the two different confinement potentials. From the middle panel of Fig. 5.6, we note that the contributions of the Hartree potential and of the exchange-correlation potential affect very little the linear dependence on l predicted by the independent-particle model [see Eq. (5.8)].

Figure 5.7 shows the single-particle spectra for the $N = 210$ dot, as a function of l and for B values in the range of 1.29–10.28 T (teslas), which correspond to the incompressible states with a reported ν value. In order to evaluate ν we used Eq. (5.23) with $\rho = 210/\pi R^2$ and $R = 160$ nm. At $B = 10.28$ T ($\nu = 1$), the system is completely spin-polarized. Note that, due to the sign of the term linear in l in the KS Eqs. (4.76), most occupied levels have a positive l, i.e. a negative angular momentum, and that the $(M, \uparrow)$ band is lower in energy than the $(M, \downarrow)$ band due

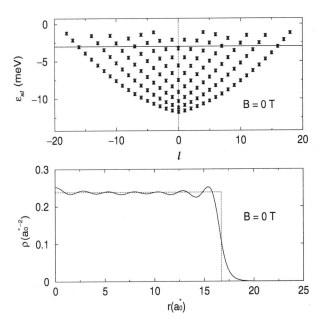

Fig. 5.5 Top: Single-particle energies as a function of l for the $N = 210$ dot at $B = 0$. The full triangles represent spin-up states, the open triangle spin-down states. The horizontal line represents the chemical potential of electrons. Bottom: Electron density ρ as a function of r. The dotted line shows the density of the jellium disk.

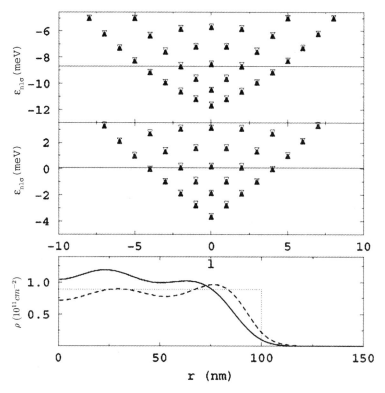

Fig. 5.6 Top: The same as the top of Fig. 5.5, but for the $N = 25$ dot with confinement produced by the jellium disk. Middle: The same as the top, but with parabolic confinement. Bottom: Corresponding electron densities ρ as a function of r. Full line: Disk confinement. Dashed line: Parabolic confinement. Dotted line: Density of the jellium disk.

to the negative value of g^* used in the calculation. For this reason, in the following we indicate by L_z the value of the total orbital angular momentum with opposite sign, and by S_z the component of the total spin in the z direction.

It is remarkable that, for the 210 electron dot, the magnetic fields corresponding to the different ν values follow the law which holds for the bulk, i.e. $B(\nu) = \frac{1}{\nu}B(1)$, up to ν values as large as $\nu = 11$. A similar behavior is observed in the $N = 25$ case, up to $\nu = 6$. Confinement potentials like those produced by a disk favor the existence of incompressible regions in the inner region of the dot, because the electrons in these regions tend to establish a density close to that of the positive background in order to screen the Coulomb potential. This can be seen, for example, in the incompressible states shown in Fig. 5.7.

Figure 5.8 shows some electronic densities and magnetizations $m(r) = \rho_+ - \rho_-$ for the 210-electron dot, together with the L_z and $2S_z$ values predicted by the calculation. From this figure we see that these states are the finite dimension analogs of the Landau incompressible states with $\nu = 1 - 4, 6$ and 8; the latter, for an electronic density of 2.61×10^{11} cm^{-2}, would be realized exactly at the same B

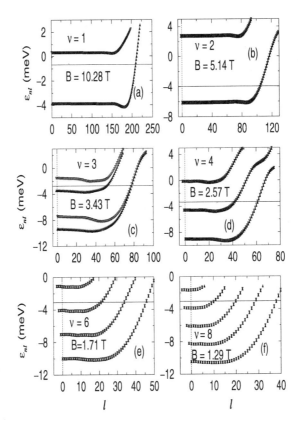

Fig. 5.7 Single-particle energies of the $N = 210$ dot as a function of l for $\nu = 1$–6, 8. The full triangles represent the $(M, \uparrow)$ bands, and the open triangles the $(M, \downarrow)$ bands. The horizontal line shows the chemical potential of the electrons.

values. Actually, the densities in Fig. 5.8 have a steplike shape, whose plateau has the value $\rho = \nu B / 2\pi c$.

Let us discuss now the results relative to the 25-electron dot, obtained by using the two different confinement potentials. In particular, let us compare the results at the same values of ν rather than B. Note that the completely polarized states $\nu = 1$ correspond to $B = 3.56$ T in the case of disk confinement, and to $B = 4.58$ T in the harmonic oscillator case. For $\nu > 1$, the B values obey the law $B(\nu) = \frac{1}{\nu} B(1)$.

Figure 5.9 shows the single-particle energies for $\nu = 1, 2$ and 3, and Fig. 5.10 the electron densities. It is interesting to note that the two confinement potentials, whose parameters were fixed so as to reproduce the experimental energy of the dipole mode at $B = 0$, yield rather different electronic densities. In general, the ground state configurations corresponding to parabolic confinement are more compact.

We will now describe some results for the 210-electron dot obtained by Pi *et al.* (1998). In the case of such a large dot, it is found that the values of the angular

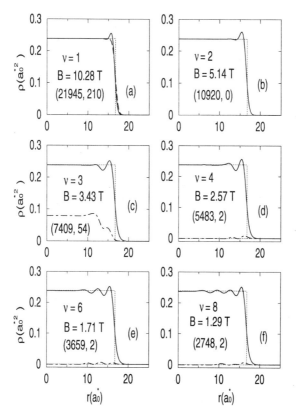

Fig. 5.8 Electron densities $\rho(r)$ (full curves), and spin magnetizations (dotted–dashed curves) corresponding to some of the configurations of Fig. 5.7. The dashed curve in the top left panel represents the density of the MDD state. The values of $(L_z, 2S_z)$ are also plotted.

momentum at different B follow the law

$$L_z = \nu^{-1} \frac{N(N-1)}{2}, \qquad (5.61)$$

and the spin is approximately given by

$$2S_z = \begin{cases} 0 & \nu \text{ even} \\ N/\nu & \nu \text{ odd} \end{cases}, \qquad (5.62)$$

and these expressions are valid in the limit of large N values for incompressible Landau states. The validity of these equations is explicit from Fig. 5.11, where we plotted the orbital and spin angular momenta as a function of B. The small plateaux found along the $L_z(B)$ curve at integer ν values, which appear in the $2S_z(B)$ curve as well, demonstrate the robustness of these states against changes in B, and their resistance toward L_z and S_z changes. In fact, because of the term $\omega_c^2 r^2$ in (4.67), when B increases starting from a given value $B(\nu)$, the electronic

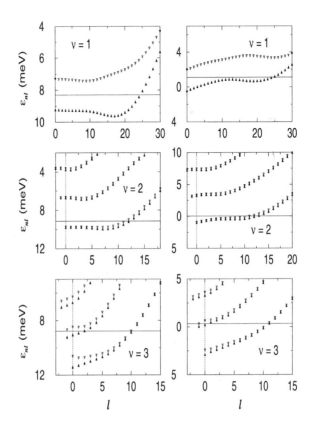

Fig. 5.9 Single-particle energies of the $N = 25$ dot as a function of l for $\nu = 1, 2, 3$. Full triangles represent the $(M, \uparrow)$ bands, and the open triangles the $(M, \downarrow)$ bands. The horizontal line shows the chemical potential of electrons. The panels on the left correspond to disk confinement; those on the right to parabolic confinement.

density is compressed, thereby increasing the electrostatic energy of the dot. This energy is abruptly compensated for by a sudden increase of the angular momentum through the population of single-particle levels with higher l values. This effect is fast because of the low density of single-particle states around the Fermi surface when ν is an integer, and is quite similar to the effect described for smaller dots (Ferconi and Vignale, 1994). On the other hand, the high density of single-particle states at the Fermi energy, when the state is compressible (i.e. $\nu > 1$ is not an integer), makes L_z increase gradually with B.

In the $\nu = 1$ panel of Fig. 5.8, the density (5.18) of the MDD state is plotted. Recall that for $N \to \infty$ this state is the incompressible state with $\nu = 1$ of the quantum Hall effect, and that for sufficiently high magnetic fields this state has an exact expression, i.e. a Slater determinant of single-particle orbitals with angular momenta $0, 1, \ldots, N - 1$, belonging to the first Landau level.

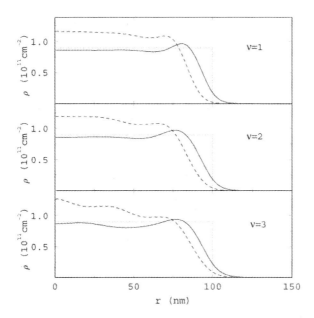

Fig. 5.10 Electron densities $\rho(r)$ for the $N = 25$ dot, for disk confinement (full lines) and for parabolic confinement (dashed lines) corresponding to the configurations of Fig. 5.9.

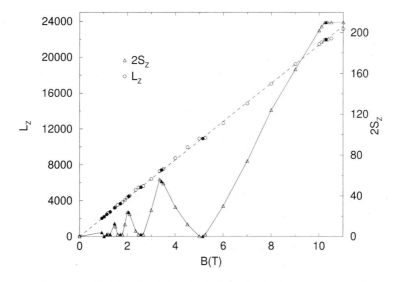

Fig. 5.11 L_z (circles and left hand scale) and $2S_z$ (triangles and right hand scale), as a function of B for the $N = 210$ dot. The values corresponding to integer filling factors from 1 to 11 are the full symbols, and the line connecting the $2S_z$ points is a guide for the eye. The dashed line is from Eq. (5.61).

The natural question, to which people have tried to give an answer in many papers (see for example Ferconi and Vignale, 1997), is whether the MDD state can be the ground state of a quantum dot. The stability region of such a completely spin-polarized state in the N–B plane is bound to the left by the line B_f, which, for a given number of electrons, represents the magnetic field for which $2S_z = N$. To the right it is bound by the line B_r, where the reconstruction of the quantum dot boundary begins, which means the process by which electrons begin to be transferred to single-particle states with $l > N - 1$. This region is very narrow because, after complete polarization of the system, the magnetic field is very effective in reconstructing the dot boundary. For the $N = 25$ dot, one finds that $B_f = 4.46 \leq B \leq B_r = 4.70$ T for harmonic confinement, and $B_f = 3.42 \leq B \leq B_r = 3.70$ T for the disk. For parabolic confinement the stability region of the MDD state narrows as N increases, and vanishes beyond a critical N value. For example, in the case of the 210-electron dot, it has been impossible to find the B value corresponding to the MDD state for parabolic confinement. These results appear to have been confirmed by the recent experimental work of Oosterkamp *et al.* (1999).

For the jellium disk confinement, and for $N = 210$, it seems that the MDD state exists, as indicated by Figs. 5.7 and 5.8. As can be seen from the top left panel of Fig. 5.8, the two densities [i.e. the one yielded by the CDFT calculation and the one from Eq. (5.18)] are very similar, with only one difference arising at the boundary. The protuberance in the CDFT density at the boundary is due both to the confinement potential, which changes abruptly at the boundary, and to the exchange-correlation energy, which also varies very fast there.

It is also very interesting to consider the energy splitting $\Delta E_{\downarrow\uparrow}$ between the Landau bands $(M, \uparrow)$ and $(M, \downarrow)$, reported for example in Fig. 5.7. This difference is very small for even values of ν, which correspond to ground states with $S_z \simeq 0$ (paramagnetic states), because the Zeeman term in (4.67) and (4.76) is also very small (for example, $\simeq 0.26$ meV at $B = 10.28$ T), and $\Delta E_{\downarrow\uparrow} \simeq |g^* \mu_0 B|$. On the contrary, the energy splitting is large for odd ν values, even when the applied magnetic field is small; compare, for example, the cases $\nu = 3$ and $\nu = 6$ in the same figure. These ground states have rather high values of S_z (they are ferromagnetic states) and the large splitting is due to the exchange-correlation potential, which depends on the magnetization: $\Delta E_{\downarrow\uparrow} \simeq 2|W_{xc}|$. W_{xc} is zero for paramagnetic states, and large for ferromagnetic ones, giving rise to an energy splitting much larger than that due to the Zeeman term. The HF, LSDFT and CFDT calculations yield this large splitting, while the Hartree approximation does not. The exchange and correlation effects are the origin of the simple structure of the MDD state, since they prevent the two $(0, \uparrow)$ and $(0, \downarrow)$ Landau bands from being close in energy at $\nu = 1$, and so from contributing to setting up the ground state. As we already stressed in Chapter 4, these spin splittings are similar to the isospin ones that occur in nuclei for which the number of neutrons differs from the number of protons. The exchange-correlation potential that produces them is the analog of the symmetry potential in the nuclei.

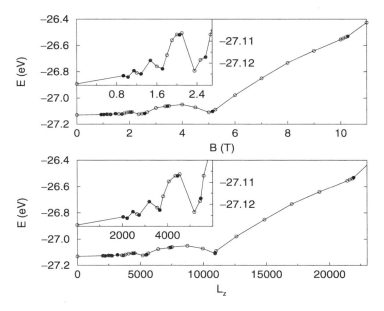

Fig. 5.12   Top: Total energy of the $N = 210$ dot as a function of B. The values corresponding to the integer filling factors in the range 1–11 are indicated by full symbols. The line connecting the points is a guide for the eye. The inset shows the low-B region with the relative scale to the right. Bottom: The same as the top, but as a function of L_z.

The B dependence of the total energy E of the 210-electron dot is shown in Fig. 5.12. Local minima of E are clearly observed at ν values corresponding to the paramagnetic states of the dot. On the other hand, local maxima occur at ν values corresponding to ferromagnetic states. The energy exhibits an inflection point at the ferromagnetic state $\nu = 1$ with $2S_z = 210$. Therefore, in this dot, which has quite a large N value, one observes several transitions from paramagnetic to ferromagnetic states. In dots with a small number of electrons this does not occur, and normally one observes just one transition between these two states. In the bottom panel of the same figure, we report the L_z dependence of E, exhibiting local minima in correspondence with the paramagnetic states. The existence of many paramagnetic states plays a fundamental role in keeping the dot energy around its value at $B = 0$. $|E|$ decreases as S_z increases, and vice versa. Only beyond the $\nu = 2$ state does E increase monotonously with S_z, and finally reaches its full-polarization value at about $\nu = 1$.

The chemical potential μ and the ionization potential $IP = E_{N-1} - E_N$ are shown in Fig. 5.13. They exhibit a saw-tooth behavior as a function of B, with an abrupt decrease in the vicinity of integer ν values. The corresponding L_z values are a sort of magic numbers, corresponding to which the dot is especially stable. Note that the large size of this dot is the cause of the great similarity between the μ and IP values, as well as of the large oscillations of the ionization potential.

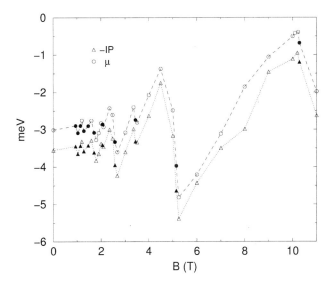

Fig. 5.13 Chemical potential of the electrons (circles) and $-IP$ (ionization potential — triangles) of the $N = 210$ dot, as a function of B. The points corresponding to integer filling factors in the range 1–11 are indicated by full symbols.

Another interesting quantity, which we shall use later on, is the electric field $\mathcal{E}(r)$ produced by the electrons of the dot. This field is plotted in Fig. 5.14 for the B values corresponding to $\nu = 2/3, 1, 4$. $\mathcal{E}(r)$ has a strong peak at the dot boundary, with features reflecting the complex structure of the dot density in that region.

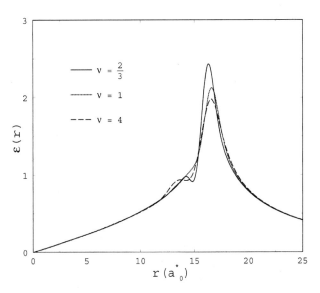

Fig. 5.14 Electric field $\mathcal{E}(r)$ produced by the electrons of the $N = 210$ dot, as a function of r for the reported filling factors.

So far we have considered mainly incompressible states with ν values that are integers ≥ 1. These are particularly simple configurations whose ground states can be described by a single Slater determinant, and which in the density functional theory are normally labeled as pure v-representable states (Dreizler and Gross, 1990). Examples of non-v-representable densities are those corresponding to degenerate ground states. Many ground states of quantum dots, corresponding to noninteger ν values, which are compressible, are found to be non-v-representable. An example of such configurations is shown in Fig. 5.15, where we plot the single-particle energies of the $N = 210$ dot, for $B = 4, 7$ and 9 T. These results were obtained at a temperature of 0.1 °K.

Contrary to the incompressible configurations described above, it can be shown that these ground states are highly degenerate, with many electronic levels whose energies are very close to the chemical potential. The panels on the left side of Fig. 5.16 show the thermal occupation numbers f_i [see Eq. (4.80)] for the same values of B, and those on the left side of Fig. 5.17 show the corresponding densities and magnetizations. The corresponding EDTF values are shown in the panels on

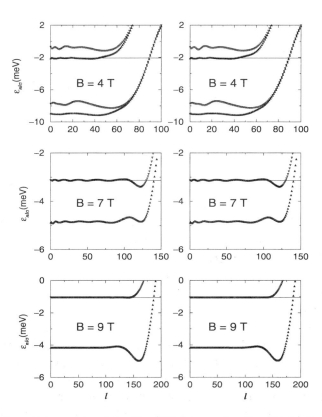

Fig. 5.15 Single-particle energies of the $N = 210$ dot as a function of l for $B = 4, 7$ and 9 T, respectively. The horizontal line is the chemical potential of the electrons. The labeling is the same as in Fig. 5.7. Panels on the left: DFT results at $T = 0.1$ °K. Panels on the right: EDFT results.

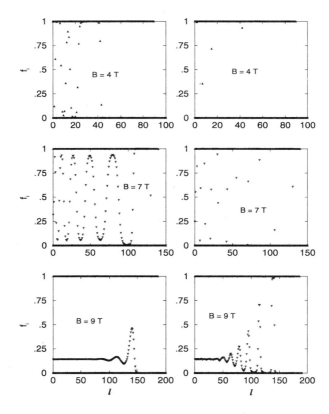

Fig. 5.16 Single-particle occupation numbers of the $N = 210$ dot as a function of l for $B = 4, 7$ and 9 T, respectively. Panels on the left: DFT results at $T = 0.1\ ^{\circ}K$. Panels on the right: EDFT results.

the right sides of the two figures in question. The L_z and $2S_z$ values predicted by the two theories are reported in Table 5.2. We recall that the EDFT is a generalization of the density functional theory, which is used to treat cases of degenerate Kohn–Sham ground states (see Subsection 4.8).

As can be seen from the table, the particle number fluctuations, given by $\Delta N/N$, are small and thus allow the use of the grand canonical ensemble to describe the dot (see Subsection 4.7).

From inspection of Figs. 5.15–5.17 and of Table 5.2, it may be concluded that the finite temperature version of the CDFT allows us to solve the technical problem that arises when the single-particle density of states is large, and it is not possible to solve the Kohn–Sham equations at zero temperature with occupation number 0 or 1. Moreover, it allows us to obtain the electron densities and other characteristics of highly correlated states that cannot be described even approximately in terms of integer occupation numbers. Of course, it is necessary that the correlations are included in the density functional.

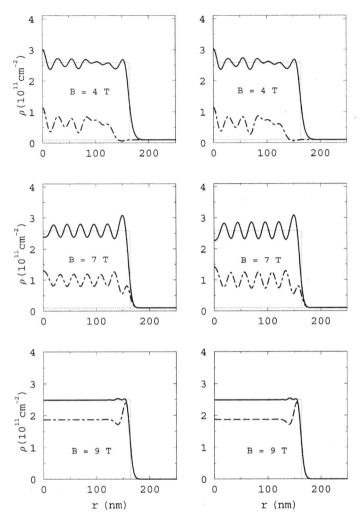

Fig. 5.17 Electronic densities $\rho(r)$ (full curves) and spin magnetizations (dashed curves) of the $N = 210$ dot as a function of r for $B = 4, 7$ and 9 T, respectively. Panels on the left: DFT results at $T = 0.1$ $^{\circ}$K. Panels on the right: EDFT results.

The finite temperature CDFT and EDFT were also applied to the study of configurations corresponding to fractional values of ν which are smaller than or equal to 1 (Heinonen *et al.*, 1995; Lubin *et al.*, 1997; Pi *et al.*, 1998). The results that were found, for example for the density, are very similar to what one obtains starting from the Laughlin wave functions.

Finally, we note that the CDFT single-particle energies for the integer filling factors of Fig. 5.7 are very similar to the results obtained in the HF approximation (Gudmundsson and Palacios, 1995). On the contrary, the energies obtained in the range $1 < \nu < 2$ are qualitatively different. In this region, and for dots with few

Table 5.2. Energies, L_z and $2S_z$ values predicted by CDFT and EDFT theories for different strengths of the magnetic field B.

B (T)	E (eV)		L_z		$2S_z$		$\Delta N/N$ (%)	
	CDFT	EDFT	CDFT	EDFT	CDFT	EDFT	CDFT	EDFT
4	-27.05	-27.05	8729	8730	30	30	0.8	0.3
7	-26.85	-26.85	14812	14811	72	72	1.6	0.8
9	-26.64	-26.64	19217	19218	164	164	2.1	1.8

electrons, the HF approximation results in a crossing of the $(0,\uparrow)$ and $(0,\downarrow)$ bands, which is not reproduced by the CDFT calculations for similar N values. It is likely that cancelations that occur between the exchange term (which is generally overestimated by HF) and the correlation term prevent such level crossing in density functional calculations.

Very recently, LSDA calculations have been carried out for dots where the electrons are confined by anisotropic potentials (Reimann *et al.*, 1998; Puente and Serra, 1999). As an example of these results, we report (Puente and Serra, 1999) in Fig. 5.18

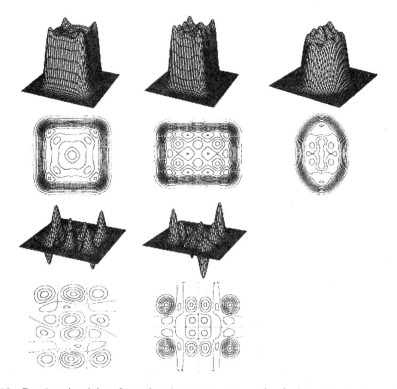

Fig. 5.18 Density $\rho(x,y)$ (top figures) and magnetizations $m(x,y)$ of the $N=20$ dot confined by a square jellium a rectangular jellium and an anisotropic parabola, respectively. The magnetization of the deformed parabola is identically zero everywhere.

the densities and magnetizations of a 20-electron dot, in a square jellium, in a rectangular jellium, and in a deformed parabola. For the square jellium one obtains a very rapidly varying density exhibiting maxima at the corners and four additional maxima arranged according to square symmetry. A similar structure appears for rectangular geometry. In the case of the deformed parabola, the density has an ellipsoidal shape, with an axis ratio close to ω_x/ω_y. In the inner part of the dot we notice three areas of local maxima, aligned along the major axis. It is interesting to note that, while for the parabola the magnetization is identically zero, for the square and the rectangle there is a magnetization wave in the ground state. The amplitude of this wave is about 15% and 25% of the density maximum height in the two cases.

5.7   Quantum Wires

The application of a negative voltage to lithographically defined gates over a GaAs–AlGaAs heterostructure allows the underlying two-dimensional electron gas (2DEG) to be electrostatically squeezed into various shapes. One of the most noteworthy successes of this technique is the experimental realization of a quasi-one-dimensional channel within a 2DEG: a quantum wire (Thortnton *et al.*, 1986, Berggren *et al.*, 1986, Wharam *et al.*, 1988, van Wees *et al.*, 1988, Thomas *et al.*, 1996, 1998). In this Subsection we study those systems in the independent and LSDA models.

Let us consider a single infinitely long quantum wire in the y direction built on a 2DEG by introducing a confining potential along the x direction. This potential is assumed to be parabolic, $\frac{1}{2}m\omega_0^2 x^2$. In the presence of an external magnetic field perpendicular to the xy plane in which the electrons are confined, and in the Landau gauge $\mathbf{A} = (0, B_z x, 0)$, the single-particle Hamiltonian of the electrons can be written as

$$h_0 = -\frac{1}{2}(\nabla_x^2 + \nabla_y^2) - i\omega_c x \nabla_y + \frac{1}{2}(\omega_c^2 + \omega_0^2)x^2 + \frac{1}{2}g^*\mu_0 B\sigma_z, \qquad (5.63)$$

where we have used, as usual, effective atomic units ($\hbar = e^2/\epsilon = m = 1$), and ϵ is the dielectric constant and m the electron effective mass $m = m^*m_e$ ($\epsilon = 12.4$, $g^* = -0.44$ and $m^* = 0.067$ for GaAs).

The single-particle Schrödinger equation for h_0 is exactly soluble and the solution is a shifted harmonic oscillator wave function in the x direction and a free particle with a renormalized mass in the y direction,

$$\epsilon_{n,\sigma}(k) = (n + 1/2)\Omega + \frac{1}{2}(\omega_0/\Omega)^2 k^2 + \frac{1}{2}g^*\mu_0 B\eta_\sigma, \qquad (5.64)$$

where $k = k_y$, $\Omega = (\omega_c^2 + \omega_0^2)^{1/2}$, $\eta_\sigma = \pm 1$ and $n = 0, 1, 2, \ldots$ is the subband index. Compared with the case of a quantum well in a perpendicular magnetic field (see Subsection 9.13), the degeneracy of Landau levels has been broken by the confining potential in the x direction. In the high magnetic field limit, however, the renormalized mass $m_r = m(\Omega/\omega_0)$ becomes infinite and one recovers the Landau degeneracy. The single particle wave function is

$$\psi_{nk\sigma}(\mathbf{r}) = (1/L)^{1/2} e^{iky} \phi_n(x - ck) \chi_\sigma, \tag{5.65}$$

where

$$\phi_n(x) = (2^n n! \pi^{1/2} \ell)^{-1/2} e^{-x^2/(2\ell^2)} H_n(x/\ell) \tag{5.66}$$

is the harmonic oscillator wave function and $H_n(x/\ell)$ the Hermite polynomial of degree n. $\ell = (1/\Omega)^{1/2}$ is the width of the wave function which becomes the magnetic length if ω_0 is zero. The center of the wave function is shifted by an amount $X = ck$ with $c = \omega_c/(\omega_c^2 + \omega_0^2)$.

Note that the above results strongly simplify if the magnetic field is in the xy plane, for example parallel to the direction of the wire (y direction), which is a situation often studied by experimentalists in their transport measurements [see for example Thomas *et al.* (1996) and Liang *et al.* (2000)]. In fact, in this case the magnetic field does not affect the orbital motion of the electrons ($\omega_c = 0$ in the above equations).

The effect of the interaction can be taken into account by means of Hartree–Fock (Gudmundsson *et al.*, 1995; Brataas *et al.*, 1997) or local spin density approximation (LSDA) (Knorr and Golby, 1994; Wang and Berggren, 1996; Camels and Gold, 1997; Wang and Berggren, 1998) calculations. Here we present a self-consistent calculation, within LSDA, of the ground state of GaAs quantum wires of infinite length and finite width, with a zero magnetic field (Malet *et al.*, 2005).

In the LSDA, the single electron wave functions are given by the solution of the Kohn–Sham (KS) equations

$$\left[-\frac{1}{2}\nabla_x^2 - \frac{1}{2}\nabla_y^2 + \frac{1}{2}\omega_0^2 x^2 + \int d\mathbf{r}' \frac{\rho(\mathbf{r}')}{|\mathbf{r} - \mathbf{r}'|} \right.$$
$$\left. + v_{\text{xc}}(\mathbf{r}) + w_{\text{xc}}(\mathbf{r})\eta_\sigma \right] \varphi_i^\sigma(\mathbf{r}) = \varepsilon_{i,\sigma}\varphi_i^\sigma(\mathbf{r}), \tag{5.67}$$

where i stands for the set of quantum numbers, but spin, which characterizes the two-dimensional single-particle wave functions, and the *two-dimensional* electronic density of the wire is

$$\rho(\mathbf{r}) = \sum_{i,\sigma} |\varphi_i^\sigma(\mathbf{r})|^2,$$

with $\mathbf{r} \equiv (x, y)$, $\eta_\sigma = 1(-1)$ if $\sigma = \uparrow (\downarrow)$ and

$$v_{\text{xc}}(\mathbf{r}) = \frac{\partial \mathcal{E}_{\text{xc}}[\rho(\mathbf{r}), m(\mathbf{r})]}{\partial \rho(\mathbf{r})}, \quad w_{\text{xc}}(\mathbf{r}) = \frac{\partial \mathcal{E}_{\text{xc}}[\rho(\mathbf{r}), m(\mathbf{r})]}{\partial m(\mathbf{r})}, \tag{5.68}$$

with $\rho(\mathbf{r}) = \rho^\uparrow(\mathbf{r}) + \rho^\downarrow(\mathbf{r})$ and $m(\mathbf{r}) = \rho^\uparrow(\mathbf{r}) - \rho^\downarrow(\mathbf{r})$.

The exchange-correlation energy per unit surface $\mathcal{E}_{xc}$ has been constructed from the results on the nonpolarized and fully polarized 2DEG in the same way as in Subsection 4.6. Translational invariance along the y direction imposes solutions to Eq. (5.67) of the form

$$\varphi_i^\sigma(\mathbf{r}) = \frac{1}{\sqrt{L}} e^{iky} \phi_n^\sigma(x), \tag{5.69}$$

where $n = 0, 1, 2, \ldots$ is the subband index. Inserting Eq. (5.69) into Eq. (5.67) one gets

$$\left[-\frac{1}{2}\nabla_x^2 + \frac{1}{2}\omega_0^2 x^2 + \int\int dx'dy' \frac{\rho(x')}{\sqrt{(x-x')^2 + (y-y')^2}} \right.$$
$$\left. + v_{xc}(x) + w_{xc}(x)\eta_\sigma \right] \phi_n^\sigma(x) = \epsilon_{n,\sigma} \phi_n^\sigma(x), \tag{5.70}$$

where we have introduced the band-head energy $\epsilon_{n,\sigma}$,

$$\epsilon_{n,\sigma} = \varepsilon_{i,\sigma} - \frac{k^2}{2}, \tag{5.71}$$

and the 2D density, which is y-independent, reads

$$\rho_\sigma(x) = \frac{1}{\pi} \sum_n \sqrt{2(\epsilon_F - \epsilon_{n,\sigma})} |\phi_n^\sigma(x)|^2. \tag{5.72}$$

The 1D electron density ρ_1 is obtained integrating over x:

$$\rho_1 = \frac{1}{\pi} \sum_{n,\sigma} \sqrt{2(\epsilon_F - \epsilon_{n,\sigma})} = \rho_1^\uparrow + \rho_1^\downarrow. \tag{5.73}$$

The Fermi energy ϵ_F fixes the number of subbands that are filled in the ground state of the wire. For each value of ρ_1, it is determined by solving the KS Eqs. (5.70) self-consistently under the condition that Eq. (5.73) is fulfilled.

For an infinite wire like the one considered here, the Hartree potential

$$V_H = \int\int dx'dy' \frac{\rho(x')}{\sqrt{(x-x')^2 + (y-y')^2}} \tag{5.74}$$

is obviously divergent. As in the case of the homogeneous electron gas in two or three dimensions, this requires the introduction of a neutralizing positive background. One possible way is to assume that this background is such that the Hartree potential is strictly canceled out, and only the exchange and correlation energy terms appear in the KS equations [see for example Reimann *et al.* (1999)]. In the following we carry out KS calculations for this case, which we call the exchange-correlation model (xc model). Other possibilities have been investigated by Gudmundsson *et al.* (1995) and Malet *et al.* (2005).

The energy per unit length of the wire can be calculated as

$$\frac{E}{L} = \sum_{n,\sigma} \left[\frac{1}{6\pi} [2(\epsilon_F - \epsilon_{n,\sigma})]^{3/2} + \frac{1}{\pi} \sqrt{2(\epsilon_F - \epsilon_{n,\sigma})} \epsilon_{n,\sigma} \right] + \int dx \mathcal{E}_{\mathrm{xc}}(x)$$

$$- \int dx \rho(x) v_{\mathrm{xc}}(x) - \int dx m(x) w_{\mathrm{xc}}(x). \tag{5.75}$$

The energy per electron E/N is obtained dividing Eq. (5.75) by ρ_1.

The KS equations have been solved for wires confined by two different values of the harmonic potential, namely $\omega_0 = 2$ and $4\,\mathrm{meV}$, and for ρ_1 densities up to filling six subbands. Values of ω_0 between 2.5 and 3.5 meV have been determined for the gate voltage close to the threshold of the second subband for long quantum wires using a magnetic depopulation technique (Tarucha *et al.*, 1995).

In Figs. 5.19 and 5.20 we have plotted E/N and the magnetization $\xi = (\rho_1^\uparrow - \rho_1^\downarrow)/\rho_1$ as a function of the one-dimensional electron density for the exchange-correlation model. The numbers along the E/N curves correspond to the number

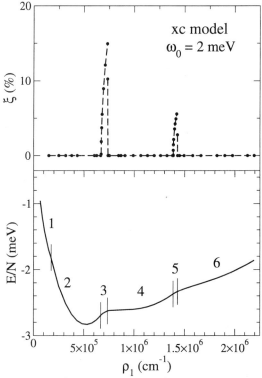

Fig. 5.19 Energy per electron (meV; bottom panel) and magnetization (top panel) in the exchange-correlation model for parabolic confinement of $\omega_0 = 2\,\mathrm{meV}$ as a function of the linear density (cm^{-1}). The regions separated by vertical lines correspond to the indicated number of occupied subbands. For one single occupied subband, the system is fully polarized.

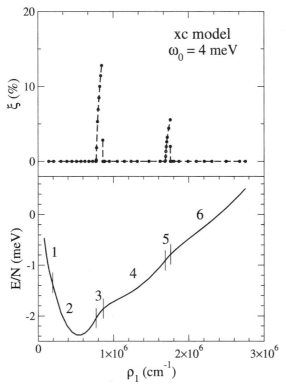

Fig. 5.20 The same as Fig. 5.19, for $\omega_0 = 4\,\mathrm{meV}$.

of occupied subbands, and the vertical lines indicate the boundary between neighbor j and $j+1$ subband regions. We have found that the transition from an even to an odd number of subband neighboring regions, like $2 \to 3$ or $4 \to 5$, is smooth, as the changes in ξ indicate. On the contrary, the transition from an odd to an even number of subband neighboring regions, like $1 \to 2$ or $3 \to 4$, is abrupt, with metastability regions not shown in the figures extending to the left and right of the crossing points. Apart from the region corresponding to the first subband, which is fully polarized, the magnetization reached in all the other odd subband regions is below the maximum value one would naively expect, i.e. $1/3$ for $j = 3$ and $1/5$ for $j = 5$. This is due to the exchange-correlation energy, which lifts the degeneracy of the $\uparrow, \downarrow$ subbands. It is also worth noting that the odd subband number regions are rather narrow, getting narrower as ρ_1 increases. The relevance of determining the boundaries of these density regions is that, in a mean field model, they fix the extension of the conductance plateaux, as we will see in Chapter 8.

In Figure 5.21 we show the density profiles of the 2D electronic density for a lateral confining potential with $\omega_0 = 4\,\mathrm{meV}$. The values of the 1D electronic densities are indicated, and correspond to configurations with 1, 3 and 6 occupied subbands. Also shown is the *local* magnetization $\xi = m(x)/\rho(x)$ for the three- subband systems.

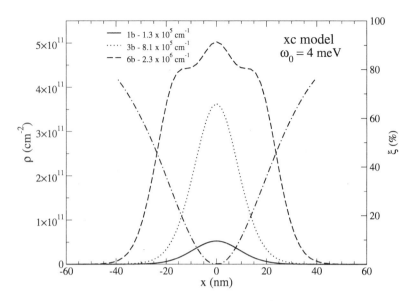

Fig. 5.21 Several density profiles of the 2D electron density as a function of x (nm) for a harmonic frequency $\omega_0 = 4\,\text{meV}$ (left scale). Also shown is the local magnetization ξ for the three-subband case (right scale). The values of the 1D electronic densities, which correspond to configurations with one, three and six occupied subbands, are indicated.

It can be seen that in this latter case, the wire presents an edge magnetization that increases as ρ decreases.

In conclusion, the LSDA calculation at a zero magnetic field shows that as the 1D electron density increases, the system can be in different phases characterized by different values of the the magnetization. These phases correspond to the filling of an increasing number of electronic subbands. Due to the exchange-correlation interaction, one finds that for odd numbers of occupied subbands larger than one, the system adquires some edge magnetization. The width of the non-paramagnetic density regions is rather narrow, and it decreases as the density increases. As we will see in Chapter 8, these results can be used to address the conductivity of quantum wires. This quantity can be measured in transport experiments and permits one to test the predictions of the theoretical models. In particular, there seems to be experimental evidence that at low density the ground state of the wire is ferromagnetic (Thomas *et al.*, 1996; Reilly *et al.*, 2001).

5.8   The Aharanov–Bohm Effect and Quantum Rings

Recently, the availability of nanoscopic semiconductor ring structures (Garcia *et al.*, 1997; Lorke and Luyken, 1998; Lorke *et al.*, 2000; Warburton *et al.*, 2000; Keyser *et al.*, 2003) has stimulated a strong interest in the properties of quantum rings, in particular those related to the fractional Aharanov–Bohm oscillations of the energy and of the persistent current in the ring. These oscillations have been observed in mesoscopic (Mailly, Chapelier and Benoit, 1993) and nanoscopic (Fuhrer *et al.*, 2001; Keyser *et al.*, 2003) rings in GaAlAs/GaAs heterostructures and studied within many theoretical approaches [see, for example, Viefers *et al.* (2000) and references therein].

The Aharanov–Bohm effect is a direct consequence of the properties of eigenfunctions of isolated one-dimensional quantum rings, which cause the periodicity of all physical quantities. The reason for this behavior is the observation (Hund, 1938; Byers and Yang, 1961; Bloch, 1970; Buttiker, Imry and Landauer, 1983; Cheung *et al.*, 1988) that in the case of an ideal one-dimensional ring without impurities, which encloses a magnetic flux Φ, the vector potential can be eliminated from the Schrödinger equation for the free motion of the electrons,

$$\frac{1}{2m} \left(-i\hbar \frac{d}{dx} + \frac{e}{c} A(x) \right)^2 \psi_\ell(x) = \epsilon_\ell \psi_\ell(x), \tag{5.76}$$

by introducing a gauge transformation. The result is that the field now enters the calculation via the flux-modified boundary condition:

$$\psi_\ell(L) = \exp\left[\frac{i2\pi\Phi}{\Phi_0}\right] \psi_\ell(0), \tag{5.77}$$

where L is the ring circumference, $\Phi_0 = hc/e$ and we have used as spatial variable $x = L\theta/2\pi$, so that x varies between 0 and L. The situation is then analogous to the one-dimensional Bloch problem, as seen by identifying $2\pi\Phi/\Phi_0$ and kL. The energy levels of the ring, given by

$$\epsilon_\ell = \frac{\hbar^2}{2m} \left[\frac{2\pi}{L} \left(\ell - \frac{\Phi}{\Phi_0} \right) \right]^2, \quad \ell = 0, \pm 1, \pm 2, \ldots, \tag{5.78}$$

form microbands as a function of Φ with period Φ_0, analogous to the Bloch electron bands in the extended k zone picture (see Fig. 5.22). The current carried by level ϵ_ℓ at zero temperature is

$$I_\ell = -\frac{e v_\ell}{L}, \quad v_\ell = \frac{1}{\hbar} \frac{\partial \epsilon_\ell}{\partial k_\ell}, \tag{5.79}$$

or, using the above analogy,

$$I_\ell = -c \frac{\partial \epsilon_\ell}{\partial \Phi}. \tag{5.80}$$

Figure 5.22 shows schematically the energies of the eigenstates as a function of flux. In the absence of impurities in the ring, the curves form intersecting parabolas. In the presence of disorder, there are gaps at the points of intersection, in the same way as band gaps form in the band structure problem. Whereas the eigenenergies

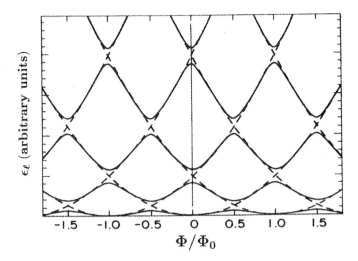

Fig. 5.22   Schematic diagram of the energy levels (5.78) as a function of the flux Φ/Φ_0 in a one-dimensional ring with (solid curve) and without impurities (dashed curve).

of the one-dimensional ring are symmetric in the flux, the current carried by an eigenstate, which is given by

$$I_\ell = \frac{2\pi e\hbar}{mL^2}\left[\ell - \frac{\Phi}{\Phi_0}\right], \quad \ell = 0, \pm 1, \pm 2, \ldots, \tag{5.81}$$

is antisymmetric in the flux. Moreover, since at an integer or half-integer flux quantum the energy is maximum or minimum, at these values of Φ the current is zero.

Up to now we have considered a single electron moving in the ideal one-dimensional ring. For many electrons, it is possible to observe new effects in the current due to the crossings of the energy levels. For example, one can place two electrons in the state $\ell = 0$. If $N = 3$, the third electron, for $\Phi/\Phi_0 \simeq 0$, can be placed in either the $\ell = 1$ state or the $\ell = -1$ state, depending on whether Φ/Φ_0 is infinitesimally negative or positive. These rearrangements, when permitted by Fermi statistics, lead to discontinuous jumps in the current, with a change of sign. Repeating the analysis for cases $N = 4\ell$, $4\ell + 1$, $4\ell + 2$, $4\ell + 3$ leads to the results given in the article by Loss and Golbart (1991). For odd numbers of electrons they found that there is an approximate half-quantum flux periodicity, which, however, improves as the number of electrons increases. For even numbers of electrons there is one quantum flux periodicity, as in the case of spinless electrons.

The electron–electron interaction can change the previous picture and yield a decrease of period and amplitude of the oscillations of the ground state energy and persistent current. This so-called "fractional Aharonov–Bohm effect" is due to electron correlations and has been studied by means of exact diagonalization (Niemela *et al.*, 1996) of the many-electron Hamiltonian (5.1) with the parabolic potential $m\omega_0^2 r^2/2$ substituted by a shifted parabolic potential,

$$v_{\text{ext}}(r) = \frac{1}{2}m\omega_0^2(r - R_0)^2,$$ (5.82)

which better models the lateral confinement potential in the case of a ring. Calculations with the LSDA theory and diffusion Monte Carlo have also been performed (Emperador, Pederiva and Lipparini, 2003), and in the following we present some results obtained by these authors. Taking $R_0 = 20$ nm, $\omega_0 = 330$ meV, an effective mass $m = 0.063\, m_e$ and a dielectric constant $\epsilon = 12.4$, the electron density turns out to extend over a region with a width that is 15% of R_0. The ring is therefore nearly equivalent to a quantum wire with periodic boundary conditions at both ends, and can be regarded as quasi-one-dimensional. The flux across the ring is $\Phi = \pi R_0^2 B$, where B is the applied magnetic field.

In Fig. 5.23 we plot the ground state energy as a function of B in the case of $N = 10$ electrons in the ring. With $R_0 = 20$ nm, a magnetic field intensity of 3.29 T corresponds to one flux quantum crossing the ring. The upper panel of the figure shows the ground state energy for the ring containing noninteracting electrons in different (L, S) configurations. The lower panel gives the energy for the same configurations as those calculated in LSDA and in diffusion Monte Carlo. The curves in the lower panel of the figure are fits to the numerical calculations obtained with the expression

$$E(\Phi) = \frac{N}{2}\hbar\omega_0 + E(0) + \frac{\hbar^2}{2mR_0^2}\sum_{\alpha}\left(\ell_\alpha - \frac{\Phi}{\Phi_0}\right)^2.$$ (5.83)

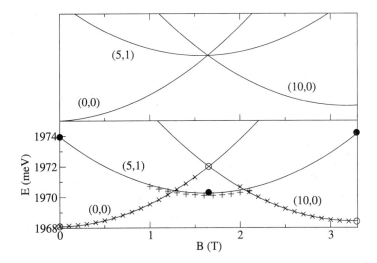

Fig. 5.23 Ground state energy as a function of B for an $N = 10$ quasi-one-dimensional ring. Upper panel: Noninteracting electrons. Lower panel: LSDA (crosses and pluses) and DFM (empty and full circles) results. The (L, S) of each configuration is indicated.

In this expression, the term including the sum running over the occupied sp levels represents the single-particle kinetic energy, and $E(0)$ is the interaction energy, extracted from the DMC calculation. It depends only on the spin of the ground state and is independent of the angular momentum. This is due to the fact that in quasi-one-dimensional rings, the only relevant difference between the single-particle wave functions with different angular momentum l, entering the trial wave function, is the phase. The radial part is always close to a Gaussian centered in R_0 due to the large strength of the lateral confining potential. For the same reason, the shape of the single-particle wave functions is very weakly affected by the magnetic field, and the interaction energy does not depend on Φ, as well. On the other hand, the spin dependence of $E(0)$ is mainly due to the exchange term, which is dominating in the $S = 1$ state.

In the case of noninteracting electrons, the three electronic configurations shown in the upper panel of the figure are degenerate at $\Phi = \Phi_0/2$, corresponding to $B = 1.645$ T. The interaction decreases the energy of the $(L, S) = (5, 1)$ configuration, in agreement with Hund's rule, and yields an alternation of spin singlet ($S = 0$) and spin triplet ($S = 1$) ground states as a function of B, which leads to a period of $\Phi_0/2$ of the Aharonov–Bohm oscillations to be compared with the period of one flux for the noninteracting case.

Increasing the radius of the ring, the density of electrons in the ring lowers and the single-particle energies become almost degenerate. The interaction mixes the quasidegenerate single-particle levels and destroys circular symmetry in the intrinsic frame of reference of the system (Koskinen *et al.*, 2001; Bormann and Harting, 2001; Puente and Serra, 2001; Pederiva *et al.*, 2002) by means of a localization of the electrons in this frame. In the case of total localization, the Wigner molecule rotates as a rigid body and all the electrons have the same angular momentum, $\ell = L/N$. In this case the spin of the electrons would not play any role and the total energy would become

$$E(\Phi) = \frac{N}{2}\hbar\omega_0 + E'(0) + \frac{\hbar^2}{2mNR_0^2}\left(L - \frac{\Phi}{\Phi_0}N\right)^2, \qquad (5.84)$$

where the interaction energy $E'(0)$ is now a constant independent of L, S and the magnetic field. Hence, in this case the period of the Aharonov–Bohm oscillations is Φ_0/N and all the curves at fixed L, given by Eq. (5.84), would have the minima at the same energy, $E_{\min} = \frac{N}{2}\hbar\omega_0 + E'_0$. The occurrence of a situation close to that described above is demonstrated by Fig. 5.24, where we have plotted as a function of Φ/Φ_0 the fixed phase DMC ground state energy for the $N = 4$ electron ring of radius 120 nm in a shifted parabolic potential with $\omega_0 = 22$ meV. With these values of parameters we still have a quasi-one-dimensional system, but this time with an electron density much lower than in the $N = 10$ case discussed previously ($r_s = \pi R_0/N$ is now nine effective atomic units, to be compared with $r_s = 0.6$ of the previous case). Furthermore, this choice of parameters models a ring similar to

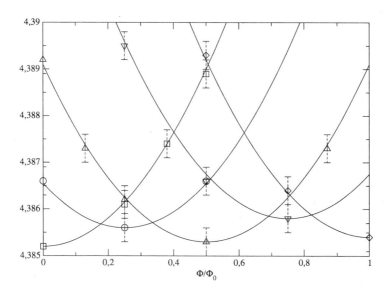

Fig. 5.24 Fixed phase DMC ground state energy as a function of Φ/Φ_0 for an $N = 4$ quasi-one-dimensional ring of very low electronic density ($r_s = 9$). Squares correspond to the $L = 0$ configuration; circles to $L = 1$; up triangles to $L = 2$; down triangles to $L = 3$; diamonds to $L = 4$.

the one realized by Keyser *et al.* (2003), where oscillations of the persistent current with a period $\Phi_0/4$ have been measured.

From the figure it is possible to observe that all the values of the angular momentum appear in the ground state band and that the period of the oscillations is $\Phi_0/4$, in agreement with the experimental findings. This fractional periodicity is due to the localization of the electrons in the intrinsic frame of reference of the system. The system minimizes its energy by maximizing the relative distance between electrons. A further indication of such localization can be found in the behavior of the angular pair distribution function

$$g(\theta) = \int \rho^{(2)}(\theta_1, \theta_2)/\rho(\theta_1)\rho(\theta_2) \, d\Theta, \qquad (5.85)$$

where $\rho^{(2)}$ is the diagonal two-body density, defined by

$$\rho^{(2)}(\theta_1, \theta_2) = \int r_1 dr_1 \, r_2 dr_2 dr_3 \cdots dr_N |\Psi(\mathbf{r}_1 \cdots \mathbf{r}_N)|^2, \qquad (5.86)$$

and $\rho(\theta)$ is the density averaged along the radial coordinate. The function $g(\theta)$ describes the probability of finding two electrons at relative angle $\theta = |\theta_1 - \theta_2|$, and is integrated in the center-of-mass angular coordinate $\Theta = (\theta_1 + \theta_2)/2$. In the case of a narrow ring this function gives essentially the same information contained in the pair distribution function $g(\mathbf{r}_1, \mathbf{r}_2)$. In fact, the density is narrow in the radial direction (so the integration in the radial coordinates does not hide essential information on the shape of the distribution), and the distribution of electrons in the

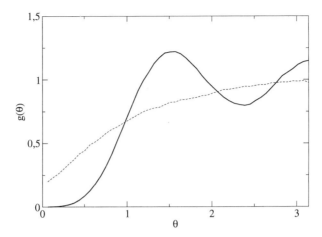

Fig. 5.25   Angular pair distribution function $g(\theta)$ for the $L = 2$, $S = 0$ ground state configuration of the $N = 4$ electron ring. Solid line: Result for $r_s = 9$. Oscillations indicate that electrons tend to keep constant distances. The maxima occur at $\theta = \pi/2$ and $\theta = \pi$, compatibly with the arrangement on a square. Dotted line: Result for $r_s = 1.5$. In the latter case no oscillations are present, and the system is not localized.

angular coordinate in the absence of external fields breaking the circular symmetry can be expected to depend only on the difference between the angular coordinates.

In Fig. 5.25 we report the results for the $L = 2$, $S = 0$ ground state configuration of the $N = 4$ ring at $\Phi = \Phi_0/2$, taken as an example. The figure shows well defined peaks at angles corresponding to $\theta = \pi/2$, π, which is in correspondence with the vertices of a square in which the first electron occupies one of the vertices. This tendency for the electrons to keep constant angular distances can be interpreted as a localization of the electrons in the two-body density. For the sake of comparison, in the same figure we report the results of a calculation for the same number of electrons, but performed at a higher density ($r_s = 1.5$). In this case it is possible to see how $g(\theta)$ does not show any structure, but for the presence of the correlation hole surrounding the electron at the origin. This means that electrons do not arrange in any particular way inside the ring. Analogous results were found in the exact diagonalization calculation by Niemela *et al.* (1996) of low density quasi-one-dimensional quantum rings containing up to four electrons.

5.9   Quantum Molecules

As we have discussed in the previous subsections, semiconductor quantum dots (QD's) have many analogies with "natural" atoms. One of the most appealing is the capability of forming molecules. Indeed, systems composed of two QD's, "artificial" quantum molecules (QM's), coupled either laterally or vertically, have recently been investigated experimentally (Kastner, 1993; Waugh *et al.*, 1995; Schmidt *et al.*, 1997; Schedelbeck *et al.*, 1997; Blick *et al.*, 1998; Luyken *et al.*, 1999; Brodsky et al., 2000; Pi *et al.*, 2001; Bayer *et al.*, 2001; Ono *et al.*, 2002) and theoretically (Yannouleas and Landman, 1999; Wensauer *et al.*, 2000; Palacios and Hawrylak, 1995; H. Imamura *et al.*, 1996; O. Mayrock *et al.*, 1997; G. Burkard *et al.*, 1999; L. Martín-Moreno *et al.*, 2000; Hu *et al.*, 1996; Oh *et al.*, 1996; Tamura, 1998; Asano, 1998; Rontani *et al.*, 1999; Partons and Peeters, 2000; Mayrock *et al.*, 1997; Tokura *et al.*, 1999; Burkard *et al.*, 2000; Pi *et al.*, 2001; Ancilotto *et al.*, 2003; Rontani *et al.* 2004).

Here we present experimental and theoretical addition energy spectra and electrochemical potentials (Pi *et al.*, 2001; Ancilotto *et al.*, 2004) of vertical diatomic QM's on going from the strong to the weak coupling limits that correspond to small and large interdot distances, b, respectively, and with and without the presence of an external magnetic field. The interpretation of the experimental results is based on the application of local spin density functional theory (LSDFT), where the ground state of the system is obtained solving the Kohn–Sham equations. The problem is simplified by the imposed axial symmetry around the z axis, which allows one to write the single-particle (sp) wave functions as $\phi_{nl\sigma}(r, z, \theta, \sigma) = u_{nl\sigma}(r, z)e^{-\iota l\theta}\chi_\sigma$, with $l = 0, \pm1, \pm2\ldots$; $-l$ is the projection of the single-particle orbital angular momentum on the symmetry axis. As usual, we use effective atomic units and in the numerical applications we consider GaAs, for which we take $\epsilon = 12.4$, and $m^* = 0.067$.

5.9.1   *The B = 0 Case*

In cylindrical coordinates the KS equations at a zero magnetic field read

$$\left[-\frac{1}{2}\left(\frac{\partial^2}{\partial r^2} + \frac{1}{r}\frac{\partial}{\partial r} - \frac{l^2}{r^2} + \frac{\partial^2}{\partial z^2} \right) + V_{cf}(r, z) \right.$$
$$\left. + V^H + V^{\mathrm{xc}} + W^{\mathrm{xc}}\eta_\sigma \right] u_{nl\sigma}(r, z) = \epsilon_{nl\sigma}u_{nl\sigma}(r, z), \tag{5.87}$$

where $\eta_\sigma=+1(-1)$ for $\sigma=\uparrow(\downarrow)$, $V_{cf}(r, z)$ is the confining potential, $V^H(r, z)$ is the direct Coulomb potential, and $V^{\mathrm{xc}} = \partial\mathcal{E}_{\mathrm{xc}}(\rho, m)/\partial\rho|_{\mathrm{gs}}$ and $W^{\mathrm{xc}} = \partial\mathcal{E}_{\mathrm{xc}}(\rho, m)/\partial m|_{\mathrm{gs}}$ are the variations of the exchange-correlation energy density $\mathcal{E}_{\mathrm{xc}}(\rho, m)$ written in terms of the electron density $\rho(r, z)$ and of the local spin magnetization $m(r, z) \equiv \rho^\uparrow(r, z) - \rho^\downarrow(r, z)$ taken at the ground state (gs). As usual, $\mathcal{E}_{\mathrm{xc}}(\rho, m) \equiv \mathcal{E}_{\mathrm{x}}(\rho, m) + \mathcal{E}_{\mathrm{c}}(\rho, m)$ has been built from 3D homogeneous electron gas calculations. This yields a simple analytical expression for the exchange contribution $\mathcal{E}_{\mathrm{x}}(\rho, m)$ (see Chapter 4). For the correlation contribution $\mathcal{E}_{\mathrm{c}}(\rho, m)$ we have used the parametrizations proposed by Perdew and Zunger (1981) based on the results of Ceperley and Alder.

For a double QD the confining potential has been taken to be parabolic in the xy plane with $\omega = 5\,\mathrm{meV}$, and a double quantum well structure $V_{cf}(z)$ in the z direction whose wells are of same width, w, and have depths $V_0 \pm \delta$ with $\delta \ll V_0$:

$$V_{cf}(r, z) = \frac{1}{2}m\omega^2 r^2 + V_{cf}(z). \tag{5.88}$$

The single-particle states which are even (odd) under reflections with respect to the $z = 0$ plane are called bonding (antibonding) states.

Owing to the axial symmetry, the direct Coulomb potential $V^H(\vec{r}) = \int d\vec{r}' \rho(\vec{r}')/|\vec{r} - \vec{r}'|$ can be written as

$$V^H(r, z) = 2 \int_0^\infty r' dr' \int_{-\infty}^{+\infty} dz' \Delta\rho(\vec{r}')[(r + r')^2 + (z - z')^2]^{1/2} \mathbf{E}(\alpha^2), \quad (5.89)$$

where $\mathbf{E}$ is the complete elliptic integral of the second kind and $\alpha^2 \equiv 4rr'/[(r + r')^2 + (z - z')^2]$. A direct integration of Eq. (5.89) is very time-consuming, and it is not advisable to get the direct Coulomb potential. Instead, we obtain $V^H(r, z)$ solving the Poisson equation

$$\Delta V^H(\vec{r}) = -4\pi\rho(\vec{r}) \quad (5.90)$$

using the conjugate gradient method. This requires a knowledge of $V^H(r, z)$ at the mesh boundary, which can be obtained by direct integration of Eq. (5.89) only for the (r, z) points which constitute the mesh boundary. For the details of the calculation, refer to the work by Pi *et al.* (2001), which uses a relaxation method to solve the partial differential equations arising from a high order discretization of the KS equations on a spatial mesh in cylindrical coordinates.

The experiment we want to analyze is illustrated in Figure 5.26, showing (a) a schematic diagram of of a submicron circular mesa, diameter D, containing two vertically coupled QD's, and (b) a scanning electron micrograph of a typical mesa after gate metal deposition. The description of the starting material, a special triple barrier resonant tunneling structure, and the processing recipe can be found in the paper by Austing *et al.* (1998). Current I_d flows through the two QD's, separated by the central barrier of thickness b, between the substrate contact and grounded top contact in response to voltage V_d applied to the substrate, and gate voltage V_g. The structures are cooled to about 100 mK and no magnetic field is applied. For the materials one typically uses, the energy splitting between the bonding and antibonding sets of single-particle (sp) molecular states, Δ_{SAS}, can be varied from

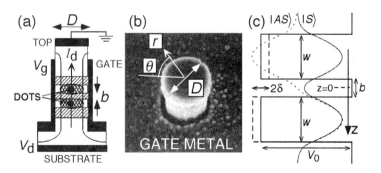

Fig. 5.26 Schematic diagrams of (a) a mesa containing two vertically coupled quantum dots and (c) double quantum well structure, and (b) scanning electron micrograph of a typical circular mesa.

about 3.5 meV for $b = 2.5$ nm (strong coupling) to about 0.1 meV for $b = 7.5$ nm (weak coupling). This is expected to have a dramatic effect on the electronic properties of QM's. The panel (c) of the figure schematically shows the double quantum well structure and its unperturbed bonding $|S\rangle$ and antibonding $|AS\rangle$ sp wave functions. We have taken $V_0 = 225$ meV and $w = 12$ nm, which are appropriate for the actual experimental devices. If δ is set to zero, the artificial molecule is symmetric ("homonuclear" diatomic QM); otherwise, it is asymmetric ("heteronuclear" diatomic QM). In the calculations here, δ is 0, or is set to a realistic value of 0.5 or 1 meV. In the homonuclear case Δ_{SAS} is well reproduced by the law $\Delta_{SAS}(b) = \Delta_0, e^{-b/b_0}$, with $b_0 = 1.68$ nm, and $\Delta_0 = 19.1$ meV. It is easy to check that in the weak coupling limit 2δ is approximately the energy splitting between the bonding and antibonding sp states which would be almost degenerate if δ is 0. For this reason we call the mismatch (offset) the quantity 2δ.

Figure 5.27(a) shows calculated addition energy spectra, $\Delta_2(N) = E(N+1) - 2E(N) + E(N-1)$, for homonuclear QM's with realistic values of b conveniently normalized as $\Delta_2(N)/\Delta_2(2)$. $E(N)$ is the total energy of the N-electron system.

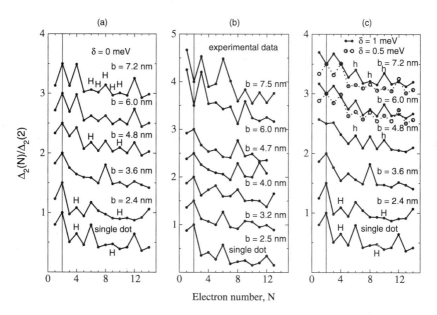

Fig. 5.27 (a) Calculated $\Delta_2(N)/\Delta_2(2)$ for homonuclear QM's with different interdot distances, b. Also shown is the calculated reference spectrum for a single QD. (b) Experimental QM addition energy spectra, $\Delta_2(N)/\Delta_2(2)$, for several interdot distances between 2.5 and 7.5 nm. Also shown is an experimental reference spectrum for a single QD. (c) Same as panel (a) but for heteronuclear QM's obtained using a $2\delta = 2$ meV mismatch (dotted lines for $b = 6.0$ and 7.2 nm are for $2\delta = 1$ meV). In each panel the curves have been vertically offset so that at $N = 2$ they are equally separated by 0.5 units for clarity. All traces in panels (a) and (c) except 3.6 and 6.0 nm: $H(h)$ marks cases where we could clearly identify Hund's first rule like filling within single dot, or bonding or antibonding states (constituent dot states).

$\Delta_2(N)$ can reveal a wealth of information about the energy required to place an extra electron into a QD or QM system. For small b ($\Delta_{\mathrm{SAS}} \gtrsim \hbar\omega$) the spectrum of a few-electron QM is rather similar to a single QD, at least for $N < 7$. At intermediate dot separation, the spectral pattern becomes more complex. However, a simple picture emerges at larger interdot distances when the molecule is about to dissociate. For example, at $b = 7.2$ nm strong peaks at $N = 2$, 4, 12 and a weaker peak at $N = 8$ appear that can be easily interpreted from the peaks appearing in the single QD spectrum. The peaks at $N = 4$, 12 in the QM are a consequence of symmetric dissociation into two closed shell (magic) $N = 2$ and 6 QD's respectively, whereas the peak at $N = 8$ corresponds to the dissociation of the QM into two identical stable QD's holding four electrons, each filled according to Hund's first rule to give maximal spin. The QM peak at $N = 2$ is related to the localization of one electron on each constituent dot, the two-electron state being a spin-singlet QM configuration.

Since the modeled QM is homonuclear, each sp wave function is shared 50% – 50% between the two constituent QD's. Electrons are completely delocalized in the strong coupling limit. As b increases, Δ_{SAS} decreases and eventually bonding, $|S\rangle$, and antibonding, $|AS\rangle$, sp molecular states become quasidegenerate. Electron localization can thus be achieved combining these states as $(|S\rangle \pm |AS\rangle)/\sqrt{2}$.

We conclude from Fig. 5.27(a) that the fingerprint of a dissociating few-electron homonuclear diatomic QM is the appearance of peaks in $\Delta_2(N)$ at $N = 2$, 4, 8, 12. This is a robust statement, as it stems from the well-understood shell structure of a single QD. If we now compare this picture with the experimental spectra shown in Fig. 5.27(b), we are led to conclude that the experimental devices are not homonuclear but heteronuclear QM's.

The origin of the mismatch is the difficulty in fabricating two perfectly identical constituent QD's in the QM's discussed here, even though all the starting materials incorporate two nominally identical quantum wells. This mismatch can clearly influence the degree of delocalization–localization, and the consequences will depend on how big 2δ is in relation to Δ_{SAS}. Here 2δ is typically 0.5–2 meV and nearly always with the upper QD (nearest the top contact of the mesa) states at higher energy than the corresponding lower QD states [see Fig. 5.26(c)].

Figure 5.27(b) shows experimental spectra, also normalized as $\Delta_2(N)/\Delta_2(2)$, for QM's with b between 2.5 and 7.5 nm, deduced accurately from peak spacings between Coulomb oscillations ($I_d - V_g$) measured by applying an arbitrarily small bias ($V_d < 100\,\mu\mathrm{V}$). Likewise, also shown is a reference spectrum for a single QD. The diameters of the mesas lie in the range of 0.5–0.6 μm, and while all mesas are circular, we cannot exclude the possibility that the QM's and QD's inside the mesas may actually be slightly noncircular, and that the confining potential is not perfectly parabolic as N increases. We emphasize the following: (i) the spectrum for the most strongly coupled QM ($b = 2.5$ nm) resembles that of the QD up to the third shell ($N = 12$); (ii) for intermediate coupling ($b = 3.2$ to 4.7 nm), the QM spectra are quite different from the QD spectrum, and a fairly noticeable peak appears at $N = 8$; (iii) for weaker coupling ($b = 6.0$ and 7.5 nm), the spectra are different again, with prominent peaks at $N = 1$ and 3.

We confirmed the heteronuclear character of the QM's by performing LSDFT calculations with a 2 meV mismatch. The results are displayed in Fig. 5.27(c). For $b = 6.0$ and 7.2 nm, spectra for a 1 meV mismatch are also given. One-to-one comparison between theory and experiment of *absolute* values is not helpful, because the QM's (QD's) actually behave in a very complex way. In particular, 2δ can vary from device to device, and probably it decreases with N. Nonetheless, the overall agreement between theory and experiment of the general spectral shape is quite good, indicating the crucial role played by the mismatch. In particular, the appearance of the spectra in the weak coupling limit for small N values is now correctly given, as well as the evolution with b of the peak appearing at $N = 8$ for intermediate coupling. A comparison between panels (a) and (c) of Fig. 5.27 reveals that for smaller values of b ($\lesssim 4.8$ nm), for a reasonable choice of parameters (ω, δ), a mismatch does not produce sizeable effects. The reason is that the electrons are still rather delocalized, and distributed fairly evenly between the two dots. Exceptions to this substantial delocalization may arise only when both the constituent single QD states are magic, as discussed below, at intermediate coupling. For larger interdot distances, a mismatch induces electron localization. The manner in which it happens is determined by the balance between interdot and intradot Coulomb repulsion, and by the degree of the mismatch between the sp energy levels, and so is difficult to predict except in some trivial cases for certain model parameters (ω, δ). For example, a large mismatch compared to $\tilde{\omega}$ will cause the QD of depth $V_0 - \delta$ to eventually "go away empty."

Finally, still assuming perfect coherency, a deeper theoretical understanding of heteronuclear QM dissociation can now be gained from analysis of the evolution with b of the sp molecular wave functions. Thus, for each sp wave function $\phi_{nl\sigma}(r, z, \theta) = u_{nl\sigma}(r, z)e^{-il\theta}\chi_\sigma$ we introduce a z probability distribution function defined as

$$\mathcal{P}(z) \equiv 2\pi \int_0^\infty dr\, r[u(r, z)]^2. \qquad (5.91)$$

Figure 5.28 shows $\mathcal{P}(z)$ for (a) $N = 6$, (b) $N = 8$ and (c) $N = 12$ (the deeper well is always in the $z > 0$ region), each for several values of b. States are labeled as $\sigma, \pm\pi, \pm\delta, \ldots$, depending on the $l = 0, \pm1, \pm2, \ldots$ sp angular momentum, and $\uparrow, \downarrow$ indicate the spins. In each subpanel, the probability functions are plotted, ordered from bottom to top, according to the increasing energies of the orbitals. For each b, the third component of the total spin and total orbital angular momentum of the ground state are also indicated by the standard spectroscopic notation $^{2S_z+1}|L_z|$ with $\Sigma, \Pi, \Delta, \ldots$ denoting $|L_z| = 0, 1, 2, \ldots$. We conclude that:

(i) QM's dissociate more easily for smaller values of b, if they yield magic number QD's, as is the case for $N = 12 \to 6 + 6$ for $b = 4.8$ nm (c) or $N = 4 \to 2 + 2$ (not shown), for example.

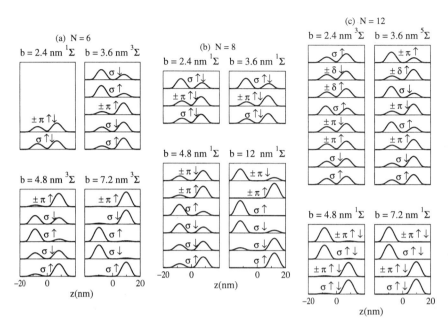

Fig. 5.28 Calculated probability distribution functions $\mathcal{P}(z)$ (arbitrary units) as a function of z for the heteronuclear $N = 6$, $N = 8$ and $N = 12$ QM's (a), (b) and (c) respectively, using a $2\delta = 2\,\text{meV}$ mismatch.

(ii) Particularly for intermediate values of b, not all orbitals contribute equally to the QM bonding, i.e. the degree of hybridization is not the same for all QD sp orbitals. See for example the π and σ states in the $b = 4.8$ nm panel of (a).

(iii) At larger b, dissociation can lead to Hund's first rule like filling in one of the QD's and full shell filling in the other dot. See for example the $b = 7.2$ nm panel in (a) for $N = 6$, which dissociates into $2 + 4$. The same happens for the $N = 10$ QM, which dissociates into $4 + 6$ (not shown). In other cases, dissociation leads to Hund's first rule like filling in each of the QD's, as shown in the $b = 12$ nm panel of (b) for $N = 8$, which breaks into $4 + 4$. In close analogy with natural molecules, atomic nuclei, or multiply charged simple metal clusters, homo- and heteronuclear QM's choose preferred dissociation channels yielding the most stable QD configurations.

(iv) Some configurations are extremely difficult to disentangle: even at very large b, there can still be orbitals contributing to the QM bonding. A good example of this is the $N = 8$ QM for $b = 12$ nm (b).

5.9.2 $B \neq 0$

In this subsection we present experimental and theoretical ground state electrochemical potentials for a diatomic QM in the intermediate coupling regime corresponding to an interdot distance, $b = 3.2$ nm ($\Delta_{\text{SAS}} \sim 3\,\text{meV}$), for magnetic fields ($B$) up to about 5 T. We assume here that the quantum-mechanical coupling is sufficiently strong that the QM can be regarded as a symmetric "homonuclear" diatomic QM. We consider two different configurations: one corresponding to an applied magnetic field parallel $(B_\parallel)$ to the drain current I_d flowing through the constituent QD's, and the other corresponding to an applied magnetic field perpendicular $(B_\perp)$ to I_d. The latter has received relatively little attention (see Tokura *et al.*, 2000; Burkard *et al.*, 2000; Sasaki *et al.*, 1998; Sànchez *et al.*, 2001; Sasaki *et al.*, 2002) Note that the QM physics we discuss in both magnetic field configurations is particularly relevant to the subject of solid state quantum computing [see, for example, Burkard *et al.* (2000)]. We also assume $V_0 = 225\,\mathrm{meV}$ and $w = 12\,\mathrm{nm}$, as in the $B = 0$ case, and a harmonic oscillator potential in the xy plane of fixed strength $\hbar\omega = 4.42\,\mathrm{meV}$. This lateral confinement energy has been determined for $N = 6$ electrons using a law that quantitatively describes the phases of QM's in the strong, intermediate and weak coupling regimes as a function of $B_\parallel$ for a number of electrons, N, between 12 and 36. Lacking a better prescription *at smaller N*, $\hbar\omega$ has been kept fixed for all N analyzed here ($N < 7$) instead of obscuring the results by further introducing an ad hoc N dependency determined by a fitting procedure.

The solution of the KS equations in the $B_\parallel$ case is greatly simplified by explicitly using the axial symmetry of the system. The additional terms in the KS Eqs. (5.87) due to the presence of an arbitrary magnetic field are given below. The inclusion of these terms crucially does not break the axial symmetry of the KS Hamiltonian in the $B_\parallel$ case.

In the symmetric gauge the vector potential $\vec{A}(\vec{r})$ corresponding to a constant magnetic field $\vec{B}$ is written as $\vec{A} = (\vec{B} \wedge \vec{r})/2$, and its contribution to the KS Hamiltonian is

$$\mathcal{H}_m = \frac{e\hbar}{2mc}\vec{B}\cdot\vec{L} + \frac{e^2}{2mc^2}\vec{A}^2 + g_s^*\mu_B\vec{B}\cdot\vec{S} \tag{5.92}$$

where g_s^* is the effective gyromagnetic factor, $\vec{L}$ and $\vec{S}$ are respectively the orbital and spin angular momentum operators, and μ_B is the Bohr magneton. Writing $\vec{B} = B(\sin\theta_B, 0, \cos\theta_B)$ and introducing the cyclotron frequency $\omega_c = eB/mc$, it can be easily checked that $\mathcal{H}_m = \mathcal{H}_{m_R} + i\mathcal{H}_{m_I}$, with

$$\mathcal{H}_{m_R} = \frac{1}{8}m\omega_c^2[x^2\cos^2\theta_B + y^2 + z^2\sin^2\theta_B - 2xz\sin\theta_B\cos\theta_B] + \frac{1}{2}g_s^*\mu_B\eta_\sigma B, \tag{5.93}$$

$$\mathcal{H}_{m_I} = -\frac{1}{2}\hbar\omega_c\left[\sin\theta_B\left(y\frac{\partial}{\partial z} - z\frac{\partial}{\partial y}\right) + \cos\theta_B\left(x\frac{\partial}{\partial y} - y\frac{\partial}{\partial x}\right)\right],$$

where $\eta_\sigma = +1(-1)$ for $\sigma=\uparrow(\downarrow)$ with respect to the direction of the applied magnetic field.

Equation(5.93) reduces to the $B_\parallel$ case when $\theta_B = 0$, and to the $B_\perp$ case when $\theta_B = \pi/2$. In the former, since $x\frac{\partial}{\partial y} - y\frac{\partial}{\partial x}$ is proportional to L_z, the problem remains axially symmetric.

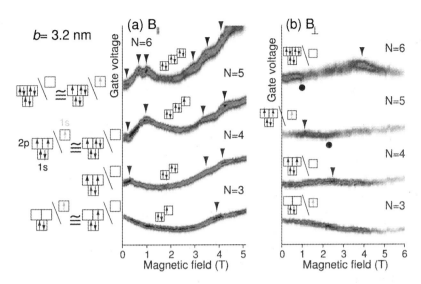

Fig. 5.29   Experimental B field dependence of the third to sixth Coulomb oscillation peaks (gs electrochemical potentials for $3 \leq N \leq 6$) in (a) the $B_{\parallel}$ case and (b) the $B_{\perp}$ case.

It is worth noting that if $B \neq 0$ then the sp wave functions $\Psi_{\sigma}(x, y, z)$ are complex, with their real and imaginary parts being coupled by $\mathcal{H}_m$.

The experimental ground state electrochemical potentials for $N = 3\text{–}6$, as a function of B, are shown in Fig. 5.29 for (a) the parallel case and (b) the perpendicular case. What is actually shown is the B field dependence of the third, fourth, fifth and sixth current (Coulomb oscillation) peaks measured by sweeping V_g in the linear conductance regime for a small $V_d \simeq 0.1$ mV. It is clearly evident that the dependencies for parallel and perpendicular cases are very different — in particular, the former is stronger than the latter. We now attempt to explain the general appearance in both cases, and particularly the features marked by the different symbols, by using the computational methods described in the previous subsection.

The $B_{\parallel}$ Case

As in single QD's, at low B fields, upward kinks (cusps) in the experimental N electron gs QM electrochemical potentials as a function of B are interpreted as changes in the N electron gs configuration of the QM which arise from sp level crossings (Tarucha *et al.*, 1996; Oosterkamp *et al.*, 1999; Kouwenhoven *et al.*, 1997). We have plotted in Fig. 5.30 the calculated gs electrochemical potential $\mu(N)$, defined as

$$\mu(N) = E(N) - E(N - 1), \tag{5.94}$$

where E is the total energy of the N electron QM gs, as a function of $B_{\parallel}$ up to $N = 6$. The superscript $+ \ (-)$, which labels the Greek letters used for the total angular momentum $(^{2S_z+1}|L_z|)$, refers to even (odd) states under reflection with respect to the $z = 0$ plane bisecting the QM. Even states are bonding (symmetric) states, and odd states are antibonding (antisymmetric) states. The subscript $g(u)$ refers to positive (negative) parity states. All these are good quantum numbers in the $B_{\parallel}$ case and can be used to label the different gs's ("phases"). Following Partons and Peeters (2000), we have also calculated the "isospin" quantum number (the bond order in molecular physics) defined as $I_z = (N_B - N_{AB})/2$, with $N_{B(AB)}$ being the number of electrons in bonding (antibonding) sp states. This is an exact quantum number for homonuclear QM's in a parallel magnetic field.

Given the complexity of real vertical QM structures and the challenge in modeling them, a comparison between Figs. 5.29(a) and 5.30 reveals a rather good agreement between theory and experiment. As a guide, and consistent with the calculated states and the observed $B_{\parallel}$ dependence, we indicate in Fig. 5.29(a) in a simple box style the dominant gs configurations at or near $B = 0$, and others at a higher field which are stable over a relatively wide range of $B_{\parallel}$. Up and down arrows indicate spin-up and spin-down electrons, and black (gray) arrows represent electrons in bonding (antibonding) sp states. For $N = 3, 5, 6$, near $B = 0$, because the gs's are close to each other, i.e. stable over a fairly narrow range, we show two configurations which in practice are hard to resolve. Some of these involve the population of the lowest antibonding state with a single electron, so isospin is nonmaximal. Above $B = 1$ T, however, all the antibonding states are depopulated, so isospin is maximal $(I_z = N/2)$, and filling of the QM resembles that of a single QD. The identifiable gs transitions in Fig. 5.29(a) are marked by black triangles. As expected, most appear as upward kinks. A couple — see the first kinks for $N = 5$ and 6 — appear as downward kinks because of the gs transitions which occur at almost the same $B_{\parallel}$ in $N = 4$ and 5 respectively.

Looking further at other details in Fig. 5.30, for $N = 2$, the singlet–triplet transition occurs at about 4.6 T, which is close to the experimental value (Amaha *et al.*, 2001) of ~ 4.2 T (not shown). We have found from the calculations an MDD configuration made up of electrons filling just bonding sp states (MDD$_B$), which has a total angular momentum $L_z = N(N - 1)/2$, and extends from ~ 4.9 to ~ 9.5 T for $N = 3$, from ~ 5.1 to ~ 9.0 T for $N = 4$, from ~ 5.4 to ~ 8.8 T for $N = 5$, and from ~ 5.6 to ~ 8.3 T for $N = 6$.

In Fig. 5.30 we can see that for the larger N values studied here, the increase in angular momentum of the QM gs as it evolves from $B = 0$ toward the MDD$_B$ is accompanied by two isospin flips caused by electrons jumping from antibonding to bonding states and vice versa. Phase transitions from $-$ to $+$ gs's involve $\Delta I_z = 1$ flips, whereas those from $+$ to $-$ gs's involve $\Delta I_z = -1$ flips, and both are clearly seen in Fig. 5.30 for $N = 5$ and 6. Interestingly, they only happen for $B_{\parallel} < 2$ T. We can see that after reaching the $\nu_B = 2$ gs (i.e. a filling factor two-QM state made up of just bonding sp states), which corresponds to the $^1\Delta_g^+$ phase for $N = 4$, to the $^2\Gamma_g^+$ phase for $N = 5$, and to the $^1I_g^+$ phase for $N = 6$, only bonding sp states

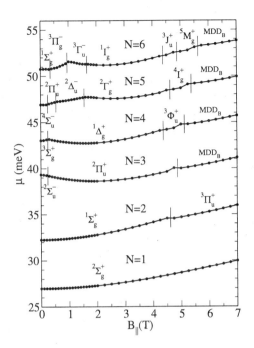

Fig. 5.30 Theoretical ground state electrochemical potentials in the $B_\parallel$ case (dots) for $N \leq 6$. The lines have been drawn as a to guide for the eye. The vertical ticks along the $\mu(N)$ lines indicate phase boundaries. The various states are identified by standard spectroscopic notation discussed in the text.

are occupied, and as a consequence the QM reaches the MDD$_B$ state in a similar way to how a single QD reaches the MDD state, namely by populating bonding sp states of higher and higher sp orbital angular momentum l values. In general, these isospin flips can produce a complex pattern in the sp spectrum as a function of $B_\parallel$. As an example of this complexity, we present in Fig. 5.31 the sp levels for $N = 6$ as a function of l for different $B_\parallel$ values. It can be seen in this figure that as B increases, the QM undergoes isospin flips. Firstly, the $l = 0\uparrow$ antibonding sp state becomes occupied, as shown in the panels corresponding to $B = 0.5$ and 1.2 T. After another isospin flip caused by the depopulation of the same sp state, the QM reaches the $\nu_B = 2$ phase corresponding to the $^1I_g^+$ configuration ($B = 3$ T panel). From this phase on, the spin polarization steadily increases until the QM reaches the MDD$_B$ phase ($B = 6$ T panel).

The $B_\perp$ Case

In the $B_\perp$ case, even if the experimental device is axially symmetric about the z axis and the constituent QD's are identical, the magnetic Hamiltonian (5.92) breaks the axial symmetry and the reflection symmetry about the $z = 0$ plane. As a consequence, the sp states no longer have a well-defined orbital angular momentum or

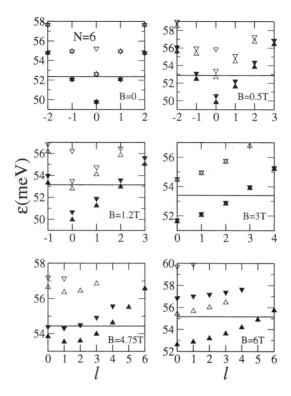

Fig. 5.31 Single-particle energy levels as a function of l for different values of $B_{\parallel}$ at $N = 6$. Upward (downward) triangles denote ↑ (↓) spin states. Open (solid) triangles correspond to antibonding (bonding) states. The horizontal lines represent the Fermi level. The value of $B_{\parallel}$ is indicated in each panel.

parity, and the bonding or antibonding labels strictly do not make sense. Crucially, within the LSDFT, the only good quantum number is the spin projection *along* the direction of the applied magnetic field, which we call $s_{\perp}$, and the gs electrochemical potentials as a function of $B_{\perp}$ are expected to be much smoother than in the $B_{\parallel}$ case.

The situation of a $B_{\perp}$ field, unlike the $B_{\parallel}$ case, lacks an analytical solution even for the case of noninteracting electrons. We show in Fig. 5.32 the calculated noninteracting sp spectrum as a function of $B_{\perp}$. At $B = 0$ (and only in this case), the energy difference between the bonding and antibonding $l = 0$ sp states is just Δ_{SAS} (likewise for the $l = 1$ states). Also, the energy difference between $l = 1$ and 0 bonding (or antibonding) states is just $\hbar\omega$. The small splitting between spin-up and -down states that originate from a common sp state with a well-defined orbital angular momentum at $B = 0$ is due to the Zeeman term.

As already noted, the even (bonding) or odd (antibonding) character of the sp levels defining a QM state is strictly lost when a magnetic field perpendicular to I_{d} is present. Intriguingly, however, the bonding/antibonding character present at

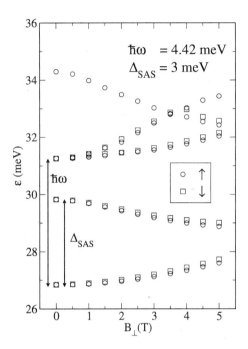

Fig. 5.32 Energies of the nine lower-lying noninteracting sp levels as a function of $B_\perp$. Δ_{SAS} and $\hbar\omega$ are marked. For each symbol, the direction of $s_\perp$ is indicated in the box.

$B = 0$ is sometimes retained to a large degree by the sp states at finite $B_\perp$ values. We have indeed found that the expectation value of the $z \rightarrow -z$ reflection operator

$$\langle \Pi_z \rangle = \int d\vec{r}\,\Psi^*(\vec{r})\Pi_z\Psi(\vec{r}) = \int d\vec{r}\,\Psi^*(x,y,z)\Psi(x,y,-z) \qquad (5.95)$$

is very close to ± 1, as it should be for bona fide bonding/antibonding states, in many cases even for relatively large values of $B_\perp$.

As an example of this, we show in Fig. 5.33 the energies of the occupied sp states as a function of $B_\perp$ for $N = 5$ and 6. Solid triangles represent "quasibonding" states with $\langle \Pi_z \rangle \geq 0.95$. Note that at 0 T for $N = 5$ the sp bonding state at $\epsilon \sim 48.2$ meV is twofold-degenerate, and likewise for two of the $N = 6$ sp bonding states at $\epsilon \sim 52$ meV. Open triangles represent "quasi-antibonding" states with $\langle \Pi_z \rangle \leq -0.95$. Actually, there is only one such occupied antibonding sp state for $N = 5$ at $B = 0$, and none for $N = 6$. All other open symbols (circles and squares) correspond to sp states with negative $\langle \Pi_z \rangle$ values larger than -0.95, i.e. they cannot really be regarded even as "quasi-antibonding" states.

The figure also shows that states that evolve from $l = 0$ sp states at $B = 0$ retain a quasibonding character up to quite high values of $B_\perp$ (at least up to 5 T), whereas other states, which at $B = 0$ are $l = 1$ sp states, do not. The quasibonding robustness of the lower-lying sp states may be due to the small effect that the

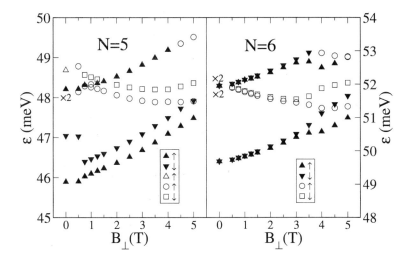

Fig. 5.33 Single-particle energy levels as a function of $B_\perp$ for $N = 5$ (left panel) and 6 (right panel). For each symbol, the direction of $s_\perp$ is indicated in the box. States that are twofold-degenerate are indicated by the ×2 symbol. Solid and open symbols are discussed in the text.

applied magnetic field has on states that are $l = 0$ sp states at $B = 0$. The $B_\perp$ evolution of what at 0 T are the $2p$ states is rather similar for $N = 5$ and 6, with a change from solid to open symbols near $B_\perp = 4$ T.

Interestingly, in spite of the lack of any spatial symmetry in the system when a perpendicular field is applied, the sp levels are still clearly distributed into shells, as in the noninteracting case. Notice also the different splitting between ↑ and ↓ states. For saturated (zero) spin ($N = 6$ case), this is essentially due to the small Zeeman term, whereas for nonsaturated spin ($N = 5$ case) the splitting is mostly due to the spin-dependent part of the exchange-correlation energy, W^{xc} term in Eq. (5.87), and this effect is larger the higher the value of the gs spin. This explains the sizeable splitting between the two lower-lying sp levels for $N = 5$ up to $B_\perp \sim 0.5$ T and the splitting of all the sp levels for $N = 6$ above $B_\perp \sim 3.5$ T (see also Fig. 5.32).

The calculated gs electrochemical potentials are shown in Fig. 5.34 as a function of $B_\perp$. Comparison with Fig. 5.29(b) shows that the agreement with experiment is good for $3 \leq N \leq 6$. We have indicated the value of the total $S_\perp$ for all the relevant gs phases. In the $B = 0\text{–}5$ T range, there are some $B_\perp$-induced changes in $S_\perp$, and these give rise to upward kinks [also marked in Fig. 5.29(b) by solid down triangles]. Some downward kinks, identified by vertical arrows in the $N = 5$ and 6 gs electrochemical potentials, do not correspond to changes in the N-electron $S_\perp$. They are associated with sp level crossings between sp states of different $\langle \Pi_z \rangle$ value of the $N - 1$ electron system. This is the case for $N = 5$ at $B_\perp \sim 1.5$ T and $N = 6$ at $B_\perp \sim 0.75$ T, as can be seen in Fig. 5.34 [also marked in Fig. 5.29(b) by solid circles]. Because of the lack of spatial symmetry in the system, we do not, in general, attempt to identify the (dominant) gs configurations. The configurations

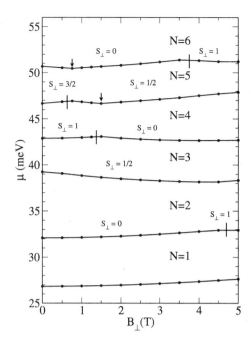

Fig. 5.34 Theoretical ground state electrochemical potentials in the $B_\perp$ case (dots) for $N \le 6$. The lines have been drawn as a guide for the eye. The vertical marks along the $\mu(N)$ lines indicate phase boundaries. The value of $S_\perp$ in each phase is given. We have indicated by vertical arrows downward kinks arising from sp level crossings in the $N-1$ electron ground state that do not produce phase transitions in the N electron ground state.

shown in a simple box style in Fig. 5.29(b) are the dominant gs configurations at 0 T and they are expected to remain so for small values of $B_\perp$. The singlet–triplet transition for $N = 2$ in Fig. 5.34 appears at ~ 4.7 T, a value comparable with that found in the $B_\parallel$ case. The $B_\perp$-induced singlet–triplet transition in the experimental data is discussed elsewhere (Sasaki *et al.*, 2002). It can also be seen that for $N = 4$ Hund's first rule like filling occurs for $B_\perp < 1.5$ T, even if the gs configuration is not strictly axially symmetric. Nonetheless, for $N = 6$ at $B = 0$ we have found an axially symmetric configuration corresponding to a 2D harmonic oscillator shell-like filling.

References

Amaha, S. *et al.*, *Sol. State Commun.* **119**, 183 (2001).
Ancilotto, F. *et al.*, *Phys. Rev.* **B67**, 205311 (2003).
Asano, Y., *Phys. Rev.* **B58**, 1414 (1998).
Ashoori, R.C. *et al.*, *Phys. Rev. Lett.* **71**, 613 (1993).
Austing, D.G. *et al.*, *Physica* **B249**, 206 (1998).
Austing, D.G. *et al.*, *Phys. Rev.* **B60**, 11514 (1999).
Barranco, M., L. Colletti, E. Lipparini, M. Pi and Ll. Serra, *Phys. Rev.* **B61**, 8289 (2000).
Bayer, M. *et al.*, *Science* **291**, 451 (2001).

Berggren, K.F., T.J. Thornton, D.J. Newson and M. Pepper, *Phys. Rev. Lett.* **57**, 11769 (1986).

Bertsch, G.F. and T. Papenbrock, *Phys. Rev. Lett.* **83**, 5412 (1999).

Bhaduri, R.K., S.M. Reimann, A.G. Choudhuri and M.K. Srivastava, cond. matt/9908210 (1999).

Blick, R.H. *et al.*, *Phys. Rev. Lett.* **80**, 4032 (1998).

Bloch, F., *Phys. Rev.* **B2**, 109 (1970).

Bormann, P. and J. Harting, *Phys. Rev. Lett.* **86**, 3120 (2001).

Brataas, A., A.G. Mal'shukov, C. Steinbach, V. Gudmundsson and K.A. Chao, *Phys. Rev.* **B55**, 13161 (1997).

Brodsky, M. *et al.*, *Phys. Rev. Lett.* **85**, 2356 (2000).

Burkard, G. *et al.*, *Phys. Rev.* **B59**, 2070 (1999).

Burkard, G. *et al.*, *Phys. Rev.* **B62**, 2581 (2000).

Buttiker, M., Y. Imry and R. Landauer, *Phys. Lett.* **A96**, 365 (1983).

Byers, N. and C.N. Yang, *Phys. Rev. Lett.* **7**, 46 (1961).

Caillol, J.M., D. Levesque, J.J. Weis and J.P. Hansen, *Stat. Phys.* **28**, 25 (1982).

Camels, L. and A. Gold, *Phys. Rev.* **B56**, 1762 (1997).

Cheung, H.F. *et al.*, *Phys. Rev.* **B37**, 6050 (1988).

Dahl, C., *Sol. State Commun.* **80**, 673 (1991).

Darwin, C.G., *Proc. Cambridge Philos. Soc.* **27**, 86 (1933).

Demel, T., D. Heitmann, P. Grambow and K. Ploog, *Phys. Rev. Lett.* **64**, 788 (1990).

Dippel, O., P. Schmelcher and L.S. Cederbaum, *Phys. Rev.* **A49**, 4415 (1994).

Dreizler, R.M. and E.K.U. Gross, *Density Functional Theory* (Springer-Verlag, Berlin, 1990).

Ebert, G., K. von Klitzing, C. Probst and K. Ploog, *Sol. State Commun.* **44**, 95 (1982).

Emperador, A., F. Pederiva and E. Lipparini, *Phys. Rev.* **B68**, 115312 (2003).

Ferconi, M. and G. Vignale, *Phys. Rev.* **B50**, 14 722 (1994); *Phys. Rev.* **B56**, 12108 (1997).

Fock, V., *Z. Phys.* **47**, 446 (1928).

Fuhrer, A. *et al.*, *Nature* **413**, 822 (2001).

Garcia, J.M. *et al.*, *Appl. Phys. Lett.* **71**, 2014 (1997).

Gudmundsson, V., A. Brataas, P. Grambow, B. Meurer, T. Kurth and D. Heitmann, *Phys. Rev.* **B24**, 17744 (1995).

Gudmundsson, V. and J.J. Palacios, *Phys. Rev.* **B52**, 11266 (1995).

Haugset, T. and H. Haugereud, *Phys. Rev.* **A57**, 3809 (1998).

Heinonen, O., M.I. Lubin and M.D. Johnson, *Phys. Rev. Lett.* **75**, 4110 (1995).

Hu, J. *et al.*, *Phys. Rev.* **B54**, 8616 (1996).

Hund, F., *Ann. Phys. (Leipzig)* **32**, 102 (1938).

Imamura, H. *et al.*, *Phys. Rev.* **B53**, 12 613 (1996).

Isihara, A., *Sol. State Phys.* **42**, 271 (1989).

Kastner, M.A., *Physics Today* **46**, 24 (1993).

Keyser, U.F. *et al.*, *Phys. Rev. Lett.* **90**, 196601 (2003).

Knorr, W. and R.W. Godby, *Phys. Rev.* **B50**, 1779 (1994).

Koskinen, M. *et al.*, *Phys. Rev.* **B63**, 205323 (2001).

Kouwenhoven, L.P. *et al.*, *Science* **278**, 1788 (1997).

Laughlin, R.B., *Phys. Rev. Lett.* **50**, 1395 (1983).

Liang, C.T., M. Pepper, M.Y. Simmons, C.G. Smith and D.A. Ritchie, *Phys. Rev.* **B61**, 9952 (2000).

Lipparini, E., N. Barberan, M. Barranco, M. Pi and Ll. Serra, *Phys. Rev.* **B56**, 12375 (1997).

Lorke, A. *et al.*, *Phys. Rev. Lett.* **84**, 2223 (2000).

Lorke, A. and R.J. Luyken, *Physica* **B256**, 424 (1998).

Loss, D. and P. Goldbart, *Phys. Rev.* **B43**, 13762 (1991).

Lubin, M.I., O. Heinonen and M.D. Johnson, *Phys. Rev.* **B56**, 10373 (1997).

Luyken, R.J. *et al.*, *Nanotechnology* **10**, 14 (1999).

Madhav, A.V. and T. Chakraborty, *Phys. Rev.* **B49**, 8163 (1994).

Mailly, D., C. Chapellier and A. Benoit, *Phys. Rev. Lett.* **70**, 2020 (1993).

Malet, F., M.Pi, M. Barranco and E. Lipparini, *Phys. Rev.* **B72**, 205326 (2005).

Manninen, M., S. Viefers, M. Koskinen and S.M. Reimann, *Phys. Rev.* **B64**, 245322 (2001).

Martín-Moreno, L. *et al.*, *Phys. Rev.* **B62**, R10 633 (2000).

Mayrock, O. *et al.*, *Phys. Rev.* **B56**, 15 760 (1997).

Niemela, K. *et al.*, *Europhys. Lett.* **36**, 533 (1996).

Oh, J.H. *et al.*, *Phys. Rev.* **B53**, R13264 (1996).

Ono, K. *et al.*, *Science* **297**, 1313 (2002).

Oosterkamp, T.H. *et al.*, *Phys. Rev. Lett.* **82**, 2931 (1999).

Oosterkamp, T.H., J.W. Janssen, L.P. Kouwenhoven, D.G. Austing, T. Honda and S. Tarucha, *Phys. Rev. Lett.* **82**, 2931 (1999).

Palacios, J.J. and P. Hawrylak, *Phys. Rev.* **B51**, 1769 (1995).

Partoens, B. and F.M. Peeters, *Phys. Rev. Lett.* **84**, 4433 (2000).

Pederiva, F., A. Emperador and E. Lipparini, *Phys. Rev.* **B66**, 165314 (2002).

Peeters, F.M., *Phys. Rev.* **B42**, 1486 (1990).

Perdew, J.P. and A. Zunger, *Phys. Rev.* **B23**, 5048 (1981).

Peredes, B., P. Fedichev, J.I. Chirac and P. Zoller, *Phys. Rev. Lett.* **87**, 402 (2001).

Petrov, D.S., M. Holzmann and G.V. Shlyapnikov, *Phys. Rev. Lett.* **84**, 2551 (2000).

Pi, M. *et al.*, *Phys. Rev. Lett.* **87**, 066801 (2001).

Pi, M. *et al.*, *Phys. Rev.* **B63**, 115316 (2001).

Pi, M., M. Barranco, A. Emperador, E. Lipparini and Ll. Serra, *Phys. Rev.* **B57**, 14783 (1998).

Pitaevskii, L.P. and A. Rosch, *Phys. Rev.* **A55**, R853 (1997).

Puente, A. and Ll. Serra, *Phys. Rev. Lett.* **83**, 3266 (1999).

Puente, A. and Ll. Serra, *Phys. Rev.* **B63**, 125334 (2001).

Recati, A., F. Zambelli and S. Stringari, *Phys. Rev. Lett.* **86**, 377 (2001).

Reilly *et al.*, *Phys. Rev.* **B63**, 121311(R) (2001).

Reimann, S.M. *et al.*, *Phys. Rev.* **B58**, 8111 (1998).

Reimann, S.M., M. Koskinen and M. Manninen, *Phys. Rev.* **B59**, 1613 (1999).

Rontani, M. *et al.*, *Phys. Rev.* **B69**, 085327 (2004).

Rontani, M. *et al.*, *Sol. State Commun.* **112**, 151 (1999); *ibid.* **119**, 309 (2001).

Sánchez, D., L. Brey and G. Platero, *Phys. Rev.* **B64**, 235304 (2001).

Sasaki, S., D.G. Austing and S. Tarucha, *Physica* **B256**, 157 (1998).

Sasaki, S., Y. Tokura, D.G. Austing, K. Muraki and S. Tarucha, unpublished (2002).

Schedelbeck, G. *et al.*, *Science* **278**, 1792 (1997).

Schmidt, T. *et al.*, *Phys. Rev. Lett.* **78**, 1544 (1997).

Schüller, C., K. Keller, G. Biese, E. Ulrichs, L. Rolf, C. Steinebach, D. Heitmann and K. Eberl, *Phys. Rev. Lett.* **80**, 2673 (1998).

Serra, Ll., A. Puente and E. Lipparini, *Physica* **E14**, 391 (2002).

Tamura, H., *Physica* **B249–251**, 210 (1998).

Tarucha, S. *et al.*, *Phys. Rev. Lett.* **77**, 3613 (1996).

Tarucha, S., T. Honda and T. Saku, *Sol. State Commun.* **94**, 413 (1995).

Thomas, K.J., J.T. Nicholls, M.Y. Simmons, M. Pepper, D.R. Mace and D.A. Ritchie, *Phys. Rev. Lett.* **77**, 135 (1996).

Thomas, K.J., J.T. Nicholls, N.J. Appleyard, M.Y. Simmons, M. Pepper, D.R. Mace, W.R. Tribe and D.A. Ritchie, *Phys. Rev.* **B58**, 4846 (1998).

Thornton, T.J. *et al.*, *Phys. Rev. Lett.* **56**, 1198 (1986).

Tokura, Y. *et al.*, *J. Phys. Condens. Matt.* **11**, 6023 (1999); Y. Tokura *et al.*, *Physica* **E6**, 676 (2000).

Tsui, D.C., H. Stormer and A.C. Gossard, *Phys. Rev. Lett.* **48**, 1559 (1982).

van Wess, B.J., H. van Houten, C.W.J. Beenakker, J.G. Williamsoon, L.P. Kouwenhoven, D.van der Marel and C.T. Foxon, *Phys. Rev. Lett.* **60**, 848 (1988).

Viefers, S. *et al.*, *Phys. Rev.* **B62**, 16777 (2000).

von Klitzing, K., G. Dorda and M. Pepper, *Phys. Rev. Lett.* **45**, 494 (1980).

Wang, C.K. and K.F. Berggren, *Phys. Rev.* **B54**, R14257 (1996).

Wang, C.K. and K.F. Berggren, *Phys. Rev.* **B57**, 4552 (1998).

Warburton, R.J. *et al.*, *Nature* **405**, 926 (2000).

Waugh, F.R. *et al.*, *Phys. Rev. Lett.* **75**, 705 (1995).

Wensauer, A. *et al.*, *Phys. Rev.* **B62**, 2605 (2000).

Wharam, D.A., T.J. Thornton, R. Newbury, M. Pepper, H. Ahmed, J.E.F. Frost, D.G. Hasko, D.C. Peacock, D.A. Ritchie and G.A.C. Jones, *J. Phys.* **C21**, L209 (1988).

Willet, R., J.P. Eisenstein, H. Stormer, D.C. Tsui, A.C. Gossard and J.H. English, *Phys. Rev. Lett.* **59**, 1776 (1987).

Yannouleas, C. and U. Landman, *Phys. Rev. Lett.* **82**, 5325 (1999).

Chapter 6

Spin–Orbit Coupling in the Confined 2D Electron Gas

A novel technology based on the use of the electron spin, as opposed to the more traditional use of the electron charge, is emerging under the name of spintronics. Several spin-based electronic devices have already proved their importance, even at a commercial level, such as the spin-valve read heads. A review of the status of this incipient field can be found in Wolf *et al.* (2001) and Zutíc, Fabian and Das Sarma (2004).

The spin–orbit (SO) coupling is an essential mechanism for most spintronic devices, since it links the spin and the charge dynamics, opening up the possibility of spin control through electric fields (Datta and Das, 1990). Indeed, recent experimental and theoretical investigations have shown that the SO coupling affects the charge transport and, more specifically, the conductance fluctuations of chaotic quantum dots in a parallel magnetic field (Folk *et al.*, 2001; Halperin *et al.*, 2001; Aleiner and Fal'ko, 2001) and the conductance of quantum wires (Moroz and Barnes, 1999; Pershin, Nesteroff and Privman, 2004; Serra, Sánchez and López, 2005). It also affects the far-infrared absorption, introducing peculiar correlations between the charge and spin oscillating densities (Manger *et al.*, 2001; Valín-Rodríguez, Puente and Serra, 2002; Chakraborty and Pietiläinen, 2005; Tonello and Lipparini, 2004; Usaj and Balseiro, 2004). A sufficiently strong SO coupling can lead (Valín-Rodríguez *et al.*, 2002; Emperador, Lipparini and Pederiva, 2003) to spin inversion, with an alternating B dependence, similar to the observations from capacitance spectroscopy experiments of both vertical (Ashoori, 1996; Ciorga *et al.*, 2000) and lateral quantum dots (Voskoboynikov *et al.*, 2001). The SO coupling also affects the electronic spin precession in a magnetic field introducing changes in the Larmor frequency (Valín-Rodríguez *et al.*, 2002; Valín-Rodríguez, Puente and Serra, 2004; Malet *et al.*, 2006), and gives rise to the spin-Hall effect (Murakami *et al.*, 2003; Sinova *et al.*, 2004).

In this chapter we analyze the combined effects of SO coupling, magnetic (B) fields and spatial deformation in fixing the spin and other ground state properties of model semiconductor nanostructures.

6.1   Spin–Orbit Coupling in Semiconductors

Similarly to what happens in atoms and nuclei, the SO interaction in semiconductors has a relativistic origin. When an electric field is present in the system, its relativistic transformation to the rest frame of the electrons originates an effective magnetic field, depending on the momentum of the electrons, which couples to their spin in a similar way to the Zeeman interaction, yielding an SO coupling of the type

$$H_{\mathrm{SO}} = \frac{e\hbar}{4m_e c^2}\sigma \cdot (\mathbf{E} \wedge \mathbf{v}), \qquad (6.1)$$

where σ is the Pauli matrix vector and $\mathbf{E} = 1/e\nabla V$ is the electric field generated by the electrostatic potential filled by the particles. Equation (6.1) is valid in a vacuum. For a slow electron, $v/c \ll 1$ and a weak electric field, H_{SO} is small because the Dirac gap in the denominator $2m_e c^2 \simeq 1$ MeV. However, in semiconductors electrons experience SO coupling that can be much stronger than in a vacuum. This is due to two basic sources (Rashba, 2004, Winkler, 2004). Bloch electrons move close to the nuclei, with velocities that are close to relativistic, and equations of the band theory of narrow gap semiconductors are similar to the Dirac equation but with a gap parameter E_g (the energetic gap between the valence and conduction bands) instead of the Dirac gap $2m_e c^2$ and an electric field $1/e\nabla V_v$ in the valence band. The symmetry of crystals, and especially the symmetry of microstructures, is essentially lower than the symmetry of a vacuum. As a result, new terms that critically change the spin dynamics appear in the electron Hamiltonian.

Therefore, low symmetry narrow gap systems formed from heavy chemical elements are the best candidates for a strong SO coupling. For electrons in semiconductors of type III–V, the SO contribution to the total Hamiltonian H is (Dresselhaus, 1955)

$$\mathcal{H}_D = \gamma \sum_{i=1}^{N}[\sigma_x p_x (p_y^2 - p_z^2) + \text{c.p.}]_i, \qquad (6.2)$$

where p_k represent the different components of the electronic momentum and, c.p. stands for cyclic permutations of all indices. This Hamiltonian is known as the Dresselhaus Hamiltonian. The constant γ is about 20 eV Å^3 for GaAs and about 150–250 eV Å^3 for InAs, InSb and GaSb. For a narrow (001) quantum well, the Hamiltonian of Eq. (6.2) reduces to the 2D Dresselhaus term (Lommer, Malcher and Rössler, 1985; Bychkov and Rashba, 1985; D'yakanov and Kachorovskii, 1986; Voskoboynikov *et al.*, 2001)

$$\mathcal{H}_D = \frac{\lambda_D}{\hbar} \sum_{i=1}^{N}[p_x \sigma_x - p_y \sigma_y]_i, \qquad (6.3)$$

where the Dresselhaus parameter λ_D is determined by the quantum well vertical width z_0 as $\lambda_D \approx \gamma(\pi/z_0)^2$. With $z_0 = 100$ Å, λ_D ranges from about 2×10^{-10} to 2×10^{-9} eV cm and decreases rapidly with z_0.

While the $\mathcal{H}_D$ term comes from the bulk inversion asymmetry (BIA) (lack of symmetry for inversion in the substrate material), there exists another contribution,

$$\mathcal{H}_R = \frac{\lambda_R}{\hbar} \sum_{i=1}^{N} [p_y \sigma_x - p_x \sigma_y]_i, \tag{6.4}$$

known as the Rashba term, which appears because of a structure inversion asymmetry (SIA) (Bychkov and Rashba, 1984). This is an asymmetry in the growth direction (z) connected to a lack of symmetry for inversion in the structure of the quantum well. This SIA can be changed by applying an electric field across the quantum well. For InAs-based quantum wells, typical values of λ_R are about $\lambda_R = 10^{-9}$ eV cm; however, values as large as $\lambda_R = 6 \times 10^{-9}$ eV cm have also been reported (Cui *et al.*, 2002). There is no simple way to calculate λ_R, because it depends both on the field E inside the quantum well and the boundary conditions at the interfaces [see, for example, Pfeffer and Zawadzki (1998)]. The importance of this interaction stems from the fact that λ_R can be controllably changed by a gate voltage, as was shown by Nitta *et al.* (1997) and Engels *et al.* (1997). This property makes SIA a prospective candidate for developing a spin transistor (Datta and Das, 1990).

6.2   Spin–Orbit Effects on Single-Particle States in Quantum Wells

In the effective mass, dielectric constant approximation, the quantum well Hamiltonian H can be written as $H = H_0 + \frac{e^2}{\epsilon} \sum_{i<j=1}^{N} \frac{1}{|\mathbf{r}_i - \mathbf{r}_j|}$, where H_0 is the one-body Hamiltonian consisting of the kinetic, Zeeman, Rashba and Dresselhaus terms:

$$H_0 \equiv \sum_{j=1}^{N} [h_0]_j = \sum_{j=1}^{N} \left[\frac{P^+ P^- + P^- P^+}{4m} + \frac{1}{2} g^* \mu_B B \sigma_z \right.$$
$$\left. + \frac{\lambda_R}{2i\hbar} (P^+ \sigma_- - P^- \sigma_+) + \frac{\lambda_D}{2\hbar} (P^+ \sigma_+ + P^- \sigma_-) \right]_j . \tag{6.5}$$

$m = m^* m_e$ is the effective electron mass in units of the bare electron mass m_e, $P^\pm = P_x \pm i P_y$, $\sigma_\pm = \sigma_x \pm i \sigma_y$, where the σ's are the Pauli matrices, and $\mathbf{P} = -i\hbar\nabla + \frac{e}{c}\mathbf{A}$ represents the canonical momentum in terms of the vector potential $\mathbf{A}$, which in the following we write in the Landau gauge, $\mathbf{A} = B(0, x, 0)$, with $\mathbf{B} = \nabla \times \mathbf{A} = B\hat{\mathbf{z}}$. The second term in Eq. (6.5) is the Zeeman energy, where $\mu_B = \hbar e/2m_e c$ is the Bohr magneton, and g^* is the effective gyromagnetic factor. The third and fourth terms are the usual Rashba and Dresselhaus interactions, respectively. Note that for bulk GaAs, taken here as an example, $g^* = -0.44$, $m^* = 0.067$, and the dielectric constant is $\epsilon = 12.4$. To simplify the expressions, in the following we will use effective atomic units $\hbar = e^2/\epsilon = m = 1$.

Introducing the operators

$$a^{\pm} = \frac{1}{\sqrt{2\omega_c}} P^{\pm} \qquad (6.6)$$

with $[a^-, a^+] = 1$ and $\omega_c = eB/mc$ being the cyclotron frequency, the single-particle (sp) Hamiltonian h_0 can be rewritten as

$$h_0/\omega_c = \frac{1}{2}(a^+ a^- + a^- a^+) - \frac{1}{2}\frac{\omega_L}{\omega_c}\sigma_z - \frac{1}{2}i\tilde{\lambda}_R(a^+ \sigma_- - a^- \sigma_+)$$

$$+ \frac{1}{2}\tilde{\lambda}_D(a^+ \sigma_+ + a^- \sigma_-), \qquad (6.7)$$

where $\omega_L = |g^* \mu_B B|$ is the Larmor frequency and $\tilde{\lambda}_{R,D} = \lambda_{R,D}\sqrt{2/\omega_c}$. For the spinor $|\phi\rangle \equiv \binom{\phi_1}{\phi_2}$ (we will use "1" for the top component and "2" for the bottom component of any spinor), the Schrödinger equation $h_0|\phi\rangle = \varepsilon|\phi\rangle$ adopts the form

$$\left[\begin{array}{cc} \frac{1}{2}(a^+ a^- + a^- a^+) - \omega_L/(2\omega_c) - \varepsilon & i\tilde{\lambda}_R a^- + \tilde{\lambda}_D a^+ \\ -i\tilde{\lambda}_R a^+ + \tilde{\lambda}_D a^- & \frac{1}{2}(a^+ a^- + a^- a^+) + \omega_L/(2\omega_c) - \varepsilon \end{array} \right] \times \binom{\phi_1}{\phi_2} = 0.$$

$$(6.8)$$

We expand ϕ_1 and ϕ_2 into oscillator states $|n\rangle$ as $\phi_1 = \sum_{n=0}^{\infty} a_n|n\rangle$, $\phi_2 = \sum_{n=0}^{\infty} b_n|n\rangle$, on which a^+ and a^- act in the usual way, i.e. $\frac{1}{2}(a^+ a^- + a^- a^+)|n\rangle = (n + \frac{1}{2})|n\rangle$, $a^+|n\rangle = \sqrt{n+1}|n+1\rangle$, $a^-|n\rangle = \sqrt{n}|n-1\rangle$, and $a^-|0\rangle = 0$. This yields the infinite system of equations

$$(n + \alpha - \varepsilon)b_n - i\tilde{\lambda}_R\sqrt{n}\,a_{n-1} + \tilde{\lambda}_D\sqrt{n+1}\,a_{n+1} = 0,$$

$$(n + \beta - \varepsilon)a_n + i\tilde{\lambda}_R\sqrt{n+1}\,b_{n+1} + \tilde{\lambda}_D\sqrt{n}\,b_{n-1} = 0, \qquad (6.9)$$

for $n \geq 0$, with $a_{-1} = 0$, $b_{-1} = 0$, and $\alpha = (1 + \omega_L/\omega_c)/2$, $\beta = (1 - \omega_L/\omega_c)/2$.

6.2.1   *The Case in Which Either $\lambda_R = 0$ or $\lambda_D = 0$*

When only the Rashba or Dresselhaus terms are considered, Eqs. (6.9) can be exactly solved (Rashba, 1960; Das, Datta and Reifenberger, 1990; Fal'ko 1992; Schliemann, Egues and Loss, 2003). In the $\lambda_D = 0$ case, combining Eqs. (6.9) one obtains

$$\left[(n + \alpha - \varepsilon)(n - 1 + \beta - \varepsilon) - n\tilde{\lambda}_R^2 \right] b_n = 0,$$

$$\left[(n + \alpha - \varepsilon)(n - 1 + \beta - \varepsilon) - n\tilde{\lambda}_R^2 \right] a_{n-1} = 0, \qquad (6.10)$$

either of which yields the energies

$$\varepsilon_n^\pm = n \pm \sqrt{\frac{1}{4}\left(1 + \frac{\omega_L}{\omega_c}\right)^2 + \frac{2}{\omega_c}\lambda_R^2 n}. \tag{6.11}$$

One also obtains

$$(n - 1 + \beta - \varepsilon_n^\pm)a_{n-1}^{\varepsilon_n^\pm} = -i\tilde{\lambda}_R \sqrt{n}\, b_n^{\varepsilon_n^\pm}, \tag{6.12}$$

which together with the normalization condition $|a_{n-1}^{\varepsilon_n^\pm}|^2 + |b_n^{\varepsilon_n^\pm}|^2 = 1$ solves exactly the problem [for $n = 0$, $a_{-1} = 0$, $b_0 = 1$, and $\varepsilon_0 = \frac{1}{2}(1 + \omega_L/\omega_c)$].

Equations (6.10) indicate that in the series expansion of the spinor $|\phi\rangle$, only one a_i and one b_i coefficient appear. Specifically,

$$|n_d\rangle = \begin{pmatrix} a_{n-1}^{\varepsilon_n^+}|n - 1\rangle \\ b_n^{\varepsilon_n^+}|n\rangle \end{pmatrix}; \quad |n_u\rangle = \begin{pmatrix} a_n^{\varepsilon_{n+1}^-}|n\rangle \\ b_{n+1}^{\varepsilon_{n+1}^-}|n + 1\rangle \end{pmatrix}. \tag{6.13}$$

In the limit of zero SO, the spinors $|n_d\rangle$ and $|n_u\rangle$ become $|n\rangle\binom{0}{1}$ and $|n\rangle\binom{0}{1}$, respectively. The exact expressions for the a_i and b_i coefficients entering Eq. (6.13) are easy to work out. Expressions valid up to $\lambda_{R,D}^2$ order are given in the next subsection.

The $\lambda_R = 0$ case can be worked out similarly. One obtains the secular equation

$$(n + \beta - \varepsilon)(n - 1 + \alpha - \varepsilon) - n\tilde{\lambda}_D^2 = 0, \tag{6.14}$$

which yields

$$\varepsilon_n^\pm = n \pm \sqrt{\frac{1}{4}\left(1 - \frac{\omega_L}{\omega_c}\right)^2 + \frac{2}{\omega_c}\lambda_D^2 n}. \tag{6.15}$$

One also obtains

$$(n - 1 + \alpha - \varepsilon_n^\pm)\, b_{n-1}^{\varepsilon_n^\pm} = -\tilde{\lambda}_D \sqrt{n}\, a_n^{\varepsilon_n^\pm}, \tag{6.16}$$

which together with the normalization condition $|a_n^{\varepsilon_n^\pm}|^2 + |b_{n-1}^{\varepsilon_n^\pm}|^2 = 1$ solves exactly the problem (in this case, for $n = 0$, $b_{-1} = 0$ and $a_0 = 1$).

Again, in the series expansion of the spinor $|\phi\rangle$, only one a_i and one b_i coefficient appear:

$$|n_d\rangle = \begin{pmatrix} a_{n+1}^{\varepsilon_{n+1}^-}|n + 1\rangle \\ b_n^{\varepsilon_{n+1}^-}|n\rangle \end{pmatrix}; \quad |n_u\rangle = \begin{pmatrix} a_n^{\varepsilon_n^+}|n\rangle \\ b_{n-1}^{\varepsilon_n^+}|n - 1\rangle \end{pmatrix}, \tag{6.17}$$

and the same comments as before apply.

6.2.2 The General Case Where $\lambda_R \neq 0$ and $\lambda_D \neq 0$

If the two terms are simultaneously considered, the SO interaction couples the states of all Landau levels, and an exact analytical solution to Eqs. (6.9) is unknown, and likely does not exist. We are going to find an approximate solution that in the $\lambda_{R,D}^2/\omega_c \ll 1$ limit coincides with the results of second order perturbation theory, i.e. it is valid up to $\tilde{\lambda}_{R,D}^2$ order, and it is quite accurate as compared with exact results obtained numerically. Combining Eqs. (6.9), one can write

$$\left[n + \alpha - \varepsilon - \tilde{\lambda}_R^2 \frac{n}{n-1+\beta-\varepsilon} - \tilde{\lambda}_D^2 \frac{n+1}{n+1+\beta-\varepsilon} \right] b_n$$
$$= -i\tilde{\lambda}_R\tilde{\lambda}_D \left[\frac{\sqrt{n(n-1)}}{n-1+\beta-\varepsilon} b_{n-2} - \frac{\sqrt{(n+1)(n+2)}}{n+1+\beta-\varepsilon} b_{n+2} \right] \quad (6.18)$$

and

$$\left[n + \beta - \varepsilon - \tilde{\lambda}_R^2 \frac{n+1}{n+1+\alpha-\varepsilon} - \tilde{\lambda}_D^2 \frac{n}{n-1+\alpha-\varepsilon} \right] a_n$$
$$= -i\tilde{\lambda}_R\tilde{\lambda}_D \left[\frac{\sqrt{n(n-1)}}{n-1+\alpha-\varepsilon} a_{n-2} - \frac{\sqrt{(n+1)(n+2)}}{n+1+\alpha-\varepsilon} a_{n+2} \right]. \quad (6.19)$$

The approximate solution is obtained by taking $a_{n-2} = a_{n+2} = b_{n-2} = b_{n+2} = 0$ in the above equations. This means that for each level $|n\rangle$, the SO interaction is allowed to couple it only with the $|n-1\rangle$ and $|n+1\rangle$ levels. This solution, which consists of a $|n_d\rangle$ and a $|n_u\rangle$ spinor, is therefore obtained by solving first the secular, cubic equation

$$(n+\alpha-\varepsilon)(n-1+\beta-\varepsilon)(n+1+\beta-\varepsilon)$$
$$= \tilde{\lambda}_R^2 n(n+1+\beta-\varepsilon) + \tilde{\lambda}_D^2(n+1)(n-1+\beta-\varepsilon). \quad (6.20)$$

Together with the equations

$$(n-1+\beta-\varepsilon)a_{n-1} = -i\tilde{\lambda}_R\sqrt{n}b_n,$$
$$(n+1+\beta-\varepsilon)a_{n+1} = -\tilde{\lambda}_D\sqrt{n+1}b_n, \quad (6.21)$$

and the normalization condition $|a_{n-1}|^2 + |a_{n+1}|^2 + |b_n|^2 = 1$, they determine the $|n_d\rangle$ solution. The solution corresponding to the $|n_u\rangle$ spinor is obtained by solving the secular equation

$$(n+\beta-\varepsilon)(n-1+\alpha-\varepsilon)(n+1+\alpha-\varepsilon) = \tilde{\lambda}_R^2(n+1)(n-1+\alpha-\varepsilon)$$
$$+ \tilde{\lambda}_D^2 n(n+1+\alpha-\varepsilon). \quad (6.22)$$

Together with the equations

$$
\begin{aligned}
(n - 1 + \alpha - \varepsilon)b_{n-1} &= -\tilde{\lambda}_D \sqrt{n}\, a_n, \\
(n + 1 + \alpha - \varepsilon)b_{n+1} &= i\tilde{\lambda}_R \sqrt{n+1}\, a_n,
\end{aligned}
\tag{6.23}
$$

and $|a_n|^2 + |b_{n-1}|^2 + |b_{n+1}|^2 = 1$, they determine the $|n_d\rangle$ solution.

Since all the estimates available in the literature [see, for example, Tonello and Lipparini (2004), Könemann *et al.* (2005), Malet *et al.* (2006)] yield $\lambda_{R,D}^2$ ($\lambda_{R,D}^2 m/\hbar^2$) values of the order of 10 μeV, and ω_c ($\hbar\omega_c$) in GaAs is of the order of the meV even at small B (~ 1 T), it is worth examining the above solutions in the $\tilde{\lambda}_{R,D}^2 = 2\lambda_{R,D}^2/\omega_c \ll 1$ limit, in which the secular equations have solutions easy to interpret.

To order $\tilde{\lambda}_{R,D}^2$, the relevant solution to Eq. (6.20) containing both SO terms is

$$
\varepsilon_n^d = n + \alpha + 2n\frac{\lambda_R^2}{\omega_c + \omega_L} - 2(n+1)\frac{\lambda_D^2}{\omega_c - \omega_L},
\tag{6.24}
$$

which corresponds to the spinor $|n_d\rangle$,

$$
|n_d\rangle = \begin{pmatrix} a_{n-1}^{\varepsilon_n^d}|n - 1\rangle + a_{n+1}^{\varepsilon_n^d}|n + 1\rangle \\ b_n^{\varepsilon_n^d}|n\rangle \end{pmatrix},
\tag{6.25}
$$

with coefficients

$$
\begin{aligned}
a_{n-1}^{\varepsilon_n^d} &= i\tilde{\lambda}_R \sqrt{n}\,\frac{\omega_c}{\omega_c + \omega_L}, \\
a_{n+1}^{\varepsilon_n^d} &= -\tilde{\lambda}_D \sqrt{n+1}\,\frac{\omega_c}{\omega_c - \omega_L}, \\
b_n^{\varepsilon_n^d} &= 1 - \frac{1}{2}\tilde{\lambda}_R^2 n\left(\frac{\omega_c}{\omega_c + \omega_L}\right)^2 - \frac{1}{2}\tilde{\lambda}_D^2(n+1)\left(\frac{\omega_c}{\omega_c - \omega_L}\right)^2.
\end{aligned}
\tag{6.26}
$$

In the following, we will refer to this solution as the quasi-spin-down (qdown) solution, since in the zero spin–orbit coupling limit $|n_d\rangle$ becomes $|n\rangle\binom{0}{1}$. Analogously, Eq. (6.22) has the solution

$$
\varepsilon_n^u = n + \beta - 2(n+1)\frac{\lambda_R^2}{\omega_c + \omega_L} + 2n\frac{\lambda_D^2}{\omega_c - \omega_L},
\tag{6.27}
$$

which corresponds to the spinor $|n_u\rangle$,

$$
|n_u\rangle = \begin{pmatrix} a_n^{\varepsilon_n^u}|n\rangle \\ b_{n-1}^{\varepsilon_n^u}|n - 1\rangle + b_{n+1}^{\varepsilon_n^u}|n + 1\rangle \end{pmatrix},
\tag{6.28}
$$

with coefficients

$$b_{n-1}^{\varepsilon_n^u} = \tilde{\lambda}_D \sqrt{n} \frac{\omega_c}{\omega_c - \omega_L},$$

$$b_{n+1}^{\varepsilon_n^u} = i\tilde{\lambda}_R \sqrt{n+1} \frac{\omega_c}{\omega_c + \omega_L},$$

$$a_n^{\varepsilon_n^u} = 1 - \frac{1}{2}\tilde{\lambda}_R^2(n+1)\left(\frac{\omega_c}{\omega_c + \omega_L}\right)^2 - \frac{1}{2}\tilde{\lambda}_D^2 n\left(\frac{\omega_c}{\omega_c - \omega_L}\right)^2. \qquad (6.29)$$

In the following, we will refer to this solution as the quasi-spin-up (qup) solution, since in the zero SO coupling limit $|n_u\rangle$ becomes $|n\rangle\binom{0}{1}$. When either λ_R or λ_D is zero, Eqs. (6.25) and (6.28) reduce to the exact equations (6.13) and (6.17), respectively, and the corresponding a_i and b_i coefficients, valid up to order $\lambda_{R,D}$, can be extracted from Eqs. (6.26) and (6.29). These equations show that a_n and b_n are of order $O(1)$, whereas $a_{n\pm 1}$ and $b_{n\pm 1}$ are of order $O(\lambda_{R,D})$, and $a_{n\pm 2}$ and $b_{n\pm 2}$ are of order $O(\lambda_{R,D}^2)$. This shows that the neglected terms in Eqs. (6.18) and (6.19) are of order $O(\lambda_{R,D}^4)$.

The sp energies obtained from Eqs. (6.24) and (6.27), valid in the $\lambda_{R,D}^2/\omega_c \ll 1$ limit, are

$$E_n^d = \left(n + \frac{1}{2}\right)\omega_c + \frac{\omega_L}{2} + 2n\lambda_R^2 \frac{\omega_c}{\omega_c + \omega_L} - 2(n+1)\lambda_D^2 \frac{\omega_c}{\omega_c - \omega_L},$$

$$\qquad (6.30)$$

$$E_n^u = \left(n + \frac{1}{2}\right)\omega_c - \frac{\omega_L}{2} - 2(n+1)\lambda_R^2 \frac{\omega_c}{\omega_c + \omega_L} + 2n\lambda_D^2 \frac{\omega_c}{\omega_c - \omega_L}.$$

These sp energies coincide with the ones that can be derived from second order perturbation theory with the standard expression

$$E_n^{(2)} = \frac{1}{4}\sum_{m \neq n} \frac{|\langle m| - i\tilde{\lambda}_R\,\omega_c(a^+\sigma_- - a^-\sigma_+) + \tilde{\lambda}_D\,\omega_c(a^+\sigma_+ + a^-\sigma_-)|n\rangle|^2}{E_n^0 - E_m^0}, \qquad (6.31)$$

where $|n\rangle = |n,\uparrow\rangle$, $|n,\downarrow\rangle$ are the spin-up and spin-down eigenstates of the sp Hamiltonian $\frac{1}{2}(a^+a^- + a^-a^+)\omega_c - \frac{1}{2}\omega_L\sigma_z$ with eigenvalues $E_n^0(\uparrow) = (n+\frac{1}{2})\omega_c - \frac{1}{2}\omega_L$ and $E_n^0(\downarrow) = (n+\frac{1}{2})\omega_c + \frac{1}{2}\omega_L$, respectively.

The approximate solution is very accurate in the high B limit (see below). It also carries an interesting piece of information in the opposite limit of vanishing B. In this limit ($\omega_L, \omega_c \ll \lambda_{R,D}^2$), Eqs. (6.20) and (6.22) yield the solutions

$$E_n^d = \sqrt{2\omega_c[n\lambda_R^2 + (n+1)\lambda_D^2]},$$

$$\qquad (6.32)$$

$$E_n^u = \sqrt{2\omega_c[(n+1)\lambda_R^2 + n\lambda_D^2]},$$

which show that, at $B \simeq 0$, to order $\lambda_{R,D}^2$, the Landau levels are not split owing to the SO interaction, as one might naively infer from Eqs. (6.30). Another merit of the approximate solution is that it shows in a transparent way the interplay between the three spin-dependent interactions, namely Zeeman, Rashba and Dresselhaus.

Such interplay has been also discussed by Malet *et al.* (2006), in relation with the violation of the Larmor theorem due to the SO couplings, and by Valín-Rodríguez and Nazmitdinov (2006), where the Zeeman and SO interplay is discussed using the unitarily transformed Hamiltonian technique. Note also that in GaAs quantum wells, due to the sign of g^*, the lowest energy level is the qup one at the energy $E_0^u = \frac{1}{2}\omega_c - \frac{1}{2}\omega_L - 2\lambda_R^2 \omega_c/(\omega_c + \omega_L)$, containing the Rashba contribution alone, whereas the following level is the qdown one at the energy $E_0^d = \frac{1}{2}\omega_c + \frac{1}{2}\omega_L - 2\lambda_D^2 \omega_c/(\omega_c - \omega_L)$, containing the Dresselhaus contribution alone. For all the other levels, both SO terms contribute to the level energies.

We have assessed the quality of the above analytical solutions, Eqs. (6.30), by comparing them with exact numerical results for some particular cases. Indeed, the exact solution to Eqs. (6.9) can be obtained in the truncated space spanned by the lower $\mathcal{N}$ oscillator levels. Mathematically, Eqs. (6.9) are then cast into a linear eigenvalue problem of the type

$$\mathbf{M} \begin{pmatrix} \mathbf{a} \\ \mathbf{b} \end{pmatrix} = \varepsilon \begin{pmatrix} \mathbf{a} \\ \mathbf{b} \end{pmatrix}, \tag{6.33}$$

where $\mathbf{M}$ is a $2\mathcal{N} \times 2\mathcal{N}$ matrix, while $\mathbf{a}$ and $\mathbf{b}$ are column vectors made with the sets of coefficients $\{a_n, n = 0, \ldots, \mathcal{N} - 1\}$ and $\{b_n, n = 0, \ldots, \mathcal{N} - 1\}$, respectively. We have diagonalized $\mathbf{M}$ using a large enough $\mathcal{N}$ to ensure good convergence in the lower eigenvalues. The top panel of Fig. 6.1 displays a comparison of numerical (symbols) and analytical (solid lines) energies as a function of the Rashba SO strength, for a fixed Dresselhaus strength, both in units of ω_c, namely $y_R = \lambda_R^2/\omega_c$ and $y_D = \lambda_D^2/\omega_c = 0.01$. The chosen values for y_D and y_R are within the expected range for a GaAs quantum well. For instance, if $m\lambda_{R,D}^2/\hbar^2 \sim 10\,\mu\text{eV}$ and $B \sim 1$ T, $(m\lambda_{R,D}^2/\hbar^2)/\hbar\omega_c \sim 10^{-2}$. There is excellent agreement between analytical and numerical results, differences starting to be visible only for strong Rashba intensities and high Landau bands. Actually, in Fig. 6.1, the largest value of the adimensional ratio between Rashba SO and cyclotron energy $y_R = \lambda_R^2/\omega_c$ is 0.05, small enough to validate the analytical expression. Notice, however, that for larger y_R values (not shown in the figure), i.e. for small enough B, Eqs. (6.30) no longer reproduce the numerical results. For GaAs this happens for magnetic fields below 0.1 T. Similarly, the bottom panel of Fig. 6.1 displays a comparison of numerical (symbols) and analytical (solid lines) energies as a function of the Dresselhaus SO strength y_D, for a fixed Rashba strength, $y_R = \lambda_D^2/\omega_c = 0.01$. For every Landau level, both panels show a crossing between the $|n_u\rangle$ state, which is at lower energy for $y_{R,D} \ll 0.01$ because $g^* < 0$, and the $|n_d\rangle$ state, which eventually lies lower in energy. This crossing is due to the interplay between the two SO terms.

Figure 6.2 compares the a_n and b_n amplitudes from the numerical diagonalization with the analytical result of Eq. (6.29). For this purpose, we have chosen the rightmost qup states of the third Landau band in both panels of Fig. 6.1. These are the states with the largest SO intensities in Fig. 6.1. Note that even for these largest

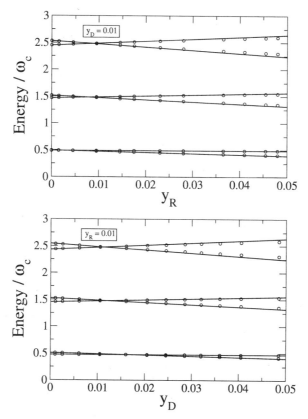

Fig. 6.1 Top panel: Lower energy levels for a GaAs quantum well as a function of the Rashba intensity $y_R = \lambda_R^2/\omega_c$ for a fixed Dresselhaus intensity, $y_D = \lambda_D^2/\omega_c = 0.01$. The solid lines show the analytical result, Eq. (6.30), while symbols correspond to the exact diagonalization, Eq. (6.33). Bottom panel: Lower energy levels for a GaAs quantum well as a function of the Dresselhaus intensity $y_D = \lambda_D^2/\omega_c$ for a fixed Rashba intensity, $y_R = \lambda_R^2/\omega_c = 0.01$. The solid lines show the analytical result, Eq. (6.30), while symbols correspond to the exact diagonalization, Eq. (6.33).

SO couplings the analytical prediction is still excellent since the a_2 numerical and analytical amplitudes are very close, and there are only small a_0 and a_4 numerical corrections. For the b_n amplitudes the comparison is also quite good and there are no relevant numerical corrections for n different from 1 and 3, as predicted by Eq. (6.29). For the qdown states similar results are found.

6.3 Single-Particle Level Transitions in Quantum Wells

We can use the preceding results to study the sp transitions induced in the system by the interaction with a left-circular polarized electromagnetic wave propagating along the z direction, i.e. perpendicular to the plane of motion of the

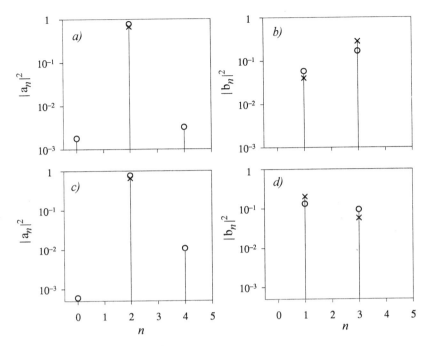

Fig. 6.2 Histograms with the a_n and b_n amplitudes for the rightmost qup state of the third Landau band of Fig. 6.1. Panels (a) and (b) are for $y_R = 0.05$ and $y_D = 0.01$, while (c) and (d) correspond to $y_R = 0.01$ and $y_D = 0.05$, respectively. The circles and crosses correspond to numerical and analytical results, i.e. to Eq. (6.33) and Eq. (6.29), respectively.

electrons, whose vector potential is $\mathbf{A}(t) = 2A(\cos\theta\,\hat{i} + \sin\theta\,\hat{j})$, with $\theta = \omega t - qz$. The sp interaction Hamiltonian $\mathbf{J} \cdot \mathbf{A}/c + g^*\mu_B\,\mathbf{s} \cdot (\nabla \times \mathbf{A})$, where $\mathbf{J} = e\mathbf{v}/\sqrt{\epsilon}$, reads

$$h_{\text{int}} = \frac{e}{c\sqrt{\epsilon}}A\left(v_-e^{i\theta} + v_+e^{-i\theta}\right) + \frac{1}{2}g^*\mu_B qA\left(\sigma_-e^{i\theta} + \sigma_+e^{-i\theta}\right), \qquad (6.34)$$

where the velocity operator $v_\pm$ is defined as $v_\pm \equiv -i[x \pm iy, H] = P^\pm \pm i\lambda_R\sigma_\pm + \lambda_D\sigma_\mp$.

The Hamiltonian h_{int} can be rewritten as

$$h_{\text{int}} = \frac{e}{c\sqrt{\epsilon}}A\sqrt{2\omega_c}\left(\alpha^-e^{i\theta} + \alpha^+e^{-i\theta}\right) + \frac{1}{2}g^*\mu_B qA\left(\sigma_-e^{i\theta} + \sigma_+e^{-i\theta}\right), \qquad (6.35)$$

where the operators α^+ and α^- acting on the spinor $|\phi\rangle$ are

$$\alpha^+ = \begin{bmatrix} a^+ & i\tilde{\lambda}_R \\ \tilde{\lambda}_D & a^+ \end{bmatrix}, \quad \alpha^- = \begin{bmatrix} a^- & \tilde{\lambda}_D \\ -i\tilde{\lambda}_R & a^- \end{bmatrix}. \qquad (6.36)$$

In the dipole approximation ($q \approx 0$), the charge-density excitation operator is $v_\pm$. We note that, even in the presence of e–e interactions, this operator satisfies the f-sum rule:

$$\sum_n \langle 0|x \mp iy|n\rangle\langle n| - iv_\pm|0\rangle = \sum_n \omega_{n0}|\langle n|x \pm iy|0\rangle|^2$$

$$= \frac{1}{2}\langle 0|[x \mp iy, [H, x \pm iy]]|0\rangle = 2N, \qquad (6.37)$$

where N is the electron number and ω_{n0} are the excitation energies.

We consider next several useful examples of sp matrix elements involving the operators α^+, which is proportional to v_+, and σ_-, and the qup and qdown sp states of Eqs. (6.25) and (6.28). For the operator α^+, we can write in general

$$\langle \psi|\alpha^+|\phi\rangle = \psi_1^* a^+ \phi_1 + i\tilde\lambda_R \psi_1^* \phi_2 + \tilde\lambda_D \psi_2^* \phi_1 + \psi_2^* a^+ \phi_2, \qquad (6.38)$$

and have to distinguish between qup–qup, qdown–qdown, qup–qdown, and qdown–qup transitions. The qup–qup and qdown–qdown transitions represent the usual cyclotron resonance (CR), and the qup–qdown and qdown–qup are related to spin-flip transitions.

Let us start with the qup–qup and qdown–qdown transitions. To the order $\lambda_{R,D}^2$ they are dominated by the transition $n \to n+1$ at the energies $E_{n+1}^d - E_n^d$ and $E_{n+1}^u - E_n^u$ with matrix elements $|\langle (n+1)_u|\alpha^+|n_u\rangle| = |\langle (n+1)_d|\alpha^+|n_d\rangle| = \sqrt{n+1}$. The energy splitting of the cyclotron resonance is

$$\Delta E_{CR} = \left|4\lambda_R^2 \frac{\omega_c}{\omega_c + \omega_L} - 4\lambda_D^2 \frac{\omega_c}{\omega_c - \omega_L}\right|. \qquad (6.39)$$

The α^+ excitation operator also induces a qup–qdown transition with energy $E_n^d - E_n^u$ and matrix element $|\langle n_d|\alpha^+|n_u\rangle| = \tilde\lambda_D \,\omega_L/(\omega_c - \omega_L)$. This is a spin-flip transition. In particular, when $n = 0$ it is related to the Larmor resonance at the energy

$$\Delta E_L = \omega_L + 2\left(\lambda_R^2 \frac{\omega_c}{\omega_c + \omega_L} - \lambda_D^2 \frac{\omega_c}{\omega_c - \omega_L}\right). \qquad (6.40)$$

Note that the transition matrix element is linear in $\tilde\lambda_D$, and that in the presence of the Rashba interaction alone, α^+ causes no spin-flip transition.

For the operator σ_- one gets $\langle\psi|\sigma_-|\phi\rangle = 2\psi_2^* \phi_1$. The dominant transition is the spin-flip excitation at energy $E_n^d - E_n^u$ with matrix element $|\langle n_d|\sigma_-|n_u\rangle| = 2$. The qup–qup and qdown–qdown cyclotron resonances at energies $E_{n+1}^d - E_n^d$ and $E_{n+1}^u - E_n^u$ are also excited with strengths $|\langle (n+1)_u|\sigma_-|n_u\rangle| = |\langle (n+1)_d|\sigma_-|n_d\rangle| = 2\tilde\lambda_D \sqrt{n+1}\,\omega_c/(\omega_c - \omega_L)$.

Other excitations that deserve some attention are those induced by the operators $\alpha^+\sigma_\pm$ and $\alpha^+\sigma_z$. They are detected in inelastic light scattering experiments as spin dipole resonances (Eriksson *et al.*, 1999) The operator $\alpha^+\sigma_z$ excites the same cyclotron states as α^+, at the energies $E_{n+1}^d - E_n^d$ and $E_{n+1}^u - E_n^u$, and with the same transition matrix element, $\sqrt{n+1}$. In contrast, the operator $\alpha^+\sigma_+$ mainly induces the transition from qdown to qup states at the energy $E_{n+1}^u - E_n^d$, whereas the operator $\alpha^+\sigma_-$ induces the transition from qup to qdown states at the energy

$E_{n+1}^d - E_n^u$. The transition matrix elements are given by $|\langle (n+1)_u | \alpha^+ \sigma_+ | n_d \rangle| = |\langle (n+1)_d | \alpha^+ \sigma_- | n_u \rangle| = 2\sqrt{n+1}$. We thus see that the dipole transitions between Landau levels $|n\rangle$ and $|n+1\rangle$ at "unperturbed" energies $E_{n+1} - E_n$ are split by the SO interaction, an effect that under some circumstances may be observed, as will be discussed in Subsection 6.8.

6.4   The Magnetic Field in Plane

If the magnetic field is parallel to the plane of the quantum well, the situation changes substantially. In this case, the orbital effect of the applied in-plane magnetic field is frozen owing to the strong confinement in the vertical direction produced by the electrostatic potential of the quantum well, and the only effect of the field appears through the Zeeman interaction. The quantum well Hamiltonian is given by

$$
H_0 \equiv \sum_{j=1}^{N} [h_0]_j = \sum_{j=1}^{N} \left[\frac{p^2}{2m} - \frac{1}{2}\omega_L (\cos\theta\sigma_x + \sin\theta\sigma_y) \right.
$$
$$
\left. + \frac{\lambda_R}{\hbar}(p_y\sigma_x - p_x\sigma_y) + \frac{\lambda_D}{\hbar}(p_x\sigma_x - p_y\sigma_y) \right]_j , \qquad (6.41)
$$

where the angle θ represents the azimuthal orientation of the magnetic field in the (x, y) plane. Hamiltonian (6.41) preserves the translational invariance and the momentum is conserved. The solution to the single-particle Schrödinger equation $h\varphi = \epsilon\varphi$ is translationally invariant in the (x, y) plane and has the form

$$
\varphi = e^{i\mathbf{k}\cdot\mathbf{r}} \begin{pmatrix} \alpha \\ \beta \end{pmatrix}, \qquad (6.42)
$$

where $\begin{pmatrix} \alpha \\ \beta \end{pmatrix}$ is a spinor to be determined. One gets (Chang, 2005; Valín-Rodríguez and Nazmitdinov, 2006)

$$
\frac{k^2}{2m}\begin{pmatrix} \alpha \\ \beta \end{pmatrix} + \lambda_R \left[k_y \begin{pmatrix} \beta \\ \alpha \end{pmatrix} + ik_x \begin{pmatrix} \beta \\ -\alpha \end{pmatrix} \right] + \lambda_D \left[k_x \begin{pmatrix} \beta \\ \alpha \end{pmatrix} + ik_y \begin{pmatrix} \beta \\ -\alpha \end{pmatrix} \right]
$$
$$
- \frac{1}{2}\omega_L \left[\cos\theta \begin{pmatrix} \beta \\ \alpha \end{pmatrix} - i\sin\theta \begin{pmatrix} \beta \\ -\alpha \end{pmatrix} \right] = \epsilon \begin{pmatrix} \alpha \\ \beta \end{pmatrix}, \qquad (6.43)
$$

from which the following spectrum is recovered:

$$
\epsilon_{\mathbf{k}\pm} = \frac{k^2}{2m} \pm \left[A^2 + B^2 \right]^{1/2},
$$

$$
A = \lambda_R k_y + \lambda_D k_x - \frac{1}{2}\omega_L \cos\theta, \qquad (6.44)
$$

$$
B = \lambda_R k_x + \lambda_D k_y + \frac{1}{2}\omega_L \sin\theta.
$$

A careful analysis of this solution has been made by Valín-Rodríguez and Nazmitdinov (2006). The interplay between the Zeeman interaction and the SO couplings induces an anisotropic effective gyromagnetic factor that would lead to the observation of a continuum of Zeeman sublevels in the spin structure of the spectrum. Depending on the relative orientation between momentum and the applied field, the value of the effective g factor can change its sign with respect to the initial bulk value.

6.5 Spin–Orbit Effects on Single-Particle States in Quantum Dots

A description of SO effects on the single-particle states of quantum dots can be simply obtained through the findings of Subsection 6.2 by adding to the quantum well Hamiltonian a harmonic confinement potential $1/2mw_0^2r^2$ and working in the symmetric gauge, $\mathbf{A} = B/2(-y, x, 0)$, with $\mathbf{B} = \nabla \times \mathbf{A} = B\hat{z}$.

Introducing the operators

$$a^{\pm} = \frac{1}{\sqrt{2\Omega}}\left(P^{\pm} \pm \frac{i}{2}\Omega(1-\gamma)Q^{\pm}\right),$$

$$b^{\pm} = \frac{1}{\sqrt{2\Omega}}\left(P^{\mp} \pm \frac{i}{2}\Omega(1+\gamma)Q^{\mp}\right),$$

(6.45)

with $Q^{\pm} = x \pm iy$, $[b^-, b^+] = 1$, $[a^-, a^+] = 1$, $\Omega = \sqrt{w_c^2 + 4w_0^2}$, $\gamma = w_c/\Omega$, $w_c = eB/mc$ being the cyclotron frequency, the sp dot Hamiltonian h_0 can be rewritten as

$$\frac{h_0}{\Omega} = \frac{1}{2} + \frac{1+\gamma}{2}a^+a^- + \frac{1-\gamma}{2}b^+b^- - \frac{1}{2}\frac{w_L}{\Omega}\sigma_z$$

$$- \frac{1}{4}i\tilde{\lambda}_R\left[(1+\gamma)(a^+\sigma_- - a^-\sigma_+) + (1-\gamma)(b^-\sigma_- - b^+\sigma_+)\right]$$

$$+ \frac{1}{4}\tilde{\lambda}_D\left[(1+\gamma)(a^+\sigma_+ + a^-\sigma_-) + (1-\gamma)(b^-\sigma_+ + b^+\sigma_-)\right], \quad (6.46)$$

where $w_L = |g^*\mu_B B|$ is the Larmor frequency and $\tilde{\lambda}_{R,D} = \lambda_{R,D}\sqrt{\frac{2}{\Omega}}$. For the spinor $|\phi\rangle \equiv \binom{\phi_1}{\phi_2}$, the Schrödinger equation $h_0|\phi\rangle = \varepsilon|\phi\rangle$ adopts the form

$$\left[-\frac{i}{2}\tilde{\lambda}_R\left((1+\gamma)a^+ + (1-\gamma)b^-\right) + \frac{1}{2}\tilde{\lambda}_D\left((1+\gamma)a^- + (1-\gamma)b^+\right)\right]\phi_1$$

$$+ \left[\frac{1}{2}(a^-a^+ + \gamma a^+a^- + (1-\gamma)b^+b^-) + w_L/(2\Omega) - \varepsilon\right]\phi_2 = 0, \quad (6.47)$$

$$\left[\frac{1}{2}(a^-a^+ + \gamma a^+a^- + (1-\gamma)b^+b^-) - w_L/(2\Omega) - \varepsilon\right]\phi_1$$

$$+ \left[\frac{i}{2}\tilde{\lambda}_R\left((1+\gamma)a^- + (1-\gamma)b^+\right) + \frac{1}{2}\tilde{\lambda}_D\left((1+\gamma)a^+ + (1-\gamma)b^-\right)\right]\phi_2 = 0.$$

(6.48)

We now expand ϕ_1 and ϕ_2 into oscillator states $|n, m\rangle$ as $\phi_1 = \sum_{n,m=0}^{\infty} a_{n,m} |n, m\rangle$, $\phi_2 = \sum_{n,m=0}^{\infty} b_{n,m} |n, m\rangle$, on which a^+, a^- and b^+, b^- act in the usual way, i.e. $a^+ |n, m\rangle = \sqrt{n+1} |n+1, m\rangle$, $a^- |n, m\rangle = \sqrt{n} |n-1, m\rangle$, and $a^- |0, m\rangle = 0$, $b^+ |n, m\rangle = \sqrt{m+1} |n, m+1\rangle$, $b^- |n, m\rangle = \sqrt{m} |n, m-1\rangle$, and $b^- |n, 0\rangle = 0$. Remember also that the angular momentum operator is given by

$$\hat{L} = a^+ a^- - b^+ b^- \tag{6.49}$$

with eigenvalues

$$\hat{L} |n, m\rangle = (n - m) |n, m\rangle. \tag{6.50}$$

Making a comparison with the representation in cylindrical coordinates (r, θ), where the motion in the angular variable separates, and the appropriate component of the angular momentum ℓ is a good quantum number taking the values $\ell = -\tilde{n}, -\tilde{n} + 2, \ldots, \tilde{n} - 2, \tilde{n}$, where $\tilde{n} = 0, 1, \ldots$ is the principal quantum number, and $n_r = (\tilde{n} - |\ell|)/2$ is the radial quantum number, the pair of quantum numbers $(\tilde{n}, \ell)$ and (n, m) are related by

$$\tilde{n} = n + m, \quad \ell = n - m. \tag{6.51}$$

We get the infinite system of equations

$$\left(\frac{1+\gamma}{2} n + \frac{1-\gamma}{2} m + \alpha - \varepsilon \right) b_{n,m} - \frac{i}{2} \tilde{\lambda}_R \left((1+\gamma)\sqrt{n} a_{n-1,m} \right.$$

$$+ (1-\gamma)\sqrt{m+1} a_{n,m+1} \right)$$

$$+ \frac{1}{2} \tilde{\lambda}_D \left((1+\gamma)\sqrt{n+1} a_{n+1,m} + (1-\gamma)\sqrt{m} a_{n,m-1} \right) = 0, \tag{6.52}$$

$$\left(\frac{1+\gamma}{2} n + \frac{1-\gamma}{2} m + \beta - \varepsilon \right) a_{n,m} + \frac{i}{2} \tilde{\lambda}_R \left((1+\gamma)\sqrt{n+1} b_{n+1,m} \right.$$

$$+ (1-\gamma)\sqrt{m} b_{n,m-1} \right) + \frac{1}{2} \tilde{\lambda}_D \left((1+\gamma)\sqrt{n} b_{n-1,m} \right.$$

$$+ (1-\gamma)\sqrt{m+1} b_{n,m+1} \right) = 0, \tag{6.53}$$

for $n, m \geq 0$, with $a_{-1} = 0$, $b_{-1} = 0$, and $\alpha = (1 + \omega_L/\Omega)/2$, $\beta = (1 - \omega_L/\Omega)/2$.

In the limits $\lambda_{R,D}^2/\Omega \ll 1$, from Eqs. (6.52) and (6.53) one can derive an approximate but very accurate analytical solution which is valid at the order $\lambda_{R,D}^2/\Omega$. Note that, owing to the confinement, even at $B = 0$ when $\Omega = \omega_0$ this approximation should be very good, $\lambda_{R,D}^2$ being of the order of the μ eV and ω_0 of the meV. This approximate solution is obtained by neglecting in these equations all the terms except those which couple, through the SO interaction, each level $|n, m\rangle$ only with the $|n-1, m\rangle$ and $|n+1, m\rangle$ or the $|n, m-1\rangle$ and $|n, m+1\rangle$ levels. In terms of the quantum numbers $\tilde{n}, \ell$, in the above approximation, one mixes each level $|\tilde{n}, \ell\rangle$ only with the $|\tilde{n}-1, \ell-1\rangle$, $|\tilde{n}+1, \ell+1\rangle$ and the $|\tilde{n}+1, \ell-1\rangle$ and $|\tilde{n}-1, \ell+1\rangle$ levels. It is possible to see that the SO coupling with all the other levels is at minimum of order $(\lambda_{R,D}^2/\Omega)^2$. In this approximation, combining Eqs. (6.52) and (6.53) we can write

$$
\left(\gamma_+ n + \gamma_- m + \alpha - \varepsilon - \tilde{\lambda}_R^2 \right.
$$

$$
\times \left[\frac{\gamma_+^2 n}{\gamma_+(n-1)+\gamma_- m+\beta-\varepsilon} + \frac{\gamma_-^2(m+1)}{\gamma_+ n+\gamma_-(m+1)+\beta-\varepsilon} \right]
$$

$$
\left. - \tilde{\lambda}_D^2 \left[\frac{\gamma_+^2(n+1)}{\gamma_+(n+1)+\gamma_- m+\beta-\varepsilon} + \frac{\gamma_-^2 m}{\gamma_+ n+\gamma_-(m-1)+\beta-\varepsilon} \right] \right) b_{n,m} = 0
$$

$$(6.54)$$

and

$$
\left(\gamma_+ n + \gamma_- m + \beta - \varepsilon - \tilde{\lambda}_R^2 \right.
$$

$$
\times \left[\frac{\gamma_+^2(n+1)}{\gamma_+(n+1)+\gamma_- m+\alpha-\varepsilon} + \frac{\gamma_-^2 m}{\gamma_+ n+\gamma_-(m-1)+\alpha-\varepsilon} \right]
$$

$$
\left. - \tilde{\lambda}_D^2 \left[\frac{\gamma_+^2 n}{\gamma_+(n-1)+\gamma_- m+\alpha-\varepsilon} + \frac{\gamma_-^2(m+1)}{\gamma_+ n+\gamma_-(m+1)+\alpha-\varepsilon} \right] \right) a_{n,m} = 0,
$$

$$(6.55)$$

where we have defined $\gamma_+ = (1+\gamma)/2$ and $\gamma_- = (1-\gamma)/2$.

To order $\tilde{\lambda}_{R,D}^2$, the relevant solution to Eq. (6.54) containing both SO terms is

$$
\varepsilon_{n_d,m_d} = \gamma_+ n + \gamma_- m + \alpha + 2\lambda_R^2 \left[\frac{\gamma_+^2 n}{\gamma_+\Omega+\omega_L} - \frac{\gamma_-^2(m+1)}{\gamma_-\Omega-\omega_L} \right]
$$

$$
- 2\lambda_D^2 \left[\frac{\gamma_+^2(n+1)}{\gamma_+\Omega-\omega_L} - \frac{\gamma_-^2 m}{\gamma_-\Omega+\omega_L} \right],
$$

$$(6.56)$$

which corresponds to the spinor $|n_d, m_d\rangle$,

$$
|n_d, m_d\rangle = \begin{pmatrix} a_{n-1,m}^{\varepsilon_{n_d,m_d}} |n-1,m\rangle + a_{n+1,m}^{\varepsilon_{n_d,m_d}} |n+1,m\rangle + a_{n,m-1}^{\varepsilon_{n_d,m_d}} |n,m-1\rangle + a_{n,m+1}^{\varepsilon_{n_d,m_d}} |n,m+1\rangle \\ b_{n,m}^{\varepsilon_{n_d,m_d}} |n,m\rangle \end{pmatrix},
$$

$$(6.57)$$

that coefficients

$$
a_{n-1,m}^{\varepsilon_{n_d,m_d}} = i\tilde{\lambda}_R \frac{\gamma_+\Omega\sqrt{n}}{\gamma_+\Omega+\omega_L},
$$

$$
a_{n+1,m}^{\varepsilon_{n_d,m_d}} = -\tilde{\lambda}_D \frac{\gamma_+\Omega\sqrt{n+1}}{\gamma_+\Omega-\omega_L},
$$

$$a_{n,m-1}^{\varepsilon_{n^d},m^d} = \tilde{\lambda}_D \frac{\gamma_- \Omega \sqrt{m}}{\gamma_- \Omega + \omega_L},$$

$$a_{n,m+1}^{\varepsilon_{n^d},m^d} = -i\tilde{\lambda}_R \frac{\gamma_- \Omega \sqrt{m+1}}{\gamma_- \Omega - \omega_L},$$

$$b_{n,m}^{\varepsilon_{n^d},m^d} = 1 - \frac{1}{2}\tilde{\lambda}_R^2 \Omega^2 \left[\frac{\gamma_+^2 n}{(\gamma_+ \Omega + \omega_L)^2} + \frac{\gamma_-^2 (m+1)}{(\gamma_- \Omega - \omega_L)^2} \right]$$
$$- \frac{1}{2}\tilde{\lambda}_D^2 \Omega^2 \left[\frac{\gamma_+^2 (n+1)}{(\gamma_+ \Omega - \omega_L)^2} + \frac{\gamma_-^2 m}{(\gamma_- \Omega + \omega_L)^2} \right]. \tag{6.58}$$

In the following, we will refer to this solution as the quasi-spin-down (qdown) solution, since in the zero spin–orbit coupling limit $|n_d, m_d\rangle$ becomes $|n, m\rangle \begin{pmatrix} 0 \\ 1 \end{pmatrix}$. Analogously, Eq. (6.55) has the solution

$$\varepsilon_{n^u,m^u} = \gamma_+ n + \gamma_- m + \beta - 2\lambda_R^2 \left[\frac{\gamma_+^2 (n+1)}{\gamma_+ \Omega + \omega_L} - \frac{\gamma_-^2 m}{\gamma_- \Omega - \omega_L} \right]$$
$$+ 2\lambda_D^2 \left[\frac{\gamma_+^2 n}{\gamma_+ \Omega - \omega_L} - \frac{\gamma_-^2 (m+1)}{\gamma_- \Omega + \omega_L} \right], \tag{6.59}$$

which corresponds to the spinor $|n_u, m_u\rangle$,

$$|n_u, m_u\rangle = \begin{pmatrix} a_{n,m}^{\varepsilon_{n^u},m^u} |n, m\rangle \\ b_{n-1,m}^{\varepsilon_{n^u},m^u} |n-1, m\rangle + b_{n+1,m}^{\varepsilon_{n^u},m^u} |n+1, m\rangle + b_{n,m-1}^{\varepsilon_{n^u},m^u} |n, m-1\rangle \\ + b_{n,m+1}^{\varepsilon_{n^u},m^u} |n, m+1\rangle \end{pmatrix}, \tag{6.60}$$

with coefficients

$$b_{n-1,m}^{\varepsilon_{n^u},m^u} = \tilde{\lambda}_D \frac{\gamma_+ \Omega \sqrt{n}}{\gamma_+ \Omega - \omega_L},$$

$$b_{n+1,m}^{\varepsilon_{n^u},m^u} = +i\tilde{\lambda}_R \frac{\gamma_+ \Omega \sqrt{n+1}}{\gamma_+ \Omega + \omega_L},$$

$$b_{n,m-1}^{\varepsilon_{n^u},m^u} = -i\tilde{\lambda}_R \frac{\gamma_- \Omega \sqrt{m}}{\gamma_- \Omega - \omega_L},$$

$$b_{n,m+1}^{\varepsilon_{n^u},m^u} = -\tilde{\lambda}_D \frac{\gamma_- \Omega \sqrt{m+1}}{\gamma_- \Omega + \omega_L}, \tag{6.61}$$

$$a_{n,m}^{\varepsilon_{n^u},m^u} = 1 - \frac{1}{2}\tilde{\lambda}_D^2 \Omega^2 \left[\frac{\gamma_+^2 n}{(\gamma_+ \Omega - \omega_L)^2} + \frac{\gamma_-^2 (m+1)}{(\gamma_- \Omega + \omega_L)^2} \right]$$
$$- \frac{1}{2}\tilde{\lambda}_R^2 \Omega^2 \left[\frac{\gamma_+^2 (n+1)}{(\gamma_+ \Omega + \omega_L)^2} + \frac{\gamma_-^2 m}{(\gamma_- \Omega - \omega_L)^2} \right].$$

In the following, we will refer to this solution as the quasi-spin-up (qup) solution, since in the zero SO coupling limit $|n_u, m_u\rangle$ becomes $|n, m\rangle \begin{pmatrix} 0 \\ 1 \end{pmatrix}$.

Note that, by neglecting in the energy denominators of Eqs. (6.56) and (6.59), ω_L with respect to $\gamma_\pm \Omega$, one gets the result

$$\epsilon_{n_r \ell s} = \frac{1}{2}(2n_r + |\ell| + 1)\sqrt{4\omega_0^2 + \omega_{cs}^2} + \frac{\omega_{cs}}{2}\ell - \frac{1}{2}\omega_L s, \qquad (6.62)$$

with $\omega_{cs} = \omega_c + 2(\lambda_R^2 - \lambda_D^2)s$ and $s = \pm 1$, obtained in the limit $\mathcal{H}_{R,D} \gg H_Z$ by Valín-Rodríguez et al. (2002) by means of unitary transformation techniques (see Appendix B). Note also that, due to the approximations made to get results (6.56) and (6.59) for the single-particle energies, an unphysical divergence can occur in these equations at a high magnetic field when $\gamma_- \Omega - \omega_L = 0$.

We have assessed the quality of the above analytical solutions, Eqs. (6.56)–(6.59), by comparing them with exact numerical results for some particular cases. Indeed, the exact solution to Eqs. (6.52) and (6.53) can be obtained numerically in the truncated space spanned by the lower $\mathcal{N}$, $\mathcal{M}$ oscillator levels. Figures 6.3–6.6

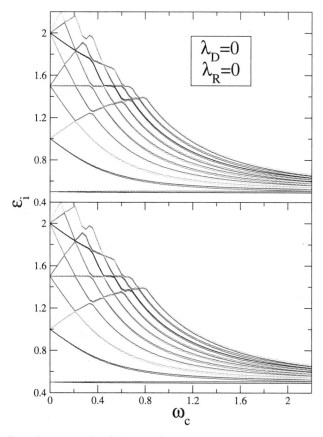

Fig. 6.3 Single-Particle energies for GaAs quantum dot with $\omega_0 = 0.3$ and $\lambda_D = \lambda_R = 0$ in dot units. Top panel: Analytical results from Eqs. (6.56) and (6.59). Lower panel: Numerical results from the exact diagonalization of Eqs. (6.52) and (6.53).

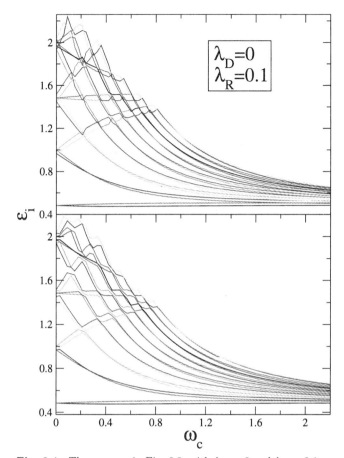

Fig. 6.4   The same as in Fig. 6.3, with $\lambda_D = 0$ and $\lambda_R = 0.1$.

display a comparison of numerical and analytical energies for some values of the Rashba and Dresselhaus strengths. The chosen values for $\lambda_{D,R}$ ($\lambda_{D,R}\sqrt{m}/\hbar$) are within the expected range for GaAs quantum dots. As one can see from the figures, there is excellent agreement between analytical and numerical results. Quite good agreement is also found among the values of the coefficients $a_{n,m}$ and $b_{n,m}$ calculated from the analitycal formulas and from the exact diagonalization.

As compared to the Fock–Darwin spectra of quantum dots without the SO coupling shown in Fig. 6.3, the most outstanding features in the energy spectra of quantum dots with the SO coupling (shown in Figs. 6.4 and 6.5) are the lifting of degeneracy at a vanishing magnetic field, rearrangement of some of the levels at small fields, and level repulsion at a higher magnetic field. When the Rashba and Dresselhaus strengths are equal, as in Fig. 6.6, the SO coupling introduces only a constant shift with respect the Fock–Darwin energies. This can be easily recovered through Eqs. (6.56)–(6.59) by just putting $\omega_L = 0$ in the energy denominators.

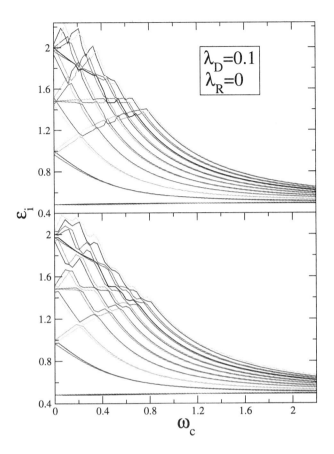

Fig. 6.5 The same as in Fig. 6.3, with $\lambda_D = 0.1$ and $\lambda_R = 0$.

In fact, in this limit one gets $(\lambda_D = \lambda_R \equiv \lambda)$

$$\varepsilon_{n^d, m^d} = \gamma_+ n + \gamma_- m + \alpha - 2\frac{\lambda^2}{\Omega}, \qquad (6.63)$$

$$\varepsilon_{n^u, m^u} = \gamma_+ n + \gamma_- m + \beta - 2\frac{\lambda^2}{\Omega}. \qquad (6.64)$$

6.5.1 *Single-Particle Transitions in Dots*

Analogously to what we have done in Subsection 6.3 for quantum wells, we can use the preceding results to study the sp transitions induced in the dot by the interaction with a left-circular polarized electromagnetic wave propagating along the z direction [see Eq. (6.34)]. By expressing the operators $P^\pm$ as a function of the $a^\pm$ and $b^\pm$ of Eq. (6.45), one gets

$$P^\pm = \sqrt{2\Omega}(\gamma_+ a^\pm + \gamma_- b^\mp), \qquad (6.65)$$

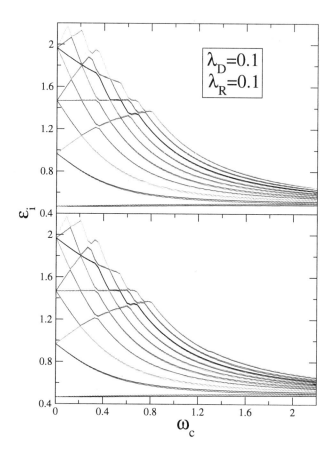

Fig. 6.6 The same as in Fig. 6.3, with $\lambda_D = 0.1$ and $\lambda_R = 0.1$.

and the Hamiltonian h_{int} can be rewritten as

$$h_{\text{int}} = \frac{e}{c\sqrt{\epsilon}} A \sqrt{2\Omega} \left(\alpha^- e^{i\theta} + \alpha^+ e^{-i\theta} \right) + \frac{1}{2} g^* \mu_B q A \left(\sigma_- e^{i\theta} + \sigma_+ e^{-i\theta} \right), \qquad (6.66)$$

where the operators α^+ and α^- which act on the spinor $|\phi\rangle$ are

$$\alpha^+ = \begin{bmatrix} \gamma_+ a^+ + \gamma_- b^- & i\tilde{\lambda}_R \\ \tilde{\lambda}_D & \gamma_+ a^+ + \gamma_- b^- \end{bmatrix}, \quad \alpha^- = \begin{bmatrix} \gamma_+ a^- + \gamma_- b^+ & \tilde{\lambda}_D \\ -i\tilde{\lambda}_R & \gamma_+ a^- + \gamma_- b^+ \end{bmatrix}.$$
$$(6.67)$$

Starting from the operators α^+ and σ_- and proceeding in the same way as in the quantum well case, one can calculate qup–qup, qdown–qdown, qup–qdown and qdown–qup transitions. As before, the qup–qup and qdown–qdown transitions represent the usual dipole resonance, and the qup–qdown and qdown–qup are related to spin-flip transitions. Let us start with the qup–qup and qdown–qdown transitions. To the order $\lambda_{R,D}^2$ they are dominated by the transitions $\Delta n = 1, \Delta m = 0$

with $\Delta\ell = 1$ and $\Delta n = 0, \Delta m = 1$ with $\Delta\ell = -1$. One gets for the energies of the qup and qdown dipole transitions

$$E_{n+1^d,m^d} - E_{n^d,m^d} = \frac{1}{2}(\Omega + \omega_c) + 2\lambda_R^2 \frac{\gamma_+^2 \Omega}{\gamma_+\Omega + \omega_L} - 2\lambda_D^2 \frac{\gamma_+^2 \Omega}{\gamma_+\Omega - \omega_L},$$

$$E_{n^d,m+1^d} - E_{n^d,m^d} = \frac{1}{2}(\Omega - \omega_c) - 2\lambda_R^2 \frac{\gamma_-^2 \Omega}{\gamma_-\Omega - \omega_L} + 2\lambda_D^2 \frac{\gamma_-^2 \Omega}{\gamma_-\Omega + \omega_L},$$

$$E_{n+1^u,m^u} - E_{n^u,m^u} = \frac{1}{2}(\Omega + \omega_c) - 2\lambda_R^2 \frac{\gamma_+^2 \Omega}{\gamma_+\Omega + \omega_L} + 2\lambda_D^2 \frac{\gamma_+^2 \Omega}{\gamma_+\Omega - \omega_L},$$

$$E_{n^u,m+1^u} - E_{n^u,m^u} = \frac{1}{2}(\Omega - \omega_c) + 2\lambda_R^2 \frac{\gamma_-^2 \Omega}{\gamma_-\Omega - \omega_L} - 2\lambda_D^2 \frac{\gamma_-^2 \Omega}{\gamma_-\Omega + \omega_L}.$$

$$(6.68)$$

The qup–qdown excitations are dominated by the transitions $\Delta n = 0, \Delta m = 0$ with $\Delta\ell = 0$. One gets for the energy of this spin-flip transition

$$E_{n^d,m^d} - E_{n^u,m^u} = \omega_L + 2\lambda_R^2 \left[\frac{\gamma_+^2 \Omega(2n+1)}{\gamma_+\Omega + \omega_L} - \frac{\gamma_-^2 \Omega(2m+1)}{\gamma_-\Omega - \omega_L} \right]$$

$$- 2\lambda_D^2 \left[\frac{\gamma_+^2 \Omega(2n+1)}{\gamma_+\Omega - \omega_L} - \frac{\gamma_-^2 \Omega(2m+1)}{\gamma_+\Omega - \omega_L} \right]. \quad (6.69)$$

The α^+ excitation operator induces both kinds of the above excitations, but with a matrix element which vanishes with λ in the case of the qup–qdown transitions with energy given by Eq. (6.69). Also, the spin-flip operator σ_- can excite both of the above transitions but in this case the matrix elements vanishing with λ are the ones relative to the qup–qup and qdown–qdown transitions given by Eqs. (6.68). Another excitation which deserves some attention in quantum dots is the one induced by the operator $\alpha^+\sigma_z$, which can be detected in Raman scattering experiments. In the absence of the Coulomb interaction, the operator $\alpha^+\sigma_z$ excites the same states as α^+ at the energies given by Eqs. (6.68); however, when the Coulomb interaction is switched on, a completely different collective low energy state, the magnon state, is originated (see Subsection 9.13.2).

When the SO coupling strength is put equal to zero, from the above equations one gets the usual results of the generalized Kohn and Larmor theorems (see Subsections 8.9.1 and 8.10):

$$\Delta E_{\Delta\ell=\pm1} = \frac{1}{2}(\Omega \pm \omega_c), \quad (6.70)$$

$$\Delta E_{\Delta\ell=0} = \omega_L. \quad (6.71)$$

As compared to the result of Eq. (6.70), the most important features in the results of Eqs. (6.68) with the SO coupling are the splitting of the two branches of excitation at a vanishing magnetic field with a subsequent crossing of spectra at a low magnetic field, and an anticrossing in the lower energy branch at a higher magnetic field, which is mainly due to the interplay between the Zeeman and SO couplings (Pietiläinen and Chakraborty, 2006). For GaAs this second effect is practically absent owing to the smallness of the Larmor frequency with respect to the values of ω_0 and ω_c considered here. In this case one can safely take $\omega_L = 0$ in the energy denominators of Eqs. (6.68) to get the simplified results

$$\Delta E_{\Delta\ell=+1} = \frac{1}{2}(\Omega + \omega_c)\left[1 \pm \frac{\lambda_R^2 - \lambda_D^2}{\Omega}\right],$$
$$\Delta E_{\Delta\ell=-1} = \frac{1}{2}(\Omega - \omega_c)\left[1 \mp \frac{\lambda_R^2 - \lambda_D^2}{\Omega}\right],$$

(6.72)

where the $\pm$ and $\mp$ signs correspond to the qdown–qdown and qup–qup transitions, respectively. Equation (6.72) explicitly shows the splitting of the two branches of excitation at $B = 0$ and the subsequent crossing of spectra at a low magnetic field. The anticrossing in the lower energy branch is shown (Pietiläinen and Chakraborty, 2006) in Figs. 6.7 and 6.8, where we plot the optical absorption spectra for noninteracting electrons confined in InAs and InSb quantum dots for $\lambda_D = 0$ and various values of λ_R. These results have been obtained starting from the exact diagonalization of Eqs. (6.52) and (6.53) and we have used for InAs and InSb the parameters $m/m_e = 0.042$, $\epsilon = 14.6$, $g^* = -14$, $m/m_e = 0.014$, $\epsilon = 17.88$, $g^* = -40$, respectively. The confinement potential strength was taken to be $\omega_0 = 7.5$ meV in all the cases.

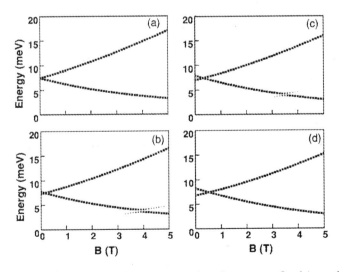

Fig. 6.7 Optical absorption spectra for noninteracting electrons confined in an InAs quantum dot for various values of the Rashba SO coupling strength (in meV nm): $\lambda_R = 0$ (a), 20 (b), 30 (c), 40 (d) and $\lambda_D = 0$.

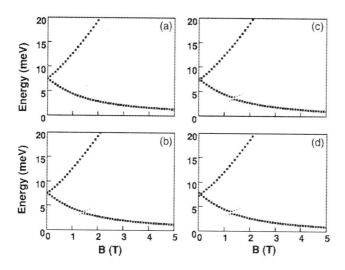

Fig. 6.8 The same as in Fig. 6.7, for an InSb quantum dot.

By taking $\omega_L = 0$ also in the energy denominators of Eq. (6.69), one then gets the simplified result (Valín-Rodríguez *et al.*, 2002) for the energies of the qup–qdown transitions,

$$\Delta E_{\Delta \ell = 0} = \omega_L + 2(\lambda_R^2 - \lambda_D^2)\ell + 2(\lambda_R^2 - \lambda_D^2)(2n_r + |\ell| + 1)\frac{\omega_c}{\Omega}, \qquad (6.73)$$

where ℓ is the angular momentum in the intrinsic frame of the precessionally active orbital, i.e. that with spin-flip allowed by the Pauli principle, normally the highest or the second-highest occupied level in an odd-N dot. Even-N systems will generally not possess a net spin at a low enough magnetic field. This equation already allows us to point out several interesting differences from the result of Eq. (6.71): (a) at $B = 0$ the SO precessional frequency does not vanish when $\ell \neq 0$, with the offset indicating the SO coupling intensity; (b) in general, positive and negative ℓ orbitals will display different B dispersions for a fixed $\lambda_R^2 - \lambda_D^2$; (c) when the precessionally active orbital changes owing to an internal rearrangement, the SO precessional frequency will display a discontinuity.

6.6 Spin–Orbit Effects on Single-Particle States in Quantum Wires

In the following we study the single-particle quantum wire Hamiltonian $h = h_0 + h_R + h_D$, where h_0 (in effective atomic units),

$$h_0 = -\frac{1}{2}(\nabla_x^2 + \nabla_y^2) - i\omega_c x \nabla_y + \frac{1}{2}(\omega_c^2 + \omega_0^2)\, x^2 - \frac{1}{2}\omega_L \sigma_z,$$

is as solved in Subsection 5.7 and h_R, h_D are the Rashba and Dresshelaus Hamiltonians

$$h_R = \lambda_R(p_y \sigma_x - p_x \sigma_y)$$

and

$$h_D = \lambda_D(p_x \sigma_x - p_y \sigma_y),$$

respectively. For the spinor $|\phi\rangle \equiv \begin{pmatrix} \phi_1(x) \\ \phi_2(x) \end{pmatrix} e^{iky}$, the Schrödinger equation $h|\phi\rangle = E|\phi\rangle$ adopts the form

$$\left[-\frac{1}{2}\frac{d^2}{dx^2} + \frac{1}{2}\Omega^2(x+x_0)^2 + \frac{1}{2}\frac{\omega_0^2}{\Omega^2}k^2 - \omega_L/2 - E \right]\phi_1 = 0,$$

$$\left[\lambda_R\left(k+\omega_c x + \frac{d}{dx}\right) + i\lambda_D\left(k+\omega_c x - \frac{d}{dx}\right) \right]\phi_2 = 0,$$

(6.74)

$$\left[\lambda_R\left(k+\omega_c x - \frac{d}{dx}\right) - i\lambda_D\left(k+\omega_c x + \frac{d}{dx}\right) \right]\phi_1 = 0,$$

$$\left[-\frac{1}{2}\frac{d^2}{dx^2} + \frac{1}{2}\Omega^2(x+x_0)^2 + \frac{1}{2}\frac{\omega_0^2}{\Omega^2}k^2 + \omega_L/2 - E \right]\phi_2 = 0,$$

(6.75)

where $\Omega = \sqrt{\omega_c^2 + \omega_0^2}$, ω_c is the cyclotron frequency and $x_0 = \omega_c k/\Omega^2$. By introducing the operators

$$a^\pm = \frac{1}{\sqrt{2\Omega}}\left(p_x \pm i\Omega(x+x_0)\right)$$

(6.76)

with $[a^-, a^+] = 1$, the sp wire Schrödinger equation can be rewritten as

$$\left[a^+a^- + \frac{1}{2} - \frac{1}{2}\frac{\omega_L}{\Omega} - \varepsilon \right]\phi_1 = 0,$$

$$\left[\tilde{\lambda}_R(\tilde{k} + i(1-\gamma)a^+ + i(1+\gamma)a^-) + i\tilde{\lambda}_D(\tilde{k} - i(1+\gamma)a^+ - i(1-\gamma)a^-) \right]\phi_2 = 0,$$

(6.77)

$$\left[\tilde{\lambda}_R(\tilde{k} - i(1+\gamma)a^+ - i(1-\gamma)a^-) - i\tilde{\lambda}_D(\tilde{k} + i(1-\gamma)a^+ + i(1+\gamma)a^-) \right]\phi_1 = 0,$$

$$\left[a^+a^- + \frac{1}{2} + \frac{1}{2}\frac{\omega_L}{\Omega} - \varepsilon \right]\phi_2 = 0,$$

(6.78)

where we have defined $\varepsilon\Omega = E - \frac{1}{2}\frac{\omega_0^2}{\Omega^2}k^2$, $\tilde{k} = \sqrt{\frac{2}{\Omega}\frac{\omega_0^2}{\Omega^2}}k$, $\gamma = \omega_c/\Omega$ and $\tilde{\lambda}_{R,D} = \sqrt{\frac{1}{2\Omega}}\lambda_{R,D}$. We now expand $\phi_1(x)$ and $\phi_2(x)$ into the oscillator states $|n\rangle$ of h_0

[see Eq. (5.66)],

$$|n\rangle = (2^n n! \pi^{1/2} b)^{-1/2} e^{-(x+x_0)^2/(2b^2)} H_n \left((x+x_0)/b\right),$$

where $b = (\hbar/m\Omega)^{1/2}$ is the width of the wave function, as $\phi_1 = \sum_{n=0}^{\infty} a_n |n\rangle$, $\phi_2 = \sum_{m=0}^{\infty} b_m |m\rangle$. The operators a^+, a^- act on $|n\rangle$ in the usual way, i.e. $a^+|n\rangle = \sqrt{n+1}|n+1\rangle$, $a^-|n\rangle = \sqrt{n}|n-1\rangle$ and $a^-|0\rangle = 0$. One finally gets the infinite system of equations

$$\begin{aligned}
(n+\beta-\varepsilon)a_n &= -\tilde{k}\tilde{\lambda}_+ b_n - i\tilde{\lambda}_1\sqrt{n}\,b_{n-1} - i\tilde{\lambda}_2\sqrt{n+1}\,b_{n+1}, \\
(n+\alpha-\varepsilon)b_n &= -\tilde{k}\tilde{\lambda}_- a_n + i\tilde{\lambda}_3\sqrt{n}\,a_{n-1} + i\tilde{\lambda}_4\sqrt{n+1}\,a_{n+1},
\end{aligned} \tag{6.79}$$

for $n \geq 0$, with $a_{-1} = 0$, $b_{-1} = 0$. In the above equations we have defined $\alpha = (1+\omega_L/\omega_c)/2$, $\beta = (1-\omega_L/\omega_c)/2$, $\tilde{\lambda}_\pm = \tilde{\lambda}_R \pm i\tilde{\lambda}_D$, $\tilde{\lambda}_1 = \tilde{\lambda}_- - \gamma\tilde{\lambda}_+$, $\tilde{\lambda}_2 = \tilde{\lambda}_- + \gamma\tilde{\lambda}_+$, $\tilde{\lambda}_3 = \tilde{\lambda}_+ + \gamma\tilde{\lambda}_-$, $\tilde{\lambda}_4 = \tilde{\lambda}_+ - \gamma\tilde{\lambda}_-$.

From Eq. (6.79) one sees that the term proportional to $\tilde{k}\tilde{\lambda}$ couples quasi-spin-up and quasi-spin-down states of the nth subband, whereas the other SO terms couple different quasispin states of the neighboring subbands. By giving B and k, the full Hamiltonian eigenvalues can be calculated by solving numerically the set of equations (6.79) (Knobbe and Schäppers, 2005; Zhang *et al.*, 2006). Note also that from Eqs. (6.79) one recovers Eqs. (6.9), valid for the quantum well by just putting $\omega_0 = 0$.

Figure 6.9 shows the energy levels of the quantum wire at a zero magnetic field for different intensities of the SO coupling. The SO effect on the spectrum is very

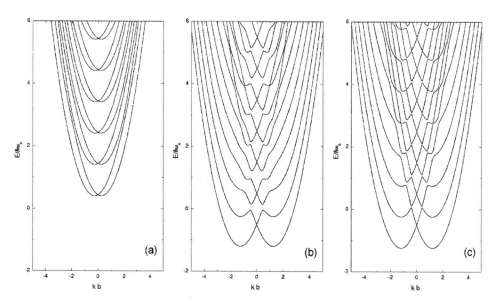

Fig. 6.9 Energy spectra of quantum wire ($b = \sqrt{\hbar/m\omega_0}$) at $B = 0$ with different values of the Rashba and Dresselhaus SO strengths. (a) $\tilde{\lambda}_R^2 = 0.01$ and $\tilde{\lambda}_D^2 = 0.005$. (b) $\tilde{\lambda}_R^2 = 0.75$ and $\tilde{\lambda}_D^2 = 0.25$. (c) $\tilde{\lambda}_R^2 = \tilde{\lambda}_D^2 = 0.5$. In (b) and (c) one can see a significant anticrossing.

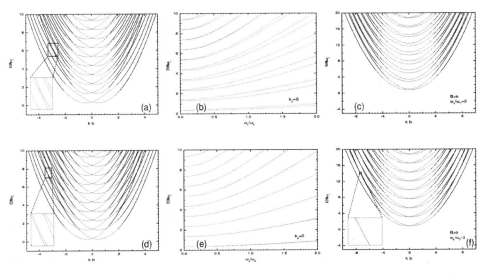

Fig. 6.10   (a) Energy dispersion at a zero magnetic field ($b = \sqrt{\hbar/m\Omega}$) with $\tilde{\lambda}_R^2 = 0.1$ and $\tilde{\lambda}_D^2 = 0.05$. The inset shows the lower subband anticrossing. (b) Energy spectrum of the same quantum wire as a function of ω_c/ω_0. The Zeeman energy splitting is chosen to be $g^*\mu_B B = -0.015\omega_c$. (c) Energy dispersion of the wire at a finite magnetic field ($\omega_c/\omega_0 = 2$). (d)–(f) The same as in (a)–(c), with $\tilde{\lambda}_R^2 = \tilde{\lambda}_D^2 = 0.075$.

weak in Fig. 6.9(a), where $\tilde{\lambda}_R^2 = 0.01$ and $\tilde{\lambda}_D^2 = 0.005$. For strong SO coupling [Fig. 6.9(b)], when $\tilde{\lambda}_R^2 = 0.75$ and $\tilde{\lambda}_D^2 = 0.25$, the situation is completly different and the coupling of neighboring subbands occurs, yielding a significant anticrossing. The anticrossing still exists in Fig. 6.9(c) when $\tilde{\lambda}_R^2 = \tilde{\lambda}_D^2 = 0.5$.

Figure 6.10 shows the energy spectra of the wire in a magnetic field for different values of the SO strenths. The magnetic field and the SO coupling lead to a rather complex energy spectrum, in which energy spacings between different subbands vary at a fixed magnetic field. For the sake of comparison, Fig. 6.10(a) shows the energy spectrum at $B = 0$ with $\tilde{\lambda}_R^2 = 0.1$ and $\tilde{\lambda}_D^2 = 0.05$, while in Fig. 6.10(d), $\tilde{\lambda}_R^2 = \tilde{\lambda}_D^2 = 0.075$. In both subbands there is anticrossing at a zero magnetic field. The energy spectrum for $k = 0$ as a function of ω_c/ω_0 is plotted in Figs. 6.10(b) and 6.10(e). One can see that, when the magnetic field increases, the subband separation becomes larger owing to the increase of Ω, and that the previous spin degeneracy at $k = 0$ is broken by the magnetic field in the case shown in Fig. 6.10(b) but not in the case of Fig. 6.10(e), owing to cancelation effects. Finally, Figs. 6.10(c) and 6.10(f) show the energy dispersion with k at a finite magnetic field. In Fig. 6.10(c) the anticrossing has disappeared with respect to the case of a zero magnetic field. On the contrary, some anticrossing persists in the case shown in Fig. 6.10(f).

Finally, we note that an analytical solution to Eqs. (6.79) can be found in the limit $h_0, h_Z \gg h_R, h_D$. In this case the energies of the quasiup and quasidown states are given by

$$
\varepsilon_n^u = n + \beta - \frac{\tilde{k}^2 \tilde{\lambda}_+ \tilde{\lambda}_- \Omega}{\omega_L} + n\frac{\tilde{\lambda}_1 \tilde{\lambda}_4 \Omega}{\Omega - \omega_L} - (n+1)\frac{\tilde{\lambda}_2 \tilde{\lambda}_3 \Omega}{\Omega + \omega_L},
$$

$$
\varepsilon_n^d = n + \alpha + \frac{\tilde{k}^2 \tilde{\lambda}_+ \tilde{\lambda}_- \Omega}{\omega_L} - (n+1)\frac{\tilde{\lambda}_1 \tilde{\lambda}_4 \Omega}{\Omega - \omega_L} + n\frac{\tilde{\lambda}_2 \tilde{\lambda}_3 \Omega}{\Omega + \omega_L},
$$

(6.80)

and are very close to the results of the numerical diagonalization taken in the same limit.

It is also interesting to study the case of a magnetic field parallel to the plane of the electron motion. In this case, the orbital effect of the applied in-plane magnetic field is frozen owing to the strong confinement in the vertical direction and the only effect of the field appears through the Zeeman interaction. The relevant Hamiltonian for the quantum wire becomes

$$
h_0 = \frac{1}{2}p^2 + \frac{1}{2}\omega_0^2 x^2 + \lambda_R(p_y\sigma_x - p_x\sigma_y) + \frac{1}{2}g^*\mu_B\sigma \cdot \mathbf{B},
$$

(6.81)

where, for simplicity, we have considered only the Rashba SO term and the magnetic field is given by $\mathbf{B} = B_x\hat{x} + B_y\hat{y}$. For the spinor $|\phi\rangle \equiv \left(\begin{smallmatrix}\phi_1(x)\\\phi_2(x)\end{smallmatrix}\right)e^{iky}$, the eigenvalue problem for h_0 can be solved analytically (Pershin, Nesteroff and Privman, 2004) by assuming that $p_x\phi(x) \ll k\phi(x)$, which is reasonable for very narrow wires and allows one to neglect the term $\lambda_R p_x\sigma_y$ in Eq. (6.81). One then obtains

$$
E_n^{\pm} = \left(n + \frac{1}{2}\right)\omega_0 + \frac{k^2}{2} \mp \left[(g^*\mu_B B/2)^2 + g^*\mu_B\lambda_R kB\sin\theta + (\lambda_R k)^2\right]^{1/2}.
$$

(6.82)

In Eq. (6.82), the up and down quasispin states are denoted by $\pm$, and θ is the angle of the magnetic field with the y axis, so that $B\sin\theta$ is equal to B_x. The energy spectrum corresponding to Eq. (6.82) is illustrated in Fig. 8.5(a) in the case of $\theta = \pi/4$. The important features of the shown subbands are the appearance of a gap at $k = 0$ between two spin-split bands and of local extrema.

6.7 The LSDA with Spin–Orbit Coupling

In the previous subsections we have studied the noninteracting model in quantum wells, dots and wires including the effects of SO coupling and the magnetic field, and we have generated a complete basis of single-particle wave functions and energies which in some cases are analytical expressions. The effect of the electron–electron interaction can be studied by performing an exact diagonalization of the many-body Hamiltonian H on this complete basis. This program has been carried out by Califano, Chakraborty and Pietiläinen (2005) in quantum wells, and by Chakraborty and Pietiläinen (2005) and Pietiläinen and Chakraborty (2006) in quantum dots. These authors have studied the influence of the Rashba SO coupling on the incompressible Laughlin state of a quantum well and on the dipole absorption spectrum of a parabolic quantum dot. The diagonalization calculations were carried out for a maximum of four electrons. Here we present a different approach which includes Coulomb interaction effects in the theory by means of a self-consistent LSDA calculation. Since $\mathcal{H}_{R,D}$ break the symmetry of a single spin quantization axis, in the following we develop a noncollinear LSDA theory to be used in calculations for quantum dots and wires (Heinonen, Kinaret and Johnson, 1999; Valín-Rodríguez et al., 2002).

Let us start with quantum dots in a perpendicular magnetic field and introduce the Hartree potential

$$V_H(\mathbf{r}) = \frac{e^2}{\epsilon} \int d\mathbf{r}' \frac{\rho(\mathbf{r}')}{|\mathbf{r}' - \mathbf{r}|}, \tag{6.83}$$

where ϵ is the semiconductor dielectric constant ($\epsilon = 12.4$ for GaAs). The Kohn–Sham orbitals will be two-component spinors of the type

$$\varphi_i(\mathbf{r}, \eta) \equiv \begin{pmatrix} \varphi_i(\mathbf{r}, \uparrow) \\ \varphi_i(\mathbf{r}, \downarrow) \end{pmatrix}, \quad i = 1, \ldots, N, \tag{6.84}$$

where η is an independent variable, similar to $\mathbf{r}$, which can only take the two values $\uparrow, \downarrow$ and, as a rule, we assume that $\uparrow$ and $\downarrow$ correspond to spin in the upward and downward z directions, respectively. We define the spin density matrix in terms of such spinors as

$$\rho_{\eta\eta'}(\mathbf{r}) = \sum_{i=1}^{N} \varphi_i^*(\mathbf{r}, \eta)\varphi_i(\mathbf{r}, \eta') \equiv \begin{pmatrix} \rho_{\uparrow\uparrow} & \rho_{\uparrow\downarrow} \\ \rho_{\downarrow\uparrow} & \rho_{\downarrow\downarrow} \end{pmatrix}. \tag{6.85}$$

The LSDA exchange-correlation functional $E_{\mathrm{xc}}[\rho_{\eta\eta'}]$ yields the following 2×2 potential matrix:

$$V_{\mathrm{xc},\eta\eta'}(\mathbf{r}) = \frac{\delta E_{\mathrm{xc}}[\rho_{\eta\eta'}]}{\delta \rho_{\eta\eta'}(\mathbf{r})}. \tag{6.86}$$

The resulting Kohn–Sham equations read

$$\left[\frac{\mathbf{P}^2}{2m} + \frac{1}{2}m\left(\omega_x^2 x^2 + \omega_y^2 y^2\right) + s_\eta \frac{1}{2}g^* \mu_B B + V_H(\mathbf{r}) \right] \varphi_i(\mathbf{r}, \eta)$$

$$+ \sum_{\eta'} \left[\frac{\lambda_R}{\hbar} \left(P_y \sigma_x - P_x \sigma_y\right)_{\eta\eta'} + \frac{\lambda_D}{\hbar} \left(P_x \sigma_x - P_y \sigma_y\right)_{\eta\eta'} \right.$$

$$\left. + V_{\mathrm{xc},\eta\eta'}(\mathbf{r}) \right] \varphi_i(\mathbf{r}, \eta') = \varepsilon_i\, \varphi_i(\mathbf{r}, \eta), \tag{6.87}$$

where $s_\eta = \pm 1$ for $\eta = \uparrow, \downarrow$ and $\mathbf{P}$ is the canonical momentum with the vector potential in the symmetric gauge.

The functional derivatives in $V_{xc,\eta\eta'}(\mathbf{r})$, can be found in the LSDA, as follows. In terms of the usual particle and magnetization densities,

$$\rho(\mathbf{r}) = \left\langle 0 \left| \sum_{i=1}^{N} \delta(\mathbf{r} - \mathbf{r}_i) \right| 0 \right\rangle = \sum_{i,\eta} |\varphi_i(\mathbf{r},\eta)|^2,$$

$$m_a(\mathbf{r}) = \left\langle 0 \left| \sum_{i=1}^{N} \delta(\mathbf{r} - \mathbf{r}_i)\sigma_a \right| 0 \right\rangle = \sum_{i,\eta,\eta'} \varphi_i^*(\mathbf{r},\eta)\sigma_a^{\eta,\eta'} \varphi_i(\mathbf{r},\eta'),$$

(6.88)

we have

$$\begin{pmatrix} \rho_{\uparrow\uparrow} & \rho_{\uparrow\downarrow} \\ \rho_{\downarrow\uparrow} & \rho_{\downarrow\downarrow} \end{pmatrix} = \frac{1}{2} \begin{pmatrix} \rho + m_z & m_x + im_y \\ m_x - im_y & \rho - m_z \end{pmatrix}.$$

(6.89)

We define a diagonal density matrix by means of a local unitary transformation U:

$$U\rho U^+ = n \equiv \begin{pmatrix} n_\uparrow & 0 \\ 0 & n_\downarrow \end{pmatrix}.$$

(6.90)

The local rotation is given by

$$U = \begin{pmatrix} e^{i\phi(\mathbf{r})/2} \cos \dfrac{\theta(\mathbf{r})}{2} & e^{-i\phi(\mathbf{r})/2} \sin \dfrac{\theta(\mathbf{r})}{2} \\ -e^{i\phi(\mathbf{r})/2} \sin \dfrac{\theta(\mathbf{r})}{2} & e^{-i\phi(\mathbf{r})/2} \cos \dfrac{\theta(\mathbf{r})}{2} \end{pmatrix}.$$

(6.91)

The local rotation angles are determined by the equations

$$\tan \phi(\mathbf{r}) = -\frac{m_y(\mathbf{r})}{m_x(\mathbf{r})},$$

(6.92)

$$\tan \theta(\mathbf{r}) = \frac{\sqrt{m_x^2(\mathbf{r}) + m_y^2(\mathbf{r})}}{m_z(\mathbf{r})}.$$

(6.93)

The diagonal local densities are then (we omit the $\mathbf{r}$ arguments)

$$n_\uparrow = \frac{1}{2}(\rho + m_z \cos \theta) + \mathrm{Re}\left\{\rho_{\uparrow\downarrow} e^{i\phi} \sin \theta\right\},$$

(6.94)

$$n_\downarrow = \frac{1}{2}(\rho - m_z \cos \theta) - \mathrm{Re}\left\{\rho_{\uparrow\downarrow} e^{i\phi} \sin \theta\right\}.$$

(6.95)

Knowing $n_\uparrow$ and $n_\downarrow$ at point $\mathbf{r}$, we can now use the familiar relations of collinear LSDA to compute the exchange-correlation potentials:

$$\begin{pmatrix} v_\uparrow & 0 \\ 0 & v_\downarrow \end{pmatrix} \equiv \begin{pmatrix} \dfrac{\delta E_{xc}[n_\uparrow, n_\downarrow]}{\delta n_\uparrow} & 0 \\ 0 & \dfrac{\delta E_{xc}[n_\uparrow, n_\downarrow]}{\delta n_\downarrow} \end{pmatrix}.$$

(6.96)

Finally, we only need to "undo" the rotation back to the laboratory frame. The resulting expression for the exchange-correlation potential V^{xc} can be written as

$$V_{\eta\eta'}^{(\text{xc})} \equiv \begin{pmatrix} v_0 + \Delta v \cos\theta & \Delta v e^{-i\phi}\cos\theta \\ \Delta v e^{i\phi}\cos\theta & v_0 - \Delta v\cos\theta \end{pmatrix}, \qquad (6.97)$$

where we have defined

$$v_0 = \frac{v_\uparrow + v_\downarrow}{2}, \qquad (6.98)$$

$$\Delta v = \frac{v_\uparrow - v_\downarrow}{2}. \qquad (6.99)$$

This scheme fully determines the 2×2 V^{xc} potential matrix in terms of the spinor orbitals and the LSDA energy functional. It allows the description of spin textures where the spin orientation can vary from one point to another of the system.

The set of equations (6.87) is solved by discretizing the xy plane in a uniform grid of points and applying an iterative scheme to reach full self-consistency in $V_H(\mathbf{r})$ and $V_{\text{xc},\eta\eta'}(\mathbf{r})$. In some cases this procedure might get trapped in a local minimum. Therefore, several calculations with different random initial conditions have to be used to ensure that the proper energy minimum is reached. The stability with the number of mesh points must also be checked.

The above formalism can be generalized to the case of the quantum wire with SO interaction and an in-plane magnetic field. The noninteracting Hamiltonian of this system was written in Eq. (6.81) of the previous subsection taking into account only the Rashba term. Here we include the Dresselhaus term and the electron–electron interaction in the LSDA. In this system the Kohn–Sham orbitals are two-component spinors of the type

$$\Psi_{nk}(\mathbf{r},\eta) \equiv |\Psi_{nk}\rangle \equiv \frac{1}{\sqrt{L}}\begin{pmatrix}\varphi_{nk}(x,\uparrow)\\ \varphi_{nk}(x,\downarrow)\end{pmatrix} e^{iky}. \qquad (6.100)$$

In Eq. (6.100) we have used the translational invariance along y to introduce a continuous k wave number, and the index $n = 1,2,3,\ldots$ is labeling the different energy bands. The quantum labels are therefore (n,k), while η, as before, is an independent variable, which can only take the two values $\uparrow,\downarrow$. This implies that we have no spin label for the bands, and each of them contains the two components.

Each band fulfills a Kohn–Sham spinorial equation:

$$h_{\text{KS}}[\rho,\mathbf{m}]\Psi_{nk} = \varepsilon_{nk}\Psi_{nk}. \qquad (6.101)$$

In Eq. (6.101) we have a functional dependence on electron density ρ and spin magnetization $\mathbf{m}$. The latter is a vector. In order to obtain these densities we need to introduce the temperature T, the chemical potential μ and the Fermi function

$$f_\mu(\varepsilon_{nk}) = \frac{1}{1 + e^{(\varepsilon_{nk}-\mu)/k_BT}}. \qquad (6.102)$$

The Fermi function is giving the occupation of the (n, k) state.

The electron density is now

$$\rho(x) = \sum_n \frac{1}{2\pi} \int dk \, \langle \Psi_{nk} | \delta(\mathbf{r}_i - \mathbf{r}) | \Psi_{nk} \rangle_{\mathbf{r}_i \eta_i} f_\mu(\varepsilon_{nk})$$

$$= \sum_n \frac{1}{2\pi} \int dk \, \left[|\varphi_{nk}(x, \uparrow)|^2 + |\varphi_{nk}(x, \downarrow)|^2 \right] f_\mu(\varepsilon_{nk}), \qquad (6.103)$$

and the one-dimensional density is given by the integral of $\rho(x)$ over x:

$$\rho_{1D} = \int dx \sum_n \frac{1}{2\pi} \int dk \, \left[|\varphi_{nk}(x, \uparrow)|^2 + |\varphi_{nk}(x, \downarrow)|^2 \right] f_\mu(\varepsilon_{nk}). \qquad (6.104)$$

The k integration is in principle from $-\infty$ to $+\infty$. In practice one truncates from $-k_{\max}$ to $+k_{\max}$. The idea is to compute Ψ_{nk} for all the chosen states on a "kgrid with N_k points" and for all the n's up to a chosen $n_{\max}$; then one performs the integration in Eq. (6.104) using a high precision method for the k domain (a Bode rule). Notice that the translational invariance implies that all densities (actually all physical variables) depend only on x, defining the transversal profile of the wire for a specific quantity.

For the $a = x, y, z$ components of the magnetization we have in a similar way

$$m_a(y) = \sum_n \frac{1}{2\pi} \int dk \langle \Psi_{nk} | \delta(\mathbf{r}_i - \mathbf{r}) \sigma_a | \Psi_{nk} \rangle_{\mathbf{r}_i \eta_i} \, f_\mu(\varepsilon_{nk}), \qquad (6.105)$$

where σ_a is the corresponding Pauli matrix. The three components then read

$$m_x(x) = \sum_n \frac{1}{2\pi} \int dk \, 2\mathrm{Re} \left[\varphi_{nk}(x, \uparrow)^* \varphi_{nk}(x, \downarrow) \right] f_\mu(\varepsilon_{nk}), \qquad (6.106)$$

$$m_y(x) = \sum_n \frac{1}{2\pi} \int dk \, 2\mathrm{Im} \left[\varphi_{nk}(x, \uparrow)^* \varphi_{nk}(x, \downarrow) \right] f_\mu(\varepsilon_{nk}), \qquad (6.107)$$

$$m_z(x) = \sum_n \frac{1}{2\pi} \int dk \, \left[|\varphi_{nk}(x, \uparrow)|^2 - |\varphi_{nk}(x, \downarrow)|^2 \right] f_\mu(\varepsilon_{nk}). \qquad (6.108)$$

The Kohn–Sham functional Hamiltonian $h_{\mathrm{KS}}[\rho, \mathbf{m}]$ can be separated in different pieces, consisting of the kinetic, confining and exchange-correlation term, plus the SO term (Rashba and the Dresselhaus contributions) and a Zeeman contribution given by a magnetic field in an arbitrary orientation in the xy plane,

$$\mathbf{B} = B(\cos \phi_B \mathbf{u}_x + \sin \phi_B \mathbf{u}_y), \qquad (6.109)$$

where ϕ_B is the azimuthal angle. The extension to include a vertical magnetic field can be done easily.

Thus, the Hamiltonian is $h_{\mathrm{KS}} = h_0 + h_Z + h_{\mathrm{SO}}$, where

$$h_0 = \frac{p_x^2 + p_y^2}{2m} + \frac{1}{2}m\omega_0^2 x^2 + V_{\eta\eta'}^{(\mathrm{xc})}(x), \qquad (6.110)$$

$$h_Z = \mathcal{E}_z(\sigma_x \cos\phi_B + \sigma_y \sin\phi_B), \qquad (6.111)$$

$$h_{\mathrm{SO}} = \frac{\lambda_R}{\hbar}(p_y\sigma_x - p_x\sigma_y) + \frac{\lambda_D}{\hbar}(p_x\sigma_x + p_y\sigma_y). \qquad (6.112)$$

We have introduced the self-consistent potential due to quantum exchange and correlation $V_{\eta\eta'}^{(\mathrm{xc})}(x)$, the Zeeman energy $\mathcal{E}_Z \equiv g^*\mu_B B$, and the Rashba and Dresselhaus parameters λ_R, λ_D. In principle one should also include the Hartree term, but we will consider that this is exactly canceled by some neutralizing background contribution. Taking into account that on our translationally invariant states p_y can be substituted by $\hbar k$, and introducing a complex SO coupling parameter $\gamma \equiv \lambda_R + i\lambda_D$, with which the full SO contributions read

$$h_R + h_D = \begin{pmatrix} 0 & \gamma k + \gamma^* \frac{d}{dx} \\ \gamma^* k - \gamma \frac{d}{dx} & 0 \end{pmatrix}, \qquad (6.113)$$

we arrive at the matrix eigenvalue equation:

$$\begin{pmatrix} -\dfrac{\hbar^2}{2m}\dfrac{d^2}{dx^2} + \dfrac{\hbar^2 k^2}{2m} + \dfrac{1}{2}m\omega_0^2 x^2 + V_{\uparrow\uparrow}^{(\mathrm{xc})} & \gamma^*\dfrac{d}{dx} + \gamma k + \mathcal{E}_Z e^{-i\phi_B} + V_{\uparrow\downarrow}^{(\mathrm{xc})} \\[2ex] -\gamma\dfrac{d}{dx} + \gamma^* k + \mathcal{E}_Z e^{i\phi_B} + V_{\downarrow\uparrow}^{(\mathrm{xc})} & -\dfrac{\hbar^2}{2m}\dfrac{d^2}{dx^2} + \dfrac{\hbar^2 k^2}{2m} + \dfrac{1}{2}m\omega_0^2 x^2 + V_{\downarrow\downarrow}^{(\mathrm{xc})} \end{pmatrix}$$
$$\times \begin{pmatrix} \varphi_{nk\uparrow} \\ \varphi_{nk\downarrow} \end{pmatrix} = \varepsilon_{nk} \begin{pmatrix} \varphi_{nk\uparrow} \\ \varphi_{nk\downarrow} \end{pmatrix}, \qquad (6.114)$$

where we have defined $\varphi_{nk\eta} \equiv \varphi_{nk}(x, \eta)$.

The potential matrix $V_{\eta\eta'}^{(\mathrm{xc})}(x)$ is related to the energy functional E_{xc} in the noncollinear framework in the same way as discussed previously for the quantum dot case.

For each (nk) we have therefore to diagonalize the problem $h_{\mathrm{KS}}\Psi_{nk} = \varepsilon_{nk}\Psi_{nk}$, keeping the lowest $n_{\max}$ eigenvalues and eigenvectors $\{\varepsilon_{nk}, \Psi_{nk}(x)\}$. In order to do this, we introduce a real space discretization of the x axis, from $-x_{\max}$ to $+x_{\max}$. This defines N_x points, and since the two components are coupled, the resulting matrix is $2N_x \times 2N_x$. Once $\{\varepsilon_{nk}, \Psi_{nk}(x)\}$ are known, we proceed to compute the density and magnetization, which allow us to start a new iteration. Self-consistency is achieved when these things do not change from iteration to iteration. For N_x's of the order of 100, the diagonalization is extremely fast and although it is repeated many times ($N_k \times n_{\max}$ times) the calculation is quite efficient.

The total energies per unit of length are easy to calculate piece by piece using the Hamiltonian separation described previously, computing

$$\int dx \sum_n \frac{1}{2\pi} \int dk \langle \Psi_{nk} | h_{\mathrm{KS}} | \Psi_{nk} \rangle. \tag{6.115}$$

For the kinetic and confining terms, we obtain

$$E_{\mathrm{kin}} = \frac{1}{4\pi} \int dx \sum_n \int dk \left\{ |\varphi'_{nk\uparrow}|^2 + |\varphi'_{nk\downarrow}|^2 \right.$$
$$\left. + k^2 \left(|\varphi_{nk\uparrow}|^2 + |\varphi_{nk\downarrow}|^2 \right) \right\} f_\mu(\varepsilon_{nk}), \tag{6.116}$$

$$E_{\mathrm{conf}} = \frac{\omega_0^2}{4\pi} \int dx \sum_n \int dk \, x^2 \left(|\varphi_{nk\uparrow}|^2 + |\varphi_{nk\downarrow}|^2 \right) f_\mu(\varepsilon_{nk}), \tag{6.117}$$

with $\varphi'_{nk\eta} \equiv d\varphi_{nk\eta}/dx$. For the SO terms, using the preceding definition of h_R and h_D in terms of the complex coupling parameter $\gamma \equiv \lambda_R + i\lambda_D$, we have

$$E_{\mathrm{SO}} = \int dx \sum_n \frac{1}{2\pi} \int dk \left(\varphi^*_{nk\uparrow} \; \varphi^*_{nk\downarrow} \right) \begin{pmatrix} 0 & \gamma k + \gamma^* \frac{d}{dx} \\ \gamma^* k - \gamma \frac{d}{dx} & 0 \end{pmatrix}$$
$$\times \begin{pmatrix} \varphi_{nk\uparrow} \\ \varphi_{nk\downarrow} \end{pmatrix} f_\mu(\varepsilon_{nk}). \tag{6.118}$$

Performing the matrix multiplications we get

$$E_{\mathrm{SO}} = \int dx \sum_n \frac{1}{2\pi} \int dk \left\{ \left(\gamma k \varphi^*_{nk\uparrow} \varphi_{nk\downarrow} + \mathrm{c.c.} \right) \right.$$
$$\left. + \gamma^* \varphi^*_{nk\uparrow} \varphi'_{nk\downarrow} - \gamma \varphi^*_{nk\downarrow} \varphi'_{nk\uparrow} \right\} f_\mu(\varepsilon_{nk}). \tag{6.119}$$

Using integration by parts we may write

$$\int dx \varphi^*_{nk\uparrow} \varphi'_{nk\downarrow} = \frac{1}{2} \int dx \left(\varphi^*_{nk\uparrow} \varphi'_{nk\downarrow} - \varphi'^*_{nk\uparrow} \varphi_{nk\downarrow} \right). \tag{6.120}$$

With this trick we can symmetrize the last two contributions in Eq. (6.119) and obtain an expression that is apparently real. The final result is

$$E_{\mathrm{SO}} = \int dx \sum_n \frac{1}{2\pi} \int dk \left\{ 2k \, \mathcal{Re} \left[\gamma^* \varphi^*_{nk\downarrow} \varphi_{nk\uparrow} \right] \right.$$
$$\left. + \mathrm{Re} \left[\gamma \left(\varphi'^*_{nk\downarrow} \varphi_{nk\uparrow} - \varphi^*_{nk\downarrow} \varphi'_{nk\uparrow} \right) \right] \right\} f_\mu(\varepsilon_{nk}). \tag{6.121}$$

Finally, we have the Zeeman term

$$E_Z = \mathcal{E}_Z \int dx \sum_n \frac{1}{2\pi} \int dk \, 2 \left\{ \cos \phi_B \, \mathcal{Re} \left[\varphi^*_{nk\downarrow} \varphi_{nk\uparrow} \right] \right.$$
$$\left. - \sin \phi_B \, \mathcal{Im} \left[\varphi^*_{nk\downarrow} \varphi_{nk\uparrow} \right] \right\} f_\mu(\varepsilon_{nk}), \tag{6.122}$$

and the exchange-correlation term

$$E_{xc} = \int dx \varepsilon_{xc}(x)\rho(x). \tag{6.123}$$

Thus, the total energy per unit of length is

$$E_{total} = E_{kin} + E_{conf} + E_{SO} + E_{xc} + E_Z. \tag{6.124}$$

In order to check the accuracy of the calculation, one can compute the total energy in an alternative way, which consists in writing the total energy as (see Subsection 4.4)

$$E'_{total} = \sum_n \frac{1}{2\pi} \int dk \varepsilon_{nk} f_\mu(\varepsilon_{nk}) + \int dx \varepsilon_{xc}(x)\rho(x) - \int dx \left(n_\uparrow v_{xc\uparrow} + n_\downarrow v_{xc\downarrow}\right). \tag{6.125}$$

One must then check that it gives the same result as Eq. (6.124) with a high precision.

In the case of quantum dots and wires where there is a lateral parabolic confinement, one can distinguish between two regimes of weak and strong SO coupling in the following way. We define a characteristic SO energy as $\Delta_{SO} = \Delta_{SO}^R + \Delta_{SO}^D$, where

$$\Delta_{SO}^{R,D} = \frac{m\lambda_{R,D}^2}{2\hbar^2}. \tag{6.126}$$

To characterize the SO regime, we then use the ratio of the SO to the confining energy:

$$\Delta_{R,D} = \frac{m\lambda_{R,D}^2}{2\hbar^3\omega_0}. \tag{6.127}$$

We have used the values $\Delta_R = 0.0037$, $\Delta_D = 0.015$ to represent a typical weak SO coupling regime, and the values $\Delta_R = 0.093$, $\Delta_D = 0.37$ to represent a typical strong SO coupling regime. The weak and strong coupling results we will discuss in Subsection 6.10 have been obtained using these parameters, except when their values are explicitly given.

6.8   Comparison with Experiments in Quantum Wells

An actual confrontation of the theoretical results we have obtained with the experiments is not an easy task, because of the smallness of the SO effects, and because of the way they are presented in the available literature, which makes it extremely difficult to carry out a quantitative analysis of such a subtle effect. Thus, we have to satisfy ourselves with a semiquantitative analysis, or to point out that these results are compatible with fairly rough estimated values of the SO coupling constants. We present now three such examples and a possible way to increase SO effects so that they could be easier to determine.

Using unpolarized far-infrared radiation, Manger *et al.* (2001) have measured the cyclotron resonance in GaAs quantum wells at different electron densities. The main finding of the experiment is a well-resolved splitting of the CR for $\nu = 3, 5$ and 7, and no significant splitting for $\nu = 1$ and for even filling factors. We have seen that the SO interaction couples charge-density and spin-density excitations yielding the SO splitting of the CR given in Eq. (6.39). However, this expression, by itself, is unable to explain the filling factor dependence of the observed splitting, for which one has to bear in mind that the SO coupling between the $\sum_i P_i^-$ and $\sum_i P_i^- \sigma_z^i$ operators is strongly enhanced when the spin gs is not zero and that the electron–electron interaction contributes to the splitting through a factor $\mathcal{K}$. [see Tonello and Lipparini (2004), Lipparini *et al.* (2006) and Subsection 8.11]. Equation (6.39) has to be generalized to include these features. One gets

$$\Delta E_{\mathrm{CR}} = \left| \frac{2S_z}{N} 4 \left(\lambda_R^2 \frac{\omega_c}{\omega_c + \omega_L} - \lambda_D^2 \frac{\omega_c}{\omega_c - \omega_L} \right) + \mathcal{K} \omega_c \right|, \qquad (6.128)$$

where the factor $2S_z/N$ takes into account the actual spin contents of the ground state. This equation embodies the theoretical explanation of the experimental findings (Manger *et al.*, 2001). In particular, it gives an appreciable splitting only for odd filling factors, for which the spin ground state S_z is not zero. The analysis of the experimental splittings using Eq. (6.128) yields values for the quantity $m|\lambda_R^2 - \lambda_D^2|/\hbar^2$ of about $30\,\mu\mathrm{eV}$, in agreement with the ones recently used to reproduce the spin splitting in quantum dots (Konemann *et al.*, 2005) and wells (Malet *et al.*, 2006). This is, in our opinion, one of the clearest evidences of a crucial SO effect on a physical observable, because its absence would imply that the physical effect does not show up.

The spin splitting of the first three Landau levels of a GaAs quantum well has been measured in a magnetoresistivity experiment by Dobers *et al.* (1988). This splitting is not influenced by the e–e interaction due to the Larmor theorem (see Subsection 8.10), and there is no spin splitting as B goes to zero [see Eq. (6.32)]. Both facts are in agreement with the analysis of the experimental data, and with previous theoretical considerations about the B dependence of the gyromagnetic factor g^*, whose determination was the physical motivation of the magnetoresistivity experiment by Dobers *et al.* (1988). These authors have derived a B- and n-dependent g^* factor $g^*(B,n) = g_0^* - c(n + \frac{1}{2})B$, where g_0^* and c are fitting constants that depend on the actual quantum well. The possibility of an SO shift was not considered, and their chosen law for g^* implies that the spin splitting energy ΔE_n does depend on the Landau level index n entering in a B^2 term, as they have $\Delta E_n = |g^* \mu_B B|$. A B dependence in g^* is crucial for explaining the experimental data, and also for reproducing them theoretically.

For the spin splitting of the Landau levels we obtain

$$\Delta E_n = \omega_L + 2(2n + 1) \left(\lambda_R^2 \frac{\omega_c}{\omega_c + \omega_L} - \lambda_D^2 \frac{\omega_c}{\omega_c - \omega_L} \right); \qquad (6.129)$$

recall that $\omega_L = |g^* \mu_B B|$, i.e. a splitting that increases with n because of the SO coupling. This SO correction has been worked out for the $n = 0$ level in the paper by Malet *et al.* (2006), using the equation of motion method (see Subsection 8.11). It is known that the experimental results by Dobers *et al.* (1988) for $n = 1$ and 2 can be reproduced if g^* depends on n and B, as already shown in that reference. We have verified that the n dependence of g^* cannot be mimicked by the n dependence introduced by the SO interaction, Eq. (6.129). Recently, the analysis of g^* has been extended to a wider magnetic field range using time-resolved Faraday rotation spectroscopy (Salis *et al.*, 2001; Sih *et al.*, 2004).

As a third example, we address the inelastic light scattering excitation of the spin dipole modes at $\nu = 2$ as measured by Eriksson *et al.* (1999). For this filling factor, in the absence of SO coupling the spin-density inter-Landau level spectrum is expected to be a triplet mode (Kallin and Halperin, 1984) excited by the three operators $\sum_i P_i^+ \sigma_{z,\pm}$ with energy splittings given by the Zeeman energy ω_L. In the presence of SO interactions, we still expect a triplet mode to appear. Indeed, for $\nu = 2$ we have $S_z = 0$ and the cyclotron and spin dipole modes excited by the operator $\sum_i P_i^+ \sigma_z^i$ are decoupled, as previously discussed. Thus, for this operator only one single mode should be detected at an average energy, $\omega = \omega_c(1 + k)$. The other operators, $\sum_i P_i^+ \sigma_\pm^i$, yield the two other spin dipole modes at the energies

$$E^\pm = \omega_c(1 + k) \pm \omega_L \pm 4 \left(\lambda_R^2 \frac{\omega_c}{\omega_c + \omega_L} - \lambda_D^2 \frac{\omega_c}{\omega_c - \omega_L} \right). \qquad (6.130)$$

The splitting is thus symmetric and depends on the SO strengths. In the experiment, triplet excitations were observed in all measured samples up to electron densities corresponding to $r_s = 3.3$ (we recall that $r_s = 1/\sqrt{\pi n_e}$). B was accordingly changed to keep the filling factor at $\nu = 2$. Only one triplet mode spectrum at $B = 2.2$ T was shown. From this spectrum, we infer that there is space for a $\sim 5-10\%$ SO effect on the splitting, assuming that at this fairly small magnetic field, g^* is that of bulk GaAs: $g^* = -0.44$. We have estimated that $m|\lambda_R^2 - \lambda_D^2|/\hbar^2 \simeq 10\,\mu eV$, in line with the previous findings. Systematic measurements, especially at high B, where the splitting is larger, are called for to allow a more quantitative analysis.

Another unequivocal signature of SO effects in quantum wells would be the detection of the Larmor state in photoabsorption experiments. The strength of this transition is given by $2\frac{\lambda_D^2}{\omega_c}(\frac{\omega_L}{\omega_c - \omega_L})^2$ and depends only on the Dresselhaus SO coupling — see the comment immediately after Eq. (6.40). In most experiments, B is perpendicularly applied to the plane of motion of the electrons, and for GaAs the strength is so small that it has never been resolved.

We finally discuss the effect of tilting the applied magnetic field using the expressions derived in the Appendix A. Equation (6.139) can be used to obtain the splitting of the cyclotron resonance which generalizes Eq. (6.39) for tilted magnetic fields:

$$\Delta E_{\text{CR}} = 4 \left[(C_R \, \mathcal{V} + C_D \, \mathcal{Z}) \frac{1}{1 + |g^*|m^* S/2} \right.$$

$$\left. - (C_R \, \mathcal{Z} + C_D \, \mathcal{V}) \frac{1}{1 - |g^*|m^* S/2} \right], \qquad (6.131)$$

where $C_{R,D} \equiv m\lambda^2_{R,D}/\hbar^2$, and the tilting angle θ enters the quantities $\mathcal{V}$, $\mathcal{Z}$, and $\mathcal{S}$ defined in Appendix A. Tilting effects might arise because of the $1 - |g^*|m^*\mathcal{S}/2$ denominator in the above equation, but sizeable effects on ΔE_{CR} should only be expected for materials such that $|g^*|m^*/2$ is large. This is not the case for GaAs, but it is, for example, for InAs and InSb, which have $|g^*|m^*/2 = 0.169$ and 0.355, respectively. For the latter case the dependence of ΔE_{CR} with the *in-of-well* field B_x, with a fixed B_z, is shown in Fig. 6.11. Notice that ΔE_{CR} is sharply increased when B_x exceeds a given value (1 T for the parameters in Fig. 6.11), which is proving the strong enhancement of SO effects introduced by the horizontal component of the tilted field configuration.

Figure 6.11 also shows the comparison with the exact diagonalization data (symbols), indicating that the analytical formula, Eq. (6.131), is accurate up to rather large tilting angles and for varying relative weights of Rashba and Dresselhaus terms. As a matter of fact, this analytical result does not depend on B_z although, for the sake of comparison with the exact diagonalization, we have used $B_z = 1$ T in Fig. 6.11. The evolution with B_x is not always monotonous, especially for $C_R > C_D$, where we find an initial decrease of ΔE_{CR} with increasing B_x, vanishing at $B_x \sim 0.8$ T, and eventually increasing again.

We want to point out that the one-band effective mass approach (EMA) used here is most appropriate for wide gap semiconductors like GaAs ($E_g = 1.43$ eV), and it is currently used for SO studies in this material; see for example Pietiläinen and Chakraborty (2006) and references therein. For narrow gap semiconductors like InSb ($E_g = 0.18$ eV), EMA can still give accurate results if corrections coming from

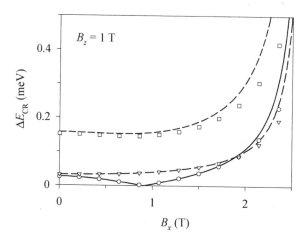

Fig. 6.11 Splitting of the cyclotron resonance for an InSb quantum well as a function of the in-of-well field B_x when $B_z = 1$ T. The lines show the results from the analytical formula, Eq. (6.131), while the symbols correspond to the exact diagonalization of Eq. (6.137). Defining $C_{R,D} \equiv m\lambda^2_{R,D}/\hbar^2$, the shown results are for: $C_R = 30\,\mu$eV and $C_D = 10\,\mu$eV — solid line and circles; $C_R = 10\,\mu$eV and $C_D = 10\,\mu$eV — long-dashed line and triangles; $C_R = 10\,\mu$eV and $C_D = 30\,\mu$eV — short-dashed line and squares.

nonparabolicity are employed (as an alternative to multiband models). However, as long as no comparison with a given experiment is carried out for InSb (we basically look for physical phenomena), the obtained results, neglecting nonparabolicity, are still meaningful but of course more qualitative than for GaAs. It is also worth recalling that the envelope function approximation implicit in EMA integrates the details described by the Bloch functions in the employed parameters (effective masses, Luttinger parameter for holes, etc.) so that only the envelope functions remain, in either the presence or the absence of an external magnetic field. We cannot discard the posibility that for InSb the underlying band structure might have some influence on the optical transitions, but discussing it in detail is beyond the scope of this book. We want also to indicate that the largest employed magnetic field for InSb, $B = 2$ T, is still fairly small, so that the magnetic confinement radius is large and no relevant effects on possible induced admixture of hole states is expected.

Finally, the tilting also affects the spin splitting of the Landau levels

$$
\frac{\Delta E_n}{\omega_c} = \frac{|g^*|m^*}{2}\mathcal{S} + 2(2n+1)\left[(y_R\,\mathcal{V} + y_D\,\mathcal{Z})\frac{1}{1+|g^*|m^*\mathcal{S}/2}\right.
$$
$$
\left. - (y_R\,\mathcal{Z} + y_D\,\mathcal{V})\frac{1}{1-|g^*|m^*\mathcal{S}/2}\right],   \tag{6.132}
$$

which generalizes Eq. (6.129) for $\theta \neq 0$. As we have commented before, in an recent experiment where spin precession frequencies in an InGaAs quantum well have been measured using electrically detected electron-spin resonances (Sih *et al.*, 2004), a strong dependence of the effective gyromagnetic factor g^{eff} on the applied tilted B has been found. In particular, at $\theta = 45°$ g^{eff} exhibits oscillations with B which indicate its sensitivity to the Landau level filling, and a coupling between spin and orbital eigenstates which is explicitly present in the SO term of Eq. (6.132). The effective g factor that can be extracted from this equation at $\theta = 45°$, by taking the ratio $2\Delta E_n/m^*\mathcal{S}\omega_c$, has the structure

$$
|g^{\mathrm{eff}}(B,n)| = |g_0^*| + \left(n+\frac{1}{2}\right)\left[c_1 B + \frac{c_2}{B}\right],   \tag{6.133}
$$

where the parametrization $g^* = g_0^* - c_1(n+\frac{1}{2})B$ of Dobers *et al.* (1988) and Sih *et al.* (2004) has been introduced, and the c_2 term is the SO contribution. For the smaller B values in the experiment, and for reasonable values of $m\lambda_{R,D}^2/\hbar^2$, of the order of 1–10 μeV, the SO contribution is important enough and should not be neglected; under these circumstances, time-resolved Faraday rotation spectroscopy could be sensible for Rashba and/or Dresselhaus SO effects.

6.9   LSDA Results and Comparison with Experiments for Quantum Dots

Figure 6.12 displays the magnetic field evolution of the single-particle energies (6.62) for the last occupied single-particle level, with $\lambda_R = 0$ and $\lambda_D \approx \gamma(\pi/z_0)^2$ for a

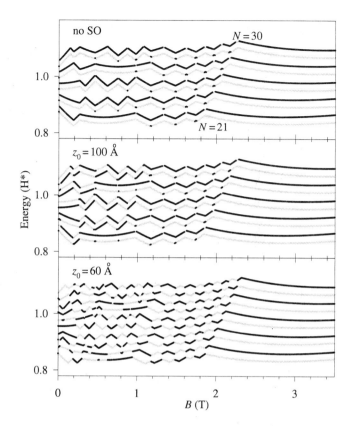

Fig. 6.12 Single-particle energies in effective atomic units of the highest occupied level for a dot with N electrons. Circular symmetry with $\omega_x = \omega_y = 1.1$ meV has been assumed. Each line has been shifted vertically by a small amount representing the charging energy. Light and dark gray tones correspond to up and down spin, respectively.

circular dot with N electrons, except for a constant representing the charging energy. These values give the chemical potential of the dot for a varying electron number, as measured for instance in capacitance spectroscopy experiments (Ashoori, 1996; Ciorga *et al.*, 2000). (Notice that, unlike the results reported in Figs. 5.2–5.4, they do not include the effect of the interaction calculated in the constant interaction model. Consequently the results reported in the top of Fig. 6.12 differ from that of Fig. 5.2 by a constant.) Actually the parabola coefficient has been taken from a fit to the experiments. The spin is indicated in Fig. 6.12 with light and dark gray tones. In the absence of SO coupling each line corresponds to a given spin, except for a very small fluctuation due to the Zeeman energy at some cusps and valleys. In this case the traces arrange themselves in parallel pairs of up and down spin. As shown in the two lower panels of Fig. 6.12, the SO coupling produces sizeable up and down fluctuations of the spin. For $z_0 = 100$ Å, i.e. weak SO coupling, the fluctuations start at low magnetic fields and they extend up to $B \approx 1$ T. An even stronger SO ($z_0 = 60$ Å) produces spin inversions up to the last level crossing, which marks the filling factor $\nu = 2$ line. Besides, in the latter case the traces are no longer paired but, instead, anticorrelated with a π phase shift, specially in the region just before $\nu = 2$. The results of Fig. 6.12 can help to interpret the experiments of Ashoori (1996) and Ciorga *et al.* (2000), which observed anticorrelated behavior in the traces and spin alternation with increasing B, respectively.

The effects of deformation and Coulomb interaction have been analyzed by Valín-Rodríguez *et al.* (2002) using the local spin density approximation within the spinorial formalism of Subsection 6.7, by taking $\lambda_R = 0$ and $\lambda_D \approx \gamma(\pi/z_0)^2$, with $\gamma \approx 20$ eV Å^3 for GaAs.

Figure 6.13 displays the LSDA density $\rho(\mathbf{r})$ and spin magnetization $\mathbf{m}(\mathbf{r})$ for $N = 9$ electrons in a circular confining potential with $\omega_x = \omega_y = 6$ meV. We have selected this electron number as a representative case in which to check the robustness of the analytically predicted spin inversions. A Dresselhaus SO coupling with $z_0 = 62$ Å and $B = 2.5$ T has been assumed. The magnetization density indicates the local orientation of the spin vector and it is related to the spin-density matrix by $m_x = 2\,\mathrm{Re}[\rho_{\uparrow\downarrow}]$, $m_y = 2\,\mathrm{Im}[\rho_{\uparrow\downarrow}]$ and $m_z = \rho_{\uparrow\uparrow} - \rho_{\downarrow\downarrow}$. We note from Fig. 6.13 that circular symmetry is conserved by both ρ and m_z, while the horizontal magnetization $\mathbf{m}_\parallel \equiv (m_x, m_y)$ shows an angle-dependent texture. This result is in good agreement with the independent particle model solution given in Subsection 6.5, which

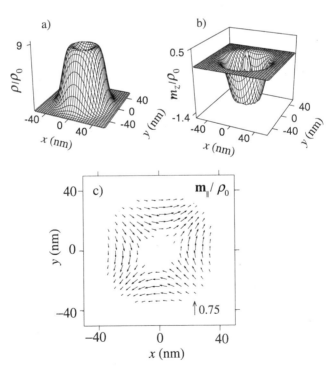

Fig. 6.13 Density ρ and magnetization $\mathbf{m}$ for a circular dot with $N = 9$ electrons, $\hbar\tilde{\omega}_x = \hbar\tilde{\omega}_y = 6$ meV, $B = 2.5$ T and $z_0 = 62$ Å. The values have been scaled by $1/9$ of the dot central density, i.e. $\rho_0 = (2.2\ 10^{-5}\ \text{Å}^{-2})/9$.

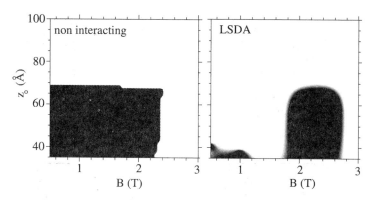

Fig. 6.14 Vertical spin evolution in the B–z_0 plane for the for a circular dot with $N = 9$ electrons, $\hbar\tilde{\omega}_x = \hbar\tilde{\omega}_y = 6$ meV. White (black) indicates upward (downward) total spin.

predicts that $\mathbf{m}_\parallel(\mathbf{r}) \sim \rho(r)\,(y, x)$. Note also that for this particular z_0 and B the vertical spin is predominantly inverted, giving a negative value for the total vertical spin $\langle S_z \rangle$.

Figure 6.14 shows $\langle S_z \rangle$ as a function of B and the intensity of the SO coupling, given by z_0, for $N = 9$ electrons in a circular confining potential with $\omega_x = \omega_y = 6$ meV. In agreement with the above discussion, for $B \leq 2.2$ T the noninteracting model predicts spin inversion when decreasing the dot width. The LSDA also yields spin-inverted regions, although with some conspicuous differences. The interaction inhibits the spin-flip at low magnetic fields and low widths, shifting the inversion region to 1.8 T $\leq B \leq 2.6$ T and leaving only a small residue for $z_0 \leq 40$ Å and $B \leq 1.2$ T. It is worth mentioning that although in the laboratory frame $\langle S_z \rangle$ is not restricted to discrete values (because of the transformation U in Appendix B), in practice its fluctuations increase with the SO strength, but they are generally small.

In Fig. 6.15 we show the vertical spin as a function of the applied magnetic field and the dot deformation, for two different values of the SO coupling. In this figure the mean value $(\omega_x + \omega_y)/2$ was kept fixed to 6 meV while the ratio $\delta = \omega_y/\omega_x$ was varied to obtain different elliptical shapes. Comparing left and right panels, we see again an interaction-induced quenching of the spin inversion at low magnetic fields. Although this occurs for the two displayed widths, at $z_0 = 48$ Å a larger region with inverted spin is found. Figure 6.15 shows that, having fixed the SO coupling strength, spin inversion can be achieved in many cases either by increasing the deformation or by increasing the magnetic field. The comparison of the $N = 9$ numerical results with the single-particle model (see Appendix B) allows us to conclude that, in spite of the differences, the spin inversions in the LSDA are qualitatively similar to the noninteracting ones. As a final piece of information we mention that the energy gap between the highest occupied and lowest unoccupied Kohn–Sham levels in the above cases stays in the range of $[0.3, 0.9]$ meV. This result provides a measure of the

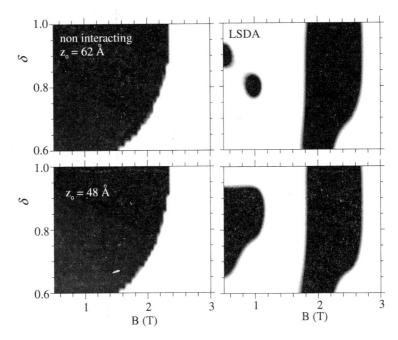

Fig. 6.15 Vertical spin evolution in the B–δ plane, where $\delta = \omega_y/\omega_x$ labels the deformation, for a fixed value of the SO coupling. The upper row corresponds to $z_0 = 62$ Å in the noninteracting model (left) and the LSDA (right). The lower row shows the same results for $z_0 = 48$ Å. We have used the color convention of Fig. 6.14.

ground state stability and, therefore, of the relative spin stiffness against thermal fluctuations.

We next discuss the spin precessional properties of electrons confined to a model GaAs quantum dot, including SO coupling. As we have anticipated in Subsection 6.5, this mechanism gives rise to a rich variety of spin precessional frequencies, depending on the orbital state of the electrons, even in the absence of a vertical magnetic field, as compared to the Larmor frequency for systems without SO coupling.

In order to excite the electronic spin precession, one needs to perturb the ground state spin configuration. The usual way to achieve this consists in applying a horizontal magnetic field for a certain time interval that rotates the spin and triggers the precessional motion. In the absence of SO coupling ($\lambda_{R,D} = 0$) the only allowed transitions are the spin-flips $(n_r\ell \uparrow) \leftrightarrow (n_r\ell \downarrow)$, leading to an in-plane spin precession at the usual Larmor frequency, $\omega_L = |g^*|\mu_B B/\hbar$. This is a well-known result valid even when spin-independent interactions are present (Larmor theorem). When SO coupling is considered, besides the pure spin-flips other transitions involving additional changes in ℓ and/or n_r are allowed. In addition to monopolar $\delta\ell = 0$, dipolar $\delta\ell = \pm 1$ and quadrupolar $\delta\ell = 2$ spin-flip transitions also contribute with different weights. As we will show below, the $\delta\ell = 0$ transitions of the pure Larmor mode are still the dominant ones in the precessional spectrum with SO coupling; the dipolar ones are weaker by more than an order of magnitude while the quadrupolar spin-flip excitations turn out to be negligible in all cases studied. It is worth pointing out that even transitions between orbitals with different n_r could contribute because of the nonorthogonality of ↑ and ↓ radial functions, although these will normally involve high energies and low strengths due to the small deviation from pure orthogonality.

The lowest $\Delta\ell = 0$ spin-flip mode, which we will call the SO precessional mode ω_P, was discussed in Subsection 6.5 in the framework of the noninteracting model with SO coupling and has a frequency [see Eq. (6.73)], in the low $\lambda_{R,D}$ limit, given by

$$\omega_P = \omega_L + 2(\lambda_R^2 - \lambda_D^2)\ell + 2(\lambda_R^2 - \lambda_D^2)(2n_r + |\ell| + 1)\frac{\omega_c}{\Omega} . \qquad (6.134)$$

This mode is the dominant excitation in the spin rotation spectrum and it can be considered in a natural way as the modification of the pure Larmor mode ω_L by the SO coupling. As we have already noted, at $B = 0$ the SO precessional frequency does not vanish when $\ell \neq 0$, with the offset indicating the SO coupling intensity. Furthermore, in general, positive and negative ℓ orbitals will display different B dispersions for a fixed $\lambda_R^2 - \lambda_D^2$ and when the precessional active orbital changes owing to an internal rearrangement the SO precessional frequency will display a discontinuity.

The effects of the higher order term in the SO coupling and Zeeman interaction and of the Coulomb interaction have been taken into account by Valín-Rodríguez *et al.* (2002) by means of a real-time simulation method within the LSDA functional theory. They have implemented in a spatial grid the solution to the time-dependent Kohn–Sham equation, labeling the discrete set of orbitals by an index j,

$$i\hbar\frac{\partial}{\partial t}\varphi_j(\mathbf{r}, \eta) = \sum_{\eta'} h_{KS}(\mathbf{r}, \eta\eta')\varphi_j(\mathbf{r}, \eta'), \qquad (6.135)$$

where $h_{KS}(\mathbf{r}, \eta\eta')$ is the Kohn–Sham Hamiltonian of Eq. (6.87).

A "real time" simulation of the precession can be performed by taking the stationary solutions to Eq. (6.135), rotating in spin space with a given horizontal axis (say, the x axis) and using the resulting perturbed spinors as the starting point for the time evolution. Figure 6.16 displays one such simulation for an $N = 7$ quantum dot without Coulomb interaction and only with Dresselhaus SO coupling. The analysis is based on the x component of the total spin, in time (lower panel) and energy (upper panel) domains. The different features discussed above in the analytical model are nicely manifested by the numerical signals. From Fig. 6.16 we can also see quantitatively the strength of the SO precessional mode with respect to the doubly split upper and lower branches of the dipole modes. The minor energy differences between the numerical and analytical peak positions can be attributed to effects beyond $O(\lambda_D^2)$ and, also, to neglecting the ω_L terms in the energy denominators of Eqs. (6.68) and (6.69) in order to get Eqs. (6.72) and (6.134), respectively.

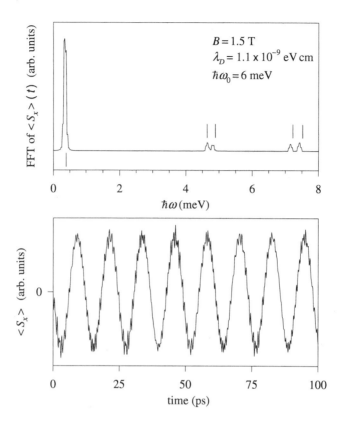

Fig. 6.16   Real time simulation of the spin evolution following an initial rotation with the x axis. Shown is the x component of total spin in time (lower) and energy (upper) domains. The vertical bars in the upper panel indicate the analytical energies of Eqs. (6.134) and (6.72).

A systematics of the lowest peak energy, i.e. the SO precessional mode, is gathered in Fig. 6.17 as a function of the magnetic field for two different λ_D's. Focusing first on the noninteracting results, we note that for the smaller λ_D the agreement between analytical and numerical values is excellent, proving the equivalence of both methods; while the slight differences for $\lambda_D = 1.1 \times 10^{-9}$ eV cm can be understood on the basis of the previous discussion. The already-mentioned offset at $B = 0$ with respect to the Larmor frequency is clearly seen in Fig. 6.17 for $N = 7$ and 11, as well as the diferent slopes for different ℓ and λ_D values.

A better understanding of the precessional mode systematics is obtained from Fig. 6.18, which displays the level scheme as a function of the magnetic field for $\lambda_D = 1.1 \times 10^{-9}$ eV cm. In this figure the active level for the same electron numbers of Fig. 6.17 is marked with thick dots and dashes. We note the clear correspondence of the discontinuities in Fig. 6.17 with the crossings in Fig. 18, which correspond to changes in the precessional level.

Figure 6.19 is the analog of Fig. 6.16 within the LSDA. The time signal has a similar large period, but the lower period modulations are manifestly different.

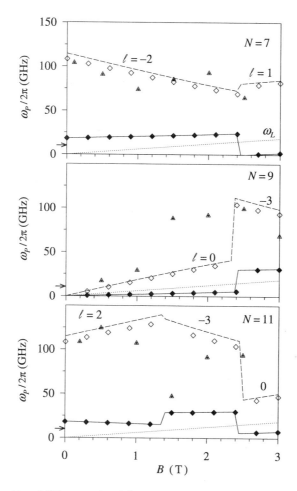

Fig. 6.17 Systematics of SO precessional frequencies $\omega_{\mathcal{P}}$ as a function of a magnetic field for two different values of the Dresselhaus parameter λ_D: in the analytical model (lines), and from the numerical calculation without (diamonds) and with Coulomb interaction (triangles with crosses). Solid lines and symbols correspond to $\lambda_D = 0.4 \times 10^{-9}$ eV cm; dashed lines, open symbols and crossed triangles to $\lambda_D = 1.1 \times 10^{-9}$ eV cm. The dotted line shows the Larmor frequency. Also indicated is the ℓ value of the precessionally active orbital in the analytical model, which at a given B is the same for the two λ_D values. The arrows on the vertical scale indicate the approximate lower frequency that can be obtained from the time simulation window of 100 ps.

Accordingly, the Fourier transform (upper panel) shows a similar low energy precessional mode but the distribution of minor peaks is rather different. The dipole peaks of Fig. 6.16 are washed out and, instead, new excitations at $\hbar\omega \approx 1.7$ and ≈ 4 meV appear. As was discussed by Valín-Rodríguez, Puente and Serra (2002) in the context of the far-infrared absorption, these excitations are collective spin oscillations known as dipole magnons. Figure 6.17 also shows the LSDA numerical calculations for the higher SO coupling constant. The characteristics of the precessional

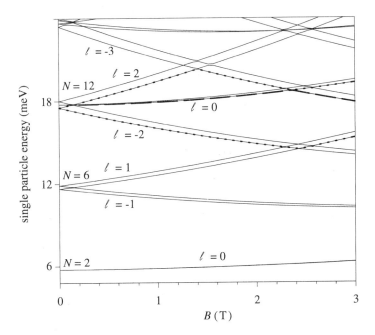

Fig. 6.18 Energy level scheme as a function of the magnetic field for the same dots of Fig. 6.17 with an SO parameter $\lambda_D = 1.1 \times 10^{-9}$ eV cm. The angular momentum for each level and the electron number at shell closures are indicated. The curves marked with thick dots and dashes indicate the precessionally active level for the electron numbers $N = 7, 9$ and 11.

mode discussed above are qualitatively retained within the LSDA, although with the important difference that the discontinuity points are changed because of the interaction-induced orbital rearrangements.

In general, along with the transverse magnetic field the system will be probed by an electric field. This modifies the relative strength of the precessional mode with respect to the plasmon and magnon peaks of Figs. 6.16 and 6.19. We have checked this numerically by using an initial charge translation, simulating the effect of the electric field at $t = 0$, simultaneously to the spinor rotation. The corresponding spectra display the same peaks of Fig. 6.16 for the noninteracting case and of Fig. 6.19 in the LSDA, but with different heights. Therefore, only the strength, not the energy of the precessional mode, depends on the coupling with the electric field.

We finally study the effects of the SO coupling on the far-infrared (FIR) optical absorption spectrum of quantum dots. We use a real-time simulation within the LSDA formalism of the same type discussed above for the precessional motion but starting from a different initial perturbation of the wave function. Physically, this corresponds for instance to the interaction with a short laser pulse or with an appropriate projectile. In the calculation it can be mimicked simply by a rigid translation of the wave functions with the operator $T(\mathbf{a}) = e^{-\mathbf{a} \cdot \nabla}$ or by an initial impulse with $\Pi(\mathbf{q}) = e^{-i\mathbf{q} \cdot \mathbf{r}}$. When either $\mathbf{a}$ or $\mathbf{q}$ is small, these perturbations induce

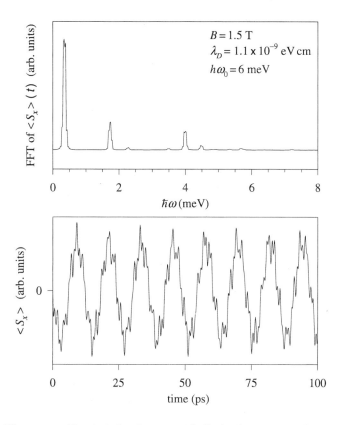

Fig. 6.19   The same as Fig. 6.16, for the case with Coulomb interaction between electrons.

dipole oscillations predominantly and the system's response is restricted in the linear regime. After the initial perturbation one keeps track of the time-dependent dipole moment d: $\langle \mathbf{e} \cdot \mathbf{r} \rangle$, where $\mathbf{e}$ corresponds to the direction of the initial perturbation given by $\mathbf{a}$ or $\mathbf{q}$.

The left panels of Fig. 6.20 show the numerical result for the FIR absorption of dots with $N = 6$ and 10 electrons with a Dresselhaus parameter for a dot width of $z_0 = 50$ Å and without Coulomb interaction. As a representative case, we consider an external parabolic confinement with energy $\hbar\omega_0 = 4.2$ meV. The splitting around $\hbar\omega_0$ nicely agrees with the analytical prediction (arrows) of Eq. (6.72) at $B = 0$, thus proving the validity of the model. For the $N = 10$ case we note the existence at low energy of a peak with much less strength than the dominant ones (notice the enlarged scale). This, too, can be explained within the analytical model realizing that $N = 10$ has a partial occupation of the third oscillator major shell and thus the possibility of quasispin-flip transitions qup–$qdown$ at an energy $4\lambda_D^2 m/\hbar^2$, indicated with an arrow in Fig. 6.20. Note that in spite of having $\Delta\ell = 0$ these transitions can manifest themselves in the dipole spectrum due to the SO coupling.

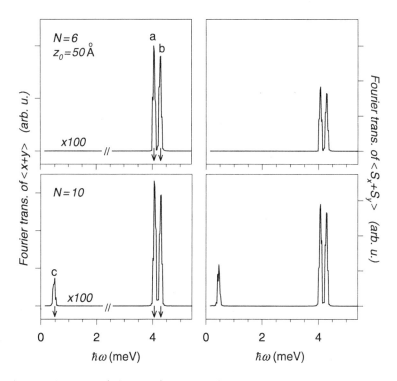

Fig. 6.20 FIR absorption (left panels) obtained from the Fourier transform of the real-time dipole oscillation for a dot width of $z_0 = 50$ Å, without Coulomb interaction. The right panels display the corresponding frequency transform of the horizontal spin signal. The arrows indicate the energies of the split Kohn modes (at ≈ 4 meV) and of the quasispin flip modes (at ≈ 0.5 meV) using the analytical model for this z_0.

This is further proved in the right panels of Fig. 6.20, where the Fourier transform of $\langle S_x + S_y \rangle$ is plotted. The same frequencies of the dipole absorption manifest themselves in the spin channels, although with different relative strengths. It is also interesting to analyze the spatial patterns of induced density $\delta\rho(\mathbf{r}, t)$, parallel spin $\delta\mathbf{S}_\parallel(\mathbf{r}, t)$ and vertical spin $\delta S_z(\mathbf{r}, t)$ for a given oscillation mode and a fixed time (Fig. 6.21), where we define the spin density as $\mathbf{S}(\mathbf{r}) = \sum_{i=1}^{N} \langle \varphi_i \,|\, \vec{\sigma}\delta(\mathbf{r} - \mathbf{r}_i) \,|\, \varphi_i \rangle$. This local-signal analysis has been performed using the method of Valín-Rodríguez, Puente and Serra (2001). Quite remarkably, the oscillation of $\delta S_z(\mathbf{r}, t)$ is always in a direction perpendicular to that of the charge density, which is explained in the analytical model as a coherent quasiup and quasidown excitation due to the SO term. On the contrary, the parallel spin patterns depend on whether the mode has a quasispin-flip character or not. For the split Kohn modes (peaks a and b), $\delta\mathbf{S}_\parallel(\mathbf{r}, t)$ is localized in the two regions perpendicular to the charge oscillation axis. However, the quasispin-flip peak (c) shows no angular dependence of the pattern. It can be shown that this behavior is in complete agreement with the prediction of the analytical model using the spinors of Subsection 6.5.

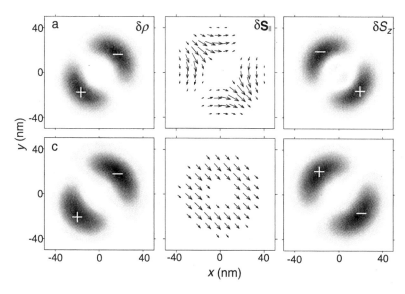

Fig. 6.21 Spatial patterns of oscillating density $\delta\rho(\mathbf{r},t)$, horizontal spin $\delta\mathbf{S}_\parallel(\mathbf{r},t)$ and vertical spin $\delta S_z(\mathbf{r},t)$ at a fixed time for the a and c dipole modes of Fig. 6.20. The gray scale of the left and right panels indicates the signal magnitude while the corresponding sign is superimposed in white.

We next consider how the features discussed above are modified by the addition of the Coulomb interaction treated in the LSDA approximation. The modifications stemming from the Coulomb interaction are twofold; first, the electron–hole transitions move in energy due to the change in the effective mean field; second, new collective peaks may appear as a result of coherent electron–hole excitations. Figure 6.22 shows the interacting results for $N = 6$ with $z_0 = 50$ and 35 Å. Note that the splitting of the Kohn-like modes disappears for the larger width and, at the same time, a collective peak begins to appear at an energy below the electron–hole states. These results clearly show the formation of the collective plasmon and magnon states lying above and below the electron–hole transitions, respectively. We stress that in the absence of SO coupling the magnon state does not contribute to the FIR absorption. The lower left panel of Fig. 6.22 indicates that a great enhancement of the magnon contribution is obtained by reducing the dot width, which is accompanied by an important Landau damping of the plasmon state into a bundle of peaks. The Coulomb interaction thus suppresses the splitting of the Kohn-like modes although it introduces the mechanism of Landau damping that ultimately, and rather abruptly with z_0, dominates for decreasing widths.

The right panels of Fig. 6.22 show the horizontal spin spectra induced by the dipole shift. These results permit one to ascertain the spin character of the different modes and, by comparison with the left panels, quantify their relevance to the FIR absorption. The relative spin or density character of each state is also seen from the amplitudes $\mathcal{A}$ of density and spin oscillation. For instance, when $z_0 = 50$ Å we find

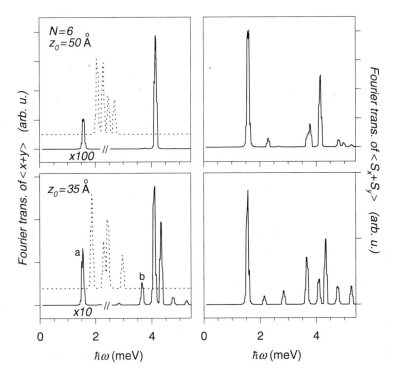

Fig. 6.22 The same as Fig. 6.20, including Coulomb interactions. The dashed line in the left panels displays the electron–hole transitions in the self-consistent mean field.

that $\mathcal{A}[\delta S_z(\mathbf{r}, t)] \approx 60\,\mathcal{A}[\delta\rho(\mathbf{r}, t)]$ for the magnon and $\mathcal{A}[\delta S_z(\mathbf{r}, t)] \approx 0.4\,\mathcal{A}[\delta\rho(\mathbf{r}, t)]$ for the plasmon, which shows the dominant spin and density type for magnon and plasmon states, respectively.

The characterization of the interacting peaks is facilitated by the comparison of the spatial patterns with the corresponding ones in the noninteracting model. Figure 6.23 displays the density and spin patterns for two of the peaks of the $z_0 = 35$ Å spectrum (lower left panel of Fig. 6.22). As in the noninteracting model, the oscillation of $\delta S_z(\mathbf{r}, t)$ is orthogonal to that of $\delta\rho(\mathbf{r}, t)$ and the horizontal spin patterns show a priviledged spin direction, i.e. they have a net induced spin. Note in particular that peak b of Fig. 6.22 has no angular dependence in the horizontal spin pattern while for peak a there are two regions with enhanced $\delta\mathbf{S}_\parallel(\mathbf{r}, t)$. We have checked that these features are not changed when the polarization of the mode, given by the direction of the density oscillation, is varied. We stress that the mechanism discussed here would permit control of the horizontal spin oscillation by tuning the frequency of the applied electric field. This effect of spin-to-charge conversion is a major requisite for hybrid spintronic devices.

Figure 6.24 presents the spectra with Coulomb interaction obtained for $N = 10$, an open shell system in the independent-particle model. At $z_0 = 50$ Å the open shell character of this dot is clearly maintained, with low energy electron–hole transitions

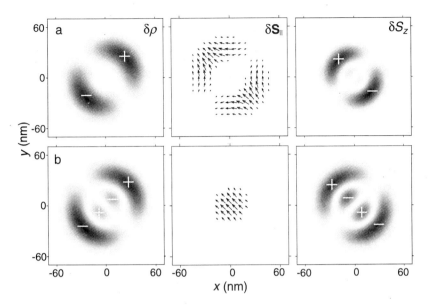

Fig. 6.23 The same as Fig. 6.21, including Coulomb interaction. The upper and lower rows correspond to the peaks labeled a and b in Fig. 6.22, respectively.

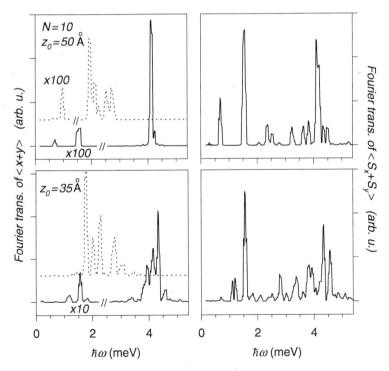

Fig. 6.24 The same as Fig. 6.22, for $N = 10$.

and an associated collective mode at $\omega \approx 0.7$ meV similar to the quasispin-flip excitation of the analytical model. For narrower dots, however, the energy of quasi spin-flip transitions increases and as a result this mode merges with the magnon and other electron–hole states. Contrary to the quasispin-flip state, both magnon and plasmon mean energies are not dependent on the Dresselhaus coupling.

In practice the control of the dot vertical extent might be quite delicate and will depend on the type of quantum dot. The feasibility of this procedure by changing the fixed charge concentrations or other device parameters in etched quantum dots can be estimated from the 3D calculations of Kumar, Laux and Stern (1990) and Bruce and Maksym (2000). An additional electric field in the vertical direction will squeeze the electron density against the $Al_xGa_{1-x}As$ barrier, effectively reducing the dot width.

Finally, in Figs. 6.25 and 6.26 we report the calculated absorption spectra (dipole-allowed) for four-electron InAs and InSb quantum dots of Pietiläinen and Chakraborty (2006). Interaction effects are included in the calculation by means of diagonalization techniques and results are given for a parabolic confinement strength $\omega_0 = 7.5$ meV and for various values of the Rashba SO coupling strength. A striking feature visible in the absorption spectra of the InAs quantum dot is the appearance of discontinuities, anticrossings, and new modes in addition to the two main ($\lambda_R = 0$) absorption lines. These optical signatures of the SO interaction are consequences of the multitude of level crossings and level repulsions that occur in the energy spectra. The latter can be attributed to an interplay between the SO and Zeeman couplings. At moderate SO coupling strengths the absorption spectra do not

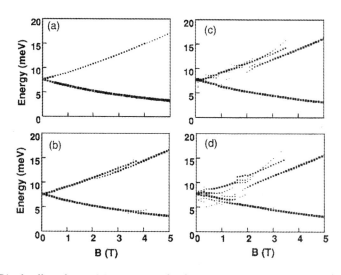

Fig. 6.25 Dipole-allowed transition energies for four interacting electrons confined in an InAs quantum dot and for various values of the SO coupling strength (in meV nm): λ_R=(a) 0, (b) 20, (c) 30 and (d) 40 and $\lambda_D = 0$. The size of the points in the figures is proportional to the calculated intensity.

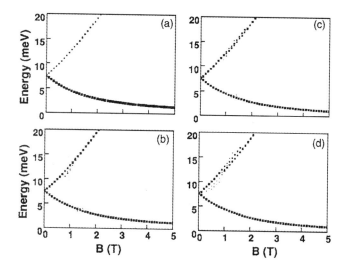

Fig. 6.26 The same as Fig. 6.25, but for a four-electron InSb quantum dot.

essentially differ from the single-particle spectra. But when the coupling strength increases, the deviations from the pure parabolic confinement also increase, which in turn implies that the lowest final states of dipole-allowed transitions are no longer achievable by adding $1/2(\Omega \pm \omega_c)$ to the initial state energies. In particular, this results in discontinuities and anticrossing as well as appearance of new modes. As compared to the spectra of InAs dot, that of the InSb four-electron dot shows a clear difference in the almost total absence of anticrossing and discontinuities. This is partly due to the very large Zeeman coupling which practically nullifies the SO interaction at the coupling strengths λ_R we are concerned with. Another reason is the large kinetic energies due to the very small electron effective mass. Because the strength of the Coulomb interaction is somewhat smaller than in InAs owing to the different dielectric constants, correlations caused by the mutual electronic interactions are effectively much smaller in InSb than in InAs.

6.10   LSDA Results for Quantum Wires

Here we present the solution of the noncollinear LSDA theory of Subsection 6.7 for a quantum wire with a harmonic confinement potential and in the presence of SO coupling for ρ_1 densities up to filling six subbands. Both the weak and strong coupling regimes are covered. To present the results, we have used the harmonic oscillator length $l_0 = \sqrt{\hbar/m\omega_0}$ and have expressed both the linear density ρ_{1D} and the wave number k in units of l_0^{-1}. For a typical energy value $\hbar\omega_0 = 4$ meV, a unit linear density $\rho_{1D} = l_0^{-1}$ is about 5.9×10^5 cm^{-1}. The energies are expressed in $\hbar\omega_0$ units.

Figure 6.27 shows the energy per electron E/N at $B = 0$ as a function of ρ_{1D} in the weak and strong SO coupling regimes. Due to the exchange-correlation energy

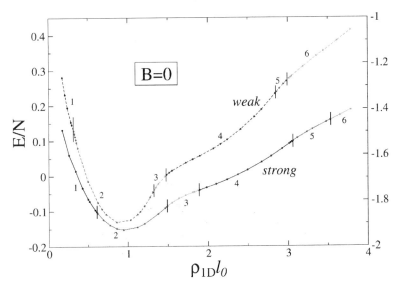

Fig. 6.27 Energy per electron (in $\hbar\omega_0$ units) as a function of the linear density at $B = 0$. The regions separated by vertical lines are characterized by the indicated number of *distinct* subbands crossed by the electron chemical potential μ, i.e. partially occupied subbands. Some of these subbands are crossed more than twice by μ, producing the anomalous steps in the conductance discussed in the text. The vertical left (right) scale corresponds to the weak (strong) SO regime. The lines have been drawn to guide the eye.

(Malet *et al.*, 2005), E/N is not a monotonous function of the linear density, and neither is the chemical potential μ. The electron–electron interaction also means that, for some ρ_{1D} values, we have found two possible configurations corresponding to a number of occupied subbands that differ in one unit. When this occurs, we choose the configuration that corresponds to the smallest E/N, the other one being metastable in some sense. Examples of the existence of these metastable configurations when the SO interaction is not taken into account can be found in Malet *et al.* (2005).

We have studied the effect of the exchange-correlation interaction in several situations involving in-plane magnetic fields and different strengths of the SO interaction, and have found it difficult to systematize, as its effect depends on the actual value of other variables that characterize the quantum wire, as ρ_{1D}, the orientation ϕ_B of the applied B, and the values of the SO coupling constants. In general, V^{xc} has a tendency to enhance magnetic field effects, which can be noticed when comparing the subband structure with and without V^{xc} at the same density, and — perhaps the most interesting feature — V^{xc} acts in some cases as an applied magnetic field, especially at low densities. Indeed, we have found that for some configurations, the subband structure at $B = 0$ when V^{xc} is taken into account turns out to be qualitatively the same as when a certain B is applied in-plane and $V^{xc} = 0$. Analogously, when a magnetic field is already applied, V^{xc} may act as if it were an additional field increasing the value of the actual B field, or contributing to create an effective in-plane magnetic field with an orientation different from that of the actually applied field. Likely, the lack of a common spin axis when SO effects are taken into account has much to do with the complex effect of V^{xc} on the subband structure.

Since in many previous works only the Rashba SO interaction has been taken into account, it is pertinent to begin with the discussion of the V^{xc} effects in the $\Delta_D = 0$ situation [see Eq. (6.127)]. As an example, Fig. 6.28 shows the results corresponding to a low density QW, $\rho_{1D}l_0 = 0.17$, for an applied B field of 20 T and $\phi_B = 0$ in a strong SO regime, namely $\Delta_R = 0.37$. One may see that when $V^{xc} = 0$ the first subband presents the symmetric double minimum structure already found by other authors (Pershin, Nesteroff and Privmann, 2004; Serra, Sanchez and Lopez, 2005), whose existence yields an anomalous step in the conductance — see for example Fig. 2(c) of the paper by Pershin, Nesteroff and Privmann (2004) and Fig. 5 (top panel), of the one by Serra, Sanchez and Lopez (2005). The effect of V^{xc} at this low density is to induce an asymmetry in the lowest subbands transforming

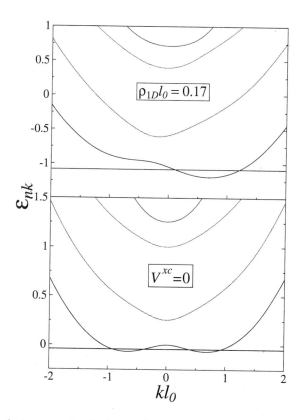

Fig. 6.28 Single electron energies (in $\tilde{\omega}_0$ units) for $\rho_{1D}l_0 = 0.17$ and $B = 20$ T, $\phi_B = 0$ in a strong SO regime characterized by $\Delta_R = 0.37$, $\Delta_D = 0$, as a function of the linear momentum kl_0. The thin horizontal line represents the chemical potential. The effect of V^{xc} has not been included in the results shown in the bottom panel.

the symmetric double minima structure, characteristic of the $V^{xc} = 0$, $\phi_B = 0$ case when only the Rashba term is considered (Pershin, Nesteroff and Privmann, 2004), into a structure rather similar to the one corresponding to $V^{xc} = 0$, $\phi_B = \pi/4$, as shown in Fig. 2(b) of the previous reference. This effect yields only one minimum below the chemical potential contributing to the conductance G (see the end of this subsection and Subsection 8.8). We should note that, for small Δ_R values, the double minimum structure is not found even when $V^{xc} = 0$, whereas in a very strong regime, for example $\Delta_R = 0.83$, this structure is also found for odd $n > 1$ values. In this case, the changes induced by V^{xc} are qualitatively similar to the ones displayed in Fig. 6.28.

When the SO interaction is present, spin is not a good quantum number and it is possible to find spin textures across the wire (Serra, Sanchez and Lopez, 2005). This is illustrated in Fig. 6.29, which corresponds to the situation displayed in Fig. 6.28. The left panel corresponds to the $V^{xc} = 0$ case. The vector plot shows the in-plane spin magnetization, and the solid line corresponds to the z component [see Eq. (6.105)]. One thus see that the effect of V^{xc} is to help produce textures that otherwise SO alone would not yield.

Exchange-correlation effects also appear when both SO contributions are taken into account. Figure 6.30 shows the energy subband structure in one of the most interesting situations for the discussion of the conductance that we will carry out at the end of the subsection. It corresponds to the strong coupling regime for a $\phi_B = \pi/2$ magnetic field. In both panels, conspicuous subband gaps and local extrema appear near $k = 0$. The interesting feature is the weak local maxima at $k > 0$ for

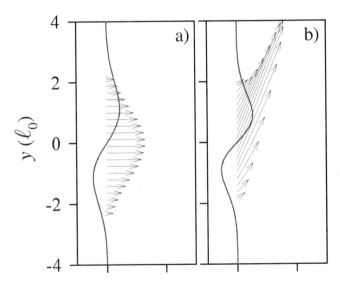

Fig. 6.29 Spin texture across the wire (y direction, in l_0 units) corresponding to the situation displayed in Fig. 6.28. The left panel corresponds to the $V^{xc} = 0$ case. The vector plot shows the in-plane spin, and the solid line corresponds to the z component.

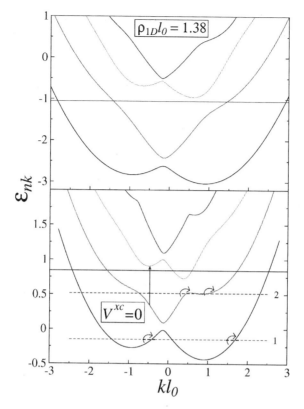

Fig. 6.30 Single electron energies (in $\tilde{\omega}_0$ units) for $\rho_{1D}l_0 = 1.38$ and $B = 20$ T, $\phi_B = \pi/2$ in a strong SO regime characterized by $\Delta_R = 0.093$, $\Delta_D = 0.37$, as a function of the linear momentum kl_0. The effect of V^{xc} has not been included in the results shown in the bottom panel. In both panels, the thin horizontal line represents the chemical potential for the linear density $\rho_{1D}l_0 = 1.38$, whereas the dashed horizontal lines in the bottom panel represent the chemical potential for two smaller linear densities chosen to show different kinds of intrasubband excitations, represented by curved arrows near the corresponding Fermi level, which contribute to the QW conductance, as discussed in Subsection 8.8. The vertical arrow in the bottom panel represents an intersubband transition.

the even subbands when V^{xc} is not considered (bottom panel). Similar structures had been found by Moroz and Barnes (1999), who addressed the $B = 0$ case for the Rashba SO interaction. The existence of these maxima is the reason for the "anomalous steps" in the conductance that appear on top of the "ordinary steps" at even e^2/h values (see the end of this subsection and Subsection 8.8). The inclusion of V^{xc} washes out these structures, as can be seen in the top panel of Fig. 6.30. This can be understood by noting the additional subband splitting induced by V^{xc}, which enhances the curvature of the bands and gives rise to local maxima. On the other hand, the well-known (Pershin, Nesteroff and Privmann, 2004; Serra, Sanchez and Lopez, 2005) local extrema present in the odd bands, responsible for the anomalous

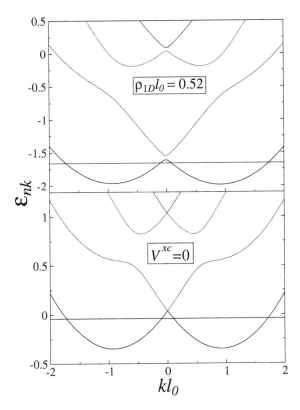

Fig. 6.31 Single electron energies (in $\tilde{\omega}_0$ units) for $\rho_{1D}l_0 = 0.52$ and $B = 0$ T in a strong SO regime characterized by $\Delta_R = 0.093$, $\Delta_D = 0.37$, as a function of the linear momentum kl_0. The thin horizontal line represents the chemical potential. The effect of V^{xc} has not been included in the results shown in the bottom panel.

structures on top of the ordinary steps in the odd e^2/h values, remain qualitatively unaffected by the inclusion of the exchange-correlation interaction.

Another situation is shown in Fig. 6.31, again in the strong SO regime, corresponding to $B = 0$ and $\rho_{1D}l_0 = 0.52$. It can be seen that V^{xc} produces a subband structure similar to that in the bottom panel of Fig. 6.28, and this is also reflected in the conductance, as will be discussed later on.

A particular situation appears when $\lambda_D = \lambda_R$ at a zero magnetic field. It has already been addressed when $V^{xc} = 0$, showing that subband anticrossings disappear when the two SO strengths are equal. We have found that the inclusion of V^{xc} does not change the crossing properties of the subbands in any SO regime, and that it induces only a small subband splitting.

Exchange-correlation effects are also found in the weak SO coupling regime, though in this case no local extrema appear. However, we will see that the effects of V^{xc} in the conductance are also apparent. As said above, all these V^{xc} features are especially apparent at low densities, becoming notably weaker or disappearing for

$n \geq 2$, in which case, only small k asymmetries are observed in odd subbands when $B = 0$ — also noticed in the conductance. When an in-plane B acts on the QW in the weak SO coupling regime, the most apparent effect of V^{xc} is to slightly enhance the B effects without producing qualitative changes in the subband structure.

The noncollinear Kohn–Sham calculation discussed in the previous subsection allows one to evaluate the KS conductance as (for the derivation see Subsection 8.8)

$$G = \frac{e^2}{h} \sum_{k_n} 1, \qquad (6.136)$$

where we have denoted with $\sum_{k_n}$ the sum over all the possible intrasubband-allowed excitations.

This amounts to counting the number of cuts (k_n) of the chemical potential with partially occupied subbands corresponding to positive slope values. Referring to Fig. 6.30, these allowed transitions are exemplified as curved arrows for two possible situations that correspond to two different values of the chemical potential.

In the $V^{\text{xc}} = 0$ case, in the absence of B and SO effects, $\epsilon_{nk} = (n+1/2)\omega_0 + k^2/2$ and the subbands are spin degenerate. Thus, only one intrasubband excitation (one single intersection k_n) contributes to G for each subband n. As a consequence, Eq. (6.136) gives the usual conductance quantization of the Landauer–Büttiker formalism, where each spin-degenerate subband contributes e^2/h to the conductance, yielding the result $G = \frac{2e^2}{h} \sum_n 1$. However, different results for G may arise due to magnetic field, SO and V^{xc} effects on the energy spectrum. This is schematically illustrated in the bottom panel of Fig. 6.30, where at certain values of the chemical potential more than one intrasubband excitation per subband (more than one intersection point k_n) contributes e^2/h to G.

Thus, the conductance of a quantum wire is particularly sensitive to its ground state structure — in particular, to symmetry-breaking effects induced by V^{xc}, like spontaneous spin polarization and Wigner crystallization, for which there seems to be some experimental evidence (Thomas *et al.*, 2004; Thomas *et al.*, 1996; Reilly *et al.*, 2001).

Figures 6.32–6.34 constitute the main result of our study, displaying the conductance in $G_0 = 2e^2/h$ units as a function of the electronic linear density in the strong and weak SO regimes, considering the possibility of having an in-plane applied magnetic field for two particular orientations with respect to the wire direction, and the cases in which the exchange-correlation energy is included or is not.

Figure 6.32 shows the $B = 0$ case for the weak and strong SO regimes when the exchange-correlation energy is taken into account and when it is not. It can be seen that when $V^{\text{xc}} = 0$ the conductance displays the usual steps of the spin-degenerate case commented on before. Contrarily, when $V^{\text{xc}} \neq 0$ the induced spin splitting in the energy subbands gives rise to steps at semi-integer multiples of G_0 for both SO coupling regimes. These steps are apparently narrower than those corresponding to integer multiples of G_0 because the splitting of the subbands due

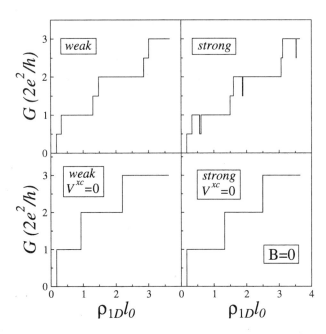

Fig. 6.32 Conductance as a function of the linear density for $B = 0$ in the strong and weak SO regimes when the exchange-correlation energy is taken into account (top panels), and when it is not (bottom panels).

to the confinement — coming from the $(n + 1/2)\omega_0$ term — is much larger than the one induced by V^{xc}.

An interesting feature appears in the strong SO regime, in which "anomalous plateaux" are found on top of the mentioned steps corresponding to semi-integer multiples of G_0. Indeed, it can be seen that G has a nonmonotonic behavior as a function of $\rho_{1D}l_0$, but presents 0.5 G_0 drops for some values of the electronic density. The origin of these plateaux can be inferred from the top panel of Fig. 6.31, in which the combination of the spin splitting induced by V^{xc} and the well-known subband k splitting induced by the (strong) SO coupling gives rise to two possible intrasubband excitations per subband (two intersection points k_n) in some small ranges of the chemical potential values (and thus of electronic densities), yielding the anomalous structure in the conductance. These anomalous structures have already been found when considering an applied in-plane magnetic field (Pershin, Nesteroff and Privman, 2004; Serra, Sanchez and Lopez, 2005). Contrarily, in our case B is zero, and thus it is a genuine exchange-correlation interaction effect which seems to mimic in some cases the effect of an applied magnetic field.

Figure 6.33 shows the conductance in the weak SO regime when the magnetic field is applied along the $\phi_B = 0$ and $\pi/2$ directions. As expected (Pershin, Nesteroff and Privman, 2004; Serra, Sanchez and Lopez, 2005), we have found plateaux at semi-integer multiples of G_0 even when $V^{xc} = 0$. The larger effect of V^{xc} at low

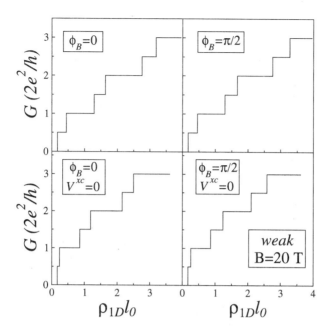

Fig. 6.33 Conductance as a function of the linear density for $B = 20$ T in the weak SO regime when the exchange-correlation energy is taken into account (top panels), and when it is not (bottom panels). The azimuthal angle of the magnetic field is indicated.

densities can be inferred from the difference in the width of the first semi-integer steps when exchange-correlation effects are included and when they are not, which shows that V^{xc} combines with B giving rise to a larger effective magnetic field. Lastly, some evidence of the B-like acting of V^{xc} stems from the comparison of the results of this figure with those displayed in the left panels of Fig. 6.32 corresponding to the $B = 0$ weak SO coupling case.

Figure 6.34 shows the conductance in the strong SO regime. When $V^{\text{xc}} \neq 0$, the conductance is qualitatively similar to that at $B = 0$ for the same SO regime — top right panel of Fig. 6.32. Only the width of the anomalous plateaux varies appreciably, especially when $\phi_B = \pi/2$. When $V^{\text{xc}} = 0$ and $\phi_B = 0$, the structure is similar to that displayed in the top panels of Fig. 6.34 but, as in the weak SO regime, the first semi-integer step is narrower than when $V^{\text{xc}} \neq 0$. As before, the same happens for most steps. New interesting structures appear when $\phi_B = \pi/2$ for the $V^{\text{xc}} = 0$ case: in addition to the just-mentioned ones at semi-integer multiples of G_0, anomalous plateaux at integer multiples of G_0 are also found. They are narrower than the semi-integer ones and their existence is due to the presence of local maxima in the subband spectrum for even values of n, which have already been discussed previously (see for example the bottom panel of Fig. 6.30). It is worth noting that these structures are not robust in the sense that the exchange-correlation energy washes them out (compare the right panels of Fig. 6.34). These additional

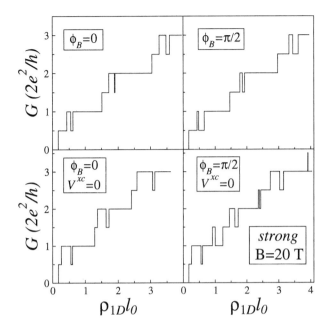

Fig. 6.34 Conductance as a function of the linear density for $B = 20$ T in the strong SO regime when the exchange-correlation energy is taken into account (top panels), and when it is not (bottom panels). The azimuthal angle of the magnetic field is indicated.

anomalous plateaux have also been found by Moroz and Barnes (1999), where neither the V^{xc} nor the Dresselhaus SO interactions were considered.

Finally, note that the above-mentioned behavior is found when $\phi_B = 0$ instead of $\pi/2$, if the values of Δ_R and Δ_D are interchanged. This is due to the particular interplay of the Rashba and Dresselhaus SO interactions and the orientation of the magnetic field.

6.11   Appendix A

In this appendix we generalize some of the expressions derived in Subsection 6.3 to the case in which B has a in-of-well component, such as $B = (B_x, 0, B_z)$. The Zeeman term then becomes $\frac{1}{2}g^*\mu_B \mathbf{B}\cdot\boldsymbol{\sigma} = -\frac{1}{2}\omega_L^z(\sigma_x \tan\theta + \sigma_z)$, where we have introduced the zenithal angle θ, $\tan\theta = B_x/B_z$, and the "z-Larmor" frequency $\omega_L^z = |g^*\mu_B B_z|$, with $\omega_L^z/\omega_c = |g^*|m^*/2$. The Schrödinger equation (6.8) then becomes

$$\begin{bmatrix} \frac{1}{2}(a^+a^- + a^-a^+) - \omega_L^z/(2\omega_c) - \varepsilon & i\tilde{\lambda}_R a^- + \tilde{\lambda}_D a^+ - [\omega_L^z/(2\omega_c)]\tan\theta \\ -i\tilde{\lambda}_R a^+ + \tilde{\lambda}_D a^- - [\omega_L^z/(2\omega_c)]\tan\theta & \frac{1}{2}(a^+a^- + a^-a^+) + \omega_L^z/(2\omega_c) - \varepsilon \end{bmatrix}$$

$$\times \begin{pmatrix} \phi_1 \\ \phi_2 \end{pmatrix} = 0. \tag{6.137}$$

The calculation proceeds as before, Eq. (6.9) becoming

$$(n + \alpha - \varepsilon)\, b_n - \frac{\alpha - \beta}{2} \tan\theta\, a_n - i\tilde{\lambda}_R \sqrt{n}\, a_{n-1} + \tilde{\lambda}_D \sqrt{n+1}\, a_{n+1} = 0,$$

$$(n + \beta - \varepsilon)\, a_n - \frac{\alpha - \beta}{2} \tan\theta\, b_n + i\tilde{\lambda}_R \sqrt{n+1}\, b_{n+1} + \tilde{\lambda}_D \sqrt{n}\, b_{n-1} = 0,$$

$$(6.138)$$

where $\alpha = (1 + \omega_L^z/\omega_c)/2$ and $\beta = (1 - \omega_L^z/\omega_c)/2$.

The sp spectrum Eq. (6.30) becomes

$$E_n^d = \left(n + \frac{1}{2} \right) \omega_c + \frac{\omega_L^z}{2}\, \mathcal{S}$$

$$+ 2n \left[\mathcal{U}\, (\lambda_R^2 + \lambda_D^2) + (\lambda_R^2\, \mathcal{V} + \lambda_D^2\, \mathcal{Z}) \frac{\omega_c}{\omega_c + \omega_L^z\, \mathcal{S}} \right]$$

$$- 2(n+1) \left[\mathcal{U}\, (\lambda_R^2 + \lambda_D^2) + (\lambda_R^2\, \mathcal{Z} + \lambda_D^2\, \mathcal{V}) \frac{\omega_c}{\omega_c - \omega_L^z\, \mathcal{S}} \right],$$

$$E_n^u = \left(n + \frac{1}{2} \right) \omega_c - \frac{\omega_L^z}{2}\, \mathcal{S} \qquad (6.139)$$

$$+ 2n \left[\mathcal{U}\, (\lambda_R^2 + \lambda_D^2) + (\lambda_R^2\, \mathcal{Z} + \lambda_D^2\, \mathcal{V}) \frac{\omega_c}{\omega_c - \omega_L^z\, \mathcal{S}} \right]$$

$$- 2(n+1) \left[\mathcal{U}\, (\lambda_R^2 + \lambda_D^2) + (\lambda_R^2\, \mathcal{V} + \lambda_D^2\, \mathcal{Z}) \frac{\omega_c}{\omega_c + \omega_L^z\, \mathcal{S}} \right].$$

where we have defined $\mathcal{S} = 1/\cos\theta$, $\mathcal{U} = \sin^2\theta/4$, $\mathcal{V} = (1 + \cos\theta)^2/4$, and $\mathcal{Z} = (1 - \cos\theta)^2/4$. When $\theta = 0$, $\mathcal{U} = \mathcal{Z} = 0$, $\mathcal{V} = 1$, and Eq. (6.139) reduces to Eq. (6.30).

6.12 Appendix B

Here we consider the noninteracting Hamiltonian $H = H_0 + H_D + H_R + H_Z$, where H_0 consists of the kinetic and confinement energies, i.e.

$$H_0 = \sum_{i=1}^{N} \left[\frac{\mathbf{P}^2}{2m} + \frac{1}{2} m(\omega_x^2 x^2 + \omega_y^2 y^2) \right]_i ;$$

the Zeeman term H_Z is given by $H_Z = g^* \mu_B B S_z$, and H_D and H_R are the Dresselhaus and Rashba Hamiltonians, respectively.

By using effective atomic units and assuming that $h_0 \gg h_{R,D} \gg h_Z$, where with h we have indicated the single-particle Hamiltonians, and expanding in powers of $\lambda_{R,D}$, an analytic diagonalization to $O(\lambda_{R,D}^3)$ in spin space is possible with a unitary transformation (Aleiner and Fal'ko, 2001) $\tilde{h} = U^+ h U$, where

$$U = \exp\left(-i\left[(\lambda_D x + \lambda_R y)\sigma_x - (\lambda_D y + \lambda_R x)\sigma_y \right] \right) . \qquad (6.140)$$

The transformed Hamiltonian is given by

$$\tilde{H} = \sum_{j=1}^{N} \left[\frac{\mathbf{P}^2}{2m} + \frac{1}{2}m(\omega_x^2 x^2 + \omega_y^2 y^2) + (\lambda_D^2 - \lambda_R^2)\mathcal{L}_z \sigma_z + \frac{1}{2}g^* \mu_B B \sigma_z \right]_j$$
$$- N(\lambda_D^2 + \lambda_R^2) + O(\lambda_{R,D}^3), \tag{6.141}$$

where we have defined the *canonical* angular momentum operator $\mathcal{L}_z = xP_y - yP_x$. Despite the spin diagonalization, the x and y degrees of freedom in (6.141) are still coupled through the vector potential in the kinetic energy and in $\mathcal{L}_z$. With a second transformation of the same type used in Subsection 5.5, each spin component can be recast in a separable form. Specifically, defining $\hat{H}_\eta = U_{2\eta}^+ \tilde{H}_\eta U_{2\eta}$, with $\eta = \uparrow, \downarrow$ and U given by Eq. (5.43), we obtain

$$\hat{H}_\eta = \sum_{j=1}^{N_\eta} \left[\frac{p_x^2}{2M_{1\eta}} + \frac{M_{1\eta}}{2}\Omega_{1\eta}^2 x^2 + \frac{p_y^2}{2M_{2\eta}} + \frac{M_{2\eta}}{2}\Omega_{2\eta}^2 y^2 + \frac{1}{2}g^* \mu_B B s_\eta \right]_j$$
$$- N_\eta(\lambda_D^2 + \lambda_R^2) + O(\lambda_{R,D}^3), \tag{6.142}$$

where $s_\eta = \pm 1$ for $\eta = \uparrow, \downarrow$. Assuming, without the loss of generality, that $\omega_x \geq \omega_y$ the masses and frequencies of the decoupled oscillators are

$$M_{k\eta} = \frac{2m\sqrt{(\omega_x^2 + \omega_y^2 + \omega_{c\eta}^2)^2 - 4\omega_x^2\omega_y^2}}{\omega_x^2 - \omega_y^2 \pm \omega_{c\eta}^2 + \sqrt{(\omega_x^2 + \omega_y^2 + \omega_{c\eta}^2)^2 - 4\omega_x^2\omega_y^2}},$$
$$\Omega_{k\eta} = \frac{1}{\sqrt{2}}\left[\omega_x^2 + \omega_y^2 + \omega_{c\eta}^2 \pm \sqrt{(\omega_x^2 + \omega_y^2 + \omega_{c\eta}^2)^2 - 4\omega_x^2\omega_y^2} \right]^{1/2}, \tag{6.143}$$

with the upper (lower) sign in $\pm$ corresponding to $k = 1$ (2). We have defined in (6.143) a spin-dependent cyclotron frequency which incorporates the SO correction,

$$\omega_{c\uparrow,\downarrow} = \frac{eB}{mc} \pm 2(\lambda_D^2 - \lambda_R^2). \tag{6.144}$$

The solution to Eq. (6.142) is given by products of x and y harmonic oscillator functions which, when transformed back to the laboratory frame, yield the desired solutions to the original Hamiltonian. The eigenvalues for each spin can be labeled by the number of quanta in the x and y oscillators:

$$\varepsilon_{N_1 N_2 \eta} = \left(N_1 + \frac{1}{2}\right)\hbar\Omega_{1\eta} + \left(N_2 + \frac{1}{2}\right)\hbar\Omega_{2\eta} + s_\eta \frac{1}{2}g^* \mu_B B - (\lambda_D^2 + \lambda_R^2). \tag{6.145}$$

By taking $\omega_x = \omega_y$ in Eq. (6.145) one gets the result reported in Eq. (6.62).

References

Aleiner, I.L. and V.I. Fal'ko, *Phys. Rev. Lett.* **87**, 256801 (2001).

Ashoori, R.C., *Nature* **379**, 413 (1996); see also references therein.

Bruce, N.A. and P.A. Maksym, *Phys. Rev.* **B61**, 4718 (2000).

Bychkov, Yu. A. and E.I. Rashba, *Proc. 17th Int. Conf. on Physics of Semiconductors,* San Francisco, 1984, (Springer, Berlin, 1985), p. 321.

Bychkov, Yu. A. and E.I. Rashba, *Sov. Phys.–JETP Lett.* **39**, 78 (1984).

Califano, M., T. Chakraborty and P. Pietiläinen, *Phys. Rev. Lett.* **94**, 246801 (2005).

Chakraborty, T. and P. Pietiläinen, *Phys. Rev. Lett.* **95**, 136603 (2005).

Chang, M.C., *Phys. Rev.* **B71**, 085315 (2005).

Ciorga, M., A.S. Sachrajda, P. Hawrylak, C. Gould, P. Zawadzki, S. Jullian, Y. Feng and Z. Wasilewski, *Phys. Rev.* **B61**, R16315 (2000).

Cui, L.J. *et al.*, *Appl. Phys. Lett.* **80**, 3132 (2002).

Das, B., S. Datta and R. Reifenberger, *Phys. Rev.* **B41**, 8278 (1990).

Datta, S. and B. Das, *Appl. Phys. Lett.* **56**, 665 (1990).

Dobers, M., K.V. Klitzing and G. Weimann, *Phys. Rev.* **B38**, 5453 (1988).

Dresselhaus, G., *Phys. Rev.* **100**, 580 (1955).

D'yakonov, M.I. and V.Y. Kachorovskii, *Sov. Phys. Semicond.* **20**, 110 (1986).

Emperador, A., E. Lipparini and F. Pederiva, *Phys. Rev.* **B70**, 125302 (2004).

Engels, G., J. Lange, T. Schäpers and H. Lüth, *Phys. Rev.* **B55**, R1958 (1997).

Eriksson, M.A., A. Pinczuk, B.S. Dennis, S.H. Simon, L.N. Pfeiffer and K.W. West, *Phys. Rev. Lett.* **82** , 2163 (1999).

Fal'ko, V.I., *Phys. Rev.* **B46**, R4320 (1992).

Folk, J.A., S.R. Patel, K.M. Birnbaum, C.M. Marcus, C.I. Duruöz and J.S. Harris, *Phys. Rev. Lett.* **86**, 2102 (2001).

Halperin, B.I., A. Stern, Y. Oreg, J.N.H.J. Cremers, J.A. Folk and C.M. Marcus, *Phys. Rev. Lett.* **86**, 2106 (2001).

Heinonen, O., J.M. Kinaret and M.D. Johnson, *Phys. Rev.* **B59**, 8073 (1999).

Kallin, C. and B.I. Halperin, *Phys. Rev.* **B30**, 5655 (1984).

Knobbe, J. and T.H. Schäpers, *Phys. Rev.* **B71**, 035311 (2005).

Könemann, J., R.J. Haug, D.K. Maude, V.I. Fal'ko and B.L. Altshuler, *Phys. Rev. Lett.* **94**, 226404 (2005).

Kumar, A., S.E. Laux and F. Stern, *Phys. Rev.* **B42**, 5166 (1990).

Lipparini, E., M. Barranco, F. Malet and M. Pi, *Phys. Rev.* **B74**, 125302 (2006).

Lommer, G., F. Malcher and U. Rössler, *Phys. Rev.* **B32**, 6965 (1985).

Malet, F., E. Lipparini, M. Barranco and M. Pi, *Phys. Rev.* **B73**, 125302 (2006).

Manger, M., E. Batke, R. Hey, K.J. Friedland, K. Köhler and P. Ganser, *Phys. Rev.* **B63**, 121203(R) (2001).

Moroz, A.V. and C.H.W. Barnes, *Phys. Rev.* **B60**, 014272 (1999).

Murakami, S., N. Nagaosa and S.C. Zhang, *Science* **301**, 1348 (2003).

Nitta, J., T. Akazaki, H. Takayanagi and T. Enoki, *Phys. Rev. Lett.* **78**, 1335 (1997).

Pershin, Y.V., J.A. Nesteroff and V. Privman, *Phys. Rev.* **B69**, 121306(R) (2004).

Pfeffer, P. and W. Zawadzki, *Phys. Rev.* **B59**, R5312 (1998).

Pietiläinen, P. and T. Chakraborty, *Phys. Rev.* **B73**, 155315 (2006).

Rashba, E.I., *Fiz. Tverd. Tela (Leningrad)* **2**, 1224 (1960) [*Sov. Phys. Solid State* **2**, 1109 (1960)].

Rashba, E.I., *Physica* **E20**, 189 (2004).

Reilly *et al.*, *Phys. Rev.* **B63**, 121311(R) (2001).

Salis, G., D.D. Awschalom, Y. Ohno and H. Ohno, *Phys. Rev.* **B64**, 195304 (2001).

Schliemann, J., J.C. Egues and D. Loss, *Phys. Rev.* **B67**, 085302 (2003).

Serra, Ll., D. Sánchez and R. López, *Phys. Rev.* **B72**, 235309 (2005).

Sih, V., W.H. Lau, R.C. Myers, A.C. Gossard, M.E. Flatté and D.D. Awschalom, *Phys. Rev.* **B70**, 161313(R) (2004).

Sinova, J. *et al.*, *Phys. Rev. Lett.* **92**, 126603 (2004).

Thomas, K.J., J.T. Nicholls, M.Y. Simmons, M. Pepper, D.R. Mace and D.A. Ritchie, *Phys. Rev. Lett.* **77**, 135 (1996).

Thomas, K.J., D.L. Sawkey, M. Pepper, W.R. Tribe, I. Farrer, M.Y. Simmons and D.A. Ritchie *J. Phys. Cond. Matt.* **16**, L279 (2004).

Tonello, P. and E. Lipparini, *Phys. Rev.* **B70**, 081201(R), (2004).

Usaj, G. and C.A. Balseiro, *Phys. Rev. B* **70**, 041301(R), (2004).

Valín-Rodríguez, M. and R. G. Nazmitdinov, *Phys. Rev.* **B73**, 235306 (2006).

Valín-Rodríguez, M., A. Puente and Ll. Serra, *Phys. Rev.* **B64**, 205307 (2001).

Valín-Rodríguez, M., A. Puente and Ll. Serra, *Phys. Rev.* **B66**, 045317 (2002).

Valín-Rodríguez, M., A. Puente and Ll. Serra, *Phys. Rev.* **B69**, 085306 (2004).

Valín-Rodríguez, M., A. Puente, Ll. Serra and E. Lipparini, *Phys. Rev.* **B66**, 165302 (2002).

Valín-Rodríguez, M., A. Puente, Ll. Serra and E. Lipparini, *Phys. Rev.* **B66**, 235322 (2002).

Voskoboynikov, O., C.P. Lee and O. Tretyak, *Phys. Rev.* **B63**, 165306 (2001).

Winkler, R., *Physica* **E22**, 450 (2004).

Wolf, S.A., D.D. Awschalom, R.A. Buhrman, J.M. Daughton, S. von Molnár, M.L. Roukes, A.Y. Chtchelkanova and D.M. Treger, *Science* **294**, 1488 (2001).

Zhang, S., R.Liang, E. Zhang and Y. Liu, *Phys. Rev.* **B73**, 155316 (2006).

Zutíc, I., J. Fabian and S. Das Sarma, *Rev. Mod. Phys.* **76**, 323 (2004).

Chapter 7

Monte Carlo Methods

7.1 Introduction

The term "Monte Carlo" was used for the first time in the 1940's by a group of scientists who were working at Los Alamos on the Manhattan Project, to indicate a class of mathematical methods for the numerical solution of integration problems by probabilistic methods. Although the ideas underlying Monte Carlo (MC) methods were already known in the 18th century (Comte de Buffon, 1777), the development of these algorithms is recent, in part due to the facility of their implementation on the computer. In many cases, Monte Carlo methods are the only efficient approach to the evaluation of many-dimensional integrals of physical interest, such as computations of averages in a statistical ensemble, or the evaluations of expectations on a quantum state. The time required to evaluate an integral with standard integration methods (for example the Simpson rule) depends exponentially on the dimensionality of the space under consideration. As we will see, the stochastic integration methods have the property that they require a computational effort which depends only algebraically on the dimensionality.

7.2 Standard Quadrature Methods

Suppose we want to compute the integral of a function defined in a space of dimension D, on a hypercube of side L:

$$I(D) = \int_0^L dx_1 \cdots \int_0^L dx_D F(x). \tag{7.1}$$

For $D = 1$, the numerical value of the integral can be approximated by subdividing the area subtended by the curve $y = F(x)$, into rectangles with fixed base h (Simpson rule) (see Fig. 7.1). The smaller h is, the smaller the error in the approximation of the subtended area. By adding all the rectangle areas, we obtain an approximate estimate which, in the limit where the base tends to zero, equals the

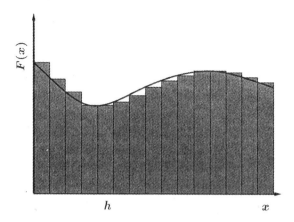

Fig. 7.1 Integration by the first Simpson rule in one dimension: the integral is approximated by the sum of the areas of rectangles with base h and height determined by the values of the function, evaluated on a grid of points.

value of the integral. By generalizing this procedure to a function defined in a space with D dimensions, we have

$$I(D) \approx h^D \sum F(x), \qquad (7.2)$$

where the summation extends over all the points of a lattice defined in the integration hypercube.

If the lattice base h is made to tend to zero, we will obtain the exact value of the integral, like in the one-dimensional case discussed previously. An upper limit of the error made in the ith interval, for a one-dimensional curve, may be obtained by the derivative of the function:

$$\Delta = h^2 \frac{d}{dx_i} F(x_i).$$

By generalizing to D dimensions,

$$\Delta = h^{D+1} \nabla F(x) \approx h^{D+1},$$

where, for the approximation on the right hand side, we have assumed that the gradient of the function is finite in the considered integration interval. Therefore, the total error may be estimated by adding the contributions of the N intervals:

$$\Delta_{\text{tot}} \approx N h^{D+1}.$$

If L is the side of the hypercube, the number of lattice points can be expressed as

$$N = \left(\frac{L}{h}\right)^D.$$

Then, the total error becomes

$$\Delta_{\text{tot}} \approx \left(\frac{L}{h}\right)^D h^{D+1} = L^D h = Vh, \tag{7.3}$$

where $V = L^D$ is the volume of the hypercube. It is possible to write the previous equation in terms of N, thus obtaining

$$\Delta_{\text{tot}} = \frac{LV}{N^{\frac{1}{D}}}.$$

This allows us to estimate the number of points required to obtain a precision of the order of ε for the numerical evaluation of the integral:

$$\varepsilon = N^{-\frac{1}{D}} \implies N = \varepsilon^{-D}.$$

In a typical case, for example the calculation of statistical averages in a system with 20 particles ($D = 60$), the requirement of an error of the order of $\varepsilon = 0.1$ entails the use of $N = 10^{60}$ points. Even with a Tflop computer (i.e. capable of performing 10^{12} operations per second), the time required for the calculation will not be less than 10^{48} seconds, which largely exceeds the age of the universe. Therefore, in general, the standard integration methods are unsuitable for many-body problems.

7.3 Random Variable Distributions and the Central Limit Theorem

In order to introduce the MC methods, it is useful to briefly recall some elementary notions of statistics. The main mathematical objects involved in MC methods are the so-called *random variables*, or *stochastic variables*. These variables may be defined by the following properties:

(1) The ensemble of the values that they may take: $\{x\}$.
(2) The probability law $P(x)$ for each value of the variable (also called the *probability density*). This function must have the following properties:

$$P(x) \geq 0 \quad \forall\, x,$$
$$\int P(x)dx = 1. \tag{7.4}$$

N stochastic variables are said to be *independent* if

$$P(x_1, x_2, \ldots, x_N) = P(x_1)P(x_2)\cdots P(x_N). \tag{7.5}$$

The MC quadrature methods rely on one of the fundamental theorems of statistics — the *central limit theorem*. Let us consider a function f with real values. Let S_N be the stochastic variable defined by

$$S_N = \frac{1}{N} \sum_{i=1}^{N} f(x_i), \tag{7.6}$$

where x_i are N independent stochastic variables, which we assume to have the same probability density. The theorem states that, whichever $P(x)$, in the limit $N \longrightarrow \infty$, the probability density of S_N is given by

$$P[S_N] \overset{N \to \infty}{=} \frac{1}{\sqrt{2\pi\sigma_N^2}} \exp\left[-\frac{(S_N - \langle f \rangle)^2}{2\sigma_N^2} \right], \tag{7.7}$$

where

$$\langle f \rangle = \int f(x)P(x)dx, \tag{7.8}$$

$$\sigma_N^2 = \frac{1}{N-1}\sigma_S^2 \tag{7.9}$$

and

$$\sigma_S^2 = \int f^2(x)P(x)dx - \langle f \rangle^2. \tag{7.10}$$

This theorem immediately suggests a way to estimate integrals written in the form (7.8), provided that we are capable of *sampling* values of the random variable x from the probability density $P(x)$. After N samples, it will be enough to compute the value of the stochastic variable S_N. How far will this estimate be from the exact value of the integral? This piece of information is included in the variance of S_N, σ_N, which can in turn be estimated using Eq. (7.10):

$$\sigma_S^2 \sim \frac{1}{N} \sum_{i=1}^{N} f^2(x_i) - S_N^2. \tag{7.11}$$

The previous considerations apply independently of the dimensionality of the space, and can be extended to the computation of more general integrals. Suppose that we want to calculate the integral of a function $F(x)$ defined in an n-dimensional space. Let us consider a probability density P from which we can extract values for the variable x (the latter, in general, will be an n-dimensional vector). The integral can be written in the following form:

$$\int F(x)dx = \int \frac{F(x)}{P(x)}P(x)dx \equiv \int f(x)P(x)dx. \tag{7.12}$$

Once more, from the central limit theorem, it follows that the stochastic variable

$$S_N = \frac{1}{N} \sum_{i=1}^{N} f(x_i) \tag{7.13}$$

has a probability density which is a Gaussian around the mean value $\langle f \rangle$ (which is precisely the integral to be computed), with a statistical error

$$\sigma_N \propto \frac{1}{\sqrt{N}}. \tag{7.14}$$

The MC integration method precisely consists in generating a sequence of points distributed according to a given probability density P, in order to estimate the integral of a given function by means of (7.13). The fact that the dependence of the statistical error on the estimate of the integral has a dependence on the number of sampled points which does not depend on dimensionality should be compared with what happens for standard integration procedures, where the number of necessary points grows exponentially. Hence, the MC method very quickly becomes computationally advantageous, as soon as the dimension of the space in which the integrand is defined exceeds a few units.

7.4 Sampling Techniques

The main problem one faces in the application of MC methods is the successful generation of sequences of points distributed according to a given probability density, which, in many cases is established *a priori* by the problem under study.

7.4.1 *The Inversion Rule*

All programming languages provide a random number generator sampling a random variable uniformly distributed in the interval $[0, 1)$. Trivial transformations can be used to generate random numbers uniformly distributed in a generic interval $[a, b)$. A useful recipe for sampling simple densities by transforming the output of such a function is the so-called inversion rule. If x is a random variate with distribution $P(x)$ and $y = f(x)$, the probability density of y will satisfy the following property:

$$P(x)dx = P(y)dy. \tag{7.15}$$

Assuming that $P(x) = 1$ in $[0, 1)$, and given a $P(y)$ to be sampled, the function f connecting x and y will be

$$P(y) = \frac{df^{-1}(y)}{dy} \Rightarrow f^{-1}(y) = \int_0^y dy' P(y'). \tag{7.16}$$

A simple example is the case of an exponential probability density. In this case we have $P(y) = e^{-y}$. Equation (7.16) promptly gives $y = f(x) = -\ln(1 - x)$.

The most commonly used implementation of the inversion rule is the Box–Muller formulas (Muller, 1958). Given two random numbers ξ_1 and ξ_2 uniformly distributed in $[0,1)$, the two numbers

$$\eta_1 = \sqrt{-\ln(1-\xi_1)}\cos(2\pi\xi_2), \qquad (7.17)$$

$$\eta_2 = \sqrt{-\ln(1-\xi_2)}\sin(2\pi\xi_1) \qquad (7.18)$$

are distributed according to

$$P(\eta) = \frac{1}{\sqrt{2\pi}}e^{-\frac{\eta^2}{2}}. \qquad (7.19)$$

A generic Gaussian of mean x_0 and standard deviation σ can be sampled by using the transformation

$$\eta' = \sigma\eta + x_0.$$

These results are commonly used in the implementation of more sophisticated algorithms.

7.4.2 The Rejection Method

This is the most elementary way to sample a generic probability density in a low-dimensional space. Suppose for simplicity that we have a one-dimensional random variable x with probability density $P(x)$. Let us define A as the maximum value assumed by $P(x)$ on the set of values that defines the random variable. It is possible to sample $P(x)$ by extracting two random numbers $x \in \{x\}$ and $y \in [0, A)$. If $y < P(x)$, then x is *accepted* as a sampled number, otherwise it is *rejected*, and one must proceed with a further extraction. It is clear that this procedure will lead to a large number of points sampled from regions where $P(x)$ is large, and vice versa.

7.4.3 Markov Chains

The probability functions to be sampled in problems of physical interest are usually very complicated, and simple methods such as inversion or acceptance/rejection are either not viable or not efficient enough.

As an example, let us assume that we need to calculate expectation values of physical observables for a quantum system of N interacting particles. We also assume that we know analytically the wave function of the system in the ground state, $\Psi(\mathbf{r_1}\cdots\mathbf{r_N})$. The calculation of the expectation value of a local observable $\hat{O}$ requires the evaluation of integrals of the following kind:

$$\langle\Psi|\hat{O}|\Psi\rangle = \int |\Psi(\mathbf{r}_1\cdots\mathbf{r}_N)|^2 O(\mathbf{r}_1\cdots\mathbf{r}_N)\,d\mathbf{r}_1\cdots d\mathbf{r}_N. \qquad (7.20)$$

In this case, the MC method has a direct intuitive meaning: by generating points distributed according to $|\Psi|^2$ (which represents the probability of finding the particles in a given configuration), and by calculating the respective values of operator $\hat{O}$, it is possible to obtain the desired expectation.

This problem has a strict formal analogy with the computation of estimates in a classical statistical ensemble of particles interacting via a potential $V(q = q_1 \cdots q_N)$, such as the canonical ensemble at temperature T:

$$\langle O(q) \rangle = \frac{\int dq_1 \cdots dq_N O(q) e^{-\frac{V(q)}{K_B T}}}{\int dq_1 \cdots dq_N e^{-\frac{V(q)}{K_B T}}}. \tag{7.21}$$

It is known that in a classical system the trajectories of the particles are the solutions to the equation of motion of the system. This fact allows us to define an algorithm which generates successive configurations for the positions and velocities of the particles, obeying Boltzmann statistics. Intuitively, it would be convenient to exploit some analogous feature, to generate configurations of a quantum system. Indeed, it is possible to define a sort of stochastic dynamics which relates, one to the other, sequentially generated configurations, in such a way that their final distribution conforms with a given probability density. Such pseudodynamics, which does not have any physical meaning, can be formalized by the theory of Markov chains.

A Markov chain is defined as a sequence of stochastic variables $\{x_i\}$, such that

$$x_1 \quad \text{has probability} \quad P_1(x),$$
$$P_k(x) = \int w_k(x_k, x_{k-1}) P_{k-1}(x_{k-1}) dx_{k-1}. \tag{7.22}$$

The function $w_k(x_k, x_{k-1})$ is often called the *transition probability* or *transition matrix*. In general, it depends on the value of index k.

A Markov chain is called *stationary* when the transition probability w_k does not depend on the index k. From this definition, it follows immediately that a stationary chain is completely defined once the distribution probability for the first variable in the chain $P_1(x)$ and the transition probability w are established. In fact, we have

$$P_2(x_2) = \int dx_1 w(x_2, x_1) P_1(x_1),$$

$$P_3(x_3) = \int dx_2 w(x_3, x_2) P_2(x_2) = \int dx_2 \int dx_1 w(x_3, x_2) w(x_2, x_1) P_1(x_1),$$

$$\vdots \tag{7.23}$$

$$P_k(x_k) = \int dx_1 \cdots \int dx_{k-1} w(x_k, x_{k-1}) \cdots w(x_2, x_1) P_1(x_1)$$

$$\equiv \int dx_1 \cdots dx_{k-1} w^{k-1} P_1(x_1).$$

In order to make the notation less clumsy, we will summarize the previous equations as

$$P_k = w^{k-1}P_1, \qquad (7.24)$$

meaning that P_k is the probability density of the element of the Markov chain obtained after sampling N times the same transition matrix, without explicitly writing the integrals.

Let us now assume that a w exists, which can be easily sampled. Given an initial configuration x_1 (which we can imagine as being sampled by a uniform probability density), a relevant question is whether the configurations which are generated in sequence do converge toward some limiting distribution, P_∞. In general, it is difficult to prove the existence of this limit. However, in the cases of physical interest, all necessary conditions are fulfilled. Therefore, in what follows we will rely on some assumptions:

- The limit

$$P_\infty = \lim_{k \to \infty} w^k P_1 \qquad (7.25)$$

  exists and corresponds to a good probability distribution.
- The transition probability w has the ergodicity property: given two points x and y in configuration space, it must be that

$$\exists\, N < \infty \quad \text{such that} \quad w^N(x, y) > 0. \qquad (7.26)$$

  This means that it is always possible to reach a given point in space, starting from any other point, in a finite number of steps.

If the limit P_∞ exists, it is an eigenvector of the matrix w. In fact,

$$w P_\infty = \lim_{k \to \infty} w^{k+1} P_1 = P_\infty. \qquad (7.27)$$

Therefore, the transition matrices have the property that all of their eigenvalues are located on the circle of unit radius in the complex plane, and this implies the uniqueness of P_∞. To prove this, we first note that since the probability distributions are normalized, there can exist no eigenvalue $\lambda > 1$. Let us suppose that we have a basis of eigenfunctions φ_n of w, and let us expand the distribution P_1 on this basis. We will have

$$w^k P_1 = \sum_n c_n w^k \varphi_n \Rightarrow w^k P_1 = \sum_n c_n \lambda^k \varphi_n, \qquad (7.28)$$

and if $\lambda < 1$, then

$$\lim_{k \to \infty} w^k P_1 = 0. \qquad (7.29)$$

The previous analysis tells us that stationary Markov chains eventually sample some probability density, P_∞. However, this property becomes interesting only if, given an arbitrary probability density to be sampled, we can find a transition matrix w that makes the chain sample exactly *that* probability density.

A further step forward can be made noting that there is a further condition that, in general, we wish that it is satisfied in a real physical system at equilibrium: the stochastic pseudodynamics should not generate configurations that are distributed in an inhomogeneous way. For example, if we compute some properties of a gas or a liquid in a container, we do not want a steady state in which the configurations fill only a part of the available volume. This requirement is reflected in a constraint imposed upon w and P, which is known as the *detailed balance* condition:

$$P_\infty(x)w(y,x) = P_\infty(y)w(x,y). \tag{7.30}$$

This means that the probability of moving the configuration from the neighborhood of point x to the neighborhood of point y should be proportional only to the values of the probability function P, and should not depend on the particular event under consideration. In other words,

$$\frac{w(x,y)}{w(y,x)} = \frac{P_\infty(x)}{P_\infty(y)}. \tag{7.31}$$

This condition is extremely important. In fact it is possible to start from Eq. (7.31) in order to find a recipe which allows one to sample the transition matrix w (which in general is unknown), which yields, as a limit, a given $P_\infty(x) \equiv P$ which we want to sample. To see this, let us factorize the matrix w into the two following factors:

(1) A "proposal" for the next position, T. This is a recipe that tells us how to generate a candidate point for the Markov chain.
(2) A probability of accepting the next position, A. This factor decides whether the proposed point has to be set as the next point in the Markov chain.

Therefore, we will have

$$w(y,x) = T(y,x)A(y,x). \tag{7.32}$$

The matrix $T(y,x)$ can be suitably chosen and, in general, it will be a distribution that we know how to sample. From the previous relations, we find that the matrix $A(y,x)$ should have the following property:

$$\frac{A(y,x)}{A(x,y)} = \frac{P(y)T(x,y)}{P(x)T(y,x)} \Rightarrow A(y,x) = F\left(\frac{P(y)}{P(x)}\frac{T(x,y)}{T(y,x)}\right), \tag{7.33}$$

where the function F should be such that

$$\frac{F(x)}{F(1/x)} = x. \tag{7.34}$$

The following are examples of functions which satisfy the above relations:

(1) $F(x) = \min(1, x)$,
(2) $F(x) = x/(1 + x)$.

We are now ready to define an algorithm for generating a Markov chain that samples an arbitrary probability distribution, P. First, it is necessary to define a transition probability T that we know how to sample. Starting then from an arbitrary point x_1 in configuration space, a subsequent configuration x_2 distributed according to

$$P(x_2) = \int_{x_1} T(x_2, x_1) P(x_1) \, dx_1$$

must be generated. This is usually obtained by adding to the initial position a vector which is sampled from T. In order to ensure detailed balance, the new configuration will be either retained or rejected with a probability $A(y, x)$ defined by (7.33), extracting a random number uniformly distributed in $[0, 1)$, and comparing it with the value of the acceptance probability. If the random number is smaller than the acceptance probability, then the forward transition is accepted, and the generated point becomes the next point in the chain. Otherwise the backward transition dominates, and the next point in the chain will become the original point. The procedure must then be repeated.

The ensemble of points in configuration space generated by such a procedure, and which constitutes a Markov chain, is often called a *random walk*, and the moving point is called the *walker*.

7.5   The Metropolis Algorithm $[M(RT)^2]$

This algorithm constitutes the simplest practical implementation of the scheme described in the previous subsections. The abbreviation by which it is indicated derives from the names of its developers (N. Metropolis, A.W. Rosenbluth, M.N. Rosenbluth, A.H. Teller, and E. Teller, 1953); however, it is often referred to as the Metropolis algorithm. In this algorithm the transition proposal $T(y, x)$ is assumed to be uniform in a D-dimensional hypercube of side Δ (see Fig. 7.2). Therefore, given a point $\mathbf{x}_n$, the next point in the Markov chain is generated in the following way:

$$\mathbf{x}_{n+1} = \mathbf{x}_n + \Delta \xi, \tag{7.35}$$

where the components of the vector ξ are extracted at random in the interval

$$-\frac{1}{2} \leq \xi \leq \frac{1}{2}.$$

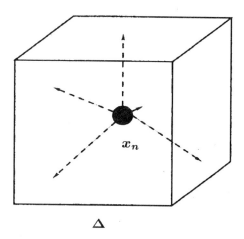

Fig. 7.2 In the $M(RT)^2$ algorithm the positions of the particles are chosen at random inside a cube of side Δ.

The transition proposal is symmetrical, i.e. $T(y, x) = T(x, y)$, and the acceptance probability $A(y, x)$ depends only on $P(x_n)$ and $P(x_{n+1})$. In the Metropolis algorithm, one makes the following choice for the function F:

$$A(x, y) = \min\left(1, \frac{P(y)}{P(x)}\right). \tag{7.36}$$

As already mentioned above, in order to decide whether a proposed move is accepted or not, one extracts a random number ξ in the interval $[0,1)$ and compares it to A. The next move is then determined according to the following criterion:

$$\begin{cases} x_{n+1} = \text{new position} & \text{if } \xi < \dfrac{P(y)}{P(x)} \\[3mm] x_{n+1} = x_n & \text{if } \xi > \dfrac{P(y)}{P(x)} \end{cases} \tag{7.37}$$

It is extremely important to note that rejecting the new configuration means that the next configuration will coincide with the previous one. This means that the estimates evaluated on the old configuration must be added once more to the cumulant.

After a number of steps, this algorithm generates a distribution of points x_i according to the probability distribution P. Therefore, after the generation of N points the estimate of an integral of the form

$$\int dx\, P(x) F(x)$$

will be given by

$$\frac{1}{N} \sum_{i=1}^{N} F(x_i).$$

7.6 Variational Monte Carlo for ^{4}He

As an example of application of the Metropolis algorithm to a real physical system, we will show how to compute the energy per particle of the helium isotope with mass 4, at the temperature $T = 0$ K. ^{4}He atoms have total spin $S = 0$, and therefore, at temperatures close to absolute zero, behave like quantum particles which obey the Bose–Einstein statistics. The interaction between He atoms is known with great accuracy (see for example Korona). However, a not-too-bad approximation for the inter-atomic potential is given by a potential of the Lennard–Jones type:

$$V(r_{12}) = 4\epsilon \left[\left(\frac{\sigma}{r_{12}} \right)^{12} - \left(\frac{\sigma}{r_{12}} \right)^6 \right], \tag{7.38}$$

where $\epsilon = 10.4$ K and $\sigma = 2.556$ Å.

The system is modeled as an ensemble of N point particles confined in a box, which is periodically replicated in space. The volume of the box is such that the ratio $\rho = N/V$ is the density of the system at which we want to perform the simulation. For ^{4}He, the zero-pressure density is known experimentally to be about $\rho_0 = 0.02186$ Å^{-3}, and the system remains fluid up to a freezing density of about $\rho_f = 0.02623$ Å^{-3}, corresponding to a pressure of about 25 atmospheres.

The Hamiltonian of the system is therefore

$$H = -\frac{\hbar^2}{2m} \sum_{i=1}^{N} \nabla^2 + \sum_{i<j} V(r_{ij}). \tag{7.39}$$

Since we do not know the wave function of the ground state, we look for an approximate solution to the time-independent Schrödinger equation:

$$H\Psi_0 = E_0 \Psi_0, \tag{7.40}$$

where E_0 is the energy of the ground state.

Let us consider the following *trial function* (McMillan, 1965):

$$\Psi_T = \prod_{i<j} e^{-\frac{1}{2}\left(\frac{b}{r_{ij}}\right)^m} = \prod_{i<j} f(r_{ij}). \tag{7.41}$$

This function has the right symmetry properties for a system of Bosons (i.e. it is symmetric under the exchange of two particles), and has the property of canceling out when $r_i = r_j$, coherently with the fact that owing to the hard core repulsion, the distance between two atoms cannot decrease arbitrarily (see Fig. 7.3).

Let us now consider the variational principle (Riesz theorem):

$$E_T = \frac{\langle \Psi_T | H | \Psi_T \rangle}{\langle \Psi_T | \Psi_T \rangle} \geq \frac{\langle \Psi_0 | H | \Psi_0 \rangle}{\langle \Psi_0 | \Psi_0 \rangle} = E_0. \tag{7.42}$$

In general, if we chose a trial function which depends on a set of parameters $\{\alpha\}$, the best representation of the ground state within that class of wave functions is given by the condition

$$\frac{\partial E_T}{\partial \{\alpha\}} = 0. \tag{7.43}$$

This derivative cannot be evaluated analytically. It is therefore necessary to compute explicitly the matrix elements $\langle \Psi_T | H | \Psi_T \rangle$ in (7.42), i.e.

$$E_T = \frac{\int dR \Psi_T^*(R) H \Psi_T(R)}{\int dR \Psi_T^*(R) \Psi_T(R)}, \tag{7.44}$$

where $R = \{\mathbf{r}_1 \cdots \mathbf{r}_N\}$ are the coordinates of the N atoms. This does not look like an integral that we can evaluate by MC. However, considering that $\Psi_T(R)$ is a positive-defined, nodeless function we can recast the integral at the numerator this way:

$$E_T = \frac{\int dR \Psi_T^2(R) \frac{H \Psi_T(R)}{\Psi_T(R)}}{\int dR \Psi_T^2(R)}. \tag{7.45}$$

The quantity $\frac{H \Psi_T(R)}{\Psi_T(R)}$ is called *local energy*, and is an expression that can be explicitly evaluated once $\Psi_T(R)$ is given. The integral is now manifestly of a form suitable for MC evaluation, and can be calculated by means of the $M(RT)^2$ algorithm by considering

$$P = \prod_{i<j} e^{-\left(\frac{b}{r_{ij}}\right)^m}. \tag{7.46}$$

Note that P is not normalized, but this is not a problem since in a Metropolis-like algorithm it appears only in the ratio $P(y)/P(x)$.

As a first step, one generates a random distribution of points in the simulation box (i.e. the distribution P_1 of the previous subsections). Subsequently, new atomic positions are generated according to (7.35), and the value of the wave function is computed in the new point of configuration space. The move is either accepted or rejected according to (7.36). Hence, the MC estimate for the energy per particle becomes

$$E_T \cong \frac{1}{N} \sum_{i=1}^{N} \frac{H \psi_T(R_i)}{\psi_T(R_i)}. \tag{7.47}$$

Some other details need to be taken into account. For instance, the fact that a finite simulation box is used imposes the truncation of both the potential and the correlation pseudopotential at a given length, which is typically the largest possible, i.e. half the side of the simulation box. However, this procedure has some drawbacks. In particular, the estimate of the potential energy will not include a sizeable contribution from the attractive tail. If the interaction is not long-ranged, it is possible

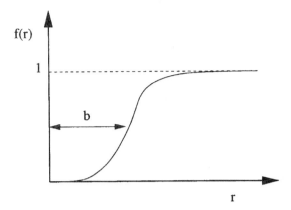

Fig. 7.3 The function $u(r_{ij})$ used in the variational Monte Carlo calculation of the energy of liquid ^{4}He. This function cancels out in correspondence with the repulsive part of the interatomic potential, and is ~ 1 for large distances. The parameter b describes the amplitude of the correlation well of the particle.

to add a *tail correction*, which can be estimated by assuming that the pair distribution function $g(r)$ is constant for $r > L/2$. We will therefore add to the estimate of the potential energy per particle a quantity ΔV,

$$\Delta V = 2\pi\rho \int_{L/2}^{\infty} dr r^2 g(r) V(r), \tag{7.48}$$

which can be evaluated in advance. If the potential is long-ranged (like in the case of Coulomb systems), more sophisticated techniques, like Ewald summations, should be used. As regards the pseudopotentials, one has to be careful about removing the contribution to the kinetic energy coming from the discontinuity of the derivative at the cutoff distance. This is usually achieved by continuously connecting the pseudopotential with a function that at the cutoff distance is zero together with all the derivatives.

The estimate that one can obtain from this procedure is $E_{\mathrm{VMC}} = -5.72\ ^\circ$K for the energy per particle (McMillan, 1965). This should be compared with the experimental value $-7.14\ ^\circ$K. With a simple model like this, it is also possible to obtain information on other properties of ^{4}He. In particular, it is possible to estimate the fraction of Bose–Einstein condensate which turns out to be about 8%, in good agreement with experimental data.

7.7   Variational Monte Carlo for Fermions

The previous example involved calculations for a many-Boson system. In the case of a many-Fermion system the variational Monte Carlo (VMC) procedure does not essentially change. Obviously it will be necessary to start from a trial wave function which has the correct symmetry properties under the permutation group, i.e. it must be antisymmetric for an odd permutation of the particles. This is usually achieved by constructing a Slater determinant of single-particle orbitals which might be obtained by a mean field calculation of the kind presented in the previous chapters. If the interaction commutes with the z component of the spin of the particles, as the Coulomb interaction, it is possible to factorize the determinant of the N particles into two smaller determinants, one for the $N_\uparrow$ particles with spin up and one for the $N_\downarrow$ particles with spin down, with $N_\uparrow + N_\downarrow = N$. The wave function will therefore have the following general form:

$$\Psi_T^A(\mathbf{r}_1 \cdots \mathbf{r}_N) = f(\mathbf{r}_1 \cdots \mathbf{r}_N) D_\uparrow(\mathbf{r}_1 \cdots \mathbf{r}_{N_\uparrow}) D_\downarrow(\mathbf{r}_{N_\uparrow+1} \cdots \mathbf{r}_N), \qquad (7.49)$$

where the function f includes quantum correlations induced by the interaction. As for the case of Bosons, in order to compute the total energy it is possible to sample the square modulus of the wave funtction, and to compute the average of the local energy:

$$E_T = \frac{\int dR (\Psi_T^A(R))^2 \frac{H\Psi_T^A(R)}{\Psi_T^A(R)}}{\int dR (\Psi_T^A(R))^2}. \qquad (7.50)$$

However, one might notice that the antisymmetric trial function must have nodes in the configuration space, and therefore the local energy $\frac{H\Psi_T^A(R)}{\Psi_T^A(R)}$ will be divergent. Some further remarks are in order. If the wave function was exact, the local energy would be constant throughout the space, but for a subset of zero measure, where it would not be defined. If the wave function is approximated, because it approaches the node at distance r as $1/r$, this will be the typical divergence of the local energy. However, the density of walkers, which is determined by $(\Psi_T^A(R))^2$, will go to zero as r^2, and therefore the integral at the numerator of Eq. (7.50) is always convergent.

An interesting collection of early VMC calculations can be found in the paper by Ceperley, Chester and Kalos (1977), where it was pointed out how the computations for a many-Fermion system do not necessarily require a CPU time proportional to N^3, as it would be required by a full computation of a determinant of an $N \times N$ matrix. This dependence can be reduced to N^2 by saving the minors $(N-1) \times (N-1)$ of the Slater matrix, provided that the orbitals depend on the coordinates of a single particle. Ceperley also produced a more refined calculation on the electron gas, in which the notion of parameter-independent Jastrow to describe two-body correlations among electrons was introduced. Later Schmidt, Lee, Kalos and Chester (1981) performed very refined variational calculations on liquid ^{3}He, showing how the quality of the variational result strongly depends not only on the correlations, but also on the single-particle wave functions used in the Slater determinant, and therefore on the nodal structure of the wave function.

The simplest wave function that can be used to model N atoms of normal (i.e. not superfluid) liquid ^{3}He is that of a correlated Fermi gas:

$$\Psi_{JS}^A = \prod_{i<j} e^{-\frac{1}{2}\left(\frac{b}{r_{ij}}\right)^5} \text{Det}_{\uparrow}(e^{-i\mathbf{k}_\alpha \cdot \mathbf{r}_l})\text{Det}_{\downarrow}(e^{-i\mathbf{k}_\alpha \cdot \mathbf{r}_m}). \tag{7.51}$$

This function is often referred to as a Jastrow–Slater (JS) wave function. There are two main simple improvements that can be made on this form. The first is to introduce explicit three-body correlations (triplets), which are induced by backflow-like correlations first dicussed by Feynman and Cohen (1956),

$$\chi_{ijk} = \exp[\xi(r_{ij})\xi(r_{jk})\mathbf{r}_{ij} \cdot \mathbf{r}_{jk}], \tag{7.52}$$

which can be conveniently rewritten as

$$\Pi_{i<j<k}\chi_{ijk}^2 = \exp\left[\lambda_T \sum_{i<j}^{N} \xi^2(r_{ij})r_{ij}^2 - \frac{\lambda_T}{2}\sum_{l=1}^{N} \mathbf{G}(l) \cdot \mathbf{G}(l)\right], \tag{7.53}$$

where the function ξ has the form

$$\xi(r) = \left(\frac{r - R_T}{R_T}\right)\exp\left[\left(\frac{r - S_T}{\omega_T}\right)^2\right], \quad r < R_T, \tag{7.54}$$

and

$$\mathbf{G}(l) = \sum_{i \neq l}^{N} \xi(r_{il})\mathbf{r}_{il}. \tag{7.55}$$

The constants λ_T, R_T, S_T and ω_T are variational parameters. The corresponding wave function is labeled Jastrow–Triplet–Slater (JTS).

A second kind of improvement that can be made in the wave function is the use of modified single-particle functions which effectively include the so-called *backflow* correlations, which account for the local modification of the density around a moving Fermion in the direction of the momentum. The modified orbitals are of the form

$$\phi_{\mathbf{k}}(\mathbf{r}_i) = \exp\left\{i\mathbf{k} \cdot \left[\mathbf{r}_i + \lambda_B \sum_{j \neq i}^{N} \eta(r_{ij})\mathbf{r}_{ij}\right]\right\}, \tag{7.56}$$

where the function η has the same form as the one used for the triplet correlations:

$$\eta(r) = \left(\frac{r - R_B}{R_B}\right)\exp\left[\left(\frac{r - S_B}{\omega_B}\right)^2\right], \quad r < R_B, \tag{7.57}$$

with λ_B, R_B, S_B and ω_B variational parameters. An interesting fact to note is that in this Jastrow–Slater backflow (JSB) wave function, the orbitals are no longer dependent on the coordinates of a single particle, but rather on the coordinates of all the atoms in the simulation box. This fact causes the necessary computational

Table 7.1. VMC Estimate of the ground state energy of ^{3}He at equilibrium density $\rho = 0.01659\,\text{Å}^{-3}$ obtained for different wave functions and compared to experiment (Schmidt 1981).

Wave function	$E/N(K)$
Mass 3 Bosons	
J	$-2.92(3)$
JT	$-3.40(2)$
^{3}He	
JS	$-1.08(3)$
JTS	$-1.61(3)$
JSB	$-1.55(4)$
JTSB	$-1.91(3)$
Experiment	$-2.47(1)$

time to scale as N^3. In Table 7.1 we report the values of the variational estimate of the energy per particle in normal ^{3}He at density $\rho = 0.01659\,\text{Å}^{-3}$ for different trial wave functions, compared with the energy per particle of a fluid of Bosons with mass 3 amu (spinless ^{3}He). The interatomic interaction used is the HFDHE2 potential of Aziz *et al.* (1979).

From the table it can be seen how the inclusion of both three-body correlations and backflow gives a very strong improvement to the variational estimate of the energy. However, the discrepancy with the experimental value is still rather large.

The addition of backflow correlation does not give a very large contribution to the energy of the electron gas. However, in a very interesting work, Kwon *et al.* (1994) showed that the use of modified orbitals is essential when trying to compute properties which are more closely related to the single-particle properties. In particular, it was shown how the MC computation of particle–hole excitations greatly benefits from the use of orbitals including backflow. In general, the issue of finding variational wave functions with a good nodal surface still remains an open question of great importance, as we will see in the following subsections.

7.8 Optimization of Variational Wave Functions

When a variational wave function contains just a few parameters, it is possible to find their optimal value repeating an MC evaluation of the expectation value of the Hamiltonian for a set of values on a grid. Sometimes, physical intuition severely limits the range of reasonable parameters, further simplifying the search.

However, more elaborate wave functions might contain up to several tenths of variational parameters. In this case a blind search does not lead to any useful result. One straightforward and interesting observation is that given a variational wave function characterized by a set of values of the variational parameters $\{\alpha\}$, if we want to slightly change the values of one or more parameters, and obtain a new set of values $\{\alpha'\}$, we do not need to revaluate the expectation of an operator with a new simulation. In fact,

$$\frac{\langle\Psi_T(R,\{\alpha'\})|\hat{O}(R)\Psi_T(R,\{\alpha'\})\rangle}{\langle\Psi_T(R,\{\alpha'\})|\Psi_T(R,\{\alpha'\})\rangle} = \frac{\int dR|\Psi_T(R,\{\alpha'\})|^2 O(R)}{\int dR|\Psi_T(R,\{\alpha'\})|^2}$$

$$= \frac{\int dR\dfrac{|\Psi_T(R,\{\alpha\})|^2}{|\Psi_T(R,\{\alpha\})|^2}|\Psi_T(R,\{\alpha'\})|^2 O(R)}{\int dR\dfrac{|\Psi_T(R,\{\alpha\})|^2}{|\Psi_T(R,\{\alpha\})|^2}|\Psi_T(R,\{\alpha'\})|^2}$$

$$= \frac{\int dR|\Psi_T(R,\{\alpha\})|^2 w(R)O(R)}{\int dR|\Psi_T(R,\{\alpha\})|^2 w(R)}. \tag{7.58}$$

The last expression can be evaluated in MC as a weighted sum:

$$\frac{\langle\Psi_T(R,\{\alpha'\})|\hat{O}(R)\Psi_T(R,\{\alpha'\})\rangle}{\langle\Psi_T(R,\{\alpha'\})|\Psi_T(R,\{\alpha'\})\rangle} \simeq \frac{\frac{1}{M}\sum_{i=1}^M w(R_i)O(R_i)}{\frac{1}{M}\sum_{i=1}^M w(R_i)}, \tag{7.59}$$

where R_i are sampled from the probability density $|\Psi_T(R,\{\alpha\})|^2$, and the weights are the quotients $w(R) = |\Psi_T(R,\{\alpha'\})|^2/|\Psi_T(R,\{\alpha\})|^2$. This means that with configurations sampled using a set of parameters $\{\alpha\}$ we can estimate the expectations of operators for other sets of parameters, provided that $w(R_i) \sim 1$, i.e. the two wave functions have a consistent overlap.

This procedure is particularly useful if we want to compute gradients of the expectation value of the Hamiltonian in the parameter space. The knowledge of gradients allows us to employ more efficient methods to locate the minimum, such as conjugate gradients, or more elaborate schemes, like the Levemberg–Marquardt algorithm.

Moroni *et al.* (1995) devised a systematic scheme (named Euler Monte Carlo) to improve the variational functions for ^{3}He and ^{4}He, in which the pseudopotentials defining Jastrow and triplet (and backflow) correlations are expanded on a set of suitable basis functions $\chi(r)$:

$$\Psi_T(R) = \exp\left\{-\frac{1}{2}\sum_{i<j} u_2(r_{ij}) + \frac{1}{2}\sum_{i<j<k} u_3(\mathbf{r}_{ij}\cdot\mathbf{r}_{jk})\right\},$$

$$u_2(r) = u_2^0(r) + \sum_m a_m\chi_m(r), \tag{7.60}$$

$$u_3(r) = u_3^0(\mathbf{r}_{ij}\cdot\mathbf{r}_{jk}) + \sum_{\text{cyc}}\sum_{m,n,l} b_{mn}^l\chi_m(r_{ij})\chi_n(r_{ij})P_l(\hat{\mathbf{r}}_{ij}\cdot\hat{\mathbf{r}}_{jk}),$$

where P_l are the Legendre polynomials. The functions $u_2^0(r)$ and $u_3^0(\mathbf{r}_{ij}\cdot\mathbf{r}_{jk})$ are the ansatzs for the Jastrow and triplet correlations for He already introduced in the previous subsections. The single-particle functions including backflow correlations for ^{3}He are also expanded in the same way. The functions $\chi(r)$ were chosen to be of the form

$$
\begin{aligned}
\chi_m(r) &= \left[1 - \cos\left(\frac{2\pi m}{L - 2r_c}\left(r - \frac{L}{2}\right)\right)\right] r^{-5} (r > r_c) \\
&= r^{-5}(r < r_c),
\end{aligned}
\tag{7.61}
$$

where L is the side of the simulation box, and r_c is an arbitrary cutoff which roughly corresponds to the width of the hard core potential. The improvement in the estimate of the energy per particle due to the optimization of the correlations is remarkable. In Tables 7.2 and 7.3 we report the results published by Moroni *et al.* by optimizing piece by piece the parameters in the expanded correlation pseudopotentials for ^{4}He at the experimental saturation density $\rho = 0.02186$ Å^{-3} and ^{3}He at the experimental saturation density $\rho = 0.01635$ Å^{-3}, respectively. The results are compared with diffusion Monte Carlo (DMC) results which represent the best possible estimate (exact for Bosons) of the energy per particle for a given potential. The label "O refers to the use of expanded pseudopotential.

The most impressive achievements in the optimization of variational wave functions have been made in the computation of the ground state of atoms and

Table 7.2. Estimate of the ground state energy of liquid ^{4}He at equilibrium density for different variational wavefunctions. DMC indicates the value obtained by projecting with Diffusion Monte Carlo (Moroni *et al.*, 1995).

^{4}He wave function	$E/N(K)$
J	$-5.702(5)$
OJ	$-6.001(16)$
OJT	$-6.862(16)$
OJOT	$-6.901(4)$
DMC	$-7.143(4)$

Table 7.3. Estimate of the ground state energy of liquid ^{3}He at equilibrium density for different variational wavefunctions. DMC indicates the value obtained by projecting with Diffusion Monte Carlo (Moroni *et al.*, 1995).

^{3}He wave function	$E/N(K)$
J	$-1.085(34)$
OJ	$-1.233(30)$
OJB	$-1.659(21)$
OJOT	$-1.709(17)$
OJTB	$-2.055(15)$
OJOTB	$-2.095(6)$
DMC	$-2.299(5)$

Table 7.4. Ground state energies of first row dimers computed by VMC and DMC (Filippi and Umriger, 1996).

Molecule	CSF,D	$E_{HF}(H)$	E_0 (H)	E_{VMC} (H)	E_{DMC} (H)	$\%_{VMC}$	$\%_{DMC}$
Li_2	1,1	-14.87152	-14.9954	$-14.97343(7)$	$-14.9911(1)$	$82.26(5)$	$96.5(1)$
	4,5			$-14.98850(4)$	$-14.9938(1)$	$94.43(4)$	$98.7(1)$
Be_2	1,1	-29.13242	$-29.33854(5)$	$-29.2782(1)$	$-29.3176(4)$	$70.70(7)$	$89.8(2)$
	5,16			$-29.3129(1)$	$-29.3301(2)$	$87.56(6)$	$95.9(1)$
B_2	1,1	-49.09088	$-49.415(2)$	$-49.3115(3)$	$-49.3778(8)$	$68.06(8)$	$88.5(2)$
	6,11			$-49.3602(2)$	$-49.3979(6)$	$83.10(7)$	$94.7(2)$
C_2	1,1	-75.40620	$-75.923(5)$	$-75.7567(5)$	$-75.8613(8)$	$67.82(9)$	$88.1(2)$
	4,16			$-75.8282(4)$	$-75.8901(7)$	$81.66(7)$	$93.6(1)$
N_2	1,1	-108.9928	$-109.542(3)$	$-109.3756(6)$	$-109.487(1)$	$69.7(1)$	$89.9(2)$
	4,17			$-109.4376(5)$	$-109.505(1)$	$80.94(8)$	$93.1(2)$
O_2	1,1	-149.6659	$-150.326(8)$	$-150.1507(6)$	$-150.268(1)$	$73.4(1)$	$91.0(2)$
	4,7			$-150.1885(5)$	$-150.277(1)$	$79.08(8)$	$92.5(2)$
F_2	1,1	-198.7701	$-199.529(9)$	$-199.3647(7)$	$-199.478(2)$	$78.26(9)$	$93.2(2)$
	2,2			$-199.4101(6)$	$-199.487(1)$	$84.23(8)$	$94.3(1)$

molecules. In this case the correlation pseudopotentials are inferred by analytic properties of the electron–electron and the electron–electron–nucleus wave functions. Moreover, multiconfiguration wave functions in which a linear combination of configurations for a state of given symmetry are used in order to systematically improve the nodal structure. A nice example is given in the paper by Filippi and Umrigar (1996), in which the ground state energies of dimers are computed. In Table 7.4 we report the results of such calculations, in which the percentage of the correlation energy (referred to as the exact, nonrelativistic, infinite mass nucleus energy E_0) recovered by the VMC and then projecting with fixed node diffusion Monte Carlo methods are shown. In the second column are reported the number of configurations and the number of Slater determinants employed in the calculations. It can be clearly seen how the improvement of the nodal structure at variational level leads to considerably better results.

Further improvements have recently been made in the field of wave function optimization by systematically minimizing a linear combination of variance and energy (Filippi and Umrigar, 2005). A recent discussion on energy minimization by means of Newton, linear and perturbative methods can be found in a work by Toulouse and Umrigar (2007).

7.9 Monte Carlo Methods and Quantum Mechanics

The most successful application of Monte Carlo methods in physics is the possibility of solving exactly the quantum many-body problem, at least in some cases. These algorithms are often generically referred to as quantum Monte Carlo (QMC) methods. However, it is possible to identify at least three great families of QMC algorithms:

(1) *Variational Monte Carlo.* We already described this method in the previous subsections. In general, VMC consists in finding an approximate solution to the Schrödinger equation, depending on some parameters, and in utilizing the variational principle for the determination of the best value of the parameters themselves.*

(2) *Projection methods.* These are methods that solve exactly the Schrödinger equation for the ground state of a system of interacting Bosons at temperature $T = 0$. They may be used to yield approximate, but extremely optimized, solutions for Fermion systems. The two main algorithms are the diffusion Monte Carlo (DMC) and the Green's function Monte Carlo (GFMC).

(3) *Path integral Monte Carlo.* This consists in evaluating averages on the density matrix of a system (and thus at finite temperature), using a path-integral-like expansion. It is exact for Boson systems, and can be used in approximate form for Fermions (Ceperley, 1995).

In the following, we will consider only projection methods, and will describe one of their possible derivations and their main characteristics.

7.10 Propagation of a State in Imaginary Time

In quantum mechanics, the propagation of a state from time t_0 to time t is given by the following expression:

$$\psi(R, t) = e^{-iH(t-t_0)/\hbar}\psi(R, t_0), \tag{7.62}$$

where $R = \{\mathbf{r}_1 \cdots \mathbf{r}_N\}$ are the coordinates of the N particles. Let us reparametrize time by means of an *imaginary time*, $\tau = it$. Formally, we can rewrite the same propagator in the following way:

$$\psi(R, \tau) = e^{-H(\tau-\tau_0)/\hbar}\psi(R, \tau_0). \tag{7.63}$$

Let us set $\hbar = 1$, so that the imaginary time has the dimension of a reciprocal energy. Moreover, let us put $\tau_0 = 0$. Next we expand the state ψ on a basis of eigenstates of H:

$$\psi(R, 0) = \sum_n c_n \phi_n(R). \tag{7.64}$$

---

*For a general discussion, see for example Ceperley and Kalos (1986).

By applying the propagator (7.63), we obtain

$$\psi(R,\tau) = e^{-H\tau}\psi(R,0) = \sum_n c_n e^{-E_n\tau}\phi_n(R), \qquad (7.65)$$

where E_n are the eigenvalues corresponding to ϕ_n. From (7.65), we note that the components $c_n e^{-\tau E_n}$ of the propagated state along the eigenstates of H decay exponentially, each with a different characteristic time which is proportional to the excitation energy.

However, it is possible to normalize the wave function in such a way as to keep the component along one of the eigenstates constant. In particular, by considering the function

$$\psi'(R,\tau) = e^{E_0\tau}\psi(R,\tau), \qquad (7.66)$$

the expansion (7.65) becomes

$$\psi'(R,\tau) = \sum_n c_n e^{-\tau(E_n-E_0)}\phi_n(R). \qquad (7.67)$$

In the $\tau \to \infty$ limit, we obtain

$$\lim_{\tau\to\infty} \psi'(R,\tau) = c_0\phi_0(R), \qquad (7.68)$$

i.e. the normalized state, propagated in imaginary time, tends to the ground state, at most, multiplied by a constant. In general, *the propagation in imaginary time of an arbitrary wave function of a system projects from this function the eigenstate of H which has the lowest energy, among those along which the initial state has a nonvanishing component.* For a many-body system, if one does not impose any constraint, H has both symmetric and antisymmetric eigenstates with respect to particle exchange. It can be shown that the lowest-energy solution, and hence the state projected by imaginary time propagation, is always symmetric. The projection algorithm can then be straightforwardly applied only to Boson systems, and hence, for simplicity we will assume for the moment that the system under consideration consists only of Bosons.

7.11   The Schrödinger Equation in Imaginary Time

Equation (7.63) represents the formal solution to a many-body Schrödinger equation in imaginary time:

$$-\frac{\partial}{\partial\tau}\psi(R,\tau) = H\psi(R,\tau), \qquad (7.69)$$

i.e.

$$-\frac{\partial}{\partial\tau}\psi(R,\tau) = -\frac{\hbar^2}{2m}\sum_{i=1}^{N}\nabla_i^2\psi(R,\tau) + V(R)\psi(R,\tau), \qquad (7.70)$$

where $R = \{r_1, r_2, \ldots, r_N\}$. Equation (7.70) is equivalent to a diffusion equation with an absorption term. To better understand its meaning, it is interesting to examine it in two important limiting cases:

(a) **The zero-mass limit**

If the mass is very small, the kinetic term becomes dominant, and the equation reduces to

$$-\frac{\partial}{\partial \tau}\psi(R,\tau) = -\frac{\hbar^2}{2m}\sum_{i=1}^{N}\nabla_i^2\psi(R,\tau).\qquad(7.71)$$

It is possible to express the exact solution of (7.71) in integral form:

$$\psi(R,\tau) = \int G(R-R',\tau)\psi(R',0)dR',\qquad(7.72)$$

with the Green function given by

$$G(R-R',\tau) = \frac{1}{(4\pi D\tau)^{\frac{dN}{2}}}\exp\left\{-\frac{(R-R')^2}{4D\tau}\right\},\qquad(7.73)$$

where d is the dimensionality of the system, N is the number of particles and $D = \hbar^2/2m$. The latter quantity plays the role of a diffusion constant.

The function $\psi(R,\tau)$ describes the probability that a particle is at point R at time τ, while the Green function describes the probability of passing from R to R' in a time interval τ. Finally, $\psi(R',0)$ is the probability of being at R' at time $\tau = 0$. Therefore, to generate system configurations distributed according to the solution of (7.71), it is sufficient to start from an arbitrary particle distribution $\psi(R',0)$, and then to generate displacements according to the Green function G, i.e. to generate a vector in $d \times N$ dimensions in which each component is a random number extracted from the Gaussian distribution G, and then sum it to the initial positions of the particles. Therefore, the pseudodynamics for the walker is given by the equation

$$r_i' = r_i + \xi_i.\qquad(7.74)$$

The positions $\{r_i'\}$ will be distributed according to $\psi(R,\tau)$.

(b) **The infinite-mass limit**

If the mass becomes very large, the dominant term is that of the potential energy and we obtain the following equation:

$$-\frac{\partial}{\partial \tau}\psi(R,\tau) = V(R)\psi(R,\tau).\qquad(7.75)$$

For a normalized state $e^{\tau E_0}\psi(R,\tau)$ the equation is modified in the following way:

$$-\frac{\partial}{\partial \tau}\psi(R,\tau) = [V(R) - E_0]\psi(R,\tau),\qquad(7.76)$$

and its solution is

$$\psi(R, \tau) = P_B = e^{-[V(R)-E_0]\tau}. \tag{7.77}$$

Equation (7.77) is interpreted as the probability that a walker, which is at R, survives in this same position after a time τ. Therefore, the walker population tends to increase where $V(R) < E_0$, and to vanish where $V(R) > E_0$. From the point of view of the algorithm, this means that we should compute the value of Eq. (7.77) at point R, and generate at the same point a number of walkers equal to $[P_B + \xi]$, where ξ is a random number uniformly distributed between 0 and 1. If the resulting number is 0, the walker is destroyed. This process is called *branching*.

In general, the situation is intermediate between these two limits, and the algorithm will have to implement both the diffusive dynamics and the branching process. For this purpose, we note that the propagator can be decomposed by using the Trotter formula:

$$e^{-\tau H} = e^{-\tau(T+V)} \approx e^{-\tau T} e^{-\tau V} + o(\tau). \tag{7.78}$$

In the $\tau \to 0$ limit, the decomposition becomes exact. This disagrees with the fact that, in order to arrive at a correct projection of the ground state, propagation should take place for long imaginary times, at least of the order of a few of the decay time $\tau_0 = 1/(E_1 - E_0)$ of the component along the first excited state. However, it is possible to split the propagation in time τ into M propagation steps $\Delta\tau = \tau/M$. If M is large, the approximation (7.78) will be accurate at order $\Delta\tau$.

The previous considerations yield a first possible structure of the projection algorithm:

(1) Generation of a walker population distributed according to $\psi(R, 0)$.
(2) Diffusion of the walkers following (7.74) for a short time $\Delta\tau$.
(3) Calculation of the survival and/or multiplication probability of a walker according to (7.77).
(4) Back to step 1, and repeat until propagation for a sufficiently long time τ is achieved.

The required ground state eigenvalue is the value of E_0 that makes the population stationary.

7.12   Importance Sampling

If the potential V has divergences (for example, in a system of electrons interacting through the Coulomb potential), the expression for the branching probability may diverge, and this causes large fluctuations in the walker population. In these cases it is convenient to employ an approximate wave function (obtained, for example, by a VMC calculation) to drive the walkers in such a way that they avoid classically forbidden regions of configuration space. For example, in the presence of a repulsive potential, the wave function should cancel or have a minimum on the hyperplanes defined by $r_i = r_j$. Let us consider, then, a trial function $\psi_T(R)$, known as *an importance function*. We sample the walkers from a density $f(R, \tau) = \psi_T(R)\phi(R, \tau)$ which evolves in imaginary time, and which in the $\tau \to \infty$ limit is proportional to $\psi_T(R)\phi_0(R)$. We are interested in finding the equation which governs the distribution $f(R, \tau)$, i.e. the analog of the Schrödinger equation in imaginary time introduced in the previous subsections. Let us define the *pseudoforce* acting on particle i:

$$F_i(R) = 2\nabla_i \ln \psi_T(R) = 2\frac{\nabla_i \psi_T(R)}{\psi_T(R)}.$$

It may be noted that

$$\begin{aligned}
\nabla\left(f(R, \tau)F(R)\right) &= 2\nabla\psi_T(R)\nabla\phi(R, \tau) + 2\phi(R, \tau)\nabla^2\psi_T(R), \\
\nabla^2 f(R, \tau) &= \psi_T(R)\nabla^2\phi(R, \tau) + 2\nabla\psi_T(R)\nabla\phi(R, \tau) \\
&\quad + \phi(R, \tau)\nabla^2\psi_T(R).
\end{aligned} \tag{7.79}$$

By combining these expressions, it is possible to obtain an expression for the Laplacian of the function $\phi(R, \tau)$:

$$\psi_T(R)\nabla^2\phi(R, \tau) = \nabla^2 f(R, \tau) - \nabla[f(R, \tau)F(R)] + f(R, \tau)\frac{\nabla^2\psi_T(R)}{\psi_T(R)}. \tag{7.80}$$

In order to obtain the desired distribution in the $\tau \to \infty$ limit, the function $\phi(R, \tau)$ must satisfy the imaginary-time-dependent Schrödinger equation:

$$\frac{\partial}{\partial\tau}\phi(R, \tau) = D\sum_{i=1}^{N}\nabla_i^2\phi(R, \tau) - V(R)\phi(R, \tau). \tag{7.81}$$

By multiplying both sides of Eq. (7.81) by $\psi_T(R)$, and inserting (7.80), we obtain the following equation for the probability density $f(R, \tau)$:

$$\begin{aligned}
\frac{\partial}{\partial\tau}f(R, \tau) = D\sum_{i=1}^{N}\nabla_i^2 f(R, \tau) - D\sum_{i=1}^{N}\nabla_i \cdot [f(R, \tau)F_i(R)] \\
- E_L(R)f(R, \tau),
\end{aligned} \tag{7.82}$$

where

$$E_L(R) = -D\sum_{i=1}^{N}\frac{\nabla_i^2\psi_T(R)}{\psi_T(R)} + V(R) = \frac{H\psi_T(R)}{\psi_T(R)}$$

is the *local energy* operator.

Equation (7.82) is a Fokker–Planck equation with a source term. To understand the walker dynamics associated with this equation, let us carry out the same kind of analysis as for the Schrödinger equation in imaginary time, i.e. let us consider the

two limits of zero mass and infinite mass. The former case corresponds to neglecting
the term depending on the local energy. In this way we obtain a simplified equation:

$$\frac{\partial}{\partial \tau} f(R, \tau) = D \sum_{i=1}^{N} \left(\nabla_i^2 f(R, \tau) - \nabla_i \cdot [f(R, \tau) F_i(R)] \right), \tag{7.83}$$

which, for $\tau \to \infty$, has the solution

$$f(R, \tau) = \psi_T^2(R).$$

This can be shown by transforming back the Fokker–Plank equation into an imaginary time Schrödinger equation, and showing that $\exp[-2\log(\psi_T(R))]$ is the eigenstate with the lowest eigenvalue, which is therefore the asymptotic distibution of
walkers for $\tau \to \infty$. This also means that, if we neglect the branching process, the
Fokker–Planck equation in imaginary time for $f(R, \tau)$ corresponds to a suitable
dynamics for the sampling of the variational probability density $\psi_T^2(R)$. The Green
function for Eq. (7.83) is a Gaussian similar to (7.73) and its center, contrary to
what happens for the free diffusion equation, is shifted along the gradient of the
wave function ψ_T with respect to the origin R. Thus, the MC dynamics has a new
term, called the *drift* term, which describes the effect of the pseudoforce on the evolution of the walker population. Therefore, starting with a walker at position R in
the $(d \times N)$-dimensional space, the new position R' is generated moving each particle according to the following expression:

$$r_i' = r_i + D \frac{2\nabla_i \psi_T(R)}{\psi_T(R)} \Delta\tau + \xi_i,$$

where ξ is a vector in which each component is a random number with Gaussian
distribution as in (7.74).

In the $m \to \infty$ limit, we have for the normalized function

$$\frac{\partial}{\partial \tau} f(R, \tau) = [E_0 - E_L(R)] f(R, \tau). \tag{7.84}$$

Now the branching probability becomes

$$P_B = e^{-\Delta\tau(E_L(R) - E_0)}.$$

Therefore, the introduction of the importance function modifies two steps of the
algorithm:

(1) Addition of the drift term in the walker dynamics.
(2) Use of the local energy in place of the potential energy in the branching term.

7.13 Importance Sampling and Green Functions

An alternative and cleaner way to introduce importance sampling in DMC passes through the solution of the imaginary time Schrödinger equation in terms of Green functions. The following derivation is due to M.H. Kalos. As already mentioned, the imaginary time propagator can be written in integral form, and the propagated solution is given by

$$\psi(R, \tau) = \int G(R, R', \tau)\psi(R', 0)dR'. \tag{7.85}$$

The basic Green function for a small increment in imaginary time $\Delta\tau$ can be written as

$$G(R, R', \Delta\tau) = \frac{1}{\sqrt{(2\pi\Delta\tau)^{3N}}} e^{\frac{(R-R')^2}{2\Delta\tau} - (V(R') - E_T)\Delta\tau}, \tag{7.86}$$

where we have posed for simplicity $\hbar = m = 1$, and therefore a diffusion constant $D = 1$. We want to sample a density of walkers which benefits of the information contained in a trial wave function which we might have determined by means of a VMC calculation. The easiest way to achieve this goal is to multiply the density of walkers by the trial function itself:

$$\Psi_T(R)\psi(R, \Delta\tau) = \int G(R, R', \Delta\tau)\Psi_T(R)\psi(R', 0)dR', \tag{7.87}$$

which can in turn be rewritten as

$$\Psi_T(R)\psi(R, \Delta\tau) = \int G(R, R', \Delta\tau)\frac{\Psi_T(R)}{\Psi_T(R')}\Psi_T(R')\psi(R', 0)dR'. \tag{7.88}$$

The propagator of the density of walkers is therefore

$$G(R, R', \Delta\tau)\frac{\Psi_T(R)}{\Psi_T(R')}. \tag{7.89}$$

The density of walkers is approximated by a sum of delta functions, one in correspondence with the position of each walker. Each delta function is smeared over an imaginary time interval $\Delta\tau$ by a propagator that, however, does not in general preserve the normalization of the density. This means that before proceeding with the propagation, we should evaluate the norm of the propagated delta function at the next step, i.e. the multiplicity of new points generated starting from R'. This means that we need to compute

$$N(R') = \int G(R, R', \Delta\tau)\frac{\Psi_T(R)}{\Psi_T(R')}dR. \tag{7.90}$$

This integral can be easily evaluated if we consider that for a small $\Delta\tau$, the displacement $|R - R'|$ will be small. We can therefore expand $\Psi_T(R)$ around R' and obtain

$$N(R') \approx \int G(R, R', \Delta\tau) \left[1 + \frac{\nabla \Psi_T(R')}{\Psi_T(R')}(R - R') \right.$$

$$\left. + \left(1 - \frac{\delta_{ij}\delta_{kl}}{2} \right) \frac{\sum_{ijkl} \frac{\partial^2 \Psi_T(R')}{\partial x'_{il} \partial x'_{jk}}(x_{il} - x'_{il})(x_{jk} - x'_{jk})}{\Psi_T(R')} \right] dR, \qquad (7.91)$$

where x_{il} denotes the lth component of the coordinates of the ith particle. This integral can be easily computed by noting that the dependence on R in the Green function $G(R, R', \Delta\tau)$ is only in the Gaussian part, which is symmetric with respect to $R - R'$. Therefore all the odd terms will cancel after integration, and we will be left with

$$N(R') \approx \left[1 + \frac{\Delta\tau}{2} \frac{\sum_k \nabla_k^2 \Psi_T(R')}{\Psi_T(R')} \right] e^{-\Delta\tau[V(R') - E_T]}$$

$$\simeq \exp\left[\left(\frac{1}{2} \frac{\nabla^2 \Psi_T(R')}{\Psi_T(R')} + E_T - V(R') \right) \Delta\tau \right]$$

$$\simeq \exp\left[-\Delta\tau \left(\frac{H\psi_T(R')}{\Psi_T(R')} - E_T \right) \right], \qquad (7.92)$$

which is the expression for the weight of the configuration derived in the previous subsection. The displaced configurations will be generated sampling the kinetic part of the Green function, which will be modified by the importance sampling as follows:

$$G(R, R', \Delta\tau) \simeq \left[1 + \frac{\nabla \Psi_T(R')}{\Psi_T(R')}(R - R') \right] \exp\left[-\frac{(R - R')^2}{2\Delta\tau} \right]$$

$$\simeq \exp\left[\nabla \log \Psi_T(R')(R - R') - \frac{(R - R')^2}{2\Delta\tau} \right]$$

$$\simeq \exp\left[\frac{1}{2\Delta\tau} (R - R' - \nabla \log \Psi_T(R')\Delta\tau) \right], \qquad (7.93)$$

which again is the previously derived expression for the drifted Gaussian. In the last step we explicitly neglected contributions to the Green function of order $o(\Delta\tau^2)$.

This Green-function-based approach is extremely useful whenever one seeks for a generalization of the standard algorithm, because it explicitly starts from the propagator, without the need for rewriting the differential equation governing the evolution of the walker distributions. In particular, it can be seen that instead of using a drifted Gaussian and a multiplicity which depend on the local energy, importance sampling could be achieved by sampling the non-importance-sampled wave function and accepting/rejecting the new configuration with probability $\Psi_T(R)/\Psi_T(R')$.

Importance sampling drastically reduces the walker population fluctuations, even in the presence of diverging potentials. In fact, a proper choice of the analytic form of $\psi_T(R)$ guarantees that $E_L(R)$ is always finite (*cusp condition*), a condition that is exactly fulfilled if $\psi_T(R)$ is an exact eigenstate of the Schrödinger equation. Another advantage of this algorithm is that it is possible to define in a natural way an energy estimator, other than the normalization constant. In the limit of propagation for an infinite imaginary time, the mean value of the local energy $E_L(R)$ computed on the distribution $f(R, \bar{\tau})$ is exactly the ground state energy. In fact,

$$E_0 = \frac{\langle \psi_T(R) | H \phi_0(R) \rangle}{\langle \psi_T(R) | \phi_0(R) \rangle}, \tag{7.94}$$

but since the operator H is Hermitian, we have

$$E_0 = \frac{\langle \phi_0(R) | H \psi_T(R) \rangle}{\langle \phi_0(R) | \psi_T(R) \rangle} = \lim_{\tau \to \infty} \frac{\int f(R, \tau) \frac{H \psi_T(R)}{\psi_T(R)} \, dR}{\int f(R, \tau) dR}. \tag{7.95}$$

Therefore, the energy eigenvalue may be obtained as the mean value of the local energy on the distribution $f(R, \tau)$. For any other estimator, one does not get mean values on the ground state, but rather matrix elements between the ground state and the trial variational state. However, let us suppose that for an operator $\hat{O}$ there exists a value $\epsilon \ll 1$ such that $|\psi_T\rangle - |\phi_0\rangle \sim \epsilon|\psi_{\text{res}}\rangle$, where $|\psi_{\text{res}}\rangle$ is a suitable combination of excited states of the system. It is easily verified that the estimator

$$2\langle \phi_0 | \hat{O} \psi_T \rangle - \langle \psi_T | \hat{O} \psi_T \rangle \tag{7.96}$$

differs from the matrix element on the ground state by an order of ϵ^2. Such a corrected estimator is named a *mixed estimator*. More complicated methods for finding pure estimators are present in the literature.

A final consideration concerns the validity of approximation (7.78). In general, the terms $o(\tau)$ which are neglected in the implementation of the algorithm have a nonnegligible effect on the energy value (*time-step error*). For this reason it is, in general, necessary to carry out a τ extrapolation of the results. By suitable modifications of the algorithm, the effects of the time-step error can be reduced. In Fig. 7.4, we see how different implementations of DMC lead to very different τ dependencies of the results.[†]

7.14 Fermion Systems and the Sign Problem

Let us suppose that we want to use a DMC-like algorithm to study a many-body Fermion (e.g. electron) system. We might consider starting from an antisymmetric variational trial function ψ_T, and then applying the algorithm as described above. From the expression (7.95), which gives the mean value of the energy, we deduce that the result is, in principle, correct if we take as energy reference E_0^A (i.e. the energy of the antisymmetric ground state), because the components on the symmetric states

---

[†]An exhaustive discussion about the dependence of the time-step error on the details of the algorithm may be found in the paper by Umrigar, Nightingale and Runge (1993).

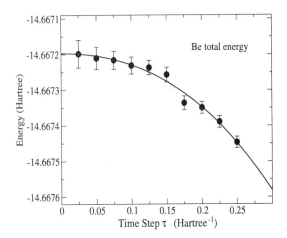

Fig. 7.4 Dependence of the total energy on the time-step τ in the calculation of the total energy of the Be atom in the DMC algorithm (courtesy C.J. Umrigar).

vanish by orthogonality. Therefore, the term in (7.67) which includes $e^{-\tau(E_0-E_0^A)}$ and would be diverging gives actually no contribution. However, if we compute the variance of the ground state energy E_0^A,

$$\sigma^2 = |\langle H \rangle^2 - \langle H^2 \rangle|, \tag{7.97}$$

then the symmetric components survive in the expression

$$\langle H^2 \rangle = \frac{\int \phi_0(R)\psi_T(R)(\frac{H\psi_T(R)}{\psi_T(R)})^2 dR}{\int \phi_0(R)\psi_T(R)dR}, \tag{7.98}$$

and the term

$$\sum_n c_n \exp\{-(E_n - E_0^A)\tau\}\phi_{0n} \tag{7.99}$$

diverges exponentially. Thus, it is possible to obtain exact mean eigenvalues, though affected by an infinitely large error.

This problem is known as the *sign problem* and, up until now, it constitutes the most severe limit to the application of the DMC method to Fermion systems.

In any case, there are a certain number of approximations that make the use of DMC possible and, in some cases, yield almost exact results. The most widespread of these is the *fixed node approximation*. Essentially, it consists in imposing artificial boundary conditions on the problem, by requiring the DMC projected wave function to have the same zeroes as the antisymmetric trial function. In this way, the problem is reduced to a standard calculation for the positive (or negative) part of the wave function. The antisymmetry and continuity of the wave function imply that it must become zero on a $(3N - 1)$-dimensional hypersurface in the $3N$-dimension space of the system under study. Such a hypersurface (the *nodal structure*) is in general

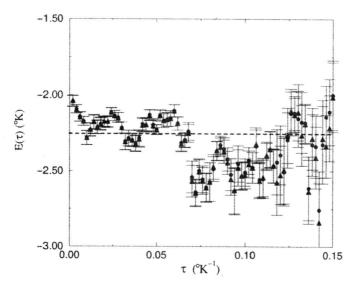

Fig. 7.5 Evolution of the mean value of energy in a DMC calculation with free nodes for 14 ^{3}He particles. Note that the error bars rapidly increase with imaginary time.

not known, but it may be optimized by means of variational methods. Though the resulting error may be estimated, it cannot be determined *a priori*. It can be proved that E_{FN} being the eigenvalue estimated within the fixed node approximation, the following inequality holds:

$$E_T^A > E_{FN} \geq E_0^A, \tag{7.100}$$

where E_T^A is the eigenvalue estimated by VMC and E_0^A is the exact eigenvalue. The last equality holds if and only if the nodal structure of the trial wave function is exact.

The results obtained in the fixed node approximation are, in general, very good for electron systems (they have been presented several times in this book), while in other cases (like ^{3}He) the method is not as satisfactory. A good review of the fixed node diffusion MC method can be found in the paper by Foulkes, Mitas, Needs and Rajagopal (2001), where a proof of the upper bound property of the fixed node estimate of the eigenvalue is presented, together with a collection of applications to solid state physics. A thorough discussion on the magnitude of fixed node errors in systems of chemical interest was published by Grossman (2002), who computed fixed node energies for a set of molecules, and compared them with exact results. From this analysis it turns out that while in certain systems on easily reaches chemical accuracy, for several molecules the energies are not so accurate. There is no systematic behavior in the observed fluctuations.

Recently, remarkable progress has been made in the development of an algorithm which is able to exactly project the ground state of Fermion systems (Kalos and Pederiva, 1999, 2000). At present, this is still an active research field.

7.15 Two Examples of DMC Calculations in Many-Body Systems

In the previous chapters we have reported many results of DMC calculations in electron systems like the electron gas in two and three dimensions, quantum dots and metallic clusters. However, DMC has been applied to many other many-body systems and in the following, as an example, we illustrate the calculations of the equation of state of a Fermi gas in the BEC–BCS crossover and of the ground state energy of a many-nucleon system.

7.15.1 *The Equation of State of Diluted Ultracold Fermi Gases in the BEC–BCS Crossover*

In this subsection, we report the calculation (Astrakharchik *et al.*, 2004) of the equation of state of a homogeneous two-component Fermi gas described by the Hamiltonian (see also Subsection 4.3.3)

$$H = \left(\sum_{i=1}^{N_\uparrow} \frac{p_i^2}{2m} + \sum_{i'=1}^{N_\downarrow} \frac{p_{i'}^2}{2m} \right) + \sum_{i,i'} V(\mathbf{r}_i - \mathbf{r}_{i'}), \qquad (7.101)$$

where m denotes the mass of the particles, i and i' label, respectively, spin-up and spin-down particles and $N_\uparrow = N_\downarrow = N/2$, N being the total number of atoms. The interspecies interatomic interaction was modeled by an attractive square well potential: $V(r) = -V_0$ for $r < R_0$, and $V(r) = 0$ otherwise. In order to ensure that the mean interparticle distance is much larger than the range of the potential (dilute condition), one uses $\rho R_0^3 = 10^{-6}$, where $\rho = k_F^3/3\pi^2$ is the gas density. By varying the depth V_0 of the potential one can change the value of the s wave scattering length, which for this potential is given by

$$a = R_0 \left[1 - \frac{\tan(K_0 R_0)}{K_0 R_0} \right], \qquad (7.102)$$

where $K_0^2 = mV_0/\hbar^2$. K_0, is varied in the range $0 < K_0 < \pi/R_0$, which yields for the range of variation of $k_F a$ $-6 \le -1/k_F a \le 6$. This includes the deep BEC and BCS regimes and the unitary limit (Bartenstein *et al.*, 2004; Bourdel *et al.*, 2004; O'Hara *et al.*, 2002) where the value of $|a|$ can be orders of magnitude larger than the inverse Fermi momentum and one enters a new strongly correlated regime. For $K_0 R_0 < \pi/2$ the potential does not support a two-body bound state and $a < 0$. For $K_0 R_0 > \pi/2$, instead, the scattering length is positive, $a > 0$, and a molecular state appears whose binding energy ϵ_b is determined by the trascendental equation

$$\sqrt{|\epsilon_b|m/\hbar^2} R_0 \frac{\tan(\bar{K}_0 R_0)}{\bar{K}_0 R_0} = -1, \qquad (7.103)$$

where $\bar{K}^2 = K_0^2 - |\epsilon_b|m/\hbar^2$. The value $K_0 = \pi/2R_0$ corresponds to the unitary limit where $|a| = \infty$ and $\epsilon_b = 0$.

The following trial wave functions were used: a BCS wave function,

$$\Psi_{\text{BCS}}(\mathbf{R}) = \mathcal{A}(\Phi(r_{11'})\Phi(r_{22'})\cdots\Phi(r_{N_\uparrow N_\downarrow})), \qquad (7.104)$$

and a Jastrow–Slater (JS) wave function of the type of Eq. (7.51),

$$\Psi_{\text{JS}} = \prod_{ii'} \phi(r_{ii'}) \left[\mathcal{A} \prod_{i\alpha} e^{-i\mathbf{k}_\alpha \cdot \mathbf{r}_i} \right] \left[\mathcal{A} \prod_{i'\alpha} e^{-i\mathbf{k}_\alpha \cdot \mathbf{r}_{i'}} \right], \qquad (7.105)$$

where $\mathcal{A}$ is the antisymmetrizer operator ensuring the correct antisymmetric properties under particle exchange. In the JS wave function, Eq. (7.105), the plane wave orbitals have wave vectors $\mathbf{k}_\alpha = 2\pi/L(\ell_{\alpha x}\hat{x} + \ell_{\alpha y}\hat{y} + \ell_{\alpha z}\hat{z})$, where L is the size of the periodic cubic box fixed by $\rho L^3 = N$, and ℓ are integer numbers. The correlation functions $\Phi(r)$ and $\phi(r)$ in Eqs. (7.104) and (7.105) are constructed from solutions to the two-body Schrödinger equation with the square well potential $V(r)$. In particular, in the region $a > 0$ one takes for the function $\Phi(r)$ the bound state solution $\Phi_{\text{bs}}(r)$ with energy ϵ_b and in the region $a < 0$ the unbound state solution corresponding to zero scattering energy: $\Phi_{\text{us}}(r) = (R_0 - a)\sin(K_0 r)/[r\sin(K_0 R_0)]$ for $r < R_0$ and $\Phi_{\text{us}}(r) = 1 - a/r$ for $r > R_0$. In the unitary limit, $|a| \to \infty$, $\Phi_{\text{bs}}(r) = \Phi_{\text{us}}(r)$.

The JS wave function Ψ_{JS}, Eq. (7.105), is used only in the region of negative scattering length, $a < 0$, with a Jastrow factor $\phi(r) = \Phi_{\text{us}}(r)$ for $r < \bar{R}$. In order to reduce possible size effects due to the long range tail of $\Phi_{\text{us}}(r)$, one uses $\phi(r) = C_1 + C_2\exp{-\alpha r}$ for $r > \bar{R}$, with $\bar{R} < L/2$ a matching point. The coefficients C_1 and C_2 are fixed by the continuity condition for $\phi(r)$ and its first derivative at $r = \bar{R}$, whereas the parameter $\alpha > 0$ is chosen in such a way that $\phi(r)$ goes rapidly to a constant. Residual size effects have been finally determined by carrying out calculations with an increasing number of particles, $N = 14, 38$ and 66.

In Fig. 7.6 we show the fixed node diffusion MC results for the energy per particle in the BEC–BCS crossover with the binding energy subtracted from E/N for $N = 66$ atoms and the potential $V(r)$ with $\rho R_0^3 = 10^{-6}$ as a function of the interaction parameter $-1/k_F a$. The numerical simulations are carried out with both the BCS and JS wave functions. For $-1/k_F a > 0.4$ one finds that Ψ_{JS} gives lower energies E/N, whereas for smaller values of $-1/k_F a$, including the unitary limit and the BEC region, the function Ψ_{BCS} is preferable. In the BCS region, $-1/k_F a > 1$, the results for E/N are in agreement with the perturbation expansion of a weakly attractive Fermi gas (see Subsection 10.3.1):

$$\frac{E_F}{N\epsilon_F} = 1 + \frac{10}{9\pi}k_F a + \frac{4(11 - 2\ln 2)}{21\pi^2}(k_F a)^2 + \cdots . \qquad (7.106)$$

In the unitary limit one finds that $E/N = \xi\epsilon_F$, with $\xi = 0.42(1)$. The value of the parameter $\beta = \xi - 1$ has been measured in experiments with trapped Fermi gases (Bartenstein *et al.*, 2004; Bourdel *et al.*, 2004; O'Hara *et al.*, 2002), but the precision is too low to make a stringent comparison with theoretical predictions. In the region of positive scattering length, $E/N - \epsilon_b/2$ decreases by decreasing $k_F a$,

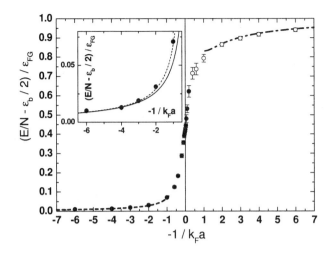

Fig. 7.6 Energy per particle in the BEC–BCS crossover with the binding energy subtracted from E/N. Solid symbols: Results with Ψ_{BCS}. Open symbols: Results with Ψ_{JS}. The dotted–dashed line is the expansion (7.106) holding in the BCS region and the dashed line corresponds to the expansion (7.107) holding in the BEC regime. Inset: Enlarged view of the BEC regime $-1/K_F a \leq -1$. The solid line corresponds to the mean field energy [first term in the expansion (7.107)]; the dashed line includes the beyond mean-field correction [Eq. (7.107)]. (Courtesy S. Giorgini.)

and for values of $-1/k_F a$ smaller than -0.3, rapidly approaches zero, indicating the formation of bound molecules. In the BEC region, $-1/k_F a < -1$, the DMC energies agree with the equation of state of a repulsive gas of molecules:

$$\frac{E/N - \epsilon_b/2}{\epsilon_F} = \frac{5}{18\pi} k_F a_m \left(1 + \frac{128}{15\sqrt{6\pi^3}} (k_F a_m)^{3/2} + \cdots \right), \qquad (7.107)$$

where the first term corresponds to the mean field energy of a gas of molecules of mass $2m$ and density $\rho/2$ interacting with the positive molecule–molecule scattering length a_m, and the second term corresponds to the RPA correlation energy (see Subsection 10.3.1). If for a_m one uses the value calculated by Petrov *et al.* (2004), $a_m = 0.6a$, one obtains the curves shown in Fig. 7.6. If, instead, one uses a_m as a fitting parameter to the DMC calculation in the region $-1/k_F a < -1$, one obtains $a_m/a = 0.62(1)$. A detailed knowledge of the equation of state of the homogeneous system is important for the determination of the frequencies of collective modes in trapped systems (see Chapter 11) which have recently been measured in the BEC–BCS crossover regime (Kinast *et al.*, 2004; Bartenstein *et al.*, 2004).

7.15.2   *Many-Nucleon Systems*

The use of quantum MC methods for computing ground state properties of nuclei and nuclear matter is made harder by the fact that realistic nucleon–nucleon potentials depend not only on the distances among the particles, but also on the relative spin–isospin state of the nucleons [see for example Wiringa, Stocks and Schiavilla (1995)], essentially owing to processes involving the exchange of one or more pions among the nucleons. An important role is also played by three-nucleon forces, which are essential for reproducing the equation of state of infinite nuclear matter (Pieper *et al.*, 2001). The first consequence of this dependence is that the structure of the correlations in the wave function should in turn contain an operatorial part depending on the spin and isospin degrees of freedom, making the whole description extremely complex.

Despite the extreme difficulty of the problem, accurate Green function MC calculations, similar to the DMC calculations described in the previous subsections, were performed for systems of $A \leq 12$ nucleons. Why such a severe limitation on the size? As already mentioned, the most recent Hamiltonians for a many-nucleon system are written as

$$H = -\frac{\hbar^2}{2m} \sum_{i=1}^{N} \nabla_i^2 + \sum_{i<j} V_2(r_{ij}) + i \sum_{i<j} V_3(\mathbf{r_{ij}} \cdot \mathbf{r_{ik}}). \tag{7.108}$$

For simplicity, let us focus only on the two-body potential. One of the most popular forms explicitly written in coordinate space is the class of the *Argonn–Urbana* potentials AVX:

$$AVX = \sum_{i<j} \sum_{p=1}^{X} v_p(r_{ij}) O^{(p)}(i,j), \tag{7.109}$$

where the operators $O^{(p)}(i,j)$ include the spin–isospin dependence. For instance, the $AV8$ potential includes the following operators:

$$O^{p=1,8}(i,j) = (1, \vec{\sigma}_i \cdot \vec{\sigma}_j, S_{ij}, \vec{L}_{ij} \cdot \vec{S}_{ij}) \otimes (1, \vec{\tau}_i \cdot \vec{\tau}_j), \tag{7.110}$$

where the operator $S_{ij} = 3\vec{\sigma}_i \cdot \hat{r}_{ij} \vec{\sigma}_j \cdot \hat{r}_{ij} - \vec{\sigma}_i \cdot \vec{\sigma}_j$ is the tensor operator, and $\vec{L}_{ij} = -i\hbar \vec{r}_{ij} \times (\vec{\nabla}_i - \vec{\nabla}_j)/2$ and $\vec{S}_{ij} = \hbar(\vec{\sigma}_i + \vec{\sigma}_j)/2$ are the relative angular momentum and the total spin for the pair ij. The potential can be further simplified into a spin/isospin-independent part and a spin/isospin-dependent part:

$$V_2 = V_{\text{SII}} + V_{\text{SID}}. \tag{7.111}$$

As seen previously, projection MC algorithms are based on the imaginary propagation of an initial state. In general a wave function for a set of N nucleons will depend on spin and isospin degrees of freedom, which must be evolved in the same way as the coordinates of the nucleons. In fact, the propagator is

$$e^{-H\Delta\tau} \sim e^{-T\Delta\tau} e^{-V_{\text{SII}}\Delta\tau} e^{-V_{\text{SID}}\Delta\tau} \tag{7.112}$$

The spin/isospin-dependent part will act on the spin/isospin degrees of freedom. The operators O_p have typically a quadratic dependence on the spin/isospin operators. The action of these operators on the corresponding degrees of freedom can be correctly treated by only summing explicitly over all the possible relative states of the nucleons. This operation is computationally extremely expansive, and the cost grows exponentially with the number of nucleons. This is the main reason for the limit on the number of nucleons that can be treated by standard QMC. In Fig. 7.7 a summary of the computations of ground and excited states of light nuclei made by GFMC using a two-body $AV18$ potential plus a three-body force $IL2$ is plotted (Pieper, 2005). As can be seen, the agreement with experimental binding energies is very good.

Recently a different approach has been suggested by Schmidt and Fantoni (1999), involving the use of auxiliary degrees of freedom (auxiliary fields). In order to simplify the description of the method, let us focus on a pure neutron system. In this case all the operators in the potential will be isoscalar, and the operators will depend on spin only. If we further limit ourselves to a potential containing only the operators $1, \vec{\sigma}_i \cdot \vec{\sigma}_j, S_{ij}$, we can write

$$V^{\mathrm{SID}} = \sum_{i,j} \sigma_{i\alpha} A_{i\alpha;j\beta} \sigma_{j\beta}, \qquad (7.113)$$

where the Latin indices in the $3N \times 3N$ matrix $A_{i\alpha;j\beta}$ indicate the nucleons, and the Greek indices indicate the Cartesian components. The matrix A has real eigenvalues and eigenvectors. It is therefore possible to diagonalize the matrix, and use the computed eigenvectors to rewrite the potential in the following form:

$$V^{\mathrm{SID}} = \frac{1}{2} \sum_n \left[\sum_{i,j} \sigma_{i\alpha} \psi_n^{i\alpha} \lambda_n \psi_n^{j\beta} \sigma_{j\beta} \right]. \qquad (7.114)$$

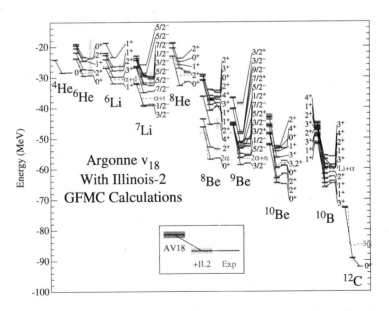

Fig. 7.7 Results of Green function MC calculations (Pieper, 2005) for light nuclei with masses up to $A = 12$, for both the ground state and low-lying excited states. The force used is a combination of a two-body force of the Argonne class ($AV18$) and a three-body force of the Illinois class ($IL2$). (Courtesy S. Pieper)

If one defines new N-body spin operators

$$O_n = \sum_i \sigma_{i\alpha} \psi_n^{i\alpha'}, \tag{7.115}$$

the spin-dependent potential becomes

$$V^{\text{SID}} = \frac{1}{2} \sum_{n=1}^{3N} \lambda_n O_n^2. \tag{7.116}$$

In the short time limit we can decompose the imaginary time propagator of the diffusion process, which projects the ground state out of a trial wave function in the following way:

$$e^{-H\Delta\tau} \sim e^{-T\Delta\tau} e^{-V_c\Delta\tau} e^{-V^{\text{SID}}\Delta\tau}, \tag{7.117}$$

where $V_c = \sum V_{\text{ext}}(r_i) + V^{\text{SII}}$ is the spin-independent part of the interaction. The propagation accounting for the kinetic and V_c operators gives rise to the usual drift–diffusion scheme of DMC shown in the previous subsections. The spin-dependent two-body potential part $e^{-V^{\text{SID}}\Delta\tau}$ is handled by making use of the so-called *Hubbard–Stratonovich transformation*:

$$e^{-\frac{1}{2}\lambda_n O_n^2 \Delta\tau}$$

$$= \frac{1}{\sqrt{2\pi}} \int_{-\infty}^{+\infty} dx_n\, e^{-\frac{x_n^2}{2} - \sqrt{-\lambda_n \Delta\tau}\, x_n O_n}, \tag{7.118}$$

with

$$e^{-V^{\text{SID}}\Delta\tau} \sim \prod_n e^{-\frac{1}{2}\lambda_n O_n^2 \Delta\tau}. \tag{7.119}$$

This transformation reduces the operators quadratic in the spin to integrals over operators which are *linear* in the spin, at the expense of introducing extra variables, which are usually named *auxiliary fields*. The action of such operators on the spin (isospin) variables amounts to a rotation of the spinors of an angle which is determined by the imaginary time step τ and by the value of the auxiliary variable, which is sampled by the Gaussian $e^{-\frac{x_n^2}{2}}$ appearing in the Hubbard–Stratonovich integral. In this way, the explicit sum over the states of the nucleons, which has an exponentially diverging computational cost, is replaced by an integration procedure which implies computations scaling with the cube of the number of nucleons considered. This speedup opened the way for computing by DMC properties of the homogeneous neutron and nuclear matter (Sarsa *et al.*, 2003; Gandolfi *et al.*, 2007), and of nuclei as heavy as ^{40}Ca.

In Table 7.5 the AFDMC energies per particle in MeV for the simplified interaction $AV6'$ interaction obtained with systems with 14, 38, 66 and 114 neutrons

Table 7.5. Binding energy of pure neutron matter computed by AFDMC at different densities [Sarsa *et al.*, 2003].

ρ (fm^{-3})	AFDMC(14)	AFDMC(38)	AFDMC(66)	AFDMC(114)	CBF	CBF+E_0
0.12	14.96(6)	13.76(9)	14.93(4)	15.5(7)*	14.3	14.9
0.16	19.73(5)	18.56(8)	20.07(5)	20.99(9)	19.5	20.4
0.20	25.29(6)	24.4(1)	26.51(6)	26.6(1)	25.2	26.4
0.32	48.27(9)	49.8(1)	53.11(9)	55.3(2)	50.4	53.2
0.40	69.9(1)	74.5(2)	79.4(2)	81.8(6)*	74.1	77.9

at various densities are displayed. The AFDMC results are compared with results obtained with the same interaction model and a correlation operator of the type v_6 by means of correlated basis functions (CBF) techniques which are representative of the value which would be obtained for an infinite number of nucleons. The results marked as CBF+E_0 include an estimate of a systematic error usually made in CBF calculations.

In Fig. 7.8 we show the equation of state of symmetric nuclear matter computed by AFDMC, compared with Brueckner–Hartree–Fock calculations and Fermi hypernetted chain (FHNC) calculations using a potential $AV6'$ (Bombaci *et al.*, 2006). It

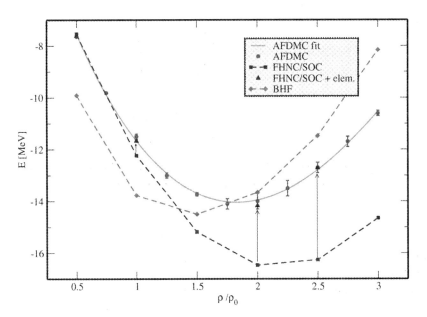

Fig. 7.8 Equation of state of symmetric nuclear matter (approximated by a periodic system of 28 nucleons) computed by means of AFDMC (Gandolfi *et al.*, 2007), and compared with Brueckner–Hartree–Fock (BHF) and Fermi hypernetted chain results (Bombaci *et al.*, 2006). The arrows indicate the approximate magnitude of the corrections to FHNC results due to the inclusion of the leading order elementary diagrams. The discrepancy with the AFDMC result is in this case largely removed.

is interesting to note how the exact AFDMC result considerably differs from both BHF and FHNC, in particular at high densities. Such differences can be removed if estimates of corrections of systematic errors made in these approximations are included (as an example, these corrections are displayed by arrows in the case of FHNC).

References

Astrakharchik, G.E., J. Boronat, J. Casurellas and S. Giorgini, *Phys. Rev. Lett.* **93**, 200404 (2004).

Aziz, R.A., V.P. Nain, J.S. Carley, W.L. Taylor and G.T. Conville, *J. Chem. Phys.* **70**, 4330 (1979).

Bartenstein, M. *et al.*, *Phys. Rev. Lett.* **92**, 203201 (2004).

Bartenstein, M. *et al.*, *Phys. Rev. Lett.* **92**, 120401 (2004).

Bombaci, I., A. Fabrocini, A. Polls and I. Vidaña, *Phys. Lett.* **B609**, 232 (2005).

Bourdel, T. *et al.*, *Phys. Rev. Lett.* **93**, 050401 (2004).

Box, G.E.P. and M.E. Muller, *Ann. Math. Stat.* **29**, 610 (1958).

Ceperley, D.M., *Rev. Mod. Phys.* **67**, 279 (1995).

Ceperley, D.M., G.V. Chester and M.H. Kalos, *Phys. Rev.* **B16**, 3081 (1977).

Ceperley, D.M. and M.H. Kalos, *Monte Carlo Methods in Statistical Physics*, ed. K. Binder (Springer-Verlag, Berlin, 1986), p. 145.

Comte de Buffon, G., *Essai d'arithmétique morale, Supplément à l' Histoire Naturelle.* Vol. 4 (1777).

Feynman, R.P. and M. Cohen, *Phys. Rev.* **102**, 1189 (1956).

Filippi, C. and C.J. Umrigar, *J. Chem. Phys.* **105**, 213 (1996).

Foulkes, W.M.C., L. Mitas, R. J. Needs and G. Rajagopal *Rev. Mod. Phys.* **73**, 33 (2001).

Gandolfi, S., F. Pederiva, S. Fantoni and K.E. Schmidt, *Phys. Rev. Lett.* **98**, 102503 (2007).

Grossman, J.C., *J. Chem. Phys.* **117**, 1434 (2002).

Kalos, M.H. and F. Pederiva, *Quantum Monte Carlo Methods in Physics and Chemistry*, eds. C.J. Umrigar and M.P. Nightingale (Kluwer, 1999), p. 263.

Kalos, M.H. and F. Pederiva, *Phys. Rev. Lett.* **85**, 3547 (2000).

Kinast, J. *et al.*, *Phys. Rev. Lett.* **92**, 150402 (2004).

Korona, T., H.L. Williams, R. Bukowski, B. Jeziorski and K. Szalewicz, *J. Chem. Phys.* **106**, 5109 (1997).

Kwon, Y., D.M. Ceperley and R.M. Martin, *Phys. Rev.* **B50**, 1684 (1994).

McMillan, W.L., *Phys. Rev.* **A138**, 442 (1965).

Metropolis, N., A.W. Rosenbluth, M.N. Rosenbluth, A.H. Teller and E. Teller, *J. Chem. Phys.* **21**, 1087 (1953).

Moroni, S., S. Fantoni and G. Senatore, *Phys. Rev.* **B52**, 013547 (1995)

O'Hara, K.M. *et al.*, *Science* **298**, 2179 (2002).

Petrov, D.S., C. Salomon and G.V. Shlyapnikov, *Phys. Rev. Lett.* **93**, 090404 (2004).

Pieper, S., *Nul. Phys.* **A751**, 516 (2005).

Pieper, S.C., V.R. Pandharipande, R.B. Wiringa and J. Carlson, *Phys. Rev.* **C64**, 014001 (2001).

Sarsa, A., S. Fantoni, K. E. Schmidt and F. Pederiva, *Phys. Rev.* **C68**, 024308 (2003).

Schmidt, K.E. and S. Fantoni, *Phys. Lett.* **B446**, 99 (1999).

Schmidt, K.E., Mi.A. Lee, M.H. Kalos and G.V. Chester, *Phys. Rev. Lett.* **47**, 807 (1981).

Toulouse, J. and C. J. Umrigar, *J. Chem. Phys.* **126**, 084102 (2007)

Umrigar, C.J. and C. Filippi, *Phys. Rev. Lett.* **94**, 150201 (2005)
Umrigar, C.J., M.P. Nightingale and K.J. Runge, *J. Chem. Phys.* **99**, 2865 (1993).
Wiringa, R.B., V.G.J. Stocks and R. Schiavilla, *Phys. Rev.* **C56**, 38 (1995).

As general references to the Monte Carlo simulation in classical and quantum systems, the following texts may be useful:

Allen, M.P. and D.J. Tildesley, *Computer Simulation of Liquids* (Clarendon, Oxford, 1987).
Hammond, B.L., W.A. Lester, Jr. and P. J. Reynolds, "Monte Carlo Methods in *Ab Initio* Quantum Chemistry," *World Scientific Lecture and Course Notes in Chemistry*, Vol. 1 (World Scientific, 1994).
Kalos, M.H. and P.A. Withlock, *Monte Carlo Methods: Basic*, Vol. 1 (J. Wiley and Sons, New York, 1986).

Chapter 8

The Linear Response Function Theory

8.1 Introduction

The second part of this book is devoted to the study of the excited states of interacting many-particle systems, which can be described by a Hamiltonian like that in (1.1). Moreover, the particles interact with an external oscillating field (responsible for the excitation of the system) through an interaction Hamiltonian H_{int}.

As we did in the first part of the book, which is devoted to the study of the ground state of a many-body system, we will consider both homogeneous and finite systems of Fermions and Bosons. In any case, the interesting quantities that can be tested experimentally are the matrix elements

$$|\langle n|F|0\rangle|^2 \tag{8.1}$$

of an observable F between the ground and excited states of H (which in turn are the solutions to the equation $H|n\rangle = E_n|n\rangle$), and the corresponding excitation energies

$$E_n - E_0. \tag{8.2}$$

In Chapter 1 we showed that these quantities can be easily computed for a one-body operator $F = \sum_{i=1}^{N} f(x_i)$, in the case of the independent-particle model (IPM), where two-body interaction is neglected in H. The relevant excited states are the one-particle–one-hole states $|i^{-1}m\rangle$, and the matrix elements of the operator F are given by (1.21) for Fermions, and (1.23) for Bosons. The corresponding excitation energies are given by the difference $\epsilon_{mi} = \epsilon_m - \epsilon_i$ between the single-particle energies of the hole and of the particle. It might be asked whether the IPM predictions are in agreement with experimental data, or fail completely. To try to give an answer to this question, let us compute, for a homogeneous system of Fermions and Bosons, the dynamic form factor

$$S(\mathbf{q},\omega) = \sum_n |\langle n|\rho_{\mathbf{q}}|0\rangle|^2 \delta(\omega - \omega_{no}), \quad \omega_{no} = E_n - E_0, \tag{8.3}$$

for the density operator

$$\rho_{\mathbf{q}} = \sum_{i=1}^{N} e^{i\mathbf{q}\cdot\mathbf{r}_i} = \int d\mathbf{r}\, e^{i\mathbf{q}\cdot\mathbf{r}} \rho(\mathbf{r}), \tag{8.4}$$

with $\rho(\mathbf{r}) = \sum_{i=1}^{N} \delta(\mathbf{r}-\mathbf{r}_i)$. This quantity, which was introduced in Subsection 1.8.1, can be directly measured by inelastic scattering experiments, and in the IPM is given by

$$S(\mathbf{q},\omega) = \sum_{mi} |\langle mi^{-1}|\rho_{\mathbf{q}}|0\rangle|^2 \delta(\omega - \epsilon_{mi}), \tag{8.5}$$

where the sum runs on all the one-particle–one-hole states. In the case of non-interacting homogeneous systems, the single-particle wave functions are plane waves.

In the 3D Fermion case, the calculation of $S(\mathbf{q},\omega)$ was carried out in Subsection 1.8.1 for small q (i.e. q much smaller than the Fermi momentum), and we obtained the analytical result

$$S(\mathbf{q},\omega) = \begin{cases} V\dfrac{m^2\omega}{2\pi^2 q} & \text{for } 0 \leq \omega \leq \dfrac{qk_F}{m} \\[2mm] 0 & \text{for } \omega > \dfrac{qk_F}{m} \end{cases}, \tag{8.6}$$

which shows that for any value of q, the dynamic form factor $S(\mathbf{q},\omega)$ tends to zero when $\omega \to 0$, as a consequence of the Pauli principle, which limits the number of low energy excitations for Fermions. The excitation spectrum is a continuum in the range

$$0 \leq \epsilon_{mi} \leq qk_F/m\,.$$

In the case of a three-dimensional homogeneous system of Bosons at zero temperature, whose IPM ground state is the product function

$$|0\rangle = \varphi_{i_o}(x_1)\varphi_{i_o}(x_2)\cdots\varphi_{i_o}(x_n) \tag{8.7}$$

[where the $\varphi_{i_o}(x)$ are all equal to each other], and is given by the plane waves $\varphi_{i_o} = e^{i\mathbf{p}\cdot\mathbf{r}}/\sqrt{V}$, characterized by the momentum $\mathbf{p}$ (which can be put equal to zero), the calculation of the matrix element (1.23) for the density operator leads immediately to the result

$$\langle mi_o^{-1}|\rho_{\mathbf{q}}|0\rangle = \sqrt{N}, \tag{8.8}$$

and to the existence of only one possible excited state at energy $\epsilon_{mi_o} = q^2/2m$. Thus, for a system of noninteracting Bosons we have the following expression for the dynamic form factor:

$$S(\mathbf{q},\omega) = N \cdot \delta\left(\omega - \frac{q^2}{2m}\right). \tag{8.9}$$

In summary, the IPM makes the following predictions for the dynamic form factor: for Fermions, the existence of a continuum of single-particle states with energy ω between 0 and qk_F/m, among which the strength is distributed, and which — for all (small) q values — goes to zero when ω goes to zero; for Bosons it is the existence of only one single-particle state at energy $q^2/2m$ which picks up all of the available strength.

In nature, the closest realization of an interacting Fermi gas in three dimensions is an alkali metal; for example, in sodium each atom has a valence electron which is nearly free because it is very far apart from the core electrons placed into closed shells. The ensemble of such valence electrons constitutes a gas of nearly-free electrons, which move in the field of the uniformly distributed ions, and under the effect of their mutual Coulomb interaction. The excitation of this valence electron gas is realized by striking an electron beam on the alkali metal, and measuring the inelastic cross-section of this process at various transferred momenta, as a function of the energy transferred to the system. The inelastic cross-section is proportional to the dynamic form factor.

The data of Fig. 8.1 show that $S(\mathbf{q},\omega)$ behaves quite differently from the prediction of the IPM in the limit of small q. The cross-section at fixed q, as a function of the energy absorbed by the system, is characterized by the presence of a peak at a well-defined energy, which does not go to zero when q tends to zero, and which collects practically all of the excitation strength of the process. It is only at high values of q that this state decays into low energy single-particle states. This excited

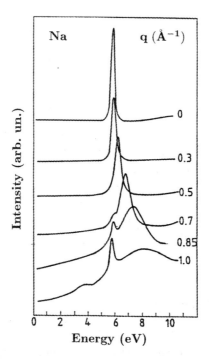

Fig. 8.1 Experimental excitation spectrum of sodium (Von Felde *et al.*, 1989) as a function of energy, for different values of the transferred momentum.

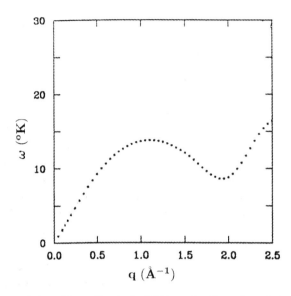

Fig. 8.2 Experimental data (Donnelly *et al.*, 1981) on the dispersion of elementary excitations in superfluid ^{4}He.

state, at small values of the transferred momentum, has nothing in common with the single-particle states of the IPM, and is a collective excitation of the system called plasmon.

As regards Bosons, if we consider superfluid ^{4}He as an example of a homogeneous system of Bosons, the experimental data of Fig. 8.2 show that at low q, these excitations have a phonon nature with a q-linear dispersion law which has nothing in common with the IPM predictions.

Therefore, the comparison with experimental data evidences the shortcomings of the IPM for the description of the excited states of a Fermion or Boson quantum liquid, and poses the problem of setting up a general theory for the description of these states. Such a theory should take into account the interaction among particles.

By "quantum liquids" we mean both the "real" quantum liquids that are found in nature, i.e. ^{4}He (Bose) and ^{3}He (Fermi), which have the property of being liquid in such a low temperature range that quantum effects connected to degeneracy and interaction cannot be neglected, and the conduction electrons in metals, semimetals and semiconductors, as well as nuclear matter and the very recently realized Bose condensates of alkali atoms.

8.2 The General Formalism

An N-particle system described by Hamiltonian H can be excited by making it interact with a beam of external particles, and subsequently studied by measuring its response to the external field that describes this interaction. If the interaction is sufficiently weak, the system response is linear and completely determined by the system's intrinsic properties (i.e. those in the absence of the external field).

The theory of the linear response function (Kubo, 1956, 1957; Pines and Noziéres, 1966) allows the study of the excited homogeneous and nonhomogeneous states of interacting systems, both at zero and at finite temperature. We will begin with the zero temperature response, and discuss the finite temperature effects in another subsection.

The interaction Hamiltonian of a system with an external field that oscillates in time with frequency ω may be written as

$$H_{\text{int}} = \lambda(G^\dagger e^{-i\omega t} + G e^{i\omega t}) e^{\eta t}. \tag{8.10}$$

The quantity λ gives the field intensity. G is an operator which depends on the space, spin, isospin, ... variables of the N particles of the system, and contains the physical information pertaining to the type of excitations to which the system is subjected. The factor $e^{\eta t}$, with η positive and small, guarantees that as $t \to -\infty$ the system is described by the unperturbed Hamiltonian H and is in the ground state $|0\rangle$.

The interaction (8.10) produces time dependence in the N-particle wave function $|\psi(t)\rangle$, which is the solution to the equation

$$(H + H_{\text{int}})|\psi(t)\rangle = i\frac{\partial}{\partial t}|\psi(t)\rangle, \tag{8.11}$$

and, consequently, the mean value of an operator F on such a state can be written as

$$\langle\psi(t)|F|\psi(t)\rangle - \langle 0|F|0\rangle = F_+ e^{-i\omega t} e^{\eta t} + F_- e^{i\omega t} e^{\eta t}. \tag{8.12}$$

Equation (8.11) is to be solved with the boundary condition that for $t \to -\infty$ the system is in the ground state $|0\rangle$.

The linear response function of the system to the external field (8.10) is defined as

$$\chi(F, G, \omega) = \lim_{\lambda \to 0} \frac{F_+}{\lambda}. \tag{8.13}$$

Note that this definition generalizes to the quantum case the classical definition of linear response. For example, consider the electrostatic polarizability α of a classical system. This is the linear response of the system to an external electric field E. If P is the induced dipole moment, we will have

$$\alpha = \lim_{E \to 0} \frac{P}{E}. \tag{8.14}$$

The linear response function to an electric field that oscillates in time is also known as dynamic polarizability.

Using the usual procedures of perturbation theory, let us look for solutions to (8.11) of the form

$$|\psi(t)\rangle = \sum_n a_n(t) e^{-iE_n t}|n\rangle,\tag{8.15}$$

where the boundary condition corresponds to

$$a_n(-\infty) = \begin{cases} 1 & \text{if } n = 0 \\ 0 & \text{if } n \neq 0 \end{cases}.\tag{8.16}$$

By retaining only the first order terms in λ, we obtain find $(n \neq 0)$

$$a_n(t) = \lambda \left[\frac{\langle n|G^\dagger|0\rangle}{\omega - \omega_{no} + i\eta} e^{(-i\omega + i\omega_{no} + \eta)t} - \frac{\langle n|G|0\rangle}{\omega + \omega_{no} - i\eta} e^{(i\omega + i\omega_{no} + \eta)t} \right].\tag{8.17}$$

Then, at first order in λ we have

$$\langle \psi(t)|F|\psi(t)\rangle - \langle 0|F|0\rangle$$

$$= \sum_n [\langle 0|F|n\rangle a_n(t) e^{-i\omega_{no} t} + \langle n|F|0\rangle a_n^\dagger(t) e^{i\omega_{no} t}].\tag{8.18}$$

By substituting (8.17) into (8.18), separating the coefficients of $e^{-i\omega t}$ from those of $e^{i\omega t}$, and bearing in mind that $\langle 0|G|n\rangle^* = \langle n|G^\dagger|0\rangle$, we obtain from the definition (8.13) the following expression for the linear response:

$$\chi(F, G, \omega) = \sum_n \left[\frac{\langle 0|F|n\rangle\langle n|G^\dagger|0\rangle}{\omega - \omega_{no} + i\eta} - \frac{\langle 0|G^\dagger|n\rangle\langle n|F|0\rangle}{\omega + \omega_{no} + i\eta} \right].\tag{8.19}$$

From Eq. (8.19) we see that the knowledge of χ means the knowledge of both the excitation energies (i.e. the poles of χ) and the matrix elements of the operators (i.e. the residues in the poles of χ). Therefore, if we succeed in calculating the linear response, we automatically achieve what we are looking for, viz. the matrix elements between ground and excited states, and the excitation energies.

If F is Hermitian and equal to G, or if $F = G$ but not Hermitian (e.g. $F = G = \rho_{\mathbf{q}}$) and the unperturbed system is time-reversal-invariant (so that $|\langle n|F|0\rangle|^2 = |\langle n|F^\dagger|0\rangle|^2$), then Eq. (8.19) takes the simplified form

$$\chi(F, \omega) = 2 \sum_n \omega_{no} \frac{|\langle n|F|0\rangle|^2}{(\omega + i\eta)^2 - \omega_{no}^2}.\tag{8.20}$$

Note that in the case of the density operator, the relation $|\langle n|\rho_{\mathbf{q}}|0\rangle|^2 = |\langle n|\rho_{\mathbf{q}}^\dagger|0\rangle|^2$ and consequently Eq. (8.20) still holds even if the time reversal symmetry is broken as long as the system retains parity invariance. This case, i.e. the density response, will be discussed in detail in Subsection 8.5. Note also that in the case of the nuclei, Eq. (8.20) cannot be derived from Eq. (8.19) starting from operators of the form

$F = G = \sum_i f(\mathbf{r}_i)\tau^+$, where τ^+ is the isospin operator which, by operating on the proton state, produces the neutron state. In fact, for nuclei with an excess of neutrons, the operators F and $F^\dagger = \sum_i f(\mathbf{r}_i)\tau^-$ (where τ^- changes the neutron state into that of the proton) excite different states and the two excitation strengths

$$|\langle n| \sum_i f(\mathbf{r}_i)\tau^+|0\rangle|^2$$

and

$$|\langle n| \sum_i f(\mathbf{r}_i)\tau^-|0\rangle|^2$$

are different.

Using the Dirac relation

$$\lim_{\eta \to 0} \frac{1}{x - a + i\eta} = \mathrm{P}\left(\frac{1}{x - a}\right) - i\pi\delta(x - a), \qquad (8.21)$$

we see that the real and imaginary parts of χ are related to the dynamic form factor

$$S(F, \omega) = \sum_n |\langle n|F|0\rangle|^2 \delta(\omega - \omega_{no}), \qquad (8.22)$$

which generalizes the definition (8.3) to any excitation operator F, by

$$\mathrm{Re}\,\chi(F, \omega) = \int_0^\infty d\omega' S(F, \omega') \mathrm{P}\left(\frac{2\omega'}{\omega^2 - \omega'^2}\right), \qquad (8.23)$$

and

$$S(F, \omega) - S(F, -\omega) = -\frac{1}{\pi}\mathrm{Im}[\chi(F, \omega)]. \qquad (8.24)$$

Note that at zero temperature, $S(F, -\omega)$ is identically zero for $\omega > 0$, because the excitation energies ω_{no} are always positive since the system is initially in the ground state. Therefore, for positive ω values, $S(F, \omega)$ and the imaginary part of χ coincide. As we will see, at finite temperature this is no longer the case.

If $F = G$ but not Hermitian, such as in the case of the operators

$$F = G = \sum_i^N x_i \sigma_i^+, \quad \sigma^+ = \frac{1}{2}(\sigma_x + i\sigma_y),$$

or

$$F = G = \sum_i^N r_i e^{i\theta},$$

and if the system is not time-reversal-invariant (i.e. it is either spin-polarized or angular-momentum-polarized), or if we have a nucleus with an excess of neutrons and we are considering charge-exchange processes described by operators like those discussed previously, then in these cases χ takes the form

$$\chi = \sum_n \left[\frac{|\langle n|F^\dagger|0\rangle|^2}{\omega - \omega_{no} + i\eta} - \frac{|\langle n|F|0\rangle|^2}{\omega + \omega_{no} + i\eta} \right]. \tag{8.25}$$

In this case the operators F and $F^\dagger$ excite different states of the system and $S(F, \omega) \neq S(F^\dagger, \omega)$, and the following relation holds:

$$-\frac{1}{\pi}\text{Im } \chi(\omega) = S(F^\dagger, \omega) - S(F, -\omega). \tag{8.26}$$

This relation allows us to extract both $S(F^\dagger)$ and $S(F)$, once the imaginary part of the linear response is known along the whole (positive and negative) ω axis.

8.3 The Linear Response Function and Sum Rules

The linear response function (8.20) can be related to the moments m_k of the dynamic form factor (8.22), defined by

$$m_k = \int_0^\infty d\omega \omega^k S(F, \omega) = \sum_n \omega_{no}^k |\langle 0|F|n\rangle|^2, \tag{8.27}$$

by expanding $\chi(F, \omega)$ as a function of ω. In the two cases of $\omega \to \infty$ and $\omega \to 0$ we obtain

$$\lim_{\omega \to \infty} \chi(F, \omega) = \frac{2}{\omega^2}\left(m_1 + \frac{1}{\omega^2}m_3 + \cdots \right) \tag{8.28}$$

and

$$\lim_{\omega \to 0} \chi(F, \omega) = -2(m_{-1} + \omega^2 m_{-3} + \cdots), \tag{8.29}$$

respectively.

By using the completeness relation $\sum_n |n\rangle\langle n| = 1$ and the equation $H|n\rangle = E_n|n\rangle$, it is possible to write the moments (8.27) with $k \geq 0$ as the mean values on the ground state of commutators [ ] and anticommutators { } of the excitation operator F and of the Hamiltonian H. For time-reversal-invariant systems, the following sum rules are derived:

$$m_0 = \frac{1}{2}\langle 0|\{F, F^\dagger\}|0\rangle - \langle 0|F|0\rangle^2,$$

$$m_1 = \frac{1}{2}\langle 0|[F, [H, F^\dagger]]|0\rangle,$$

$$m_2 = \frac{1}{2}\langle 0|\{[F, H], [H, F^\dagger]\}|0\rangle, \tag{8.30}$$

$$m_3 = \frac{1}{2}\langle 0|[[F, H], [H, [H, F^\dagger]]]|0\rangle.$$

Note that in the odd sum rules only commutators appear, while in the even ones anticommutators appear as well, and that in the expansion (8.28) only odd moments with $k > 0$ appear. The moments with $k < 0$ may also be expressed through commutators and anticommutators. For example, we can write

$$m_{-1} = \frac{1}{2}\langle 0|[[X^\dagger, H], X]|0\rangle,$$

$$m_{-2} = \frac{1}{2}\langle 0|\{X^\dagger, X\}|0\rangle, \tag{8.31}$$

where the operator X is the solution to the equation

$$[H, X] = F. \tag{8.32}$$

In the expansion (8.29), only moments with negative odd k enter. Among these, the most important is certainly m_{-1}, which is connected to the static polarizability $\chi(F, 0)$ of the system by the relation

$$m_{-1} = \sum_n \frac{|\langle 0|F|n\rangle|^2}{\omega_{no}} = -\frac{\chi(F, 0)}{2}. \tag{8.33}$$

As an example of the calculation of a sum rule, let us compute m_1 for the electric dipole operator

$$F = D = e \sum_{i=1}^N z_i,$$

in the case where the system is a neutral atom with N electrons and Hamiltonian given by

$$H = \sum_{i=1}^N \left(\frac{p^2}{2m} - \frac{Ne^2}{r}\right)_i + \sum_{i<j} \frac{e^2}{|\mathbf{r}_i - \mathbf{r}_j|}.$$

The dynamic form factor for the operator D is directly observable — for example, in photoabsorption experiments where the cross-section $\sigma(\omega)$ is related to $S(D, \omega)$ by the relationship

$$\sigma(\omega) = 4\pi^2 \omega S(D, \omega). \tag{8.34}$$

The evaluation of the double commutator of Eq. (8.30) is very simple, because the dipole operator commutes with both the one-body nuclear potential and the two-body Coulomb interaction, since both are local terms, and the only contribution comes from the kinetic energy. The result is

$$m_1(D) = \frac{N\hbar^2 e^2}{2m}, \tag{8.35}$$

which is independent of the model and known as the Thomas–Reich–Kuhn sum rule. Using Eq. (8.34) we have

$$\int \sigma(\omega)d\omega = 4\pi^2 m_1(D) = 2\pi^2 \frac{N\hbar^2 e^2}{m}, \tag{8.36}$$

which means that the area underlying the photoabsorption cross-section plotted as a function of the photon energy is proportional to the atomic charge.

The sum rules (8.30) may be generalized to any operator F and G, starting from the following mixed expression:

$$m_k^{\pm} = \frac{1}{2} \sum_n \omega_{no}^k (\langle 0|F|n\rangle\langle n|G^{\dagger}|0\rangle \pm \langle 0|G^{\dagger}|n\rangle\langle n|F|0\rangle)$$

$$= \frac{1}{2}(\langle 0|F(H - E_0)^k G^{\dagger}|0\rangle \pm \langle 0|G^{\dagger}(H - E_0)^k F|0\rangle). \tag{8.37}$$

For $k = 0, 1, 2$ we obtain

$$m_0^+ = \frac{1}{2}\langle 0|\{F, G^{\dagger}\}|0\rangle,$$

$$m_1^+ = \frac{1}{2}\langle 0|[F, [H, G^{\dagger}]]|0\rangle, \tag{8.38}$$

$$m_2^+ = \frac{1}{2}\langle 0|\{[F, H], [H, G^{\dagger}]\}|0\rangle,$$

and

$$m_0^- = \frac{1}{2}\langle 0|[F, G^{\dagger}]|0\rangle,$$

$$m_1^- = \frac{1}{2}\langle 0|\{F, [H, G^{\dagger}]\}|0\rangle, \tag{8.39}$$

$$m_2^- = \frac{1}{2}\langle 0|[[F, H], [H, G^{\dagger}]]|0\rangle.$$

Clearly, when $G = G^{\dagger} = F$, we will have m_0^+, m_1^+, m_2^+ coinciding with the sum rules of Eq. (8.30), while m_0^-, m_1^-, m_2^- vanish. This holds also in the $F = G = \rho_{\mathbf{q}}$ case, provided the unperturbed system is parity- or time-reversal-invariant.

The sum rules m_0^-, m_1^+, $m_2^-, \ldots$ enter the $\omega \to \infty$ expansion of the linear response $\chi(F, G, \omega)$ of (8.19):

$$\lim_{\omega \to \infty} \chi(F, G, \omega) = \frac{2}{\omega}\left(m_0^- + \frac{1}{\omega}m_1^+ + \frac{1}{\omega^2}m_2^- \cdots\right). \tag{8.40}$$

As an example of application of the mixed sum rules (8.37), let us consider a quantum dot under the effect of a static magnetic field perpendicular to the electron motion (which is confined on the x, y plane), with Hamiltonian H given by Eq. (5.1), and let us compute the sum rules m_0^-, m_1^+, m_2^- and m_3^+ for the operators

$$F = F^- = \sum_{j=1}^{N} r_j e^{-i\theta_j}$$

and

$$G^\dagger = F^+ = \sum_{j=1}^{N} r_j e^{+i\theta_j}.$$

These are dipole operators with angular momentum -1 and $+1$, respectively, and if the dot in its ground state is angular-momentum-polarized and has $\langle 0|L_z|0\rangle = -L_0$, with $L_0 = \sum_i^{\text{occ}} l_i$, then they excite states with $\Delta L_z = \mp 1$, respectively. For a high external magnetic field, these two excitations — at IPM level — are one-particle–one-hole excitations among single-particle states belonging to one Landau level in the first case (edge excitations), and among states belonging to two different Landau levels (bulklike) in the second case.

The calculation of the commutators is very simple because the two-body interaction commutes both with $F^\pm$ and with the results of the commutation between H and $F^\pm$, which is given by

$$[H, F^\pm] = \pm\frac{\omega_c}{2} F^\pm - iP^\pm, \quad P^\pm = \sum_{j=1}^{N}(p_x \pm ip_y)_j.$$

We can thus find (in effective atomic units) the following:

$$\begin{aligned} m_0^- &= 0, \\ m_1^+ &= N, \\ m_2^- &= \omega_c N, \\ m_3^+ &= (\omega_0^2 + \omega_c^2)N. \end{aligned} \tag{8.41}$$

The results for the above moments may be used to compute the excited state frequencies of the operators $F^\pm$, if one makes the further hypothesis that such moments are saturated by only two modes with frequencies $\omega_\pm$, which are excited respectively by the operators $F^\pm$:

$$S^\pm(\omega) = \sigma^\pm \delta(\omega - \omega_\pm), \tag{8.42}$$

where $\sigma^\pm = |\langle \omega_\pm|F^\pm|0\rangle|^2$ are the corresponding strengths. The hypothesis (8.42), in conjunction with the result $m_0^- = 0$, implies that $\sigma^+ = \sigma^-$. In this case, then, one immediately finds that

$$\omega_+ - \omega_- = \frac{m_2^-}{m_1^+} = \omega_c, \tag{8.43}$$

and

$$\frac{\omega_+^3 + \omega_-^3}{\omega_+ + \omega_-} = \frac{m_3^+}{m_1^+} = \omega_0^2 + \omega_c^2, \tag{8.44}$$

from which it follows that

$$\omega_\pm = \sqrt{\omega_0^2 + \frac{1}{4}\omega_c^2} \pm \frac{\omega_c}{2}, \tag{8.45}$$

which is an exact result for the Hamiltonian (5.1) of the quantum dot (generalized Kohn theorem; see also Subsection 8.9.1).

Review papers on sum rules, where many applications of this method are discussed, are those of Leonardi and Rosa-Clot (1971); Bohigas, Lane and Martorell (1979); Lipparini and Stringari (1989); and Orlandini and Traini (1990).

8.4 Finite Temperature

The formalism of the linear response function at finite temperature is very similar to that at zero temperature developed above. However, there are some differences worthy of being stressed.

At zero temperature the excitation operator F induces a transition between the ground state $|0\rangle$ and one of the excited states $|n\rangle$ with $n \neq 0$. If the system is at equilibrium at a finite temperature T, then the ground state is a statistical mixture defined by the density matrix (2.76):

$$D = \sum_k p_k |k\rangle\langle k|,$$

where p_k is the probability of finding the system in the state $|k\rangle$, given by

$$p_k = \frac{e^{-\beta E_k}}{Z}, \tag{8.46}$$

with $Z = \sum_k e^{-\beta E_k}$ equal to the partition function. The dynamic form factor associated with the excitation operator F is generalized from Eq. (8.22) to

$$S_T(F,\omega) = \sum_{k \neq n} p_k |\langle n|F|k\rangle|^2 \delta(\omega - \omega_{nk}), \tag{8.47}$$

where $\omega_{nk} = E_n - E_k$.

During the interaction of the system with a test particle at finite temperature, it is possible to transfer energy both from the particle to the system, and from the system to the test particle, because negative excitation energies ω_{nk} are admissible. At thermal equilibrium, the two probabilities [which are proportional to $S(F,\omega)$ and $S(F,-\omega)$, respectively] are not independent. The relationship existing between

$S(F, \omega)$ and $S(F, -\omega)$ is known as the principle of detailed balance. To derive it, it is sufficient to interchange the indices n and k in (8.47), obtaining

$$S_T(F, \omega) = Z^{-1} \sum_{k \neq n} e^{-\beta E_n} |\langle k|F|n \rangle|^2 \delta(\omega + \omega_{nk})$$

$$= Z^{-1} \sum_{k \neq n} e^{-\beta(E_n - E_k)} e^{-\beta E_k} |\langle n|F^\dagger|k \rangle|^2 \delta(\omega + \omega_{nk})$$

$$= e^{\beta \omega} S_T(F^\dagger, -\omega). \tag{8.48}$$

Assuming time reversal invariance so that $S_T(F^\dagger, -\omega) = S_T(F, -\omega)$, we finally find that

$$S_T(F, \omega) = e^{\beta \omega} S_T(F, -\omega). \tag{8.49}$$

At zero temperature the imaginary part of the response function and the dynamic form factor coincide for positive ω, and only processes by which energy is transferred to the system are allowed. At finite temperature these two functions are quite different for small $\beta \omega$ values. This can be easily seen starting from the Kubo result for the linear response function at finite temperature:

$$\chi(F, \omega) = Z^{-1} \sum_{kn} e^{-\beta E_k} |\langle n|F|k \rangle|^2 \frac{2\omega_{nk}}{(\omega + i\eta)^2 - \omega_{nk}^2}, \tag{8.50}$$

which holds for time-reversal-invariant systems. Using relation (8.21), the imaginary part of χ can be written as

$$\mathrm{Im}\, \chi(\omega) = -\pi [S_T(F, \omega) - S_T(F, -\omega)], \tag{8.51}$$

and using the detailed balance condition (8.49) we obtain the relation

$$\mathrm{Im}\, \chi(\omega) = -\pi [1 - e^{-\beta \omega}] S_T(F, \omega), \tag{8.52}$$

known as the fluctuation–dissipation theorem. Relation (8.52) shows that the imaginary part of the linear response function and $S_T(F, \omega)$ can be different at finite temperature. In general, $S_T(F, \omega)$ is much more temperature-dependent than $\mathrm{Im}\, \chi(\omega)$, which, for this reason, is a more fundamental quantity from the point of view of many-body theory. It is also known as the dissipative component of the response function of the system. In fact, by means of second order perturbation theory, it is possible to connect $\mathrm{Im}\, \chi(\omega)$ to the energy transferred per unit time by the external field to the system (see Subsection 8.5).

The moments of the dynamic form factor (8.47) are given by the following generalization of (8.27):

$$m_p = \int_{-\infty}^{+\infty} S_T(F, \omega) \omega^p d\omega = \sum_{k \neq n} p_k |\langle n|F|k \rangle|^2 \omega_{nk}^p. \tag{8.53}$$

Note that these moments, contrary to the $T = 0$ case, get contributions from negative ω as well. It can be shown that the moments defined as in Eq. (8.53) obey the usual sum rules, i.e. the m_p with $p > 0$ are the mean values on the statistical mixture D. For example, in the case of the energy-weighted sum rule m_1 for the Hermitian operator F, we have

$$m_1 = \frac{1}{2}\text{Tr}\{D[F,[H,F]]\}$$

$$= \frac{1}{2}\text{Tr}\{D(2FHF - HF^2 - F^2H)\}$$

$$= \frac{1}{2}\sum_{k \neq n}|\langle n|F|k\rangle|^2(2E_n - E_k - E_k)p_k$$

$$= \sum_{k \neq n}|\langle n|F|k\rangle|^2(E_n - E_k)p_k. \tag{8.54}$$

In what follows, for the sake of simplicity, we will indicate the mean value of an operator A on the statistical mixture D by $\langle A \rangle$, which will stand for*

$$\text{Tr}(DA) = \sum_n p_n\langle n|A|n\rangle.$$

8.5   The Density Response

The density response is a quantity that allows one to study the dynamic behavior of an interacting many-body system, at both zero and finite temperature. It can be measured by inelastic scattering reactions and can be used to provide important information both on the collective states that are excited at small transferred momentum, and on the momentum distribution which characterizes the system at high transferred momentum.

As we have already discussed in Subsection 8.2, the density response corresponds to the case $F = G = \rho_{\mathbf{q}}$ in (8.19). If the unperturbed system is parity-invariant or time-reversal-invariant, it is possible to write the density response $\chi(\mathbf{q}, \omega)$ as

$$\chi(\mathbf{q}, \omega) = 2\sum_{kn} p_k \omega_{nk} \frac{|\langle n|\rho_{\mathbf{q}}|k\rangle|^2}{(\omega + i\eta)^2 - \omega_{nk}^2}. \tag{8.55}$$

It can be written as a function of the dynamic form factor (8.47), as

$$\chi(\mathbf{q}, \omega) = \int_{-\infty}^{+\infty} d\omega'\left[\frac{S(\mathbf{q}, \omega)}{\omega - \omega' + i\eta} - \frac{S(-\mathbf{q}, \omega)}{\omega + \omega' + i\eta}\right], \tag{8.56}$$

*See Eq. (2.77).

and the real and imaginary parts of χ can be expressed as

$$\text{Re }\chi(\mathbf{q},\omega) = \int_{-\infty}^{+\infty} d\omega' \left[S(\mathbf{q},\omega)\text{P}\frac{1}{\omega-\omega'} - S(-\mathbf{q},\omega)\text{P}\frac{1}{\omega+\omega'} \right] \quad (8.57)$$

and

$$S(\mathbf{q},\omega) - S(-\mathbf{q},-\omega) = -\frac{1}{\pi}\text{Im}[\chi(\mathbf{q},\omega)]. \quad (8.58)$$

If the system interacts weakly with a test particle, and $v(\mathbf{q})$ is the Fourier transform of the interaction, the probability per unit time that the particle transfers momentum $\mathbf{q}$ and energy ω to the system, $P(\mathbf{q},\omega)$, is related to the dynamic form factor $S(\mathbf{q},\omega)$ by

$$P(\mathbf{q},\omega) = 2\pi|v(\mathbf{q})|^2 S(\mathbf{q},\omega). \quad (8.59)$$

As was discussed in Subsection 8.4, at finite temperature the test particle can either transfer momentum $\mathbf{q}$ and energy ω to the system, or extract them from the system. The energy transferred per unit time between the external field and the system is given by

$$\frac{dE}{dt} = 2\pi|v(\mathbf{q})|^2\omega(S(\mathbf{q},\omega) - S(-\mathbf{q},-\omega)) = -2|v(\mathbf{q})|^2\omega\text{Im}[\chi(\mathbf{q},\omega)]. \quad (8.60)$$

Equations (8.53) and (8.58), together with the identity $S(\mathbf{q},\omega) = S(-\mathbf{q},\omega)$, which holds for parity- or time-reversal-invariant systems, allow us to write the following relation among the odd moments of $S(\mathbf{q},\omega)$ and the imaginary part of χ:

$$m_{2p+1} = -\frac{1}{\pi}\int_0^\infty \omega^{2p+1}\text{Im }[\chi(\mathbf{q},\omega)]d\omega, \quad (8.61)$$

independent of temperature. For example, the sum rule m_1, also known as the f-sum rule in the case of local potentials, for which

$$[H,\rho_\mathbf{q}] = \mathbf{q}\cdot\mathbf{j_q}, \quad (8.62)$$

where

$$\mathbf{j_q} = \frac{1}{2m}\sum_{i=1}^N (\mathbf{p}_i e^{i\mathbf{q}\cdot\mathbf{r}_i} + e^{i\mathbf{q}\cdot\mathbf{r}_i}\mathbf{p}_i), \quad (8.63)$$

takes the value

$$m_1 = \frac{N\mathbf{q}^2}{2m}. \quad (8.64)$$

The result (8.64) holds for any q and at all temperatures, and does not depend on the model used to derive it. For example, the IPM, which for small q and zero temperature yields the result (8.6) for $S(\mathbf{q}, \omega)$, reproduces the f-sum rule:

$$m_1 = \int_0^{\frac{qk_F}{m}} \frac{m^2 V}{2\pi^2 q} \cdot \omega^2 d\omega = \frac{q^2 V}{2m} \cdot \frac{k_F^3}{3\pi^2} = \frac{q^2}{2m} \cdot V\rho = \frac{Nq^2}{2m}. \tag{8.65}$$

Another important odd sum rule is the m_3, given by

$$m_3 = \frac{1}{2}\langle [[\rho_{\mathbf{q}}, H], [H, [H, \rho_{-\mathbf{q}}]]] \rangle. \tag{8.66}$$

For local potentials, for which (8.62) holds, we see that the m_3 is determined by the double commutator of the Hamiltonian with the component of the current (8.63) parallel to $\mathbf{q}$.

Among the odd negative moments let us consider m_{-1}, which is related to the static polarizability $\chi(\mathbf{q}, 0)$ by

$$m_{-1} = -\frac{1}{2}\chi(\mathbf{q}, 0). \tag{8.67}$$

In the $\mathbf{q} \to 0$ limit and for homogeneous systems, it is easy to show (Pines and Nozières, 1966) that the static polarizability of neutral systems is related to the isothermal compressibility K of the system by the relation (compressibility sum rule)

$$\lim_{q \to 0} \chi(\mathbf{q}, 0) = -2 \lim_{q \to 0} \int_{-\infty}^{+\infty} S(\mathbf{q}, \omega) \frac{1}{\omega} d\omega = -N\rho K, \tag{8.68}$$

where ρ is the density of the homogeneous system and the compressibility K is defined by

$$\frac{1}{K\rho} = \frac{\partial P}{\partial \rho}, \tag{8.69}$$

and the pressure P can be obtained from the energy of the system

$$P = \rho^2 \frac{\partial(E/N)}{\partial \rho}. \tag{8.70}$$

The isothermal compressibility can also be expressed through the isothermal sound velocity v as $(K\rho)^{-1} = mv^2$.

The even moments are more difficult to evaluate and, as we will see, their relation with $\mathrm{Im}(\chi)$ depends explicitly on temperature. Let us consider, for example, the moment m_0, given by

$$m_0 = \int_{-\infty}^{+\infty} S(\mathbf{q}, \omega) d\omega = N S(\mathbf{q}), \tag{8.71}$$

where $S(\mathbf{q})$ is the static form factor:

$$S(\mathbf{q}) = \frac{1}{N} \sum_{k \neq n} p_k |\langle n|\rho_\mathbf{q}|k\rangle|^2$$

$$= \frac{1}{N} \left(\sum_{k,n} p_k |\langle n|\rho_\mathbf{q}|k\rangle|^2 - \sum_k p_k |\langle k|\rho_\mathbf{q}|k\rangle|^2 \right)$$

$$= \frac{1}{N} \left(\sum_{k,n} p_k \langle k|\rho_{-\mathbf{q}}|n\rangle \langle n|\rho_\mathbf{q}|k\rangle - \sum_k p_k |\langle k|\rho_\mathbf{q}|k\rangle|^2 \right)$$

$$= \frac{1}{N} (\langle \rho_{-\mathbf{q}}\rho_\mathbf{q}\rangle - |\langle \rho_\mathbf{q}\rangle|^2), \tag{8.72}$$

which, as is obvious from (8.72), is determined by the density fluctuations.

In the limit of zero temperature, $S(\mathbf{q})$ can be expressed through the ground state two-body density matrix as

$$S(\mathbf{q}) = 1 + \frac{1}{N} \int d\mathbf{r}_1 d\mathbf{r}_2 [\rho^{(2)}(\mathbf{r}_1,\mathbf{r}_2) - \rho^{(1)}(\mathbf{r}_1)\rho^{(1)}(\mathbf{r}_2)] e^{i\mathbf{q}\cdot(\mathbf{r}_1-\mathbf{r}_2)}. \tag{8.73}$$

If the system is homogeneous, $\rho^{(2)}(\mathbf{r}_1,\mathbf{r}_2)$ can only depend on $r = |\mathbf{r}_1 - \mathbf{r}_2|$ owing to translational invariance, and we have

$$\rho^{(2)}(\mathbf{r}_1,\mathbf{r}_2) = \rho^2 g(r), \tag{8.74}$$

where

$$g(r) = \frac{1}{4} \sum_{\sigma,\sigma'} g_{\sigma,\sigma'}(r)$$

is the spin-averaged probability of finding two particles at relative distance r, while ρ is the one-body diagonal density which is a constant of the system. Therefore, we obtain

$$S(\mathbf{q}) = 1 + \rho \int d\mathbf{r}(g(r) - 1)e^{i\mathbf{q}\cdot\mathbf{r}}. \tag{8.75}$$

This formula allows us to derive $g(r)$ from $S(\mathbf{q})$ by taking the Fourier transform. Note that for large $\mathbf{q}$, $S(\mathbf{q}) \to 1$. In this limit, only incoherent processes are important, in which the particle is scattered by the individual constituents of the system, and only the $i = j$ term in the sum $\rho_{-\mathbf{q}}\rho_\mathbf{q} = \sum_{i,j} e^{i\mathbf{q}\cdot(\mathbf{r}_i-\mathbf{r}_j)}$ contributes to the static structure factor. In the same limit, the dynamic form factor can be written as

$$S(\mathbf{q},\omega) = \int n(\mathbf{p})\delta\left(\omega - \left(\frac{q^2}{2m} - \frac{\mathbf{q}\cdot\mathbf{p}}{m}\right)\right) d\mathbf{p}, \tag{8.76}$$

and is entirely determined by the momentum distribution $n(\mathbf{p})$ of the system (P. Hohenberg and P. Platzman, 1966). The situation is completely different at small $\mathbf{q}$, where coherent processes dominate.

By using the relation (8.52), together with the antisymmetry property of $\text{Im}(\chi)$ under the exchange of ω into $-\omega$ [see Eq. (8.51)], it is possible to put $S(\mathbf{q})$ in the form

$$NS(\mathbf{q}) = \int_{-\infty}^{+\infty} S(\mathbf{q},\omega)d\omega = -\frac{1}{2\pi}\int_{-\infty}^{+\infty} \text{Im } \chi(\mathbf{q},\omega)\coth\frac{\beta\omega}{2}d\omega. \tag{8.77}$$

When $\mathbf{q}$ is small, the integral in (8.77) is dominated by the region where $\omega \ll KT$ and it is possible to replace the cotangent by the reciprocal of its argument. Then, if we use the result of Eq. (8.68), we find that

$$\lim_{q\to 0} S(\mathbf{q}) = \frac{\rho K}{\beta}. \tag{8.78}$$

This shows that at low transferred momentum the density fluctuations have a thermal origin. In any case, at very low temperature thermal fluctuations are quenched and the validity of (8.78) is limited to very low momenta.

8.6   The Density Response for Noninteracting Homogeneous Systems

The density response in the case of noninteracting homogeneous systems (both Bosons and Fermions) can be computed analytically. Though the contribution of the interaction is essential for the correct calculation of the density response, and especially so at low $\mathbf{q}$, it is important to know the free response because it is an essential ingredient in theories which take interaction into account. At zero temperature, the free response function [hereafter indicated as $\chi_0(\mathbf{q},\omega)$] is given by the expression

$$\chi_0(\mathbf{q},\omega) = \sum_{mi} 2\epsilon_{mi}\frac{|\langle mi^{-1}|\rho_{\mathbf{q}}|0\rangle|^2}{(\omega+i\eta)^2 - \epsilon_{mi}^2}, \tag{8.79}$$

where the summation runs over all one-particle–one-hole states.

In the case of Bosons, and using the result (8.8), the calculation of the matrix element of the density operator in (8.79) leads immediately to the result

$$\chi_0^B(\mathbf{q},\omega) = \frac{\frac{q^2}{m}\cdot N}{(\omega+i\eta)^2 - (\frac{q^2}{2m})^2}, \tag{8.80}$$

from which it is seen that $\chi_0^B(\mathbf{q},\omega)$ has only one pole at energy $q^2/2m$. From Eq. (8.80), we can very easily obtain the result (8.9), which holds for positive ω. Then, by taking the limit $\omega \to \infty$, we have

$$\lim_{\omega\to\infty} \chi_0^B(\mathbf{q},\omega) = \frac{Nq^2}{m\omega^2}\left(1 + \frac{1}{\omega^2}\left(\frac{q^2}{2m}\right)^2 + \cdots\right) \tag{8.81}$$

and so the correct value for the m_1 sum rule is $m_1 = Nq^2/2m$, and for m_3 the prediction is $m_3 = m_1(q^2/2m)^2$. The latter reflects the fact that in the ideal Boson gas there is only one excited state at energy $q^2/2m$.

Moreover, from Eq. (8.80) it follows that for $\mathbf{q} \to 0$ the static response diverges as $1/q^2$, a result that is due to the infinite compressibility of an ideal Boson gas.

The Fermion case is somewhat more complex, because in the Fermi gas there is a continuum of one-particle–one-hole excited states, with excitation energies given by

$$\omega_{mi} = \epsilon_m - \epsilon_i = \frac{(\mathbf{p} + \mathbf{q})^2}{2m} - \frac{\mathbf{p}^2}{2m}. \tag{8.82}$$

Using the results of Subsection 1.8.1, it is possible to write the free response function of Fermions as

$$\chi_0^F(\mathbf{q}, \omega) = 2 \sum_{\substack{p < k_F \\ |\mathbf{p}+\mathbf{q}| > k_F}} \frac{\frac{2\mathbf{p}\cdot\mathbf{q}}{m} + \frac{q^2}{m}}{\omega^2 - \left(\frac{\mathbf{p}\cdot\mathbf{q}}{m} + \frac{q^2}{2m}\right)^2}, \tag{8.83}$$

where the factor 2 takes into account the sum over the spin index. Therefore, the calculation of the response function is reduced to an integration over the region of momentum space such that $p < k_F$ and $|\mathbf{p} + \mathbf{q}| > k_F$, out of which one-particle–one-hole pairs with total momentum $\mathbf{q}$ may be excited. This region depends on the dimensionality of the system. It is a portion of a sphere in 3D, of the Fermi circle in 2D, and of the Fermi segment in 1D. The free response function of Fermions is known as the Lindhard function.

Introducing the distributions (1.76), Eq. (8.83) can be rewritten as

$$\chi_0^F(\mathbf{q}, \omega) = 2L^D \int \frac{d\mathbf{p}}{(2\pi)^D} n_{\mathbf{p}}(1 - n_{\mathbf{p}+\mathbf{q}}) \frac{\frac{2\mathbf{p}\cdot\mathbf{q}}{m} + \frac{q^2}{m}}{\omega^2 - \left(\frac{\mathbf{p}\cdot\mathbf{q}}{m} + \frac{q^2}{2m}\right)^2}. \tag{8.84}$$

By means of the change of variable $\mathbf{p}' = -\mathbf{p} - \mathbf{q}$, it is easy to show that the second integral in (8.84),

$$I_2 = \int \frac{d\mathbf{p}}{(2\pi)^D} n_{\mathbf{p}} n_{\mathbf{p}+\mathbf{q}} \frac{\frac{2\mathbf{p}\cdot\mathbf{q}}{m} + \frac{q^2}{m}}{\omega^2 - \left(\frac{\mathbf{p}\cdot\mathbf{q}}{m} + \frac{q^2}{2m}\right)^2},$$

vanishes because $I_2 = -I_2$. Therefore, we are left with the calculation of

$$\chi_0^F(\mathbf{q}, \omega) = 2L^D \int \frac{d\mathbf{p}}{(2\pi)^D} n_{\mathbf{p}} \frac{\frac{2\mathbf{p}\cdot\mathbf{q}}{m} + \frac{q^2}{m}}{\omega^2 - \left(\frac{\mathbf{p}\cdot\mathbf{q}}{m} + \frac{q^2}{2m}\right)^2}$$

$$= 2L^D \int \frac{d\mathbf{p}}{(2\pi)^D} n_{\mathbf{p}} \left(\frac{1}{\omega - \left(\frac{\mathbf{p}\cdot\mathbf{q}}{m} + \frac{q^2}{2m}\right)} - \frac{1}{\omega + \left(\frac{\mathbf{p}\cdot\mathbf{q}}{m} + \frac{q^2}{2m}\right)} \right). \tag{8.85}$$

In the one-dimensional case, taken as an example, we have

$$\chi_0^{1D}(\mathbf{q},\omega) = \frac{L}{\pi} \cdot \int_{-k_F}^{k_F} dp \left[\frac{1}{\omega - (\frac{pq}{m} + \frac{q^2}{2m})} - \frac{1}{\omega + (\frac{pq}{m} + \frac{q^2}{2m})} \right],$$

so that

$$\chi_0^{1D}(\mathbf{q},\omega) = \frac{L}{\pi} \cdot \left[-\frac{m}{q} \ln \left(\omega - \frac{pq}{m} - \frac{q^2}{2m} \right) - \frac{m}{q} \ln \left(\omega + \frac{pq}{m} + \frac{q^2}{2m} \right) \right]_{-k_F}^{k_F}$$

$$= -\frac{m\,L}{q\,\pi} \left[\ln \left(\frac{\omega - \frac{k_F q}{m} - \frac{q^2}{2m}}{\omega + \frac{k_F q}{m} - \frac{q^2}{2m}} \right) + \ln \left(\frac{\omega + \frac{k_F q}{m} + \frac{q^2}{2m}}{\omega - \frac{k_F q}{m} + \frac{q^2}{2m}} \right) \right]$$

$$= -\frac{m\,L}{q\,\pi} \ln \left[\frac{\omega^2 - (\frac{k_F q}{m} + \frac{q^2}{2m})^2}{\omega^2 - (\frac{k_F q}{m} - \frac{q^2}{2m})^2} \right]. \tag{8.86}$$

Note that for $\mathbf{q}$ fixed and smaller than $2k_F$, we find that $\chi_0^{1D}(\mathbf{q},\omega)$ has a finite negative value for $\omega = 0$. This free response function has a continuum of poles which correspond to the continuum of single-particle excited states (see also Subsection 1.8.1) in the range

$$\frac{k_F q}{m} - \frac{q^2}{2m} \le \omega \le \frac{k_F q}{m} + \frac{q^2}{2m},$$

and it tends to zero starting from positive ω when ω tends to infinity. If q is small, the region of the poles gets narrower and the one-dimensional Fermi gas resembles a Boson gas, because in the $q \to 0$ limit the continuum of excited states tends to only one single-particle excited state at energy $\omega = k_F q/m$ (see Fig. 8.3).

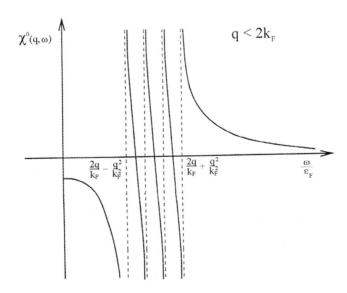

Fig. 8.3   Free response function of the one-dimensional Fermi gas, at fixed q.

From Eq. (8.86), the following result is obtained for the static polarizability of the one-dimensional Fermi gas:

$$
\begin{aligned}
\frac{\chi_0^{1D}(\mathbf{q}, 0)}{L} &= -\frac{2m}{q\pi} \ln\left(\frac{\frac{k_F q}{m} + \frac{q^2}{2m}}{\frac{k_F q}{m} - \frac{q^2}{2m}} \right) \\
&= -\frac{2m}{q\pi} \ln\left(\frac{1 + \frac{q}{2k_F}}{1 - \frac{q}{2k_F}} \right) \\
&= -\frac{2m}{q\pi} \left[\ln\left(1 + \frac{q}{2k_F}\right) - \ln\left(1 - \frac{q}{2k_F}\right) \right],
\end{aligned}
\tag{8.87}
$$

and thus

$$
\lim_{q \to 0} \frac{\chi_0^{1D}(\mathbf{q}, 0)}{N} = -\frac{2m}{q\pi\rho}\left[\frac{q}{2k_F} + \frac{q}{2k_F} \right] = -\rho\frac{4m}{\pi^2 \rho^3},
\tag{8.88}
$$

from which we derive the value $K = 4m/\pi^2\rho^3$ for the system compressibility. This value coincides with the one that can be obtained starting from the expression for the energy of the free one-dimensional gas given in (1.79), using relation (1.82).

For homogeneous systems in two and three dimensions, the free response functions in terms of the dimensionless quantities

$$
\hat{q} = \frac{q}{k_F}, \quad \hat{\omega} = \frac{m\omega}{k_F^2}, \quad \hat{\chi}_0(\hat{q}, \hat{\omega}) = \frac{1}{\nu_0}\chi_0(q, \omega),
\tag{8.89}
$$

where ν_0 are the densities of states at the Fermi surface, which are in turn given by

$$
\nu_0 = \begin{cases} L^3 \dfrac{m k_F}{\pi^2} & \text{for } D = 3 \\[2mm] L^2 \dfrac{m}{\pi} & \text{for } D = 2 \\[2mm] L \dfrac{2m}{\pi k_F} & \text{for } D = 1 \end{cases},
\tag{8.90}
$$

turn out to have the form ($\omega = \omega_1 + i\omega_2$)

$$
\begin{aligned}
\hat{\chi}_{3D}^0(\hat{q}, \hat{\omega}) \\
= \frac{1}{4\hat{q}} &\left[\left(\frac{1 - a^2 + c^2}{2} - iac \right) \ln \frac{(1+a)^2 + c^2}{(1-a)^2 + c^2} \right. \\
&- \left(\frac{1 - b^2 + c^2}{2} - ibc \right) \ln \frac{(1+b)^2 + c^2}{(1-b)^2 + c^2} \\
&+ (-2ac - i(1 + c^2 - a^2)) \left(\tan^{-1}\frac{1+a}{c} + \tan^{-1}\frac{1-a}{c} \right) \\
&\left. + (2bc + i(1 + c^2 - b^2)) \left(\tan^{-1}\frac{1+b}{c} + \tan^{-1}\frac{1-b}{c} \right) - 4b \right] \quad \text{(in 3D),}
\end{aligned}
\tag{8.91}
$$

where

$$a = \hat{\omega}_1/\hat{q} - \hat{q}/2, \quad b = \hat{\omega}_1/\hat{q} + \hat{q}/2, \quad c = \hat{\omega}_2/\hat{q},$$

and

$$\hat{\chi}_{2D}^0(\hat{q}, \hat{\omega}) = -1 - \frac{\sqrt{(a+ic)^2 - 1} - \sqrt{(b+ic)^2 - 1}}{\hat{q}} \quad \text{(in 2D)}, \qquad (8.92)$$

where the square root is defined by the choice of the cut in the upper semiplane. Finally, the one-dimensional response (8.86) in terms of dimensionless quantities is written as

$$\hat{\chi}_{1D}^0(\hat{q}, \hat{\omega}) = \frac{1}{2\hat{q}} \ln \left[\frac{\hat{\omega}^2 - (\hat{q} - \hat{q}^2/2)^2}{\hat{\omega}^2 - (\hat{q} + \hat{q}^2/2)^2} \right].$$

The behavior of the free response functions in 2D and 3D, as a function of ω_1, is analogous to the one-dimensional case. The only difference is in the range of the continuum of poles, which for $q \leq 2k_F$ is limited between 0 and $k_F q/m + q^2/2m$. The linear response function in the 3D case is schematically shown in Fig. 8.4. The one suitable for the 2D case is absolutely similar. The generalization of the Lindhard response function to finite temperature has been discussed by Khanna and Glyde (1976). Finally, note that in the case of nuclear matter, in the computation of the above quantities one should take into account the isospin degeneracy, which has the effect that many of such quantities should be multiplied by 2.

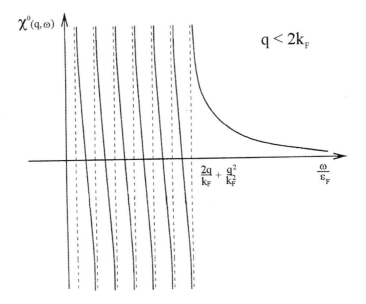

Fig. 8.4   Free response function for the three-dimensional Fermi gas, at fixed q.

8.7 The Current Response to an Electromagnetic Field

The current response to an electromagnetic field is the quantity that allows the study of the conductivity and diamagnetic properties of an interacting charged many-body system.

Of particular interest is the transverse current response function which is obtained through the general expression (8.19) by setting $F = G = j_{\mathbf{q},\perp}$:

$$\chi_\perp(\mathbf{q},\omega) = 2 \sum_n \omega_{n0} \frac{|\langle n|j_{\mathbf{q},\perp}|0\rangle|^2}{(\omega + i\eta)^2 - \omega_{n0}^2}, \tag{8.93}$$

where $j_{\mathbf{q}}$ is the current operator,

$$\mathbf{j}_{\mathbf{q}} = \frac{1}{2m} \sum_{i=1}^N (\mathbf{p}_i e^{i\mathbf{q}\cdot\mathbf{r}_i} + e^{i\mathbf{q}\cdot\mathbf{r}_i} \mathbf{p}_i),$$

and $j_{\mathbf{q},\perp}$ is an arbitrary component of $\mathbf{j}_{\mathbf{q}}$ perpendicular to $\mathbf{q}$. $\chi_\perp(\mathbf{q},\omega)$ is related [see, for example, Pines and Nozières (1966)] to the transverse conductivity by

$$\sigma_\perp(\mathbf{q},\omega) = \frac{ie^2}{\omega} \left(\chi_\perp(\mathbf{q},\omega) + \frac{N}{m} \right), \tag{8.94}$$

and its static part $\chi_\perp(\mathbf{q},0)$ is connected to the magnetic permeability μ_M by

$$\mu_M = \lim_{q\to 0} \frac{1}{1 + \frac{4\pi e^2}{c^2 q^2}(\chi_\perp(\mathbf{q},0) + \frac{N}{m})}. \tag{8.95}$$

The corresponding diamagnetic susceptibility χ_D is then given by

$$\chi_D = \frac{\mu_M - 1}{4\pi}. \tag{8.96}$$

The longitudinal current response function, defined as

$$\chi_\|(\mathbf{q},\omega) = \frac{2}{q^2} \sum_n \omega_{n0} \frac{|\langle n|\mathbf{q}\cdot\mathbf{j}_{\mathbf{q}}|0\rangle|^2}{(\omega + i\eta)^2 - \omega_{n0}^2}, \tag{8.97}$$

can be related to the density response $\chi(\mathbf{q},\omega)$ of the previous subsection using the identity

$$\chi_\|(\mathbf{q},\omega) + \frac{N}{m} = \frac{\omega^2}{q^2}\chi(\mathbf{q},\omega), \tag{8.98}$$

which immediately follows from Eqs. (8.62) and (8.64). The result (8.98) expresses the gauge invariance of the theory. The relevant excited states in (8.97) involve single-pair, collective states (plasmons) and multipair excitations. On the other hand, plasmons do not contribute to the transverse current response function (8.93), since transverse excitations do not involve density changes which are at the origin of plasmons (see Subsection 9.4).

From Eq. (8.98) it follows that

$$\chi_\|(\mathbf{q}, 0) = -\frac{N}{m}. \tag{8.99}$$

This result is directly equivalent to the f-sum rule. In the opposite limit, $\mathbf{q} \equiv 0$ and $\omega \neq 0$, the longitudinal and transverse current responses coincide, and the relevant matrix elements entering their expressions are also the ones determining absorption of light on the system [see Eq. (8.34)] since the dipole operator $\mathbf{D} = \sum_i \mathbf{r}_i$ is connected to the momentum operator by $[H, \mathbf{D}] = -(i/m)\mathbf{P}$, where $\mathbf{P} = \sum_i \mathbf{p}_i$. For translationally invariant systems for which the identity $[H, \mathbf{P}] = 0$ holds, one obtains

$$\chi_\|(0, \omega) = \chi_\perp(0, \omega) = 0, \tag{8.100}$$

as follows immediately from Eqs. (8.93) and (8.97) and the result

$$\omega_{n0} \langle n|\mathbf{P}|0\rangle = \langle n|[H, \mathbf{P}]|0\rangle = 0. \tag{8.101}$$

In the $q \equiv 0$ limit the conductivity is given by

$$\sigma(0, \omega) = \frac{iNe^2}{m\omega}, \tag{8.102}$$

and hence

$$\mathrm{Re}\,[\sigma(0, \omega)] = 0. \tag{8.103}$$

The DC conductivity which is obtained from the real part of the conductivity by taking the $\mathbf{q} \to \mathbf{0}$ limit and subsequently the $\omega \to 0$ limit, is zero for translationally invariant systems. The homogeneous electron gas does not absorb any light. The possibility of nearly free particle systems to absorb light and to conduct is due therefore to its imperfections or deviations from homogeneity.

In the general case (nontranslationally invariant systems) one has

$$\mathrm{Re}\,[\sigma(0, \omega)] = -e^2 \mathrm{Im}\left[\frac{\chi(P_\alpha/m, \omega)}{\omega}\right] = \pi e^2 \frac{S(P_\alpha/m, \omega)}{\omega}, \tag{8.104}$$

where $S(P_\alpha/m, \omega)$ is the strength of the operator P_α/m,

$$S(P_\alpha/m, \omega) = \sum_n |\langle n|P_\alpha/m|0\rangle|^2 \delta(\omega - \omega_{n0}), \tag{8.105}$$

and P_α/m is an arbitrary component of $\mathbf{P}/m$. It is straightforward to demonstrate that the conductivity (8.104) fulfills the sum rule:

$$\int \mathrm{Re}\,[\sigma(0, \omega)]d\omega = \int \pi e^2 \frac{S(P_\alpha/m, \omega)}{\omega}d\omega = \frac{\pi e^2 N}{2m}. \tag{8.106}$$

Using the formalism introduced in the previous two subsections, it is then easy to generalize result (8.104) to the case of finite temperature. The result is

$$\text{Re}\,[\sigma(0,\omega)] = \pi e^2 \frac{1}{\omega}[1 - e^{-\beta\omega}]S_T(P_\alpha/m,\omega), \tag{8.107}$$

where

$$S_T(P_\alpha/m,\omega) = \sum_{kn} p_k |\langle n|P_\alpha/m|k\rangle|^2 \delta(\omega - \omega_{nk}), \tag{8.108}$$

with

$$p_k = e^{-\beta E_k}/Z, \quad Z = \sum_k e^{-\beta E_k}.$$

For nontranslationally invariant systems the Hamiltonian can be written as $H = H_0 + V$, where, as an example, H_0 is the homogeneous electron gas part, while V is the potential which causes other forces besides electron–electron interactions. For example, V could be the sum of the interactions with the crystalline potential, with impurities of density operator $\rho_i(\mathbf{q})$, or with phonons. In this case one has

$$\left[H, \frac{P_\alpha}{m}\right] = \left[V, \frac{P_\alpha}{m}\right] = \frac{i}{m}\nabla_\alpha V = -\frac{i}{m}F_\alpha, \tag{8.109}$$

where F_α is the force due to the potential V. By using the result

$$\left\langle n\left|\left[H, \frac{P_\alpha}{m}\right]\right|0\right\rangle = \omega_{n0}\left\langle n\left|\frac{P_\alpha}{m}\right|0\right\rangle = -\frac{i}{m}\langle n|F_\alpha|0\rangle, \tag{8.110}$$

Eq. (8.104) yields the following expression for conductivity:

$$\text{Re}\,[\sigma(0,\omega)] = \frac{\pi e^2}{m^2\omega^3}S(F_\alpha,\omega), \tag{8.111}$$

where

$$S(F_\alpha,\omega) = -\frac{1}{\pi}\text{Im}\,[\chi(F_\alpha,\omega)] = \sum_n |\langle n|F_\alpha|0\rangle|^2\delta(\omega - \omega_{no}). \tag{8.112}$$

Equations (8.111) and (8.112) are easily generalized to the case of finite temperature and serve as another possible starting point for the evaluation of the conductivity. Applications of these formulas for several different kinds of potentials V can be found in the book by Mahan (1981).

We now discuss the diamagnetic properties of a many-body system. They are obtained from the transverse current response (8.93) by taking first $\omega = 0$ and then the $\mathbf{q} \to 0$ limit.

In a homogeneous system, the main contribution to the static transverse current response, in the $\mathbf{q} \to 0$ limit, comes from single-pair excitations (the longitudinal plasmon mode does not contribute for reasons of symmetry, while multipair excitations are negligible in the long wavelength limit). The static response is given by

$$\chi_\perp(\mathbf{q},0) = -\frac{2}{m^2} \sum_p \frac{n_\mathbf{p}(1 - n_{\mathbf{p}+\mathbf{q}})(\mathbf{p} \cdot n_\perp)^2}{\omega_{\mathbf{pq}}}, \tag{8.113}$$

where $n_\mathbf{p}$ is the Fermi distribution (1.76), $\omega_{\mathbf{pq}}$ is the excitation energy (1.103) and $n_\perp$ is a unit vector perpendicular to $\mathbf{q}$. It is easy to show, by a suitable interchange of indices, that the Pauli exclusion principle plays no role in the summation over p in (8.113). We then obtain

$$\chi_\perp(\mathbf{q},0) = -\frac{2}{m} \sum_{p \leq p_F} \frac{(\mathbf{p} \cdot n_\perp)^2}{\mathbf{p} \cdot \mathbf{q} + q^2/2}. \tag{8.114}$$

Replacing the sum with an integral and carrying out the angular integrations, one gets

$$\chi_\perp(\mathbf{q},0) = -\frac{N}{2m} + \frac{1}{2\pi^2 mq} \int_0^{p_F} dp\, p\, (p^2 - q^2/4) \ln\left(\frac{p - q/2}{p + q/2}\right). \tag{8.115}$$

Expanding the logarithm in powers of $q/2p$, one finally obtains

$$\lim_{q \to 0} \chi_\perp(\mathbf{q},0) = -\frac{N}{m} + \frac{N}{m} \frac{q^2}{4p_F^2}, \tag{8.116}$$

where the term of order q^2 is responsible for the weak diamagnetism first calculated by Landau. The magnetic permeability is then given by

$$\mu_M = \frac{1}{1 + N\pi e^2/mc^2 p_F^2}, \tag{8.117}$$

and the corresponding diamagnetic susceptibility is given by

$$\chi_D = \frac{\mu_M - 1}{4\pi} \simeq -\frac{Ne^2}{4mc^2 p_F^2} = -\frac{1}{3}\chi_P, \tag{8.118}$$

where χ_P is the paramagnetic spin susceptibility for the noninteracting electron gas ($\chi_P = -\mu_0\chi_\sigma$ of Chapter 1).

In a finite size many-particle system, the static transverse current response can be calculated exactly at order q^2 for the many-body Hamiltonian (1.1). In order to reach the goal, it is convenient to write $j_{\mathbf{q},\perp}$ as

$$j_{\mathbf{q},\perp} = \frac{i}{2}\left([H, O] - \frac{q}{m}Q\right), \tag{8.119}$$

where

$$O = \sum_{i=1}^N (x_i e^{iqy_i} + e^{iqy_i} x_i)$$

and

$$Q = \sum_{i=1}^{N} (x_i p_i^y e^{iqy_i} + e^{iqy_i} x_i p_i^y).$$

Using the result of (8.119) one can write

$$\chi_\perp(\mathbf{q}, 0) = -2 \sum_n \frac{|\langle n|j_{\mathbf{q},\perp}|0\rangle|^2}{\omega_{n0}}$$

$$= -\frac{1}{2} \sum_n \omega_{n0} |\langle n|O|0\rangle|^2$$

$$+ \frac{q}{2m} \langle 0|Q^\dagger O + O^\dagger Q|0\rangle - \frac{q^2}{2m^2} \sum_n \frac{|\langle n|Q|0\rangle|^2}{\omega_{n0}}. \tag{8.120}$$

This expression, truncated at order q^2, gives

$$\chi_\perp(\mathbf{q}, 0) = -\frac{N}{m} + \frac{Nq^2}{4m} \langle x^2 + y^2 \rangle - \frac{q^2}{2m^2} \sum_n \frac{|\langle n|L_z|0\rangle|^2}{\omega_{n0}}, \tag{8.121}$$

where

$$N \langle x^2 + y^2 \rangle = \left\langle 0 \left| \sum_i (x_i^2 + y_i^2) \right| 0 \right\rangle$$

and L_z is the z component of the angular momentum. Using the results

$$\Theta = 2 \sum_n \frac{|\langle n|L_z|0\rangle|^2}{\omega_{n0}}$$

and

$$\Theta_{\text{rig}} = mN \langle x^2 + y^2 \rangle,$$

where Θ is the moment of inertia of the system (see Subsection 11.1.5) and Θ_{rig} the rigid value of Θ, one gets for the magnetic permeability

$$\mu_M = \frac{1}{1 + \frac{\pi e^2}{m^2 c^2} (\Theta_{\text{rig}} - \Theta)}, \tag{8.122}$$

and for the diamagnetic susceptibility

$$\chi_D = \frac{\mu_M - 1}{4\pi} \simeq -\frac{e^2}{4m^2 c^2} \Theta_{\text{rig}} \left(1 - \frac{\Theta}{\Theta_{\text{rig}}} \right). \tag{8.123}$$

In the case of spherical systems $\Theta = 0$, since the eigenstates of the Hamiltonian are also the eigenstates of the angular momentum, χ_D takes the usual value,

$$\chi_D = -\frac{e^2}{4mc^2} N \langle x^2 + y^2 \rangle.$$

However, for deformed systems, Θ is different from zero and may give a large contribution to the diamagnetic susceptibility.

8.8   The Conductivity of Quantum Wires

In Subsections 5.7 and 6.6 we have considered a single infinitely long quantum wire in the y direction built on a two-dimensional electron gas by introducing a parabolic confining potential $1/2m\omega_0^2 x^2$ along the x direction, and studied its ground state properties by means of self-consistent Kohn–Sham calculations. We have considered different situations in which a magnetic field and a spin–orbit coupling can be present and affect the motion of the electrons in the system. In all these cases, the quantum wire is characterized by an energy spectrum with a subband structure in which the energies $\epsilon_{n,k}$ depend on a subband index n and on the momentum in the y direction. In fact, when the spin–orbit coupling is in, we have no spin label for the bands, and each of them contains the two components. The energies $\epsilon_{n,k}$ are plotted as a function of k in Fig. 8.5(a), in the case of a very narrow wire in the presence of an in-plane magnetic field and of spin–orbit coupling and neglecting the electron–electron interaction [see Pershin, Nesteroff and Privman (2004)]. In this figure we have stressed the two possible types of excitations — interband (between different bands) and intraband (inside a band) — with arrows.

In the calculations we are considering, the electrons are treated within the effective mass model in two dimensions, with the motion restricted to the xy plane, as usual. The SO terms forbid the formation of good spin states. Therefore, in general, the orbitals will be two-component spinors of the type

$$\Psi_{nk}(\mathbf{r},\eta) \equiv \frac{1}{\sqrt{L}}\begin{pmatrix}\varphi_{nk}(x,\uparrow)\\ \varphi_{nk}(x,\downarrow)\end{pmatrix}e^{iky}, \tag{8.124}$$

where η is an independent variable, which can only take the two values $\uparrow$, $\downarrow$. The intraband excitations, which exist owing to the confinement which breaks the translational invariance of the system along the x direction and are gapless, are responsible for the DC conductivity of the wire. This quantity can be obtained starting from Eqs. (8.104) and (8.105), by taking $P_\alpha/m = P_y/m = \sum_i v_y^i$ along the wire direction:

$$\mathrm{Re}\,[\sigma(0,\omega)] = \frac{\pi e^2}{\omega}\sum_\ell |\left\langle \ell\left|\sum_i v_y^i\right|0\right\rangle|^2 \delta(\omega - \omega_{\ell 0}). \tag{8.125}$$

Note that along the x direction, the DC conductivity is zero owing to the fact that in this case Eqs. (8.109)–(8.112) hold and the dipole interband excitations induced by the $F_x = -\sum_i \omega x_i$ operator have a gap. As a consequence, the system can absorb light and conduct only at finite frequency (see also Subsection 8.9.2).

a)

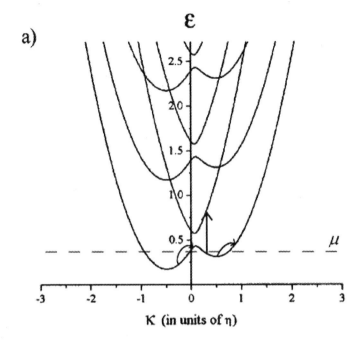

b)

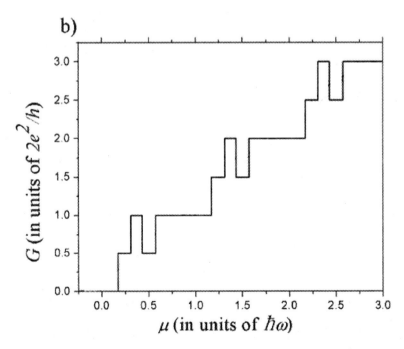

Fig. 8.5 (a) Typical energy spectrum (Pershin, Nesteroff and Privman, 2004) ϵ in units of $\hbar\omega_0$ of a quantum wire in an in-plane magnetic field and with spin–orbit interaction, as a function of the momentum k in units of $\eta = (2m\hbar\omega_0)^{1/2}$ — see Eq. (6.82); (b) conductance G as a function of chemical potential μ.

From Eq. (8.125), by using

$$v_y = -i[y, H], \tag{8.126}$$

and working in the $q \to 0$ limit where one can assume that $\sum_i y_i = \sum_i e^{iqy_i}/iq$ (the identity operator does not excite the system), one can write

$$\mathrm{Re}\left[\frac{\sigma(q,\omega)}{L}\right] = \frac{\pi e^2}{L} \sum_\ell \frac{\omega_{\ell 0}}{q^2} |\langle \ell| \sum_i e^{iqy_i}|0\rangle|^2 \delta(\omega - \omega_{\ell 0}). \tag{8.127}$$

[Note that this result also follows immediately from Eq. (8.98).] In the $q \to 0$ limit, the operator $\sum_i e^{iqy_i}$ induces intraband excitations between the states $|\Psi_{nk}\rangle$ and $|\Psi_{nk+q}\rangle$ of Eq. (8.124), which we are going to use in the following to evaluate the matrix elements, with excitation energies given by

$$\omega_{\ell 0} = \epsilon_{n,k+q} - \epsilon_{n,k} = \left.\frac{\partial \epsilon_{n,k}}{\partial k}\right|_{k=k_n} q \equiv \alpha_{k_n} q, \tag{8.128}$$

where k_n are the k values corresponding to the intersections of the n subband with the chemical potential μ for positive slopes α_{k_n}. Indeed, it is crucial to realize that for $q > 0$, only intraband excitations with $k + q > k$ are allowed. Referring to Fig. 8.5, some of these transitions are exemplified as curved arrows. Since in the $q \to 0$ limit the matrix elements of the operator $\sum_i e^{iqy_i}$ can be taken equal to the unity, as follows from

$$\langle \ell| \sum_i e^{iqy_i}|0\rangle = \sum_\eta \int \Psi^*_{nk+q}(\mathbf{r}, \eta) e^{iqy} \Psi_{nk}(\mathbf{r}, \eta) d\mathbf{r}$$

$$= \frac{1}{L} \sum_\eta \int dx\, dy\, \varphi^*_{nk+q}(x, \eta) e^{-i(k+q)y}\, e^{iqy}\, \varphi_{nk}(x, \eta)\, e^{iky}$$

$$= 1 + O(q) \tag{8.129}$$

and from $\sum_\ell = \sum_{k_n} \frac{L}{2\pi} \int dk = \sum_{k_n} \frac{L}{2\pi} q$, valid in the same limit, one finally gets

$$\mathrm{Re}\left[\frac{\sigma(q,\omega)}{L}\right] = \frac{\pi e^2}{q^2} \sum_{k_n} \frac{q}{2\pi} \alpha_{k_n} q \delta(\omega - \alpha_{k_n} q), \tag{8.130}$$

where we have denoted with $\sum_{k_n}$ the sum over all the possible intrasubband allowed excitations.

From Eq. (8.130), taking the cosine Fourier transform, it can be easily shown that

$$\mathrm{Re}\left[\frac{\sigma(y,\omega)}{L}\right] = \frac{e^2}{2\pi} \sum_{k_n} \cos\left(\frac{\omega\, y}{\alpha_{k_n}}\right). \tag{8.131}$$

Thus in the limit $\omega \to 0$ and restoring $\hbar$, we get for the conductance G (defined as the $q \to 0$, $\omega \to 0$ limit of Re $[\frac{\sigma(y,\omega)}{L}]$)

$$G = \frac{e^2}{h} \sum_{k_n} 1. \qquad (8.132)$$

In the noninteracting model and in the absence of magnetic and spin–orbit effects, where $\epsilon_{n,k} = (n + 1/2)\omega + k^2/2$ and the subbands are degenerate in spin, only one intraband excitation (one intersection point, $k = k_n$) contributes to G for each band (see Fig. 8.6). As a consequence Eq. (8.132) gives the usual conductance quantization of the Landauer–Büttiker formalism [see for example Kelly (1995), Beenakker and van Houten (1991), Datta (1995)], where each of the spin-degenerate subbands contributes e^2/h to the conductance, yielding the result $G = \frac{2e^2}{h} \sum_n 1$. However, different results for G can arise owing to magnetic field, spin–orbit and interaction effects on the energy spectrum. For example, this is the case of the energy spectrum of Fig. 8.5(a), where at certain values of the chemical potential more than one intraband excitation per band (more intersection points, $k = k_n$) contributes e^2/h to the conductance, yielding for G as a function of μ the values shown in Fig. 8.5(b). Hence, we can conclude that the conductance is a quantity particularly sensitive to the ground state properties of quantum wires. In particular, it is very sensitive to symmetry-breaking effects of the ground state which are induced by the electron–electron interaction, like spontaneous spin polarization and Wigner crystalization for which there seems to be experimental evidence (Thomas *et al.*, 2004; Thomas *et al.*, 1996; Reilly *et al.*, 2001). Kohn–Sham calculations yielding

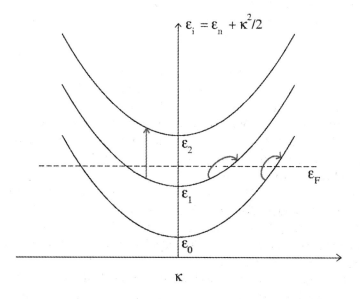

Fig. 8.6 Energy spectrum of a noninteracting quantum wire.

spin-polarized ground states at low density were presented in Subsection 5.7. In these calculation one fixes the 1D density in the wire and solves self-consistently the Kohn–Sham equations, which yields the energy spectrum and chemical potential. At low density, in the absence of external magnetic fields and spin–orbit couplings, the system is ferromagnetic, and in this region of density the conductance takes the value e^2/h, in agreement with some experiments. Inclusion of magnetic field and spin–orbit effects in the calculation can yield the occurrence of peaks of the conductance as the ones shown and discussed in Subsection 6.10.

Finally, we make some considerations about the universality of result (8.132). The derivation of Eq. (8.132) made above is in the mean field scheme since it assumes for the excitation energies in the starting formula (8.125) for σ, the gapless intraband excitations of the mean field energy spectrum. Now the question is whether this mean field scheme yields an exact result for G (a universal law), as stated by some authors in the literature, or, on the contrary, must be corrected by interaction effects.

We know (see for example Pines and Noziéres) that in the limit where the wave vector q and energy ω related to the excitations fulfill the conditions $qv_F, \omega \ll \epsilon_F$, only 1p–1h excitations contribute correctly to the dynamic response of the system, and therefore to σ, which is a response of this type to be calculated in that limit. In this sense one can conclude that the Landau–Fermi theory for σ would be the exact theory for this quantity, since it is well known that in this limit the Landau theory for the dynamic response of Fermi liquids is exact (see Subsection 9.1). But the Landau theory is not a mean field theory. Rather, it is analogous to the time-dependent local density (TDLDA) theory in the limit where the parameters F_1 and F_2 are negligible, which seems to be a good approximation for electrons. In fact the parameter F_0 (the only important one, together with the direct Coulomb) is related to the compressibility, like it is the exchange correlation particle–hole residual interaction in the TDLDA.

The other question is whether one has to use for the response the full one or the screened one. In other words if one has to calculate $\sigma = \langle j \rangle / E_{ext}$ or $\sigma' = \langle j \rangle / (E_{ext} + E_{pol})$, where E_{ext} is the applied field and $E_{ext} + E_{pol}$ is an effective field internal to the wire. In my opinion and according to Pines and Noziéres (1966) and other authors [see for example A. Kawabata (1995), D. Maslov and M. Stone (1995)], one has to use the screened one, which means that one has to calculate σ', the conductivity, as the current response to an electric field which is the sum of the external and the polarization field (see Subsection 9.12).

If this is true, Eq. (8.132) is not universal and must be corrected for exchange-correlation effects which affect the TDLDA screened response. Use of the screened response in the TDLDA, which takes care of 1p–1h excitations practically like in the Landau theory, makes the TDLDA calculation of σ', obtained in the $q, \omega \to 0$ limits, practically an exact calculation.

In the following we report on such a calculation (Malet *et al.*, 2005) in the case of the quantum wire in the absence of magnetic and spin–orbit effects, which was described in Subsection 5.7. For simplicity, we address only the lowest density cases, namely when the quantum wire is ferromagnetic and only one subband is occupied, and when it is paramagnetic and only two subbands (spin-up and spin-down) are occupied, being degenerate and equally populated. These two situations are the more interesting ones from an experimental viewpoint. The generalization of the method to the case with more than two subbands is straightforward but requires a more elaborate calculation.

Our starting point is the relation (Izuyama, 1961; Pines and Noziéres, 1966)

$$\frac{\sigma'(q,\omega)}{L} = i\frac{\omega e^2}{q^2}\frac{\chi_{\mathrm{sc}}(q,\omega)}{L} ,\qquad(8.133)$$

where $\chi_{\mathrm{sc}}(q,\omega)$ is the screened response function. This relation substitutes Eq. (8.127), discussed above. Note that (8.127) can be obtained through Eq. (8.133) by using the density response function

$$\chi(q,\omega) = \sum_\ell 2\omega_{\ell 0}\frac{|\langle \ell| \sum_i e^{iqy_i}|0\rangle|^2}{\omega^2 - \omega_{\ell 0}^2} ,$$

at the place of the screened one. Equation (8.133) is derived by calculating σ as the linear current response to the effective field $E = E_{\mathrm{ext}} + E_{\mathrm{pol}}$. $\chi_{\mathrm{sc}}(q,\omega)$ is related to $\chi(q,\omega)$ by (see Subsection 9.12)

$$\chi(q,\omega) = \frac{\chi_{\mathrm{sc}}(q,\omega)}{1 - v(q)\chi_{\mathrm{sc}}(q,\omega)},\qquad(8.134)$$

where

$$v(q) = 2\int dx\, dx'|\phi_0(x)|^2\, |\phi_0(x')|^2 K_0[q(x - x')]\qquad(8.135)$$

is the quasi-1D Fourier transform of the Coulomb potential [see Subsection 5.7 and Malet *et al.*, (2005)], $\phi_0(x) = \phi_0^\uparrow(x)$ for the ferromagnetic case and $\phi_0(x) = \phi_0^\uparrow(x) = \phi_0^\downarrow(x)$ for the paramagnetic case. As we will show in Subsection 9.12, in the RPA (time-dependent Hartree, without exchange effects), the electron response to the effective field (the sum of the external field and the local induced field), $\chi_{\mathrm{sc}}(q,\omega)$, is approximated by the mean field response

$$\chi_{\mathrm{sc}}(q,\omega) = \chi_0(q,\omega) ,\qquad(8.136)$$

yielding

$$\chi^{\mathrm{RPA}}(q,\omega) = \frac{\chi_0(q,\omega)}{1 - v(q)\chi_0(q,\omega)} .\qquad(8.137)$$

Use of the screened response in the RPA approximation then yields for the conductivity the result of Eq. (8.132).

To go beyond the RPA, exchange-correlation terms must be taken into account by modifying Eq. (8.136) as

$$\chi_{sc}(q,\omega) = \frac{\chi_0(q,\omega)}{1 + \frac{v(q)}{L}\mathcal{G}(q,\omega)\chi_0(q,\omega)} , \tag{8.138}$$

where the dynamic local field correction $\mathcal{G}(q,\omega)$ has been introduced (see Subsection 10.4). Equation (8.138) is the most general way to express $\chi_{sc}(q,\omega)$ in terms of the free response function $\chi_0(q,\omega)$, and yields for $\chi(q,\omega)$

$$\chi(q,\omega) = \frac{\chi_0(q,\omega)}{1 - \frac{v(q)}{L}[1 - \mathcal{G}(q,\omega)]\chi_0(q,\omega)}. \tag{8.139}$$

In the following we are interested only in the $\omega = 0$ limit, so the frequency dependence of the local field correction is suppressed:

$$\chi_{sc}(q,\omega) = \frac{\chi_0(q,\omega)}{1 + \frac{v(q)}{L}\mathcal{G}(q)\chi_0(q,\omega)}. \tag{8.140}$$

In this form, the frequency dependence of $\chi_{sc}(q,\omega)$ comes only from $\chi_0(q,\omega)$. An important property of the local field correction is (Iwamoto and Pines, 1984)

$$\lim_{q\to 0} v(q)\mathcal{G}(q) = v(0)\mathcal{G}(0) = \frac{1}{\rho_1^2 K_0}\left(1 - \frac{K_0}{K}\right), \tag{8.141}$$

where K is the compressibility of the system and K_0 its free value. In the situations we are considering, we have $K_0 = 1/\pi^2\rho_1^3$ for the ferromagnetic case, and $K_0 = 4/\pi^2\rho_1^3$ for the paramagnetic case, and the compressibility K can be calculated from the standard — thermodynamical — expression

$$\frac{1}{K} = \rho_1^2\left(\rho_1\frac{\partial^2 E/N}{\partial\rho_1^2} + 2\frac{\partial E/N}{\partial\rho_1}\right), \tag{8.142}$$

where ρ_1 is the one-dimensional density (see Subsection 5.7).

For the locally neutral system, the energy functional contains only the kinetic and exchange-correlation pieces plus the confining potential in the x direction. One gets

$$\frac{E}{N} = \frac{\pi^2}{c}\rho_1^2 + \frac{1}{\rho_1}\int dx\,\mathcal{E}_{xc}(x) + \text{const}, \tag{8.143}$$

where $c = 6$ for the ferromagnetic case, and $c = 24$ for the paramagnetic case.

Hence

$$\frac{1}{K} = \frac{1}{K_0} + \rho_1^2 I, \tag{8.144}$$

where we have used (see Subsection 5.7)

$$\frac{\partial\mathcal{E}_{xc}}{\partial\rho_1} = \frac{\partial\mathcal{E}_{xc}}{\partial\rho}\frac{\partial\rho}{\partial\rho_1} = \frac{\partial\mathcal{E}_{xc}}{\partial\rho}|\phi_0(x)|^2,$$

and defined

$$I = \int dx \, \frac{\partial^2 \mathcal{E}_{\text{xc}}}{\partial \rho^2} |\phi_0(x)|^4. \tag{8.145}$$

From Eqs. (8.140), (8.128) and (8.129), and by the definition of $\chi_0(q,\omega)$ with $\sum_\ell = \sum_\sigma \frac{L}{2\pi} \int dk = \sum_\sigma \frac{L}{2\pi} q$, valid in the small q limit, one gets

$$\frac{L}{\chi_{\text{sc}}(q,\omega)} = \frac{L}{\chi_0(q,\omega)} + v(0)\mathcal{G}(0) = \frac{\omega^2 - (\alpha_{k_0}q)^2}{c'\alpha_{k_0}q^2/\pi} + v(0)\mathcal{G}(0), \tag{8.146}$$

with $c' = 1$ for the ferromagnetic case ($n = 0, \uparrow$ band occupied in the ground state), and with $c' = 2$ for the paramagnetic case ($n = 0, \uparrow$ and $n = 0, \downarrow$ bands occupied in the ground state). Hence

$$\frac{1}{L} \chi_{\text{sc}}(q,\omega) = \gamma c' \frac{\alpha_{k_0}q^2}{\pi\gamma} \frac{1}{\omega^2 - (\alpha_{k_0}q/\gamma)^2} = \gamma \frac{1}{L} \chi_0(q,\omega,\alpha_{k_0}/\gamma), \tag{8.147}$$

where

$$\gamma = \left(1 - c' \frac{v(0)\mathcal{G}(0)}{\pi\alpha_{k_0}}\right)^{-1/2} = \sqrt{\frac{K}{K_0}}. \tag{8.148}$$

It is now possible to calculate the real part of the conductivity and thus the conductance along the line of Eqs. (8.127)–(8.132), using $\chi_{\text{sc}}(q,\omega)/L = \gamma \chi_0(q,\omega,\alpha_{k_0}/\gamma)/L$ at the place of the mean field response $\chi_0(q,\omega,\alpha_{k_0})/L$. One easily gets

$$G = \frac{e^2}{h} \sqrt{\frac{K}{K_0}} \tag{8.149}$$

for the ferromagnetic case, and

$$G = \frac{2e^2}{h} \sqrt{\frac{K}{K_0}} \tag{8.150}$$

for the paramagnetic case. In both situations the ratio K/K_0 is calculated from the expression

$$\frac{K}{K_0} = \frac{1}{1 + \frac{2c'}{\pi^2 \rho_1} I}, \tag{8.151}$$

where I is as defined in Eq. (8.145) with $\phi_0(x) = \phi_0^\uparrow(x)$ for the ferromagnetic case, and $\phi_0(x) = \phi_0^\uparrow(x) = \phi_0^\downarrow(x)$ for the paramagnetic case.

The ratio $\gamma = \sqrt{K/K_0}$ is plotted in Fig. 8.7 as a function of ρ_1 for different values of the frequency of the lateral confining potential ω_0. One sees from this figure that when the wire is in the first — polarized — subband, γ approximately ranges from 1.5 to 1 for all the confinements considered here, yielding for this subband a conductance which goes from $G \simeq 0.7(2e^2/h)$ to $G \simeq 0.5(2e^2/h)$ with increasing density, in agreement with the experimental data of Reilly *et al.* (2001). The discontinuity of γ and K when passing from one to two subbands reflects the phase

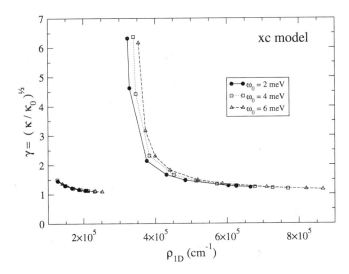

Fig. 8.7 Ratio $\gamma = \sqrt{K/K_0}$ as a function of the linear density ρ_1 (cm^{-1}) for the xc model of the quantum wire of Subsection 5.7 and three values of the harmonic frequency ω_0. The lines have been drawn to guide the eye.

transition occurring in the system. After that, the paramagnetic state has a conductance that, starting from values slightly larger than $2e^2/h$, decreases with increasing density to the measured value $G \simeq 2e^2/h$. Despite the fact that the present model for an infinite quantum wire is obviously an oversimplification of the actual experimental device, it yields a qualitative agreement with measurements, especially the observed density dependence of the conductance. Note that at values of ρ_1 lower than the ones considered here, γ, as calculated from Eqs. (8.148)–(8.151), becomes imaginary because the compressibility is negative. This result for the compressibility is well known (Camels and Gold, 1997), and it appears also for the electron gas in two and three dimensions, reflecting the failure at low densities of the jellium model in which the background of positive ions is kept indefinitely rigid.

8.9   Magnetoconductivity

In this subsection we show that the magnetoconductivity of an interacting two-dimensional electron gas submitted to a perpendicular magnetic field is a universal property of the system that follows from the validity of Kohn's theorem.

8.9.1   *The Kohn Theorem*

By using the creation $(O_\lambda^\dagger)$ and annihilation operators (O_λ) of the excited states [see Eqs. (9.149) and (9.150)], one can rewrite the Schrödinger equation as

$$[H, O_\lambda^\dagger]|0\rangle = (E_\lambda - E_0)O_\lambda^\dagger|0\rangle, \tag{8.152}$$

where $|0\rangle$ is the system ground state. This equation (equation of motion) is used, in general, as a starting point to set up approximate theories for the excited states, and the relative excitation energies, of a many-body system. However, there are cases where these equations provide exact solutions for such quantities.

Kohn (1961) considered the case of a two-dimensional system of N electrons with charge $-e$, in a uniform magnetic field perpendicular to the plane where the particles move (i.e. the x–y plane), with the Hamiltonian ($\hbar = 1$) given by

$$H = \frac{1}{2m} \sum_{i=1}^{N} \mathbf{P}_i^2 + g\mu_0 B S_z + \sum_{i<j}^{N} V(|\mathbf{r}_i - \mathbf{r}_j|) = H_0 + V, \tag{8.153}$$

where we have chosen the vector potential $\mathbf{A}$ in such a way as to give a constant magnetic field along the z axis [e.g. $\mathbf{A} = B(0, x, 0)$], g is the gyromagnetic factor of electrons, and μ_0 is Bohr's magneton. In Eq. (8.153) S_z is the total spin of the electrons and

$$\mathbf{P} = \sum_{i=1}^{N} \mathbf{P}_i = \sum_{i=1}^{N} \left(\mathbf{p}_i + \frac{e}{c} \mathbf{A}(\mathbf{r}_i) \right)$$

is their total kinetic momentum ($\mathbf{p}_i = -i\nabla_i$ is the canonical momentum of the ith electron).

Starting from Eq. (8.153), it is easy to verify that

$$[H, P_+] = \omega_c P_+, \tag{8.154}$$

where $\omega_c = \frac{eB}{mc}$ is the cyclotron frequency, and

$$P_+ = P_x + iP_y, \tag{8.155}$$

because the total momentum $\mathbf{P}$ commutes with the interaction term:

$$\left[\sum_{i<j}^{N} V(|\mathbf{r}_i - \mathbf{r}_j|), \mathbf{P} \right] = 0. \tag{8.156}$$

Equation (8.154), together with the equation of motion (8.152), shows that the state $P_+|0\rangle$ is an exact eigenstate of H with energy $E_0 + \omega_c$. This eigenstate is known as the cyclotron resonance of the system, and its frequency cannot be directly affected by any effect caused by interparticle interaction. This result is known in the literature as the Kohn theorem.

The state $P_+|0\rangle$ is the only state of the system that is excited if we immerse the system in a weak microwave field at frequency ω_c. In fact, describing such a field at frequency ω by a vector potential $\left(\mathbf{E} = -\frac{1}{c} \frac{\partial \mathbf{A}'}{\partial t} \right)$,

$$\mathbf{A}' = c \left(\frac{E_x}{i\omega} \hat{x} + \frac{E_y}{\omega} \hat{y} \right) e^{-i\omega t},$$

and choosing $E_x = E_y = E$, we find that the interaction Hamiltonian, $H' = -\frac{1}{c}\mathbf{j}\cdot\mathbf{A}'$, with $\mathbf{j} = -e\sum_{i=1}^{N}\mathbf{v}_i$ and $\mathbf{v} = -i[\mathbf{r}, H]$, between the system and the homogeneous microwave field is given by

$$H' = \frac{e}{2mc}\sum_{i=1}^{N}(\mathbf{P}_i\cdot\mathbf{A}'_i + \mathbf{A}'_i\cdot\mathbf{P}_i) = \frac{eE}{i\omega m}P_+ e^{-i\omega t}. \qquad (8.157)$$

This operator induces a transition from the ground state $|0\rangle$ to the excited state $P_+|0\rangle$, absorbing the energy ω_c. As a result, a narrow absorption peak emerges at frequency $\omega = \omega_c$.

In the case of a system confined by a potential of the parabolic type, such as in quantum dots, it is necessary to add to Hamiltonian (8.153) the confinement potential

$$V_{\text{ext}} = \frac{1}{2}m\sum_{i=1}^{N}\omega_0^2 r_i^2.$$

By using in this case the symmetric gauge $\mathbf{A} = \frac{B}{2}(-y, x, 0)$ and defining $P_\pm = P_x \pm iP_y$, $Q_\pm = Q_x \pm iQ_y$, with

$$Q_x = \sum_{i=1}^{N}x_i, \quad Q_y = \sum_{i=1}^{N}y_i,$$

and using the commutation relations

$$\left[\sum_{i=1}^{N}\frac{P_i^2}{2m}, Q_\pm\right] = -\frac{i}{m}P_\pm, \quad \left[\sum_{i=1}^{N}\frac{P_i^2}{2m}, P_\pm\right] = \pm\omega_c P_\pm,$$

$$[V_{\text{ext}}, P_\pm] = im\omega_0^2 Q_\pm, \qquad (8.158)$$

together with (8.156), it is easy to prove that the following result holds:

$$\left[H, Q_\pm - \frac{i\omega_\pm}{m\omega_0^2}P_\pm\right] = \omega_\pm\left(Q_\pm - \frac{i\omega_\pm}{m\omega_0^2}P_\pm\right), \qquad (8.159)$$

where $\omega_\pm = \sqrt{\omega_0^2 + 1/4\omega_c^2} \pm \omega_c/2$.

Equation (8.159), together with the equation of motion (8.152), proves that the states $(Q_\pm - i(\omega_\pm/m\omega_0^2)P_\pm)|0\rangle$ are exact eigenstates of H with respective energies $E_0+\omega_\pm$. These eigenstates are the bulk and edge plasmon resonances of the quantum dots under a magnetic field, whose frequencies cannot be directly affected by any effect caused by electron interactions. This result is known in the literature as the generalized Kohn theorem. Such plasmon states are the only ones which are excited in photoabsorption reactions with long wavelength photons (far-infrared absorption), and have been observed in several experiments (see Subsection 9.13).

A generalized Kohn theorem holds also in the case of a quantum wire for which one has to add to Hamiltonian (8.153) the confinement potential

$$V_{\text{ext}} = \frac{1}{2}m\sum_{i=1}^{N}\omega_0^2 x_i^2,$$

which confines the 2D electron gas in the x direction. In this case it is straightforward to prove that the following result holds:

$$\left[H, \sum_{i=1}^{N}\left(-m\Omega x + i\left(p_x + i\frac{\omega_c}{\Omega}p_y\right)\right)\right] = \Omega\sum_{i=1}^{N}\left(-m\Omega x + i\left(p_x + i\frac{\omega_c}{\Omega}p_y\right)\right),$$

(8.160)

where $\Omega = \sqrt{\omega_0^2 + \omega_c^2}$.

Equation (8.160) proves that the state

$$\frac{1}{\sqrt{2m\Omega N}}\sum_{i=1}^{N}\left(-m\Omega x + i\left(p_x + i\frac{\omega_c}{\Omega}p_y\right)\right)|0\rangle$$

is an exact eigenstate of the wire Hamiltonian with energy $\Omega = \sqrt{\omega_0^2 + \omega_c^2}$. This state is the plasmon resonance of the quantum wire under a magnetic field and is the only one that is excited in far-infrared experiments (Demel *et al.*, 1988).

The generalized Kohn theorem applies to all systems undergoing harmonic confinement. In the absence of a magnetic field, it implies the existence of a collective state at energy ω_0. This state has been observed in cold and dilute Boson gases (see Subsection 9.6) and in metal clusters (Subsection 9.8.2). The Kohn theorem is not fulfilled by the mean field theories, like the Hartree, HF and Kohn–Sham theories. However, as we will see in the next chapter, it is satisfied by the RPA and TDLDA theories.

8.9.2 *Magnetoconductivity and the Quantum Hall Effect*

By using the results of the previous subsection we now evaluate the transverse conductivity of a two-dimensional electron gas in a magnetic field perpendicular to the plane of motion of the electrons, with the Hamiltonian given by (8.153). In the presence of an electric field in the x direction, giving rise to the interaction Hamiltonian $H' = -\frac{Ej_x}{i\omega}e^{-i\omega t}$, the xx conductivity (we use linear response theory as discussed in Subsection 8.2) is given by

$$\sigma_{xx}(0,\omega) = \frac{\langle j_x\rangle}{E} = \frac{i}{\omega}\frac{\sum_n 2\omega_{n0}\,|\langle n\,|j_x|\,0\rangle|^2}{\omega^2 - \omega_{n0}^2},$$

(8.161)

so that its real part is given by

$$\text{Re}\,[\sigma_{xx}(0,\omega)] = \frac{\pi e^2}{\omega}S\left(\sum_{i=1}^{N}(p_x/m)_i,\omega\right),$$

(8.162)

where we have used

$$j_x = -e \sum_{i=1}^{N} p_x^i/m, \tag{8.163}$$

and the strength S is given by

$$S\left(\sum_{i=1}^{N}(p_x/m)_i, \omega\right) = \sum_n \left|\left\langle n \left| \sum_{i=1}^{N}(p_x/m)_i \right| 0 \right\rangle\right|^2 \delta(\omega - \omega_{n0})$$

$$= \left|\left\langle 0 \left| \left[\sum_{i=1}^{N}(p_x/m)_i, O_{\omega_c}^\dagger\right] \right| 0 \right\rangle\right|^2 \delta(\omega - \omega_c), \tag{8.164}$$

$$O_{\omega_c}^\dagger = \sqrt{\frac{1}{2Nm\omega_c}} \sum_{i=1}^{N}(-m\omega_c x + ip_x - p_y)_i, \tag{8.165}$$

and we have used the results (8.154) and (8.155) and normalized the state $|\omega_c\rangle = O_{\omega_c}^\dagger|0\rangle$ to unity.

From Eqs. (8.164) and (8.165) one gets for the xx conductivity per unit of surface

$$\text{Re}\left[\sigma_{xx}(0,\omega)\right] = \frac{\pi e^2}{2m}\rho\delta(\omega - \omega_c). \tag{8.166}$$

The yx conductivity is then given by

$$\sigma_{yx}(0,\omega) = \frac{\langle j_y \rangle}{E} = \frac{i}{\omega} \sum_n \left(\frac{\langle 0|j_y|n\rangle\langle n|j_x|0\rangle}{\omega - \omega_{n0}} - \frac{\langle 0|j_x|n\rangle\langle n|j_y|0\rangle}{\omega + \omega_{n0}}\right), \tag{8.167}$$

where

$$j_y = -e \sum_{i=1}^{N}(p_y/m + \omega_c x)_i. \tag{8.168}$$

Once more, by using the fact that only the cyclotron state $|\omega_c\rangle = O_{\omega_c}^\dagger|0\rangle$, with $O_{\omega_c}^\dagger$ given by (8.165), can be excited in (8.167), it is easy to show that the real part of the yx conductivity is given by

$$\text{Re}[\sigma_{yx}(0,\omega)] = -\frac{e^2 \rho \omega_c}{m} P\left(\frac{1}{\omega^2 - \omega_c^2}\right). \tag{8.169}$$

The DC yx conductivity, which is obtained through (8.169) taking the $\omega = 0$ limit, coincides with the classical Hall conductivity. We stress, however, that the result (8.169) is an exact result of quantum mechanics. Therefore, if ν Landau levels are filled by electrons with density $\rho = \nu\frac{eB}{ch}$, we find that

$$\text{Re}[\sigma_{yx}] = \frac{ec\rho}{B} = \nu\frac{e^2}{h}. \tag{8.170}$$

At the same time, the DC xx conductivity in the direction of the electric field is zero. It is important to note that even though ν is an integer, the Hall conductivity is actually proportional to the ground state density ρ, which is a continuous variable. That is, the second equality in (8.170) holds only at a particular ρ.

If the magnetic field is so strong that only the first Landau level is filled with a fractional filling factor, as happens for the Laughlin states, then Eq. (8.170) holds also for $\nu = 1/m$. The integer quantized Hall effect has been observed in 1980 by von Klitzing *et al.*, and the fractional quantized Hall effect has been measured by Tsui *et al.* (1982) and Willet *et al.*, (1987).

The calculation of the Hall conductivity is also very instructive in the case of a system confined by a potential of the parabolic type for which the exact solutions (8.159) hold. In the presence of an electric field in the x direction, the real part of the xx conductivity is given by (we use now the symmetric gauge)

$$\mathrm{Re}[\sigma_{xx}(0,\omega)] = \frac{\pi e^2}{\omega} S\left(\sum_{i=1}^{N}\left(p_x/m - \frac{\omega_c}{2}y\right)i, \omega\right), \tag{8.171}$$

where

$$
S\left(\sum_{i=1}^{N}\left(p_x/m - \frac{\omega_c}{2}y\right)_i, \omega\right)
$$

$$
= \sum_{n}\left|\langle n|\sum_{i=1}^{N}\left(p_x/m - \frac{\omega_c}{2}y\right)i|0\rangle\right|^2 \delta(\omega - \omega_{n0})
$$

$$
= \left|\langle 0|\left[\sum_{i=1}^{N}\left(p_x/m - \frac{\omega_c}{2}y\right)i, O_+^\dagger\right]|0\rangle\right|^2 \delta(\omega - \omega_+)
$$

$$
+ \left|\langle 0|\left[\sum_{i=1}^{N}\left(p_x/m - \frac{\omega_c}{2}y\right)i, O_-^\dagger\right]|0\rangle\right|^2 \delta(\omega - \omega_-), \tag{8.172}
$$

$$
O_\pm^\dagger = \frac{1}{2}\sqrt{\frac{m\bar{\omega}}{N}}\sum_{i=1}^{N}\left(x \pm iy - \frac{i}{m\bar{\omega}}(p_x \pm ip_y)\right)i, \tag{8.173}
$$

and we have used the result (8.159), defined as $\bar{\omega} = \sqrt{\omega_0^2 + \omega_c^2/4}$ so that $\omega_\pm = \bar{\omega} \pm \omega_c/2$, and normalized the states $|\omega_\pm\rangle = O_\pm^\dagger|0\rangle$ to unity. Performing the commutators in (8.172), we get for the conductivity per unit surface area

$$\mathrm{Re}[\sigma_{xx}(0,\omega)] = \frac{\pi e^2}{4m}\frac{\rho}{\bar{\omega}}(\omega_+\delta(\omega - \omega_+) + \omega_-\delta(\omega - \omega_-)). \tag{8.174}$$

From this equation, in the limit $\omega_0 \equiv 0$, one recovers immediately Eq. (8.166), since in this limit $\omega_- = 0$ and $\omega_+ = \omega_c$.

The real part of the yx conductivity can be calculated in an analogous way. One starts from (8.167) and uses as the excited states of this equation the two states $|\omega_\pm\rangle = O_\pm^\dagger |0\rangle$. After a simple calculation one finds that

$$\text{Re}\,[\sigma_{yx}] = -\frac{e^2 \rho}{2m\bar{\omega}} P \left(\frac{\omega_+^2}{\omega^2 - \omega_+^2} - \frac{\omega_-^2}{\omega^2 - \omega_-^2} \right). \tag{8.175}$$

Once again, we see that from this equation, in the limit $\omega_0 \equiv 0$, one recovers immediately Eq. (8.169). However, for $\omega_0 \neq 0$, differently from the bulk case, the DC conductivity of the dot vanishes since the two excitation branches in the dot contribute to the DC conductivity in an equal and opposite way.

Finally, we consider the case of a quantum wire. In analogy with the case of the quantum well, we can use the creation operator

$$O^\dagger = \frac{1}{\sqrt{2m\Omega N}} \sum_{i=1}^{N} \left(-m\Omega x + i \left(p_x + i\frac{\omega_c}{\Omega} p_y \right) \right) \tag{8.176}$$

of Eq. (8.160) to calculate the matrix elements entering the general expressions (8.161) and (8.167) for $\sigma_{xx}(0, \omega)$ and $\sigma_{yx}(0, \omega)$, respectively. We then easily gets

$$\text{Re}\,[\sigma_{xx}(0, \omega)] = \frac{\pi e^2}{2m} \rho \delta(\omega - \Omega) \tag{8.177}$$

and

$$\text{Re}\,[\sigma_{yx}(0, \omega)] = -\frac{e^2 \rho \omega_c}{m} P \left(\frac{1}{\omega^2 - \Omega^2} \right). \tag{8.178}$$

From Eqs. (8.177) and (8.178) it follows that the DC σ_{xx} conductivity is zero, i.e. the wire can conduct and absorb light in the x direction only at the finite frequency $\omega = \Omega$, whereas the DC σ_{yx} conductivity is given by

$$\text{Re}\,[\sigma_{yx}] = \frac{e^2 \rho}{m} \frac{\omega_c}{\sqrt{\omega_0^2 + \omega_c^2}}, \tag{8.179}$$

and goes to zero when $\omega_c \to 0$.

For the quantum wire, the calculation of the σ_{yy} conductivity is also important. In the presence of an electric field in the y direction (along the wire), giving rise to the interaction Hamiltonian $H' = -\frac{E j_y}{i\omega} e^{-i\omega t}$, the yy conductivity is given by

$$\sigma_{yy}(0, \omega) = \frac{\langle j_y \rangle}{E} = \frac{i}{\omega} \frac{\sum_n 2\omega_{n0} |\langle n | j_y | 0 \rangle|^2}{\omega^2 - \omega_{n0}^2}, \tag{8.180}$$

so that its real part is given by

$$\text{Re}\,[\sigma_{yy}(0, \omega)] = \frac{\pi e^2}{\omega} \sum_n \left| \langle n | \sum_{i=1}^{N} v_y^i | 0 \rangle \right|^2 \delta(\omega - \omega_{n0}). \tag{8.181}$$

Differently from the xx and yx conductivities which get contributions from the interband excitations at energy Ω and are consequences of the generalized Kohn theorem, the yy conductivity gets contributions from the intraband excitations which exist in the wire due to the confinement in the x direction which breaks the translational invariance of the system and curves the single-particle energy levels of Fig. 8.6. (Note that by putting $\omega_0 = 0$ this curvature disappears and one remains with the Landau levels for which intraband excitations do not exist.) The calculation of σ_{yy} yields the Landauer–Buttiker type formulas which have been discussed in Subsection 8.8.

8.10   The Larmor Theorem

Starting from Eq. (8.153) (with or without confinement potential), it is also easy to verify that the following result is found:

$$[H, S_\mp] = \pm\omega_L S_\mp, \qquad (8.182)$$

where $S_\mp = S_x \mp iS_y$ and $\omega_L = |g\mu_0 B|$ is the Larmor frequency. Equation (8.182), together with the equation of motion (8.152), proves that for a system with spin-polarized ground state ($\langle 0|S_z|0\rangle = (N_\uparrow - N_\downarrow)/2$ for negative g or g^*, as in the case of electrons in GaAs semiconductors), the state $S_-|0\rangle$ is an exact eigenstate of H with energy $E_0 + \omega_L$. This eigenstate is known as the Larmor state of the system, and its frequency cannot be directly affected by any effect caused by interparticle interaction. Equation (8.182), together with the result $[H, S_z] = 0$, allows us to write

$$\frac{d\mathbf{S}}{dt} = i[H, \mathbf{S}] = |g|\mu_0 \mathbf{S} \wedge \mathbf{B}. \qquad (8.183)$$

This result is known in the literature as the Larmor theorem, and is the equation for a precessing spin. In the exact theory, precession takes place at the angular frequency $\omega = \omega_L$.

8.10.1   *Electron Spin Precession with Rashba Spin–Orbit Coupling*

Spin precession of electrons can be induced in semiconductors not only by the Zeeman term in the Hamiltonian but also by SO coupling terms. Particularly important is the Rashba SO term, since its intensity can be controlled by applying a vertical electric field to the heterostructure, as proved experimentally by Nitta *et al.* (1997). For the Rashba Hamiltonian

$$H_R = \sum_{i=1}^{N} \left[\frac{p_x^2 + p_y^2}{2m} + \lambda_R(p_y\sigma_x - p_x\sigma_y) \right]_i = \sum_{i=1}^{N} h_i, \qquad (8.184)$$

where λ_R is a strength parameter, one gets

$$\frac{d\mathbf{S}}{dt} = i[H_R, \mathbf{S}] = 2\lambda_R \mathbf{S} \wedge \boldsymbol{\Omega}(\mathbf{p}), \qquad (8.185)$$

where

$$\boldsymbol{\Omega}(\mathbf{p}) = \hat{z} \wedge \mathbf{p}. \qquad (8.186)$$

Therefore, the evolution under H_R leads again to a spin precession of frequency ω_P given by the energy difference between the corresponding "up" and "down" eigenstates of h_R [see Eq. (1.125)]. Namely, $\omega_P = 2\lambda_R|\mathbf{p}|$. In Eq. (8.185), $\boldsymbol{\Omega}$ is the precession vector analogous to $\mathbf{B}$ in the Larmor precession, but with the difference that now the plane of spin precession depends on the spatial orientation of $\mathbf{p}$. Note also that the precession axis of $\boldsymbol{\Omega}$ always lies in the plane (x, y) of the heterostructure. This precessional motion is at the basis of the spin field effect transistor proposed by Datta and Das (1990). The device is based on spin injecton and spin detection by a ferromagnetic source and drain, and on spin precession about the built-in field $\boldsymbol{\Omega}$ in the quasi-one-dimensional channel of an ordinary field effect transistor. Since the precession axis of $\boldsymbol{\Omega}$ always lies in the channel plane (x, y), the results are insensitive to the relative orientation of $\hat{z}$ and the principal crystal axes. From Eq. (8.185), one gets for the evolution of the expectation value for a spin perpendicular to the plane, s_z, and a spin parallel to the in-plane $\mathbf{p}$, $s_{\parallel} = \mathbf{s} \cdot \mathbf{p}/p$,

$$\frac{ds_z}{dt} = 2\lambda_R p s_{\parallel}, \qquad \frac{ds_{\parallel}}{dt} = -2\lambda_R p s_z . \qquad (8.187)$$

The average spin component along $\boldsymbol{\Omega}$, $s_{\perp} = \mathbf{s} \cdot (\mathbf{p} \wedge \hat{z})/p$ is constant. As a result, $s_{\parallel} = s_{\parallel}(0) \cos(\omega_P t)$, and the injected spin at the source is labeled with (0). If ϕ is the angle between $\mathbf{p}$ and the source-drain axis, the electron will reach the drain at time $t' = Lm/(p \cos \phi)$, with the spin $s_{\parallel}$ precessing at the angle $\Phi = 2\lambda_R m L$, where L is the source-drain separation. The average spin at the drain in the direction of magnetization is $s_{\parallel}(t') \cos \phi + s_{\perp}(0) \mathbf{m} \cdot (\mathbf{p} \wedge \hat{z})$, so the current is modulated by $1 - \cos^2 \phi \sin^2(\Phi/2)$, the probability of finding the spin in the direction of magnetization $\mathbf{m}$. Note that Φ does not depend on the momentum of the carriers. As the spread ϕ in the momenta increases, the modulation effect decreases. The largest effect is seen for $\phi = 0$, where the current modulation factor is $\cos^2(\Phi/2)$. It was therefore proposed that ϕ be limited by further confining the electron motion along $\phi = 0$ using a one-dimenssional channel as a wave guide. Spin modulation of the current becomes ineffective if transport is diffusive. Taking typical values for $\lambda_R \simeq 1 \times 10^{-11}$ eV m, and $m = 0.1 m_e$, current modulation should be observable at source-drain separations of $L \geq 1 \, \mu$m, setting the scale for ballistic transport.

Other proposals of spin-based devices relying on the spin–orbit coupling mechanism are spin filters (Koga *et al.*, 2002) and spin guides (Valín-Rodríguez, Puente and Serra, 2003).

8.11 Deviations from Kohn and Larmor Theorems Due to Spin–Orbit Coupling

In this Subsection we show that the SO interaction violates the Kohn and Larmor theorems and therefore strongly affects the optical and magnetic properties of semiconductor quantum wells by inducing a strong coupling between charge density, spin density and spin-flip excitations in the long wavelength limit. We show that the energy splitting of the cyclotron resonance and the dispersion relation of the Larmor resonance are a clear and quantitative signature of SO coupling in these systems.

8.11.1 *Spin–Orbit Splitting of Cyclotron Resonance*

The influence of impurity and band structure effects on the cyclotron resonance (CR) is an important topic that has been investigated in many experiments on space charge layers (Küblbeck and Kotthaus, 1975; Ensslin *et al.*, 1987; Batke *et al.*, 1988; Summers *et al.*, 1993; Besson *et al.*, 1992; Hu *et al.*, 1995). These effects break the translational invariance of the system and as a consequence invalidate the Kohn theorem, according to which in the CR experiments a single line at the cyclotron energy $\omega_c = eB/mc$ should be observed. Indeed, in a recent CR investigation of high mobility electron space charge layers in GaAs (Manger *et al.*, 2001), a line splitting of the CR resonance due to band structure influences was clearly observed. The main features of the experiment are: a well-resolved splitting of CR for filling factors $\nu = 3$ and 5 $\left(\nu = 2\pi \ell^2 \rho, \ell = \sqrt{\frac{\hbar}{eB}}\right)$ which increases with the electron carrier density ρ; a similar behavior, but less pronounced for $\nu = 7$; a gain in strength of the line with lower transition frequency with increasing ρ; and finally no significant splitting for the even filling factors.

In the following we show that the SO interaction is also the main cause of the observed CR splitting. Moreover, assuming the presence of an additional small, but notnegligible, nonlocal electron–electron interaction, it can explain all the features observed in the experiment by Manger *et al.* For simplicity we develop the calculation only for the Dresselhaus SO interaction; inclusion of the Rashba term is straightforward.

Adding the Dresselhaus term

$$H_D = \frac{\lambda}{2} \sum_{i=1}^{N} [P_+\sigma_+ + P_-\sigma_-]_i \tag{8.188}$$

to the Hamiltonian $H = H_0 + V$ of Eq. (8.153), and using the basic commutation rule $[P^-, P^+] = 2\omega_c$, one gets

$$\left[H_0 + V + H_D, \sum_{i=1}^{N} P_i^+\right] = \omega_c \sum_{i=1}^{N} P_i^+ + \omega_c\lambda \sum_{i=1}^{N} \sigma_-^i \tag{8.189}$$

and violates the Kohn theorem for which $[H_0 + V, \sum_{i=1}^{N} P_i^+] = \omega_c \sum_{i=1}^{N} P_i^+$. From Eq. (8.189), one sees that the SO coupling mixes density excitations induced by the operator $\sum_{i=1}^{N} P_i^+$ with the spin-flip excitations induced by $\sum_{i=1}^{N} \sigma_-^i$. Next, the spin-flip operator $\sum_{i=1}^{N} \sigma_-^i$, at the order λ^2, mixes the density excitations with the spin-density ones induced by the operator $\sum_{i=1}^{N} P_i^+ \sigma_z^i$, since one has

$$\left[H_0 + V + H_D, \sum_{i=1}^{N} \sigma_-^i \right] = \omega_L \sum_{i=1}^{N} \sigma_-^i + 2\lambda \sum_{i=1}^{N} P_i^+ \sigma_z^i . \qquad (8.190)$$

One can then try to solve the equations of motion

$$[H_0 + V + H_D, O^\dagger]| = \omega O^\dagger, \qquad (8.191)$$

at the order λ^2, with the creation operator $O^\dagger = a \sum_{i=1}^{N} P_i^+ + b \sum_{i=1}^{N} \sigma_-^i + c \sum_{i=1}^{N} P_i^+ \sigma_z^i$, which mixes density, spin-flip and spin-density excitations. To do this it is necessary to consider explicitly the effects of the electron–electron interaction V, since differently from the operators $\sum_{i=1}^{N} P_i^+$ and $\sum_{i=1}^{N} \sigma_-^i$, the spin density operator $\sum_{i=1}^{N} P_i^+ \sigma_z^i$ does not commute with V. These effects will be treated in the following in the Brueckner–Hartree–Fock (BHF) approximation of Chapter 3. The reason to use BHF is that the usual theories, like RPA, time-dependent Hartree–Fock and time-dependent local density approximations (see next chapter), do not give any effect on the splitting of the CR resonance, whereas non local effective theories, like the Landau theory and BHF, do.

In BHF, the relevant nonlocal term in the energy functional for studying the cyclotron resonance is given by [see also Eq. (3.90)]

$$\int v_0(\rho)(\rho\tau - \mathbf{j}^2)d\mathbf{r} - \int v_1(\rho)\mathbf{j}_1^2 d\mathbf{r}, \qquad (8.192)$$

where ρ and τ are the one-body diagonal and kinetic energy densities, respectively, and the current densities $\mathbf{j}$ and $\mathbf{j}_1$ are given by

$$\mathbf{j} = \langle \Psi | \frac{1}{2} \sum_{i=1}^{N} (\mathbf{P}_i \delta(\mathbf{r} - \mathbf{r}_i) + \text{H.c.}) | \Psi \rangle$$

and

$$\mathbf{j}_1 = \langle \Psi | \frac{1}{2} \sum_{i=1}^{N} (\mathbf{P}_i \delta(\mathbf{r} - \mathbf{r}_i) \sigma_z^i + \text{H.c.}) | \Psi \rangle ,$$

respectively. By taking in Eq. (8.192) $v_0(\rho) = k_0/2\rho$ and $v_1(\rho) = k_1/2\rho$, one gets a BHF potential:

$$V = \sum_{i=1}^{N} \left[k_0 \frac{P_i^2}{2} - k_0 \mathbf{j} \cdot \mathbf{P}_i - k_1 \mathbf{j}_1 \cdot \mathbf{P}_i \sigma_z^i \right] . \qquad (8.193)$$

The first term of this equation, added to the kinetic energy term of Eq. (8.153), gives rise to a constant effective mass (coming from the electron–electron interaction),

$1/m^*_{ee} = 1 + k_0$. A self-consistent vibrating dipole–dipole interaction $\delta V(\mathbf{r}, t)$ (see also Subsection 11.2.6), to be used in time-dependent BHF calculations, is derived from the last two terms of Eq. (8.193) by imposing irrotational currents $\delta\mathbf{j} = \beta(t)\rho\nabla f$, where $f = x, y$, on both $\mathbf{j}$ and $\mathbf{j}_1$ (Lipparini and Stringari, 1981; Stringari and Dalfovo, 1990; Serra *et al.*, 1999). The irrotational nature of currents in collective motion follows naturally from the assumption that the collective state exhausts the excitation strength, as is the case for the CR resonance and the spin density mode in the long wavelength limit. By using self-consistency to determine $\beta(t)$, one gets for $\delta V(\mathbf{r}, t)$

$$\delta V(\mathbf{r}, t) = \sum_{i=1}^{N} \left[-\frac{k_0}{2N} \left\langle \sum_{j=1}^{N} P_j^- \right\rangle P^+ \right.$$
$$\left. -\frac{k_1}{2N} \left\langle \sum_{j=1}^{N} \left(P_j^- \sigma_z^j - \frac{2S_z}{N} P_j^- \right) \right\rangle \left(P^+ \sigma_z - \frac{2S_z}{N} P^+ \right) \right]_i + \text{H.c.},$$

(8.194)

where $2S_z = N_\uparrow - N_\downarrow$ and $N_\uparrow$ ($N_\downarrow$) is the number of spin-up electrons (down). Note that the time dependence of $\delta V(\mathbf{r}, t)$ is in the $\langle \ldots \rangle$ spatial foldings with the densities induced by a time-dependent external field.

The total Hamiltonian $H = H_0^* + H_D + \delta V(\mathbf{r}, t)$, where H_0^* includes the effective mass $1/m^*_{ee} = 1 + k_0$ in the kinetic energy term of Eq. (8.153), can now be solved analytically within the RPA by finding the operators O^+ solution to the equation of motion (8.191). We have used the methods illustrated in Subsection 9.10 to compute the commutators of a one-body operator F with the Hamiltonian H as

$$[H, F] = [H_0^* + H_D, F] + \delta V(F),$$
(8.195)

where $H_0^* + H_D$ is the static Hamiltonian with the effective mass $1/m^*_{ee} = 1 + k_0$, and $\delta V(F)$ is the change (linear in F) induced in the time-dependent potential by the unitary transformation e^{iF}. For the potential of Eq. (8.194) one gets

$$\delta V(F) = -\frac{k_0}{2N} \langle 0| \left[\sum_{j=1}^{N} P_j^-, F \right] |0\rangle \sum_{i=1}^{N} P_i^+$$
$$-\frac{k_1}{2N} \langle 0| \left[\sum_{j=1}^{N} P_j^- \sigma_z^j - \frac{2S_z}{N} P_j^-, F \right] |0\rangle \sum_{i=1}^{N} \left(P_i^+ \sigma_z^i - \frac{2S_z}{N} P_i^+ \right),$$
(8.196)

where $|0\rangle$ is the static BHF ground state and analogously for the Hermitian conjugate term of Eq. (8.194). The two terms of Eq. (8.195) have a different physical meaning: the commutator $[H_0^* + H_D, F]$ originates from the static, one-body properties of the Hamiltonian, while the term $\delta V(F)$ originates from the renormalization of the self-consistent potential. The latter contribution is essential for taking into account the RPA correlations.

By using Eqs. (8.195) and (8.196) it is now easy to get the result

$$\left[H_0 + V, \sum_{i=1}^{N} P_i^+ \sigma_i^z\right] = \omega_c(1+k) \sum_{i=1}^{N} P_i^+ \sigma_i^z - 2\frac{S_z}{N}\omega_c k \sum_{i=1}^{N} P_i^+ , \qquad (8.197)$$

where $k = k_0 - k_1(1 - (2S_z/N)^2)$, which allows us to get the solution to the equation of motion (8.191), at the order λ^2, with the ansatz $O^\dagger = \sum_{i=1}^{N}[aP^+ + b\sigma_- + cP^+\sigma_z]_i$.

From Eq. (8.197), one sees that the electron–electron interaction mixes spin-density excitations induced by the operator $\sum_{i=1}^{N} P_i^+ \sigma_i^z$ with the density excitations induced by $\sum_{i=1}^{N} P_i^+$. It is also important to note that the above theory is equivalent to the Landau theory for the two-dimensional electron gas in the long wave-length limit (see Subsection 9.1), if one identifies the strengths k_0 and k_1 with the combinations of Landau parameters $-F_1^s/(2 + F_1^s)$ and $-F_1^a/(2 + F_1^s)$. This allows us to give an estimate of the strength k by using for F_1^s and F_1^a the available MC calculation (Kwon, Ceperley and Martin, 1994). This calculations shows that k is negative, strongly density-dependent (it decreases with increasing density ρ) and equal to $\simeq -2 \times 10^{-2}$ at $\rho = 3.32 \times 10^{-11}$ cm^{-2}, which is the highest value of the density reported. The CR experiment (Manger *et al.*, 2001) we are going to analyze covers the density regime from 2 to 13×10^{-11}cm^{-2}, the energy ω_c lies in some range around 100 cm^{-1}, yielding the estimate $-k\omega_c \simeq 2$ cm^{-1}, and the observed splitting is in the range $\simeq 1-4$ cm^{-1}. $k\omega_c$ is a key quantity of the model which scales as m^*/ϵ^2. Under the same conditions of density and magnetic field, one can then vary this quantity by changing the material.

By using Eqs. (8.189), (8.190) and (8.197) it is now possible to solve the equations of motion (8.191) with the operator $O^+ = \sum_{i=1}^{N}[aP^+ + b\sigma_- + cP^+\sigma_z]_i$ yielding a homogeneous system of linear equations for the coefficients a, b and c, from which the energies ω_ρ, ω_σ and ω_P of the three coupled modes are obtained by solving the secular equation (valid at the order λ^2):

$$(\omega - \omega_c)(\omega - \omega_c(1+k))(\omega - \omega_L) = -\frac{4S_z}{N}\lambda^2 k\omega_c^2. \qquad (8.198)$$

For each energy solution, the homogeneous linear system, supplemented with the normalization condition $\langle 0|[(O^+)^\dagger, O^+]|0\rangle = 1$, gives the coefficients a, b and c.

In the cases of $\lambda = 0$ the three modes are uncoupled and the solutions to Eq. (8.191) are given by

$$O_\rho^+ = \sqrt{\frac{1}{2N\omega_c}} \sum_{i=1}^{N} P_i^+ ,$$

$$O_\sigma^+ = \sqrt{\frac{1}{2N\omega_c(1 - (2S_z/N)^2)}} \sum_{i=1}^{N} \left(P_i^+ \sigma_z^i - \frac{2S_z}{N} P_i^+\right), \qquad (8.199)$$

$$O_P^+ = \sqrt{\frac{1}{8S_z}} \sum_{i=1}^{N} \sigma_-^i ,$$

and

$$\omega_\rho = \omega_c , \quad \omega_\sigma = \omega_c(1+k), \quad \omega_P = \omega_L, \qquad (8.200)$$

where the subscripts ρ, σ and P refer to density, spin-density and spin-flip excitations in the long wavelength limit, respectively. The dipole strength is distributed among the above states as follows ($Q_\rho = \sum_{i=1}^{N}(x+iy_i)$):

$$|\langle 0|Q_\rho|\omega_\rho\rangle|^2 = \frac{2\rho}{\omega_c}, \quad |\langle 0|Q_\rho|\omega_\sigma\rangle|^2 = 0, \quad |\langle 0| \quad Q_\rho|\omega_P\rangle|^2 = 0, \qquad (8.201)$$

so that the Kohn and Larmor theorems are fulfilled, and according to them the spin-density and spin-flip modes are not excited by the density operator Q_ρ, and the corresponding matrix elements vanish.

When λ, k and S_z are different from zero, dipole absorption gives rise to three excitation modes at the energies

$$\omega_\rho = \omega_c\left[1 + \frac{1}{2}\left(k - \sqrt{k^2 - 16\frac{S_z}{N}k\lambda^2\frac{1}{\omega_c - \omega_L}}\right)\right]$$

$$\simeq \omega_c + \lambda^2\frac{4S_z}{N}\frac{\omega_c}{\omega_c - \omega_L} \;,$$

$$\omega_\sigma = \omega_c\left[1 + \frac{1}{2}\left(k + \sqrt{k^2 - 16\frac{S_z}{N}k\lambda^2\frac{1}{\omega_c - \omega_L}}\right)\right]$$

$$\simeq \omega_c(1+k) - \lambda^2\frac{4S_z}{N}\frac{\omega_c}{\omega_c - \omega_L} \;,$$

$$\omega_P = \omega_L - k\lambda^2\frac{4S_z}{N}\frac{\omega_c^2}{(\omega_c - \omega_L)(\omega_c(1+k) - \omega_L)}, \qquad (8.202)$$

which are created by the operators

$$O_\rho^+ = a\sum_{i=1}^{N}\left[P^+ + \frac{\lambda\omega_c}{\omega_c - \omega_L}\sigma_- - \frac{\omega_\rho - \omega_c}{2(S_z/N)k\omega_c}P^+\sigma_z\right]_i \;,$$

$$O_\sigma^+ = c\sum_{i=1}^{N}\left[P^+\sigma_z - 2S_z/N\frac{k\omega_c}{\omega_\sigma - \omega_c}P^+ - 2S_z/N\lambda\frac{\omega_c}{\omega_c - \omega_L}\sigma_-\right]_i \;,$$

$$O_P^+ = b\sum_{i=1}^{N}\left[\sigma_- - \frac{4k\omega_c\lambda S_z/N}{(\omega_c - \omega_L)(\omega_c(1+k) - \omega_L)}P^+\right.$$

$$\left. - \frac{2\lambda}{\omega_c(1+k) - \omega_L}P^+\sigma_z\right]_i . \qquad (8.203)$$

All these modes get dipole strengths as measured in far-infrared experiments. These are given by $|\langle 0|Q_\rho|\omega_\rho\rangle|^2 = |\langle 0|[Q_\rho, O_\rho^+]|0\rangle|^2$, $|\langle 0|Q_\rho|\omega_\sigma\rangle|^2 = |\langle 0|[Q_\rho, O_\sigma^+]|0\rangle|^2$ and $|\langle 0|Q_\rho|\omega_P\rangle|^2 = |\langle 0|[Q_\rho, O_P^+]|0\rangle|^2$ and are all different from zero. The above results are able to explain all the features observed in the CR experiment of Manger *et al.* (2001) and in particular the measured energy splitting ΔE of the cyclotron resonance that has been interpreted (Tonello and Lipparini, 2004) as being due to the SO and interaction couplings between the density and spin-density modes. From Eq. (8.202) one gets for this energy splitting the result

$$\Delta E = \left| -\omega_c k + \frac{8S_z}{N} \lambda^2 \frac{\omega_c}{\omega_c - \omega_L} \right| . \tag{8.204}$$

In fact, the estimates for $-k\omega_c$ and $2\lambda^2$ which we have done before give for ΔE the right order of magnitude of the observed splitting. Moreover, at fixed density, both $\frac{8S_z}{N}\lambda^2$ and $k\omega_c$ decrease for increasing filling factors since $\frac{2S_z}{N} = \frac{1}{\nu}$ and also ω_c goes as $\frac{1}{\nu}$, explaining why the splitting is much better experimentally resolved at $\nu = 3$ and 5, than at for $\nu = 7$. Finally, the strength $|\langle 0|Q_\rho|\omega_\sigma\rangle|^2$ vanishes [see Tonello and Lipparini (2004)] when $\frac{2S_z}{N} = 1$, explaining why at filling factor $\nu = 1$ no splitting is observed. It is also interesting to compare the result (8.204) for CR splitting with that of the nonparabolicity model for the GaAs conduction band. This single-particle model predicts a splitting proportional to B^2 and dipole strengths which do not reproduce the experimental results. In the above theory the interaction enters in a natural and crucial way for reproducing energy splittings and strengths. In particular, the splitting is linear in B.

One should also note that when the Rashba term is added to the previous theory one gets for the energy splitting the result of Eq. (6.128), which generalizes that of Eq. (8.204).

Finally, we stress that recently (Syed *et al.*, 2003; Henriksen *et al.*, 2006) a large splitting [up to 2 meV, about an order of magnitude larger than the one measured by Manger *et al.*, and interpreted with Eq. (8.204)] has been measured in AlGaN/GaN and AlGaAs/GaAs heterostructures by means of far-infrared transmission experiments. Though the origin of this large splitting remains uncertain, Eqs. (8.202) and (8.203) predict that three lines can be excited in far-infrared trasmission experiments on quantum wells, and therefore that besides that of Eq. (8.204) another energy splitting of the CR should be observed at the energy $\omega_\rho - \omega_P$, which is much large than the previous one.

8.11.2 *Spin Splitting of Landau Levels*

Starting from the same Hamiltonian, $H = H_0 + V + H_D$ of the previous subsection, and using the basic commutation rule $[\sigma_+, \sigma_-] = 4\sigma_z$ and $[P^-, P^+] = 2\omega_c$, one gets

$$\left[H_0 + V + H_D, \sum_{i=1}^{N} \sigma^i_- \right] = \sum_{i=1}^{N} [\omega_L \sigma_- + 2\lambda P^+ \sigma_z]_i \tag{8.205}$$

and violates the Larmor theorem for which $[H_0 + V, \sum_{i=1}^{N} \sigma^i_-] = \omega_L \sum_{i=1}^{N} \sigma^i_-$. From Eq. (8.205), one sees that the SO coupling mixes spin-flip excitations induced by the operator $\sum_{i=1}^{N} \sigma^i_-$ with the spin-density excitations induced by $\sum_{i=1}^{N} P^+ \sigma^i_z$. Next, the spin-density operator $\sum_{i=1}^{N} P^+ \sigma^i_z$, at the order λ^2, again mixes itself with the spin-flip operator due to the entrance in the game of the monopole operator $\sum_{i=1}^{N} t_i \sigma_-^i$, $t = (P^- P^+ + P^+ P^-)$:

$$\left[H_0 + V + H_D, \sum_{i=1}^{N} P_i^+ \sigma_i^z \right] = \sum_{i=1}^{N} \left[\omega_c (1+k) P^+ \sigma^z + \frac{1}{2} \lambda_D t \sigma_- \right]_i$$

$$= \sum_{i=1}^{N} \left[\omega_c (1+k) P^+ \sigma^z + \lambda_D (2n+1) \omega_c \sigma_- + \frac{1}{2} \lambda_D (t - (t)) \sigma_- \right]_i, \qquad (8.206)$$

where n is the Landau level index and we have used $(t)/2 = (2n+1)\omega_c$ and neglected some terms which are irrelevant to the final results. One can then try to solve the equations of motion

$$[H_0 + V + H_D, O^\dagger]| = \omega O^\dagger,$$

at the order λ^2, with the creation operator $O^\dagger = \sum_{i=1}^{N} [a \sigma_-^i + b P^+ \sigma_z]_i$, which mixes spin-flip and spin-density excitations. The SO mixing with the spin-density excitation renormalizes the energy of the spin-flip transition which is equal to the spin splitting of the Landau levels, and is now given by

$$\omega_{L,\lambda} = \frac{1}{2} \left[\omega_c (1+k) + \omega_L - \sqrt{(\omega_c (1+k) - \omega_L)^2 + 8\lambda^2 (2n+1)\omega_c} \right]$$

$$\simeq \omega_L - 2(2n+1)\lambda^2 \frac{\omega_c}{\omega_c - \omega_L}. \qquad (8.207)$$

When the Rashba term is taken into account, one gets for the energy of the quasi-Larmor state the result

$$\omega_{L,\lambda} = \omega_L + 2(2n+1) \left[\lambda_R^2 \frac{\omega_c}{\omega_c + \omega_L} - \lambda_D^2 \frac{\omega_c}{\omega_c - \omega_L} \right],$$

which coincides with that of Eq. (6.129). For a comparison with experiments, refer to Subsection 6.8.

8.12 Spin-Hall Conductivity

The spin-Hall effect predicted by Murakami *et al.* (2003) and Sinova *et al.* (2004) has produced an intense theoretical activity (Shen *et al.*, 2004; Adagideli and Bauer, 2005; Rashba, 2004; Raimondi and Schwab, 2005; Chang, 2005; Dimitrova, 2005; Erlingsson and Loss, 2005; Bernevig and Zhang, 2006; Bellucci and Onorato, 2006). This spin-Hall effect is called "intrinsic" — as it arises from the band structure, even in the absence of scattering — to distinguish it from that caused by asymetries in scattering for up and down spins, which is called the extrinsic spin-Hall effect (D'yakanov and Perel, 1971; Hirsch, 1999; Zhang, 2000; Inoue, Bauer and Molenkamp, 2004; Mishchenko, Shytov and Halperin, 2004; Chalaev and Loss, 2005; Tse and Das Sarma, 2006). The intrinsic spin-Hall effect has been experimentaly observed in n-doped semiconductors (Kato *et al.*, 2004; Sih *et al.*, 2005), and the extrinsic one in 2D hole gases (Wunderlich *et al.*, 2005).

Spin-Hall effects, as considered in these references, result from SO coupling in which an electrical current flowing through a sample may lead to spin transport in a perpendicular direction and spin accumulation at lateral boundaries, and does not require an applied magnetic field (B) (Sinova *et al.*, 2006; Engel, Rashba and Halperin, 2006). The starting points for the description of this spin-Hall effect are the Fermi gas model in two dimensions with Rashba interaction, illustrated in Subsection 1.8.3, and the following expression for the spin conductivity Σ_{zyx}:

$$\Sigma_{zyx}(\omega) = \frac{\mathcal{J}_{zy}}{E_x} = -2\frac{ie}{S}\sum_n \frac{\langle 0|\sum_{i=1}^N v_x^i|n\rangle\langle n|\sum_{j=1}^N \mathcal{J}_{zy}^j|0\rangle}{\omega^2 - \omega_{n0}^2}, \qquad (8.208)$$

obtained in linear response theory as a current response to an electric field in the x direction (see Subsections 8.2 and 8.7). In Eq. (8.208), S is the area of the quantum well, the spin current $\mathcal{J}_{zy}^j$ is defined as

$$\mathcal{J}_{zy}^j = \frac{1}{2}(v_y\sigma_z + \sigma_z v_y)_j, \qquad (8.209)$$

with $v_k^j = -i[r_k^j, H]$, and $|0\rangle$ and $|n\rangle$ are the exact gs and excited states of the model Hamiltonian H and $\omega_{n0} = E_n - E_0$ are the corresponding excitation energies.

Sinova *et al.* (2004) have calculated the dc ($\omega = 0$) spin conductivity with the Rashba Hamiltonian of Subsection 1.8.3:

$$H_0^R = \sum_{i=1}^N \left[\frac{p_x^2 + p_y^2}{2m} + \lambda_R(p_y\sigma_x - p_x\sigma_y)\right]_i = \sum_{i=1}^N h_i,$$

which can be solved analytically, yielding the eigenenergies and eigenstates of Eq. (1.125). The corresponding energy spectrum is illustrated in Fig. 1.7 and shows two branches of excitation, ϵ_- and ϵ_+. These authors considered only interbranch transitions [from the lower (ϵ_-) to the upper (ϵ_+) branch] and got the result

$$\Sigma_{zyx}^{inter}(0) = 2\frac{ie}{S}\frac{S}{(2\pi)^2}$$

$$\times \int_{k_+^F}^{k_-^F} k\,dk\,d\psi \frac{\int d\mathbf{r}\varphi_-^*(p_x/m - \lambda_R\sigma_y)\varphi_+ + \int d\mathbf{r}\varphi_-(p_y/m\sigma_z)\varphi_+^*}{4\lambda_R^2 k^2}.$$

$$(8.210)$$

By using the results $p_x\varphi_\pm = k\cos\psi\varphi_\pm$, $p_y\varphi_\pm = k\sin\psi\varphi_\pm$, $\varphi_-^*\sigma_z\varphi_+ = 1$, $\varphi_-^*\sigma_y\varphi_+ = i\sin\psi$, one then gets

$$\Sigma_{zyx}^{inter}(0) = \frac{e}{8m\pi\lambda_R}(k_+^F - k_-^F) = -\frac{e}{4\pi}, \qquad (8.211)$$

independent of λ_R! This universal result for the spin conductivity produced great enthusiasm and a big amount of papers. However, it was later pointed out by many authors that the result was incorrect. In fact, if one includes in the calculation the contribution of intrabranch transitions, one sees that this contribution exactly cancels out the previous one, and one then gets as the final result $\Sigma_{zyx}(0) = \Sigma_{zyx}^{\text{inter}}(0) + \Sigma_{zyx}^{\text{intra}}(0) = 0$. $\Sigma_{zyx}^{\text{intra}}(0)$ can be readily calculated by using in Eq. (8.208) $v_x = -i[x, H]$ and $\mathcal{J}_{zy}^j = (-i/2m\lambda_R)[H, \sigma_y]$ to write the general formula

$$\Sigma_{zyx}(0) = \frac{ie}{m\lambda_R S} \sum_n \langle 0| \sum_{i=1}^N x_i |n\rangle \langle n| \sum_{j=1}^N \sigma_y^j |0\rangle \ . \tag{8.212}$$

Then, working in the $q \to 0$ limit, where one can assume that $\sum_i x_i = \sum_i e^{iqx_i}/iq$ (the identity operator does not excite the system), $(\varphi_\pm|e^{iqx}|\varphi_\pm) = 1$ and using $\sum_n = S/(2\pi)^2 \int k \ dk \ d\psi = S/(2\pi)^2 \int k^F q \cos\psi \ d\psi$, valid in the same limit, and $\varphi_\mp^* \sigma_y \varphi_\mp = \pm \cos\psi$, one easily gets

$$\Sigma_{zyx}^{\text{intra}}(0) = \frac{e}{m\lambda_R S} \frac{S}{(2\pi)^2} \left[\int_{-\pi/2}^{+\pi/2} k_-^F q \cos^2\psi \ d\psi \frac{1}{q} \right.$$

$$\left. - \int_{-\pi/2}^{+\pi/2} k_+^F q \cos^2\psi \ d\psi \frac{1}{q} \right]$$

$$= \frac{e}{8m\pi\lambda_R}(k_-^F - k_+^F) = -\Sigma_{zyx}^{\text{inter}}(0). \tag{8.213}$$

However, it was pointed out that there exists a finite and small range of frequency $1/\tau < \omega < 2\lambda_R k^F$, with $1/\tau$ giving the transport rate of a clean sample and $2\lambda_R k^F$ the SO splitting, where

$$\Sigma_{zyx}^{\text{inter}}(\omega) \simeq -\frac{e}{4\pi},$$

$$\Sigma_{zyx}^{\text{intra}}(\omega) \to 0 \ as \ \left(\frac{1}{\tau\omega}\right)^2, \tag{8.214}$$

and therefore $\Sigma_{zyx}(\omega) = -\frac{e}{4\pi}$, independent of the details of the impurity scattering.

8.12.1 *The Spin-Hall Effect at a Finite Magnetic Field*

We discuss here a different situation — to some extent more similar to the charge-Hall effect — that involves spin currents. We are going to show that when a two-dimensional interacting electron gas is submitted to a perpendicular B, owing to the existence of incompressible quantum Hall effect states and to the applicability of Kohn's theorem, an in-plane applied electric field E induces a spin current perpendicular to it, whose conductivity is quantized (Lipparini and Barranco, 2006).

8.12.2 *Spin-Hall Conductivity Without SO Coupling*

We consider a two-dimensional system made up of N electrons with charge $-e$, in a uniform magnetic field perpendicular to the plane where the particles move — the $(x - y)$ plane — with the usual Hamiltonian (in effective atomic units),

$$H = \frac{1}{2m} \sum_{i=1}^{N} \mathbf{P}_i^2 - \frac{1}{2} \omega_L \sigma_z + \sum_{i<j}^{N} V(|\mathbf{r}_i - \mathbf{r}_j|), \qquad (8.215)$$

where the $-$ sign in front of the Zeeman energy corresponds to negative g^* values, and has some relevance in the discussion of the resonant spin-Hall conductivity which we will carry out below. Our choice is motivated by the fact that for most materials of frequent study, like GaAs, InAs or InSb, $g^* < 0$, which implies that within a given Landau level, spin-up electrons have lower energies than spin-down electrons. The SO interaction may alter this situation.

Since the total momentum $\sum_{j=1}^{N} \mathbf{P}_j$ commutes with the interaction term, the Kohn theorem holds (see Subsection 8.9.1) and the normalized state

$$|\omega_c\rangle = \frac{1}{\sqrt{2N\omega_c}} \sum_{j=1}^{N} P_j^+ |0\rangle \qquad (8.216)$$

is an exact eigenstate of H with energy $E_0 + \omega_c$, where E_0 is the ground state (gs) energy. This eigenstate is the cyclotron resonance state of the system. By using this result, we now evaluate the real part of the spin conductivity given by Eqs. (8.208) and (8.209). Since the velocity operator $\sum_{i=1}^{N} v_x^i$ has the same quantum numbers as the cyclotron operator $\sum_{j=1}^{N} P_j^+$, as stated by the Kohn theorem, it can excite only the cyclotron state, Eq. (8.216). As a consequence, in Eq. (8.208) only $|\omega_c\rangle$ contributes to the sum over the excited states $|n\rangle$, and one obtains for the spin conductivity the simpler expression

$$\Sigma_{zyx}(\omega) = -2 \frac{ie}{S} \frac{\langle 0| \sum_{i=1}^{N} v_x^i |\omega_c\rangle \langle \omega_c| \sum_{j=1}^{N} \mathcal{J}_{zy}^j |0\rangle}{\omega^2 - \omega_c^2}. \qquad (8.217)$$

By using Eq. (8.216) together with $\sum_{j=1}^{N} P_j^- |0\rangle = 0$, it is straightforward to obtain the matrix elements

$$\langle 0| \sum_{i=1}^{N} v_x^i |\omega_c\rangle = (1/\sqrt{2N\omega_c}) \langle 0| \left[\sum_{i=1}^{N} v_x^i, \sum_{j=1}^{N} P_j^+ \right] |0\rangle,$$

$$\langle \omega_c| \sum_{j=1}^{N} \mathcal{J}_{zy}^j |0\rangle = (1/\sqrt{2N\omega_c}) \langle 0| \left[\sum_{i=1}^{N} P_i^-, \sum_{j=1}^{N} \mathcal{J}_{zy}^j \right] |0\rangle,$$

and from them the final expression for the spin conductivity,

$$\Sigma_{zyx}(\omega) = -\frac{e}{m} \omega_c \frac{\xi}{\omega^2 - \omega_c^2}, \qquad (8.218)$$

where $\xi = \rho_\uparrow - \rho_\downarrow$ is the spin magnetization of the system, and we have restored normal units. The result (8.218) is exact for the Hamiltonian (8.215). It shows that the spin conductivity increases resonantly at the cyclotron frequency.

If ν Landau levels are filled by electrons with density $\rho = \nu eB/(ch)$ and spin magnetization $\xi = \rho/\nu$ for ν odd ≥ 1, $\xi = 0$ for ν even, and $\xi = \rho$ for $\nu \leq 1$, one obtains for the dc spin conductivity

$$\Sigma_{zyx}(\omega = 0) = \begin{cases} e/h & \text{for } \nu \text{ odd} \geq 1 \\ 0 & \text{for } \nu \text{ even} \\ e\nu/h & \text{for } \nu \leq 1 \end{cases} . \tag{8.219}$$

Note that by using the same method, just changing the spin current $\mathcal{J}_{zy}^j$ [Eq. (8.209)] with the charge current $j_y^i = e v_y^i$, we have derived the corresponding result for the real part of the charge conductivity and obtained the dc charge-Hall conductivity, namely $\sigma_{yx}(\omega = 0) = e^2 \nu/h$ of Subsection 8.9.2. Both the intrinsic charge and spin quantum Hall effects are based on the validity of Kohn's theorem and on the existence of incompressible quantum Hall effect states, and do not result from any SO coupling; only the existence of a perpendicularly applied magnetic field plays a fundamental role.

Differently from the charge current, the spin current is not directly measurable. However, it can lead to spin accumulation near the sample boundaries that can be detected by means of Faraday or Kerr experiments (Kato *et al.*, 2004). This accumulation is in the transverse direction with respect to that of the applied electric field. Indeed, it can be easily checked that the longitudinal spin conductivity

$$\Sigma_{zxx}(\omega) = -\frac{ie}{\omega S} \sum_n \left(\frac{\langle 0| \sum_{i=1}^N v_x^i |n\rangle \langle n| \sum_{j=1}^N \mathcal{J}_{zx}^j |0\rangle}{\omega - \omega_{n0}} - \frac{\langle 0| \sum_{j=1}^N \mathcal{J}_{zx}^j |n\rangle \langle n| \sum_{i=1}^N v_x^i |0\rangle}{\omega + \omega_{n0}} \right) \tag{8.220}$$

is

$$\Sigma_{zxx}(\omega) = \xi \frac{e\pi}{2m} \delta(\omega - \omega_c), \tag{8.221}$$

thus yielding a zero dc longitudinal spin conductivity.

8.12.3 *Influence of the SO Coupling on the Spin-Hall Conductivity*

Addition of the Bychkov–Rashba SO interaction

$$H_R = \frac{\lambda_R}{\hbar} \sum_{j=1}^N [P_y \sigma_x - P_x \sigma_y]_j$$

to the Hamiltonian (8.215) introduces only minor changes in the spin- and charge-Hall conductivity, as we will discuss in the following. If we add the SO interaction to H, we have

$$\left[H + H_R, \sum_j P_j^+ \right] = \omega_c \sum_j \left(P^+ + i\lambda_R \sigma^+ \right)_j,$$ (8.222)

with $\sigma^+ = \sigma_x + i\sigma_y$, showing, as previously discussed, that the Kohn theorem is violated by the SO interaction. However, this violation is fairly small, since it can be shown (see also Subsection 8.11.1) that the equation of motion

$$[H + H_R, O^+] |0\rangle = \omega O^+ |0\rangle$$

is solved, to first order in λ_R, by the operator

$$O_\lambda^+ = \frac{1}{\sqrt{2N\omega_c}} \sum_j \left(P^+ + i\lambda_R \frac{\omega_c}{\omega_c + \omega_L} \sigma^+ \right)_j,$$ (8.223)

which, acting on the ground state, yields the quasicyclotron excited state $|\omega_c, \lambda\rangle = O_\lambda^+ |0\rangle$ at the excitation energy $\omega = \omega_c$. The crucial point here is that, as when the SO coupling is neglected, for $g^* < 0$ the velocity operator $\sum_{j=1}^N v_x^j$ can only excite the quasicyclotron state $|\omega_c, \lambda\rangle$. Interestingly, the quasicyclotron state can also be excited by the spin-flip operator $\sum_j \sigma_j^+$, owing to the σ^+ term introduced in the operator (8.223) by the SO coupling.

We can now proceed as before and obtain the dc spin Hall conductivity starting from Eq. (8.208). We obtain

$$\Sigma_{zyx}(\omega = 0) = 2\frac{e}{S} \frac{\langle 0| \sum_{j=1}^N x_j |\omega_c, \lambda\rangle \langle \omega_c, \lambda| \sum_{j=1}^N P_y \sigma_z^j |0\rangle}{\omega_c},$$ (8.224)

where we have used $\sum_{j=1}^N v_x^j = -i[\sum_{j=1}^N x_j, H + H_R]$ to eliminate one energy difference ω_{n0} in the denominator of Eq. (8.208), and have taken into account that only the quasicyclotron state enters the sum over the excited states. By using Eq. (8.223) is it straightforward to show that

$$\Sigma_{zyx}(\omega = 0) = \frac{e\xi}{m\omega_c} \left[1 + O(\lambda_R^2/\omega_c) \right],$$ (8.225)

which coincides with the result previously obtained without SO effects apart from corrections of order λ_R^2/ω_c. These corrections can be explicitly worked out using the techniques developed in Subsection 8.11.1. For any reasonable value of the λ_R parameter these corrections are very small as compared to 1.

However, if the Dresselhaus Hamiltonian,

$$H_D = \frac{\lambda_D}{2\hbar} \sum_{j=1}^N \left[P^+ \sigma^+ + P^- \sigma^- \right]_j,$$

is taken into account, either alone or together with the Rashba one, the conclusion may change. Indeed, H_D introduces a coupling of the quasicyclotron state to the quasi-Larmor state $|\omega_L, \lambda\rangle$ (see Subsections 6.2, 6.3 and 8.11.1) which, for the materials we are considering, is mainly excited by the σ^- spin-flip operator, $|\omega_L, \lambda\rangle \simeq \sum_j \sigma_j^- |0\rangle$. This state contributes to the n sum in Eq. (8.208) as it yields the term

$$2\frac{e}{S}\frac{\langle 0| \sum_{j=1}^{N} x_j |\omega_L, \lambda\rangle \langle \omega_L, \lambda| \sum_{j=1}^{N} P_y \sigma_z^j |0\rangle}{\omega_{L,\lambda}}, \tag{8.226}$$

which has to be added to Eq. (8.224). The matrix element is proportional to λ_D^2 and the energy denominator is the energy of the quasi-Larmor state [see Eq. (8.207) and the following one]:

$$\omega_{L,\lambda} = \omega_L + 2(2n+1)\lambda_R^2 \frac{\omega_c}{\omega_c + \omega_L} - 2(2n+1)\lambda_D^2 \frac{\omega_c}{\omega_c - \omega_L},$$

where n is the Landau level index of the last occupied level. This changes the dc spin-Hall conductivity Eq. (8.225) into

$$\Sigma_{zyx}(\omega = 0) = \frac{e\xi}{m\omega_c}\left[1 + O(\lambda_R^2/\omega_c) + O(\lambda_D^2/\omega_{L,\lambda})\right]. \tag{8.227}$$

The $\omega_{L,\lambda}$ energy denominator gives rise to a resonant behavior of the spin conductivity for materials with $g^* < 0$, if the strength of the Dresselhaus SO term is larger than the strength of the Rashba one, so that at some value of B, $\omega_{L,\lambda}$ is zero, which corresponds to a crossing of Landau levels with different spin values (see Fig. 6.1). Clearly, for this resonance to appear a fine-tuning of the sample-dependent SO coupling constants $\lambda_{R,D}$ has to be performed, which points toward a nonuniversal behavior. The resonant spin-Hall conductance has been discussed by Shen *et al.* (2004) in a somewhat infrequent situation, namely for materials with $g^* > 0$ and only taking into account the Rashba SO coupling.

It is worth seeing that, whereas they introduce small corrections to the dc charge-Hall conductivity, SO couplings *cannot* produce a resonant behavior of the dc charge-Hall conductivity. Indeed, by using $v_k^j = -i[r_k^j, H + H_{SO}]$, we can write

$$\sigma_{yx}(\omega = 0) = \frac{2ie}{S}\sum_n \langle 0| \sum_{j=1}^{N} x_j |n\rangle \langle n| \sum_{j=1}^{N} y_j |0\rangle, \tag{8.228}$$

showing that the energy denominators, responsible for any potential resonant behavior, no longer appear.

Finally, note that the dc spin conductivity can also be obtained starting from the observation made by Rashba and Dimitrova that, *when the Zeeman term in H is neglected*, the spin conductivity can be written as in Eq. (8.212) (Rashba, 2005). Since the dipole operator $\sum_{i=1}^{N} x^i$ can only excite the $|\omega_c, \lambda\rangle$ state, we have again $\Sigma_{zyx}(\omega = 0) = e\xi/m\omega_c$, showing that the spin-Hall conductivity Eq. (8.218) is a robust result. The Zeeman term is crucial for having a nonzero spin magnetization ξ, and hence a nonzero spin conductivity. Note, however, that when the Zeeman term is restored, one has to add to Eq. (8.212) a term containing an energy denominator which is responsible for the possible resonant behaviors discussed above.

8.13 Hall Conductivity in Graphene

The 2D graphene is a honeycomb lattice of carbon atoms with two sublattices. A unit cell contains two carbon atoms with one electron per π orbital (half-filled band). Its energy band can be calculated by the tight-binding model (Haldane, 1988; Zheng and Ando, 2002; Peres, Castro Neto and Guinea, 2006), and an intrinsic graphene is a semimetal with the Fermi energy located at the inequivalent K and K' points at opposite corners of its hexagonal Brillouin zone. In the effective mass approximation, the low energy phisics of a clean undoped graphene crystal is described by a four-band envelope function Hamiltonian,

$$h = v \left(P_x \tau_z \sigma_x + P_y \sigma_y \right), \tag{8.229}$$

where v is a band parameter, $\mathbf{P} = \mathbf{p} + e\mathbf{A}/c$, with $\mathbf{p}$ being the electron momentum operator, and $\mathbf{A}$ is vector potential given by $\mathbf{A} = (0, Bx, 0)$ in the Landau gauge. $\tau_z = \pm$ is a valley label that specifies one of the two inequivalent K and K' points near which low energy states occur, and σ_i are Pauli matrices representing a pseudospin degree of freedom corresponding to the two sites per primitive cell of a hexagonal lattice. Hamiltonian (8.229) defines four spin-degenerate gapless bands in which the pseudospin orientation lies in the (x, y) plane and winds around the z axis, either clockwise or counterclockwise, with a 2π planar wave vector rotation. $s_z = \pm$, the electron spin component perpendicular to the graphene plane, commutes with τ_z, and we assume that s_z and τ_z are good quantum numbers, allowing us to consider the cases $\tau_z, s_z = \pm 1$ independently.

The 2D graphene Hamiltonian in the spin-↑ K valley is

$$h = v \left(P_x \sigma_x + P_y \sigma_y \right). \tag{8.230}$$

In the absence of a magnetic field and taking into account that $[h, \mathbf{p}] = 0$, the solution to the Schrödinger equation $h\varphi = \epsilon\varphi$ is translationally invariant in the (x, y) plane and has the form

$$\varphi = \frac{1}{\sqrt{S}} e^{i\mathbf{k}\cdot\mathbf{r}} \begin{pmatrix} \alpha \\ \beta \end{pmatrix}, \tag{8.231}$$

where $\begin{pmatrix} \alpha \\ \beta \end{pmatrix}$ is a spinor to be determined. One gets

$$v \left[k_x \begin{pmatrix} \beta \\ \alpha \end{pmatrix} - i k_y \begin{pmatrix} \beta \\ -\alpha \end{pmatrix} \right] = \epsilon \begin{pmatrix} \alpha \\ \beta \end{pmatrix}, \tag{8.232}$$

from which the following solutions are recovered:

$$\epsilon_\pm = \pm v|k|,$$

$$\varphi_\pm = \frac{1}{\sqrt{2S}} e^{i\mathbf{k}\cdot\mathbf{r}} \begin{pmatrix} \pm 1 \\ e^{i\psi} \end{pmatrix}, \tag{8.233}$$

where $\psi = \arctan\frac{k_y}{k_x}$ and the $+$ solution applies to electrons in the conduction band and the $-$ one to the holes in the valence band.

In the presence of a magnetic field one gets

$$h = v\left(p_x\sigma_x + p_y\sigma_y + \omega_c x\sigma_y\right), \tag{8.234}$$

where $\omega_c = eB/c$ is the cyclotron frequency. Instead of solving the Schrödinger equation $h\varphi = \epsilon\varphi$, it is much more convenient to solve the one, $h^2\varphi = \epsilon^2\varphi$, for the square Hamiltonian. In fact a simple calculation yields

$$h^2 = 2v^2\left[\frac{p_x^2 + p_y^2}{2} + \omega_c x p_y + \frac{\omega_c^2}{2}x^2 + \frac{\omega_c}{2}\sigma_z\right], \tag{8.235}$$

which is quite similar to the single-particle Hamiltonian of a quantum well in a perpendicular magnetic field (see Subsections 5.7, 8.9.1 and 9.13.1). One then gets immediately

$$\epsilon_{n,\sigma}^2 = 2v^2\left((n+1/2)\omega_c + \frac{\omega_c}{2}\sigma\right),$$

$$\varphi_{nk\sigma} = \frac{1}{\sqrt{L}} e^{iky} \psi_n(x - \ell^2 k)\chi_\sigma, \tag{8.236}$$

where $\ell = \sqrt{1/\omega_c}$ is the magnetic length, $\psi_n(x) = (2^n n! \sqrt{\pi}\ell)^{-1/2} e^{-x^2/2\ell^2} H_n(x/\ell)$ and H_n are the Hermite polynomials. From the previous equation one then gets

$$\epsilon_{n,\sigma} = \pm\frac{v}{\ell}\sqrt{2n+1+\sigma}, \tag{8.237}$$

and the complete expression of eigenfunctions of h is

$$\varphi_{nX} = \frac{C_n}{\sqrt{L}} e^{iXy/\ell^2} \begin{pmatrix} \text{sgn}(n)i^{|n|-1}\psi_{|n|-1}(\frac{x-X}{\ell}) \\ i^{|n|}\psi_{|n|}(\frac{x-X}{\ell}) \end{pmatrix}, \tag{8.238}$$

with $X = k\ell^2$ and

$$C_n = \begin{cases} 1 & \text{if } n = 0 \\ 1/\sqrt{2} & \text{if } n \neq 0 \end{cases}, \tag{8.239}$$

$$\text{sgn}(n) = \begin{cases} 1 & \text{if } n > 1 \\ 0 & \text{if } n = 0, \\ -1 & \text{if } n < 0 \end{cases} \tag{8.240}$$

and corresponding eigenvalues

$$\epsilon_n = \operatorname{sgn}(n)\frac{v}{\ell}\sqrt{2|n|}, \tag{8.241}$$

with $n = 0, \pm 1, \pm 2, \ldots$, and $\operatorname{sgn}(n) = 1(-1)$ labels the electron (hole) levels.

Starting from Eqs. (8.238)–(8.240) one can calculate the electron density

$$\rho(x,y) = \sum_{nk}|\varphi_{nk}|^2, \tag{8.242}$$

where the sum covers all the occupied states nk and the sum over k is calculated as $\sum_k = \frac{L}{2\pi\ell^2}\int dX$. A simple calculation yields for the electrons in the spin-↑ K valley

$$\rho = \frac{1}{2\pi\ell^2}(|n| + 1/2), \tag{8.243}$$

with $|n|=0,1,2,\ldots$. The above density must then be multiplied for the degeneracy factor $g_s = 4$, accounting for spin degeneracy and sublattice degeneracy.

The Hamiltonian $\bar{H} = \sum_{i=1}^{N} h_i^2$, with h_i given by Eq. (8.235), has the cyclotron state $|\omega_c\rangle = O_{\omega_c}^\dagger|0\rangle$, with $O_{\omega_c}^\dagger$ given by Eq. (8.165) as an exact eigenstate, since $[\bar{H}, O_{\omega_c}^\dagger]|0\rangle = 2v^2\omega_c O_{\omega_c}^\dagger|0\rangle$. It follows that the initial Hamiltonian $H = \sum_{i=1}^{N} h_i$ also has the cyclotron state as an eigenstate, since $\bar{H}$ and H have a common set of eigenstates.

On the other hand, the general expression for the DC yx conductivity per unit of surface of Eq. (8.167) can be put in the form

$$\sigma_{yx} = \frac{-2ie^2}{S}\sum_n \langle 0|\sum_{i=1}^{N} y_i|n\rangle\langle n|\sum_{i=1}^{N} x_i|0\rangle \ , \tag{8.244}$$

where we have used $j_y = ie[H, \sum_{i=1}^{N} y_i]$, $j_x = ie[H, \sum_{i=1}^{N} x_i]$. Since the center-of-mass operators $\sum_{i=1}^{N} y_i$ and $\sum_{i=1}^{N} x_i$ can excite only the cyclotron state, one gets

$$\sigma_{yx} = \frac{-2ie^2}{S}\langle 0|\left[\sum_{i=1}^{N} y_i, O_{\omega_c}^\dagger\right]|0\rangle\langle 0|\left[O_{\omega_c}, \sum_{i=1}^{N} x_i\right]|0\rangle, \tag{8.245}$$

from which there immediately follows the result

$$\sigma_{yx} = e^2\frac{\rho}{\omega_c}, \tag{8.246}$$

valid for the graphene Hamiltonian as well. By using the result (8.243) one then gets (restoring $\hbar$)

$$\sigma_{yx} = g_s\frac{e^2}{h}(|n| + 1/2). \tag{8.247}$$

Plateaux in the graphene conductivity have been experimentally detected at the above sequence by Novoselov *et al.* (2005) and Zhang *et al.* (2005).

References

Adagideli, I. and G.E.W. Bauer, *Phys. Rev. Lett.* **95**, 256602 (2005).

Batke, E., H.L. Störmer, A.C. Gossard and J.H. English, *Phys. Rev.* **B37**, 3093 (1988).

Beenakker, C.W. and H. van Houten, *Quantum Transport in Semiconductor Nanostructures*, Vol. 44 of *Solid State Phys.* (Academic, New York, 1991).

Bellucci, S. and P. Onorato, *Phys. Rev.* **B73**, 045329 (2006).

Bernevig, B.A. and S.-C. Zhang, *Phys. Rev. Lett.* **96**, 106802 (2006).

Besson, W.M., E. Gornik, C.M. Engelhardt and G. Weimann, *Semicond. Sci. Technol.* **7**, 1274 (1992).

Bohigas, O., A.M. Lane and J. Martorell, *Phys. Rep.* **51**, 267 (1979).

Camels, L. and A. Gold, *Phys. Rev.* **B56**, 1762 (1997).

Chalaev, O. and D. Loss, *Phys. Rev.* **B71**, 245318 (2005).

Chang, M.-C. *Phys. Rev.* **B71**, 085315 (2005).

D'yakonov, M.I. and V.I. Perel, *Phys. Lett.* **A35**, 459 (1971).

Datta, S., *Electronic Transport in Mesoscopic Systems* (Cambridge University Press, 1995).

Datta, S. and B. Das, *Appl. Phys. Lett.* **56**, 665 (1990).

Demel, T. *et al.*, *Phys. Rev.* **B38**, 12732(R) (1988).

Dimitrova, O.V., *Phys. Rev.* **B71**, 245327 (2005).

Donnelly, R.J., J.A. Donnelly and R.N. Hills, *J. Low Temp. Phys.* **44**, 471 (1981).

Engel, H.-A., E.I. Rashba and B.I. Halperin, *Phys. Rev. Lett.* **95**, 166605 (2005).

Ensslin, K., D. Heitmann, H. Sigg and K. Ploog, *Phys. Rev.* **B36**, 8177 (1987).

Erlingsson, S.I. and D. Loss, *Phys. Rev.* **B72**, 121310(R) (2005).

Haldane, F.D.M., *Phys. Rev. Lett.* **61**, 2015 (1988).

Henriksen, E.A. *et al.*, *Physica* **E34**, 318 (2006).

Hirsch, J.E., *Phys. Rev. Lett.* **83**, 1834 (1999).

Hohenberg, P. and P. Platzman, *Phys. Rev.* **152**, 198 (1966).

Hu, C.M., T. Fridrich, E. Batke, K. Köhler and P. Ganser, *Phys. Rev.* **B52**, 12090 (1995); C.M. Hu, E. Batke, K. Köhler and P. Ganser, *Phys. Rev. Lett.* **75**, 918 (1995).

Inoue, J.I., G.E.W. Bauer and L.W. Molenkamp, *Phys. Rev.* **B70**, 041303(R) (2004).

Iwamoto, N. and D. Pines, *Phys. Rev.* **B29**, 3924 (1984).

Izuyama, T., *Prog. Theor. Phys.* **25**, 964 (1961).

Kato, Y.K., R.C. Myers, A.C. Gossard and D.D. Awschalom, *Science* **306**, 1910 (2004).

Kawabata, A., *J. Phys. Soc. Japan* **65**, 30 (1995).

Kelly, M.J., *Low Dimensional Semiconductors: Material, Physics, Technology, Devices* (Oxford University Press, 1995).

Khanna, F.C. and H.R. Glyde, *Can. J. Phys.* **54**, 648 (1976).

Koga, T., J. Nitta, H. Takayanagi and S. Datta, *Phys. Rev. Lett.* **88**, 126601 (2002).

Kohn, W., *Phys. Rev.* **123**, 1242 (1961).

Küblbeck, H. and J.P. Kotthaus, *Phys. Rev. Lett.* **35**, 1019 (1975).

Kubo, R., *Can. J. Phys.* **34**, 1274 (1956); *J. Phys. Soc. Japan* **12**, 570 (1957).

Kwon, Y., D.M. Ceperley and R.M. Martin, *Phys. Rev.* **B50**, 1684 (1994).

Leonardi, R. and M. Rosa-Clot, *Riv. Nuovo Cimento* **1**, 1 (1971).

Lipparini, E. and M. Barranco, *Physica* **E36**, 190 (2007).

Lipparini, E. and S. Stringari, *Nucl. Phys.* **A371**, 430 (1981).

Lipparini, E. and S. Stringari, *Phys. Rep.* **175**, 103 (1989).

Mahan, G.D., *Many-Particle Physics* (Plenum, New York and London, 1981).

Malet, F., M. Pi, M. Barranco and E. Lipparini, *Phys. Rev.* **B72**, 205326 (2005).

Manger, M., E. Batke, R. Hey, K.J. Friedland, K. Köhler and P. Ganser, *Phys. Rev.* **B63**, 121203R (2001).

Maslov, D.L. and M. Stone, *Phys. Rev.* **B52**, R5539 (1995).

Mishchenko, E.G., A.V. Shytov and B.I. Halperin, *Phys. Rev. Lett.* **93**, 226602 (2004).

Murakami, S., N. Nagaosa and S.C. Zhang, *Science* **301**, 1348 (2003).

Nitta, J., T. Akazaki, H. Takayanagi and T. Enoki, *Phys. Rev. Lett.* **78**, 1335 (1997).

Novoselov, K.S. *et al.*, *Nature (London)* **438**, 197 (2005).

Orlandini, G. and M. Traini, *Rep. Prog. Phys.* **54**, 257 (1991).

Peres, N.M., A.H. Castro Neto and F. Guinea, *Phys. Rev.* **B73**, 241403(R) (2006).

Pershin, Y.V., J.A. Nesteroff and V. Privman, *Phys. Rev.* **B69**, 121306(R) (2004).

Pines, D. and P. Nozières, *The Theory of Quantum Liquids* (Benjamin, New York, 1966).

Raimondi, R. and P. Schwab, *Phys. Rev.* **B71**, 033311 (2005).

Rashba, E.I., *Phys. Rev.* **B70**, 201309(R) (2004).

Reilly *et al.*, *Phys. Rev.* **B63**, 121311(R) (2001).

Serra, Ll., M. Barranco, A. Emperador, M. Pi and E. Lipparini, *Phys. Rev.* **B59**, 15290 (1999).

Shen, S.-Q., M. Ma, X.C. Xie and F.C. Zhang, *Phys. Rev. Lett.* **92**, 256603 (2004).

Sih, V., R.C. Myers, Y.K. Kato, W.H. Lau, A.C. Gossard and D.D. Awschalom, *Nature Phys.* **1**, 31 (2005).

Sinova, J., D. Culcer, Q. Niu, N.A. Sinitsyn, T. Jungwirth and A.H. MacDonald, *Phys. Rev. Lett.* **92**, 126603 (2004).

Sinova, J., S. Murakami, S.-Q. Shen and M.-S. Choi, *Solid State Commun.* **138**, 214 (2006).

Stringari, S. and F. Dalfovo, *J. Low Temp.* **78**, 1 (1990).

Summers, G.M., R.J. Warburton, J.G. Michels, R.J. Nicholas, J.J. Harris and C.T. Foxon, *Phys. Rev. Lett.* **70**, 2150 (1993).

Syed, S. *et al.*, *Phys. Rev.* **B67**, 241304R (2003).

Thomas, K.J., J.T. Nicholls, M.Y. Simmons, M. Pepper, D. R. Mace and D.A. Ritchie, *Phys. Rev. Lett.* **77**, 135 (1996).

Thomas, K.J., D.L. Sawkey, M. Pepper, W.R. Tribe, I. Farrer, M.Y. Simmons and D.A. Ritchie, *J. Phys. Cond. Matt.* **16**, L279 (2004).

Tonello, P. and E. Lipparini, *Phys. Rev.* **B70**, 081201R (2004).

Tse, W.-K. and S. Das Sarma, *Phys. Rev. Lett.* **96**, 056601 (2006).

Tsui, D.C., H. Stormer and A.C. Gossard, *Phys. Rev. Lett.* **48**, 1559 (1982).

Valín-Rodriguez, M., A. Puente and Ll. Serra, *Nanotechnology* **14**, 882 (2003).

Von Felde, A., A. Sprosser-Prou and J. Fink, *Phys. Rev.* **B40**, 10181 (1989).

von Klitzing, K., G. Dorda and M. Pepper, *Phys. Rev. Lett.* **45**, 494 (1980).

Willet, R., J.P. Eisenstein, H. Stormer, D.C. Tsui, A.C. Gossard and J.H. English, *Phys. Rev. Lett.* **59**, 1776 (1987).

Wunderlich, J., B. Kaestner, J. Sinova and T. Jungwirth, *Phys. Rev. Lett.* **94**, 047204 (2005).

Zhang, S., *Phys. Rev. Lett.* **85**, 393 (2000).

Zhang, Y. *et al.*, *Nature (London)* **438**, 201 (2005).

Zheng, Y. and T. Ando, *Phys. Rev.* **B65**, 245420 (2002).

Chapter 9

The Linear Response Function in Different Models

9.1 The Linear Response Function in Landau Theory

In this subsection we will discuss the linear response of a homogeneous Fermi liquid to an external field which oscillates at a given frequency and wavelength (Pines and Nozières, 1966). In the limit where the wave vector q and energy ω related to the excitation fulfill the conditions $qv_f, \omega \ll \epsilon_F$, which guarantee that only $1p$–$1h$ excitations contribute to the dynamic response, the system response is correctly described by the Landau theory of normal Fermi liquids. This theory allowed Landau to predict the existence of the "zero sound" in liquid ^{3}He, before its experimental observation. The propagation of zero sound, unlike that of normal (first) sound, which takes place in the hydrodynamic regime (see Subsection 1.9), is related to a collisionless (elastic) regime, and is produced by the mean self-consistent field.

In the presence of an external perturbation, which interacts with the system through an interaction potential H_{int} of the form of Eq. (8.10) with

$$G^{s,a} = \rho_{\mathbf{q}}^{s,a} = \sum_{j=1}^{N_\uparrow} e^{-i\mathbf{q}\cdot\mathbf{r}_j} \pm \sum_{j=1}^{N_\downarrow} e^{-i\mathbf{q}\cdot\mathbf{r}_j}, \qquad (9.1)$$

it is necessary to add to the Landau equations (1.136) an external field term describing the flow of particles in phase space; this flow is induced by the force $\mathcal{F}$ corresponding to H_{int}, i.e.

$$\mathcal{F}^{s,a} = -i\lambda\mathbf{q}(\rho_{\mathbf{q}}^{s,a})^* e^{-i\omega t} + \text{c.c.} \qquad (9.2)$$

The external field term, corresponding to the force $\mathcal{F}$, has the form (see Subsection 1.9)

$$\mathcal{F}^{s,a} \cdot \mathbf{v_k} \frac{\partial n_{\mathbf{k}}^0}{\partial \epsilon_{\mathbf{k}}}. \qquad (9.3)$$

In the above equations, the suffixes s, a indicate the symmetric and antisymmetric parts, respectively, of the densities and the external forces [in Eq. (9.1), the $+$ sign corresponds to G^s, and the $-$ sign to G^a], and the term (9.3) should be added in its symmetric (antisymmetric) part to the Landau equation for the symmetric (antisymmetric) distribution function. Moreover, for simplicity, in the following we will assume that the system has zero initial spin, so that the density and spin-density response functions are uncoupled. The effects due to coupling of the two responses will be discussed in one of the forthcoming subsections. Therefore, in the absence of the collision term (elastic regime), the Landau equations under an external field become

$$\frac{\partial}{\partial t}\delta n_{\mathbf{k}}^{s,a}(\mathbf{r},t) + \mathbf{v_k} \cdot \nabla_{\mathbf{r}}\delta n_{\mathbf{k}}^{s,a}(\mathbf{r},t)$$

$$- \nabla_{\mathbf{k}}n_{\mathbf{k}}^0 \cdot \sum_{\mathbf{k'}} f_{\mathbf{k},\mathbf{k'}}^{s,a} \nabla_{\mathbf{r}}\delta n_{\mathbf{k'}}^{s,a}(\mathbf{r},t) + \mathcal{F}^{s,a} \cdot \mathbf{v_k}\frac{\partial n_{\mathbf{k}}^0}{\partial \epsilon_{\mathbf{k}}} = 0. \qquad (9.4)$$

The external field in (9.4) induces oscillations of the distribution function, which may be written as

$$\delta n_{\mathbf{k}}^{s,a}(\mathbf{r},t) = \delta n_{\mathbf{k}}^{s,a}(\mathbf{q},\omega)e^{i\mathbf{q}\cdot\mathbf{r}}e^{-i\omega t} + \text{c.c.} \qquad (9.5)$$

Next, the linear response function is obtained by evaluating the density fluctuations:

$$\langle \rho_{\mathbf{q}}^{s,a} \rangle = \frac{1}{V}\int \rho^{s,a}(\mathbf{r},t)e^{-i\mathbf{q}\cdot\mathbf{r}}d\mathbf{r}$$

$$= \frac{1}{(2\pi)^3}\int d\mathbf{k}\delta n_{\mathbf{k}}^{s,a}(\mathbf{r},t)e^{-i\mathbf{q}\cdot\mathbf{r}}d\mathbf{r}$$

$$= \sum_{\mathbf{k}}\delta n_{\mathbf{k}}^{s,a}(\mathbf{q},\omega)e^{-i\omega t}. \qquad (9.6)$$

Using Eqs. (8.13) and (9.6), we finally obtain the response functions:

$$\chi^s(q,\omega) = \chi(\rho_{\mathbf{q}}^s, \rho_{\mathbf{q}}^s, \omega) = \frac{1}{\lambda}\sum_{\mathbf{k}}\delta n_{\mathbf{k}}^s(\mathbf{q},\omega),$$

$$\chi^a(q,\omega) = \chi(\rho_{\mathbf{q}}^a, \rho_{\mathbf{q}}^a, \omega) = \frac{1}{\lambda}\sum_{\mathbf{k}}\delta n_{\mathbf{k}}^a(\mathbf{q},\omega). \qquad (9.7)$$

It is convenient to put 0the quantity $\delta n_{\mathbf{k}}^{s,a}(\mathbf{q},\omega)$ in the form

$$\delta n_{\mathbf{k}}^{s,a}(\mathbf{q},\omega) = \delta(\epsilon_k - \epsilon_F)v_F u_{\mathbf{k}}(\mathbf{q},\omega),$$

where $u_{\mathbf{k}}$ is the normal displacement of the Fermi surface at point $\mathbf{k}$ [see also Eq. (1.149)], $v_F = k_F/m^*$ is the Fermi velocity, and ϵ_F is the Fermi energy. The equation for $u_{\mathbf{k}}$ is attained by substituting (9.5) into (9.4):

$$(\mathbf{q}\cdot\mathbf{v_k} - \omega)u_{\mathbf{k}}^{s,a} + \mathbf{q}\cdot\mathbf{v_k}\sum_{\mathbf{k'}} f_{\mathbf{k},\mathbf{k'}}^{s,a}\delta(\epsilon_{k'} - \epsilon_F)u_{\mathbf{k'}}^{s,a} + \lambda\frac{\mathbf{q}\cdot\mathbf{v_k}}{v_F} = 0, \qquad (9.8)$$

where we have used (1.146). By taking, without the loss of generality, $\mathbf{q}$ parallel to the z axis of momentum space, where the directions of $\mathbf{k}$ and $\mathbf{k}'$ are identified by the (θ, ϕ) and (θ', ϕ') angles, respectively, Eq. (9.8) may be rewritten as

$$q v_F (\cos\theta - s) u^{s,a}(\theta, \phi)$$
$$+ q v_F \cos\theta \, \nu(0) \int \frac{d\Omega'_{\mathbf{k}'}}{4\pi} f^{s,a}(k\hat{k}') u^{s,a}(\theta', \phi') + \lambda q \cos\theta = 0, \quad (9.9)$$

where $\nu(0)$ is the density of states at the Fermi surface (1.140) and $s = \omega/q v_F$. [Note that in the case of nuclear matter, $\nu(0) = 2V m^* k_F/\pi^2$, owing to isospin degeneracy.] In what follows, we will look for solutions to (9.9) in the form

$$u^{s,a}(\theta, \phi) = \sum_{l,m} u^{s,a}_{l,m} Y_{l,m}(\cos\theta).$$

If we use expansion (1.138) for the interaction, we see that Eqs. (9.9) are equations in which m is conserved, i.e. the nature of the solutions is determined by m, and not by l, which is mixed. The solutions with $m = 0$ are solutions in which the density is changed [$\int d\Omega u(\theta, \phi) \neq 0$], and give rise to propagation of longitudinal waves. Those with $m \neq 0$ do not cause a density change and are responsible for transverse waves ($m = 1$), quadrupolar waves ($m = 2$), etc. In what follows we will consider the longitudinal case ($m = 0$), which is the only one that involves density fluctuations and, thus, represents the high frequency counterpart of ordinary sound. For the longitudinal mode, the following expansion holds:

$$u^{s,a}(\theta, \phi) = \sum_l u^{s,a}_l P_l(\cos\theta), \quad (9.10)$$

where

$$u^{s,a}_l \equiv u^{s,a}_{l,0} \sqrt{\frac{2l+1}{4\pi}}.$$

Then we obtain

$$(\cos\theta - s) \sum_l u^{s,a}_l P_l(\cos\theta) + \cos\theta \sum_{l'} \frac{u^{s,a}_{l'}}{2l'+1} P_{l'}(\cos\theta) F^{s,a}_{l'} + \lambda \frac{\cos\theta}{v_F} = 0, \quad (9.11)$$

where we have used (1.139), $P_l(\cos k\hat{k}') = P_l(\cos\theta) P_l(\cos\theta')$ and

$$\int_{-1}^1 P_l(\cos\theta) P_{l'}(\cos\theta) d\cos\theta = \frac{2}{2l+1} \delta_{l,l'}.$$

Furthermore, from Eq. (9.11) we have

$$\frac{u^{s,a}_l}{2l+1} + \sum_{l'} \Omega_{l,l'} F^{s,a}_{l'} \frac{u^{s,a}_{l'}}{2l'+1} = -\frac{\lambda}{v_F} \Omega_{l,0}, \quad (9.12)$$

with

$$\Omega_{l,l'} = \Omega_{l',l} = \frac{1}{2}\int_{-1}^{1} dx P_l(x)\frac{x}{x-s}P_{l'}(x), \tag{9.13}$$

from which we obtain

$$\Omega_{0,0} = 1 + \frac{s}{2}\ln\frac{s-1}{s+1} = 1 + \frac{s}{2}\ln\left|\frac{s-1}{s+1}\right| + i\frac{\pi}{2}s\Theta(1-s),$$

$$\Omega_{l,1} = s\Omega_{l,0} + \frac{1}{3}\delta_{l,1},$$

$$\Omega_{2,0} = \frac{1}{2} + \frac{3s^2-1}{2}\Omega_{0,0}, \tag{9.14}$$

$$\Omega_{2,2} = \frac{1}{5} + \frac{3s^2-1}{2}\Omega_{2,0}.$$

In what follows we will use the following limits:

$$\Omega_{0,0}|_{s\to\infty} = -\frac{1}{3s^2} - \frac{1}{5s^4} - \frac{1}{7s^6} + \cdots,$$

$$\Omega_{0,0}|_{s\to1+} = 1 + \frac{1}{2}\ln\frac{s-1}{2} + \cdots. \tag{9.15}$$

From Eq. (9.7), we then obtain the response function

$$\chi^{s,a}(q,\omega) = \frac{1}{\lambda}v_F\nu(0)\frac{1}{4\pi}\int u^{s,a}(\theta,\phi)d\Omega = \frac{1}{\lambda}v_F\nu(0)u_0^{s,a}. \tag{9.16}$$

In order to derive $u_0^{s,a}$ from Eqs. (9.11) and (9.12), taking into account the three first Landau parameters (F_0, F_1, F_2), it is convenient to proceed in the following way. From Eq. (9.11), taking $\frac{1}{2}\int_{-1}^{1}d\cos\theta$, we obtain the relation

$$su_0^{s,a} = \frac{1}{3}\left(1 + \frac{1}{3}F_1^{s,a}\right)u_1^{s,a}, \tag{9.17}$$

which represents quasiparticle number conservation. Then, taking

$$\frac{1}{2}\int_{-1}^{1}\cos\theta d\cos\theta$$

we obtain the relation

$$su_1^{s,a} - (1 + F_0^{s,a})u_0^{s,a} - \frac{2}{5}\left(1 + \frac{1}{5}F_2^{s,a}\right)u_2^{s,a} = \frac{\lambda}{v_F}, \tag{9.18}$$

which expresses quasiparticle current conservation. Eliminating $u_1^{s,a}$ from the two above equations, we find that

$$\left[s^2 - \frac{1}{3}\left(1 + \frac{1}{3}F_1^{s,a}\right)(1 + F_0^{s,a})\right]u_0^{s,a} - \frac{2}{15}\left(1 + \frac{1}{3}F_1^{s,a}\right)\left(1 + \frac{1}{5}F_2^{s,a}\right)u_2^{s,a}$$

$$= \frac{1}{3}\left(1 + \frac{1}{3}F_1^{s,a}\right)\frac{\lambda}{v_F}. \tag{9.19}$$

From Eq. (9.12), taking $l = 0$ we have the relation

$$u_0^{s,a} + \Omega_{0,0}F_0^{s,a}u_0^{s,a} + \Omega_{0,1}F_1^{s,a}\frac{u_1^{s,a}}{3} + \Omega_{0,2}F_2^{s,a}\frac{u_2^{s,a}}{5} = -\frac{\lambda}{v_F}\Omega_{0,0}, \tag{9.20}$$

and taking $l = 2$ we have

$$\frac{u_2^{s,a}}{5} + \Omega_{2,0}F_0^{s,a}u_0^{s,a} + \Omega_{2,1}F_1^{s,a}\frac{u_1^{s,a}}{3} + \Omega_{2,2}F_2^{s,a}\frac{u_2^{s,a}}{5} = -\frac{\lambda}{v_F}\Omega_{2,0}. \tag{9.21}$$

By dividing these two relations side by side, and using the identities $\Omega_{0,1} = s\Omega_{0,0}$ and $\Omega_{2,1} = s\Omega_{2,0}$, we finally obtain the relation

$$\frac{u_2^{s,a}}{5} = \frac{\Omega_{2,0}u_0^{s,a}}{\Omega_{0,0} + F_2^{s,a}(\Omega_{2,2}\Omega_{0,0} - \Omega_{2,0}^2)}. \tag{9.22}$$

From Eqs. (9.19) and (9.22) we arrive at the desired equation for $u_0^{s,a}$ and, thus, to the dynamic polarizabilities, we have

$$\chi^{s,a}(q,\omega) = \frac{\frac{1}{3}\nu(0)(1 + \frac{F_1^{s,a}}{3})}{s^2 - \frac{1}{3}(1 + \frac{F_1^{s,a}}{3})(1 + F_0^{s,a}) - \frac{2}{3}\frac{(1 + \frac{1}{3}F_1^{s,a})(1 + \frac{1}{5}F_2^{s,a})\Omega_{2,0}}{\Omega_{0,0} + F_2^{s,a}(\Omega_{2,2}\Omega_{0,0} - \Omega_{2,0}^2)}}. \tag{9.23}$$

From Eq. (9.23), taking the $\omega \to 0$ and $\omega \to \infty$ limits [see Eqs. (8.28) and (8.29)], and taking account of Eqs. (9.14) and (9.15), it is possible to derive the following analytic expressions for the sum rules:

$$m_{-1}^{s,a} = \frac{N}{2}\frac{3}{1 + F_0^{s,a}}\frac{m^*}{k_F^2}, \tag{9.24}$$

$$m_1^{s,a} = \frac{N}{2}\frac{q^2}{m^*}\left(1 + \frac{F_1^{s,a}}{3}\right), \tag{9.25}$$

$$m_3^{s,a} = \frac{N}{2}\frac{q^4k_F^2}{m^{*3}}\left(1 + \frac{F_1^{s,a}}{3}\right)^2\left(\frac{3}{5} + \frac{1}{3}F_0^{s,a} + \frac{4}{75}F_2^{s,a}\right). \tag{9.26}$$

The m_{-3} sum rule diverges owing to the linear ω dependence of the strength [Im (χ)] for small ω. In Eqs. (9.24)–(9.26), the effective mass is related to the F_1^s Landau parameter through Eq. (1.147):

$$\frac{m^*}{m} = 1 + \frac{F_1^s}{3}. \tag{9.27}$$

The m_{-1}^s sum rule is known as the hydrodynamic sum rule, because it is directly related to ordinary sound velocity $v_1^2 = k_F^2(1 + F_0^s)/3mm^*$ [see Eq. (1.152)]. The m_1^s sum rule is the f-sum rule; as a consequence of Eq. (9.27) it does not depend on the interaction, in agreement with the direct calculation of m_1^s from Eq. (8.30) [see also Eq. (8.64)]. The m_3^s sum rule has been less discussed in the literature. As remarked on by Lipparini and Stringari (1989), and as we will see later on, it is related to the elastic properties of the Fermi systems.

In the case of the m_k^a sum rules for the spin-density operator $\rho_{\mathbf{q}}^a$, the following remarks apply. The m_{-1}^a sum rule can be expressed in terms of the system magnetic susceptibility [see Eq. (1.153)], through the relation

$$m_{-1}^a = -\frac{1}{2\mu_0}\chi_\sigma. \qquad (9.28)$$

Unlike the symmetric case, the m_1^a sum rule gets a contribution from the interaction. In fact, in the case of the spin channel, there is no cancelation between the combination $1 + F_1^a/3$ and the one that characterizes the effective mass, $m^*/m = 1 + F_1^s/3$.

The spin channel has deep analogies with the isospin channel of nuclear matter, with N neutrons and Z protons and $A = N + Z$ nucleons. In this case, which is the most interesting for nuclear matter because it leads to the excitation of strongly collective isovector modes (giant resonances), we have

$$G^a = \rho_{\mathbf{q}}^a = \sum_{j=1}^{N} e^{-i\mathbf{q}\cdot\mathbf{r}_j} - \sum_{j=1}^{Z} e^{-i\mathbf{q}\cdot\mathbf{r}_j}, \qquad (9.29)$$

and there is total analogy between spin and isospin operators. As remarked on before, in the case of nuclear matter one should bear in mind that the density of states at the Fermi energy is twice the amount with respect to other cases due to isospin degeneracy. In the isovectorial case, the m_{-1}^a sum rule can be expressed in terms of the b_{sym} coefficient of the volume-symmetry-energy that appears in the semi-empirical mass formula of nuclei (Weizsacker, 1935; Bethe and Bacher, 1936):

$$E = b_{\text{vol}}N - b_{\text{surf}}N^{2/3} - \frac{1}{2}b_{\text{sym}}\frac{(N-Z)^2}{A} - \frac{3}{5}\frac{Z^2e^2}{R_c}, \qquad (9.30)$$

through the relation

$$m_{-1}^a = \frac{A}{2b_{\text{sym}}}, \qquad (9.31)$$

where the parameters b_{sym} and F_0^a are connected by

$$b_{\text{sym}} = \frac{2}{3}\epsilon_F(1 + F_0^a). \qquad (9.32)$$

Using the values $b_{\text{sym}} = 50$ MeV and $\epsilon_F \simeq 37$ Mev, corresponding to a nuclear mass density $\rho = 0.17$ nucleons fm^{-3}, we have $F_0^a \simeq 1$ for nuclear matter from Eq. (9.32). Let us recall that for ^{3}He it is $F_0^a = -0.7$ and $F_0^s = 9.15$. The Landau parameter

F_0^s of nuclear matter is connected to the compressibility χ_{nm} of nuclear matter, by the relation $\chi_{nm}/9 = 2\epsilon_F(1 + F_0^s)/3$. The value of χ_{nm} is not very well determined experimentally. Using the value $\chi_{nm} \simeq 200$ MeV, we find that $F_0^s \simeq -0.1$. The Landau parameters for nuclear matter can be deduced by the effective, Skyrme-like forces, for example the SIII described in Chapter 4. In Table 9.1 we report the values of the parameters $F_0^s, F_0^a, F_1^s, F_1^a$ for some of these forces (Garcia-Recio *et al.*, 1992) (k_F in fm^{-1}):

In what follows, as an example of the application of Eq. (9.23), we will discuss the dynamic polarizability in the simplified case where only the $F_0^{s,a}$ Landau parameter is assumed to be nonzero. Using relations (9.14) and putting $F_1 = F_2 = 0$ in (9.23), we find the result

$$\chi^{s,a}(q,\omega) = \frac{-\nu(0)\Omega_{0,0}}{1 + F_0^{s,a}\Omega_{0,0}}. \tag{9.33}$$

The poles of Eq. (9.33) are given by the solutions to

$$\frac{1}{F_0^{s,a}} = \frac{s}{2}\ln\frac{s+1}{s-1} - 1, \tag{9.34}$$

where we have used the result (9.14) for $\Omega_{0,0}$. The dynamic form factor (strength function) is given by

$$S^{s,a}(q,\omega) = -\frac{1}{\pi}\text{Im }\chi^{s,a}(q,\omega).$$

First of all, we notice that if $F_0 = 0$, we find the result

$$S^{s,a}(q,\omega) = \frac{\nu(0)}{\pi}\text{Im }\Omega_{0,0} = \frac{Vm^2}{2\pi^2}\frac{\omega}{q}\Theta\left(1 - \frac{\omega}{qv_F}\right), \tag{9.35}$$

using Eq. (9.14), which is the same as that of (1.108) for the single-particle strength of the Fermi gas. However, if $F_0 > 0$, the interaction produces a discrete peak at $\omega > qv_F$. In fact, in this case (i.e. $F_0^{s,a} > 0$, corresponding to a repulsive interaction among the quasiparticles) Eq. (9.34) has a real solution for $s = \omega/qv_F$, such that $s > 1$. This solution (collective) corresponds to an excitation mode of the system with no attenuation, whose phase velocity $v = \omega/q$ is greater than v_F. However, if $-1 < F_0^{s,a} < 0$, the solution of (9.34) has imaginary components and the corresponding

Table 9.1. Landau parameters for nuclear matter as deduced by three different effective Skyrme forces.

	SIII($k_F = 1.29$)	SkM($k_F = 1.33$)	SGII($k_F = 1.33$)
F_0^s	0.299	−0.235	−0.235
F_0^a	0.865	0.975	0.733
F_1^s	−0.708	−0.630	−0.652
F_1^a	0.489	0.574	0.524

mode is damped (Landau damping). Finally, if the attraction among quasiparticles is so strong that $F_0^{s,a} < -1$, then the solution is unstable.

In the ^{3}He case, the first case ($F_0^{s,a} > 0$) is realized for the density channel (symmetric spin), and the second one for the spin-density channel (antisymmetric spin). For the isovectorial modes of nuclear matter we are again in the first case.

We now discuss the undamped solution for ^{3}He, i.e. zero sound. In this case, $F_0^s = 9.15$ is large enough that we are justified in looking for solutions to (9.34) in the limit of large s. Therefore, using the expansion of Eq. (9.15), we obtain

$$s^2 = \frac{F_0^s}{3}\left(1 + \frac{9}{5F_0^s} + \cdots\right) \tag{9.36}$$

and a wave phase velocity:

$$v_0^2 = \frac{\omega^2}{q^2} = v_F^2 \frac{F_0^s}{3}\left(1 + \frac{9}{5F_0^s} + \cdots\right). \tag{9.37}$$

If we compare this with the ordinary sound wave, as obtained from the hydrodynamic equations (1.152), taken in the same limit ($F_1^s = 0$), and which are given by

$$v_1^2 = v_F^2\left(\frac{F_0^s}{3}\right)\left(1 + \frac{1}{F^s}_0 + \cdots\right),$$

we find that the two velocities are different. The difference increases if we take into account the other Landau parameters as well. For example, the ordinary sound velocity v_1 does not depend on F_2, while the velocity v_0 of zero sound does. This mode has an elastic nature and its frequency is well described by the energy

$$\omega^2 = \frac{m_3^s}{m_1^s} = \frac{m^*}{m}q^2 v_F^2\left(\frac{3}{5} + \frac{F_0^s}{3} + \frac{4}{75}F_2^s\right). \tag{9.38}$$

The difference between the energies of the hydrodynamic (with collisions) and elastic (collisionless) modes can be detected experimentally by studying the velocity of sound in ^{3}He as a function of temperature, i.e. of the collision time. Two regimes are found: the low temperature one, in which sound propagation takes place at velocity v_0 [see Eq. (9.37)], and the high temperature one, where sound propagation occurs at ordinary sound velocity v_1. In these regimes, there is no sound attenuation. However, there is an intermediate range in which sound attenuation is not zero and varies with temperature.

The sound attenuation can be studied in the Landau theory using the "single time" approximation (Khalatninov and Abrikosov, 1958) for the collision integral:

$$I(\delta n^s)_{\text{coll}} = -\frac{1}{\tau}[\delta \tilde n_{\mathbf{k}}^s - \delta \tilde n_{\mathbf{k}}^s(l = 0, l = 1)], \tag{9.39}$$

where

$$\delta \tilde{n}_{\mathbf{k}}^s = \delta n_{\mathbf{k}}^s - \frac{\partial n_{\mathbf{k}}^0}{\partial \epsilon_{\mathbf{k}}} \sum_{\mathbf{k}'} f_{\mathbf{k},\mathbf{k}'}^s \delta n_{\mathbf{k}'}^s \qquad (9.40)$$

is the local deformation of the distribution function, while $\delta \tilde{n}_{\mathbf{k}}^s (l = 0, l = 1)$ corresponds to the components of $\delta \tilde{n}_{\mathbf{k}}^s$ which have $l = 0$ and $l = 1$ angular momentum in momentum space, and ensures mass and momentum conservation during collisions. The use of $\delta \tilde{n}_{\mathbf{k}}^s$ in place of $\delta n_{\mathbf{k}}^s$ ensures local energy conservation of quasiparticles. τ is the characteristic time of collision. The collision integral (9.39) changes the Landau equation under an external field (9.11) into the following:

$$(\cos\theta - \xi)u^s(\theta) + \cos\theta \left(\frac{\lambda}{v_F} + F_0^s u_0 + \frac{1}{3} F_1^s u_1 \cos\theta \right.$$

$$\left. + \frac{1}{5} F_2^s u_2 (3\cos^2\theta - 1) \right)$$

$$= \frac{1}{\sigma}(u_0 + u_1 \cos\theta), \qquad (9.41)$$

where $\xi = s + i/\tau q v_F$ and $\sigma = i\tau q v_F$, and we have retained Landau parameters up to F_2. From (9.41) it is possible to derive the system linear response in the presence of the collision term. The equation for the poles of the response function describes sound dispersion in a Fermi liquid in the presence of the collision term. We are going to describe the viscoelastic model (Bedell and Pethick, 1982) for sound attenuation. This model is based on a truncated solution to (9.41). This equation yields the following linear system of equations:

$$\begin{vmatrix} -s & \dfrac{G_1}{\sqrt{3}} & 0 & 0 & \cdots \\[2mm] \dfrac{G_0}{\sqrt{3}} & -s & \dfrac{2G_2}{\sqrt{15}} & 0 & \cdots \\[2mm] 0 & \dfrac{2G_1}{\sqrt{15}} & -\xi\left(1 - \dfrac{F_2}{5\sigma\xi}\right) & \dfrac{3G_3}{\sqrt{35}} & \cdots \\[2mm] 0 & 0 & \dfrac{3G_2}{\sqrt{35}} & -\xi\left(1 - \dfrac{F_3}{7\sigma\xi}\right) & \cdots \\[2mm] 0 & 0 & 0 & \dfrac{4G_3}{\sqrt{63}} & \cdots \\[2mm] \cdots & \cdots & \cdots & \cdots & \cdots \end{vmatrix} \cdot \begin{vmatrix} u_0 \\ u_1 \\ u_2 \\ u_3 \\ \cdots \\ \cdots \end{vmatrix} = -\frac{\lambda}{v_F} \begin{vmatrix} 0 \\ 1 \\ 0 \\ 0 \\ 0 \\ \cdots \end{vmatrix},$$

$$(9.42)$$

where $G_l = 1 + F_l/(2l+1)$. The structure of these equations shows that, for $l > 2$, we have $u_{l+1} \simeq u_l/s$, so that the contributions from high multipolarities are less and less important, provided that $s > 1$. This suggests that the set of equations be truncated at a given order in l. The viscoelastic model consists in truncating the system of equations at order $l \leq 2$, which means considering only the three first lines and columns of the matrix Eq. (9.42). By requiring that the determinant of the coefficients of the homogeneous system be zero, we find that

$$-\xi \left(1 - \frac{F_2}{5\sigma\xi} \right) (s^2 - s_1^2) + \frac{4}{15} sG_1G_2 = 0, \tag{9.43}$$

where $s_1^2 = (v_1/v_F)^2$. Assuming that ω is real, Eq. (9.43) yields two equations for the real q_R and imaginary q_I parts of $q(\omega)$, from which we get the sound velocity $v = \omega/q_R$ and the sound attenuation coefficient $\gamma = q_I$. Assuming further that the imaginary part of s is much smaller than the real part, we find that

$$v^2 = v_1^2 + \frac{4}{15} v_F^2 G_1 G_2 \frac{(\omega\tau)^2}{G_2 + (\omega\tau)^2}, \tag{9.44}$$

and

$$\gamma = \frac{2}{15} \frac{v_F^2}{v^3} G_1 G_2 \frac{\omega^2\tau}{G_2 + (\omega\tau)^2}. \tag{9.45}$$

Equations (9.44) and (9.45) describe the transition from the collisionless regime ($\omega\tau \gg 1$) to the hydrodynamic one ($\omega\tau \ll 1$). In the first case the sound velocity is given by

$$v_0^2 = v_1^2 + \frac{4}{15} v_F^2 G_1 G_2,$$

and the attenuation coefficient becomes

$$\gamma_0 = 2v_F^2 G_1 G_2 / 15 v_0^3 \tau.$$

In the second case the sound velocity is the ordinary one and the attenuation coefficient behaves like $\omega^2\tau$. The above results, in the case of ^{3}He, reproduce rather well the experimental data (Abel, Anderson and Wheatley, 1966). They were also applied to the study of nuclear resonances at finite temperature (Denisov, 1989; Baran *et al.*, 1995). Finally, we note that the viscoelastic model yields an expression for the dynamic polarizability which is identical to the one obtained previously, provided that all the Landau parameters other than F_0 and F_1 are neglected, and a $1/s^2$ expansion is performed.

The Landau theory has been generalized to charged Fermi systems by Silin (Silin, 1957). Silin was the first to realize that the problems due to the long range Coulomb interactions (which are singular in the $\mathbf{q} \to 0$ limit) might be overcome by partitioning the total interaction among charged quasiparticles into a regular part $f_{\mathbf{k},\mathbf{k}'}$ (which is the equivalent of the total interaction among neutral quasiparticles), and a singular part given by the Fourier transform of the Coulomb interaction. The latter is to be treated in terms of a mean polarization field acting on the quasiparticles. Such a mean polarization field $\mathcal{E}(\mathbf{r}, t)$ is connected to the density fluctuations by the equation

$$\text{div } \mathcal{E}(\mathbf{r}, t) = 4\pi e \delta n_{\mathbf{k}}^s(\mathbf{r}, t), \tag{9.46}$$

and appears in the Landau Eqs. (9.4) for $\delta n_{\mathbf{k}}^s(\mathbf{r}, t)$, through an additional external force term of the form (9.3), i.e. $-e\mathcal{E} \cdot \mathbf{v}_{\mathbf{k}} \partial n^0 / \partial \epsilon_{\mathbf{k}}$, while it does not change the Landau equation for the spin density fluctuations. $\mathcal{E}(\mathbf{r}, t)$ can be interpreted as an additional applied field, which screens the interaction between two quasiparticles due to the presence of all the other quasiparticles. Using the result

$$\mathcal{E}(q, \omega) = -iq \frac{4\pi e}{q^2} \sum_{\mathbf{k}} \delta n_{\mathbf{k}}^s(q, \omega), \tag{9.47}$$

one obtains, in charged systems, the following equation for $u_{\mathbf{k}}^s$:

$$(\mathbf{q} \cdot \mathbf{v}_{\mathbf{k}} - \omega) u_{\mathbf{k}}^s + \mathbf{q} \cdot \mathbf{v}_{\mathbf{k}} \sum_{\mathbf{k}'} \left(f_{\mathbf{k}, \mathbf{k}'}^s + \frac{4\pi e^2}{q^2} \right) \delta(\epsilon_{k'} - \epsilon_F) u_{\mathbf{k}'}^s + \lambda \frac{\mathbf{q} \cdot \mathbf{v}_{\mathbf{k}}}{v_F} = 0. \tag{9.48}$$

The solution of such an equation goes like in the case of neutral systems. The final formulas for the dynamic polarizability $\chi^s(q, \omega)$, and for the sum rules, are obtained from Eqs. (9.23)–(9.26) by simply replacing the Landau parameter F_0^s with $F_0^s + \nu(0) 4\pi e^2 / q^2$.

Much effort has been devoted to deriving the Landau parameters for the 3D electron gas, starting from the microscopic interaction [see for example Pines and Nozières (1966), Hedin and Lundqvist (1969)]. What emerges clearly from these calculations is that for the 3D electron gas the ratio m^*/m is practically equal to 1, so that the value of the Landau parameter F_1^s might be zero. The same can be assumed to hold for the corresponding parameter in the spin channel: $F_1^a \simeq 0$. The only parameters which are definitely nonzero are F_0^s and F_0^a. The most reliable values for these parameters can be deduced from the Monte Carlo values of the compressibility and magnetic susceptibility (Vosko, Wilk and Nusair, 1980; Iwamoto and Pines 1984), and are reported in Table 9.2 for different values of the density. The excited state of the 3D electron gas which corresponds to the zero sound in

Table 9.2. Landau parameters for the 3D electron gas as deduced by Monte Carlo calculations at different densities.

r_s	F_0^s	F_0^a
1	−0.17	−0.13
2	−0.36	−0.23
3	−0.55	−0.32
4	−0.74	−0.38
5	−0.95	−0.44
10	−2.03	−0.65

[3]He is called a plasmon, and has been observed in numerous experiments. One of the most complete among these experments is the one published by vom Felde, Sprosser-Prou and Fink (1989). We will describe it in Subsection 9.4.

To conclude, let us discuss the linear response in the two-dimensional Landau theory. In 2D the equation corresponding to Eq. (9.9) is

$$(\cos\theta - s)u^{s,a}(\theta) + \cos\theta\nu(0)\int\frac{d\theta'}{2\pi}f_{2D}^{s,a}(\theta,\theta')u^{s,a}(\theta') = -\frac{\lambda}{v_F}\cos\theta, \qquad (9.49)$$

where the 2D level density $\nu(0)$ at the Fermi surface is given by (1.155). Employing the expansion (1.54) for the 2D interaction, and the analog for the displacements u,

$$u^{s,a}(\theta) = \sum_l u_l^{s,a}\cos l\theta, \qquad (9.50)$$

we find that

$$(\cos\theta - s)\sum_l u_l^{s,a}\cos l\theta + \cos\theta\left(u_0^{s,a}F_0^{s,a} + \sum_{l>0}u_l^{s,a}\frac{F_l^{s,a}}{2l}\cos l\theta\right) = -\frac{\lambda}{v_F}\cos\theta. \qquad (9.51)$$

From this equation, dividing by $\cos\theta - s$ and then multiplying by $\cos m\theta$, and then performing the integration $\int_0^{2\pi}\frac{d\theta}{2\pi}$, we obtain

$$u_0^{s,a} + F_0^{s,a}u_0^{s,a}\Lambda_{0,0} + \sum_{l>0}\Lambda_{0,l}\frac{F_l^{s,a}}{2l}u_l^{s,a} = -\frac{\lambda}{v_F}\Lambda_{0,0},$$

$$\frac{u_m^{s,a}}{2} + F_0^{s,a}u_0^{s,a}\Lambda_{m,0} + \sum_{l>0}\Lambda_{m,l}\frac{F_l^{s,a}}{2l}u_l^{s,a} = -\frac{\lambda}{v_F}\Lambda_{m,0}, \quad m \neq 0, \qquad (9.52)$$

with

$$\Lambda_{m,l} = \int_0^{2\pi}\frac{d\theta}{2\pi}\cos m\theta\frac{\cos\theta}{\cos\theta - s}\cos l\theta.$$

The Λ functions have the properties

$$\Lambda_{0,0} = \begin{cases} 1 + i\dfrac{s}{\sqrt{1-s^2}} & \text{if } s \leq 1 \\[3mm] 1 - \dfrac{s}{\sqrt{s^2-1}} & \text{if } s > 1 \end{cases}, \qquad (9.53)$$

and

$$\Lambda_{1,0} = s\Lambda_{0,0}, \quad \Lambda_{2,0} = 1 + (2s^2-1)\Lambda_{0,0},$$

$$\Omega_{2,1} = s\Lambda_{2,0}, \quad \Lambda_{2,2} = \frac{1}{2} + (2s^2-1)\Lambda_{2,0}. \qquad (9.54)$$

From Eq. (9.52), performing the integration $\int_0^{2\pi} \frac{d\theta}{2\pi}$ we have

$$su_0^{s,a} = \frac{u_1^{s,a}}{2}\left(1 + \frac{F_1^{s,a}}{2}\right),\qquad(9.55)$$

and performing the integration $\int_0^{2\pi}\cos\theta\,\frac{d\theta}{2\pi}$,

$$s\frac{u_1^{s,a}}{2} - \frac{u_0^{s,a}}{2}(1+F_0^{s,a}) - \frac{u_2^{s,a}}{4}\left(1+\frac{F_2^{s,a}}{4}\right) = \frac{\lambda}{2v_F}.\qquad(9.56)$$

From the two previous equations we find that

$$\left[s^2 - \frac{1}{2}(1+F_0^{s,a})\left(1+\frac{F_1^{s,a}}{2}\right)\right]u_0^{s,a} - \frac{u_2^{s,a}}{4}\left(1+\frac{F_1^{s,a}}{2}\right)\left(1+\frac{F_2^{s,a}}{4}\right)$$
$$= \left(1+\frac{F_1^{s,a}}{2}\right)\frac{\lambda}{2v_F}.\qquad(9.57)$$

Moreover, from (9.52) we consider those with $m=0$ and $m=2$, and divide them side by side to obtain

$$u_2^{s,a} = \frac{2u_0^{s,a}\Lambda_{2,0}}{\Lambda_{0,0} + \frac{F_2^{s,a}}{2}(\Lambda_{2,2}\Lambda_{0,0} - \Lambda_{2,0}^2)}.\qquad(9.58)$$

By substituting this equation into (9.57), we arrive at the expression

$$u_0^{s,a} = \frac{\frac{1}{2}(1+\frac{F_1^{s,a}}{2})\frac{\lambda}{v_F}}{s^2 - \frac{1}{2}(1+F_0^{s,a})(1+\frac{F_1^{s,a}}{2}) - \frac{1}{2}\frac{(1+\frac{F_1^{s,a}}{2})(1+\frac{F_2^{s,a}}{4})\Lambda_{2,0}}{\Lambda_{0,0}+\frac{F_2^{s,a}}{2}(\Lambda_{2,2}\Lambda_{0,0}-\Lambda_{2,0}^2)}},\qquad(9.59)$$

and for the linear response $\chi^{s,a}(q,\omega) = v_F\nu(0)u_0^{s,a}/\lambda$, we find the expression

$$\chi_{2D}^{s,a}(q,\omega) = \frac{\frac{1}{2}\nu(0)(1+\frac{F_1^{s,a}}{2})}{s^2 - \frac{1}{2}(1+\frac{F_1^{s,a}}{2})(1+F_0^{s,a}) - \frac{1}{2}\frac{(1+\frac{F_1^{s,a}}{2})(1+\frac{F_2^{s,a}}{4})\Lambda_{2,0}}{\Lambda_{0,0}+\frac{F_2^{s,a}}{2}(\Lambda_{2,2}\Lambda_{0,0}-\Lambda_{2,0}^2)}}.\qquad(9.60)$$

From the latter, it is possible to derive the following expressions for the sum rules:

$$m_{-1,2D}^{s,a} = \frac{N}{2}\frac{1}{1+F_0^{s,a}}\frac{2m^*}{k_F^2},\qquad(9.61)$$

$$m_{1,2D}^{s,a} = \frac{N}{2}\frac{q^2}{m^*}\left(1+\frac{F_1^{s,a}}{2}\right),\qquad(9.62)$$

$$m_{3,2D}^{s,a} = \frac{N}{2}\frac{q^4k_F^2}{m^{*3}}\left(1+\frac{F_1^{s,a}}{2}\right)^2\left(\frac{3}{4}+\frac{1}{2}F_0^{s,a}+\frac{1}{16}F_2^{s,a}\right).\qquad(9.63)$$

The frequency of ordinary sound in 2D is well represented by the energy

$$\omega^2 = \frac{m_1^s}{m_{-1}^s} = q^2 \epsilon_F (1 + F_0^{s,a}), \tag{9.64}$$

while that of the elastic mode (zero sound in 2D) is well represented by the energy

$$\omega^2 = \frac{m_3^s}{m_1^s} = \frac{m^*}{m} q^2 v_F^2 \left(\frac{3}{4} + \frac{1}{2} F_0^{s,a} + \frac{1}{16} F_2^{s,a} \right). \tag{9.65}$$

Like in the three-dimensional case, the 2D Landau theory can be generalized to charged Fermi systems by introducing a mean polarization field $\mathcal{E}(\mathbf{r}, t)$, related to the density fluctuations by

$$\mathcal{E}(q, \omega) = -i\mathbf{q} \frac{2\pi e}{q} \sum_{\mathbf{k}} \delta n_{\mathbf{k}}^s(q, \omega), \tag{9.66}$$

which is the Fourier transform of the Coulomb interaction in 2D, and by adding to the 2D Landau equation the term

$$-e\mathcal{E} \cdot \mathbf{v_k} \frac{\partial n_{\mathbf{k}}^0}{\partial \epsilon_{\mathbf{k}}}. \tag{9.67}$$

From this point on, the calculation of the linear response function proceeds as in the case of neutral systems. The final formulas for the dynamic polarizability $\chi^s(q, \omega)$ and for the sum rules are obtained from Eqs. (9.60)–(9.63), simply by replacing the Landau parameter F_0^s with $F_0^s + \nu(0) 2\pi e^2/q$. For the 2D electron gas, a variational Monte Carlo calculation for the Landau parameters is available (Kwon, Ceperley and Martin, 1994). The results of such calculation for some density values are reported in Table 9.3.

The 2D electron gas can be realized at the interfaces of heterostructures of GaAs/Al$_x$Ga$_{1-x}$As (Ando, Fowler and Stern, 1982) and of other structures. The response function of such systems has been studied by means of inelastic scattering of polarized light (Pinczuk *et al.*, 1989; Luo *et al.*, 1993; Eriksson *et al.*, 1999). These experiments, which evidence the existence of "zero sound" — like collective states — will be discussed later.

Table 9.3. Landau parameters for the 2D electron gas as deduced by Monte Carlo calculations at different densities.

	$r_s = 1$	$r_s = 2$	$r_s = 3$	$r_s = 5$
F_0^s	−0.60	−0.99	−1.63	−3.70
F_0^a	−0.34	−0.41	−0.49	−0.5
F_1^s	−0.14	−0.10	−0.03	0.12
F_1^a	−0.19	−0.24	−0.26	−0.27
F_2^s	−0.07	−0.16	−0.27	−0.50
F_2^a	0.01	0.07	0.14	0.32

9.2 Time Dependent Hartree (TDH) for Homogeneous Systems: The RPA

The time-dependent Hartree–Fock (HF) equations are the time-dependent version of the static HF Eqs. (2.7) and are written as ($\hbar = 1$)

$$i\frac{\partial}{\partial t}\varphi_i(\mathbf{r}, \sigma, t) = H^{\mathrm{HF}}\varphi_i(\mathbf{r}, \sigma, t), \tag{9.68}$$

where the single-particle wave functions $\varphi_i(\mathbf{r}, \sigma, t)$, and the one-body densities $\rho(\mathbf{r}, t)$ (diagonal) and $\rho^{(1)}(\mathbf{r}, \sigma; \mathbf{r}', \sigma', t)$ (nondiagonal), upon which H^{HF} depends, are functions of time. Equations (9.68) can be obtained by requiring that the action integral

$$I = \int_{t_1}^{t_2} \langle \phi_{\mathrm{SD}}(t)|H - i\frac{\partial}{\partial t}|\phi_{\mathrm{SD}}(t)\rangle dt, \tag{9.69}$$

computed for a generic Slater determinant, be stationary against an arbitrary variation of ϕ_{SD} in the space of Slater determinants:

$$\frac{\delta I}{\delta \phi_{\mathrm{SD}}} = 0. \tag{9.70}$$

The time-dependent Hartree (TDH) equations, are obtained by replacing the HF Hamiltonian with the Hartree Hamiltonian (which depends only on the diagonal, one-body density) in (9.68):

$$H^H = \frac{-\nabla^2}{2m} + v_{\mathrm{ext}}(\mathbf{r}) + \int \rho(\mathbf{r}', t)v(\mathbf{r} - \mathbf{r}')d\mathbf{r}'. \tag{9.71}$$

The TDH equations in the presence of a time-oscillating external field of the form (8.10) are written as

$$i\frac{\partial}{\partial t}\varphi_i(\mathbf{r}, t) = \left[\frac{-\nabla^2}{2m} + v_{\mathrm{ext}}(\mathbf{r}) + \int \rho(\mathbf{r}', t)v(\mathbf{r} - \mathbf{r}')d\mathbf{r}' \right.$$
$$\left. + \lambda(g^\dagger e^{-i\omega t} + h.c.) \right]\varphi_i(\mathbf{r}, t), \tag{9.72}$$

and for a homogeneous system have the same form for both Fermions and Bosons.

In the following, we will discuss the solutions to these equations for a homogeneous system in the case where

$$G = \sum_{i=1}^N g(r_i) = \sum_{i=1}^N e^{-i\mathbf{q}\cdot\mathbf{r}_i},$$

and subsequently we will compute the Hartree response function $\chi(F, G, \omega)$ for the operator

$$F = \sum_{i=1}^{N} f(r_i) = \sum_{i=1}^{N} e^{i\mathbf{q}\cdot\mathbf{r}_i},$$

i.e. the density response. For a homogeneous system, the static external field $v_{\text{ext}}(\mathbf{r})$ in (9.72) will be equated to zero in the cases of ^{3}He, nuclear matter and Bosons, and equal to the potential generated by a jellium of positive charge (ρ_+) in the case of electrons. In the latter case, Eqs. (9.72) have the form

$$i\frac{\partial}{\partial t}\varphi_i(\mathbf{r}, t) = \left[\frac{-\nabla^2}{2m} + e^2 \int \frac{\rho(\mathbf{r}', t) - \rho_+}{|\mathbf{r} - \mathbf{r}'|} d\mathbf{r}' \right.$$

$$\left. + \lambda \cdot \left(e^{i(\mathbf{q}\cdot\mathbf{r} - \omega t)} + e^{-i(\mathbf{q}\cdot\mathbf{r} - \omega t)}\right)\right]\varphi_i(\mathbf{r}, t). \tag{9.73}$$

In the homogeneous case, owing to translational invariance, Eq. (9.72) for $\rho(\mathbf{r}, t)$ will yield solutions of the form

$$\rho(\mathbf{r}, t) = \rho_0 + \delta\rho(\mathbf{r}, t), \tag{9.74}$$

where ρ_0 is the (constant) density of the (unperturbed) ground state, and

$$\delta\rho(\mathbf{r}, t) = \delta\rho(e^{i(\mathbf{q}\cdot\mathbf{r} - \omega t)} + e^{-i(\mathbf{q}\cdot\mathbf{r} - \omega t)}), \tag{9.75}$$

where $\delta\rho$ is a constant to be determined. Therefore, the fluctuation of the operator

$$F = \sum_{i=1}^{N} e^{i\mathbf{q}\cdot\mathbf{r}_i}$$

is given by

$$\delta F(G, \omega) = \langle\psi(t)|F|\psi(t)\rangle - \langle 0|F|0\rangle$$

$$= \int d\mathbf{r} e^{i\mathbf{q}\cdot\mathbf{r}}(\rho(\mathbf{r}, t) - \rho_0)_G = L^D \delta\rho e^{i\omega t}, \tag{9.76}$$

where L^D is the volume of the system in dimension D. Moreover, the dynamic polarizability [see Eq. (8.13)] is given by

$$\chi(q, \omega) = \frac{L^D \delta\rho}{\lambda}. \tag{9.77}$$

In order to determine $\delta\rho$, we will insert $\rho(\mathbf{r}, t)$ of Eqs. (9.74) and (9.75) into (9.72), and linearize the equations. This means writing the self-consistent Hartree mean field, i.e. $U(\mathbf{r}, \rho(\mathbf{r}, t)) = \int \rho(\mathbf{r}', t)v(\mathbf{r} - \mathbf{r}')d\mathbf{r}'$, as

$$U(\mathbf{r}, \rho(\mathbf{r}, t)) = U(\mathbf{r}, \rho_0) + \left.\frac{\partial U}{\partial\rho(\mathbf{r}, t)}\right|_{\rho=\rho_0} \delta\rho(\mathbf{r}, t). \tag{9.78}$$

Therefore, from Eqs. (9.72) and (9.78) we obtain

$$i\frac{\partial}{\partial t}\varphi_i(\mathbf{r},\sigma,t) = \left[\frac{-\nabla^2}{2m} + v_{\text{ext}}(\mathbf{r}) + U(\mathbf{r},\rho_0) + (\delta\rho v(q) + \lambda)\cdot(e^{i(\mathbf{q}\cdot\mathbf{r}-\omega t)}\right.$$

$$\left. + e^{-i(\mathbf{q}\cdot\mathbf{r}-\omega t)})\right]\varphi_i(\mathbf{r},\sigma,t), \tag{9.79}$$

where $v(q)$ is the Fourier transform of the interaction potential.

For example, if the interacting potential, is the Coulomb potential, then $v(q)$ is given by

$$v(q) = \begin{cases} \text{(1D)} & 2e^2 K_0(|qa|) \\ \text{(2D)} & 2\pi e^2/q \\ \text{(3D)} & 4\pi e^2/q^2 \end{cases}, \tag{9.80}$$

where for the "one-dimensional" case we have taken a Coulomb interaction of the form $1/\sqrt{(x-x')^2 + a^2}$, and K_0 is a modified Bessel function of the second kind [see for example Gradshteyn and Ryzhik (1980)] for which, in the limit $|qa| \to 0$, we have

$$K_0(|qa|) \sim |\ln qa|. \tag{9.81}$$

Equation (9.79) can be rewritten as

$$i\frac{\partial}{\partial t}\varphi_i(\mathbf{r},\sigma,t) = \left[\frac{-\nabla^2}{2m} + v_{\text{ext}}(\mathbf{r}) + U(\mathbf{r},\rho_0)\right.$$

$$\left. + \lambda'\cdot(e^{i(\mathbf{q}\cdot\mathbf{r}-\omega t)} + e^{-i(\mathbf{q}\cdot\mathbf{r}-\omega t)})\right]\varphi_i(\mathbf{r},\sigma,t), \tag{9.82}$$

with

$$\lambda'(q) = \lambda + \delta\rho v(q). \tag{9.83}$$

In the homogeneous systems we are considering, the static Hartree potential $U(\mathbf{r},\rho_0)$ is a constant which, in the case of the electron gas, cancels exactly the one originating from v_{ext} (for the other systems, $v_{\text{ext}} = 0$) and, apart from this inessential constant, Eq. (9.82) coincides with that of a noninteracting system coupled to an external time-oscillating field, with a coupling constant λ' given by Eq. (9.83). For such a system, the density response function is the (single-particle) free response $\chi_0((\mathbf{q},\omega)$, which we studied in Subsection 8.6. Therefore, in the TDH theory, the particles of the N-body interacting system respond like noninteracting particles to the combined effect of the external field (λ) plus the polarization field induced by the density fluctuations in the self-consistent Hartree mean field. From Eq. (9.77) and from the analogous relation for the free response function,

$$\chi^0(q,\omega) = \frac{L^D \delta\rho}{\lambda'(q)}, \qquad (9.84)$$

we then obtain

$$\lambda\chi(q,\omega) = \lambda'(q)\chi^0(q,\omega) = L^D \delta\rho, \qquad (9.85)$$

which are known as the RPA equations. The solution to these equations is given by

$$\chi^{\mathrm{RPA}}(q,\omega) = \frac{\chi^0(q,\omega)}{1 - \frac{v(q)}{L^D}\chi^0(q,\omega)}, \qquad (9.86)$$

and is called the RPA response function.

The energies of the RPA excited states of the system are given by the poles of Eq. (9.86), and are obtained from the solutions to

$$1 - \frac{v(q)}{L^D}\chi^0(q,\omega) = 0. \qquad (9.87)$$

The graphical solution to this equation is obtained by plotting the free response $\chi^0(q,\omega)$ for fixed q, as a function of ω, and finding the intersections of $\chi^0(q,\omega)/L^D$ with the straight line $1/v(q)$, parallel to the ω axis.

The RPA matrix elements of the density operator $\rho_{\mathbf{q}} = F = \sum_{i=1}^{N} e^{i\mathbf{q}\cdot\mathbf{r}_i}$, between the ground state and the RPA excited states with energies ω, given by the solutions (9.87) for the poles of $\chi^{\mathrm{RPA}}(q,\omega)$, are found by calculating the residues of $\chi^{\mathrm{RPA}}(q,\omega)$ in such poles. To perform such a calculation, let us put Eq. (9.86) in the form

$$\frac{1}{\chi^{\mathrm{RPA}}(q,\omega)} = \frac{1}{\chi^0(q,\omega)} - \frac{v(q)}{L^D}, \qquad (9.88)$$

and let us expand it around the generic pole of energy $\bar\omega$, keeping in mind that in the pole $(\chi^{\mathrm{RPA}}(q,\bar\omega))^{-1} = 0$, so that $(\chi^0)^{-1} = v(q)/L^D$. As a result, we obtain

$$\frac{1}{\chi^{\mathrm{RPA}}(q,\omega)} \sim \frac{\partial}{\partial\omega}\left(\frac{1}{\chi^0}\right)\Big|_{\omega=\bar\omega}(\omega-\bar\omega) = -\left(\frac{v(q)}{L^D}\right)^2 \frac{\partial\chi^0}{\partial\omega}\Big|_{\omega=\bar\omega}(\omega-\bar\omega), \qquad (9.89)$$

i.e.

$$\chi^{\mathrm{RPA}}(q,\omega) \sim -\left(\frac{L^D}{v(q)}\right)^2 \frac{1}{\chi'_0(\omega=\bar\omega)}\frac{1}{\omega-\bar\omega}, \qquad (9.90)$$

where $\chi'_0 = \frac{\partial\chi^0}{\partial\omega}$. Therefore, the desired matrix element is given by

$$|\langle 0|\rho_{\mathbf{q}}|\bar\omega\rangle|^2 = -\left(\frac{L^D}{v(q)}\right)^2 \frac{1}{\chi'_0(\omega=\bar\omega)}. \qquad (9.91)$$

Equations (9.86), (9.87) and (9.91) hold for both Fermion and Boson systems. As discussed in Subsection 8.6, what is quite different in the two cases is the form of the free response $\chi^0(q,\omega)$. As will appear later, such a difference leads to a great diversity in the RPA solutions in the two cases.

Another important difference between Fermion and Boson systems concerns the range of validity of the RPA Eq. (9.86) at finite temperature. In these equations, the thermal effects show themselves only through the free response function, which includes the occupation numbers of the single-particle states. Whereas for Fermions the use of the Fermi–Dirac thermal occupation numbers is appropriate and the ensuing RPA equations at finite temperature make sense [see for example Braghin and Vautherin (1994), Hernandez *et al.* (1995)], in the Boson case the use of the thermal Bose–Einstein occupation numbers leads to equations which, when solved, produce wrong results.

This behavior is due to the different nature of low energy elementary excitations for the two systems, i.e. *p–h* excitations for Fermions and phonons for Bosons. While the starting point for the time-dependent theory that leads to the RPA equations, i.e. the HF ground state, in the case of finite temperature is a good one for Fermions, this is not the case for Bosons and leads to wrong results. It is only at zero temperature that the RPA equations for Bosons do provide a good solution for the excited states of the system. This fact will be discussed in Subsection 9.5. Subsequently, in Subsection 9.6, we will briefly discuss a finite temperature approach for Bosons, known as the Popov approximation, which produces correct results for these systems.

9.3 TDH for the Density Matrix and the Landau Equations

From the TDH equations

$$i\hbar\frac{\partial}{\partial t}\varphi_i(\mathbf{r},t) = \left[\frac{-\hbar^2\nabla^2}{2m} + \int \rho(\mathbf{r}',t)v(\mathbf{r}-\mathbf{r}')d\mathbf{r}'\right]\varphi_i(\mathbf{r},t), \qquad (9.92)$$

where for simplicity we have omitted the irrelevant external field term, by multiplying on the left hand side by $\varphi_i^*(\mathbf{r}',t)$ and performing a summation on occupied states, and recalling the definition of the nondiagonal, one-body density matrix, i.e.

$$\rho^{(1)}(\mathbf{r},\mathbf{r}',t) = \sum_i \varphi_i^*(\mathbf{r},t)\varphi_i(\mathbf{r}',t),$$

it is easy to obtain the equation

$$i\hbar\frac{\partial}{\partial t}\rho^{(1)}(\mathbf{r},\mathbf{r}',t) = \left[\frac{-\hbar^2}{2m}(\nabla_{\mathbf{r}}^2 - \nabla_{\mathbf{r}'}^2) + (U(\mathbf{r},\rho(\mathbf{r},t)) - U(\mathbf{r}',\rho(\mathbf{r}',t)))\right]\rho^{(1)}(\mathbf{r},\mathbf{r}',t),$$

$$(9.93)$$

where

$$U(\mathbf{r},\rho(\mathbf{r},t)) = \int \rho(\mathbf{r}',t)v(\mathbf{r}-\mathbf{r}')d\mathbf{r}'$$

is the Hartree self-consistent mean field. Then, introducing the relative and center-of-mass variables, $\mathbf{s} = \mathbf{r} - \mathbf{r}'$, $\mathbf{R} = (\mathbf{r} + \mathbf{r}')/2$, Eq. (9.93) can be rewritten as

$$\left\{ i\hbar \frac{\partial}{\partial t} + \frac{\hbar^2}{2m} \nabla_{\mathbf{s}} \cdot \nabla_{\mathbf{R}} - \left[U\left(\mathbf{R} - \frac{\mathbf{s}}{2}, t\right) - U\left(\mathbf{R} + \frac{\mathbf{s}}{2}, t\right) \right] \right\} \rho^{(1)}(\mathbf{R}, \mathbf{s}, t) = 0. \quad (9.94)$$

Let us now introduce the distribution function

$$f(\mathbf{R}, \mathbf{p}, t) = \frac{1}{(2\pi)^D} \int ds e^{\frac{i}{\hbar} \mathbf{p} \cdot \mathbf{s}} \rho^{(1)}(\mathbf{R}, \mathbf{s}, t), \quad (9.95)$$

which gives the probability of finding a particle at position $\mathbf{R}$ with momentum $\mathbf{p}$ at time t. Apart from a constant, the distribution function f coincides with the Landau distribution function $n(\mathbf{R}, \mathbf{p}, t)$, as defined in Subsection 1.9. This can easily be seen by computing f starting from the density expression $\rho^{(1)}(\mathbf{R}, \mathbf{s})$ built up by means of a static, plane wave Slater determinant. In this case, apart from a constant, f coincides with the ground state distribution (1.76). By transforming Eq. (9.94), the following equation for f is obtained:

$$\frac{\partial}{\partial t} f(\mathbf{R}, \mathbf{p}, t) + \frac{\mathbf{p}}{m} \cdot \nabla_{\mathbf{R}} f(\mathbf{R}, \mathbf{p}, t)$$

$$- \frac{1}{i\hbar} \int ds e^{\frac{i}{\hbar} \mathbf{p} \cdot \mathbf{s}} \left[U\left(\mathbf{R} - \frac{\mathbf{s}}{2}, t\right) - U\left(\mathbf{R} + \frac{\mathbf{s}}{2}, t\right) \right] \rho^{(1)}(\mathbf{R}, \mathbf{s}, t) = 0. \quad (9.96)$$

By expanding, for small $\mathbf{s}$, the difference

$$U\left(\mathbf{R} - \frac{\mathbf{s}}{2}, t\right) - U\left(\mathbf{R} + \frac{\mathbf{s}}{2}, t\right),$$

which appeared in the last term of Eq. (9.96), and using the relation

$$-\frac{1}{i\hbar} \int ds e^{\frac{i}{\hbar} \mathbf{p} \cdot \mathbf{s}} \mathbf{s} \cdot (\nabla_{\mathbf{R}} U(\mathbf{R}, t)) \rho^{(1)}(\mathbf{R}, \mathbf{s}, t) = \nabla_{\mathbf{R}} U(\mathbf{R}, t) \cdot \nabla_{\mathbf{p}} f(\mathbf{R}, \mathbf{p}, t),$$

we obtain

$$\frac{\partial}{\partial t} f(\mathbf{R}, \mathbf{p}, t) + \frac{\mathbf{p}}{m} \cdot \nabla_{\mathbf{R}} f(\mathbf{R}, \mathbf{p}, t) - \nabla_{\mathbf{R}} U(\mathbf{R}, t) \cdot \nabla_{\mathbf{p}} f(\mathbf{R}, \mathbf{p}, t) + O(\hbar^2) = 0. \quad (9.97)$$

Eq. (9.97) is the semiclassical limit ($\hbar \to 0$) of the TDH equation, and holds in the long wavelength limit, where the momentum transferred to the system is small, and the expansion of the Hartree potential used for its derivation is correct. By writing $f(\mathbf{R}, \mathbf{p}, t) = f_{\mathbf{p}}^0 + \delta f(\mathbf{R}, \mathbf{p}, t)$, and using the relations

$$\nabla_{\mathbf{R}} U(\mathbf{R}, t)) = \left. \frac{\partial U}{\partial \rho(\mathbf{R}, t)} \right|_{\rho(\mathbf{R}, t) = \rho_0} \nabla_{\mathbf{R}} \delta\rho(\mathbf{R}, t),$$

$$\sum_{\mathbf{p}} \delta f(\mathbf{R}, \mathbf{p}, t) = \delta\rho(\mathbf{R}, t),$$

$$(9.98)$$

we can immediately identify such an equation with the Landau–Vlasov Eq. (1.136) in the absence of collisions, and with the only nonzero Landau parameter F_0 given by

$$F_0 = \nu(0) \frac{\partial U}{\partial \rho}\bigg|_{\rho=\rho_0}, \qquad (9.99)$$

where $\nu(0)$ is the density of states at the Fermi surface [see Eqs. (1.140) and (1.155)]. The Landau–Vlasov equations have been extensively used to study the collective modes of nuclei [see for example Brink, Dellafiore and Di Toro (1986)].

The connection between the TDH theory and the Landau theory can also be evidenced in a direct fashion starting from the RPA response of Eq. (9.86). In fact, in the $q \to 0$, $\omega \to 0$ limit the free responses $\chi^0(q,\omega)$ of Subsection 8.6 have the limits

$$\chi_0^{3D}(q,\omega) = -\nu^{3D}(0)\Omega_{0,0}(s), \quad \chi_0^{2D}(q,\omega) = -\nu^{2D}(0)\Lambda_{0,0}(s), \qquad (9.100)$$

where $\Omega_{0,0}(s)$ and $\Lambda_{0,0}(s)$, with $s = \omega/qv_F$, are the functions which appear in the Landau theory, and are as defined in (9.14) and (9.53), respectively. The Fourier transform of the interaction potential, in the same limit, tends to

$$v(0) = L^D \frac{\partial U}{\partial \rho}\bigg|_{\rho=\rho_0}.$$

Hence, it is immediate to verify that in the $q \to 0$, $\omega \to 0$ limit, Eq. (9.86) tends to the results of Eqs. (9.23) and (9.60), with $F_1 = F_2 = 0$ and with F_0 given by (9.99).

The identification of the RPA and Landau theories is feasible only in the $\hbar \to 0$ limit, i.e. long wavelength. The RPA theory incorporates quantum effects, which are absent in the Landau theory; as we will see, such terms may be important. Moreover, the RPA theory can be generalized to nonhomogeneous systems like nuclei, metal clusters, quantum dots and free surfaces.

9.4   The RPA for the Electron Gas in Different Dimensions: The Plasmon

As an example of a Fermion system, let us consider the electron gas in different dimensions, and let us study the RPA solutions. The energies of the RPA states are given by the solutions to Eq. (9.87) where $v(q)$ is given, in the different cases, in (9.80), and the free response functions were studied in Subsection 8.6. The graphical solution of such equations in 2D and 3D is outlined in Fig. 9.1, where we plot $\chi^0(q,\omega)$, for q fixed and much smaller than $2k_F$, as a function of ω, as well as the straight line for constant $1/v(q)$. Given the form of $v(q)$ for the electron gas, such straight line lies in the upper semiplane, and the smaller q is, the nearer it is to the ω axis.

From this figure, we notice that the RPA energies, in the region of the poles of $\chi^0(q,\omega)$, are practically identical to the single-particle energies ϵ_{mi}, but besides these solutions, a new one emerges. This solution is, in general, known as the RPA

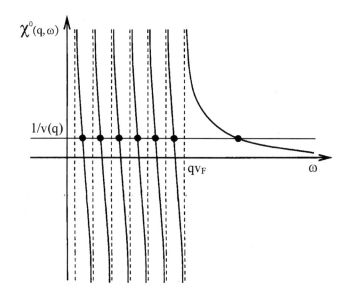

Fig. 9.1 Graphical solution of Eq. (9.87), for the 2D and 3D electron gas.

"collective solution" (zero sound ^{3}He, plasmon for the electron gas, giant resonances for nuclei), in order to differentiate it from single-particle-like solutions. The smaller q is, the farther away the energy of such a solution is from the single-particle continuum. Therefore, in order to study this new (with respect to the single-particle excitations) type of excitation, let us consider the limit of small q. In this limit we have $\omega \gg \epsilon_{mi}$, so that for $\chi^0(q,\omega)$ we can use the expansion

$$\chi^0(q,\omega) = \frac{2}{\omega^2}\left(m_1^{\mathrm{sp}} + \frac{1}{\omega^2}m_3^{\mathrm{sp}} + \cdots\right), \tag{9.101}$$

where

$$m_1^{\mathrm{sp}} = \frac{1}{2}\langle\mathrm{SD}|[[\rho_{\mathbf{q}}^\dagger,[T,\rho_{\mathbf{q}}]]]|\mathrm{SD}\rangle = \frac{Nq^2}{2m} \tag{9.102}$$

and

$$m_3^{\mathrm{sp}} = \frac{1}{2}\langle\mathrm{SD}|[[\rho_{\mathbf{q}}^\dagger,T],[T,[T,\rho_{\mathbf{q}}]]]|\mathrm{SD}\rangle = \frac{NT_0^D}{m^2}q^4 \tag{9.103}$$

are the sum rules for the density operator of the noninteracting system (T is the kinetic energy operator, and $|\mathrm{SD}\rangle$ is the ground state of the Fermi gas), and where T_0^D is the kinetic energy per particle of the Fermi gas:

$$T_0^D = \frac{D}{D+2}\frac{k_F^2}{2m}. \tag{9.104}$$

Inserting these results into Eq. (9.87), we find for the energy of the RPA collective state

$$\omega_{\text{RPA}}^2 = \omega_p^2 + \frac{2T_0^D}{m} q^2, \qquad (9.105)$$

which is valid at order q^2, and where ω_p is the plasma frequency given by

$$\omega_p = \begin{cases} (1\text{D}) & \sqrt{\dfrac{2\lambda e^2}{m}} q^2 |\ln qa| \\[12pt] (2\text{D}) & \sqrt{\dfrac{2\pi e^2 q\sigma}{m}} \\[12pt] (3\text{D}) & \sqrt{\dfrac{4\pi e^2 \rho_0}{m}} \end{cases} , \qquad (9.106)$$

and $\lambda = N/L$ is the linear density, $\sigma = N/L^2$ the surface density, and $\rho_0 = N/L^3$ the bulk density.

Now let us compute the strength of the collective solution (9.105), using Eq. (9.91) in the limit of small q, in which the expansion of Eqs. (9.101)–(9.103) for $\chi^0(q,\omega)$ holds. Using

$$\chi^0(q,\omega) \simeq \frac{Nq^2}{m\omega^2} \left(1 + \frac{2T_0^D}{m\omega^2} q^2 \right),$$

a simple calculation leads to the result

$$|\langle 0|\rho_{\mathbf{q}}|\omega_{\text{RPA}}\rangle|^2 = \left(\frac{L^D}{v(q)} \right)^2 \frac{m\omega_{\text{RPA}}^3}{2Nq^2 \left(1 + \frac{4T_0^D q^2}{m\omega_{\text{RPA}}^2} \right)}, \qquad (9.107)$$

from which we immediately have

$$\omega_{\text{RPA}} |\langle 0|\rho_{\mathbf{q}}|\omega_{\text{RPA}}\rangle|^2 = \frac{Nq^2}{2m}, \qquad (9.108)$$

which is valid at order q^2, which means that the collective state exhausts the f-sum rule in the limit of small q. Therefore, in this limit, the whole strength is in this one state, and the dynamic form factor $S(\mathbf{q},\omega)$, in the RPA approximation and for small q, can be written as

$$S^{\text{RPA}}(\mathbf{q},\omega) \sim \frac{Nq^2}{2m\omega_p} \delta(\omega - \omega_p). \qquad (9.109)$$

If we compare this result with that of the independent-particle model, which in the 3D case (which we took as an example; see Subsection 1.8.1) yielded for $q \ll k_F$

$$S_{3D}(\mathbf{q}, \omega) = V \frac{m^2 \omega}{2\pi^2 q} \qquad \text{for } \omega < q v_F,$$

$$S_{3D}(\mathbf{q}, \omega) = 0 \qquad \text{for } \omega > q v_F,$$

we immediately see that the interaction has crucial effects, which lead to the prediction of a new kind of excited state, the plasmon, which was observed experimentally in the 3D case, by inelastic electron scattering on alkali metals (see Fig. 8.1) at an energy value that is in very good agreement with the RPA prediction (for $q \simeq 0$). For example, in the case of Sodium which has an experimental density corresponding to $r_s = 3.93$, one finds that $(\omega_p = 4\pi e^2 \rho_0 / m)^{1/2} = 5.93$ eV, which is a value very close to the peak energy of Fig. 8.1 for $q \simeq 0$. Moreover, the q dispersion of the RPA energy is positive, in agreement with the experimental data of Fig. 8.1 for sodium.

The RPA strength taken by the collective state can be computed, not only in the small q limit, but for any q. The fraction $F(q)$ of the f-sum rule exhausted by the collective state as a function of q is plotted in Fig. 9.2, as calculated in the TDLDA theory (see Subsection 9.9.1), which takes into account the effect of exchange and correlations in some approximate way. From such a figure we see that starting from a given q value (critical q), as q increases, the strength taken by the collective state quickly drops to zero. The critical q is the q value at which the collective solution joins the continuum of one-particle–one-hole excited states, and

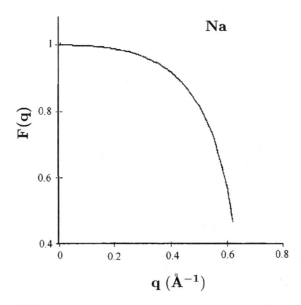

Fig. 9.2 Fraction $F(q)$ of the m_1 f-sum rule for the density operator, concentrated in the volume plasmon for sodium ($r_s = 3.93$).

the so-called Landau fragmentation of the strength takes place, or Landau damping
of the collective state into single-particle states takes place. The Landau damping
can be observed in the experimental curve of Fig. 8.1 as a broadening of the collective
state with increasing q, which corresponds to the fragmentation of the strength into
many states, rather than just into a collective one.

However, the agreement between the RPA theory and finite-q experiment is
only qualitative. From a quantitative standpoint, there are important discrepancies,
which are highlighted in Fig. 9.3. In this figure we plot the dispersion coefficient α
of the experimental bulk plasmon dispersion, defined by the equation

$$E(q) = E(0) + \frac{\alpha}{m}q^2,\qquad(9.110)$$

and normalized to the RPA dispersion coefficient, $\alpha_{\text{RPA}} = 3\epsilon_F/5\omega_p$, as a function
of the Wigner–Seitz radius r_s. The experimental values were obtained by inelastic
scattering experiments on Na, K, Rb and Cs (vom Felde *et al.*, 1989), to which
correspond the r_s values of 3.93, 4.86, 5.20 and 5.62, respectively. From this figure
we notice that the experiment predicts a strong variation of $\alpha/\alpha_{\text{RPA}}$ with r_s, which
disagrees with the RPA theory. Even negative dispersion coefficients are observed
for the r_s value corresponding to cesium. The divergences between the RPA results
for the dispersion of the bulk plasmon and the experimental results will form the
subject of further discussion in forthcoming Subsections.

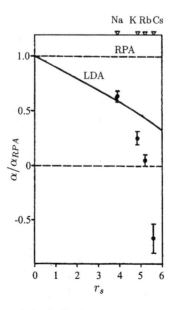

Fig. 9.3 Dispersion coefficient of the bulk plasmon, normalized to the RPA coefficient, as a
function of r_s. The experimental data are from vom Felde *et al.*, (1989). The constant dashed line
at value 1 is the RPA prediction; the full line is the TDLDA prediction.

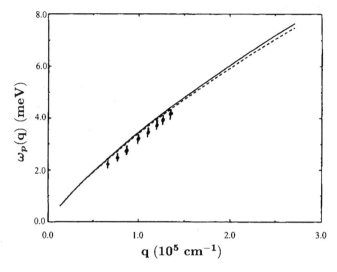

Fig. 9.4 Experimental results for the one-dimensional plasmon (Goi *et al.*, 1991), compared with the RPA (full line) and TDLDA (dashed line) theories.

Unlike the 3D case, the plasmon frequency in 2D and 1D vanishes in the limit of small q. This important feature of the low-dimensionality plasmon was observed in 2D by Allen *et al.* (1986) in inversion layers, by Grimes and Adams (1976) for electrons on liquid helium, and in 1D by Goni *et al.* (1991) in quantum wires of GaAs. In Fig. 9.4 we report the experimental dispersion of the plasmon in the GaAs quantum wire, together with the RPA prediction of Quiang Li and Das Sarma (1989, 1991) and the TDLDA ones (see Subsection 9.9). As seen from the figure, there seems to be good agreement between theory and experiment.

As regards the dimensionality effects on the collective nature of the plasmon, we notice what follows. In 2D the behavior of the RPA strength as a function of transferred momentum q is quite analogous to the 3D one, shown in Fig. 9.2. However, the situation is very different in one dimension. The fraction $F(q)$ of the f-sum rule m_1 of the density operator, which is taken by the quasi-one-dimensional plasmon, and which specifies the relative importance of the plasmon and single-particle excitations, is plotted as a function of q in Fig. 9.5 for GaAs quantum wires of varying width and density.

From the figures, it can be seen that the plasmon collective state in 1D dominates the excitation process in a very wide range of q values, much wider than in 3D and 2D. Moreover, unlike the 3D and 2D cases, in 1D the continuum of single-particle states has a gap [see Eq. (1.105) and Fig. 8.3]. This has the consequence that there exist (undamped) collective states for the 1D system even for attractive interaction for which $v(q)$ is negative, and that for a wide range of q values the excitation spectrum in 1D, unlike the 2D and 3D ones, is always dominated by collective states. As we will see in the next subsection, this is analogous to what happens for Boson

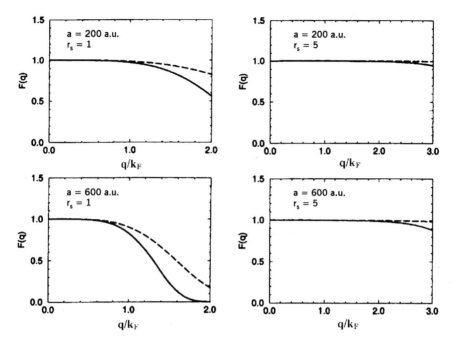

Fig. 9.5 The fraction $F(q)$ of the f-sum rule m_1 for the density operator taken by the quasi-one-dimensional plasmon, for quantum wires of various widths and densities. The dashed lines are the RPA results; the full lines are the TDLDA results (Agosti *et al.*, 1998).

systems. As a result, the 1D system has more resemblance to a Tomonaga–Luttinger liquid (Tomonaga, 1950; Luttinger, 1963), where it is assumed that the excitations have a Boson nature, rather than to a Fermi liquid, in which the excitations are of the Fermion type.

9.5 The RPA for Bosons

For systems of Bosons, the energies of the RPA states are still given by the solutions to Eq. (9.87), where $v(q)$ is now the Fourier transform of the interaction among Boson atoms, and the free response function was studied in Subsection 8.6 and is given in (8.80):

$$\chi_0^B(\mathbf{q},\omega) = \frac{\frac{Nq^2}{m}}{\omega^2 - (\frac{q^2}{2m})^2}.$$

Using the result of (9.87), we obtain for the energy of the RPA solution

$$\omega = \sqrt{\frac{\rho q^2 v(q)}{m} + \left(\frac{q^2}{2m}\right)^2}. \tag{9.111}$$

In the limit of small transferred momentum q, Eq. (9.111) leads to the phonon-like dispersion relation

$$\omega \simeq cq, \quad c = \sqrt{\frac{\rho v(0)}{m}}, \tag{9.112}$$

which is linear in q. Therefore, at small q the interaction changes the q dependence of the RPA excitations with respect to the q^2 term of the independent-particle model. The elementary excitations of the Bose systems are different from the single-particle excitations, and are of the phonon type. This fact has a very important consequence on the thermodynamics of Boson systems. In fact, thermodynamic properties based on phonon-like elementary excitations are completely different from those based on single-particle-like elementary excitations. For example, in a system of Boson atoms, the specific heat is no longer linear with the temperature as in (1.134) (which holds for Fermions), but rather grows like T^3 [see for example Huang (1987)].

The graphical solution of the RPA equations for Bosons is outlined in Fig. 9.6, where we plot $\chi^0(q, \omega)$ as a function of ω, as well as the straight line of constant value $1/v(q)$.

From this figure we see that the single-particle state at $q^2/2m$ disappears completely, and what is left is only the collective phonon state with a dispersion law given by (9.111). Such a dispersion law is plotted in Fig. 9.7 for superfluid ^{4}He,

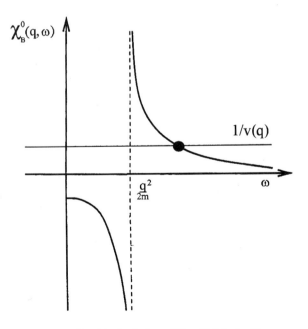

Fig. 9.6   Graphical solution of Eq. (9.87), for Bosons.

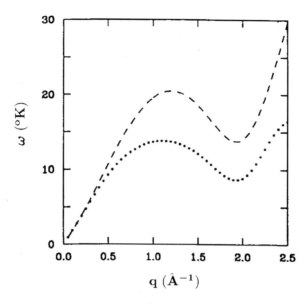

Fig. 9.7 Dispersion of the elementary excitations in superfluid ^{4}He. The dashed line is the RPA result (9.111). The dots are the experimental data by Donnelly *et al.* (1981).

together with the experimental data of Donnelly *et al.* (1981). The function $v(q)$ is fixed, through the equation

$$\chi(q) = \frac{\chi^0(q)}{1 - \frac{v(q)}{L^D}\chi^0(q)}, \qquad (9.113)$$

by the static response function, which is experimentally known in superfluid ^{4}He and is plotted in Fig. 9.8.

As can be observed from Fig. 9.7, the RPA prediction explains quite well all the features of the experimental excitation spectrum, i.e. the low-q phonon linear regime, the single-particle quadratic regime at high q when the interaction term in Eq. (9.111) becomes negligible, and lastly the roton minimum at intermediate q. However, from a quantitative point of view, it overestimates the experimental result. Such a discrepancy between theory and experiment will be further discussed later on, together with the discrepancy found in the case of bulk plasmon dispersion.

Recalling that in the $q \to 0$ limit the static polarizability $\chi(q)$ tends to $-N\rho K$, where K is the isothermal compressibility of the system, and that for Bosons the compressibility of the noninteracting system is infinite, by starting from Eq. (9.113) it is possible to connect the interaction $v(q)$ (for $q \to 0$) to the compressibility:

$$v(0) = \frac{1}{K\rho^2}. \qquad (9.114)$$

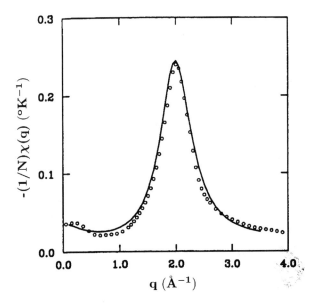

Fig. 9.8 Static response function of superfluid ^{4}He. The full line is the density functional result of Dalfovo *et al.* (1995). The dots are the experimental results by Woods and Cowley (1973).

Therefore, the coefficient c of the phonon dispersion relation (9.112) is determined by the system compressibility

$$c = \sqrt{\frac{1}{mK\rho}}, \tag{9.115}$$

and gives the propagation velocity of sound in the system. In the case of a cold and dilute Boson gas, for which it is possible to replace the two-body interaction with an effective interaction given by $v(r) = g\delta(r)$, where the constant g is determined by the scattering length a through $g = 4\pi a/m$, we have $v(q) = g$, and in this case the RPA frequency dispersion is given by

$$\omega = \sqrt{\frac{\rho q^2 g}{m} + \left(\frac{q^2}{2m}\right)^2}. \tag{9.116}$$

Note that for attractive interactions ($a < 0$), the frequency ω becomes imaginary at small q, evidencing a system instability (see Subsection 2.4).

From an unrestricted point of view, the following general considerations apply to quantum liquids with the elementary-excitation dispersion law having a form as in Fig. 9.7:

(i) At thermal equilibrium, most of the elementary excitations of the system have energies close to the minima of the curve $\omega(q)$, i.e. near $\omega = 0$ and $\omega(q_0)$, where q_0 is the momentum value where the roton minimum lies. Therefore, such minima have a paramount importance for the properties of the quantum liquid. In particular, in the neighborhood of the roton minimum it is possible to expand $\omega(q)$ as a function of $q - q_0$, thus obtaining, at second order, $\omega(q) = \Delta + (q - q_0)^2/2m^*$. Starting from this result it is possible to compute the contribution of rotons (i.e. the elementary excitations with this energy) to the thermodynamic functions of the system; the roton contribution may be greater than that of the phonon at sufficiently high temperatures.

(ii) These systems have the property of superfluidity, i.e. of flowing through narrow capillaries or fissures with no viscosity.

For a detailed description of these phenomena, see for example the books by Lifchitz and Pitaevskii (1981) and Pines and Nozières (1990).

9.6 The Time-Dependent Gross–Pitaevskii Theory

The Hamiltonian of N interacting Bosons can be written in second-quantization formalism (see Appendix 2.8), as

$$\hat{H} = \int d\mathbf{r}\,\hat{\Phi}^\dagger(\mathbf{r}) \left(\frac{-\nabla^2}{2m} + v_{\text{ext}}(\mathbf{r}) \right) \hat{\Phi}(\mathbf{r})$$

$$+ \frac{1}{2} \int d\mathbf{r}\,d\mathbf{r}'\,\hat{\Phi}^\dagger(\mathbf{r})\hat{\Phi}^\dagger(\mathbf{r}')v(\mathbf{r} - \mathbf{r}')\hat{\Phi}(\mathbf{r})\hat{\Phi}(\mathbf{r}'), \qquad (9.117)$$

where $\hat{\Phi}(\mathbf{r})$ and $\hat{\Phi}^\dagger(\mathbf{r})$ are the Boson field operators which annihilate and create a particle, respectively, at $\mathbf{r}$. Note that using in (9.117) the expansion $\hat{\Phi}(\mathbf{r}) = \sum_\alpha \varphi_\alpha(\mathbf{r})a_\alpha$, where the $\varphi_\alpha(\mathbf{r})$ are single-particle wave functions and a_α are the corresponding annihilation operators, we obtain for Bosons the equivalent of Eqs. (2.65)–(2.67), which were derived in Subsection 2.5 in the case of N interacting Fermions.

For a homogeneous gas in a volume V, we have

$$\hat{\Phi}(\mathbf{r}) = \frac{1}{\sqrt{V}} \sum_{\mathbf{p}} e^{i\mathbf{p}\cdot\mathbf{r}} a_{\mathbf{p}}, \quad N = \int d\mathbf{r}\,\hat{\Phi}^\dagger(\mathbf{r})\hat{\Phi}(\mathbf{r}) = \sum_{\mathbf{p}} a_{\mathbf{p}}^\dagger a_{\mathbf{p}}. \qquad (9.118)$$

Bose–Einstein condensation (BEC) occurs when the number n_0 of atoms that occupy a particular single-particle state becomes very large, $n_0 \equiv N_0 \gg 1$, and the ratio N_0/N stays finite in the thermodynamic limit $N \to \infty$. BEC occurs in a single-particle state $\varphi_0 = 1/\sqrt{V}$ with momentum $\mathbf{p} = 0$, and the operators relative to such a state, a_0 and $a_0^\dagger$, can be treated as numbers: $a_0 = a_0^\dagger = \sqrt{N_0}$. Therefore, the field operator $\hat{\Phi}(\mathbf{r})$ can be decomposed in the form

$$\hat{\Phi}(\mathbf{r}) = \sqrt{\frac{N_0}{V}} + \hat{\Phi}'(\mathbf{r}). \qquad (9.119)$$

Treating the operator $\hat{\Phi}'(\mathbf{r})$ as a perturbation, Bogoliubov (1947) set up a first-order theory for the elementary excitations of Bose gases which, as we will see, yields the same RPA results as in Subsection 9.5.

The generalization of the Bogoliubov theory for nonuniform systems and time-dependent configurations is given by

$$\hat{\Phi}(\mathbf{r}, t) = \Psi(\mathbf{r}, t) + \hat{\Phi}'(\mathbf{r}, t), \qquad (9.120)$$

where we have used the Heisenberg representation for the field operators. $\Psi(\mathbf{r}, t)$ is a complex function defined as the expectation value of the field operator:

$$\Psi(\mathbf{r}, t) = \langle \hat{\Phi}(\mathbf{r}, t) \rangle.$$

Its modulus identifies the density of the condensate: $\rho_0 = |\Psi(\mathbf{r}, t)|^2$. The function $\Psi(\mathbf{r}, t)$ is a classical field which has the meaning of an order parameter, and is often named the "wave function of the condensate" [for further considerations, see Dalfovo *et al.* (1999)]. The decomposition (9.120) is especially useful when $\hat{\Phi}'(\mathbf{r}, t)$ is small, i.e. when the number of particles outside the condensate is small. In this case, it is possible to derive an equation for the order parameter by expanding the theory at first order in $\hat{\Phi}'(\mathbf{r}, t)$, like in the case of homogeneous systems. Unlike the latter, we obtain a nontrivial theory at zero order for $\Psi(\mathbf{r}, t)$. To obtain the theory, we start from the Heisenberg equation for the Hamiltonian (9.117) ($\hbar = 1$):

$$i\frac{\partial}{\partial t}\hat{\Phi}(\mathbf{r}, t) = [\hat{\Phi}, \hat{H}]$$

$$= \left(-\frac{\nabla^2}{2m} + v_{\text{ext}}(\mathbf{r}) + \int d\mathbf{r}' \hat{\Phi}^\dagger(\mathbf{r}', t) v(\mathbf{r} - \mathbf{r}') \hat{\Phi}(\mathbf{r}', t)\right) \hat{\Phi}(\mathbf{r}, t),$$

$$(9.121)$$

and we replace the operator $\hat{\Phi}$ by the classical field Ψ. This approximation turns out to be a very good one in the case of a cold and dilute atom gas, for which only low energy two-body interactions are important, which can be described by the effective interaction of Subsection 2.4: $v(\mathbf{r} - \mathbf{r}') = g\delta(\mathbf{r} - \mathbf{r}')$, with $g = 4\pi a/m$ and a the scattering length in the s wave. The use of such an effective interaction in (9.121) is consistent with the replacement of $\hat{\Phi}$ by the classical field Ψ, and leads to the time-dependent Gross–Pitaevskii (GP) equations (Gross, 1961, 1963; Pitaevskii, 1961):

$$i\frac{\partial}{\partial t}\Psi(\mathbf{r}, t) = \left(\frac{-\nabla^2}{2m} + v_{\text{ext}}(\mathbf{r}) + g|\Psi(\mathbf{r}, t)|^2\right) \Psi(\mathbf{r}, t). \qquad (9.122)$$

Equations (9.122) hold in the limit where the scattering length is much smaller than the average interatomic distance, and when the number of atoms in the condensate is much greater than 1 [for more details, see for example Dalfovo *et al.* (1999)].

Due to the assumption $\Phi' = 0$, the above equations strictly hold only in the limit of zero temperature, when all particles are in the condensate. Before discussing the dynamic behavior and the finite temperature generalization, we will consider their stationary solution at zero temperature, which, as will be shown, coincides with the HF theory discussed in Subsection 2.4. In order to obtain the ground state, we write the wave function of the condensate as $\Psi(\mathbf{r}, t) = \Psi(\mathbf{r})\exp(-i\mu t)$, where μ is the chemical potential, and Ψ is real and normalized to the total number of particles, $\int d\mathbf{r}\Psi^2 = N_0 = N$. Then, the Gross–Pitaevskii Eq. (9.122) becomes

$$\left(\frac{-\nabla^2}{2m} + v_{\text{ext}}(\mathbf{r}) + g\Psi^2(\mathbf{r})\right)\Psi(\mathbf{r}) = \mu\Psi(\mathbf{r}), \qquad (9.123)$$

and is completely equivalent to the HF equations (2.60) with $\Psi(\mathbf{r}) = \sqrt{N}\varphi(\mathbf{r})$. The solutions to these equations were discussed in Subsections 2.4 and 4.3.2.

In the dynamic case and in the low temperature limit, where the properties of excitations do not depend on temperature, the excited states may be found from the frequencies ω of the GP equations. We look for solutions of the type

$$\Psi(\mathbf{r}, t) = e^{-i\mu t}[\Psi(\mathbf{r}) + u(\mathbf{r})e^{-i\omega t} + v^*(\mathbf{r})e^{i\omega t}], \qquad (9.124)$$

which correspond to small oscillations of the order parameter around the ground state value. Retaining terms linear in the complex functions u and v, the GP Eqs. (9.122) become

$$\begin{aligned}
\omega u(\mathbf{r}) &= [H_0 - \mu + 2g\Psi^2(\mathbf{r})]u(\mathbf{r}) + g\Psi(\mathbf{r})^2 v(\mathbf{r}), \\
-\omega v(\mathbf{r}) &= [H_0 - \mu + 2g\Psi^2(\mathbf{r})]v(\mathbf{r}) + g\Psi(\mathbf{r})^2 u(\mathbf{r}),
\end{aligned} \qquad (9.125)$$

where $H_0 = -\nabla^2/2m + v_{\text{ext}}(\mathbf{r})$. These coupled equations allow the calculation of the energies ω of the excitations. In a homogeneous gas, the amplitudes u and v are plane waves and $\mu = g\Psi^2(\mathbf{r})$. The resulting dispersion law is (Bogoliubov, 1947)

$$\omega^2 = \frac{q^2}{2m}\left(\frac{q^2}{2m} + 2g\rho\right), \qquad (9.126)$$

where $\rho = \Psi^2(\mathbf{r})$ is the gas density, and is the same as the RPA result of (9.116). For large momentum q the spectrum coincides with the single-particle one, while for low q values the dispersion is phonon-like.

In the case of harmonic confinement of a finite number of atoms, the GP equations yield either single-particle or phonon-like solutions according to whether the ratio Na/a_{ho} ($a_{\text{ho}} = \sqrt{1/m\omega_{\text{ho}}}$) is much smaller or much larger than 1, respectively.

The coupled Eqs. (9.125) were solved by several authors in order to compute the excited states of confined gases (Edwards *et al.*, 1996; Ruprecht *et al.*, 1996; Singh and Rokhsar, 1996; Dalfovo *et al.*, 1997; Esry, 1997; Hutchinson and Zaremba, 1997; Hutchinson, Zaremba and Griffin, 1997; You, Hoston and Lewenstein, 1997). Equations. (9.125) also allow the calculation of the density of particles flowing out of the condensate at zero temperature (quantum depletion), by summing the squared modulus of the "hole" amplitude v over all excited states: $\rho_{\text{out}} = \sum_j |v_j(\mathbf{r})|^2$ (Fetter, 1972; Dalfovo *et al.*, 1997; Hutchinson, Zaremba and Griffin, 1997).

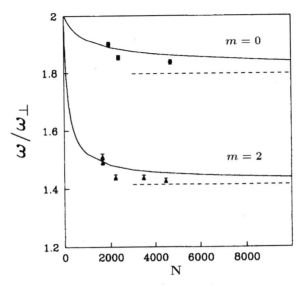

Fig. 9.9   Frequencies (in units of $\omega_\perp$) of collective excitation modes of even parity with $m = 0$ and $m = 2$, for N rubidium atoms in a symmetric axial trap (see text). The dots are experimental data of Jin *et al.* (1996). The full lines are the predictions of the Gross–Pitaevskii equations with $a/a_\perp$ equal to 3.3710^{-3} ($a_\perp = \sqrt{1/m\omega_\perp}$). The dashed lines are the asymptotic results for $Na/a_\perp \to \infty$ (Stringari, 1996).

For spherical traps the solutions to (9.125) are characterized by quantum numbers n_r, l, and m, where n_r the number of radial nodes, l the angular momentum of the excitation, and m its z component. For traps with axial symmetry, m is still a good quantum number. In Fig. 9.9 we show the lowest energy solutions of even parity with $m = 0$ and $m = 2$, for a gas of rubidium atoms confined in an axial symmetry trap ($\omega_x = \omega_y = \omega_\perp$). The deformation parameter of the trap, $\omega_z/\omega_\perp = \sqrt{8}$, corresponds to the experimental conditions of Jin *et al.* (1996), and values of N up to 10^4 are considered. The agreement between theory and experiment is very good, and evidences the importance of the interaction. In fact, without interaction the two modes would be degenerate at energy $\omega = 2\omega_\perp$.

The spectrum resulting from the numerical solution of the GP Eqs. (9.125) is shown in Fig. 9.10 for a condensate of 10^4 rubidium atoms in a spherical trap (Dalfovo *et al.*, 1997). Each state with energy ω and angular momentum l is represented by a full bar. For fixed angular momentum, the number of radial nodes n_r increases with energy. Looking at the high energy and high multipolarity eigenstates, we note that the splitting between even and odd states is approximately equal to ω_0, and that the spectrum is similar to that of a 3D harmonic oscillator. However, the states with the same value of $2n_r + l$ are not exactly degenerate like in the case of the harmonic oscillator; rather, the states with lower angular momentum are shifted to high energy due to the mean field produced by the condensate in the central region of the trap. The high energy range of the spectrum is well reproduced by a mean field description. Such a description can be obtained from the GP equations by neglecting the coupling between the positive energy (u) and negative energy (v) components of the order parameter (9.124), which is responsible for the collective nature of the solutions. This corresponds to putting $v = 0$ in the first Eq. (9.125), which reduces to the eigenvalue problem $(H_{\rm sp} - \mu)u = \omega u$, for the single-particle Hamiltonian

$$H_{\rm sp} = \frac{-\nabla^2}{2m} + v_{\rm ext}(\mathbf{r}) + 2g\rho(\mathbf{r}). \tag{9.127}$$

In this case, the eigenfunctions u fulfill the normalization conditions

$$\int d\mathbf{r} u_i^*(\mathbf{r})u_j(\mathbf{r}) = \delta_{ij}.$$

In many-body theory, this approximation is identical to the one known as the Tamm–Dancoff approximation (see next Subsection), and may also be considered as the zero temperature limit of the HF theory at finite temperature [see the second of Eqs. (2.98)]. For this reason, in the literature the mean field spectrum of Fig. 9.10 is sometimes referred to as the HF spectrum. Once we know how to compute the chemical potential and the condensate density through the static GP Eq. (9.123), it is easy to obtain the mean field-like spectrum for Hamiltonian (9.127). The mean field energies are shown in Fig. 9.10 by dashed horizontal bars. We notice that the general structure of the spectrum is very similar to the one yielded by the GP equations, apart from the low energy, low multipolarity states. In fact, the latter are the

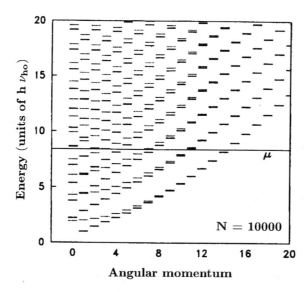

Fig. 9.10 Excitation spectrum of a dilute gas of rubidium atoms in a spherical trap with $a_{\rm ho} = 0.791 \mu m$. Full lines: Solutions to the time-dependent GP equations. Dashed lines: Single-particle spectrum for the Hamiltonian (9.127). The horizontal line is the chemical potential in units of $\omega_{\rm ho}$.

collective states that we discussed previously, which cannot be described by single-particle theories. In principle, one would expect that such low energy states might influence the Boson system thermodynamics at low temperature, thereby making the finite temperature mean field theory incorrect in this range (i.e. around the critical temperature), while the theory should work well at high temperature where single-particle excitations play the prominent role. In practice, for confined systems the thermodynamics is governed by surface excitations, which are very well reproduced by the mean field theory even at low energy (Giorgini, Pitaevskii and Stringari, 1997). Thus, and unlike what happens in homogeneous systems where phonons dominate the thermodynamic behavior of the system, and the finite temperature HF theory is unphysical at low temperature, in confined systems HF at finite temperature may be used to describe the thermodynamics of the system even at very low temperature.

A theory that describes correctly both the high temperature and low temperature regimes is the so-called Popov approximation (Popov, 1965, 1987; Griffin, 1996). This approximation is based on the one hand on the extension to temperature T of the GP equation [see the first of Eqs. (2.98)] in which the interaction among atoms of the condensate and outside the condensate is explicitly considered, and on the other hand on Bogoliubov type equations for the system excitations. The equations have the form

$$\left(\frac{-\nabla^2}{2m} + v_{\text{ext}}(\mathbf{r}) + g[\rho_0 + 2\rho_T]\right)\Psi = \mu\Psi, \tag{9.128}$$

and

$$
\begin{aligned}
\epsilon_i u_i(\mathbf{r}) &= [H_0 - \mu + 2g\rho(\mathbf{r})]u_i(\mathbf{r}) + g\rho_0(\mathbf{r})v_i(\mathbf{r}), \\
-\epsilon_i v_i(\mathbf{r}) &= [H_0 - \mu + 2g\rho(\mathbf{r})]v_i(\mathbf{r}) + g\rho_0(\mathbf{r})u_i(\mathbf{r}),
\end{aligned} \tag{9.129}
$$

where $H_0 = -\nabla^2/2m + v_{\text{ext}}(\mathbf{r})$ and $\rho_0 = |\Psi(\mathbf{r})|^2$ is the density of the condensate, while ρ_T is the density of thermal particles, given by

$$\rho_T = \sum_j (|u_j|^2 + |v_j|^2)[\exp(\beta\epsilon_j) - 1]^{-1} \quad (\beta = 1/KT), \tag{9.130}$$

with u_j, v_j and ϵ_j solutions to Eqs. (9.129). Now, these quantities depend on temperature. The sum $\rho = \rho_0 + \rho_T$ is the total density. The functions u_j, v_j are normalized according to the condition

$$\int d\mathbf{r}[u_i^*(\mathbf{r})u_j(\mathbf{r}) - v_i^*(\mathbf{r})v_j(\mathbf{r})] = \delta_{ij}. \tag{9.131}$$

In the Popov approximation, the thermal component is treated as a thermal bath that produces a static external field in the condensate equation. One ignores quantum depletion at zero temperature, which is given by $\rho_{\text{out}} = \sum_j |v_j|^2$, and which is very small for the trapped gases. Then, at very low temperature, when ρ_T is negligible with respect to ρ_0, Eq. (9.128) coincides with the stationary GP equation (9.123), and Eqs. (9.129) coincide with (9.125). Results obtained by solving the Popov equations are found in Dalfovo *et al.* (1999) and Giorgini *et al.* (1997). By comparing the thermodynamic properties of the Boson systems as obtained by the Popov theory with those of the finite temperature HF theory (see Subsection 2.6), we note that while for homogeneous gases the phonons (i.e. the low energy collective states that are taken into account only in the Popov theory) are always relevant to the calculation of thermodynamic properties, in the case of confined gases the HF approximation turns out to be excellent even at relatively low temperatures. As already discussed, this is due to the fact that at such temperatures the main contribution to thermodynamic properties comes from surface excitations, which are well reproduced by the HF theory.

9.7   Time-Dependent Hartree–Fock (TDHF) and the Matrix RPAE

In this subsection we look for solutions to the equation

$$\frac{\delta I}{\delta \phi} = 0,$$

where, in the case of Fermions, the action I is given by

$$I = \int_{t_1}^{t_2} \langle \phi_{\mathrm{SD}}(t) | H - i \frac{\partial}{\partial t} | \phi_{\mathrm{SD}}(t) \rangle dt, \qquad (9.132)$$

where $H = H_{\mathrm{HF}} + V_{\mathrm{res}}$, and $\phi_{\mathrm{SD}}(t)$ is a Slater determinant. We recall that (see Chapter 2) the HF Hamiltonian H_{HF} and the residual interaction V_{res} have the following properties:

$$\langle \mathrm{HF} | H_{\mathrm{HF}} | \mathrm{HF} \rangle = E_{\mathrm{HF}}, \quad H_{\mathrm{HF}} | i^{-1}, m \rangle = (E_{\mathrm{HF}} + \varepsilon_{m,i}) | i^{-1}, m \rangle,$$

where $\varepsilon_{m,i} = \varepsilon_m - \varepsilon_i$, $|\mathrm{HF}\rangle$ is the HF ground state and $|i^{-1}, m\rangle = a_m^\dagger a_i |\mathrm{HF}\rangle$, and

$$\langle \mathrm{HF} | V_{\mathrm{res}} | \mathrm{HF} \rangle = 0, \quad \langle \mathrm{HF} | V_{\mathrm{res}} | i^{-1}, m \rangle = 0,$$

$$V_{m,j,i,n} = \langle i^{-1}, m | V_{\mathrm{res}} | j^{-1}, n \rangle \neq 0,$$

$$V_{i,j,m,n} = \langle \mathrm{HF} | V_{\mathrm{res}} | i^{-1}, m, j^{-1}, n \rangle \neq 0,$$

$$V_{m,n,i,j} = \langle i^{-1}, m, j^{-1}, n | V_{\mathrm{res}} | \mathrm{HF} \rangle \neq 0.$$

In order to derive the TDHF equations, in the limit of small oscillations around the equilibrium configuration, let us write the generic Slater determinant $\phi_{\mathrm{SD}}(t)$ of Eq. (9.132) by the use of the Thouless theorem [see Eq. (2.69)]:

$$|\phi_{\mathrm{SD}}(t)\rangle = e^{\sum_{m,i} C_{m,i}(t) a_m^\dagger a_i} |\mathrm{HF}\rangle. \qquad (9.133)$$

Then, the equations for the coefficients $C_{m,i}$ are obtained from Eqs. (9.70), (9.132) and (9.133) by expanding (9.132) up to second order in $C_{m,i}$, and using the above properties. We find that

$$\langle \phi_{\mathrm{SD}}(t) | H - i\frac{\partial}{\partial t} | \phi_{\mathrm{SD}}(t) \rangle$$

$$= E_{\mathrm{HF}} + \sum_{m,i} \langle \mathrm{HF} | \left(H_{\mathrm{HF}} + V_{\mathrm{res}} - i\frac{\partial}{\partial t} \right) C_{m,i} | i^{-1}, m \rangle$$

$$+ \sum_{m,i} \langle i^{-1}, m | C_{m,i}^{*} \left(H_{\mathrm{HF}} + V_{\mathrm{res}} - i\frac{\partial}{\partial t} \right) | \mathrm{HF} \rangle$$

$$+ \sum_{m,i,n,j} \langle i^{-1}, m | C_{m,i}^{*} \left(H_{\mathrm{HF}} + V_{\mathrm{res}} - i\frac{\partial}{\partial t} \right) C_{n,j} | j^{-1}, n \rangle$$

$$+ \frac{1}{2} \sum_{m,i,n,j} \langle i^{-1}, m, j^{-1}, n | C_{m,i}^{*} C_{n,j}^{*} \left(H_{\mathrm{HF}} + V_{\mathrm{res}} - i\frac{\partial}{\partial t} \right) | \mathrm{HF} \rangle$$

$$+ \frac{1}{2} \sum_{m,i,n,j} \langle \mathrm{HF} | \left(H_{\mathrm{HF}} + V_{\mathrm{res}} - i\frac{\partial}{\partial t} \right) C_{m,i} C_{n,j} | i^{-1}, m, j^{-1}, n \rangle$$

$$= E_{\mathrm{HF}} + \sum_{m,i} |C_{m,i}|^{2} \varepsilon_{m,i} + \sum_{m,i,n,j} C_{m,i}^{*} C_{n,j} V_{m,j,i,n} - i\sum_{m,i} C_{m,i}^{*} \dot{C}_{m,i}$$

$$+ \frac{1}{2} \sum_{m,i,n,j} C_{m,i}^{*} C_{n,j}^{*} V_{m,n,i,j} + \frac{1}{2} \sum_{m,i,n,j} C_{m,i} C_{n,j} V_{i,j,m,n}. \qquad (9.134)$$

Bt requiring $\delta I = \frac{\delta I}{\delta C_{m,i}^{*}} \delta C_{m,i}^{*} = 0$, we find that

$$\delta I = \int_{t_1}^{t_2} \left[\sum_{m,i} \left(C_{m,i} \varepsilon_{m,i} + \sum_{n,j} C_{n,j} V_{m,j,i,n} - i\dot{C}_{m,i} \right. \right.$$

$$\left. \left. + \sum_{n,j} C_{n,j}^{*} V_{m,n,i,j} \right) \delta C_{m,i}^{*} \right] dt = 0,$$

and since the variations $\delta C_{m,i}^{*}$ are arbitrary, the final equations for the coefficients $C_{m,i}$ are

$$i\dot{C}_{m,i} = C_{m,i} \varepsilon_{m,i} + \sum_{n,j} (C_{n,j} V_{m,j,i,n} + C_{n,j}^{*} V_{m,n,i,j}). \qquad (9.135)$$

Let us now look for solutions to Eq. (9.135) that oscillate in time with frequency ω, of the form

$$C_{m,i}(t) = Y_{m,i} e^{-i\omega t} + Z_{m,i}^{*} e^{i\omega t}. \qquad (9.136)$$

We obtain

$$\varepsilon_{m,i}(Y_{m,i}e^{-i\omega t} + Z_{m,i}^*e^{i\omega t}) + \sum_{n,j}[V_{m,j,i,n}(Y_{n,j}e^{-i\omega t} + Z_{n,j}^*e^{i\omega t})$$

$$+V_{m,n,i,j}(Y_{n,j}^*e^{i\omega t} + Z_{n,j}e^{-i\omega t})] = \omega Y_{m,i}e^{-i\omega t} - \omega Z_{m,i}^*e^{i\omega t},$$

(9.137)

and by equating the coefficients of the exponentials with $\pm\omega$, we find a system of coupled equations for $Y_{m,i}$ and $Z_{m,i}$:

$$\varepsilon_{m,i}Y_{m,i} + \sum_{n,j}[V_{m,j,i,n}Y_{n,j} + V_{m,n,i,j}Z_{n,j}] = \omega Y_{m,i},$$

$$\varepsilon_{m,i}Z_{m,i} + \sum_{n,j}[V_{i,n,m,j}Z_{n,j} + V_{i,j,m,n}Y_{n,j}] = -\omega Z_{m,i}.$$

(9.138)

[Note that in order to derive these equations, we used $V_{i,j,m,n} = (V_{m,n,i,j})^*$.] Hereafter, Eqs. (9.138) will be denoted as the RPAE equations, in order to distinguish them from the RPA equations in which exchange effects are neglected both in single-particle energies and in the matrix elements. These equations, which are particularly suited for studying the excited states of nonhomogeneous systems like metal clusters and nuclei, are eigenvalue equations whose solution yields the excitation energies ω of the RPAE states, as well as the matrix elements of an excitation operator F between the ground and the excited states: $\langle 0|F|n\rangle$.

Linear RPA–TDHF equations, of the kind of Eqs. (9.138), can be derived also for Bosons systems, such as the dilute and cold gas of alkali atoms treated in the previous subsection by means of the Gross–Pitaevskii theory. In this case, the HF wave function of the condensate is written as $|\mathrm{HF}\rangle = \phi_i(\mathbf{r_1})\phi_i(\mathbf{r_2})\cdots\phi_i(\mathbf{r_N})$, and the one-particle–one-hole states which appear in the Thouless theorem (9.133) can be obtained starting from the complete set of single-particle wave functions which are eigenstates of the Hamiltonian

$$H_{\mathrm{sp}} = -\frac{\nabla^2}{2m} + v_{\mathrm{ext}}(\mathbf{r}) + g\rho_0(\mathbf{r}),$$

where ρ_0 is the density of the condensate. Following the track of calculation performed above for Fermions, one easily gets the result

$$\varepsilon_{m,i}Y_{m,i} + g\sum_n\left[\int d\mathbf{r}\phi_m^*(\mathbf{r})\phi_n(\mathbf{r})\rho(\mathbf{r})Y_{n,i} + \int d\mathbf{r}\phi_m^*(\mathbf{r})\phi_n(\mathbf{r})\rho(\mathbf{r})Z_{n,i}\right]$$

$$= \omega Y_{m,i},$$

$$\varepsilon_{m,i}Z_{m,i} + g\sum_n\left[\int d\mathbf{r}\phi_n^*(\mathbf{r})\phi_m(\mathbf{r})\rho(\mathbf{r})Z_{n,i} + \int d\mathbf{r}\phi_n^*(\mathbf{r})\phi_m(\mathbf{r})\rho(\mathbf{r})Y_{n,i}\right]$$

$$= -\omega Z_{m,i}.$$

(9.139)

The solutions of these equations are identical to those of Eqs. (9.125), as can be easily verified by expanding u and v in the latter equations on the complete set of eigenfunctions of the above Hamiltonian, $H_{\rm sp}$.

In general, the RPAE equations are written in matrix form as

$$\begin{pmatrix} A & B \\ -B^* & -A^* \end{pmatrix} \begin{pmatrix} Y(\omega) \\ Z(\omega) \end{pmatrix} = \omega \begin{pmatrix} Y(\omega) \\ Z(\omega) \end{pmatrix}, \qquad (9.140)$$

where the matrices A and B are defined in terms of the single-particle wave functions, as

$$A_{mi,nj} = \varepsilon_{mi}\delta_{mn}\delta_{ij} + \langle mj|V_{\rm res}|in\rangle_a, \quad B_{mi,nj} = \langle mn|V_{\rm res}|ij\rangle_a, \qquad (9.141)$$

and the matrix elements should be antisymmetrized by including exactly the exchange contributions:

$$\langle ab|V_{\rm res}|cd\rangle_a = \langle ab|V_{\rm res}|cd\rangle - \langle ab|V_{\rm res}|dc\rangle. \qquad (9.142)$$

This is the main difference between the microscopic HF–RPAE calculation and the RPA calculations based on the LDA theory (see Subsection 9.9), where the exchange contribution is only approximately treated in the energy functional and not in the wave function of the ground state.

To solve the RPAE equations, the following procedure is to be followed. First, the static HF problem is solved, from which we get the single-particle energies ε_α, the single-particle wave functions φ_α, and the HF ground state $|{\rm HF}\rangle$. Subsequently, all the one-particle–one-hole states $|i^{-1}, m\rangle$, with their respective excitation energies $\varepsilon_{m,i}$, are set up according to the kind of excitation to be studied. Let us assume, for example, that we want to study the photoabsorption cross-section of a metal cluster of sodium atoms, in the limit of long wavelength in which $kR \ll 1$ ($k \simeq 1/\lambda$, with R the metal cluster radius). In this limit, the photoabsorption cross-section as a function of the frequency ω of the absorbed photon is given by

$$\sigma(\omega) = 4\pi e^2 \omega S(F, \omega), \qquad (9.143)$$

where

$$F = D = \sum_{i=1}^{N} r_i \cos\vartheta_i$$

is the electric dipole operator, and the sum runs over the valence electrons of the cluster, and

$$S(D,\omega) = \sum_n |\langle 0|D|n\rangle|^2 \delta(\omega - \omega_{n,0}) \qquad (9.144)$$

is the excitation "strength" for the dipole operator.

Then, one selects the one-particle–one-hole states $|i^{-1}, m\rangle$ such that

$$F_{i,m} = D_{i,m} = \langle HF|D|i^{-1}, m\rangle \neq 0$$

(i.e. those which fulfill the selection rule $\Delta l = \pm 1$, valid for the dipole), and cuts the particle–hole space at some energy $\bar{\omega}$ such that $\varepsilon_{m,i} \leq \bar{\omega}$. For each particle–hole pair i^{-1}, m in the model space, we compute all the matrix elements of the residual interaction that appear in the RPAE Eqs. (9.138), using the states j^{-1}, n of the model space, and solve the system of coupled equations to obtain the solutions ω and $Y(\omega)_{m,i}, Z_{m,i}(\omega)$. Thus, we have as many solutions as the states i^{-1}, m of the model space, with energies ω different from the energies $\varepsilon_{m,i}$. The RPAE matrix elements $\langle 0|F|\omega\rangle$ are obtained starting from the solutions $Y(\omega)_{m,i}, Z_{m,i}(\omega)$, as

$$\langle 0|F|\omega\rangle = \sum_{m,i}(Y_{m,i}(\omega)F^*_{m,i} + Z_{m,i}(\omega)F_{m,i}), \qquad (9.145)$$

with $F_{i,m} = \langle HF|F|i^{-1}, m\rangle$. The matrix element (9.145) allows us to calculate the excitation strength $S(F, \omega)$ and, in the case taken as the example, where $F = D$, the RPAE photoabsorption cross-section as well. Clearly, the RPAE strength differs from the analogous quantity as computed in HF, given by

$$S_{HF}(F, \omega) = \sum_{m,i}|\langle i^{-1}, m|F|HF\rangle|^2\delta(\omega - \varepsilon_{m,i}). \qquad (9.146)$$

Eq. (9.145) can be derived starting from the variance

$$\langle \phi_{SD}(t)|F|\phi_{SD}(t)\rangle - \langle HF|F|HF\rangle.$$

Using for $\phi_{SD}(t)$ Eq. (9.133) expanded to first order in the coefficients $C_{m,i}(t)$ [which in turn are given by $C_{m,i}(t) = Y_{m,i}(\omega)e^{-i\omega t} + Z^*_{m,i}(\omega)e^{i\omega t}$], we obtain

$$\langle \phi_{SD}(t)|F|\phi_{SD}(t)\rangle - \langle HF|F|HF\rangle$$
$$= \sum_{m,i}[(Y_{m,i}F^*_{m,i} + Z_{m,i}F_{m,i})e^{-i\omega t} + (Z^*_{m,i}F^*_{m,i} + Y^*_{m,i}F_{m,i})e^{i\omega t}].$$

$$(9.147)$$

On the other hand, by writing $|\phi(t)\rangle = |0\rangle + \alpha(t)|\omega\rangle$, where $\alpha(t) = \alpha_0 e^{-i\omega t}$, we obtain

$$\langle \phi(t)|F|\phi(t)\rangle - \langle 0|F|0\rangle = \alpha_0(\langle 0|F|\omega\rangle e^{-i\omega t} + \langle \omega|F|0\rangle e^{i\omega t}) \qquad (9.148)$$

for the variance, so that by comparing Eqs. (9.147) and (9.148), we find (9.145), where the Y and the Z are defined apart from a normalization constant.

The normalization constant for Y and Z can be found in a simple way by introducing the creation and annihilation operators for the RPA excited states starting from (9.145) and the definitions

$$O^\dagger_\omega|0\rangle = |\omega\rangle \qquad (9.149)$$

and

$$O_\omega|0\rangle = 0, \tag{9.150}$$

for all ω's. Using the result

$$\langle 0|F|\omega\rangle = \langle 0|FO_\omega^\dagger|0\rangle = \langle 0|[F,O_\omega^\dagger]|0\rangle \approx \langle \mathrm{HF}|[F,O_\omega^\dagger]|\mathrm{HF}\rangle, \tag{9.151}$$

and comparing it to (9.145), we immediately get for the creation operator

$$O_\omega^\dagger = \sum_{m,i}(Y_{m,i}(\omega)a_m^\dagger a_i - Z_{m,i}(\omega)a_i^\dagger a_m). \tag{9.152}$$

Moreover, the annihilation operator O_ω is the adjoint operator of $O_\omega^\dagger$. Then, the normalization condition is obtained from

$$1 = \langle \omega \mid \omega \rangle = \langle 0|O_\omega O_\omega^\dagger|0\rangle = \langle 0|[O_\omega,O_\omega^\dagger]|0\rangle \approx \langle \mathrm{HF}|[O_\omega,O_\omega^\dagger]|\mathrm{HF}\rangle. \tag{9.153}$$

Using the result (9.152) and the anticommutation relations (2.62) for the operators a and $a^\dagger$, we obtain at last

$$\sum_{m,i}(|Y_{m,i}|^2 - |Z_{m,i}|^2) = 1. \tag{9.154}$$

Different solutions may be orthogonalized in order to fulfill the relation

$$\langle \mathrm{HF}|[O_\omega,O_\lambda^\dagger]|\mathrm{HF}\rangle = \delta_{\omega,\lambda}.$$

The RPAE equations can also be derived using the method of equations of motion, in which the Schrödinger equation is rewritten using the excited state creation and annihilation operators (9.149) and (9.150) as

$$[H,O_\lambda^\dagger]|0\rangle = (E_\lambda - E_0)O_\lambda^\dagger|0\rangle \equiv \omega_\lambda O_\lambda^\dagger|0\rangle. \tag{9.155}$$

From this equation, it is possible to derive the variational equation for the O and $O^\dagger$ operators:

$$\langle 0|[\delta O_\lambda,[H,O_\lambda^\dagger]]|0\rangle = \omega_\lambda\langle 0|[\delta O_\lambda,O_\lambda^\dagger]|0\rangle. \tag{9.156}$$

This equation is exact in the limit when $|0\rangle$ is the true ground state of the system. It is also very useful for us to note that in the limit in which the right hand and left hand side commutators of the equation are numbers, it turns out to be completely independent of the approximation employed for the ground state, and in the limit in which the commutators are one-body operators their mean values are well reproduced by a mean field ground state, because for the one-body operators F one expects that $\langle 0|F|0\rangle \approx \langle \mathrm{HF}|F|\mathrm{HF}\rangle$, even if $|0\rangle \neq |\mathrm{HF}\rangle$. These considerations are also the basis for the last steps of Eqs. (9.151) and (9.153).

In the RPAE approximation the excitation operators are taken in the form (9.152), and in the equations of motion (9.156) we replace the mean values of the commutators on the true state by their mean values on the HF state.

By inserting expression (9.152) into the equation of motion (9.156) and allowing δO_λ to span the ensemble of particle–hole creation $(a_m^\dagger a_i)$ and annihilation $(a_i^\dagger a_m)$ operators, one obtains Eqs. (9.138). In such a procedure there is a lack of self-consistency, because by including in $O_\lambda^\dagger$ both the creation and the annihilation operators of a particle–hole pair, one automatically excludes the possibility that the RPAE ground state is the HF one, which, however, is used to solve the equations of motion. To be consistent, one should use for the ground state in the equations of motion the solution to

$$O_\lambda|\text{RPAE}\rangle = 0, \tag{9.157}$$

for all the states λ which define the RPAE ground state. However, the solution of such a system is extremely complex, and in practice, to solve the equations of motion one computes the commutators on the $|\text{HF}\rangle$ state.

Lastly, recall the Tamm–Dancoff approximation, which was widely used in past years before the appearance of computers made it possible to solve the RPAE equations. It consists in approximating the ground state $|0\rangle$ by the HF state, and in writing the operators $O_\omega^\dagger$ as creation operators for one-particle–one-hole states:

$$O_\omega^\dagger = \sum_{m,i} Y_{m,i}(\omega) a_m^\dagger a_i. \tag{9.158}$$

In this case the equation for the ground state

$$O_\omega|\text{HF}\rangle = \sum_{m,i} Y_{m,i}^*(\omega) a_i^\dagger a_m |\text{HF}\rangle = 0 \tag{9.159}$$

is satisfied for all particle–hole states. By inserting (9.158) into the equations of motion (9.156), and considering arbitrary variations $\delta Y_{m,i}$, we obtain

$$\sum_{n,j} \langle\text{HF}|[a_i^\dagger a_m, [H, a_n^\dagger a_j]]|\text{HF}\rangle Y_{m,j} = \omega_\lambda \langle\text{HF}|[a_i^\dagger a_m, a_n^\dagger a_j]|\text{HF}\rangle Y_{m,i}. \tag{9.160}$$

Using the fact that the HF state obeys Eq. (9.159), we arrive at the Tamm–Dancoff equations

$$\sum_{n,j} (\delta_{mn}\delta_{ij}\varepsilon_{m,i} + V_{m,j,i,n}) Y_{n,j} = \omega Y_{m,i}. \tag{9.161}$$

Compared to the RPAE solutions, the Tamm–Dancoff approximation completely ignores the dynamic effects due to the presence of correlations in the ground state, which are taken into account in RPAE by the term $Z_{m,i}$ in (9.152). This term makes the RPAE state different from the HF one.

In the case of a cold dilute gas of Bosons in a magnetic trap, the Tamm–Dancoff equations lead to solutions which coincide with those of the mean field, obtained with the single-particle Hamiltonian (9.127). They coincide also with the second HF equation (2.98) at finite temperature only in the zero temperature limit.

9.8   Examples of Application of the RPA Theory

9.8.1   *The RPA with Separable Interactions*

The physics described by the matrix RPAE theory can be illustrated in a simple and clear way by a separable interaction of the kind

$$V_{\text{res}} = k \sum_{i<j} Q_i Q_j, \tag{9.162}$$

where k is a constant. This interaction emerges in a natural way in the vibrating potential method (Chapter 11), and should be used without including the exchange terms in the particle–hole matrix elements, which thus are given by

$$
\begin{aligned}
V_{m,j,i,n} &= \langle mi^{-1}|V_{\text{res}}|nj^{-1}\rangle = k Q_{m,i} Q_{n,j}^*, \\
V_{m,n,i,j} &= \langle mi^{-1}nj^{-1}|V_{\text{res}}|\text{HF}\rangle = k Q_{m,i} Q_{n,j},
\end{aligned}
\tag{9.163}
$$

where

$$Q_{m,i} = \langle m|Q|i\rangle = \sum_{\sigma} \int \varphi_m^*(\mathbf{r},\sigma) q \varphi_i(\mathbf{r},\sigma) d\mathbf{r} \quad \left(Q = \sum_i q_i \right),$$

and φ_i are the solutions to the static HF problem.

Eqs. (9.138) then become

$$
\begin{aligned}
(\varepsilon_{m,i} - \omega) Y_{m,i} + k Q_{m,i} \sum_{n,j} [Q_{n,j}^* Y_{n,j} + Q_{n,j} Z_{n,j}] &= 0, \\
(\varepsilon_{m,i} + \omega) Z_{m,i} + k Q_{m,i}^* \sum_{n,j} [Q_{n,j} Z_{n,j} + Q_{n,j}^* Y_{n,j}] &= 0,
\end{aligned}
\tag{9.164}
$$

with the solution

$$Y_{m,i} = -\frac{N Q_{m,i}}{\varepsilon_{m,i} - \omega}, \quad Z_{m,i} = -\frac{N Q_{m,i}^*}{\varepsilon_{m,i} + \omega}. \tag{9.165}$$

The constant N, which is given by

$$N = k \sum_{n,j} [Q_{n,j}^* Y_{n,j} + Q_{n,j} Z_{n,j}], \tag{9.166}$$

can be evaluated by the normalization condition (9.154). By eliminating $Y_{m,i}$ and $Z_{m,i}$ from Eqs. (9.165) and (9.166), we obtain the following dispersion relation for the excitation energies ω:

$$\frac{1}{k} = -2 \sum_{m,i} \frac{\varepsilon_{m,i} |Q_{m,i}|^2}{\varepsilon_{m,i}^2 - \omega^2}. \tag{9.167}$$

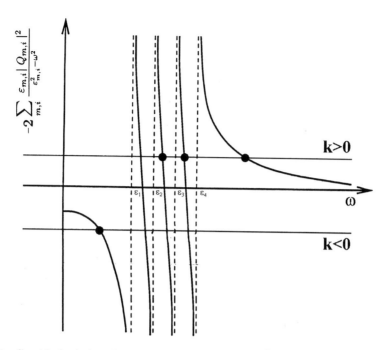

Fig. 9.11 Graphical solution of the dispersion relation (9.167) for the RPA theory with separable interactions.

Equation (9.167) is completely similar to (9.87), which yields the RPA excitation energies for homogeneous systems. It is enough to identify the free response function χ^0 with $-2\sum_{m,i} \varepsilon_{m,i} \mid Q_{m,i} \mid /(\varepsilon_{m,i}^2 - \omega^2)$ and k with $v(q)/L^D$. The graphical solution of the equation is plotted in Fig. 9.11, and clearly shows that a collective solution emerges apart from the single-particle-like ones. Note that, unlike the homogeneous case, the single-particle spectrum of a finite system like the one considered here has a gap and is discrete.

The RPA matrix elements

$$\langle \omega | F | \mathrm{HF} \rangle = \sum_{m,i} (Y_{m,i}^* F_{m,i} + Z_{m,i}^* F_{m,i}^*)$$

of Eq. (9.145) are found by computing the normalization constant and using solutions (9.165) for the Y and Z. With a simple calculation we find that

$$N = \left[\sum_{m,i} \left(\frac{4\varepsilon_{m,i}\omega \mid Q_{m,i} \mid^2}{(\varepsilon_{m,i}^2 - \omega^2)^2} \right) \right]^{-1/2} \tag{9.168}$$

and

$$\langle \omega | F | \mathrm{HF} \rangle = -N \sum_{m,i} \left(\frac{Q_{m,i}^* F_{m,i}}{\varepsilon_{m,i} - \omega} + \frac{Q_{m,i} F_{m,i}^*}{\varepsilon_{m,i} + \omega} \right). \tag{9.169}$$

An analytic solution for the collective energy can be found in the degeneration limit, in which all the particle–hole energies $\varepsilon_{m,i} = \varepsilon$ are equal. In this limit, all the RPA solutions except one are trapped at the unperturbed energy ε. The untrapped (collective) solution is obtained from Eq. (9.167) as

$$\omega^2 = \varepsilon^2 + 2\varepsilon k \sum_{m,i} |Q_{m,i}|^2, \tag{9.170}$$

and assuming that the excitation operator F coincides with Q, we find that

$$N^2 = \frac{(\varepsilon^2 - \omega^2)^2}{4\varepsilon\omega \sum_{m,i} | Q_{m,i} |^2}, \tag{9.171}$$

and

$$|\langle\omega|Q|\mathrm{HF}\rangle|^2 = \frac{\varepsilon}{\omega} \sum_{m,i} | Q_{m,i} |^2 . \tag{9.172}$$

The above equations show that the collective state energy is either higher or lower than the unperturbed energy, according to whether the residual interaction is attractive or repulsive, and that the energy-weighted sum rule (f-sum rule)

$$m_1 = \sum_n E_n|\langle 0|Q|n\rangle|^2 = \frac{1}{2}\langle 0|[F,[\mathrm{H},F]]|0\rangle$$

is conserved in the sense that

$$\omega|\langle\omega|F|\mathrm{HF}\rangle|^2 = \varepsilon \sum_{m,i} | F_{m,i} |^2 = m_1^{\mathrm{RPA}}(\omega). \tag{9.173}$$

Therefore, the same result is obtained when computing m_1^{RPA} either with the RPA matrix element or taking the commutator $\frac{1}{2}\langle\mathrm{HF}|[Q,[H,Q]]|\mathrm{HF}\rangle$ on the HF state. This is not the case for the $m_0 = \sum_n |\langle 0|Q|n\rangle|^2$ sum rule, which, when computed in RPA, is either larger or smaller than the unperturbed one, depending on whether ε/ω is larger or smaller than 1. For a general description of the properties of the sum rules mentioned here, see Subsection 9.10.

9.8.2   *The RPAE for metal clusters*

In this Subsection we will study the response of closed shell sodium clusters to the electric dipole excitation operator, and use the RPAE theory based on the HF and BHF approximations for the energies of the single-particle wave functions. We will employ both the jellium model for the metal cluster, and a model which uses pseudo-Hamiltonians (Bachelet, Ceperley and Chiocchetti, 1989) to describe the ionic potential acting on the valence electrons of the cluster (see Subsection 2.2.1).

The electric dipole excitations for sodium metal clusters were measured by several authors by means of photoabsorption reaction on a cluster beam (Knight *et al.*, 1985; Pollack *et al.*, 1991; Selby *et al.*, 1991; Borgreen *et al.*, 1993; Ellert *et al.*,

1995; Reiners *et al.*, 1995); the calculations we report here are those by Guet and Johnson (1992) and Lipparini, Serra and Takayanagi (1994). The results are compared with the RPA calculations based on the LDA theory of Ekardt (1985) and Yannouleas *et al.* (1989).

The RPAE calculations based on the HF method with Coulomb interaction will henceforth be referred to as the Coulomb–HF (CHF) approximation, while those based on the Brueckner–HF method with the effective interaction (3.89), which is used to determine both the single-particle states and the residual interaction, will be indicated as the BHF approximation.

Thus, the Hamiltonian that describes the motion of the valence electrons of sodium clusters is given in BHF by [in the following, we use atomic units (a.u.)]

$$H = \sum_i \frac{p_i^2}{2} + \sum_{i,j} \left(\frac{1}{|\mathbf{r}_i - \mathbf{r}_j|} + g^c(r_{i,j}, \nabla_{i,j}) \right) + \sum_i \hat{V}_I. \tag{9.174}$$

From this, we obtain the CHF Hamiltonian by simply dropping the correlation term g^c.

For the electronic potential

$$V = \sum_{i,j} \left(\frac{1}{|\mathbf{r}_i - \mathbf{r}_j|} + g^c(r_{i,j}, \nabla_{i,j}) \right), \tag{9.175}$$

of Eq. (9.174), and for the dipole operator $D = \sum_{i=1}^{N} z_i$, the following property holds:

$$[V, D] = 0. \tag{9.176}$$

This condition (9.176) is known as (Bohigas, Lane and Martorell, 1979; Lipparini and Stringari, 1989) the Galilean invariance of the force, and stems from the translational invariance of the electronic potential (9.175), and from the fact that the dipole operator coincides with the coordinate of the center of mass of the electrons. Then, in the spherical jellium model (JM), where the ionic potential $\hat{V}_I$ is the local operator (2.24) ($\hat{V}_I = V_+$), one finds that

$$[H, D] = -\frac{i}{2} P_z, \tag{9.177}$$

where $P_z = \sum_i p_i^z$ is the z component of the total momentum of the valence electrons. Thus, using Eq. (9.177) one obtains for the f-sum rule m_1

$$m_1^{\mathrm{JM}} = \frac{N}{2}, \tag{9.178}$$

and for the sum rule cubic in energy

$$m_3^{\mathrm{JM}} = \frac{2\pi}{3} \int_0^\infty \rho(r) \nabla^2 V_+(r) r^2 dr = \frac{2\pi}{r_s^3} \int_0^R \rho(r) r^2 dr, \tag{9.179}$$

where $\rho(r)$ is the electron density of the ground state, and in order to obtain the last expression of (9.179) we have used the Poisson equation. The m_3 sum rule depends on the ionic potential V_+ because this external field breaks the translational invariance of the system and thus does not commute with the total momentum operator. In the pseudojellium model (PJM), the ionic potential $\hat{V}_I$ of (9.174) is nonlocal and is given by the expression (Lipparini, Serra and Takayanagi, 1994)

$$\hat{V}_I = -\frac{1}{2}\nabla\alpha(r)\cdot\nabla + \mathbf{L}\beta(r)\cdot\mathbf{L} + u(r), \qquad (9.180)$$

where $\mathbf{L}$ is the angular momentum operator and the functions α, β and u are computed from a pseudo-Hamiltonian that takes into account the ionic structure (core electrons) in the interaction between valence electrons and ions. These functions are shown in Fig. 9.12 for the cluster Na_{20}.

The nonlocal character of V_I makes V_I contribute to the commutator $[H, D]$. Taking this contribution into consideration, we get the following result for the f-sum rule:

$$m_1^{\mathrm{PJM}} = 2\pi \int_0^\infty \left(1 + \alpha(r) + \frac{4}{3}r^2\beta(r)\right)\rho(r)r^2 dr, \qquad (9.181)$$

which differs from the result (9.178) and is valid in the jellium model.

The above results for m_1 and m_3 hold in the RPAE approximation, provided one uses the self-consistent mean field solution for the electron density that appears

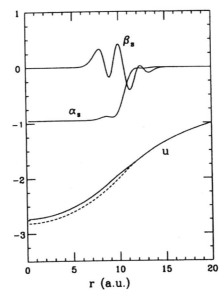

Fig. 9.12   Radial functions α, β and u for the Na_{20} cluster, in atomic units. The functions were scaled in the following way: $\alpha_s = 25\alpha$, $\beta_s = 2 \times 10^4\beta$. The local jellium potential of Eq. (2.24) is represented by the dashed line, for comparison.

in expressions (9.179) and (9.181). In fact, it can be shown that the evaluation of the mean value of the commutators

$$m_1 = 1/2\langle 0|[D,[H,D]]|0\rangle,$$

$$m_3 = 1/2\langle 0|[[D,H],[H,[H,D]]]|0\rangle$$

on the HF state is equivalent to computing the moments m_1 and m_3 with the RPAE energies and matrix elements (Bohigas, Lane and Martorell, 1979; Lipparini and Stringari, 1989; see also Subsection 9.10). As regards the inversely energy-weighted sum rule m_{-1}, which is connected to the static polarizability α of the system by the relation

$$m_{-1} = \frac{1}{2}\alpha, \tag{9.182}$$

it is possible to show in a similar way that the evaluation of the moment

$$m_{-1} = \sum_n E_n^{-1}|\langle n|D|0\rangle|^2$$

using the RPAE energies and matrix elements is equivalent to calculating the polarizability α in a static HF calculation with a dipole constraint. The RPAE calculations that we are going to report in what follows, satisfy these theorems.

In Fig. 9.13 we show the distributions of the dipole strength in the Na_{20} and Na_{92} clusters, computed in CHF and BHF with the spherical jellium model. The strengths are in units of the f-sum rule m_1 of (9.178). As a general feature, the BHF spectrum has an average excitation energy slightly higher than the CHF spectrum, which is reflected in a lower polarizability value. This is shown in Fig. 9.14, where we plot the electric-dipole polarizability per particle in CHF and BHF (we recall that for the jellium clusters the classical value of α/N is r_s^3). This behavior can be understood in terms of the role played by electron correlations. Inclusion of these correlations in BHF makes the cluster more rigid than in CHF and, as a consequence, less polarizable by an external electric field and resonant at higher energy.

In Na_{20} the excitation spectrum exhibits two main peaks, in both CHF and BHF. This agrees with the experimental data on the photoabsorption cross-sections of Pollack *et al.* (1991), which exhibit two peaks at 2.4 eV and 2.75 eV. However, the relative intensities of these two peaks are not well reproduced, because in the experiments the second peak is weaker than the first (see Fig. 9.15) and, moreover, the effect of correlations in BHF is to increase the difference between theoretical and measured peak energies, which is already present in CHF, i.e. to increase the blue shift with respect to experiment.

As the size of the cluster increases, the fragmentation of the strength increases as well, and the average excitation energy shifts to higher energies, but in any case staying well below the classical Mie frequency $\omega_{\text{Mie}} = \sqrt{4\pi\rho} = \sqrt{3/r_s^3}$, which, for sodium ($r_s \simeq 4$), is 3.4 eV. This is shown in the Na_{92} spectrum of Fig. 9.13(b).

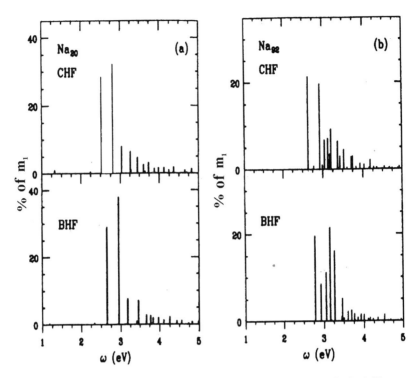

Fig. 9.13 Dipole-energy distribution for the Na$_{20}$ and Na$_{92}$ clusters with the jellium model, in units of the f-sum rule m_1. In all cases the top panel shows the calculation under the CHF+RPA approximation, while the bottom one corresponds to the BHF+RPA approximation.

As for the comparison with the TDLSDA theory (see Subsection 9.9), the BHF spectra are very close to the TDLSDA ones, and the polarizabilities are practically coincident. The disagreement with experiment cannot be explained within the HF framework, even by taking into account short range correlations like in BHF; one should, rather, incorporate effects due to the ionic structure into the model. To prove this statement, we show in Fig. 9.15 the dipole strengths as computed in the BHF+PJM approximation, for the Na$_{20}$, Na$_{21}^+$ and Na$_{35}^+$ clusters, and compare them with the experimental data. Then, in Fig. 9.16, we report the theoretical calculation in (neutral and positively charged) clusters with 40, 58 and 92 electrons. The strength is given in units of the f-sum rule of (9.181), whose value in the different cases is reported in Table 9.4. In this table, we report the values of $2m_1$ in a.u., which for the jellium equals the number of electrons [see Eq. (9.178)].

As is seen from the table, in sodium clusters m_1^{PJM} is very close to the value of the jellium model, $N/2$ [see Eq. (9.178)]. However, in other cases (e.g. lithium clusters), m_1^{PJM} is much smaller than m_1^{JM}. This effect, which is due to the non-locality of the pseudo-Hamiltonian, is the important one in order to reproduce the experimental photoabsorption results in lithium clusters (Brechignac *et al.*,

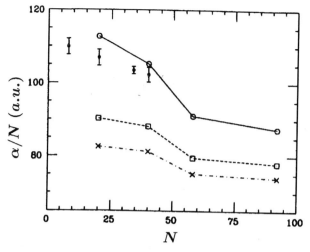

Fig. 9.14 Static dipole polarizability in some closed shell sodium clusters. The squares and the crosses correspond to the CHF and BHF approximations, respectively, in the jellium model. The circles are the results of the BHF method with the pseudojellium potential.

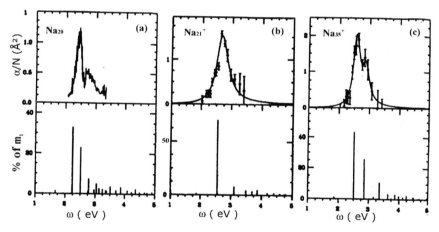

Fig. 9.15 Dipole strength distributions, in units of the f-sum rule m_1, in the pseudojellium model. The results for the Na_{20}, Na_{21}^+ and Na_{35}^+ clusters are reported in (a), (b) and (c), respectively. In all cases the top panel shows the experimental photoabsorption cross-section [Pollack *et al.*, 1991, for (a); Borggreen *et al.*, 1993, for (b) and (c)], while the bottom panel shows the theoretical results (Lipparini, Serra and Takayanagi, 1994).

1993), which in the optical region do not saturate the f-sum rule of the jellium model. This was explicitly shown by RPA calculations with pseudopotentials in lithium clusters (Serra *et al.*, 1993; Blundell and Guet, 1993; Alasia *et al.*, 1995).

From Fig. 9.15 we may conclude that, once the ionic structure effects are included, the spectra predicted by the BHF + PJM theory are in very good

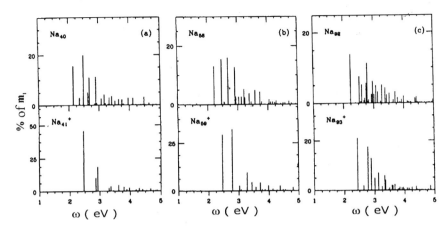

Fig. 9.16 The same as Fig. 9.15, but for the theoretical results in clusters with 40 (a), 58 (b) and 92 (c) electrons. In all cases the top panel reports the results for neutral clusters, and the bottom panel for positively charged ones.

Table 9.4. Energy-weighted sum rule within the pseudo-jellium model for different closed-shell clusters. We list the value of $2m_1$ in a.u., that for jellium is just the number of electrons.

	Na_{20}	Na_{40}	Na_{58}	Na_{92}	
$2m_1$ (a.u.)	19.41	38.76	56.12	88.95	
	Na_{21}^+	Na_{35}^+	Na_{41}^+	Na_{59}^+	Na_{93}^+
$2m_1$ (a.u.)	19.38	32.91	38.74	56.09	88.93

agreement with experiment. Especially remarkable is the degree to which it reproduces the fragmentation of the strength (Landau damping) in Na_{20} and Na_{35}^+, and the relative weighs of the peaks. For the Na_{21}^+ cluster the model foresees one dominant state with more than 70% of the strength, in good agreement with the experimental results reported in the figure (Borggreen *et al.*, 1993), as well as with those of Ellert *et al.* (1995) and Reiners *et al.* (1995). The spectrum of the Na_{41}^+ cluster also compares very well with the experiments, which evidence a dominant peak at about 2.6 eV, and a tail at around 3 eV. The high degree of fragmentation predicted for Na_{40} is also observed in the measurements of Selby *et al.* (1991).

The dipole polarizability computed in BHF + PJM is shown in Fig. 9.14. From this figure we see that the computed values are very close to the experimental values of Knight *et al.* (1985).

9.9 The Adiabatic Time-Dependent LSDA (TDLSDA)

The time-dependent version of the LSDA approximation (see Subsection 4.6), known as adiabatic TDLSDA (time-dependent local spin density approximation) [see for example Gross and Kohn (1990)], can be derived through a variational principle on the action integral

$$I = \int \langle \varphi_{\mathrm{SD}}(t)|H^{\mathrm{LSDA}} - i\frac{\partial}{\partial t}|\varphi_{\mathrm{SD}}(t)\rangle dt, \qquad (9.183)$$

where $\varphi_{\mathrm{SD}}(t)$ is a Slater determinant of single-particle, time-dependent wave functions, and

$$\langle \varphi_{\mathrm{SD}}(t)|H^{\mathrm{LSDA}}|\varphi_{\mathrm{SD}}(t)\rangle = E(\rho(\mathbf{r},t), m(\mathbf{r},t))$$

$$= T_0(t) + \int d\mathbf{r}\rho(\mathbf{r},t)\left[v_{\mathrm{ext}}(\mathbf{r}) + \frac{1}{2}U(\mathbf{r},t)\right]$$

$$+ E_{\mathrm{xc}}(\rho(\mathbf{r},t), m(\mathbf{r},t)), \qquad (9.184)$$

where

$$U(\mathbf{r},t) = \int d\mathbf{r}'\rho(\mathbf{r}',t)v(|\mathbf{r} - \mathbf{r}'|)$$

is the Hartree potential (for electrons, it is the classical Coulomb potential). If $\varphi_{i,\sigma}$ are the single-particle states that appear in the Slater determinant, for the density appearing in the above equation we have

$$\rho(\mathbf{r},t) = \sum_{\sigma} \rho_{\sigma}(\mathbf{r},t),$$

$$m(\mathbf{r},t) = \rho_{\uparrow}(\mathbf{r},t) - \rho_{\downarrow}(\mathbf{r},t), \qquad (9.185)$$

$$\rho_{\sigma}(\mathbf{r},t) = \sum_{i} |\varphi_{i,\sigma}(\mathbf{r},t)|^2,$$

and

$$T_0(t) = \int d\mathbf{r}\frac{\tau(\mathbf{r},t)}{2m}, \qquad (9.186)$$

where

$$\tau(\mathbf{r},t) = \sum_{i\sigma} |\nabla\varphi_{i,\sigma}(\mathbf{r},t)|^2.$$

By making the stationary condition $\delta I = 0$ to Eq. (9.183), we obtain the time-dependent Kohn–Sham equations for $\varphi_i(\mathbf{r},t)$:

$$i\frac{\partial}{\partial t}\varphi_{i,\sigma}(\mathbf{r},t) = \left(-\frac{\nabla^2}{2m} + v_{\mathrm{ext}}(\mathbf{r}) + U(\mathbf{r},t) + V_{\mathrm{xc}}(\mathbf{r},t) + W_{\mathrm{xc}}(\mathbf{r},t)\sigma_z\right)\varphi_{i,\sigma}(\mathbf{r},t),$$

$$(9.187)$$

where σ_z is the z component of the Pauli spin operator and, in the adiabatic approximation,

$$V_{\mathrm{xc}}(\mathbf{r},t) = \frac{\partial E_{\mathrm{xc}}(\rho, m)}{\partial\rho(\mathbf{r},t)}, \quad W_{\mathrm{xc}}(\mathbf{r},t) = \frac{\partial E_{\mathrm{xc}}(\rho, m)}{\partial m(\mathbf{r},t)} \qquad (9.188)$$

are the exchange-correlation potentials. An exhaustive discussion about the justification of the theory and the validity of the TDLSDA is found in Dobson *et al.* (1998).

Starting from Eq. (9.187), in the following we will study the TDLSDA response functions of a many-body system to an external, time-oscillating field, by employing a formalism which applies to both homogeneous and nonhomogeneous systems, and which also takes account of coupling among different responses which are due to possible spin polarization in the ground state of the system. We will separate the study of the response into the longitudinal case, in which the system excitations take place without varying the third component of the spin ($\Delta S_z = 0$), and the transverse case, in which the excitations involve a change of spin ($\Delta S_z = \pm 1$).

9.9.1 *The TDLSDA Longitudinal Response Function*

Let us put the system [Eq. (9.187)] in an external, time-oscillating field like that of Eq. (8.10), where the operator G depends on the spin coordinates and is of the form

$$G = G^{s,a} = \sum_{i=1}^{N_\uparrow} g(\mathbf{r}_i) \pm \sum_{i=1}^{N_\downarrow} g(\mathbf{r}_i) \equiv \sum_{i\sigma} g_\sigma(\mathbf{r}_i)|\sigma\rangle\langle\sigma|, \qquad (9.189)$$

with $N_\uparrow, N_\downarrow$ the numbers of system particles with spin up and spin down, respectively. The external potential causes variations in the spin densities of the system (with respect to the static densities), which can be written in the following way:

$$\delta\rho_\sigma(\mathbf{r},\omega) = \sum_{\sigma'} \int \alpha_{\sigma,\sigma'}(\mathbf{r},\mathbf{r}',\omega) g_{\sigma'}(\mathbf{r}')\, d\mathbf{r}', \qquad (9.190)$$

where $\alpha_{\sigma,\sigma'}$ is the correlation function that connects the density variation to the external field. Such density variations give rise to changes in the effective LSDA potential of (9.187), which depends on the densities, so that we have $V_{\text{eff}} = V_{\text{eff}}^{\text{gs}} + \delta V_{\text{eff}}$. To first perturbation order in δV_{eff}, the effective potential variation, in turn, causes a density variation which is given by

$$\delta\rho_\sigma(\mathbf{r},\omega) = \sum_{\sigma'} \int \alpha_{\sigma,\sigma'}^0(\mathbf{r},\mathbf{r}',\omega)\delta V_{\text{eff}}^{\sigma'}\, d\mathbf{r}', \qquad (9.191)$$

where the correlation function of the noninteracting particles $\alpha_{\sigma,\sigma'}^0$ (built up through the solutions to the stationary KS equations) is given by the expression

$$\alpha_{\sigma,\sigma'}^0(\mathbf{r},\mathbf{r}',\omega) = \delta_{\sigma,\sigma'} \sum_{mi} \left[\frac{\varphi_{i\sigma}(\mathbf{r})\varphi_{i\sigma}^*(\mathbf{r}')\varphi_{m\sigma}^*(\mathbf{r})\varphi_{m\sigma}(\mathbf{r}')}{\omega + \epsilon_{i\sigma} - \epsilon_{m\sigma}} \right.$$
$$\left. - \frac{\varphi_{i\sigma}^*(\mathbf{r})\varphi_{i\sigma}(\mathbf{r}')\varphi_{m\sigma}(\mathbf{r})\varphi_{m\sigma}^*(\mathbf{r}')}{\omega - \epsilon_{i\sigma} + \epsilon_{m\sigma}} \right], \qquad (9.192)$$

where the indices i, m refer to hole and particle states, respectively. Noting that at first order in $\delta\rho$ the variation of the effective potential is given by

$$\delta V_{\text{eff}}^{\sigma}(\mathbf{r}) = g_{\sigma}(\mathbf{r}) + \sum_{\sigma'} \int d\mathbf{r'} \left(v(|\mathbf{r} - \mathbf{r'}|) + \frac{\partial^2 E_{\text{xc}}}{\partial\rho_{\sigma}\partial\rho_{\sigma'}} \right) \delta\rho_{\sigma'}(\mathbf{r'}),\tag{9.193}$$

from Eqs. (9.190), (9.191) and (9.193) we obtain a Dyson-like integral equation,

$$\alpha_{\sigma,\sigma'}(\mathbf{r}, \mathbf{r'}, \omega) = \alpha_{\sigma,\sigma'}^{0}(\mathbf{r}, \mathbf{r'}, \omega)$$

$$+ \sum_{\sigma_1\sigma_2} \int \alpha_{\sigma,\sigma_1}^{0}(\mathbf{r}, \mathbf{r_1}, \omega) K_{\sigma_1\sigma_2}(\mathbf{r_1}, \mathbf{r_2}) \alpha_{\sigma_2\sigma'}(\mathbf{r_2}, \mathbf{r'}, \omega)\, d\mathbf{r_1}\, d\mathbf{r_2},$$

$$\tag{9.194}$$

which defines the LSDA correlation function $\alpha_{\sigma,\sigma'}$ in terms of the free correlation functions $\alpha_{\sigma,\sigma'}^{0}$. The kernel of the integral equation, which is commonly known as particle–hole residual interaction, is given by

$$K_{\sigma_1\sigma_2}(\mathbf{r_1}, \mathbf{r_2}) = v(|\mathbf{r_1} - \mathbf{r_2}|) + \left. \frac{\partial^2 E_{\text{xc}}(\rho, m)}{\partial\rho_{\sigma_1}\partial\rho_{\sigma_2}} \right|_{\text{gs}} \delta(\mathbf{r_1} - \mathbf{r_2}),\tag{9.195}$$

where

$$\left. \frac{\partial^2 E_{\text{xc}}}{\partial\rho_{\sigma}\partial\rho_{\sigma'}} \right|_{\text{gs}} = \left. \frac{\partial^2 E_{\text{xc}}}{\partial\rho^2} \right|_{\text{gs}} + (\eta_{\sigma} + \eta_{\sigma'}) \left. \frac{\partial^2 E_{\text{xc}}}{\partial\rho\partial m} \right|_{\text{gs}} + \eta_{\sigma}\eta_{\sigma'} \left. \frac{\partial^2 E_{\text{xc}}}{\partial m^2} \right|_{\text{gs}}$$

$$\equiv \mathcal{K}(r) + (\eta_{\sigma} + \eta_{\sigma'})\mathcal{M}(r) + \eta_{\sigma}\eta_{\sigma'}\mathcal{I}(r),\tag{9.196}$$

with $\eta_{\uparrow} = 1, \eta_{\downarrow} = -1$. The latter expression is the definition of the functions $\mathcal{K}, \mathcal{M}$ and $\mathcal{I}$.

The dynamic polarizability is given by

$$\chi(F, G, \omega) = \sum_{\sigma'} \int f_{\sigma'}^{*}(\mathbf{r})\delta\rho_{\sigma'}(\mathbf{r}, \omega)\, d\mathbf{r},\tag{9.197}$$

with $\delta\rho_{\sigma}$ given by (9.190). Then, we obtain

$$\chi(F^s, G^s, \omega) = \int (f_{\uparrow}^{*}(\mathbf{r})\delta\rho_{\uparrow} + f_{\downarrow}^{*}(\mathbf{r})\delta\rho_{\downarrow})\, d\mathbf{r}$$

$$= \int (f_{\uparrow}^{*}(\mathbf{r})\alpha_{\uparrow,\uparrow}g_{\uparrow}(\mathbf{r'}) + f_{\uparrow}^{*}(\mathbf{r})\alpha_{\uparrow,\downarrow}g_{\downarrow}(\mathbf{r'})$$

$$+ f_{\downarrow}^{*}(\mathbf{r})\alpha_{\downarrow,\uparrow}g_{\uparrow}(\mathbf{r'}) + f_{\downarrow}^{*}(\mathbf{r})\alpha_{\downarrow,\downarrow}g_{\downarrow}(\mathbf{r'}))d\mathbf{r}d\mathbf{r'}.\tag{9.198}$$

Note that if we substitute $\delta\rho_{\sigma}^{0}$, as given by

$$\delta\rho_{\sigma}^{0}(\mathbf{r}, \omega) = \sum_{\sigma'} \int \alpha_{\sigma,\sigma'}^{0}(\mathbf{r}, \mathbf{r'}, \omega)g_{\sigma'}(\mathbf{r'})d\mathbf{r'},\tag{9.199}$$

into (9.197), and use for $\alpha^0_{\sigma,\sigma'}$ the expression (9.192), we obtain the free response function:

$$\chi^0(F,G,\omega) = \sum_{\sigma'mi} \left(\frac{\langle m|f^*_{\sigma'}|i\rangle\langle i|g_{\sigma'}|m\rangle}{\omega - (\epsilon_m - \epsilon_i)} - \frac{\langle m|g_{\sigma'}|i\rangle\langle i|f^*_{\sigma'}|m\rangle}{\omega + (\epsilon_m - \epsilon_i)} \right), \qquad (9.200)$$

where i and m are the single-particle KS states, occupied and nonoccupied, respectively, in the ground state of the system. It is also important to note that, in order to obtain the dynamic polarizability of the system, it is not necessary to go through the solution of the integral Eq. (9.194), which, from a technical point of view, implies the difficult task of inverting large matrices. In fact, starting from Eq. (9.194), it is possible to obtain an integral equation for the density variation that enters directly into the definition of χ:

$$\begin{pmatrix} \delta\rho_\uparrow \\ \delta\rho_\downarrow \end{pmatrix} = \begin{pmatrix} \delta\rho_\uparrow^{(0)} \\ \delta\rho_\downarrow^{(0)} \end{pmatrix} + \begin{pmatrix} \alpha_{\uparrow\uparrow}^{(0)} & 0 \\ 0 & \alpha_{\downarrow\downarrow}^{(0)} \end{pmatrix} \otimes \begin{pmatrix} K_{\uparrow\uparrow} & K_{\uparrow\downarrow} \\ K_{\downarrow\uparrow} & K_{\downarrow\downarrow} \end{pmatrix} \otimes \begin{pmatrix} \delta\rho_\uparrow \\ \delta\rho_\downarrow \end{pmatrix},$$

$$(9.201)$$

where the symbols $\otimes$ indicate an implicit spatial integration on the common variable $\mathbf{r}$ of the operands. In this way, the problem of determining the dynamic polarizability is reduced to that of solving the system of $2N_p$ linear Eqs. (9.201), where N_p is the number of points used to solve the Kohn–Sham equations.

The above formalism can be quickly generalized to the finite temperature case. The only element which changes is the correlation function (9.192), which becomes

$$\alpha^0_{\sigma,\sigma'}(\mathbf{r},\mathbf{r}',\omega) = -\delta_{\sigma,\sigma'} \sum_{\alpha\beta} \varphi^*_\alpha(\mathbf{r})\varphi_\beta(\mathbf{r}) \frac{f_\alpha - f_\beta}{\epsilon_\beta - \epsilon_\alpha + \omega} \varphi^*_\beta(\mathbf{r}')\varphi_\alpha(\mathbf{r}'), \qquad (9.202)$$

where the indices α (β) refer to single-particle levels with spin σ (σ') and thermal occupation numbers f_α (f_β) (see Subsections 2.6 and 4.7). The fact that, in the heated system, thermal effects manifest themselves in the RPA or TDLDA equations only through the free correlation function and the density dependence of the residual interaction, which contain the appropriate Fermi–Dirac occupation numbers, has been discussed by many authors [see for example Braghin and Vautherin (1994), Hernandez et al. (1996)]. One can thus realize that the structure of both the RPA and TDLDA equations and its solutions are preserved.

From Eq. (9.197), taking for F and $G = F$ the symmetric and antisymmetric combinations defined in Eq. (9.189), it follows that

$$\chi(F^s, F^s, \omega) = \int f^*(\mathbf{r})f(\mathbf{r}')\alpha_{ss}(r, r', \omega) \, d\mathbf{r}d\mathbf{r}',$$

$$\chi(F^a, F^a, \omega) = \int f^*(\mathbf{r})f(\mathbf{r}')\alpha_{aa}(r, r', \omega) \, d\mathbf{r}d\mathbf{r}',$$

$$\chi(F^s, F^a, \omega) = \int f^*(\mathbf{r})f(\mathbf{r}')\alpha_{sa}(r, r', \omega) \, d\mathbf{r}d\mathbf{r}'$$

$$\chi(F^a, F^s, \omega) = \int f^*(\mathbf{r})f(\mathbf{r}')\alpha_{as}(r, r', \omega) \, d\mathbf{r}d\mathbf{r}',$$

(9.203)

where

$$\alpha_{ss} = \alpha_{\uparrow\uparrow} + \alpha_{\downarrow\downarrow} + \alpha_{\uparrow\downarrow} + \alpha_{\downarrow\uparrow},$$

$$\alpha_{aa} = \alpha_{\uparrow\uparrow} + \alpha_{\downarrow\downarrow} - \alpha_{\uparrow\downarrow} - \alpha_{\downarrow\uparrow},$$

$$\alpha_{sa} = \alpha_{\uparrow\uparrow} - \alpha_{\downarrow\downarrow} - \alpha_{\uparrow\downarrow} + \alpha_{\downarrow\uparrow},$$

$$\alpha_{as} = \alpha_{\uparrow\uparrow} - \alpha_{\downarrow\downarrow} + \alpha_{\uparrow\downarrow} - \alpha_{\downarrow\uparrow}.$$

(9.204)

The simplest physical situation to which it is possible to apply the TDLSDA theory is the paramagnetic case, in which the magnetization $m_0(r)$ of the ground state is zero, i.e. initially the system is not spin-polarized. In this case we have $\alpha_{\uparrow\uparrow}^{(0)} = \alpha_{\downarrow\downarrow}^{(0)}$, and the function $\mathcal{M}(r)$ of Eq. (9.196) vanishes identically. It can be easily verified that in such a limiting case α_{sa} and α_{as} are zero, and that α_{ss} and α_{aa} obey the Dyson equations

$$\alpha_{ss} = \alpha^{(0)} + \alpha^{(0)} \otimes K_{ss} \otimes \alpha_{ss},$$

$$\alpha_{aa} = \alpha^{(0)} + \alpha^{(0)} \otimes K_{aa} \otimes \alpha_{aa},$$

(9.205)

where the kernels are given by $K_{ss} = v(r_{12}) + \mathcal{K}\delta(r_{12})$, $K_{aa} = \mathcal{I}\delta(r_{12})$, and the free correlation function $\alpha^{(0)} = \alpha_{\uparrow\uparrow}^{(0)} + \alpha_{\downarrow\downarrow}^{(0)}$ is the same in the two channels because $\alpha_{\uparrow\uparrow}^{(0)} = \alpha_{\downarrow\downarrow}^{(0)}$. In the paramagnetic limit of longitudinal response, the charge channel (symmetric) and the spin channel (antisymmetric) are uncoupled, and the kernel of the integral equation consists of the Hartree potential term, plus an exchange-correlation term in the former case, and only an exchange-correlation term in the latter. When the system is spin-polarized in the ground state, $\alpha_{\uparrow\uparrow}^{(0)} \neq \alpha_{\downarrow\downarrow}^{(0)}$, $\mathcal{M}(r)$ is not zero, and the two other independent correlation functions α_{sa} and α_{as} produce a (symmetric) density response at F_a, and a (antisymmetric) spin density response at F_s.

Lastly, we note that it can be shown that using the residual interaction given by Eq. (9.195) in the RPAE equations (9.138), the RPAE solutions for the energies and matrix elements coincide with the TDLSDA results of the present subsection.

In homogeneous systems and when F and $G = F$ are given by the density operator $\rho_q = \sum_i e^{i\mathbf{q}\cdot\mathbf{r}_i}$ and the magnetization density operator $m_q = \sum_i e^{i\mathbf{q}\cdot\mathbf{r}_i}\sigma_i^z$, and in the paramagnetic case, the integral equation (9.205) yield uncoupled algebraic equations for the dynamic polarizabilities in the density and magnetization channels, which are written, respectively, as

$$\chi^{\rho\rho}(q,\omega) = \frac{\chi^0(q,\omega)}{1 - \frac{1}{L^D}\left(v(q) + \frac{\partial^2 E_{xc}}{\partial\rho^2}\big|_{gs}\right)\chi^0(q,\omega)}, \qquad (9.206)$$

$$\chi^{mm}(q,\omega) = \frac{\chi^0(q,\omega)}{1 - \frac{1}{L^D}\frac{\partial^2 E_{xc}}{\partial m^2}\big|_{gs}\chi^0(q,\omega)}. \qquad (9.207)$$

If we compare these equations with the RPA solution (9.86), which was derived in the density channel, we see that the TDLDA theory introduces an explicit dependence on the exchange-correlation energy in the density response, through its second derivative with respect to density. This term changes the dispersion relation of the collective state. For example, in the case of the electron gas, the TDLDA plasmon frequency in the limit of small q becomes

$$\omega_{\text{TDLDA}}^2(\mathbf{q}) = \omega_p^2 + \left(\frac{2T_0^D}{m} + \frac{\rho_0}{m}\frac{\partial^2 E_{xc}^D}{\partial\rho^2}\bigg|_{\rho=\rho_0}\right)q^2, \qquad (9.208)$$

where ω_p is given by (9.106). The TDLDA dispersion coefficient, normalized to the RPA dispersion coefficient, is plotted in Fig. 9.3. We see that the agreement with experimental data, though improved with respect to RPA, is still unsatisfactory at high r_s. In the spin-density channel, the TDLDA response is fixed by the second derivative of the exchange-correlation energy with respect to the magnetization, calculated at the magnetization value of the ground state. This quantity is different from zero also for ground states that are not spin-polarized, and is negative. The particle–hole interaction in the magnetic channel is attractive. This has the effect that (see Fig. 9.1) in 2D and 3D the TDLDA solution, given by the intersection of χ^0/L^D with the (negative) straight line $1/\frac{\partial^2 E_{xc}}{\partial m^2}\big|_{gs}$, falls into the continuum of single-particle states, so that there is no collective state in the magnetic channel. The situation is different in Q1D, where there exists a gap (see Fig. 8.3) between the continuum of single-particle excited states and the energy zero, so that the TDLDA solution in the Q1D case yields a low energy collective solution in the magnetic channel.

Finally, we note that in symmetric nuclear matter (i.e. with the same number of protons and neutrons) it is of special relevance to study the isospin-density response in which

$$F^a = G^a = \rho_{1,q} = \sum_i e^{i\mathbf{q}\cdot\mathbf{r}_i}\tau_i^z,$$

where τ_z is the third isospin component. By exploiting the analogy between spin and isospin, and that between the equations (4.60) and (4.115), it is possible to write immediately the isospin-density response function as

$$\chi^{\rho_1 \rho_1}(q,\omega) = \frac{\chi^0(q,\omega)}{1 - \frac{1}{L^D}\frac{\partial W}{\partial \rho_1}|_{gs}\chi^0(q,\omega)}, \tag{9.209}$$

where W is the symmetry potential of (4.117), and $\rho_1 = \rho_n - \rho_p$ is the isovector density. Thus, for nuclear matter (and neglecting the small nonlocal a_1 term) we have $\frac{\partial W}{\partial \rho_1}|_{gs} = b_1 + c_1\rho^\gamma$, which, with the parameter values of Tables 4.3, 4.4 and 4.10, leads to the numerical value $\frac{\partial W}{\partial \rho_1}|_{gs} = 28.7$ MeV fm^3. Thus, unlike the magnetic channel, the isospin channel has a repulsive interaction, so that the TDLDA solution yields a collective solution. Such a collective solution was experimentally observed in nuclei, and is known as giant resonance [see for example Lipparini and Stringari (1989)]. The response function of nuclear matter and liquid ^{3}He was studied in detail, starting from density functionals like those described in Subsection 4.9, by Garcia-Recio *et al.* (1992) and Alberico *et al.* (1991); that of finite nuclei by Bertsch and Tsai (1975), Liu and Brown (1976) and Liu and N. Van Giai (1983). For the response of ^{3}He and ^{4}He clusters one can then refer to the papers by Serra *et al.* (1991), Barranco *et al.* (1997) and Casas *et al.* (1995).

9.9.2 *The TDLSDA Transverse Response Function*

Assuming that the spin magnetization is along a fixed direction, which we will take as the z direction, it is possible to derive the transverse spin response in a simple way. The method was developed by Rajagopal (1978), and was applied to quantum dots by Lipparini *et al.* (1998, 1999). It turns out that the longitudinal and transverse spin responses do not couple. In what follows, we define the spin-flip operators $\sigma_\pm = 1/2(\sigma_x \pm i\sigma_y)$ in such a way that

$$\sigma_+ |\uparrow\rangle = \sigma_- |\downarrow\rangle = 0,$$

$$\sigma_+ |\downarrow\rangle = |\uparrow\rangle,$$

$$\sigma_- |\uparrow\rangle = |\downarrow\rangle.$$

Let us introduce now a spin magnetization vector $\mathbf{m}$ and assume that the exchange-correlation energy depends only on ρ and $|\mathbf{m}|$, i.e. $E_{xc} = E_{xc}[\rho, |\mathbf{m}|]$. Then, the spin-dependent exchange-correlation potential W_{xc} in Eq. (9.187)) can be written as

$$\mathbf{W}^{xc}[\rho, |\mathbf{m}|] = W_{xc}[\rho, |\mathbf{m}|]\mathbf{m}/|\mathbf{m}|,$$

with

$$W_{xc}[\rho, |\mathbf{m}|] = \partial E_{xc}[\rho, |\mathbf{m}|]/\partial|\mathbf{m}|.$$

At this point, in that equation one performs the change

$$W_{xc}\sigma_z \to \mathcal{F}_{xc}[\rho, |\mathbf{m}|]\mathbf{m} \cdot \boldsymbol{\sigma}, \tag{9.210}$$

where $\mathcal{F}_{xc}[\rho, |\mathbf{m}|] \equiv W_{xc}[\rho, |\mathbf{m}|]/|\mathbf{m}|$. Defining the spherical components $\pm$ of the vectors $\mathbf{m}$ and $\boldsymbol{\sigma}$, we have

$$W_{xc}\sigma_z \to \mathcal{F}_{xc}[\rho, |\mathbf{m}|] \, [m_z\sigma_z + 2(m_+\sigma_- + m_-\sigma_+)]. \tag{9.211}$$

In the static case, the inclusion of the densities m_+ and m_- makes no difference since they vanish identically. The situation is different when the system interacts with a time-dependent field that couples to the electron spin through

$$\mathbf{G} \cdot \boldsymbol{\sigma} = G_z\sigma_z + 2(G_+\sigma_- + G_-\sigma_+).$$

If the time dependence is harmonic, the interaction Hamiltonian causing transverse spin excitations may be written as

$$H_{int} \sim \sum_i [g(\mathbf{r}_i)\sigma_-^i e^{-\imath\omega t} + g^*(\mathbf{r}_i)\sigma_+^i e^{\imath\omega t}]. \tag{9.212}$$

H_{int} causes nonvanishing variations in the densities m_+ and m_-, which in turn generate induced potentials through first-order perturbation variations in the mean field of (9.211). This equation shows that the induced interaction is $2\mathcal{F}_{xc}\delta(\mathbf{r}_1 - \mathbf{r}_2)\sigma_-$ for m_+, and $2\mathcal{F}_{xc}\delta(\mathbf{r}_1 - \mathbf{r}_2)\sigma_+$ for m_-, computed for the ground state values of ρ and m. If we define the transverse correlation function through the equation

$$\delta m^- = \alpha_{-+} \otimes g, \tag{9.213}$$

where the symbol $\otimes$ denotes space integration on the space variable common to the operands, and if we require the self-consistency for the density variations like we did in the previous subsection, it is easy to reach the following integral equation for α_{-+}:

$$\alpha_{-+} = \alpha_{-+}^{(0)} + 2\alpha_{-+}^{(0)} \otimes \mathcal{F}_{xc}\delta(r_{12}) \otimes \alpha_{-+}, \tag{9.214}$$

in terms of the free correlation function

$$\alpha_{-+}^{(0)}(\mathbf{r}, \mathbf{r}'; \omega) = -\sum_{\alpha\beta}(f_\alpha - f_\beta)\frac{\langle\alpha|\sigma_-|\beta\rangle_\mathbf{r}\langle\beta|\sigma_+|\alpha\rangle_{\mathbf{r}'}}{\epsilon_\beta - \epsilon_\alpha + \omega}. \tag{9.215}$$

The dynamic polarizability is further given by

$$\chi_-(\omega) = f^* \otimes \delta m_-. \tag{9.216}$$

The expressions for the other transverse channel are

$$\delta m^+ = \alpha_{+-} \otimes g, \tag{9.217}$$

$$\alpha_{+-} = \alpha_{+-}^{(0)} + 2\alpha_{+-}^{(0)} \otimes \mathcal{F}_{xc}\delta(r_{12}) \otimes \alpha_{+-}, \tag{9.218}$$

with

$$\alpha^{(0)}_{+-}(\mathbf{r}, \mathbf{r}'; \omega) = \sum_{\alpha\beta}(f_\alpha - f_\beta)\frac{\langle\alpha|\sigma_+|\beta\rangle_\mathbf{r}\langle\beta|\sigma_-|\alpha\rangle_{\mathbf{r}'}}{\epsilon_\alpha - \epsilon_\beta + \omega}. \tag{9.219}$$

And finally

$$\chi_+(\omega) = f^* \otimes \delta m_+. \tag{9.220}$$

In the homogeneous case where G and F are given by the operators

$$G^- = \sum_i \exp(-i\mathbf{q} \cdot \mathbf{r})_i \sigma_i^-,$$

$$F^+ = \sum_i \exp(i\mathbf{q} \cdot \mathbf{r})_i \sigma_i^+,$$

$$G^+ = \sum_i \exp(-i\mathbf{q} \cdot \mathbf{r})_i \sigma_i^+,$$

$$F^- = \sum_i \exp(i\mathbf{q} \cdot \mathbf{r})_i \sigma_i^-,$$

it is easy to obtain the TDLSDA transverse linear response at zero temperature:

$$\chi_t(q, \omega) = \frac{\chi_t^0(q, \omega)}{1 - \frac{2}{L^D}\mathcal{F}_{xc}\chi_t^0(q, \omega)}, \tag{9.221}$$

with the free transverse linear response given by

$$\chi_t^0(q, \omega) = \sum_{mi}\left(\frac{|\langle m \downarrow|\exp(-i\mathbf{q} \cdot \mathbf{r})\sigma^-|i\uparrow\rangle|^2}{\omega - \epsilon_{mi}} - \frac{|\langle m \uparrow|\exp(i\mathbf{q} \cdot \mathbf{r})\sigma^+|i\downarrow\rangle|^2}{\omega + \epsilon_{mi}}\right). \tag{9.222}$$

In the case of an electron gas in a constant magnetic field that spin-polarizes the system and induces negligible diamagnetic effects, the collective solutions to Eq. (9.221) are not damped, even in the case of attractive p–h interaction ($\mathcal{F}_{xc} < 0$), and are known as spin waves. The difference with the longitudinal case, where the collective spin states are always damped for p–h negative interaction, lies in the presence of the magnetic field, which induces magnetization in the system and may generate a gap in the single-particle excitation spectrum for the 2D and 3D cases like the one shown in Fig. 8.3; for one-dimensional systems such a gap is present also for $B = 0$, but not for two-dimensional and three-dimensional systems (see Fig. 8.4).

Spin waves were predicted theoretically by Silin (1958, 1959) and Platzman and Wolf (1967), and observed in alkali metals by Schultz and Dunifer (1967).

In the TDLSDA formalism these spin waves can be obtained as poles of Eq. (9.221), using for the transverse free-particle wave function the expression

$$\chi_t^0(q, \omega) = -\frac{3}{4}\frac{\rho}{\epsilon_F}\left(1 + \frac{\omega}{2qv_F}\ln\frac{\omega - \omega_a - qv_F}{\omega - \omega_a + qv_F}\right), \tag{9.223}$$

which is valid in the $qv_F \ll \epsilon_F$ limit, and where

$$\omega_a = \frac{\omega_L}{1 + \frac{3\rho\mathcal{F}_{\mathrm{xc}}}{2\epsilon_F}},$$

with ω_L equal to the Larmor frequency $\omega_L = g\mu_0 B$. In the high magnetic field limit, $qv_F/\omega_a < 1$, such a free response function is quite different from the $B = 0$ one, and exhibits a gap in the single-particle excitation spectrum. The dispersion of the spin waves, in this case, is given by

$$\omega(q) = \omega_L - \frac{1}{3}\frac{q^2 v_F^2}{\omega_a}(1 + F_0^a), \qquad (9.224)$$

with $F_0^a = 3\rho\mathcal{F}_{\mathrm{xc}}/2\epsilon_F$. This solution is a collective excitation ($|\mathrm{col}\rangle$) induced by the operator F^-, since it exhausts the m_0^- sum rule of Subsection 8.3, yielding $|\langle\mathrm{col}|F^-|0\rangle|^2 = N_\uparrow - N_\downarrow$; in fact, the strength $|\langle\mathrm{col}|F^+|0\rangle|^2$ is equal to zero owing to the blocking due to the Pauli exclusion principle in the $q \to 0$ limit. To conclude, note that the solution (9.224) obeys the Larmor theorem (see Subsection 8.10).

The above formalism can be applied to the case of liquid ^{3}He (Stringari and Dalfovo, 1990), and can be easily generalized to the case of asymmetric nuclear matter ($N \neq Z$), where the transverse collective solution is known as an isospin wave (Friman, 1980; Lipparini and Stringari, 1987). In this latter case the role of the constant magnetic field B is played by the proton Coulomb potential, which, in first approximation, produces in the nuclear Hamiltonian a term of the type $-\frac{1}{2}E_C\tau_z$, with E_C equal to the Coulomb energy shift between near nuclei, i.e. the analog of the Zeeman term present in magnetic-field-polarized systems. Then, for nuclear matter, the Larmor frequency turns into the frequency $\omega_C = E_C$.

In the cases of helium and nuclear matter, one can take the interaction parameters F_0^a, which appear in the dispersion (9.224), as the Landau parameters, whose values were given in Subsection 9.1.

9.10   RPA and TDLSDA Commutators and Symmetry Restoration

In this subsection we will discuss a series of results concerning the commutators of a generic, Hermitian, one-body operator, $F = \sum_i f(i)$, with the system Hamiltonian, within the frameworks of the RPA and TDLSDA theories. In other words, we wish to evaluate the commutator $[H, F]_{\mathrm{RPA}} = [H_{\mathrm{RPA}}, F]$ where $H_{\mathrm{RPA}} = \sum_k \omega_k O_k^\dagger O_k$ and $O_k^\dagger$ (O_k) are the creation (annihilation) operators of the excited RPA states k (at frequency ω_k) of Eq. (9.152). Using the expansion of F in terms of the complete set of RPA excitations, and the Boson commutation rules for the operators $O_k^\dagger$ and O_k, we can directly evaluate the commutator $[H_{\mathrm{RPA}}, F]$. The TDLSDA case can be considered as a particular case of the RPA, in which the RPA Eqs. (9.138) are solved with the residual particle–hole interaction of Eq. (9.195), and with the single-particle wave functions and energies obtained as solutions to the Kohn–Sham equations instead of the HF ones. In general, the RPA commutator differs from the true one and, most of all, from the commutator with the static HF (or Kohn–Sham) Hamiltonian, which does not take account of dynamic correlations. The evaluation of the RPA commutator is relevant to the problem of the restoration of symmetries which are broken in the static HF or LSDA theories (see Subsection 4.11), and for the calculation of the sum rules defined in Subsection 8.3, taking into account RPA correlations. The results that we are going to show in the following were proven rigorously by Lipparini and Stringari (1989). For the RPA commutator we obtain the result

$$[H, F]_{\mathrm{RPA}} = [H_0^{\mathrm{HF}}, F] + \delta H^{\mathrm{HF}}(iF), \qquad (9.225)$$

where H_0^{HF} is the static HF Hamiltonian and

$$\delta H^{\mathrm{HF}}(iF) = \sum_i \delta H_i^{\mathrm{HF}}$$

is the change (linear in F) induced in the HF Hamiltonian by the unitary transformation e^{iF}.

The two terms of (9.225) have different physical meaning: the commutator $[H_0^{\mathrm{HF}}, F]$ originates from the static, one-body properties of the Hamiltonian, while the term $\delta H^{\mathrm{HF}}(F)$ originates from the renormalization of the self-consistent potential. The latter contribution is essential for taking into account the RPA correlations and, in general, for restoring the symmetries that are broken by the static theory. In order to expound this point, let us consider the Hamiltonian

$$H = \sum_i \frac{p_i^2}{2m} + g \sum_{i<j} \delta(\mathbf{r}_i - \mathbf{r}_j), \qquad (9.226)$$

which is of relevance for the case of a cold and dilute condensate of Bosons (see Subsection 2.4). Clearly, this Hamiltonian is invariant under translations, and so it commutes with the total momentum operator $P_z = \sum_i p_i^z$. However, the HF self-consistent Hamiltonian,

$$H^{\mathrm{HF}} = \sum_i \left(\frac{p_i^2}{2m} + g\rho(\mathbf{r}_i) \right), \qquad (9.227)$$

where ρ is the Boson density, does not commute with P_z when considered from a static point of view but yields the result

$$[H_0^{\mathrm{HF}}, iP_z] = -g \sum_i \nabla_z \rho. \qquad (9.228)$$

Nevertheless, if we consider it from the dynamical point of view, it changes owing to the variations that take place in its density-dependent terms during the collective oscillation. The variation

$$\delta H^{\mathrm{HF}}(iP_z) = g \sum_i \delta\rho(\mathbf{r}_i) \qquad (9.229)$$

is easily evaluated because the density change, which is linear in P_z and induced by the transformation e^{iP_z}, is given by

$$\delta\rho(\mathbf{r}) = \langle\mathrm{HF}| \left[\sum_i \delta(\mathbf{r}-\mathbf{r}_i), i\sum_j p_j^z \right] |\mathrm{HF}\rangle = \nabla_z \rho(\mathbf{r}). \qquad (9.230)$$

Then we find that the two contributions of Eq. (9.225) cancel out exactly, and that $[H, P_z]_{\mathrm{RPA}} = 0$, thus showing that the RPA theory restores translational symmetry. As regards the sum rules, the following equalities can be shown:

$$m_1^{\mathrm{RPA}} = \frac{1}{2}\langle\mathrm{HF}|[F, [H, F^\dagger]_{\mathrm{RPA}}]|\mathrm{HF}\rangle$$

$$\qquad (9.231)$$

$$m_3^{\mathrm{RPA}} = \frac{1}{2}\langle\mathrm{HF}|[[F, H]_{\mathrm{RPA}}, [H, [H, F^\dagger]_{\mathrm{RPA}}]_{\mathrm{RPA}}]|\mathrm{HF}\rangle,$$

where $|\mathrm{HF}\rangle$ is the HF (or LSDA) ground state, and the moments

$$m_k = \sum_n \omega_{no}^k |\langle n|F|0\rangle|^2$$

are computed by using excitation energies and matrix elements given by the solutions to the RPA (TDLSDA) equations.

The results (9.231) can be extended to the case of mixed sum rules which involve two different kinds of excitation operators F and G (see Subsection 8.3):

$$(m_0^-)^{\mathrm{RPA}} = \frac{1}{2}\langle\mathrm{HF}|[F, G^\dagger]|\mathrm{HF}\rangle,$$

$$(m_1^+)^{\mathrm{RPA}} = \frac{1}{2}\langle\mathrm{HF}|[F, [H, G^\dagger]_{\mathrm{RPA}}]|\mathrm{HF}\rangle, \qquad (9.232)$$

$$(m_2^-)^{\mathrm{RPA}} = \frac{1}{2}\langle\mathrm{HF}|[[F, H]_{\mathrm{RPA}}, [H, G^\dagger]_{\mathrm{RPA}}]|\mathrm{HF}\rangle,$$

where, like before, the moments

$$(m_k^\pm)^{\mathrm{RPA}} = \sum_n \omega_{no}^k \left(\langle 0|F|n\rangle\langle n|G^\dagger|0\rangle \pm \langle 0|G^\dagger|n\rangle\langle n|F|0\rangle \right)$$

are computed in the RPA approximation, and the mean values of the RPA commutators, on the HF ground state.

Eqs. (9.231) and (9.233) show that the calculation of some of the moments of the excitation strength in the RPA (TDLSDA) does not require the explicit solution of the RPA (TDLSDA) equations, but only the evaluation of mean values on the static HF (LSDA) state of the RPA commutators that appear in such equations, by using the rule (9.225).

As an example of the application of the above relations, let us compute, starting from the Pauli Hamiltonian (5.1), the m_1^+ sum rule for the operators

$$F = F^+ = \sum_i \exp(i\mathbf{q} \cdot \mathbf{r})_i \sigma_i^+,$$

$$G^\dagger = F^- = \sum_i \exp(-i\mathbf{q} \cdot \mathbf{r})_i \sigma_i^-.$$

This sum rule is of relevance in the calculation of the transverse response function of a two-dimensional electron gas confined in a quantum dot, in the presence of an external magnetic field perpendicular to the plane of motion of the electrons. Explicit calculation of

$$m_1^+ = \sum_n \omega_{n0} |\langle n|F^-|0\rangle|^2 + \sum_n \omega_{n0} |\langle n|F^+|0\rangle|^2$$

yields the result

$$m_1^+ = \frac{Nq^2}{2} - g^* \mu_0 B (N_\uparrow - N_\downarrow) \tag{9.233}$$

where N is the total number of Fermions in the system, and $N_\uparrow$ $(N_\downarrow)$ is the number of Fermions with spin up (down), and $N_\uparrow - N_\downarrow$ is the magnetization of the ground state.

The result (9.233) for m_1^+ is also reproduced by the TDLSDA theory, if one uses (9.225) to evaluate the TDLSDA commutator in (9.233). Therefore, the transverse spectrum as computed by the TDLSDA theory is such that it satisfies the mixed sum rule m_1^+. On the other hand, the spectra of single-particle models, like those obtained by in the Hartree, HF or Kohn–Sham approximation, violate such a sum rule. For example, the KS spectrum that one uses to build up the single-particle response function χ^0 leads to the result

$$(m_1^+)^{\text{KS}} = \frac{Nq^2}{2} - [g^* \mu_0 B + 2W_{\text{xc}}](N_\uparrow - N_\downarrow), \tag{9.234}$$

where W_{xc} is the spin-dependent, exchange-correlation potential that appears in the KS Eqs. (4.60).

In the TDLSDA theory the effect of the induced particle–hole interaction [i.e. the term δH^{HF} of Eq. (9.225)] is crucial in restoring the exact value of m_1^+: in fact, it exactly cancels out the $2W_{\text{xc}}$ contribution in (9.234).

9.11   The Linear Response Based on the Green Functions; RPAE

In this subsection we will compare the previously discussed RPA and TDLSDA theories with the approach to the linear density response function based on the particle–hole (*p–h*) Green functions. The density response function $\chi(\mathbf{q}, \omega)$ is connected to the Green function by the relation (in the following, for simplicity, we ignore spin sums)

$$\chi(\mathbf{q}, \omega) = \int \frac{d\mathbf{k}}{(2\pi)^D} G_D^{ph}(\mathbf{k}, \mathbf{q}, \omega), \tag{9.235}$$

where D indicates, as usual, the dependence on dimensionality and G_D^{ph} is the *p–h* Green function for the transition [induced by the external field $(\mathbf{q}, \omega)$] from the initial hole (*h*) state $\mathbf{k}$, to the final particle (*p*) state $\mathbf{k}+\mathbf{q}$. $G_D^{ph}(\mathbf{k}, \mathbf{q}, \omega)$ is a solution to the Dyson integral equation:

$$G_D^{ph}(\mathbf{k}, \mathbf{q}, \omega) = G_{0,D}^{ph}(\mathbf{k}, \mathbf{q}, \omega)$$
$$+ G_{0,D}^{ph}(\mathbf{k}, \mathbf{q}, \omega) \int \frac{d\mathbf{p}}{(2\pi)^D} V(\mathbf{k}, \mathbf{p}, \mathbf{q}) G_D^{ph}(\mathbf{p}, \mathbf{q}, \omega), \tag{9.236}$$

where $G_{0,D}^{ph}(\mathbf{k}, \mathbf{q}, \omega)$ is the HF Green function:

$$G_{0,D}^{ph}(\mathbf{k}, \mathbf{q}, \omega) = \frac{(1 - n_{\mathbf{k}+\mathbf{q}})n_{\mathbf{k}}}{\omega - \varepsilon_{\mathbf{kq}} + i\eta} - \frac{(1 - n_{\mathbf{k}})n_{\mathbf{k}+\mathbf{q}}}{\omega - \varepsilon_{\mathbf{kq}} - i\eta}, \tag{9.237}$$

with $n_{\mathbf{k}}$ given by (1.76) and

$$\varepsilon_{\mathbf{kq}} = \frac{(\mathbf{k} + \mathbf{q})^2}{2m} + \Sigma(\mathbf{k} + \mathbf{q}) - \frac{\mathbf{k}^2}{2m} - \Sigma(\mathbf{k})$$

[$\Sigma(\mathbf{k})$ is the Fock self-energy (2.34)], and $V(\mathbf{k}, \mathbf{p}, \mathbf{q})$ is the *p–h* interaction given by the matrix element

$$V(\mathbf{k}, \mathbf{p}, \mathbf{q}) = \langle \mathbf{k}, \mathbf{k} + \mathbf{q} | V | \mathbf{p} + \mathbf{q}, \mathbf{p} \rangle. \tag{9.238}$$

For example, for the Coulomb interaction the direct and exchange matrix elements are respectively $4\pi e^2/q^2$ and $4\pi e^2/|\mathbf{k} - \mathbf{p}|^2$ in 3D, and $2\pi e^2/q$ and $2\pi e^2/|\mathbf{k} - \mathbf{p}|$ in 2D. The response function obtained from the solution of the Dyson equation with includes exchange effects is the RPAE response function that we introduced in Subsection 9.7 starting from the TDHF equations, and that differs from the RPA response because the latter takes into account only the direct matrix elements of the interaction.

The exchange matrix elements make the solution of the integral Dyson Eq. (9.236) in homogeneous systems a very difficult task — indeed, much more difficult than in finite systems that we considered in Subsection 9.8, where spherical symmetry was assumed for the problem. To the best of the author's knowledge, such an equation was solved only in the case of the 2D interacting electron gas

(Takayanagi and Lipparini, 1995; Schulze *et al.*, 2000). For its solution, it is opportune to introduce the "vertex function" $f(\mathbf{k}, \mathbf{q}, \omega)$, defined by

$$G_D^{ph}(\mathbf{k}, \mathbf{q}, \omega) = f(\mathbf{k}, \mathbf{q}, \omega) G_{0,D}^{ph}(\mathbf{k}, \mathbf{q}, \omega), \qquad (9.239)$$

which satisfies the integral equation

$$f(\mathbf{k}, \mathbf{q}, \omega) = 1 + \int \frac{d\mathbf{p}}{(2\pi)^D} V(\mathbf{k}, \mathbf{p}, \mathbf{q}) f(\mathbf{k}, \mathbf{q}, \omega) G_{0,D}^{ph}(\mathbf{p}, \mathbf{q}, \omega). \qquad (9.240)$$

Equation (9.240) is still a singular integral equation, like the original Dyson equation, but it has the advantage now that the singularities are in the known function
$G_{0,D}^{ph}(\mathbf{p}, \mathbf{q}, \omega)$ of Eq. (9.237). The numerical results show that the exchange contributions to the static and dynamic responses are important, and that such quanti-
ties differ much from the RPA predictions, and agree with the Monte Carlo calculations for the static response. The plasmon dispersion and the single-particle transition range for the 2D electron gas are reported in Fig. 9.17.

From the curves we observe that in 2D the RPAE predicts Landau damping (the
plasmon energy touches the border of the single-particle excitations) at q values
which are relatively low with respect to the RPA predictions.

The TDLDA can be obtained from Eq. (9.236), by omitting the exchange contributions in the matrix element (9.238), and the Fock self-energy term in the single-
particle energies, and identifying the p–h interaction in momentum space with

$$V(\mathbf{k}, \mathbf{p}, \mathbf{q}) = v(q) + \left. \frac{\partial V_{\text{xc}}}{\partial \rho} \right|_{\rho=\rho_0}. \qquad (9.241)$$

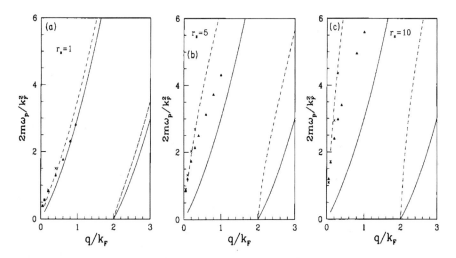

Fig. 9.17 Plasmon dispersion and single-particle transition range for the 2D electron gas. Crosses
and triangles are the plasmon dispersion in RPAE and RPA, respectively. Dashed and full lines
give the HF and Hartree single-particle transition ranges, respectively. (a) $r_s = 1$; (b) $r_s = 5$; (c)
$r_s = 10$.

Then, $V(\mathbf{k}, \mathbf{p}, \mathbf{q})$ as given by Eq. (9.241) can be taken out of the integral in (9.236) and, integrating such an equation in $d\mathbf{k}$, we obtain an algebraic equation for the response function. Solving the latter yields the TDLDA solution to (9.206). The RPA solution to Eq. (9.86) is then obtained from (9.236) by taking $V(\mathbf{k}, \mathbf{p}, \mathbf{q}) = v(q)$. Note, further, that in coordinate space the p–h interaction (9.241) becomes

$$V_{ph}(|\mathbf{r}_1 - \mathbf{r}_2|) = v(|\mathbf{r}_1 - \mathbf{r}_2|) + \frac{\partial V_{xc}}{\partial \rho}\bigg|_{\rho=\rho_0} \delta(|\mathbf{r}_1 - \mathbf{r}_2|), \qquad (9.242)$$

and that the quantity $\frac{\partial V_{xc}}{\partial \rho}|_{\rho=\rho_0}$ is connected to the compressibility K of the system by the relation

$$\frac{\partial V_{xc}}{\partial \rho}\bigg|_{\rho=\rho_0} = \frac{L^D}{\nu(0)_D}\left(\frac{K_0}{K} - 1\right), \qquad (9.243)$$

where K_0 is the compressibility of the noninteracting system, and $\nu(0)_D$ is the density of states at the Fermi surface. The "vertex function," defined in (9.240), has a very simple expression in the TDLDA theory, given by

$$f^{\text{TDLDA}}(\mathbf{q}, \omega) = \frac{1}{1 - (v(q) + \frac{\partial V_{xc}}{\partial \rho}|_{\rho=\rho_0})\chi^0(\mathbf{q}, \omega)}. \qquad (9.244)$$

In the RPA the vertex function is obtained through the above expression by setting to zero the derivative of the exchange-correlation potential.

9.12   The Screened Response Function and the Dielectric Constant

In charged systems, it is convenient to define the screened response function χ^{sc}. This is the response of the system to the sum of the external field and the Coulomb polarization field. The relation between χ and χ^{sc} in the linear response theory is easily found from the equation

$$\lambda\chi = [\delta\rho v(q) + \lambda]\chi^{\text{sc}} = \delta\rho L^D, \qquad (9.245)$$

which follows from the definitions [see Eq. (9.77)] $\chi = \frac{\delta\rho L^D}{\lambda}$, $\chi^{\text{sc}} = \frac{\delta\rho L^D}{\lambda'}$, with $\lambda' = \delta\rho v(q) + \lambda$ equal to the sum of the external field λ and the Coulomb polarization field $\delta\rho v(q)$. In this way we find that

$$\chi(\mathbf{q}, \omega) = \frac{\chi^{\text{sc}}(\mathbf{q}, \omega)}{1 - \frac{v(q)}{L^D}\chi^{\text{sc}}(\mathbf{q}, \omega)}, \qquad (9.246)$$

which leads to the following results for χ^{sc} in the RPA and the TDLDA:

$$\chi^{\text{sc}}_{\text{RPA}}(\mathbf{q}, \omega) = \chi^0(\mathbf{q}, \omega), \qquad (9.247)$$

$$\chi^{\text{sc}}_{\text{TDLDA}}(\mathbf{q}, \omega) = \frac{\chi^0(\mathbf{q}, \omega)}{1 - \frac{1}{L^D}\frac{\partial V_{xc}}{\partial \rho}|_{\rho=\rho_0}\chi^0(\mathbf{q}, \omega)}. \qquad (9.248)$$

The screened response has the following property [equivalent to Eq. (8.68), which is valid for neutral systems)]:

$$\lim_{q \to 0} \frac{\chi^{\text{sc}}(\mathbf{q}, 0)}{L^D} = -\rho^2 K, \tag{9.249}$$

where K is the system compressibility. Moreover, it is related to the dielectric constant of the system by the relation

$$\epsilon(\mathbf{q}, \omega) = 1 - v(q)\chi^{\text{sc}}(\mathbf{q}, \omega), \tag{9.250}$$

which follows from the definition

$$\chi^{\text{sc}}(\mathbf{q}, \omega) = \epsilon(\mathbf{q}, \omega)\chi(\mathbf{q}, \omega). \tag{9.251}$$

The dielectric constant $\epsilon(\mathbf{q}, \omega)$ allows us to define a screened static Coulomb potential,

$$V_{\text{eff}}(\mathbf{q}) = \frac{v(\mathbf{q})}{e\epsilon(\mathbf{q}, 0)}, \tag{9.252}$$

which gives the Fourier transform of the electrostatic potential $v(\mathbf{r})$ felt by an electron, owing to its interaction with a charge e placed at the origin of the plasma. Such an external charge polarizes the electrons of the gas in its neighborhood, so that a distant electron responds to both the external charge and the induced polarization charge. In the Thomas–Fermi approximation (Mott and Jones, 1936), and in 3D, we find that

$$V_{\text{eff}}(\mathbf{q}) = \frac{4\pi e}{q^2 + q_{\text{FT}}^2}, \tag{9.253}$$

where

$$q_{\text{FT}} = \sqrt{\frac{6\pi e^2 \rho_0}{\epsilon_F}}, \tag{9.254}$$

is the Thomas–Fermi momentum. Its reciprocal $\lambda_{\text{FT}} = q_{\text{FT}}^{-1}$ is known as the "screening length." The ratio of this quantity to the Wigner–Seitz radius is proportional to the square root of the ratio between the mean kinetic and potential energies. For an electron gas at metallic densities (i.e. r_s between 2 and 6), such a ratio is approximately 1, so that the screening length and the interparticle distance are comparable. Thus, it turns out that for the electron gas of physical interest screening is important.

The dielectric constant is also related to the longitudinal conductivity $\sigma(\mathbf{q}, \omega)$. For example, in 3D one has [see Pines and Noziéres (1966)]

$$\epsilon(\mathbf{q}, \omega) = 1 + \frac{4\pi i}{\omega}\sigma(\mathbf{q}, \omega). \tag{9.255}$$

From Eq. (9.250) one then sees that the longitudinal conductivity is simply related to $\chi^{sc}(\mathbf{q}, \omega)$ by

$$\sigma(\mathbf{q}, \omega) = i\frac{\omega e^2}{q^2}\chi^{sc}(\mathbf{q}, \omega), \qquad (9.256)$$

which is exactly the result one would get directly by evaluating σ as the current linear response to an effective electric field which is the sum of the external field and the polarization field (see Subsections 8.7 and 8.8).

9.13   Examples of Application of the TDLSDA Theory

Recently (Eriksson *et al.*, 1999), it has become possible to produce a GaAs/Al$_x$Ga$_{1-x}$As quantum well containing a high quality, two-dimensional electron gas, and to vary the density of this gas by changing the density of electron donors in the heterojunction. This experimental realization opened the way for the study of interaction effects in two-dimensional electron systems in the presence of strong magnetic fields, and for the possibility of comparing these studies with the ones carried out on quantum dots where the finite size effects are important, and different excitation modes (edge modes) are present. In particular, at present the data on inelastic scattering of photons (Raman scattering) are available for both systems (Eriksson *et al.*, 1999; Schuller *et al.*, 1998). These data allow comparison to be made between the excitation spectra of quantum dots and quantum wells, within the framework of the same reaction process. In the next two subsections we will study the excitation spectra of these systems using the TDLSDA.

9.13.1   *Quantum Wells under a Very High External Magnetic Field*

In this subsection we will study the collective states of the two-dimensional electron gas (2DEG) under a very high perpendicular magnetic field which, in the ground state of the system, causes full occupation of an integer number of Landau levels. The excited states of the 2DEG under a very strong magnetic field have been studied in the past by several authors. Of particular relevance are the calculations by Kallin and Halperin (1984) and MacDonald (1985), who studied, respectively, the dispersion with transferred momentum q and with the magnetic field B, of the density excitations (CDE) and of the magnetization density excitations (SDE) in the 2DEG, within the RPAE and TDHF approximations. The results of the two approaches are fully equivalent. Their agreement with the results of inelastic photon scattering experiments on the dispersion of SDE and CDE modes in the 2DEG at filling factor $\nu = 2$, is good only in the limit of high density. At large r_s, large discrepancies between theory and experiment are observed. Instead of only one roton minimum (see Fig. 9.18), and of the instability predicted by TDHF–RPA calculations

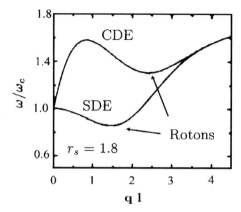

Fig. 9.18 Density excitations (CDE) and magnetization density excitations (SDE), as a function of the transferred momentum q, in the TDHF approximation. $l = \sqrt{c/eB}$ is the magnetic length.

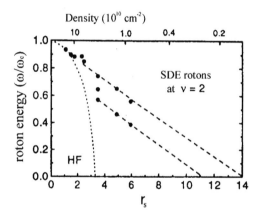

Fig. 9.19 Energies of the rotonic minima in the spin channel SDE, as a function of r_s. Dots: Experimental results. Note that for $r_s \geq 3.3$ more than one rotonic minimum is observed. Dotted line: Energy of the rotonic minimum computed in TDHF. Dashed line: Linear extrapolation of the data.

at $r_s \simeq 3$, the experimental dispersion exhibits two roton minima in the spin channel, and a possible instability only at much larger r_s values (≥ 11) (see Fig. 9.19).

In the following, we will study the dispersion of longitudinal and transverse density modes in the 2DEG, by means of the TDLSDA described in the previous subsections (Ghizzi, 2000), which, unlike the TDHF–RPA approximation, includes the effects of correlations in the local density approximation. For the longitudinal modes we will limit the calculation to the $\nu = 2$ case, which is of interest and has been studied experimentally. In the transverse channel we will later consider the case $\nu = 1$ as well, in order to evidence the effects of the spin polarization of the system, which in this channel are of paramount importance. The generalization to other filling factors is direct.

A. Longitudinal Response

In Subsection 9.9.1 we derived an integral equation for the TDLSDA correlation function $\alpha_{\sigma\sigma'}(\mathbf{r},\mathbf{r}',\omega)$ [see Eqs. (9.194)–(9.196)], whose main ingredients are the free correlation function, and the residual particle–hole interaction, which were set up starting from the solutions to the KS equations and from the exchange-correlation potential, respectively. In the case of a homogeneous gas we have $v_{\text{ext}} + U = 0$ and the exchange-correlation potentials, V_{xc} and W_{xc}, are constant, so that it is only the kinetic energy operator that determines the solutions to the Kohn–Sham equations (apart from the spin-dependent terms which split the single-particle energy levels with different spin). In the case of a very intense magnetic field, directed perpendicularly to the plane of motion of the electrons with charge $-e$, it is convenient to choose the Landau gauge for the vector potential $\mathbf{A}$, for which we have $\mathbf{A} = (0, B_z x, 0)$. Thus, we need to solve the KS equation ($\hbar = 1$),

$$\frac{1}{2m}\left(\mathbf{p} + \frac{e}{c}\mathbf{A}\right)^2 \varphi_i(x,y) = \epsilon_i \varphi_i(x,y), \tag{9.257}$$

i.e.

$$\left[-\frac{1}{2m}\left(\frac{\partial^2}{\partial x^2} + \frac{\partial^2}{\partial y^2}\right) - i\omega_c x \frac{\partial}{\partial y} + \frac{1}{2}m\omega_c^2 x^2\right]\varphi_i(x,y) = \epsilon_i \varphi_i(x,y), \tag{9.258}$$

where $\omega_c = \frac{eB}{mc}$ is the cyclotron frequency. Looking for the general solution in the form

$$\varphi_i(x,y) = \frac{1}{\sqrt{L_y}} e^{-ik_y y} \psi_M(x),$$

where k_y obeys the periodicity condition $k_y = \frac{2\pi}{L_y}n$, for $n = 0, \pm1, \pm2, \ldots$, we obtain for $\psi_M(x)$ the equation

$$\left(-\frac{1}{2m}\frac{\partial^2}{\partial x^2} + \frac{1}{2}m\omega_c^2(x - X)^2\right)\psi_M(x) = \epsilon_M \psi_M(x), \tag{9.259}$$

which is the equation of a harmonic oscillator whose equilibrium position is $X = k_y/m\omega_c$, and whose eigenstates are the Hermite polynomials. Therefore, the eigenfunctions for the KS equation are

$$\varphi_{M,X,\sigma}(x,y) = \frac{e^{-iXy/l^2}}{\sqrt{L_y \pi^{1/2} 2^M M! l}} e^{-\frac{(x-X)^2}{2l^2}} H_M\left(\frac{x - X}{l}\right)\chi_\sigma, \tag{9.260}$$

where χ_σ is eigenfunction of the spin operator, while $l = (m\omega_c)^{-1/2}$ is the magnetic length and $H_M(\frac{x-X}{l})$ is the Hermite polynomial of degree M defined by

$$H_M(\xi) = (-1)^M e^{\xi^2} \frac{d^M}{d\xi^M} e^{-\xi^2}. \tag{9.261}$$

The eigenvalues of the KS equation, taking account of the spin-dependent (constant) terms as well, are given by $(M = 0, 1, 2, \ldots)$

$$\epsilon_{M,\sigma} = \omega_c \left(M + \frac{1}{2} \right) + \left(W_{\text{xc}} + \frac{1}{2} g^* \mu_0 B \right) \sigma. \qquad (9.262)$$

We see that the eigenvalues depend only on M, which indicates the different Landau levels, and on the spin $\sigma = \pm 1$, which splits each of these levels into two sublevels set apart by the constant amount $2W_{\text{xc}} + g^* \mu_0 B$. For fixed values of M and σ, the quantum number $X = (2\pi/L_y m\omega_c)n$, for $n = 0, \pm 1, \pm 2, \ldots$, can take on infinite numbers of values; this means that each Landau level is infinitely degenerate. The spin-splitting term is negative, since both W_{xc} and g^* are negative, so that once M is fixed, the states whose spin is parallel to the magnetic field, $\sigma = 1$, have lower energies than those with $\sigma = -1$.

The electron density is given by

$$\rho(x, y) = \sum_\sigma \rho_\sigma = \sum_{Mk_y\sigma} |\varphi_{M,X,\sigma}(x, y)|^2,$$

where the sum covers all the occupied states $MX\sigma$, and it is easy to verify that in the limit $N \to \infty$ it is a constant given by

$$\rho = \frac{eB}{2\pi c}\nu \quad \nu = 1, 2, 3, \ldots, \qquad (9.263)$$

where $\nu = \nu_\uparrow + \nu_\downarrow$ is the number of Landau levels occupied by the electrons in the ground state. The density (9.263) is consistent with the initial statement

$$v_{\text{ext}} + U = 0,$$

which allowed us to simplify the KS equations.

The zero temperature ground state is determined solely by the filling factor $\nu = 2\pi\rho c/eB$. For example, $\nu = 1$ means that all electrons are in the first Landau level $M = 0$, and have spin up, while $\nu = 2$ means that the electrons occupy the two Landau levels: $M = 0$, $\sigma = \uparrow$ and $M = 0$, $\sigma = \downarrow$. These states are easily obtained experimentally, because the density and the magnetic field are two parameters that can be controlled independently of each other. The LSDA theory cannot treat the fractional regime in which $\nu < 1$. However, it can be generalized by the CDFT (see Subsection 4.7) to include this regime as well; which, however, will not be treated in this subsection. Here we will be concerned with $\nu = 2$ and $\nu = 1$ filling.

For $\nu = 2$, the system is not spin-polarized ($S_z = 0$, $W_{\text{xc}} = 0$), and in the paramagnetic ground state two Landau levels with $M = 0$ are occupied, and states with spin-up and spin-down are shifted in energy only by the Zeeman term. In this case, as was shown in Subsection 9.9.1, only the two correlation functions α^{ss} and α^{aa} of Eq. (9.205) are different from zero, and they describe the density response of the density operator $\rho(q) = \sum_i \exp(i\mathbf{q}\cdot\mathbf{r}_i)$ and the magnetization density response of the spin-density operator $m(q) = \sum_i \exp(i\mathbf{q}\cdot\mathbf{r}_i)\sigma_i^z$, respectively. Since in the integral equations for α^{ss} and α^{aa} the free correlation function is the same, the difference between the TDLSDA responses in the two channels is only due to the different kernels (residual interactions). The longitudinal TDLSDA response is obtained from the correlation functions as

$$\chi_{AA}(q,\omega) = \int d\mathbf{r}_1 d\mathbf{r}_2 \exp(i\mathbf{q}\cdot(\mathbf{r}_1-\mathbf{r}_2))\alpha_{AA}(r_1,r_2;\omega), \qquad (9.264)$$

with A equal to ρ or m. As discussed in a paper by Horing (1965), in the case of a two-dimensional electron gas under a magnetic field, the correlation functions that appear in the Dyson integral equation are functions only of the module of the spatial distance, as in the case with no magnetic field. Then, the response functions $\chi(q,\omega)$ are obtained from the correlation functions as Fourier transforms, and in this case as well, the integral Eqs. (9.205) yield algebraic equations for the dynamic polarizabilities which in 2D are written as

$$\chi_{\rho\rho}(q,\omega) = \frac{\chi^{(0)}(q,\omega)}{1-\frac{1}{S}(\frac{2\pi}{q}+\mathcal{K})\chi^{(0)}(q,\omega)}, \qquad (9.265)$$

$$\chi_{mm}(q,\omega) = \frac{\chi^{(0)}(q,\omega)}{1-\frac{1}{S}\mathcal{I}\chi^{(0)}(q,\omega)}, \qquad (9.266)$$

where $\mathcal{K} = \frac{\partial^2 E_{xc}}{\partial\rho^2}$ and $\mathcal{I} = \frac{\partial^2 E_{xc}}{\partial m^2}$. Eqs. (9.265) and (9.266) are then completely similar to those [i.e. Eqs. (9.206) and (9.207)] for the response functions of the electron gas in the absence of a magnetic field. The presence of the field, however, affects the calculation of the free response function substantially. For $\nu = 2$ we have $\chi^{(0)} = \chi^0_{\uparrow\uparrow} + \chi^0_{\downarrow\downarrow}$ and $\chi^0_{\uparrow\uparrow} = \chi^0_{\downarrow\downarrow}$. Since single-particle transitions are possible only between states with the same spin, and for which the Landau index M in the final state differs from the one in the initial state (which is zero; see Fig. 9.20), we have

$$\chi^0_{\uparrow\uparrow}(q,\omega) = \sum_{M=1}^{\infty}\sum_{k_y,k'_y}\frac{2(\epsilon_{M\uparrow}-\epsilon_{0\uparrow})|\langle M\uparrow X|e^{i\mathbf{q}\cdot\mathbf{r}}|0\uparrow X'\rangle|^2}{\omega^2-(\epsilon_{M\uparrow}-\epsilon_{0\uparrow})^2}. \qquad (9.267)$$

From Eq. (9.262), we have $\epsilon_{M\uparrow} - \epsilon_{0\uparrow} = M\omega_c$. Taking $\mathbf{q}$ along the x direction, we find for $\langle M\uparrow X|e^{i\mathbf{q}\cdot\mathbf{r}}|0\uparrow X'\rangle \equiv F_{MX,0X'}$

$$F_{MX,0X'} = \int \varphi^*_{MX}(x,y)e^{iqx}\varphi_{0X'}(x,y)\,dx\,dy$$

$$= \frac{1}{L_y l\sqrt{\pi 2^M M!}}\int e^{+i\frac{X-X'}{l^2}y}e^{iqx}e^{-\frac{(x-X)^2}{l^2}}H_M\left(\frac{x-X}{l}\right)dx\,dy, \qquad (9.268)$$

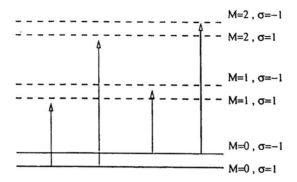

Fig. 9.20 Possible excitations for the first three occupied Landau levels in the case $\nu = 2$. The full line marks the occupied Landau levels in the ground state; the dashed line marks the nonoccupied levels.

and using

$$\int e^{+i\frac{X-X'}{l^2}y}\, dy = L_y \delta_{X,X'},$$

$$H_M\left(\frac{x-X}{l}\right) = (-1)^M e^{\frac{(x-X)^2}{l^2}} l^M \frac{d^M}{dx^M} e^{-\frac{(x-X)^2}{l^2}}, \tag{9.269}$$

we obtain

$$F_{MX,0X'} = \frac{\delta_{X,X'}}{l\sqrt{\pi 2^M M!}} \int (-1)^M l^M \frac{d^M}{dx^M}\left(e^{-\frac{(x-X)^2}{l^2}}\right) e^{iqx}\, dx. \tag{9.270}$$

The latter integral is easily solved through the well-known property of the Fourier transform (which we denote by T),

$$T\left[\frac{d^n}{dx^n} f(x)\right] = (-ik)^n T[f(x)],$$

and bearing in mind that

$$T[e^{-\frac{x^2}{a^2}}] = a\sqrt{\pi}\, e^{-\frac{k^2 a^2}{4}}.$$

At last, we obtain

$$F_{MX,0X'} = \delta_{X,X'} \frac{e^{iqX}}{\sqrt{M!}} \left(\frac{iql}{\sqrt{2}}\right)^M e^{-\frac{q^2 l^2}{4}}, \tag{9.271}$$

so that

$$|F_{MX,0X'}|^2 = \delta_{X,X'} \frac{1}{M!}\left(\frac{q^2 l^2}{2}\right)^M e^{-\frac{q^2 l^2}{2}}. \tag{9.272}$$

Therefore, we can write the free response function of the system explicitly:

$$\chi^0(q,\omega) = 2N \sum_{M=1}^{\infty} \frac{M\omega_c(\frac{q^2 l^2}{2})^M e^{-\frac{q^2 l^2}{2}}}{M!(\omega^2 - M^2\omega_c^2)}. \qquad (9.273)$$

The excitation energies $\omega(q)$ of the collective density and spin-density modes, as a function of the wave vector q, in the TDLSDA may be derived by solving the equations

$$\frac{\chi^{(0)}(q,\omega)}{S} = \left(\frac{2\pi}{q} + \mathcal{K}\right)^{-1}, \quad \frac{\chi^{(0)}(q,\omega)}{S} = \mathcal{I}^{-1}; \qquad (9.274)$$

which yield the poles of $\chi_{\rho\rho}(q,\omega)$ and $\chi_{mm}(q,\omega)$, respectively.

By bounding the model space to the single-particle transitions with energy $1\omega_c$, $2\omega_c$ and $3\omega_c$, in the calculation of the free response function of (9.273), we have

$$\frac{\chi^0}{S} = \frac{2\rho x e^{-x}}{\omega_c}\left(\frac{1}{\lambda-1} + \frac{x}{\lambda-4} + \frac{\frac{x^2}{2}}{\lambda-9}\right), \qquad (9.275)$$

where we have set $x = q^2/2l^2$ and $\lambda = \omega^2/\omega_c^2$. Inserting this result into (9.274), we then find that

$$\lambda^3 - \lambda^2\left(14 + A(x)\left(1 + x + \frac{x^2}{2}\right)\right) + \lambda\left(49 + A(x)\left(13 + 10x + \frac{5}{2}x^2\right)\right)$$

$$- 36 - A(x)(36 + 9x + 2x^2) = 0, \qquad (9.276)$$

where

$$A(x) = \frac{2\rho x e^{-x}}{\omega_c}\left(\frac{2\pi}{q} + K\right)$$

for the density channel, and

$$A(x) = \frac{2\rho x e^{-x}}{\omega_c} I$$

if we consider the spin-density channel.

In Fig. 9.21 we report the numerical solutions of (9.276) for ω/ω_c, for both density and magnetization channels, as a function of ql and for different values of the density expressed through the Wigner–Seitz radius r_s ($\rho = \pi r_s^2$).

At small r_s, the TDLSDA dispersion relation is quite similar to the TDHF–RPA results of Kallin and Halperin (1984) and MacDonald (1985). The only relevant difference is found in the $ql \gg 1$ region, where the effect of the residual interaction is negligible and the collective TDLSDA energy tends to $M\omega_c$, while those of the TDHF–RPA always tend to a value larger than $M\omega_c$ owing to the self-energy (Fock) shift among the unperturbed single-particle energies. This shift is absent in the LSDA, where the single-particle energies always differ by integer multiples of ω_c. By increasing the value of r_s, the correlation effects in the residual interaction,

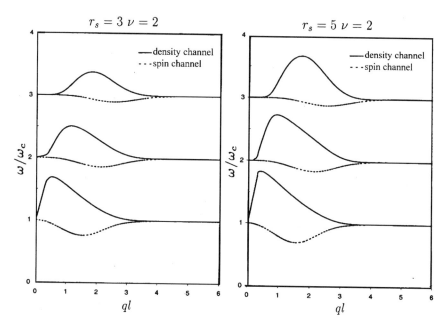

Fig. 9.21 Energies of the collective states in the density and spin-density channels in the case $\nu = 2$, as a function of ql for $r_s = 3, 5$.

which are found in the TDLSDA only, make the difference between TDLSDA and TDHF–RPA more and more important. While the TDHF–RPA approach predicts an instability in the spin mode at $r_s = 3.3$, the TDLSDA calculation takes care of such an instability by means of the correlations. The roton minimum decreases with r_s, but does not show any instability, at least up to $r_s = 10$, which is the maximum value of the considered calculation. The TDLSDA reproduces well the experimental values of Fig. 9.19 for the first roton minimum, while it is unable to reproduce the second roton minimum observed in the spin channel at large ql values.

The strengths relative to the collective states in the density and magnetization channels, whose excitation energies we indicate by ω_ρ and ω_m, respectively, are given by

$$|\langle \omega_\rho(q)|\hat{\rho}(q)|0\rangle|^2 = \frac{1}{\frac{\partial 1/\chi^{(0)}}{\partial \omega}\big|_{\omega=\omega_\rho(q)}} \qquad (9.277)$$

and

$$|\langle \omega_m(q)|\hat{m}(q)|0\rangle|^2 = \frac{1}{\frac{\partial 1/\chi^{(0)}}{\partial \omega}\big|_{\omega=\omega_m(q)}}. \qquad (9.278)$$

The strength distributions for the first three collective states of density and magnetization excitation are reported in Figs. 9.22 and 9.23, for different values of r_s and ql. From these figures we notice that the low-ql spin state is very collective,

excitation strengths

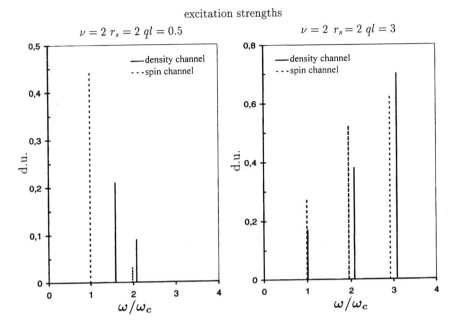

Fig. 9.22 Distribution of strengths in effective atomic units (d.u.) for GaAs samples for the first three collective excitation states in the density and spin-density channels, for various values of ql, in the case of $\nu = 2$, $r_s = 2$.

excitation strengths

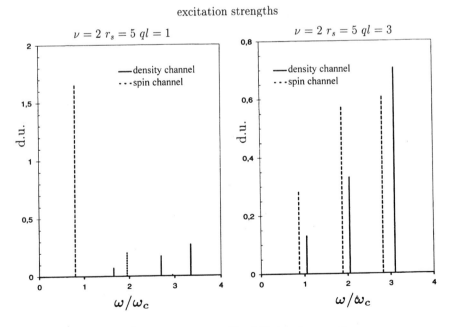

Fig. 9.23 The same as Fig. 9.22, but for $r_s = 5$.

unlike the density state, and that the strength breaks up among the possible states with increasing ql.

We further note that the TDLSDA energies and strengths saturate the f-sum rule, given by

$$m_1(\rho\rho, mm) = \frac{Nq^2}{2}. \tag{9.279}$$

This can be immediately understood by making the limit $\omega \to \infty$ in the expressions of $\chi_{\rho\rho}$ and χ_{mm} [see Eq. (8.28)], and using the result $(x = q^2l^2/2 = q^2/2\omega_c)$

$$\sum_{M=1}^{\infty} \frac{Mx^M e^{-x}}{M!} = xe^{-x} \sum_{M=1}^{\infty} \frac{x^{M-1}}{(M-1)!} = x. \tag{9.280}$$

The TDLSDA gives the following results for the sum rule cubic in energy:

$$m_3(\rho\rho) = N\left[\frac{q^6}{8} + \frac{q^2}{2}\omega_c^2 + \frac{3q^4}{4}\omega_c + \frac{\rho q^4}{2}\left(\frac{2\pi}{q} + \mathcal{K}\right)\right] \tag{9.281}$$

and

$$m_3(mm) = N\left[\frac{q^6}{8} + \frac{q^2}{2}\omega_c^2 + \frac{3q^4}{4}\omega_c + \frac{\rho q^4}{2}\mathcal{I}\right]. \tag{9.282}$$

The results (9.281), (9.282) were obtained using the result (8.28) of Chapter 8, and

$$\sum_{M=1}^{\infty} \frac{M^3 x^M e^{-x}}{M!} = x + 3x^2 + x^3. \tag{9.283}$$

The results (9.281), (9.282) may also be obtained by directly evaluating the (TDLDA) commutators of the density and magnetization operators with the Kohn–Sham Hamiltonian (see Subsection 9.10). The inversely energy-weighted sum rule $m_{-1} = -\chi(q,0)/2$ can be easily computed using the result

$$\frac{\chi^{(0)}(q,0)}{S} = \frac{-2\rho}{\omega_c}\exp\left(-\frac{q^2l^2}{2}\right)\sum_{M=1}^{\infty} \frac{1}{MM!}\left(\frac{q^2l^2}{2}\right)^M. \tag{9.284}$$

Finally, we note that the energies which are solutions to (9.274), and the mean energies $E_1 = \sqrt{m_1/m_{-1}}$ and $E_3 = \sqrt{m_3/m_1}$, tend to the limit ω_c when $q \to 0$, in accordance with the Kohn theorem.

B. Transverse Response

As in the longitudinal case, the transverse response function for the two-dimensional electron gas under a strong magnetic field is given by the same expression, (9.221), derived in Subsection 9.9.2 in the $B = 0$ case:

$$\chi_t(q,\omega) = \frac{\chi_t^0(q,\omega)}{1 - 2\mathcal{F}_{xc}\chi_t^0(q,\omega)}. \tag{9.285}$$

In the cases $\nu = 1$ and $\nu = 2$, which we are considering in the following, only the $M = 0$ Landau level is filled in the ground state (spin-up and spin-down levels for $\nu = 2$, and only the spin-up level for $\nu = 1$). The free response function is given by

$$\chi_t^0(q,\omega) = \sum_{M'=0}^{\infty} \sum_{X,X'} \left(\frac{|\langle M' \downarrow X'|F^-|0 \uparrow X\rangle|^2}{\omega - (M'\omega_c + \omega_a)} - \frac{|\langle M' \uparrow X'|F^+|0 \downarrow X\rangle|^2}{\omega + (M'\omega_c - \omega_a)} \right),$$

(9.286)

with $\omega_a = -2\mathcal{F}_{xc}m_0 - g^*\mu_0 B$, where m_0 is the ground state magnetization. In the $\nu = 1$ case, the one-particle–one-hole transitions for the first Landau level are schematized in Fig. 9.24.

Using the results found in the longitudinal case, it is easy to obtain the following expression for the free response function:

$$\chi_t^0(q,\omega) = \sum_{M'=0}^{\infty} \frac{1}{M'!} \left(\frac{q^2l^2}{2} \right)^{M'} e^{-\frac{q^2l^2}{2}}$$

$$\times \left(\frac{N_\uparrow}{\omega - (M'\omega_c + \omega_a)} - \frac{N_\downarrow}{\omega + (M'\omega_c - \omega_a)} \right),$$

(9.287)

where $N_\uparrow = N_\downarrow = N/2$ and hence $m_0 = 0$ for $\nu = 2$ and $N_\uparrow = N$ and $N_\downarrow = 0$ for $\nu = 1$.

Note that when the system is spin-polarized, the TDLSDA transverse response always gives the two sum rules

$$m_0^- = N_\uparrow - N_\downarrow, \quad m_1^+ = \frac{Nq^2}{2} + \omega_L(N_\uparrow - N_\downarrow),$$

(9.288)

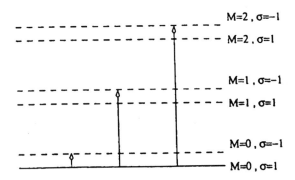

Fig. 9.24 Possible spin-flip excitations for the first Landau levels in the case $\nu = 1$. The full line indicates the Landau levels which are occupied in the ground state; the dashed line indicates unoccupied levels.

where $\omega_L = -g^*\mu_0 B$ is the generalized Larmor frequency. Equations. (9.288) follow immediately from the $\omega \to \infty$ limit of the above equations for $\chi_t(q,\omega)$ and $\chi_t^0(q,\omega)$ [see Eq. (8.40)], and from the properties

$$\sum_{M'=0}^{\infty} \frac{1}{M'!} \left(\frac{q^2 l^2}{2}\right)^{M'} e^{-\frac{q^2 l^2}{2}} = 1$$

and

$$\sum_{M'=0}^{\infty} M' \frac{1}{M'!} \left(\frac{q^2 l^2}{2}\right)^{M'} e^{-\frac{q^2 l^2}{2}} = \frac{q^2}{2\omega_c}.$$

The sum rule $m_0^- = \langle 0|[F^+, F^-]|0\rangle$ is model-independent and determined by the magnetization of the system, while m_1^+ is model-dependent and involves an explicit assumption for the Hamiltonian. Use of the Pauli Hamiltonian in the equation $m_1^+ = \langle 0|[F^+, [H, F^-]]|0\rangle$ leads exactly to the result (9.288), so that it is possible to conclude that m_1^+ is satisfied by the TDLSDA. The same does not happen for the free response, which leads to the result $m_1^+(KS) = Nq^2/2 + \omega_a(N_\uparrow - N_\downarrow)$, in which ω_a is present instead of ω_L. Also in the $q \to 0$ limit there is an important difference between the TDLSDA result and the free response one. In fact, from Eqs. (9.285) and (9.287), it immediately follows that

$$\chi_t^0(q \to 0, \omega) = \frac{m_0}{\omega - \omega_a}, \qquad (9.289)$$

$$\chi_t(q \to 0, \omega) = \frac{m_0}{\omega - \omega_L}, \qquad (9.290)$$

showing that at the lowest order in q, the TDLSDA pole is given correctly by the Larmor frequency ω_L and not by the renormalized frequency ω_a, which gives the poles of $\chi_t^0(q \to 0, \omega)$ in the same limit.

In the case $\nu = 2$, where the system has zero total spin, the lowest excitation energies are of the order of $1\omega_c$, and at $q = 0$ are given by

$$\omega_\pm = \omega_c \pm \omega_L. \qquad (9.291)$$

This result differs from the one found in the case where the system is spin-polarized. In fact, at $\nu = 1$, the $\Delta S_z = +1$ channel, excited by the operator $F^+ = \sum_i \exp(+i\mathbf{q} \cdot \mathbf{r})_i \sigma_i^+$, is blocked by the Pauli exclusion principle, and the $\Delta S_z = -1$ channel has the lowest-energy state at $\omega = \omega_L$ and a state "$1\omega_c$" at energy (in the $q \to 0$ limit)

$$\omega_- = \omega_c + \omega_L - 2\mathcal{F}_{xc}. \qquad (9.292)$$

The poles of Eq. (9.285) are solutions to the equation

$$1 - 2\mathcal{F}_{xc}\chi_t^0(q,\omega) = 0, \qquad (9.293)$$

which gives the ql dispersion of the transverse collective states. The only available experimental result concerning the measurement of the Zeeman splitting for $\nu = 2$ at small r_s is that of Eriksson *et al.* (1999). Such a splitting is well reproduced by the calculation.

9.13.2 *Quantum Dots Under Magnetic Field*

In this subsection we will present and discuss the TDLSDA results for the longitudinal and transverse dipole excitations in quantum dots under an external magnetic field (Serra *et al.*, 1999; Lipparini *et al.*, 1999; Steffens and Suhrke, 1999). These excitations were observed in photoabsorption experiments in the far-infrared region (FIR), where one observes the spectrum excited by the operator $D_s = \sum_{i=1}^{N} x_i$ (Sikorsky and Merkt, 1989; Demel *et al.*, 1990), and more recently in Raman scattering experiments (Strenz *et al.*, 1994; Schuller *et al.*, 1996), where, by the same experiment, it is possible to observe single-particle and collective excitations belonging to both the density and spin-density, and for the latter in both the longitudinal and the transverse channel. In the dipolar case, these modes are excited by the operators $D_a = \sum_{i=1}^{N} x_i \sigma_i^z$ and $D_{\pm} = \sum_{i=1}^{N} x_i \sigma_i^{\pm}$, respectively. As mentioned previously, when the dot is spin-polarized (which happens when the dot either has an odd number of electrons or is under the action of a magnetic field), the spin response is coupled to the density response, so that the external operator D_s also excites the spin modes, as D_a excites the density ones. When the system is completely spin-polarized, the two modes coincide, while at zero polarization they are uncoupled.

In the following we will assume that the quantum dots under consideration have spherical symmetry. In order to exploit such symmetry, it is suitable to consider the following dipolar external fields, which correspond to the operator G in Eq. (9.189):

$$D_s^{(\pm 1)} = re^{\pm i\theta}\begin{pmatrix}1\\1\end{pmatrix}, \quad D_a^{(\pm 1)} = re^{\pm i\theta}\begin{pmatrix}1\\-1\end{pmatrix}, \qquad (9.294)$$

with $D_s = (D_s^{(+1)} + D_s^{(-1)})/2$, and likewise for D_a.

In the longitudinal case, the integral Eqs. (9.194) for the correlation functions may be solved as a system of matrix equations in the coordinate space, after performing an angular decomposition on $\alpha_{\sigma\sigma'}$ and on the residual interaction.*

As can be easily seen by performing the angular integration in (9.190), it is only the modes with $l = \pm L$ that couple to the external fields $G_s^{(\pm L)}$ and $G_a^{(\pm L)}$. Therefore, in the dipolar case the only multipoles which appear in the integral

---

*For the residual interaction $K_{\sigma\sigma'}$, one gets

$$K_{\sigma\sigma'}(\mathbf{r}, \mathbf{r}') = \sum_l K_{\sigma\sigma'}^{(l)}(r, r')e^{il(\theta-\theta')}. \qquad (9.295)$$

equations for the correlation functions are those with $L = \pm 1$, and the corresponding response functions $\chi_{AB}^{(\pm 1)}(r, r'; \omega)$, with $A, B = s, a$, are given by

$$\chi_{AB}(\omega) = \pi^2 \int dr_1 dr_2 r_1^2 r_2^2 [\alpha_{AB}^{(+1)}(r_1, r_2; \omega) + \alpha_{AB}^{(-1)}(r_1, r_2; \omega)]$$

$$\equiv \chi_{AB}^{(+1)}(\omega) + \chi_{AB}^{(-1)}(\omega). \tag{9.296}$$

The imaginary part of $\chi_{AB}(\omega)$ is related to the dynamic structure function by Eq. (8.26), and the following relations hold:

$$\text{Re}[\chi_{AB}^{(-1)}(\omega)] = \text{Re}[\chi_{AB}^{(1)}(-\omega)],$$
$$\text{Im}[\chi_{AB}^{(-1)}(\omega)] = -\text{Im}[\chi_{AB}^{(1)}(-\omega)]. \tag{9.297}$$

These relations are important from a practical standpoint, because they allow us to determine the structure functions of both components ± 1, by using the $+1$ component over a frequency range $(-\omega_{\min}, \omega_{\max})$.

In order to perform a check on the numerical accuracy of the calculation of the structure functions, one can use the f-sum rules for the corresponding excitation operators. One finds (in effective atomic units) that

$$m_1^{(ss)} = \int S_{ss}(\omega)\omega d\omega = \frac{1}{2}\langle 0|[D_s, [H, D_s]]|0\rangle = \frac{N}{2},$$

$$m_1^{(aa)} = \int S_{aa}(\omega)\omega d\omega = \frac{1}{2}\langle 0|[D_a, [H, D_a]]|0\rangle = \frac{N}{2}, \tag{9.298}$$

$$m_1^{(as)} = m_1^{(sa)} = \int S_{as}(\omega)\omega d\omega + \int S_{sa}(\omega)\omega d\omega = \langle 0|[D_a, [H, D_s]]|0\rangle = 2S_z.$$

In the transverse case, and again to exploit the circular symmetry, one considers the operators

$$D_{\pm 1, \pm} = \sum_j r_j e^{\pm i\theta_j} \sigma_\pm^j, \tag{9.299}$$

with

$$\sum_j x^j \sigma_\pm^j = \frac{1}{2}(D_{+1,\pm} + D_{-1,\pm}), \tag{9.300}$$

keeping in mind that such operators, unlike the longitudinal case where states are excited without varying the third spin component (i.e. $\Delta S_z = 0$), now induce variations with $\Delta S_z = \pm 1$. With the use of such operators one then solves the integral Eqs. (9.214) and (9.218), and computes the polarizabilities (9.216) and (9.220).

Figures 9.25–9.27 show the dipolar structure functions of the $N = 5$, 25 and 210 dots for some selected values of the magnetic field B. The full curves correspond to the density response to D_s, and the dotted curves mark the spin-density response

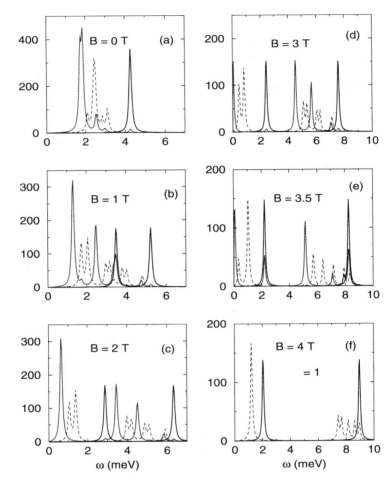

Fig. 9.25 Dipole structure function in arbitrary units of the $N = 5$ dot, as a function of frequency. The thick and dotted lines correspond to the density response to D_s and to the spin-density response to D_a, respectively. The dashed lines represent the single-particle structure function.

to D_a, i.e. to S_{ss} and to S_{aa}, respectively. The dashed curves represent the single-particle structure function.

For the dot with five electrons the parabolic confinement potential $v_{\text{ext}}(r) = m\omega_0^2 r^2/2$ was used, with $\omega_0 = 4.28$ meV, while for the other two dots the confinement potential was the one generated by a disk of positive charge, as in Subsection 5.6 [see Eq. (5.60)]. As will be shown later, the experimental results of Demel *et al.* are better reproduced for the $N = 25$ dot, using a parabolic confinement potential. This is why in the $N = 25$ case, we present in Fig. 9.28 the structure function obtained with the parabolic potential as well.

These figures show that in both the density and spin-density channels, the response at $B = 0$ is concentrated in a small energy range, either with only one

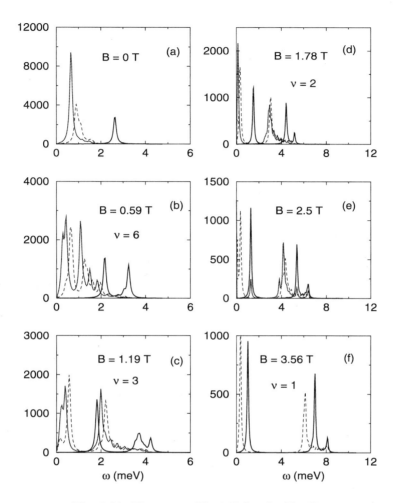

Fig. 9.26 The same as Fig. 9.25, but for $N = 25$.

peak or with several very close peaks, which exhaust most of the f-sum rule. The peak energy is lower in the spin channel than in the density one. This is due to the character of the residual interaction $K_{\sigma\sigma'}$, which is attractive in the spin channel and repulsive in the charge one, and shifts the TDLSDA response with respect to the single-particle one in opposite directions.

Moreover, the residual interaction in the spin channel is weaker than it is in the density channel, where it is not only the exchange-correlation energy that contributes, but also the direct Coulomb term. As a result, the spin response is closer to the free response. Therefore, it is difficult to distinguish the spin longitudinal collective mode from the single-particle spectrum. In the dots with a large number of electrons, this also induces a much larger Landau damping in the spin channel than it does in the density one (Schuller *et al.*, 1998).

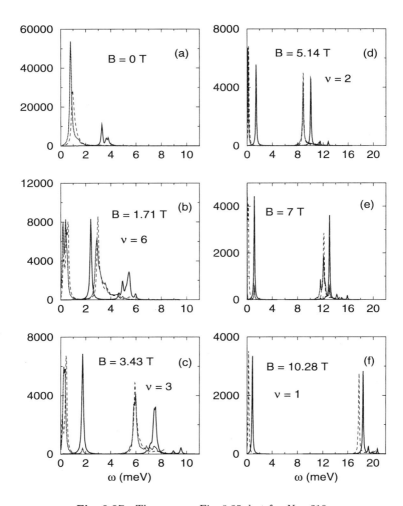

Fig. 9.27   The same as Fig. 9.25, but for $N = 210$.

At $B = 0$, as a consequence of the generalized Kohn theorem (see Subsection 8.9.1), and if the confinement potential is harmonic with frequency ω_0, the excitation energy of the dipole density mode is equal to ω_0 irrespective of the number of electrons in the dot. Otherwise (for other kinds of confinement potential), the excitation energy depends on N [see for example Serra and Lipparini (1997) and Gudmundsson and Gerhardts (1991)]. In the spin channel this theorem does not hold; moreover one observes N dependence also for parabolic confinement potentials.

When B is different from zero, the dipole mode in both channels is split into two branches — one with negative B dispersion and the other with positive B dispersion. The splitting originates from the breaking of the l degeneracy of the single-particle levels, in the presence of a magnetic field. Therefore, several new phenomena occur.

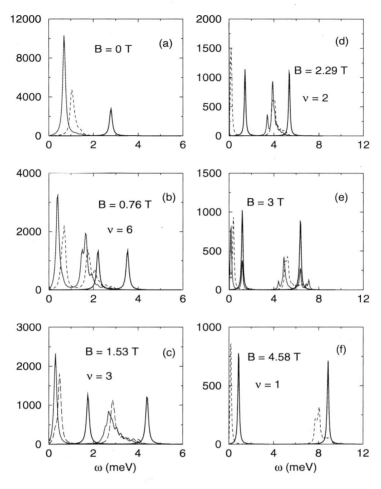

Fig. 9.28 The same as Fig. 9.26, but using the parabolic confinement potential with $\omega_0 = 2.78$ meV, instead of the disk potential.

First, we note that for B values such that the ground state spin is not zero, the two excitation modes (i.e. density and spin) are coupled. This can be seen particularly in the dot with $N = 210$ electrons. In fact, at $B = 1.71$ T ($\nu = 6$) and 5.14 T ($\nu = 2$), the system is practically paramagnetic, having $2S_z = 2$ and 0, respectively (see Fig. 5.8 of Subsection 5.6). As a consequence of that, the two modes are uncoupled, as can be seen in Fig. 9.27. Contrary to these cases, at $B = 3.43$ T ($\nu = 3$) and 7 T we have $2S_z = 54$ and 74: the system possesses a large spin magnetization in the ground state, and the two modes are clearly coupled. This is shown in Fig. 9.27, where we can observe a separate peak in the spin response at the energy of the density mode. This effect was observed experimentally by Schuller *et al.* (1998). The strength of this peak increases with S_z, and when the system is completely spin-polarized the whole strength is transferred from the spin channel to the density one. On the other hand, the spin mode can be observed in the density channel with some intensity. This effect is somewhat masked because the Kohn theorem forbids it for parabolic potentials, and for the disk potential it is of the order of $(2S_z/N)^2$.

The coupling between density and spin modes is especially evident in the mixed channel which is described by the density response to the spin-dipole excitation operator D_a, or by the spin response to the density dipole operator D_s. This is shown in Fig. 9.29 for the $N = 210$ quantum dot at $\nu = 3$. One clearly observes two peaks at the energy of the density mode, and two more peaks at the energy of the spin mode. This can be well understood by dividing the mixed structure function by a sum over the "spin-dipole states" $|a\rangle$, plus a sum over the "charge-dipole states" $|s\rangle$:

$$S_{as}(\omega) = S_{sa}(\omega) = \sum_i \langle 0|D_s|i\rangle\langle i|D_a|0\rangle\delta(\omega - \omega_{i0})$$

$$= \sum_s \langle 0|D_s|s\rangle\langle s|D_a|0\rangle\delta(\omega - \omega_{s0})$$

$$+ \sum_a \langle 0|D_s|a\rangle\langle a|D_a|0\rangle\delta(\omega - \omega_{a0}). \qquad (9.301)$$

For the disk confinement potential, the matrix element $\langle 0|D_s|a\rangle$ is different from zero and there is a contribution to S_{as} from the spin modes. For a parabolic confinement potential $\langle 0|D_s|a\rangle$ vanishes, and only the density modes contribute to S_{as} through the sum over s in (9.301).

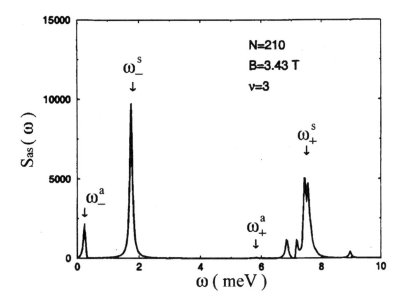

Fig. 9.29 Mixed structure function $S_{as}(\omega)$ (in effective atomic units) of the $N = 210$ dot, at $\nu = 3$.

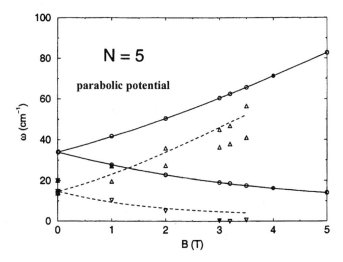

Fig. 9.30 B dispersion of the major peaks of the $N = 5$ spectrum. The circles correspond to density modes, the triangles to spin modes. The lines indicate the B dispersion of the vibrating potential model of Subsection 11.2.6, with the values fitted at $B = 0$.

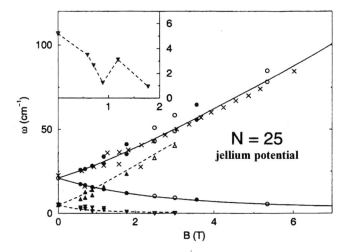

Fig. 9.31 B dispersion of the major peaks of the $N = 25$ spectrum for a disk confinement potential. The circles correspond to density modes and the triangles to spin modes. The crosses mark the experimental points from Demel *et al.* (1990). The lines represent the B dispersion of the VPM of Subsection 11.2.6, with the $B = 0$ values fitted. The inset shows the negative B dispersion branch of the spin mode. From the left to the right the solid symbols correspond to $\nu = 6$ to 1.

In Figs. 9.30–9.33 we report the dispersion with B of the major peaks of the two spectra. The density modes are represented by circles, the spin modes by triangles. The full symbols correspond to integer filling factors, and the insets show the branch with negative B dispersion of the spin mode. In the figure, also present are lines starting from the frequency at $B = 0$, and following the B dispersion

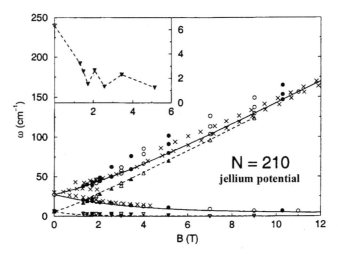

Fig. 9.32 B dispersion of the major peaks of the $N = 210$ spectrum, for a disk confinement potential. The circles correspond to density modes, the triangles to spin modes. The crosses are the experimental points from Demel *et al.* (1990). The lines represent the B dispersion of the VPM of Subsection 11.2.6, with values fitted at $B = 0$. The inset shows the negative dispersion branch of the spin mode. From left to right the solid symbols correspond to ν values from 8 to 1.

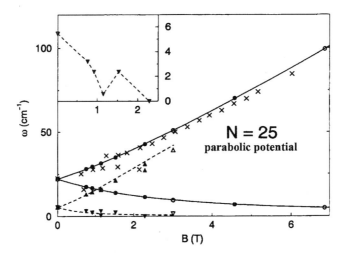

Fig. 9.33 The same as Fig. 9.31, but using a parabolic confinement potential with $\omega_0 = 2.78$ meV in place of the disk potential.

laws predicted by the vibrating potential model (VPM), which will be described in Subsection 11.2.6.

The figures evidence several interesting features that deserve discussion. As concerns the spin modes, we first note that at low B their energy is much lower than that of density modes, in agreement with the experimental findings of Schuller *et al.* At higher B values, the quantum dot is eventually completely spin-polarized, and the longitudinal density and spin modes coincide, as in the case of the 2D electron gas (Kallin and Halperin, 1984). This effect is not explicitly shown by the figure. Moreover, the branch with negative B dispersion of the spin mode exhibits a clearly oscillating behavior with ν, similar to the one observed experimentally for the density response by Bollweg *et al.* (1996), which reveals that the paramagnetic configurations of the dots have spin modes which are softer than in the ferromagnetic configurations.

The TDLSDA predicts the existence of a spin instability when the lowest-energy branch of the spin mode goes to zero at some critical value of B, between $\nu = 1$ and 2. Such an instability appears also in the static spin polarizability $\mathrm{Re}[\chi_{aa}^{(-1)}(0)]$, which becomes negative at these values of B, indicating that the ground state of the dot is no longer stable. The lack of experimental observation of collective spin states at these values of the magnetic field might be the mark of the instabilities, but it cannot be ruled out that the lack of data may be due to the Landau damping, which, in this energy range, is particularly strong. Moreover, it might be that in this region, correlations which are not included in the TDLSDA are important and quench the spin instability.

As regards the density dipole mode, the following considerations apply. For the parabolic confinement potential, Figs. 9.30 and 9.33 show that the density response follows the law

$$\omega_\pm^s = \Omega \pm \frac{\omega_c}{2},$$

where $\Omega^2 = \omega_0^2 + \omega_c^2/4$. This is so because the TDLSDA which we are using obeys the generalized Kohn theorem (see Subsection 8.9.1). For the disk confinement potential, this law is also approximately satisfied, especially by the branch with negative B dispersion. As concerns the other branch, it is systematically fragmented, especially for the $N = 210$ dot. Comparing the calculations for the $N = 25$ dot with the experimental results of Demel *et al.*, we may conclude that the parabolic potential works better than the disk one, in reproducing the experimental situation. On the contrary, the confinement potential of the $N = 210$ dot is not parabolic, and this originates the second high energy branch observed in experiment. The nonharmonic character of the confinement potential was indicated as the origin of such a branch in calculations of the Hartree + RPA type with $N \leq 30$ dots by Gudmundsson and Gerhardts (1991) and Ye and Zaremba (1994). The TDLSDA calculations support this interpretation.

To complete the description of the TDLSDA dipole response in quantum dots, we turn now to transverse spin modes. We will consider the $N = 210$ dot, which allows us to present very clear results for several ground states corresponding to integer filling factors. This will emphasize the sensitivity of the transverse response to the applied magnetic field B, which shows itself by having large oscillations of the excitation energies as a function of B, and by a nearly collapsed, low energy excitation mode $x\sigma_+$ at $\nu = 2$ (Lipparini *et al.*, 1999).

Owing to the rotational invariance in spin space, the longitudinal ($\Delta S_z = 0$) and transverse ($\Delta S_z = \pm 1$) spin excitations (SDE) are degenerate at $B = 0$ if $S_z = 0$ in the ground state of the dot. When the magnetic field is not zero, rotational invariance is broken by the Zeeman term. Then, if the ground state is paramagnetic (or quasi-paramagnetic), the SDE are split in a simple way:

$$\omega_{\pm}^{\mathrm{tr}} = \omega^{\mathrm{long}} \pm g^* \mu_0 B, \qquad (9.302)$$

where the apex *long(tr)* indicates the longitudinal (transverse) character of the mode, and the $\pm$ sign corresponds to the two possible transitions with the change of spin. If the ground state has a large value of S_z, i.e. a large spin magnetization, then Eq. (9.302) does not hold any more, and the longitudinal and transverse SDE excitations become very different due to the spin dependence of the particle–hole residual interaction (vertex correction).

In the following we will study the structure functions corresponding to the operators

$$D_a = \sum_{i=1}^{N} x_i \sigma_i^z, \quad D_{\pm} = \sum_{i=1}^{N} x_i \sigma_i^{\pm},$$

which will be indicated by $x\sigma_z$, $x\sigma_+$, $x\sigma_-$, as well as of the combination $x\sigma_x = x(\sigma_+ + \sigma_-)$.

Some of the characteristics of the structure functions can be easily understood in terms of the single-particle excitations, employed to set up the corresponding correlation function [see Eqs. (9.215) and (9.219)], whose basic ingredients are the single-particle energies $\epsilon_{nl\sigma}$ and wave functions $\phi_{nl\sigma}(\mathbf{r})$, provided by the solution of the Kohn–Sham equations. However, the particle–hole residual interaction, which has the effect of the interaction on the excited states, can drastically change the situation as represented by the single-particle response.

Figure 9.34 shows the structure function $S(\omega)$ (full lines) corresponding to $x\sigma_x$, which was subsequently decomposed into its components, $x\sigma_+$ and $x\sigma_-$, which are plotted in Figs. 9.35 and 9.36.

The corresponding single-particle responses (SPE) are represented by dashed lines; all functions are expressed in effective atomic units, and frequencies are in meV.

We will start by commenting on the results corresponding to paramagnetic ground states, in which the two single-particle states with spin-up and spin-down tend to be equally populated, leading to a ground state with a low value of S_z. As a consequence, the particle–hole residual interaction $\mathcal{F}_{\mathrm{xc}}$ (see Subsection 9.9.2), which is proportional to the magnetization m of the ground state, is weak, and the SDE collective excitations are very close in energy to the single-particle ones. In both the $x\sigma_+$ and $x\sigma_-$ components, the strength exhibits a structure at high energy very close to the single-particle one, and a low energy structure. For paramagnetic ground states, the low energy excitation is *a transverse edge spin excitation* built

$x\sigma_x$

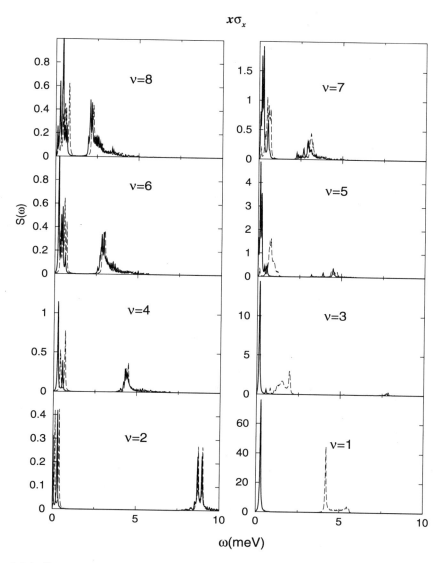

Fig. 9.34 Structure function of the $N=210$ dot, corresponding to $x\sigma_x$ (full lines). The dashed lines represent the single-particle structure function. $S(\omega)$ is in effective atomic units divided by 10^5.

up starting from the particle–hole pairs near to the Fermi level. These pairs can be easily identified in Fig. 5.7 (even values of ν), since they are at the intersection of the chemical potential with the Landau levels. The band structure of the single-particle levels also explains why the edge mode is more fragmented at low magnetic field values. For example, at $\nu = 8$ four particle–hole pairs (each with single-particle angular momenta completely different from the other pairs) contribute to the $x\sigma_-$ strength, while just one pair contributes at $\nu = 2$. These pairs are only weakly

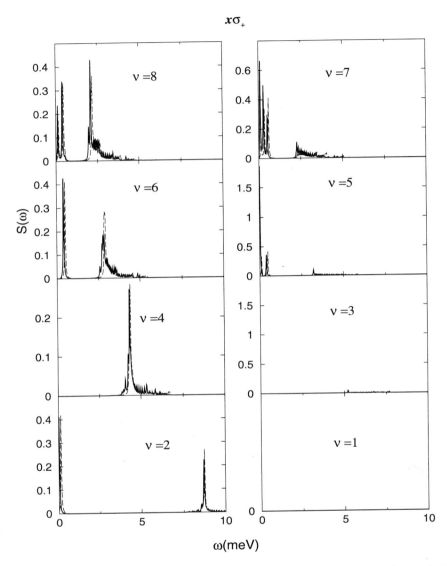

Fig. 9.35 Structure function of the $N = 210$ dot, corresponding to $x\sigma_+$ (full lines). The dashed lines represent the single-particle structure function. $S(\omega)$ is in effective atomic units divided by 10^5.

mutually correlated, and this results in an edge mode which is fourfold fragmented at $\nu = 8$, threefold at $\nu = 6$, and so on. This corresponds to the number of level crossings between the Fermi level and the $(M, \uparrow\downarrow)$ levels in Fig. 5.7.

In the $x\sigma_+$ case, the edge mode is less fragmented because some of the one-particle–one-hole spin-flip transitions with $\Delta S_z = +1$ are forbidden by the Pauli principle, owing to our arbitrary choice of directing B along the positive z axis (recall that it must be $\Delta L_z = \pm 1$, which cannot be always simultaneously fulfilled for the spin and edge conditions). The lack of the edge state in the $\nu = 4$ case is due to the special single-particle structure around the Fermi level at $B = 2.57$ T. This accidental occurrence is not relevant to the general discussion.

As can be seen in Fig. 9.35, when the dot is completely spin-polarized at $\nu = 1$, $x\sigma_+$ does not excite it any more owing to the Pauli blocking. Also, at $\nu = 3$ ($B = 3.43$ T), its strength is very small. On the contrary, the excitation produced by $x\sigma_-$, shown in Fig. 9.36, is appreciably red-shifted with respect to the single-particle response. The difference between the two strengths represents the intensity of the residual interaction (vertex correction) due to exchange and correlation terms of the

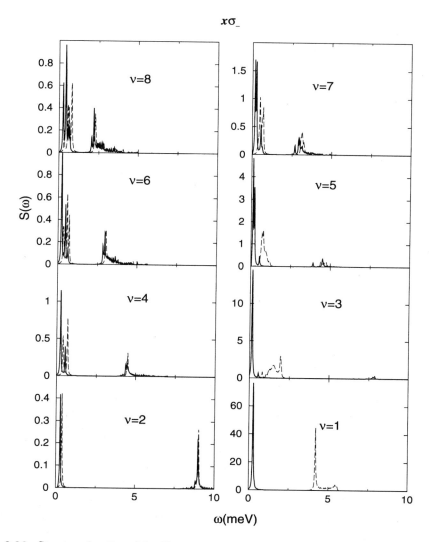

Fig. 9.36 Structure function of the $N = 210$ dot, corresponding to $x\sigma_-$ (full lines). The dashed lines represent the single-particle structure function. $S(\omega)$ is in effective atomic units divided by 10^5.

electron–electron interaction; these terms, in the time-dependent density functional theory, are the only ones contributing to the dressing of the free particle–hole vertex in the spin channel. This effect is more pronounced at $\nu = 1$, when the system is completely magnetized.

The low energy peak excited by $x\sigma_-$ exhausts almost all of the strength. Note that the SDE excitations in ferromagnetic (spin-polarized) ground states caused by both $x\sigma_+$ and $x\sigma_-$ *are not edge but bulk excitations.* Again, Fig. 5.7 helps us to understand this feature. For these ground states with odd filling factor ν, the Fermi level lies between the Landau levels with $(M_{\max}, \uparrow)$ and $(M_{\max}, \downarrow)$, the former being occupied and the latter empty. Even though finite size effects distort the bands at the dot edge, which is built up with single-particle high-l levels, it is evident that the low energy transitions with a change of spin involve single-particle states whose energy difference is exactly the energy difference between the $(M_{\max}, \uparrow\downarrow)$ levels. Such excitations exist in the two-dimensional electron gas as well (see previous subsection). The role played by the residual interaction is evident in Fig. 9.36, if we compare, for instance, the single-particle and TDLSDA strengths at $\nu = 5$ and $\nu = 3$. It can be seen that the SPE energies have nothing in common with the SDE ones at low energy. Thus, the IPM turns out to be completely useless for a quantitative analysis of the SDE excitations in spin-polarized quantum dots, and the situation becomes worse and worse as the polarization degree increases.

In Fig. 9.37 we show in more detail the transverse structure function at high B, excited by $x\sigma_x$ in the region $2 > \nu \geq 1$. In this range of filling factors, which correspond to $5.14 \text{ T} < B \leq 10.28 \text{ T}$, $2S_z$ increases from zero to 210 (see Fig. 5.11), so that the strength $x\sigma_x$ practically coincides with that of $x\sigma_-$, which we already discussed. The new and interesting feature in Fig. 9.37 is the structure of the high energy peaks. These are two orders of magnitude less intense than the low energy ones, which saturate most of the strength. Among the high energy peaks, the most energetic ones are excited by $x\sigma_-$, and the least energetic ones by $x\sigma_+$ (note that the high energy transitions induced by $x\sigma_+$ are blocked by the Pauli principle only when the system is completely spin-polarized). It can be shown that such high energy peaks are not collective, and that the SPE and SDE excitations are very similar; moreover, the centroid of the peaks excited by $x\sigma_+$ and $x\sigma_-$ follows, approximately, the same B evolution as the cyclotron frequency. The value of ω_c is marked in Fig. 9.37 by vertical arrows. When both peaks are very visible in the structure function, for example at $B = 7$ T, their splitting provides a quantitative measure of the spin-dependent exchange-correlation term W_{xc}, and its measurement, when feasible, may be considered as a spectroscopic complement for the measurement of the same quantity by conductivity experiments (Usher *et al.*, 1990), which are directly connected to the single-particle energies. In fact, since the correlation effects are negligible, such a splitting is given by the IPM result $4W_{\mathrm{xc}} + 2g^*\mu_0 B$ and is twice as large as the energy difference between the Landau levels $(M, \uparrow)$ and $(M, \downarrow)$ around the Fermi level [see Eq. (9.262)]. Such correspondence is well confirmed by

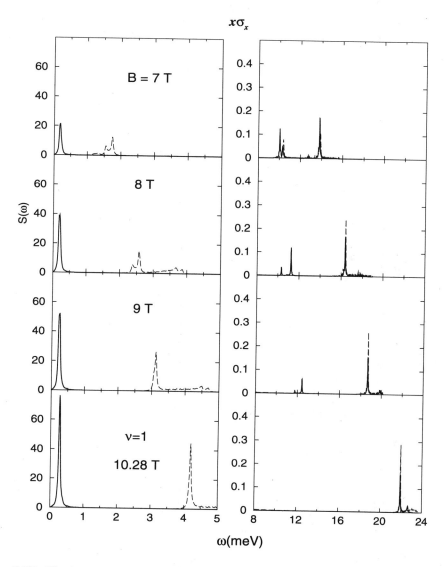

Fig. 9.37 The same as Fig. 9.34, but for $2 > \nu \geq 1$. The vertical arrows mark the value of ω_c.

the numerical LSDA calculation in the cases $B = 7$ T and 9 T, so that an explicit comparison can be made between the splitting of the TDLSDA states excited at high energy by $x\sigma_+$ and $x\sigma_-$, and the energy difference of the static Landau levels (see Fig. 5.15).

The high energy peak of the longitudinal structure function lies at $\omega^{\text{long}} \sim \omega_c$ (Serra *et al.*, 1999; see also Fig. 9.39), and it is possible to write an expression similar to Eq. (9.302):

$$\omega_{\pm}^{\text{tr}} \sim \omega^{\text{long}} \pm 2 W_{\text{xc}}. \tag{9.303}$$

The validity of (9.303) is a consequence of the weakness of the particle–hole residual interaction in the situation where such an expression holds. Note that W_{xc} is
negative.

Figure 9.38 shows the longitudinal ($x\sigma_z$, dotted lines) and transverse ($x\sigma_x$, full
lines) spin structure functions. It can be seen that in the ferromagnetic states the
strength is dominated by transverse modes. It can also be noted that for $\nu \geq 2$,

$$x\sigma_x \ ; \ x\sigma_z$$

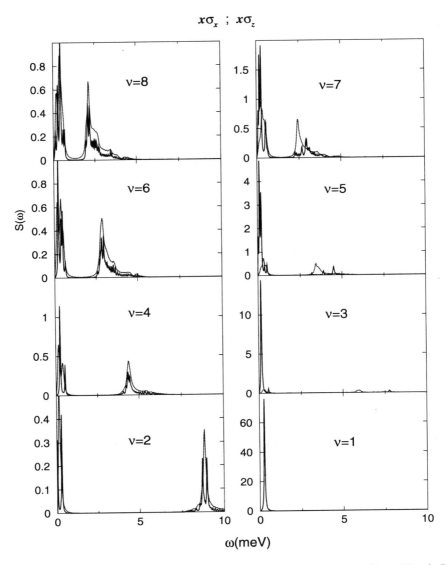

Fig. 9.38 Structure functions corresponding to $x\sigma_x$ (full lines) and to $x\sigma_z$ (dotted lines). The
strengths are in effective atomic units divided by 10^5 in the transverse case, and by 5×10^4 in the
longitudinal case, to make them clearly discernible.

i.e. low B, apart from fine-structure effects, the longitudinal and transverse spin responses have their main peaks at very similar energies.

The paramagnetic configuration at $\nu = 2$ ($B = 5.14$ T) shows the situation in which at zero spin the ground state produces the result that was obtained in (9.302): the transverse modes are shifted with respect to the longitudinal ones by the Zeeman energy (note that g^* is negative). It is also interesting to note that the low energy mode $x\sigma_+$ has practically collapsed. This hints at the existence of spin instabilities in this transverse channel, similar to the ones found for the longitudinal spin channel in the region $2 \geq \nu > 1$.

When the system is completely spin-polarized, the longitudinal dipole strengths in the density and spin channels coincide, and among the spin modes only the transverse one makes sense.

The energies of the most intense high energy peaks which are observed in the structure functions $x\sigma_+$ and $x\sigma_-$ are shown in Fig. 9.39 as a function of B. In the figure we also report the cyclotron frequency. The full symbols correspond to even filling factors, while the open ones correspond to odd filling factors in the range of $\nu = 10$ to $\nu = 1$. We also report (crosses) the values of the peaks of the longitudinal spin response.

In agreement with the previous discussion, we see that for even filling factors the energies of the transverse SDE are $\sim \omega_c \pm g^* \mu_0 B$, i.e. very close to ω_c (since the Zeeman term is very small), while for odd filling factors they are quite distant from ω_c, owing to the large contribution of the spin-dependent potential W_{xc}. In all cases, these peaks correspond to bulk modes since they involve one-particle–one-hole

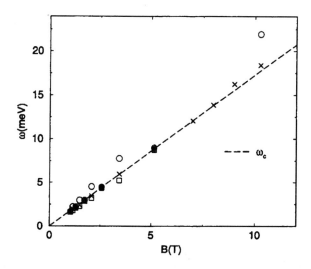

Fig. 9.39 Energies of the most intense high energy peaks excited by $x\sigma_+$ (squares) and $x\sigma_-$ (circles), as a function of B. Solid symbols correspond to even filling factors, open ones to odd filling factors for values of ν ranging from 10 ($B = 1.03$ T) to $\nu = 1$ ($B = 10.28$ T). Also reported in the figure are the energies of the longitudinal, high energy spin peaks (crosses).

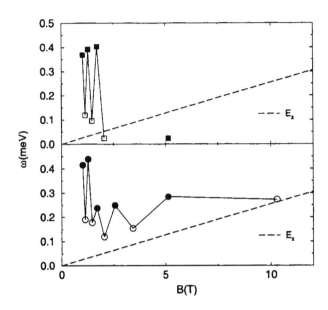

Fig. 9.40 Bottom panel: Energies of the most intense low energy peaks excited by $x\sigma_-$ as a function of B, for the same configurations as in Fig. 9.39. In the figure we also report the Zeeman energy (dashed line). Top panel: The same as the bottom panel, but for $x\sigma_+$. Some configurations are lacking because of the Pauli blocking.

transitions made up of single-particle states belonging to different Landau levels.

In a similar fashion, Fig. 9.40 shows the energies of the most intense low energy peaks of the structure functions $x\sigma_+$ (top) and $x\sigma_-$ (bottom). The full and open symbols have the same meaning as in Fig. 9.39. In the figure we also report the Zeeman energy $E_z = -g^*\mu_0 B > 0$. In order to evidence the zigzag behavior of the energies, the points corresponding to consecutive ν values were connected by a line. As we already discussed, these are edge spin modes for the paramagnetic ground states, and bulk modes for the ferromagnetic ground states.

The SDE excitation corresponding to $x\sigma_-$ is the most interesting one. In fact, this is the only transverse spin mode that shows itself at high magnetization, because the one produced by $x\sigma_+$ is blocked by the Pauli principle. For ferromagnetic states this is an undamped excitation because it is well separated from the single-particle excitations (SPE) (see Fig. 9.36). Note that the transverse SDE energy tends to the Zeeman energy E_z as B increases. When the system is completely polarized ($\nu = 1$), the energy is close to E_z, but does not coincide with it owing to finite size effects which behave like ω_c/N.

We have shown that among the SDE excitations, transverse ones are of particular relevance; in the longitudinal channel the residual interaction is very weak, and the SDE excitations undergo Landau damping because they are very close to the SPE ones (the same happens for transverse modes at low B). In the transverse channel, when the dot has an appreciable magnetization, the SDE excitations are shifted with respect to the SPE ones by the p–h vertex correction, which originates from the electron–electron interaction. As a consequence, a very collective and dispersionless SDE excitation emerges. At large spin magnetization, the Pauli exclusion principle plays a determinant role by blocking the $x\sigma_+$ component of the transverse structure function, which therefore becomes simpler. The sensitivity of the transverse response to the applied magnetic field B is revealed by strong oscillations of the collective energies with B. Such oscillations are also a consequence of the different strengths of the vertex corrections in the ferromagnetic and paramagnetic ground states.

Lastly, we note that there is a strong analogy between SDE excitations in quantum dots, and isospin excitations in nuclei with a different number of neutrons N and protons Z. In nuclei the isospin polarization given by $(N-Z)/2$ is the analog of the spin polarization of the dots. The ingredients that produce it are the Coulomb potential, which is the analog of the applied magnetic field for the dots, and the symmetry potential, which is the analog of the spin-dependent potential W_{xc}. The excitations which are the analog of those of the SDE discussed here are induced by the operators

$$D_a = \sum_{i=1}^{N} x_i \tau_i^z, \quad D_\pm = \sum_{i=1}^{N} x_i \tau_i^\pm,$$

where τ indicates the isospin operator analogous to the spin operator σ. These excitations, known as giant resonances in the isospin channels, have been studied in the framework of the time-dependent LDA theory, based on the density functional described in Subsection 4.10, by Auerbach and Klein (1983).

References

Abel, W.R., A.C. Anderson and J.C. Wheatley, *Phys. Rev. Lett.* **17**, 74 (1966).

Agosti, D., F. Pederiva, E. Lipparini and K. Takayanagi, *Phys. Rev.* **B57**, 14869 (1998).

Alasia, F., R.A. Broglia, N. Van Giai, E. Lipparini, H.E. Roman and Ll. Serra, *Phys. Rev.* **B52**, 8488 (1995).

Alberico, W.M., A Depace, A. Drago and A. Molinari, *R.N.C.* **14**, 1 (1991).

Allen, S.J., D.C. Tsui and R.A. Logan, *Phys. Rev. Lett.* **38**, 980 (1977).

Ando, T., A.B. Fowler and F. Stern, *Rev. Mod. Phys.* **54**, 437 (1982).

Auerbach, N. and A. Klein, *Nucl. Physi.* **A395**, 77 (1985).

Bachelet, G.B., D.M. Ceperley and M.G.B. Chiocchetti, *Phys. Rev. Lett.* **62**, 2088 (1989).

Baran, V., M. Colonna, M. Di Toro, A. Guarnera, V.N. Kondratyev and A. Smerzi, *Nucl. Phys.* **A599**, C 29 (1996).

Barranco, M. *et al.*, *Phys. Rev.* **B56**, 8997 (1997).

Bedell, K. and C.J. Pethick, *J. Low Temp. Phys.* **49**, 213 (1982).

Bertsch, G.F. and S.F. Tsai, *Phys. Rep.* **C18**, 126 (1975).

Bethe, H.A. and R.F. Bacher, *Rev. Mod. Phys.* **8**, 82 (1936).

Blundell, S.A. and C. Guet, *Z. Phys.* **D28**, 81 (1993).

Bogoliubov, N., *J. Phys. (Moscow)* **11**, 23 (1947).

Bohigas, O., A.M. Lane and J. Martorell, *Phys. Rep.* **51**, 267 (1979).

Bollweg, K., T. Kurth, D. Heitmann, V. Gudmundsson, E. Vasiliadou, P. Grambow and K. Eberl, *Phys. Rev. Lett.* **76**, 2774 (1996).

Borggreen, J., P. Chowdhury, N. Kebaïli, L. Lundsberg-Nielsen, K. Lützenkirchen, M.B. Nielsen, J. Pedersen and H.D. Rasmussen, *Phys. Rev.* **B48**, 17507 (1993).

Braghin, F.L. and D. Vautherin, *Phys. Lett.* **B333**, 289 (1994).

Bréchignac, C., Ph. Cahuzac, J. Leygnier and A. Sarfati, *Phys. Rev. Lett.* **70**, 2036 (1993).

Brink, D.M., A. Dellafiore and M. Di Toro, *Nucl. Phys.* **A456**, 205 (1986).

Casas, M. *et al.*, *Z. Phys.* **D35**, 67 (1995).

Dalfovo, F. *et al.*, *Phys. Rev.* **A56**, 3840 (1997).

Dalfovo, F., S. Giorgini, L. Pitaevskii and S. Stringari, *Rev. Mod. Phys.* **71**, 3 (1999).

Dalfovo, F., A. Lastri, L. Pricaupenko, S. Stringari and J. Treiner, *Phys. Rev.* **B52**, 1193 (1995).

Demel, T. *et al.*, *Phys. Rev.* **B38**, 12732(R) (1988).

Demel, T., D. Heitmann, P. Grambow and K. Ploog, *Phys. Rev. Lett.* **64**, 788 (1990).

Donnelly, R.J., J.A. Donnelly and R.N. Hills, *J. Low. Temp. Phys.* **44**, 471 (1981).

Edwards, M.R. *et al.*, *Phys. Rev. Lett.* **77**, 1671 (1996).

Ekardt, W., *Phys. Rev.* **B31**, 6360 (1985).

Ellert, C. *et al.*, *Phys. Rev. Lett.* **75**, 1731 (1995).

Eriksson, M.A., A. Pinczuk, B.S. Dennis, S.H. Simon, L.N. Pfeiffer and K.W. West, *Phys. Rev. Lett.* **82**, 2163 (1999).

Esry, B.D., *Phys. Rev.* **A55**, 1147 (1997).

Fetter, A.L., *Ann. Phys. (N.Y.)* **70**, 67 (1972).

Garcia-Recio, C., J. Navarro, N. Van Giai and L. Salcedo, *Ann. Phys.* **214**, 293 (1992).

Ghizzi, A., *Tesi di Laurea in Fisica* (Università di Trento, A.A. 1999–2000).

Goi, A.R. *et al.*, *Phys. Rev. Lett.* **67**, 3298 (1991).

Gradshteyn, I.S. and I.M. Ryzhik, *Table of Integrals, Series and Products* (Academic, New York, 1980).

Griffin, A., *Phys. Rev.* **B53**, 9341 (1996).

Grimes, C.C. and G. Adams, *Phys. Rev. Lett.* **36**, 145 (1976).

Gross, E.P., *Nuovo Cimento* **20**, 454 1961; *J. Math. Phys.* **4**, 195 (1963).

Gross, E.K.U. and W. Kohn, *Adv. Quant. Chem.* **21**, 255 (1990).

Gudmundsson, V. and R.R. Gerhardts, *Phys. Rev.* **B43**, 12098 (1991).

Guet, C. and W.R. Johnson, *Phys. Rev.* **B45**, 11283 (1992).

Hedin, L. and B.I. Lundqvist, *Solid State Phys.* **23**, 1 (1969).

Hernandez, E.S., J. Navarro, A. Polls and J. Ventura, *Nucl. Phys.* **A597**, 1 (1996).

Huang, K., *Statistical Mechanics*, second edition (John Wiley and Sons, New York, 1987).

Hutchinson, D.A. and E. Zaremba, *Phys. Rev.* **A57**, 1280 (1997).

Hutchinson, D.A., E. Zaremba and A. Griffin, *Phys. Rev. Lett.* **78**, 1842 (1997).

Iwamoto, N. and D. Pines, *Phys. Rev.* **B29**, 3924 (1984).

Jin, D.S. *et al.*, *Phys. Rev. Lett.* **77**, 420 (1996).

Kallin, C. and B. Halperin, *Phys. Rev.* **B30**, 5655 (1984).

Khalatninov, I.M. and A.A. Abrikosov, *Sov. Phys. JEPT* **6**, 84 (1958).

Knight, W.D., K. Clemenger, W.A. de Heer and W.A. Saunders, *Phys. Rev.* **B31**, 445 (1985).

Kwon, Y., D.M. Ceperley and R.M. Martin, *Phys. Rev.* **B50**, 1684 (1994) and references therein.

Landau, L.D. and E.M. Lifchitz, *Fisica Statistica*, Teoria dello Stato Condensato, Editori riuniti (Edizioni Mir, 1981).

Lipparini, E., M. Barranco, A. Emperador, M. Pi and Ll. Serra, *Phys. Rev.* **B60**, 8734 (1999).

Lipparini, E. and Ll. Serra, *Phys. Rev.* **B57**, R6830 (1998).

Lipparini, E., Ll. Serra and K. Takayanagi, *Phys. Rev.* **B49**, 16733 (1994).

Lipparini, E. and S. Stringari, *Phys. Rep.* **175**, 103 (1989).

Liu, K.F. and G.E. Brown, *Nucl. Phys.* **A265**, 385 (1976).

Liu, K.F. and N. Van Giai, *Phys. Lett.* **B65**, 23 (1976).

Luo, M.S.C., S.L. Chuang, S. Schmitt-Rink and A. Pinczuk, *Phys. Rev.* **B48**, 11086 (1993).

Luttinger, J.M., *J. Math. Phys.* **4**, 1154 (1963).

MacDonald, A.H., *J. Phys.* **C18**, 1003 (1985).

Mott, N.F. and H. Jones, *Theory of Metals and Alloys* (Dover, New York, 1936).

Nozières, P. and D. Pines, *The Theory of Quantum Liquids, Superfluid Bose Liquids* (Addison-Wesley, 1990).

Pinczuk, A. *et al.*, *Phys. Rev. Lett.* **63**, 1633 (1989).

Pines, D. and P. Nozières, *The Theory of Quantum Liquids* (Benjamin, New York, 1966), Vol. I.

Pitaevskii, L.P., *Sov. Phys. JEPT* **13**, 451 (1961).

Platzman, P.M. and P.A. Wolff, *Phys. Rev. Lett.* **18**, 280 (1967).

Pollack, S., C.R.C. Wang and M.M. Kappes, *J. Chem. Phys.* **94**, 2496 (1991).

Popov, V.N., *Sov. Phys. JEPT* **20**, 1185 (1965).

Popov, V.N., *Functional Integrals and Collective Excitations* (Cambridge University Press, 1987).

Quiang Li and S. Das Sarma, *Phys. Rev.* **B40**, 5860 (1989); **43**, 11768 (1991).

Rajagopal, A.K., *Phys. Rev.* **B17**, 2980 (1978).

Reiners, Th. *et al.*, *Phys. Rev. Lett.* **75**, 1558 (1995).

Ruprecht, P.A. *et al.*, *Phys. Rev.* **A54**, 4178 (1996).

Schüller, C. G. Biese, K. Keller, C. Steinebach, D. Heitmann, P. Grambow and K. Eberl, *Phys. Rev.* **B54**, R17304 (1996).

Schüller, C., K. Keller, G. Biese, E. Ulrichs, L. Rolf, C. Steinebach, D. Heitmann and K. Eberl, *Phys. Rev. Lett.* **80**, 2673 (1998).

Schultz, S. and G. Dunifer, *Phys. Rev. Lett.* **18**, 283 (1967).

Schulze, H.J., P. Schuck and N. Van Giai, *Phys. Rev.* **B61**, 8026 (2000).

Selby, K., V. Kresin, J. Masui, M. Vollmer, W.A. de Heer, A. Scheidemann and W. D. Knight, *Phys. Rev.* **B43**, 4565 (1991).

Serra, Ll., G.B. Bachelet, N. Van Giai and E. Lipparini, *Phys. Rev.* **B48**, 14708 (1993).

Serra, Ll., M. Barranco, A. Emperador, M. Pi and E. Lipparini, *Phys. Rev.* **B59**, 15290 (1999).

Serra, Ll. and E. Lipparini, *Europhys. Lett.* **40**, 667 (1997).

Serra, Ll., J. Navarro, M. Barranco and N. Van Giai, *Phys. Rev. Lett.* **67**, 2311 (1991).

Sikorski, Ch. and U. Merkt, *Phys. Rev. Lett.* **62**, 2164 (1989).

Silin, V.P., *Sov. Phys. JEPT* **6**, 387 (1957); **6**, 985 (1957).

Silin, V.P., *Sov. Phys. JEPT* **6**, 945 (1958); **8**, 870 (1959).

Singh, K.G. and D.S. Rokhsa, *Phys. Rev. Lett.* **77**, 1667 (1996).

Steffens, O. and M. Suhrke, *Phys. Rev. Lett.* **82**, 3891 (1999).

Strenz, R., U. Bockelmann, F. Hirler, G. Abstreiter, G. Böhm and G. Weimann, *Phys. Rev. Lett.* **73**, 3022 (1994).

Stringari, S. and F. Dalfovo, *J. Low Temp.* **78**, 1 (1990).

Takayanagi, K. and E. Lipparini, *Phys. Rev.* **B52**, 1738 (1995); **B54**, 8122 (1996).

Tomonaga, S., *Prog. Theor. Phys. (Kyoto)* **5**, 544 (1950).

Usher, A. *et al.*, *Phys. Rev.* **B41**, 1129 (1990).

vom Felde, A., J. Sprosser-Prou and J. Fink, *Phys. Rev.* **B40**, 10181 (1989).

von Weizsacker, C.F., *Z. Phys.* **96**, 431 (1935).

Vosko, S.H., L. Wilk and M. Nusair, *Can. J. Phys.* **58**, 1200 (1980).

Woods, A.D.B. and R.A. Cowley, *Rep. Prog. Phys.* **36**, 1135 (1973).

Yannouleas, C., R.A. Broglia, M. Brack and P.F. Bortignon, *Phys. Rev. Lett.* **63**, 255 (1989).

Ye, Z.L. and E. Zaremba, *Phys. Rev.* **B50**, 17217 (1994).

You, L., W. Hoston and M. Lewenstein, *Phys. Rev.* **A55**, R1581 (1997).

Yu Denisov, V., *Sov. J. Nucl. Phys.* **49**, 38 (1989).

Chapter 10

Dynamic Correlations and the Response Function

10.1 Introduction

As we saw in the previous chapter, the formulation of the many-body problem in terms of the density response function of the system produces many very interesting results for several systems; above all, for the degenerate electron gas in a background of positive charge, which is a model of basic importance for studying many properties of metals and of systems realized recently in the laboratory, such as quantum wells, metal clusters and quantum dots. This approach is very useful, in general, for all Bosonic and Fermionic quantum liquids. With this formulation, not only is it possible to study the excitation spectrum of many-body systems, but it is also possible to express in a rigorous fashion the ground state energy and the related static properties, such as the pair correlation function and the compressibility.

In the present chapter, we will study this formulation, and present the different approximations that have been used in the literature and compare them with one other and, whenever possible, with the results of "exact" calculations provided, for example by the Monte Carlo method. All the calculations are performed in the zero temperature limit.

10.2 Interaction Energy and Correlation Energy

When we evaluate the ground state energy of a many-body system, we must take into account two contributions, i.e. the kinetic one and the one due to interaction:

$$E = T + E_{\text{int}}, \tag{10.1}$$

where

$$T = \langle 0| \sum_{i=1}^{N} \frac{p_i^2}{2m} |0\rangle, \quad E_{\text{int}} = \frac{1}{2} g \langle 0| \sum_{i \neq j} v(|\mathbf{r}_i - \mathbf{r}_j|) |0\rangle. \tag{10.2}$$

In Eq. (10.2), g is the coupling constant of the interaction, for example the squared charge in the case of the Coulomb interaction; in this case $v(|\mathbf{r}_i - \mathbf{r}_j|) = |\mathbf{r}_i - \mathbf{r}_j|^{-1}$. Introducing the Fourier transform $v_{\mathbf{q}}$ of $v(|\mathbf{r}_i - \mathbf{r}_j|)$, the interaction term may be rewritten as

$$E_{\text{int}} = g\langle 0| \int \frac{d\mathbf{q}}{(2\pi)^D} \frac{v_{\mathbf{q}}}{2} (\rho_{-\mathbf{q}}\rho_{\mathbf{q}} - N)|0\rangle, \qquad (10.3)$$

where $\rho_{\mathbf{q}} = \sum_{i=1}^{N} e^{i\mathbf{q}\cdot\mathbf{r}_i}$ is the density operator. If we introduce the static form factor as defined in Eqs. (8.71) and (8.72), we obtain

$$E_{\text{int}} = g \int \frac{d\mathbf{q}}{(2\pi)^D} \frac{v_{\mathbf{q}}}{2} \left(\int_0^\infty d\omega \, S(\mathbf{q},\omega) - N + N^2\delta_{\mathbf{q}0} \right)$$

$$= -g \int \frac{d\mathbf{q}}{(2\pi)^D} \frac{v_{\mathbf{q}}}{2} \left(N - N^2\delta_{\mathbf{q}0} + \frac{1}{\pi} \int_0^\infty d\omega \, \text{Im} \, \chi(\mathbf{q},\omega) \right). \qquad (10.4)$$

Starting from (10.4) it is easy to show that, using the free response functions of Subsection 8.6 in place of χ, we can obtain for the interaction energy of homogeneous systems the Hartree–Fock energy in the Fermion case, and the Gross–Pitaevskii mean field energy for Bosons. The direct Hartree–Fock term results from the term proportional to $N^2\delta_{\mathbf{q}0}$, and in the case of the electron gas this is canceled by the field produced by the positive background charge. We recall that

$$\delta_{\mathbf{q}0} = 1/L^D \int d\mathbf{r} e^{i\mathbf{q}\cdot\mathbf{r}}.$$

It is clear that any theory which, for the response function, yields an approximation beyond the free response one, will also provide a result for the interaction energy which takes into account correlations other than the statistical ones. However, dynamic correlations contribute to the kinetic energy term as well, and in order to obtain the total contribution of dynamic correlations to the ground state energy, starting from the density response function, it is necessary to proceed further with the theory. It is necessary to use a theorem, due to Pauli, which allows the kinetic energy of the system to be computed by means of the interaction energy of Eq. (10.4). We define

$$E_0^\alpha = \langle 0|\hat{H}^\alpha|0\rangle, \qquad (10.5)$$

where

$$\hat{H}^\alpha = \hat{T} + \alpha \sum_{i\neq j} v(|\mathbf{r}_i - \mathbf{r}_j|), \qquad (10.6)$$

so that, if the coupling constant α is set equal to g, we obtain the usual interparticle interaction, while if it is set to zero we are back in the noninteracting system.

By taking into account the normalization condition of the ground state $|0\rangle$ of the system, we have

$$\frac{\partial E_0^\alpha}{\partial \alpha} = \langle 0|\frac{\partial \hat{H}^\alpha}{\partial \alpha}|0\rangle + \langle 0|\hat{H}^\alpha\left|\frac{\partial 0}{\partial \alpha}\right\rangle + \left\langle\frac{\partial 0}{\partial \alpha}\right|\hat{H}^\alpha|0\rangle$$

$$= \langle 0|\sum_{i \neq j} v(|\mathbf{r}_i - \mathbf{r}_j|)|0\rangle = \frac{E_{\text{int}}}{\alpha}, \tag{10.7}$$

and by integrating this equation with respect to α, up to the value g, we obtain

$$\int_0^g \frac{\partial E_0^\alpha}{\partial \alpha} d\alpha = \int_0^g \frac{E_{\text{int}}}{\alpha} d\alpha = E_0(\alpha = g) - E_0(\alpha = 0), \tag{10.8}$$

where

$$E_0(\alpha = 0) = \begin{cases} \dfrac{D}{D+2} N \dfrac{k_F^2}{2m} & \text{for Fermions} \\[2mm] 0 & \text{for Bosons} \end{cases} \tag{10.9}$$

Equations (10.4), (10.7) and (10.8) allow us to evaluate the ground state energy of the homogeneous system, starting from the density response function, through the equation

$$E_0 = T_0 + \int_0^g \frac{E_{\text{int}}}{\alpha} d\alpha, \tag{10.10}$$

where T_0 is the kinetic energy of the free gas.

Knowledge of the density response function of the system also allows us to calculate the contribution from dynamic correlations to the pair correlation function as defined in (8.74). In fact, from the relation between such a quantity and the static form factor (see Subsection 8.5)

$$S(\mathbf{q}) = 1 + \rho \int d\mathbf{r}(g(r) - 1)e^{i\mathbf{q}\cdot\mathbf{r}},$$

we obtain, after taking its inverse transform,

$$g(r) = 1 + \frac{1}{\rho} \int \frac{d\mathbf{q}}{(2\pi)^D}(S(\mathbf{q}) - 1)e^{-i\mathbf{q}\cdot\mathbf{r}}, \tag{10.11}$$

which allows the numerical calculation of $g(r)$.

10.3   The RPA Correlation Energy

In this subsection, we will apply the formalism described above to the case where the density response is that obtained from the RPA theory (see Subsection 9.2):

$$\chi^{\text{RPA}}(q,\omega) = \frac{\chi^0(q,\omega)}{1 - g\frac{v_{\mathbf{q}}}{L^D}\chi^0(q,\omega)},$$

with $\chi^0(q,\omega)$ the free response function of Subsection 8.6. Then, we obtain the following relation for the ground state energy in the RPA approximation:

$$E_0^{\text{RPA}} = T_0 - \int \frac{d\mathbf{q}}{(2\pi)^D}\frac{v_{\mathbf{q}}}{2}\int_0^g d\alpha$$

$$\times \left(N - N^2\delta_{\mathbf{q}0} + \frac{1}{\pi}\int_0^\infty d\omega \,\text{Im }\chi^{\text{RPA}}(\mathbf{q},\omega)\right). \tag{10.12}$$

The RPA correlation energy is subsequently obtained by subtracting from the above expression the HF energy which, as discussed previously, in the framework of the density response function, is obtained using the free response function for χ. In this way, we have for the correlation energy per particle

$$\frac{E_c^{\text{RPA}}}{N} = \frac{1}{2\pi N}\int \frac{d\mathbf{q}}{(2\pi)^D}v_{\mathbf{q}}\int_0^g d\alpha \int_0^\infty d\omega \,(\text{Im }\chi^{\text{RPA}} - \text{Im }\chi_0). \tag{10.13}$$

The RPA correlation energy has been computed for the electron gas immersed in a background of positive charge, in several dimensions, by many authors [see for example Pines and Nozières (1966), Jonson (1976), Agosti *et al.* (1998)]. The numerical results one obtains are in agreement with those derived by Gell-Mann and Brueckner (1957) in 3D, and by Isihara and Toyoda (1977) and Rajagopal and Kimball (1977) in 2D, using perturbation theory in the limit of high density $(r_s \ll 1)$. In this limit, the leading terms in the correlation energy are a series of diagrams (ring diagrams) corresponding to the most diverging direct terms which, when summed up, lead to a rigorous and converging result. Therefore, the RPA is the proper theory for the correlation energy of the electron gas in the limit of high density. As can be seen from Table 10.1, where the RPA correlation energy is compared with the results of other theories in the case of the 2D electron gas (taken as an example), the RPA theory works well only at very low values of r_s. Even at $r_s = 0.5$ there is a large discrepancy between the RPA correlation energy and the one obtained from Monte Carlo calculations (Tanatar and Ceperley, 1989).

The direct connection between the RPA theory as discussed in this subsection and the high density perturbation theory can be established by starting from the expression for the RPA correlation energy (10.13), and assuming that in

$$\chi^{\text{RPA}}(q,\omega) = \frac{\chi^0(q,\omega)}{1 - e^2\frac{v_{\mathbf{q}}}{L^D}\chi^0(q,\omega)}$$

the term $e^2 v_{\mathbf{q}} \chi^0 / L^D$ represents a small parameter. In this case, it is possible to write a series expansion of the type

$$\chi^{\text{RPA}} = \chi^0 \left(1 + e^2 \frac{v_{\mathbf{q}}}{L^D} \chi^0 + \left(e^2 \frac{v_{\mathbf{q}}}{L^D} \right)^2 (\chi^0)^2 + \cdots \right), \tag{10.14}$$

and verify that the terms proportional to $\text{Im}(\chi_0)^2$, $\text{Im}(\chi_0)^3$, and so on, give the ring diagrams of perturbation theory.

Finally, note that the RPA correlation energy can also be calculated in finite size systems. To reach this goal one has to solve both the RPAE and Tamm–Dancoff (TD) equations, (9.138) and (9.161), respectively, taking care of all the possible p–h excitations up to some cutoff energy which fixes the model space of the calculation. The RPA correlation energy is then given by the sum,

$$E_{\text{RPAE}} - E_{\text{HF}} = \frac{1}{2} \sum_\lambda (\omega_\lambda^{\text{RPA}} - \omega_\lambda^{\text{TD}}), \tag{10.15}$$

of all differences in the excitation energies between the RPAE and TD approaches [for the derivation of Eq. (10.15), see for example Ring and Schuck (1980)]. Clearly, in the calculation of the sum (10.15), one must reach convergence of the result with the cutoff. A calculation of this kind has been done, for example, in metal clusters by Reinhard (1992) and in quantum dots by Serra, Nazmitdinov and Puente (2003).

10.3.1 *The RPA Correlation Energy for the Cold and Dilute Gas of Bosons and Fermions*

As an example of analytic calculation of RPA correlations, we consider the case of a cold and dilute gas of Bosons and Fermions, for which it is possible to replace the two-body interaction by the effective interaction $v(r) = g\delta(r)$, where g is connected to the scattering length a for the low energy scattering of the gas atoms. As was shown in Subsection 9.5, in the Boson case and using the following expression for the free response,

$$\chi_0^B(\mathbf{q}, \omega) = \frac{\frac{Nq^2}{m}}{\omega^2 - (\frac{q^2}{2m})^2},$$

we obtain for the RPA response the expression

$$\chi_{\text{RPA}}^B(\mathbf{q}, \omega) = \frac{\frac{Nq^2}{m}}{\omega^2 - \frac{\rho q^2 g}{m} - (\frac{q^2}{2m})^2}, \tag{10.16}$$

which has only one pole at energy

$$\omega_{\text{RPA}}^B = \sqrt{\frac{\rho q^2 g}{m} + \left(\frac{q^2}{2m} \right)^2}.$$

The imaginary part of $\chi_{\text{RPA}}^{B}(\mathbf{q}, \omega)$ is, therefore, given by

$$\text{Im} \, \chi_{\text{RPA}}^{B}(\mathbf{q}, \omega) = -\frac{Nq^2}{m} \frac{\pi\delta(\omega - \omega_{\text{RPA}}^{B})}{2\omega_{\text{RPA}}^{B}}. \tag{10.17}$$

By inserting this result in (10.12), it is easy to obtain the following result for the RPA energy in the ground state of a dilute Boson gas:

$$\frac{E_{B}^{\text{RPA}}}{N} = \frac{\rho g}{2} + \frac{1}{2\rho} \int \frac{d\mathbf{q}}{(2\pi)^3} \left[\frac{q^2}{2m} \sqrt{1 + \frac{4m\rho g}{q^2}} - \frac{q^2}{2m} - g\rho \right], \tag{10.18}$$

where (ρ is the gas density) the first term is the (Gross–Pitaevskii) mean field energy which derives from the term proportional to $N^2 \delta_{\mathbf{q}0}$ in (10.12), and the second term is the RPA correlation energy. By expanding the integral in the correlation term, it is possible to write

$$\frac{E_{B}^{\text{RPA}}}{N} = \frac{\rho g}{2} + \frac{8}{15\pi^2} g m^3 \left(\frac{g\rho}{m} \right)^{3/2} - \frac{1}{2} m g^2 \rho \int \frac{d\mathbf{q}}{(2\pi)^3} \frac{1}{q^2}. \tag{10.19}$$

The last term in (10.19) diverges in the limit of large q. Such ultraviolet divergence can be removed by a suitable renormalization of the bare coupling constant g which appears in the δ-like interaction, by using the ladder theory of Subsection 3.3. In fact, when applied to such an interaction, the ladder theory produces the result $\tilde{g}$ [see Eq. (3.9)] for the renormalized coupling constant [with $V(q) = g$]:

$$\tilde{g} = g \left(1 - \tilde{g} \int \frac{d\mathbf{k}}{(2\pi)^3} \frac{m}{k^2} \right), \tag{10.20}$$

which, at second order in $\tilde{g}$, yields

$$g = \tilde{g} \left(1 + \tilde{g} \int \frac{d\mathbf{k}}{(2\pi)^3} \frac{m}{k^2} \right). \tag{10.21}$$

For a dilute and cold gas $\tilde{g} = 4\pi a/m$, where a is the scattering length, and finally from Eq. (10.21), we get for the renormalization of the bare coupling constant

$$g = \frac{4\pi a}{m} \left(1 + \frac{4\pi a}{m} \int \frac{d\mathbf{k}}{(2\pi)^3} \frac{m}{k^2} \right), \tag{10.22}$$

which is valid to second order. The substitution of this result into (10.19) leads to the following expression for the ground state energy per particle:

$$\frac{E_{B}^{\text{RPA}}}{N} = \frac{2\pi a \rho}{m} \left(1 + \frac{128}{15\sqrt{\pi}} \sqrt{a^3 \rho} \right). \tag{10.23}$$

This result coincides, at lowest order, with the exact result derived by Lee, Huang and Yang (1957) by expansion in powers of $\sqrt{a^3 \rho}$ of the ground state energy of a homogeneous and dilute Bose gas. A rigorous derivation of the exact energy per particle of a homogeneous and dilute Bose gas can be found in Lieb and Yngvason

(1998). These authors also derived in a rigorous way the energy of the same gas in the two-dimensional case (Lieb and Yngvason, 2000):

$$\frac{E_B^{2D}}{N} = \frac{2\pi\rho}{m} \, |\ln(\rho a^2)|^{-1}, \tag{10.24}$$

which holds at lowest order in ρa^2. This result had been derived previously by Schick and Hines (1971), Frankel and Mitchell (1978), and extended to finite temperature by Popov (1977) and Fisher and Hohenberg (1988).

In the case of a diluted Fermion gas, an analytic calculation of the RPA ground state energy, analogous to the one carried out for Bosons, is not possible. In fact, the Fermion free response function is much more complicated than that for Bosons, and the integrals which appear in the expression for the RPA ground state energy (10.12) are not analytical. However, we do obtain an analytical result if in (10.12) we use for χ^{RPA} the expansion (10.14) truncated at second order, and with $e^2 v_{\mathbf{q}} = g$, for the dilute Fermion gas. The result one obtains, by adding the second order exchange term (which, for the δ-like interaction, is a half of the direct one with opposite sign), then coincides with that of second order perturbation theory [see, for example Landau and Lifchitz 1981]:

$$\frac{E_F^{3D}}{N} = \frac{3k_F^2}{10m} \left[1 + \frac{10}{9\pi} k_F a + \frac{4(11 - 2\ln 2)}{21\pi^2}(k_F a)^2 \right], \tag{10.25}$$

where k_F is the Fermi momentum and a is the scattering length for the Fermion gas. Note that, in deriving this result, and as in the Boson case, we take into account the renormalization of the bare coupling constant. Such renormalization cancels exactly the ultraviolet divergence of the correlation energy term. Expression (10.25) gives the first terms of the expansion of the energy in powers of the parameter $k_F a$ (Huang and Yang, 1957). The above calculation also allows the derivation of quasiparticle properties of the dilute Fermion gas. For example, the effective mass turns out to be given by

$$\frac{m^*}{m} = 1 + \frac{8}{15\pi^2}(7\ln 2 - 1)(k_F a)^2. \tag{10.26}$$

The above results for the dilute Fermion gas are valid for both repulsion and attraction among Fermions, i.e. for both positive and negative scattering lengths. In fact, in the BCS regime holding in the case of attraction ($a < 0$), the corrections to the above equations are exponentially small (see Subsection 2.7).

10.4   Theories Beyond the RPA

In general, the linear response χ of the system can always be expressed in terms of the free response function χ^0, as

$$\chi(\mathbf{q},\omega) = \frac{\chi^0(\mathbf{q},\omega)}{1 - \frac{v_{\mathbf{q}}}{L^D}(1 - G(\mathbf{q},\omega))\chi^0(\mathbf{q},\omega)}, \tag{10.27}$$

where $G(\mathbf{q}, \omega)$ is the so-called field factor which is to be determined through a microscopic calculation. When $G(\mathbf{q}, \omega) = G(\mathbf{q})$, one talks about static field theories. In general, G depends on ω as well, and in this case one talks about dynamic field theories. The RPA theory has $G = 0$, while the adiabatic TDLDA theory is a static theory with

$$G^{\text{TDLDA}}(\mathbf{q}) = -\frac{1}{v_{\mathbf{q}}} \frac{\partial V_{\text{xc}}}{\partial \rho}\bigg|_{\rho=\rho_0}, \tag{10.28}$$

where V_{xc} is the exchange-correlation potential. Another static theory is the Hubbard theory (Hubbard, 1957) for the electron gas, where one assumes that

$$G^{H}(\mathbf{q}) = \frac{q^2}{2(q^2 + k_F^2)} \quad \text{(in 3D)} \tag{10.29}$$

and

$$G^{H}(\mathbf{q}) = \frac{q}{2\sqrt{q^2 + k_F^2}} \quad \text{(in 2D)}. \tag{10.30}$$

This theory takes into account in an approximate way the exchange effects which are neglected in the RPA.

$\quad$ $G(\mathbf{q}, \omega)$ has been computed numerically for the 2D electron gas, starting from the RPAE results (see Subsection 9.11) for $\chi(\mathbf{q}, \omega)$ of Takayanagi and Lipparini (1995) and Schulze *et al.* (2000), which treat the exchange exactly. At $\omega = 0$, $G(\mathbf{q}, 0)$ exhibits a linear behavior in the low-q limit, a peak around $q = 2k_F$, and tends to $1/2$ for $q \rightarrow \infty$. This is shown in Fig. 10.1, where the RPAE results for $G(\mathbf{q}, 0)$ are compared with those of other models (Pederiva *et al.*, 1997). From this figure, one notices that for $q < 2k_F$ the RPAE result is in very good agreement with the Monte Carlo (MC) results at $r_s = 1$ and $r_s = 5$, while for high momentum values one observes a substantial difference between the RPAE and MC results. In fact, the MC calculation (Moroni *et al.*, 1992) predicts a nearly linear increase of $G(\mathbf{q}, 0)$ with q, as expected (Likos *et al.*, 1997), while the RPAE yields $G(\mathbf{q}, 0)$ tending to $1/2$ for large q. As was shown Takayanagi and Lipparini (1996), a large part of the difference existing between RPAE and MC results can be accounted for by taking into consideration electron–electron correlations due to rescattering processes of the ladder diagrams, and then using the effective Brueckner–Hartree–Fock interaction (g matrix; see Chapter 3) instead of the bare interaction to solve the integral Dyson equation. The results of such calculation (RPAEG) for the static response function $\chi(q)$ are shown in Fig. 10.2, together with the results of the MC calculation. The peak at $q = 2k_F$ in $\chi(q)$ is a consequence of the similar peak in $G(\mathbf{q}, 0)$. For reasons of numerical accuracy, it is not possible to draw a definite conclusion as to the behavior of such quantities in the neighborhood of $q = 2k_F$, and in this q region there is still some ambiguity in the determination of the static field factor.

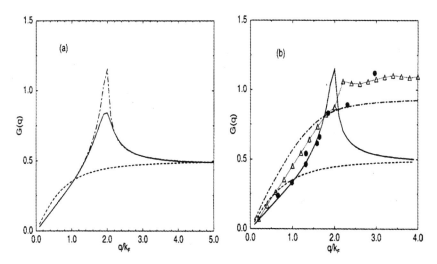

Fig. 10.1 (a) Static field factor $G(q)$ of the 2D electron gas, computed in the RPAE as a function of q/k_F for two different values of the density. Full curve: $r_s = 1$, dashed–dotted curve $r_s = 5$. Dashed curve: Hubbard approximation. (b) $G(q)$ for $r_s = 5$ in different models. Dashed curve: Hubbard approximation. Full curve: RPAE. Dashed–dotted curve: STLS approximation. Triangles: RPAEG. Full dots: QMC results.

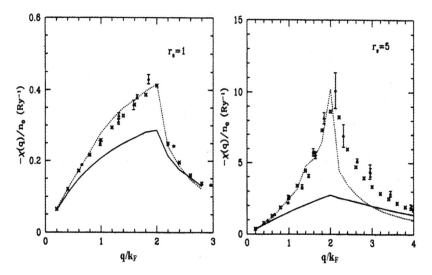

Fig. 10.2 Static linear response of the 2D electron gas as a function of q/k_F at $r_s = 1$ and $r_s = 5$. The full and dashed lines give the RPA and RPAE results, respectively. The stars indicate the RPAEG results (including the ladder correlations), and the triangles the QMC results.

10.5   The STLS Theory

The STLS (Singwi, Tosi, Land and Sjolander, 1968) model is based on an approximate solution to the equation of motion for the one-particle distribution function
$f(\mathbf{r}, \mathbf{p}, t)$, which gives the probability of finding a particle at position $\mathbf{r}$ with momentum $\mathbf{p}$ at time t. This quantity, which is related to the spatial density of the particle $\rho(\mathbf{r}, t)$ by the relationship

$$\rho(\mathbf{r}, t) = \int d\mathbf{p} f(\mathbf{r}, \mathbf{p}, t), \qquad (10.31)$$

fulfills, in the presence of an external field $v_{\text{ext}}(\mathbf{r}, t)$, the kinetic equation [see, for
example Cohen (1962)]

$$\frac{\partial}{\partial t} f(\mathbf{r}, \mathbf{p}, t) + \frac{\mathbf{p}}{m} \cdot \nabla_\mathbf{r} f(\mathbf{r}, \mathbf{p}, t) - \nabla_\mathbf{r} v_{\text{ext}}(\mathbf{r}, t) \cdot \nabla_\mathbf{p} f(\mathbf{r}, \mathbf{p}, t)$$

$$- \int \nabla_\mathbf{r} v(\mathbf{r} - \mathbf{r}') \cdot \nabla_\mathbf{p} f(\mathbf{r}, \mathbf{p}, \mathbf{r}'; \mathbf{p}', t) d\mathbf{r}' d\mathbf{p}' = 0, \qquad (10.32)$$

where $v(\mathbf{r} - \mathbf{r}')$ is the interparticle interaction potential, and $f(\mathbf{r}, \mathbf{p}, \mathbf{r}'; \mathbf{p}', t)$ is the
two-particle distribution function. The equation for the two-particle distribution
function, in turn, contains the three-particle distribution function, and so on. STLS
truncated such an infinite hierarchy of equations by the ansatz

$$f(\mathbf{r}, \mathbf{p}, \mathbf{r}'; \mathbf{p}', t) = f(\mathbf{r}, \mathbf{p}, t) f(\mathbf{r}', \mathbf{p}', t) g(\mathbf{r} - \mathbf{r}'), \qquad (10.33)$$

where $g(\mathbf{r} - \mathbf{r}')$ is the static pair correlation function [see Eq. (8.74)]. Such an
ansatz considers in an approximate fashion the short range correlations among
particles, through the function $g(\mathbf{r} - \mathbf{r}')$ which tends to 1 in the limit of large
separation between particles, while for small separation it is affected by the dynamic
correlations among particles due to interaction. Assuming that $g(\mathbf{r} - \mathbf{r}') = 1$ for all
distances, it is easy to obtain from Eqs. (10.32) and (10.33) the result of (9.97),
i.e. the semiclassical limit of the time-dependent Hartree equations, which in turn
coincide with the Landau–Vlasov equations (1.136) in the absence of collisions. If
we write

$$f(\mathbf{r}, \mathbf{p}, t) = f^0(\mathbf{p}) + \delta f(\mathbf{r}, \mathbf{p}, t), \qquad (10.34)$$

where $\delta f(\mathbf{r}, \mathbf{p}, t)$ is the deviation from the equilibrium distribution function $f^0(\mathbf{p})$,
induced by the (weak) external field, and if we linearize Eqs. (10.32), we obtain an
equation for $\delta f(\mathbf{r}, \mathbf{p}, t)$ which, once solved, allows us to write the response function
of the system to the external field

$$v_{\text{ext}}(\mathbf{r}, t) = v_{\text{ext}}(\mathbf{q}, \omega) e^{i(\mathbf{q} \cdot \mathbf{r} - \omega t)} + \text{c.c.} \qquad (10.35)$$

in the following way:

$$\chi^{\text{STLS}}(\mathbf{q}, \omega) = \frac{\chi^0(\mathbf{q}, \omega)}{1 - \frac{g v_{\mathbf{q}}}{L^D}(1 - G^{\text{STLS}}(\mathbf{q}))\chi^0(\mathbf{q}, \omega)}. \tag{10.36}$$

In Eq. (10.36), $v_{\mathbf{q}}$ is the Fourier transform of the interparticle interaction and g is the coupling constant, $\chi^0(\mathbf{q}, \omega)$ is the classical free response function, and the field factor $G^{\text{STLS}}(\mathbf{q})$ is given by

$$G^{\text{STLS}}(\mathbf{q}) = -\frac{1}{\rho_0} \int \frac{d\mathbf{q}'}{(2\pi)^D} \frac{\mathbf{q} \cdot \mathbf{q}'}{q'^2}(S(|\mathbf{q} - \mathbf{q}'|) - 1), \tag{10.37}$$

where the static form factor $S(\mathbf{q})$ is given by

$$S(\mathbf{q}) = -\frac{1}{\pi} \int_0^\infty d\omega \, \text{Im} \, \chi(\mathbf{q}, \omega). \tag{10.38}$$

Equations (10.36) and (10.37), together with (10.38), constitute a set of equations that are to be solved in a self-consistent way. The derivation of such a set of equations is completely classical. The application to quantum systems is carried out just by replacing the classical response function and field factor with their quantum counterparts. In this case, Eq. (10.36) reduces to the RPA response function provided we neglect field corrections, and Eq. (10.37) for $G^{\text{STLS}}(\mathbf{q})$, in the case of the electron gas, reproduces the Hubbard expression (10.29), (10.30) provided that we use for $S(\mathbf{q})$ the static form factor of the Fermi gas [see Eqs. (1.110) and (1.111)]. Once the self-consistent solution for all values of the density is obtained, it is possible to derive the pair correlation function through relation (10.11) from the knowledge of the static form factor. The correlation energy in the STLS approximation can be derived by integrating on the coupling constant of the interaction energy per particle:

$$\frac{E_{\text{int}}(\alpha)}{N} = \frac{\alpha}{2} \int \frac{d\mathbf{q}}{(2\pi)^D} v_{\mathbf{q}}[S(\mathbf{q}) - 1 + N\delta_{\mathbf{q}0}]. \tag{10.39}$$

This calculation was carried out for the 3D electron gas by Singwi, Tosi, Land and Sjolander (1968), in 2D by Jonson (1976), and in quasi-one-dimensional systems by Freser and Bergensen (1980) and Camels and Gold (1995, 1997) (for these systems the term $N\delta_{\mathbf{q}0}$ is canceled by the positive background).

Several authors attempted to improve the STLS theory, in order to better reproduce the high density limit ($r_s \ll 1$), and the compressibility sum rule of Eqs. (8.67)–(8.70) and (9.249). In particular, people abandoned the ansatz (10.33), by taking into account three-body correlations in the kinetic equation (10.32), and truncating the infinite series of such equations at this order [see Ichimaru (1982)]. In this way, for the local field factor one obtains the expression

$$G(\mathbf{q}) = -\frac{1}{\rho_0} \int \frac{d\mathbf{q}'}{(2\pi)^D} \frac{\mathbf{q} \cdot \mathbf{q}'}{q'^2} S(q')(S(|\mathbf{q} - \mathbf{q}'|) - 1), \tag{10.40}$$

by which one should then solve the triad of equations for χ, G and S. The result of such calculations appreciably improves the STLS approximation.

Lastly, we mention the "hypernetted chain" approximation (HNC), which was developed in the physics of classic liquids (van Leeuwen, Groeneveld and De Boer, 1959; Morita, 1960), and applied to the study of several physical systems, such as the electron gas, liquid helium and nuclear matter (Fantoni and Rosati, 1972, 1975; Ripka, 1979). Such a theory leads to the following result for $G(\mathbf{q})$:

$$G(\mathbf{q}) = -\frac{1}{\rho_0} \int \frac{d\mathbf{q}'}{(2\pi)^D} \frac{\mathbf{q} \cdot \mathbf{q}'}{q'^2} (S(|\mathbf{q} - \mathbf{q}'|) - 1)$$
$$\times \{1 + [1 - G(\mathbf{q}')][S(\mathbf{q}') - 1]\}. \tag{10.41}$$

Note that the STLS scheme is obtained from the above expression by putting $G(\mathbf{q}') = 1$ in the right hand side, while the convolution formula (10.40) is obtained by assuming that $G(\mathbf{q}') = 0$. The HNC approximation turns out to be the best approximation for static field theories, and is discussed in detail and compared to STLS and other theoretical schemes in Chihara (1973), Choquard (1978) and Chihara and Sasaki (1979). For a review of works on HNC applied to the electron gas, see Ichimaru (1982).

10.6   Comparison of Different Theories for the Electron Gas in 2D

In this subsection we will compare the results of the calculations carried out in the framework of the models discussed previously, for the case of the two-dimensional electron gas, which is taken as an example of many Fermion homogeneous systems, and for which in the literature one finds a great deal of results.

In Tables 10.1 and 10.2 we show (Pederiva, Lipparini and Takayanagi, 1997) the correlation energies and compressibilities per particle as a function of the Wigner–Seitz radius r_s computed by: Monte Carlo (QMC, Tanatar and Ceperley, 1989), RPA, Hubbard approximation (HA), RPA with the exchange term treated in an exact way (RPAE; Takayanagi and Lipparini, 1995), STLS approximation (Jonson,

Table 10.1. Correlation energy per particle of the 2D electron gas as a function of r_s in different theories.

r_s	ϵ_c^{MC}	ϵ_c^{RPA}	ϵ_c^{HA}	ϵ_c^{RPAE}	ϵ_c^{STLS}	ϵ_c^{RPAEG}
0.5	−3.64	−6.28	−4.26	−3.65	−3.40	−3.92
1	−2.99	−5.40	−3.70	−3.14	−2.87	−3.11
2	−2.26	−4.42	−3.10	−2.56	−2.11	−2.54
4	−1.55	−3.43	−2.46	−1.99	−1.47	−1.50
5	−1.34	−3.10	−2.21	−1.76	−1.25	−1.15
8	−0.98	−2.56	−1.86	−1.56	−0.90	−0.89

Table 10.2. Isothermal compressibility K_0/K of the 2D electron gas as a function of r_s in different theories.

r_s	K_0/K^{MC}	K_0/K^{RPA}	K_0/K^{HA}	K_0/K^{RPAE}	K_0/K^{STLS}	K_0/K^{RPAEG}
0.5	0.77	0.77	0.77	0.77	0.77	0.77
1	0.54	0.54	0.54	0.54	0.54	0.54
2	0.06	0.05	0.07	0.05	0.06	0.05
4	−0.96	−1.02	−0.94	−0.98	−0.96	−0.99
5	−1.49	−1.58	−1.46	−1.51	−1.49	−1.52
8	−3.12	−3.38	−3.13	−3.11	−3.09	−3.16

1976), and RPAE including ladder correlations (RPAEG; Takayanagi and Lipparini, 1996).

From these tables, comparing the different results with Monte Carlo (which should be considered the best possible calculation), we see that the exact treatment of Coulomb exchange in RPAE greatly improves the results obtained by RPA and HA, and that inclusion of the short range correlations as performed in RPAEG and in the STLS calculation is crucial for correctly reproducing the correlation energies. This trend can also be seen in Figs. 10.3 and 10.4, where, in the $r_s = 5$ case, we compare the results of different calculations for the static form factor and the pair correlation function.

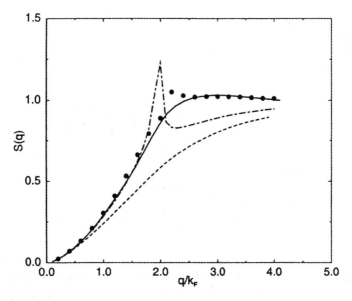

Fig. 10.3 Static form factor of the electron gas in 2D at $r_s = 5$. The dashed and dashed–dotted curves give the RPA and RPAE results, respectively. The circles give the RPAEG results (including the ladder correlations), and the full line shows the QMC result.

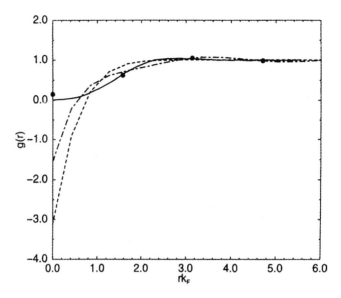

Fig. 10.4 Pair correlation function of the 2D electron gas at $r_s = 5$. The dashed and dotted–dashed curves give the RPA and RPAE results, respectively. The circles are the RPAEG results (including the ladder correlations), and the full line is the QMC result.

10.7   Quasiparticle Properties

In the RPA theory, and in local field theories such as the Hubbard theory and the STLS model, the quasiparticle properties can be calculated in a rigorous way starting from the expression

$$E_0 = T_0 - \int \frac{d\mathbf{q}}{(2\pi)^D} \frac{v_\mathbf{q}}{2} \int_0^g \frac{d\alpha}{\alpha}$$

$$\times \left(N - N^2 \delta_{\mathbf{q}0} + \frac{1}{2\pi} \int_{-\infty}^{\infty} dw \frac{\alpha \chi^0(\mathbf{q}, iw)}{1 - \frac{\alpha v_\mathbf{q}}{L^D}(1 - G(\mathbf{q}))\chi^0(\mathbf{q}, iw)} \right) \qquad (10.42)$$

for the ground state energy, and by performing functional differentiation with respect to the occupation number $n_\mathbf{k}$ (Ting, Lee and Quinn, 1975). Equation (10.42) can be derived from the results of Subsection 10.2 by means of the relation

$$\int_0^\infty dw \, \text{Im} \, \chi(\mathbf{q}, \omega) = \frac{1}{2} \int_{-\infty}^{\infty} dw \, \chi(\mathbf{q}, iw), \qquad (10.43)$$

with $\omega = iw$, and writing χ in terms of the free response function as in (10.27). In the integral that appears in (10.42), the only term which depends on the occupation number is the one involving the free response function χ^0, given by

$$\chi^0(\mathbf{q}, iw) = \int \frac{d\mathbf{k}}{(2\pi)^D} G_0^{ph}(\mathbf{k}, \mathbf{q}, iw), \qquad (10.44)$$

where $G_0^{ph}(\mathbf{k}, \mathbf{q}, iw)$ is the free p–h Green function:

$$G_0^{ph}(\mathbf{k}, \mathbf{q}, iw) = \frac{(1 - n_{\mathbf{k}+\mathbf{q}})n_{\mathbf{k}}}{iw - \varepsilon_{\mathbf{kq}} + i\eta} - \frac{(1 - n_{\mathbf{k}})n_{\mathbf{k}+\mathbf{q}}}{iw - \varepsilon_{\mathbf{kq}} - i\eta}, \tag{10.45}$$

with

$$\varepsilon_{\mathbf{kq}} = \frac{(\mathbf{k}+\mathbf{q})^2}{2m} - \frac{\mathbf{k}^2}{2m}.$$

Note that the local field factor $G(q)$ can also be a functional of $\alpha\chi^0$. The kinetic energy is given by $T_0 = \sum_{\mathbf{k}} n_{\mathbf{k}}\varepsilon_{\mathbf{k}}$ with $\frac{\delta T_0}{\delta n_{\mathbf{k}}} = \varepsilon_{\mathbf{k}}$, and the self-energy is given by

$$\Sigma(\mathbf{k}, ik) = \frac{\delta E_0}{\delta n_{\mathbf{k}}} - \varepsilon_{\mathbf{k}}. \tag{10.46}$$

The integrand (denoted by $\mathcal{F}$) in the interaction term of Eq. (10.42) is a function of $\alpha\chi^0$, and the functional derivative of $\mathcal{F}$ with respect to the occupation number is written as

$$\frac{\delta\mathcal{F}}{\delta n_{\mathbf{k}}} = \frac{\delta\mathcal{F}}{\delta\chi^0}\frac{\delta\chi^0}{\delta n_{\mathbf{k}}}, \tag{10.47}$$

where

$$\frac{\delta\chi^0}{\delta n_{\mathbf{k}}} = \mathcal{G}_0(\mathbf{k}+\mathbf{q}, iw + ik) + \mathcal{G}_0(\mathbf{k}+\mathbf{q}, -iw + ik), \tag{10.48}$$

and $\mathcal{G}_0$ is the free single-particle Green function:

$$\mathcal{G}_0(\mathbf{k}, iw) = \frac{n_{\mathbf{k}}}{iw - \varepsilon_{\mathbf{k}} - i\eta} + \frac{1 - n_{\mathbf{k}}}{iw - \varepsilon_{\mathbf{k}} + i\eta}. \tag{10.49}$$

Note that the two terms of (10.48) give the same contribution, since $\mathcal{F}$ is a symmetric function of the complex frequency iw. Moreover, the integral on the coupling constant and the functional derivative of $\mathcal{F}$ with respect to χ^0 may be arranged in the following way by introducing the variable $y = \alpha\chi^0$:

$$\int_0^g \frac{d\alpha}{\alpha}\frac{\delta\mathcal{F}(\alpha\chi^0)}{\delta\chi^0} = \frac{1}{\chi^0}\int_0^{g\chi^0} dy\frac{\delta\mathcal{F}(y)}{\delta y} = \frac{\mathcal{F}(g\chi^0)}{\chi^0}. \tag{10.50}$$

Using all these results, it is easy to obtain the following expression for the interaction self-energy:

$$\Sigma(\mathbf{k}, ik) = -g\int\frac{d\mathbf{q}}{(2\pi)^D}v_{\mathbf{q}}\frac{1}{2\pi}\int_{-\infty}^{\infty}dw\frac{\mathcal{G}_0(\mathbf{k}+\mathbf{q}, iw + ik)}{1 - \frac{gv_{\mathbf{q}}}{L^D}(1 - G(\mathbf{q}))\chi^0(\mathbf{q}, iw)}. \tag{10.51}$$

This expression may be rewritten in a more general form by introducing the screened vertex function Γ as

$$\Sigma(\mathbf{k}, ik) = -g\int\frac{d\mathbf{q}}{(2\pi)^D}v_{\mathbf{q}}\frac{1}{2\pi}\int_{-\infty}^{\infty}dw\mathcal{G}_0(\mathbf{k}+\mathbf{q}, iw + ik)\frac{\Gamma(\mathbf{q}, iw)}{\epsilon(\mathbf{q}, iw)}, \tag{10.52}$$

where

$$\Gamma(\mathbf{q}, iw) = \epsilon(\mathbf{q}, iw) f(\mathbf{q}, iw). \tag{10.53}$$

Here $\epsilon(\mathbf{q}, iw)$ is the dielectric constant defined in Eqs. (9.250) and (9.251), and $f(\mathbf{q}, iw)$ is the vertex function defined in Subsection 9.11. In general, $f(\mathbf{q}, iw)$ also depends on the variable k, and neglecting such dependence is an approximation that leads to the result

$$f(\mathbf{q}, iw) = \frac{\chi(\mathbf{q}, iw)}{\chi^0(\mathbf{q}, iw)}, \tag{10.54}$$

and hence, for static field theories, to expression (10.51).

Starting from expression (10.51), the quasiparticle properties of the electrons in the homogeneous gas were computed, in the different static field approximations. The results of these computations have been discussed in detail, for example by Rice (1965) and in the book by Mahan (1981). It is important to note that the self-energy as computed in the RPA theory [i.e. putting $G(q) = 0$ in the above expressions] has the following important characteristic: the reciprocal effective mass does not diverge any more at the Fermi surface. In other words, the divergence affecting the reciprocal mass at the Fermi surface in the HF theory [see Eq. (2.37)] is removed by the RPA theory which takes into account the dynamic electron–electron correlations.

As a general remark, in the case of electrons the quasiparticle properties are little affected by the effects due to the electron–electron interactions. Other effects, such as those due to the band structure in real metals, and to electron–phonon interactions, are more important.

The quasiparticle properties and the nonlocal effects (see next subsection) are of more relevance for other Fermion systems, and are discussed in detail in the review paper by Mahaux *et al.* (1985).

10.8   Nonlocal Effects

The local theories for the linear response function discussed above assume that the field factor $G(q, \omega)$ does not depend on frequency, but only on the transferred momentum q. Clearly, this is an approximation. In the parlance of the Landau theory for Fermi liquids, this is equivalent to assuming that the Landau parameters F_ℓ are equal to zero when $\ell \geq 1$. Only in this approximation, and in the limit of small q and ω, do the local field theories and the Landau theory coincide.

Are nonlocal effects important? The answer to this question is, in general, positive for strongly interacting liquids, like ^{3}He, where the Landau parameters F_1 and F_2 play a fundamental role in determining the frequency difference between first sound and "zero sound" (see Subsection 9.1). As we will see in the following, these effects are relevant also for the electron gas, when the density is low and correlations are more important.

In the parlance of density functional theory, this means that one is not allowed to neglect nonlocal (current) terms in the energy functional if one wants to reproduce correctly the dynamic response of the system. Current terms in the energy functional, which have no effects on the static properties of the system, may nevertheless be crucial for the dynamic properties.

In what follows we will discuss two approaches which take account of nonlocal effects in the effective interaction, through the introduction of a velocity-dependent term. The effects of such a term will be studied in connection with the problem of correctly reproducing the dispersion of phonon–roton collective states in superfluid ^{4}He, and of the plasmon ones in the electron gas.

The importance of nonlocal effects for obtaining such dispersions can be easily understood by studying the moments

$$m_k(q) = \int_0^\infty d\omega\, \omega^k S(q,\omega) \tag{10.55}$$

of the dynamic structure function $S(q,\omega)$. In the case of Bosons, such a function, at small ω, is determined primarily by collective modes; in Fermi systems by collective modes and one-particle–one-hole excitations; and in the region of high ω by the many-particle–many-hole excitations (mp–mh). The situation is schematized in Fig. 10.5. The mean field theories discussed so far account only for single-particle

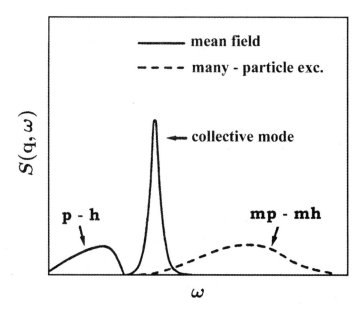

Fig. 10.5 Schematic representation of $S(q,\omega)$. For Bosons, the part corresponding to one-particle–one-hole (p–h) excitations should be ignored.

and collective excitations, but not for excitations due to many-particle, and hence they provide an incomplete description of $S(q, \omega)$. This can be seen explicitly by comparing, for example in the case of liquid ^{3}He, the f-sum rule

$$m_1(q) = \int_0^\infty d\omega \, \omega S(q, \omega) = \frac{Nq^2}{2m}, \tag{10.56}$$

which holds for interaction potentials that do not depend on velocity (for both density and spin-density excitations), with the analogous. sum rule for the spin-density operator in the Landau theory [see Eq. (9.25)]:

$$m_1^a(q) = \frac{Nq^2}{2m} \frac{1 + F_1^a/3}{1 + F_1^s/3}.$$

The result of the Landau theory ($F_1^a = -0.55$, $F_1^s = 6$ for ^{3}He) is much smaller than the exact result of (10.56), and this means that the elementary excitations considered by the Landau theory do not saturate the f-sum rule even at small q. The many-particle excitations also contribute to m_1^a at small q, thanks to the factor in ω in the f-sum rule, which amplifies the contribution of such high energy excitations. A different situation is realized in the case of the inversely energy-weighted sum rule:

$$m_{-1}(q) = \int_0^\infty d\omega \frac{S(q, \omega)}{\omega} = \frac{-\chi(q)}{2}, \tag{10.57}$$

where $\chi(q)$ is the static response of the system. Due to the factor ω^{-1} in the integral (10.57), which quenches the contribution of many-particle excitations, such a sum rule is saturated at low q by the collective mode. In the case of superfluid ^{4}He, this is the case even at rather large q values, as is evident from Fig. 9.8, where the experimental results for $\chi(q)$ are well reproduced by the density functional theory of Dalfovo *et al.* (1995), where $\chi(q)$ is given by the second functional derivative of energy in momentum space (V is the volume):

$$-\frac{N}{\chi(q)} = \frac{q^2}{4m} + \frac{\rho}{V} \int d\mathbf{r} d\mathbf{r}' \frac{\delta^2 E}{\delta\rho(\mathbf{r})\delta\rho(\mathbf{r}')} e^{i\mathbf{q}\cdot(\mathbf{r}-\mathbf{r}')}. \tag{10.58}$$

For Fermions, such as the electron gas, the static response can be computed starting from the expression

$$-\frac{V}{\chi(q)} = f_{xc}(q) - \frac{V}{\chi_{RPA}(q)} = f_{xc}(q) - \frac{V}{\chi_0(q)} + v(q), \tag{10.59}$$

where the exchange-correlation factor $f_{xc}(q)$ is connected to the local field factor $G(q)$ by

$$f_{xc}(q) = -v(q)G(q), \tag{10.60}$$

and is given by the Fourier transform of the second functional derivative of the exchange-correlation energy with respect to density:

$$f_{\text{xc}}(q) = \frac{1}{V} \int d\mathbf{r} d\mathbf{r}' \frac{\delta^2 E_{\text{xc}}}{\delta\rho(\mathbf{r})\delta\rho(\mathbf{r}')} e^{i\mathbf{q}\cdot(\mathbf{r}-\mathbf{r}')}. \tag{10.61}$$

The m_1 and m_{-1} sum rules can be used to study the dispersion of the collective modes in both Boson and Fermion systems, through their ratio:

$$\omega^2(q) = \frac{m_1(q)}{m_{-1}(q)}. \tag{10.62}$$

In the following we will study such a ratio in the case of the phonon–roton density modes of superfluid ^{4}He, and of the plasmon mode of the electron gas and, in this framework, we will discuss the relevance of nonlocal effects in the energy functional.

In the case of superfluid ^{4}He, employing the exact results for m_1 and m_{-1}, (10.56) and (10.57), we obtain

$$\omega^2(q) = \frac{Nq^2}{m|\chi(q)|}, \tag{10.63}$$

which may be considered as a rigorous upper bound for the collective mode dispersion. Note that, since the following inequality holds,

$$\sqrt{\frac{m_1(q)}{m_{-1}(q)}} \leq \frac{m_1(q)}{m_0(q)} = \frac{q^2}{2mS(q)}, \tag{10.64}$$

where $S(q)$ is the static structure factor, expression (10.63) is a better approximation to the exact dispersion than the more renowned Feynman approximation

$$\omega^F(q) = \frac{q^2}{2mS(q)}. \tag{10.65}$$

Using for $\chi(q)$ the result of the density functional theory of (10.58), which reproduces the experimental value up to high enough q values, we obtain for the dispersion of the phonon–roton mode the dashed curve in Fig. 10.6. While the phonon dispersion $\omega = cq$, with $c^2 = -(m\chi(0))^{-1}$, is well reproduced at low q, the roton one at higher q values is largely overestimated. From the previous discussion, it is clear that the estimate (10.63) can be improved by including velocity-dependent terms in the energy functional, which change the energy-weighted sum rule m_1^{col} with respect to the exact result (10.56). The difference with respect to the exact result $m_1 - m_1^{\text{col}}$ is due to the contribution of many-particle excitations, which is correctly subtracted in the computation of the energy-weighted sum rule, owing to the collective excitations alone. Since the many-particle excitations do not affect the

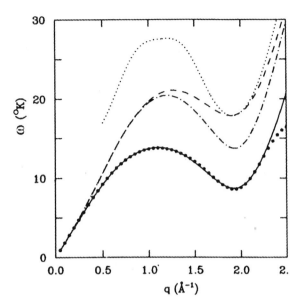

Fig. 10.6 Phonon–roton dispersion in superfluid ^{4}He. The dots are the experimental data of Donnelly *et al.* (1981); the dotted line is the Feynman approximation with the experimental value for the static form factor $S(q)$ (Svensson *et al.*, 1980); the dashed line is the result of Eq. (10.63) using, for χ, the result of the static density functional; the dotted–dashed line is the result of the dynamic theory without the backflow term, while the full line includes the backflow term.

inversely energy-weighted sum rule, the ratio $m_1^{\text{col}}/m_{-1}^{\text{col}}$ will turn out to be a better approximation for the exact dispersion of the collective mode of superfluid ^{4}He. Dalfovo *et al.* (1995) added the velocity-dependent term to the energy functional

$$E_v = -\frac{m}{4}\int d\mathbf{r}\,d\mathbf{r}'V_J(|\mathbf{r}-\mathbf{r}'|)\rho(\mathbf{r})\rho(\mathbf{r}')(\mathbf{v}(\mathbf{r})-\mathbf{v}(\mathbf{r}'))^2, \qquad (10.66)$$

where V_J is a finite range current–current effective interaction, which plays the role of a nonlocal kinetic energy, and gives rise to hydrodynamic equations (see Chapter 11) with a backflow current term. From a microscopic point of view, an effective interaction with these characteristics (i.e. finite range and velocity-dependent) emerges naturally in the Brueckner–Hartree–Fock theory described in Chapter 3 and applies to the electron gas.

The interaction (10.66) lowers the value of the energy-weighted sum rule with respect to the exact value and yields, for the homogeneous system, the result

$$m_1^{\text{col}}(q) = \frac{Nq^2}{2m}(1 - \rho(\hat{V}_J(0) - \hat{V}_J(q))), \qquad (10.67)$$

where $\hat{V}_J(q)$ is the Fourier transform of the current–current interaction $V_J(r)$. However, it does not change the value of the static response, which is fixed by the part of the energy functional which depends on density and not on velocity. The dispersion of the phonon–roton collective mode is thus given by

$$\omega^2(q) = \frac{Nq^2}{m|\chi(q)|}(1 - \rho(\hat{V}_J(0) - \hat{V}_J(q))). \qquad (10.68)$$

It is possible to fix $V_J(r)$ in order to reproduce both the phonon–roton dispersion and the energy-weighted sum rule as derived from neutron scattering experiments in the energy range where only the collective contribution to the sum rule is important. The results are shown in Figs. 10.6 and 10.7. The fact that the new functional does not satisfy the exact f-sum rule, clearly evidences the fact that the density functional theory does not take account of many-particle excitations, but only one-particle ones, and that in order to reproduce the dispersion of such excitations it is crucial to introduce current–current interaction terms in the functional. Note that the new current–current term in the dispersion (10.68) does not alter the hydrodynamic limit $q \to 0$, given by $\omega = cq$, with $c = (m|\chi(0)|)^{-1}$ the sound velocity. This is an important fact, ensured by Galilean invariance.

The current–current term is also very important for reproducing correctly the dynamics of nonhomogeneous systems, such as free surfaces, thin films and helium drops. Calculations for such systems are found in Dalfovo *et al.* (1995).

In the case of the electron gas, the situation is similar to the above-described one, but with the further complication of one-particle–one-hole excitations which, together with the collective ones, are present in the low energy spectrum of Fermions. The situation for the plasmon dispersion is the one already discussed in

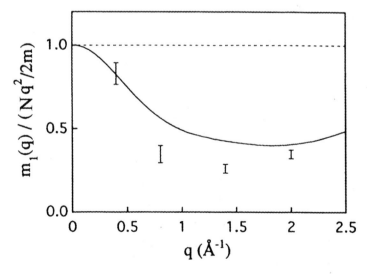

Fig. 10.7 Energy-weighted sum rule. The dashed line is the exact sum rule; the bars give the mean field contribution to the sum rule, as measured by neutron scattering (Woods and Cowley, 1973); the full line is the result of Eq. (10.67).

Subsections 9.4 and 9.9.1, and reported in Fig. 9.3. The experimental results are obtained by inelastic electron scattering experiments on alkali metals ranging from sodium to caesium. These metals represent, in nature, the physical system closest to the interacting electron gas in 3D, with a value of r_s varying from 4 to about 6. The local field theories predict for the α coefficient of the dispersion relation

$$E(q) = E(0) + \alpha q^2,$$

values which are in serious disagreement with the experimental ones for $r_s > 4$. In the following, using the sum rules discussed above, we will show that current-interaction effects (Lipparini, Stringari and Takayanagi, 1994) can improve the agreement between theory and experiment, even though other effects, such as those due to the band structure of alkali metals (Aryasetiawan and Karlsson, 1994; Taut and Sturm, 1992), appear to be also relevant for establishing a quantitative comparison with observation.

The dependence on the transferred momentum q of the density excitation strength $|\rho_{n0}|^2 = |\langle n|\rho_{\mathbf{q}}|0\rangle|^2$, of the excitation frequencies and of the sum rules, due to the possible excitations of the electron gas, can be deduced on the basis of simple arguments based on conservation laws and sum rules (Pines and Nozières, 1966), and is reported in Table 10.3, in the limit of small q, with the main contributions. In this table v_F and ϵ_F are the Fermi velocity and energy, respectively, and $\omega_p = \sqrt{4\pi\rho e^2/m}$ is the plasma frequency.

If, as we did in the case of superfluid helium, we neglect the contribution of many-particle (mp) excitations to the inversely energy-weighted sum rule, up to terms in q^4 we can write

$$m_{-1} = m_{-1}^{\text{pl}} + m_{-1}^{\text{sp}} \qquad m_1 = m_1^{\text{pl}} + m_1^{\text{mp}}. \tag{10.69}$$

If then we want to calculate the plasmon dispersion through the ratio $m_1^{\text{pl}}/m_{-1}^{\text{pl}}$, we clearly need to subtract from the exact results for m_{-1} and m_1 the single-particle and many-particle contributions, respectively. The exact result for m_1 is given by

Table 10.3. Matrix elements, excitation frequencies, and sum rule contributions of density excitations in the long wavelength limit.

	Single-particle	Plasmon	Many-particle
$\sum_n |\langle n|\rho_{\mathbf{q}}|0\rangle|^2$	q^5	$\dfrac{Nq^2}{2m\omega_p}$	q^4
ω_{n0}	qv_F	ω_p	ω_{mp}
m_{-1}	$\dfrac{4}{15}\dfrac{N\epsilon_F}{m^2\omega_p^4}q^4$	$\dfrac{Nq^2}{2m\omega_p^2}$	q^4
m_1	q^6	$\dfrac{Nq^2}{2m}$	q^4
m_3	q^8	$\dfrac{N\omega_p^2 q^2}{2m}$	q^4

the f-sum rule (10.56), and that for the inversely energy-weighted sum rule can be easily derived starting from Eq. (10.59), by performing the limit for small q and recalling the result (9.249). In this way we obtain the result, valid up to terms in q^4,

$$m_{-1} = \frac{Nq^2}{2m\omega_p}\left(1 - \frac{v^2 q^2}{\omega_p^2}\right),\tag{10.70}$$

where v^2 is the sound velocity. The single-particle contribution to m_{-1} can be evaluated rigorously (Lipparini, Stringari and Takayanagi, 1994) and is given by

$$m_{-1}^{\text{sp}} = \frac{4}{15}\frac{N\epsilon_F}{m^2\omega_p^4}q^4.\tag{10.71}$$

As regards the many-particle contribution to m_1, considering the exact result (10.56) as well as Eq. (10.69) and Table 10.3, one concludes that the corrections in q^4 due to the plasmon and many-particle excitations must cancel each other exactly. The plasmon q^4 contribution can be obtained only in nonlocal theories with finite range and velocity-dependent effective interactions. One such theory is the Brueckner–HF theory discussed in Chapter 3, in the separable approximation of Eq. (3.48), in which the effective interaction is given by $g^c = c(p)v(q)$. For the plasmon contribution to m_1 up to order q^4, it gives the result

$$m_1^{\text{pl}}(q) = \frac{Nq^2}{2m} - K\frac{Nq^4}{2m},\tag{10.72}$$

with

$$K = \frac{9}{8r_s^3}\left(v''(0)\int\left(\frac{j_1(k_F r/2)}{k_F r/2}\right)^2 c(r)r^4 dr\right.$$

$$\left. - \frac{1}{2}c''(0)\int\left(\frac{j_1(k_F r)}{k_F r}\right)^2 v(r)r^4 dr\right),\tag{10.73}$$

where $j_1(x) = 1/x(\sin x/x - \cos x)$ and $v''(0) = \frac{d^2}{dq^2}v(q)|_{q=0}$, $c''(0) = \frac{d^2}{dp^2}c(p)|_{p=0}$ and $v(r)$ and $c(r)$ are the Fourier transforms of $v(q)$ and $c(p)$, respectively. The values of K, together with those of the Fermi energy and plasmon frequency, are reported in atomic units in Table 10.4 for some values of r_s.

Table 10.4. Values of the coefficient K and of the muon multi-pair excitation energy ω_{mp} for different densities r_s. The various quantities are in atomic units.

r_s	ϵ_F	ω_p	K	ω_{mp}
2	0.460	0.613	0.05	1.52
3	0.205	0.333	0.3	0.68
4	0.115	0.217	0.5	0.48
5	0.074	0.155	1.0	0.30
6	0.051	0.118	2.0	0.20

The plasmon dispersion can then be calculated by means of the equation

$$E^{\mathrm{pl}}(q) = \sqrt{\frac{m_1^{\mathrm{pl}}}{m_{-1}^{\mathrm{pl}}}} = \omega_p \left(1 + \frac{q^2}{2m\omega_p^2} \left(mv^2 + \frac{8}{15}\epsilon_F - Km\omega_p^2 \right) \right), \qquad (10.74)$$

obtained using the results of (10.70)–(10.72). Note that putting $K = 0$ in this equation, and expressing the sound velocity through the second derivative of the LDA exchange-correlation energy with respect to the density, one obtains the TDLDA dispersion of (9.208).

From Eq. (10.74) we then obtain the following result for the coefficient α normalized to its RPA value $\alpha_{\mathrm{RPA}} = \frac{3\epsilon_F}{5\omega_p}$:

$$\frac{\alpha}{\alpha_{\mathrm{RPA}}} = \frac{5}{6\epsilon_F} \left(mv^2 + \frac{8}{15}\epsilon_F - Km\omega_p^2 \right). \qquad (10.75)$$

The predictions of such an equation are plotted in Fig. 10.8. The general trend indicated by the experiments is rather well reproduced. The term in K, which is due

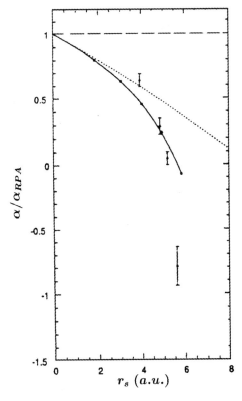

Fig. 10.8 Coefficient of the bulk plasmon dispersion, normalized to its RPA value, as a function of density. The full line is the result from Eq. (10.75). The dotted line gives the dispersion of local field theories. The experimental data are taken from Von Felde *et al.* (1989).

to nonlocal effects in the effective interaction, and which accounts for the plasmon q^4 contribution to the m_1 sum rule, plays a crucial role. In fact, in the absence of such ($K = 0$) we obtain the dotted curve, which is the prediction of local field theories.

10.9 Mean Energy of Many-Particle Excitations

Using the method of sum rules, in some cases it is possible to evaluate explicitly the value of the mean energy of the many-particle excitations through the ratio

$$\omega^{\mathrm{mp}} = \sqrt{\frac{m_3^{\mathrm{mp}}}{m_1^{\mathrm{mp}}}}, \tag{10.76}$$

where m_3^{mp} is the contribution of many-particle excitations to the sum rule weighted by the cube of energy: $m_3 = \int S(q,\omega)\omega^3 d\omega$.

In the case of the electron gas in 3D, m_3^{mp} can be computed using the result (see Table 10.3)

$$m_3^{\mathrm{mp}} = m_3 - m_1^{\mathrm{pl}}(E^{\mathrm{pl}}(q))^2, \tag{10.77}$$

which holds to order q^4, where m_3 is given by (Puff, 1965; Iwamoto, Krotscheck and Pines, 1984)

$$m_3 = \frac{Nq^2}{2m}\left(\omega_p^2 + \frac{2q^2}{m}\left(E_{\mathrm{kin}} + \frac{2}{15}E_{\mathrm{pot}}\right)\right), \tag{10.78}$$

and the second term on the right hand side of Eq. (10.77) is the plasmon contribution. In Eq. (10.78), E_{kin} and E_{pot} are the kinetic and potential energies per particle of the electron gas, whose values were computed by Monte Carlo. Using the results of the previous subsection, we can then write the mean energy of the many-particle excitations:

$$\omega^{\mathrm{mp}} = \sqrt{\frac{m_3^{\mathrm{mp}}}{m_1^{\mathrm{mp}}}}$$

$$= \sqrt{\frac{2}{mK}\left(\left(E_{\mathrm{kin}} + \frac{2}{15}E_{\mathrm{pot}}\right) - \left(\frac{1}{2}mv^2 + \frac{4}{15}\epsilon_F - Km\omega_p^2\right)\right)}. \tag{10.79}$$

The values predicted by this expression are reported in Table 10.4. As expected, ω^{mp} is systematically larger than E^{pl} and ϵ_F.

As a second example we consider the contribution of many-particle excitations to the spin-density response function of liquid ^{3}He. The sum rule cubic in energy for the spin-density operator $\rho_q^\sigma = \sum_j e^{i\mathbf{q}\cdot\mathbf{r}_j}\sigma_j^z$ is given by

$$m_3^\sigma = N\left(\frac{q^6}{8m^3} + \frac{q^4}{m^2}E_{\mathrm{kin}} + \frac{\rho q^2}{2m^2}\int[g(r) - g^\sigma(r)]\cos qz]\nabla_z^2 V(r)d\mathbf{r}\right), \tag{10.80}$$

where

$$g(r) = \rho^{-2} \sum_{\sigma_1, \sigma_2} \rho^{(2)}(1, 2; 1, 2),$$

$$g^{\sigma}(r) = \rho^{-2} \sum_{\sigma_1, \sigma_2} \sigma_1 \sigma_2 \rho^{(2)}(1, 2; 1, 2),$$

and $r = r_1 - r_2$, and $\rho^{(2)}(1, 2; 1, 2)$ is the diagonal two-body density matrix. $V(r)$ is the helium–helium interaction potential.

In the limit of small q, the m_3 sum rule becomes

$$\lim_{q \to 0} m_3^{\sigma} = N \frac{\rho q^2}{m^2} \int g_{+-}(r) \nabla_z^2 V(r) d\mathbf{r}, \qquad (10.81)$$

with $g_{+-} = 1/2(g(r) - g^{\sigma}(r))$, and in this limit is a pure effect of correlations in the ground state (Dalfovo and Stringari, 1989). In fact, the mean field theories based on the RPA, time-dependent density functional theory, or the Landau theory, only account for Pauli correlations in the ground state (a Slater determinant), and hence (see Subsection 1.8) they predict a constant value for g_{+-} ($= 2$). Therefore, the q^2 term in m_3^{σ} vanishes, consistent with the result (9.26) of the Landau theory. This result proves that in liquid ^{3}He, in the limit of small q, the m_3^{σ} sum rule is fully determined by many-particle effects.

The contribution of many-particle excitations to the f-sum rule can, then, be easily calculated in the same limit, through the difference

$$m_1^{\sigma}(\text{mp}) = m_1^{\sigma} - m_1^{\sigma}(\text{mf}) \to_{q \to 0} N \frac{\rho q^2}{m^2} \frac{1}{3} \frac{F_1^s - F_1^a}{1 + 1/3 F_1^s}, \qquad (10.82)$$

where for the mean field contribution to m_1^{σ} ($m_1^{\sigma}(\text{mf})$) in the limit of small q, we have taken the result of the Landau theory of (9.25). Combining the results of Eqs. (10.81) and (10.82) into (10.76), we can finally give an estimate of the mean energy of many-particle excitations in the spin-density channel, which turns out to be $\omega^{\text{mp}} \simeq 50$ K at saturated vapor pressure.

10.10   The Polarization Potential Model

A separation between mean-field and many-particle excitations in the framework of the linear response theory, of the same type as the one discussed above, has been developed by Pines (1985) for quantum liquids, and is known in the literature as the polarization potential model. In this theory, the linear responses in the density and spin-density channels are written as

$$\chi^s(q,\omega) = \frac{\chi^{sc}(q,\omega)}{1 - (v_c(q) + f_0^s(q) + \frac{\omega^2}{q^2} f_1^s(q))\chi^{sc}(q,\omega)}, \tag{10.83}$$

$$\chi^a(q,\omega) = \frac{\chi^{sc}(q,\omega)}{1 - (f_0^a(q) + \frac{\omega^2}{q^2} f_1^a(q))\chi^{sc}(q,\omega)}, \tag{10.84}$$

where $v_c(q)$ is nonvanishing only for charged systems (in which case it is equal to the Fourier transform of the Coulomb field), and $\chi^{sc}(q,\omega)$ is the screened response function which describes the response of density fluctuations to the combined action of the external and induced polarization fields. $\chi^{sc}(q,\omega)$ includes contributions from both single- and many-particle excitations. If we neglect many-particle contributions, then $\chi^{sc}(q,\omega) = \chi^0(q,\omega)$ (i.e. the free response) and the local mean field theories become a particular case of the Pines theory in which $f_1 = 0$.

The frequency-dependent term, which is proportional to f_1^s in (10.83), is connected to an effective mass depending on q, which characterizes the single-particle or pair excitations with momentum q, with which $\chi^{sc}(q,\omega)$ is built up. In fact, from Eq. (10.83) in the large ω limit (see Subsection 8.3), and from the results

$$\chi^s(q,\omega) = \frac{\rho q^2}{m\omega^2}, \qquad \chi^{sc}(q,\omega) = \frac{\rho q^2}{m_q^*\omega^2},$$

which hold in this limit, we obtain

$$m_q^* = m + \rho f_1^s(q). \tag{10.85}$$

On the other hand, in the spin channel, performing the high ω limit in (10.84), we obtain for the energy-weighted sum rule

$$m_1^a = \frac{\rho q^2}{2m^*} \frac{1}{1 - \rho f_1^a(q)/m^*}. \tag{10.86}$$

The polarization potential model reduces to the Landau theory in the limit $q \to 0$, $\omega \to 0$, provided the following identifications are made:

$$f_0^s(0) = \frac{F_0^s \pi}{m^*}, \qquad f_0^a(0) = \frac{F_0^a \pi}{m^*}, \tag{10.87}$$

$$f_1^s(0) = \frac{F_1^s}{1 + F_1^s/2} \frac{m^*\pi}{k_F^2}, \qquad f_1^a(0) = \frac{F_1^a}{1 + F_1^a/2} \frac{m^*\pi}{k_F^2}, \tag{10.88}$$

in the 2D case, and

$$f_0^s(0) = \frac{F_0^s \pi^2}{m^* k_F}, \qquad f_0^a(0) = \frac{F_0^a \pi^2}{m^* k_F}, \tag{10.89}$$

$$f_1^s(0) = \frac{F_1^s}{1 + F_1^s/3} \frac{m^*\pi^2}{k_F^3}, \qquad f_1^a(0) = \frac{F_1^a}{1 + F_1^a/3} \frac{m^*\pi^2}{k_F^3}, \tag{10.90}$$

in the 3D case, and provided that in the same limit we identify $\chi^{sc}(q,\omega)$ with $\Omega_{0,0}$ and $\Lambda_{0,0}$, respectively. Here $\Omega_{0,0}$ and $\Lambda_{0,0}$ are the long wavelength limits of the free response function in 3D and 2D, to excite single–quasiparticle–quasihole states of effective mass m^* around the Fermi surface (see Subsection 9.1). In Eqs. (10.87)–(10.90), F are the usual dimensionless Landau parameters. In particular, it can be remarked that from Eqs. (10.85)–(10.86) one obtains the usual results of the Landau theory, $m^*/m = 1 + F_1^s/2$ for the effective mass, and

$$m_1^a = \frac{\rho q^2}{2m^*}(1 + F_1^a/2) \qquad (10.91)$$

for the energy-weighted sum rule in the spin channel. Comparison of Eqs. (10.86) and (10.91) with the exact result $m_1^a = \rho q^2/2m$ shows once again that in the spin channel many-particle excitations contribute to m_1^a at order q^2.

One can notice that the adiabatic TDLSDA theory, which we developed in Subsection 9.9, can be generalized to include nonlocal components of the type

$$\frac{1}{2}t_1\int(\rho\tau - j^2)d\mathbf{r} \qquad (10.92)$$

in the energy functional, which yield a response that is analogous to that of the polarization potential theory in the long wavelength limit. In Eq. (10.92),

$$\tau = \left\langle \sum_i \mathbf{p}_i^2 \delta(\mathbf{r} - \mathbf{r}_i) \right\rangle$$

is the kinetic energy density, and j is the current density, defined by

$$\mathbf{j}(\mathbf{r}) = 1/2\left\langle\left(\sum_i \delta(\mathbf{r} - \mathbf{r}_i)\mathbf{p}_i + \text{H.c.}\right)\right\rangle. \qquad (10.93)$$

The form (10.92) ensures translational invariance. The coefficient t_1 is connected to $f_1^s(0)$, and hence to the Landau parameter F_1^s. For example, in 2D we have

$$t_1 = -\frac{1}{\rho}\frac{F_1^s/2}{1 + F_1^s/2},$$

as follows from the fact that the term of Eq. (10.92) produces an effective mass given by $m^*/m = (1 + t_1\rho)^{-1}$. The term of Eq. (10.92) gives rise (Stringari and Dalfovo, 1990) to a density–current response,

$$\chi_{sj} = \sum_n \left(\frac{(\rho_q)_{0n}(j_q^\dagger)_{n0}}{\omega - \omega_{n0}} - \frac{(\rho_q^\dagger)_{0n}(j_q)_{n0}}{\omega + \omega_{n0}}\right),$$

which is coupled to the density–density response χ_s by

$$\chi_s = [(v_c(q) + \mathcal{K})\chi_s + 1]\chi_s^0 - t_1\chi_{sj}\chi_{js}^0, \qquad (10.94)$$

where $\mathcal{K}$ is the second derivative of the exchange-correlation energy with respect to the density [see Eq. (9.196)], and χ_s^0 and χ_{js}^0 are the single-particle (with effective mass) density–density and density–current responses, respectively. Using the relations (Pines and Nozières, 1966)

$$\chi_{sj}^0 = \frac{m^*}{q}\omega\chi_s^0, \qquad (10.95)$$

$$\chi_{sj} = \frac{1}{q}\omega\chi_s, \qquad (10.96)$$

which link the density–density and density–current responses, we can establish an equation of the kind (10.83) with f_0^s and f_1^s independent of q and related to $\mathcal{K}$ and t_1. An analogous calculation can be carried out in the case of the spin-density–spin-density and spin-current–spin-density responses, which leads to an equation of the kind (10.84). In this case, all the quantities that appear in the above expressions have a spin dependence. For example, the spin current density $\mathbf{j}^\sigma$ becomes

$$\mathbf{j}^\sigma(\mathbf{r}) = 1/2\left\langle\left(\sum_i \delta(\mathbf{r}-\mathbf{r}_i)\mathbf{p}_i\sigma_i + \text{H.c.}\right)\right\rangle \qquad (10.97)$$

and the coefficient t_1^σ of the equation analogous to (10.92) is related to the Landau parameter F_1^a. For example, in 2D, we have

$$t_1^\sigma = -\frac{1}{\rho}\frac{F_1^a/2}{1+F_1^a/2}.$$

The polarization potential model has been applied, particularly, to the study of elementary excitations of ^{3}He and ^{4}He. An extensive review of the applications of such a model, together with a list of the most important references, can be found in Pines (1985).

10.11 The Gross–Kohn Model

In Subsection 9.9 we developed the time-dependent density functional theory in the so-called adiabatic approximation, in which one assumes for the exchange and correlation functional the same static form hypothesized to calculate the properties of the ground state of the system. In general, one expects this to be a good approximation only for processes that depend very slowly on time, i.e. one implicitly assumes the adiabatic approximation. This is reflected also in the fact that the exchange-correlation kernel of Eqs. (9.195) and (9.196) is frequency-independent. For homogeneous systems, and in the density channel, this kernel is given by

$$\lim_{q\to 0} f_{\text{xc}}^{\text{hom}}(q,\omega=0) = \frac{d^2}{d\rho^2}(\rho\epsilon_{\text{xc}}(\rho)) \equiv f_0(\rho), \qquad (10.98)$$

where $\epsilon_{xc}(\rho)$ is the exchange-correlation energy per particle of the homogeneous electron gas. $f_{xc}^{hom}(q, \omega = 0)$ was defined in (10.60) and (10.61). In general, f_{xc}^{hom} is a quantity that depends on both q and ω, and is related to the field factor $G(q, \omega)$ of (10.27) by

$$f_{xc}^{hom}(q, \omega) = -v_c(q)G(q, \omega). \tag{10.99}$$

This $f_{xc}^{hom}(q, \omega)$ has the following properties [see, for example Gross and Kohn, 1990]:

(1) $$\lim_{q \to 0} f_{xc}^{hom}(q, \omega = \infty) = -\frac{4}{5}\rho^{2/3}\frac{d}{d\rho}\left(\frac{\epsilon_{xc}(\rho)}{\rho^{2/3}}\right) + 6\rho^{1/3}\frac{d}{d\rho}\left(\frac{\epsilon_{xc}(\rho)}{\rho^{1/3}}\right) \tag{10.100}$$

$$\equiv f_\infty(\rho).$$

This follows from the sum rule cubic in energy.

(2) $$f_0(\rho) < f_\infty(\rho) < 0. \tag{10.101}$$

This follows from Monte Carlo calculations on $\epsilon_{xc}(\rho)$.

(3) $$\lim_{q \to \infty} f_{xc}^{hom}(q, \omega = 0) = -\frac{4\pi}{q^2}(1 - g(0)). \tag{10.102}$$

Here $g(r)$ is the pair correlation function (Shaw, 1970).

(4) $$\lim_{q \to \infty} f_{xc}^{hom}(q, \omega \neq 0) = -\frac{2}{3}\frac{4\pi}{q^2}(1 - g(0)). \tag{10.103}$$

This was shown by Niklasson (1974).

(5) $f_{xc}^{hom}(q, \omega)$ is a complex function that fulfills the symmetry relations

$$\text{Re } f_{xc}^{hom}(q, \omega) = \text{Re } f_{xc}^{hom}(q, -\omega),$$
$$\text{Im } f_{xc}^{hom}(q, \omega) = -\text{Im } f_{xc}^{hom}(q, -\omega). \tag{10.104}$$

Here, $f_{xc}^{hom}(q, \omega)$ is an analytic function of ω in the upper part of the complex plane and tends to a real function, $f_\infty(q)$ for $\omega \to \infty$ (Kugler, 1975). Hence, the function $f_{xc}^{hom}(q, \omega) - f_\infty(q)$ obeys the standard Kramers–Kronig relations. The imaginary part of $f_{xc}^{hom}(q, \omega)$ exhibits the high frequency behavior

$$\lim_{\omega \to \infty} \text{Im } f_{xc}^{hom}(q, \omega) = -\frac{c}{\omega^{3/2}} \tag{10.105}$$

for all values of $q < \infty$ (Holas and Singwi, 1989). As for c, its value in the limit of high density is known: $c = 23\pi/15$ (Holas and Singwi, 1989; Glick and

Long, 1971). In the same limit, the real part of $f_{\text{xc}}^{\text{hom}}(q,\omega)$ behaves like (Gross and Kohn, 1985)

$$\lim_{\omega \to \infty} \text{Re}\, f_{\text{xc}}^{\text{hom}}(q,\omega) = f_\infty(q) + \frac{c}{\omega^{3/2}}. \tag{10.106}$$

Since $c > 0$, the infinite frequency value f_∞ is approached from above. Due to relation (10.101), this implies that $\text{Re}\, f_{\text{xc}}^{\text{hom}}(q = 0, \omega)$ cannot grow monotonically from f_0 to f_∞.

All of the above considerations hold in 3D. Analogous results have been obtained in 2D by Holas and Singwi (1989) and Iwamoto (1984).

Gross and Kohn (1985) advanced a local approximation with respect to the space coordinates ($q = 0$) for the exchange-correlation kernel, which, however, includes an explicit frequency dependence. In the density channel, such an approximation for the kernel of the integral Eq. (9.205) is written as

$$K_{ss}^{\text{xc}}(\mathbf{r}_1, \mathbf{r}_2; \omega) = \delta(\mathbf{r}_1 - \mathbf{r}_2) f_{\text{xc}}^{\text{hom}}(q = 0, \omega; \rho)|_{\rho = \rho_0(\mathbf{r})}. \tag{10.107}$$

The frequency dependence of $\text{Im}\, f_{\text{xc}}^{\text{hom}}(q = 0, \omega)$ is approximated with a Padè-like interpolation between the high and low frequency limits:

$$\text{Im}\, f_{\text{xc}}^{\text{hom}}(q = 0, \omega) = \frac{a(\rho)\omega}{(1 + b(\rho)\omega^2)^{5/4}}, \tag{10.108}$$

where

$$\begin{aligned}
a(\rho) &= -c(\gamma/c)^{5/3}(f_\infty(\rho) - f_0(\rho))^{5/3}, \\
b(\rho) &= (\gamma/c)^{4/3}(f_\infty(\rho) - f_0(\rho))^{4/3},
\end{aligned} \tag{10.109}$$

where $\gamma = (\Gamma(1/4))^2/4\sqrt{2\pi}$, and f_0 and f_∞ are as given in (10.98) and (10.100), and $c = 23\pi/15$. For the exchange-correlation energy $\epsilon_{\text{xc}}(\rho)$, the parametrization of Vosko, Wilk and Nusair (1980) is employed. Subsequently, from the Kramers–Kronig relation, we get for the real part of $f_{\text{xc}}^{\text{hom}}(q = 0, \omega)$ the following result:

$$\begin{aligned}
\text{Re}\, f_{\text{xc}}^{\text{hom}}(q = 0, \omega) = f_\infty &+ \frac{a}{\pi s^2}\sqrt{\frac{8}{b}}\left(2E\left(\frac{1}{\sqrt{2}}\right) - \frac{1+s}{2}\Pi\left(\frac{1-s}{2}, \frac{1}{\sqrt{2}}\right)\right. \\
&\left. - \frac{1-s}{2}\Pi\left(\frac{1+s}{2}, \frac{1}{\sqrt{2}}\right)\right), \tag{10.110}
\end{aligned}$$

where E and Π are the complete second and third type elliptical integrals in the standard notation of Byrd and Friedman (1954), and $s^2 = 1 + b\omega^2$.

In Figs. 10.9 and 10.10 we show the real and imaginary parts of $f_{\text{xc}}^{\text{hom}}(q = 0, \omega)$, for the two density values corresponding to $r_s = 2$ and $r_s = 4$. From these figures, we note that frequency dependence grows with increasing r_s. In the $r_s \to 0$ limit, we find that the difference $f_\infty(\rho) - f_0(\rho)$ goes to zero like r_s^2, and that the depth of the minimum of $\text{Im}\, f_{\text{xc}}^{\text{hom}}(q = 0, \omega)$ still decreases proportionally to r_s^2.

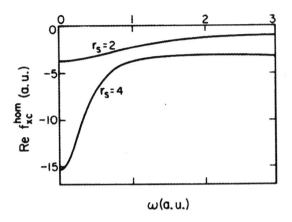

Fig. 10.9   Real part of the parametrization for $f_{\mathrm{xc}}^{\mathrm{hom}}(q,\omega)$.

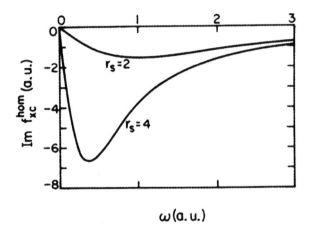

Fig. 10.10   Imaginary part of the parametrization for $f_{\mathrm{xc}}^{\mathrm{hom}}(q,\omega)$.

An extension of the parametrization (10.108) to the finite q case has been carried out by Dabrowski (1986), and the 2D case has been treated by Holas and Singwi (1989).

The Gross–Kohn theory, with the frequency-dependent parametrization described above, has been applied extensively to the calculation of the density response of a metallic surface to a uniform external electric field perpendicular to the surface itself [see for example Liebsch (1987), Kempa and Schaich (1988)]. With respect to RPA calculations, in which exchange and correlation effects are neglected, modifications are generally of the order of 10%.

Another application of the theory concerns the study of plasmon dispersion in the 3D and 2D interacting electron gas. An extensive discussion on such a problem, and on the connection between $f_{\mathrm{xc}}^{\mathrm{hom}}(q=0,\omega)$ and the spectrum of two-particles–two-holes excitations at lowest order in q, can be found in Bohm, Conti and Tosi (1996) and Nifosi, Conti and Tosi (1998), together with an extensive review of references on the problem.

Finally, we recall that, contrary to the adiabatic TDLDA, the theory of Gross and Kohn with the frequency-dependent parametrization does not fulfill the generalized Kohn theorem discussed in Subsection 8.9.1 (Dobson, 1994) when applied to the interacting electron systems confined in an external parabolic potential. This theorem guarantees the existence of a collective state at the same frequency as the harmonic potential, which corresponds to the rigid oscillation of the many-body wave function around the center of the external potential. This problem was further addressed by Vignale (1995) and by Vignale and Kohn (1996), who proposed a new theory for the exchange-correlation kernel. Such a theory has been described in detail by Vignale and Kohn (1997).

Another theory with the frequency-dependent parametrization, which satisfies the generalized Kohn theorem, is the one described at the end of Subsection 10.10 and synthetically represented by the Eqs. (10.92)–(10.96). This approach gives the result

$$f_{\text{xc}}^{\text{hom}}(q,\omega) = \mathcal{K} - \frac{m^*}{m} t_1 \frac{\omega^2}{q^2}.$$

10.12   The Method of Lorentz Transforms

Recently (Efros, Leideman and Orlandini, 1994), a new method has been proposed for the calculation of the dynamic form factor

$$S(F,\omega) = \sum_n |\langle n|F|0\rangle|^2 \delta(\omega - (E_n - E_0)).$$

This method uses the integral Lorentz transforms. It allows the calculation of $S(F,\omega)$ with no explicit knowledge of the set of excited states $|n\rangle$ (both discrete and continuous), which, in general, are very complicated.

The Lorentz transform of $S(F,\omega)$ is written as

$$\Phi(\sigma_R, \sigma_I, F) = \int_{\omega_{\min}}^{\infty} d\omega \frac{S(F,\omega)}{(\omega - \sigma_R)^2 + \sigma_I^2}. \tag{10.111}$$

Its advantage over the direct calculation of $S(F,\omega)$ lies in the resonant form of the kernel, which allows one to solve even very complex forms of $S(F,\omega)$ by choosing sufficiently small values of σ_I.

The method goes forward in two steps. First one computes $\Phi(\sigma_R, \sigma_I, F)$ by making use of the closure relationship $\sum_n |n\rangle\langle n| = 1$:

$$\Phi(\sigma_R, \sigma_I, F) = \langle 0|F^\dagger (H - E_0 - \sigma_R - i\sigma_I)^{-1} (H - E_0 - \sigma_R + i\sigma_I)^{-1} F|0\rangle$$

$$\equiv \langle \Psi|\Psi\rangle, \tag{10.112}$$

where E_0 is the energy of the ground state of the Hamiltonian H of the system, and $|\Psi\rangle$ is determined by the equation

$$(H - E_0 - \sigma_R + i\sigma_I)|\Psi\rangle = F|0\rangle. \tag{10.113}$$

Note that owing to (10.112), the norm of $|\Psi\rangle$ exists and so, contrary to the wave functions of states in continuum, $|\Psi\rangle$ cancels out at large distances like the wave functions of bounded states. This makes the solution of (10.113) much simpler than the solution of the equation for the excited states of the continuum. The solution of (10.113) is unique, since the corresponding homogeneous equation has no solution known owing to the complex nature of $\sigma = -\sigma_R + i\sigma_I$.

The second step consists in inverting the Lorentz transform (10.111), to obtain the dynamic form factor. In order that the dynamic form factor is obtained with good accuracy even at rather high frequencies, one needs an ad hoc procedure that has discussed in detail in Efros, Leideman and Orlandini (1999).

So far, the Lorentz transform method has been applied only in nuclear physics and for light nuclei up to a maximum of six nucleons. The results of such calculations for the photo disintegration and inelastic electron scattering cross-sections, as well as their comparison with experimental data, can be found in Efros, Leideman and Orlandini (1997).

References

Agosti, D., F. Pederiva, E. Lipparini and K. Takayanagi, *Phys. Rev.* **B57**, 14869 (1998).

Aryasetiawan, F. and K. Karlsson, *Phys. Rev. Lett.* **73**, 1679 (1994).

Bohm, H.M., S. Conti and M.P. Tosi, *J. Phys. Cond. Matter* **8**, 781 (1996).

Byrd, P.F. and M.D. Friedman, *Handbook of Elliptical Integrals for Engineers and Physicists* (Springer-Verlag, Berlin, 1954).

Camels, L. and A. Gold, *Phys. Rev.* **B52**, 10841 (1995); **P56**, 1762 (1997).

Chihara, J., *Prog. Theor. Phys.* **50**, 409 (1973).

Chihara, J. and K. Sasaki, *Prog. Theor. Phys.* **62**, 1533 (1979).

Choquard, Ph., in *Strongly Coupled Plasmas*, ed. G. Kalman (Plenum, New York, 1978), pp. 347.

Cohen, E.G.D., *Fundamental Problems in Statistical Mechanics* (North-Holland, Amsterdam, 1962).

Dabrowski, B., *Phys. Rev.* **B34**, 4989 (1986).

Dalfovo, F., A. Lastri, L. Pricaupenko, S. Stringari and J. Treiner, *Phys. Rev.* **B52**, 1193 (1995).

Dalfovo, F. and S. Stringari, *Phys. Rev. Lett.* **63**, 532 (1989).

Dobson, J.F., *Phys. Rev. Lett.* **73**, 2244 (1994).

Donnelly, R.J., J.A. Donnelly and R.N. Hills, *J. Low Temp. Phys.* **44**, 471 (1981).

Efros, V.D., W. Leidemann and G. Orlandini, *Phys. Lett.* **B338**, 130 (1994); *Few-Body Systems* **26**, 251 (1999); *Phys. Rev. Lett.* **78**, 432 (1997), 4015 (1997).

Fantoni, S. and S. Rosati, *Nuovo Cimento* **A10**, 145 (1972); **A25**, 593 (1975).

Fisher, D.S. and P.C. Hohenberg, *Phys. Rev.* **B37**, 4936 (1988).

Freser, W.L. and J. Bergersen, *J. Phys.* **C13**, 6627 (1980).

Gell-Mann, M. and K.A. Brueckner, *Phys. Rev.* **106**, 364 (1957).

Glick, A.J. and W.F. long, *Phys. Rev.* **B4**, 3455 (1971).

Gross, E.K.U. and W. Kohn, *Phys. Rev. Lett.* **55**, 2850 (1985).

Gross, E.K.U. and W. Kohn, *Adv. Quant. Chem.* **21**, 255 (1990).

Hines, D.F., N.E. Frankel and D.J. Mitchell, *Phys. Lett.* **A68**, 12 (1978).

Holas, A. and S. Singwi, *Phys. Rev.* **B40**, 158 (1989).

Huang, K. and C.N. Yang, *Phys. Rev.* **105**, 767 (1957).

Ichimaru, S., *Rev. Mod. Phys.* **54**, 1017 (1982).

Isihara, A. and T. Toyoda, *Ann. Phys. (N.Y.)* **106**, 394 (1977); **114**, 497 (1978).

Iwamoto, N., *Phys. Rev.* **A30**, 2597 (1984); 3289 (1984).

Iwamoto, N. and W. Gross, *Phys. Rev.* **B35**, 3003 (1987).

Iwamoto, N., E. Krotscheck and D. Pines, *Phys. Rev.* **B29**, 3936 (1984).

Jonson, M., *J. Phys.* **C9**, 3055 (1976).

Kempa, K. and W.L. Schaich, *Phys. Rev.* **B37**, 6711 (1988).

Kugler, A.A., *J. Stat. Phys.* **12**, 35 (1975).

Landau, L.D. and E.M. Lifchitz, *Fisica Statistica*, Teoria dello Stato Condensato (Editori Riuniti, Edizioni Mir, 1981).

Lee, T.D., K.W. Huang and C.N. Yang, *Phys. Rev.* **106**, 1135 (1957).

Lieb, E.H. and J. Yngvason, *Phys. Rev. Lett.* **80**, 2504 (1998).

Lieb, E.H. and J. Yngvason, *J. Stat. Phys.* **103**, 509 (2001).

Liebsch, A., *Phys. Rev.* **B36**, 7378 (1986).

Likos, C.N., S. Moroni and G. Senatore, *Phys. Rev.* **B55**, 8867 (1997).

Lipparini, E., S. Stringari and K. Takayanagi, *J. Phys. Cond. Matt.* **6**, 2025 (1994).

Mahan, G.D., *Many-Particle Physics* (Plenum, New York, London, 1981).

Mahaux, C., P.F. Bortignon, R.A. Broglia and C.H. Dasso, *Phys. Rep.* **120**, 1 (1985).

Morita, T., *Prog. Theor. Phys.* **23**, 829 (1960).

Moroni, S., D.M. Ceperley and G. Senatore, *Phys. Rev. Lett.* **69**, 1837 (1992).

Nifosi, R., S. Conti and M.P. Tosi, *Phys. Rev.* **B58**, 12758 (1998).

Niklasson, G., *Phys. Rev.* **B10**, 3052 (1974).

Pederiva, F., E. Lipparini and K. Takayanagi, *Europhys. Lett.* **40**, 607 (1997).

Pines, D., in *Pro. Int. Sch. Physics (E. Fermi)*, Course LXXXIX, eds. F. Bassani, F. Fumi and M. Tosi (North-Holland, Amsterdam, 1985).

Pines, D. and P. Nozières, *The Theory of Quantum Liquids* (Benjamin, New York, 1966).

Popov, V.N., *Theor. Math. Phys.* **11**, 565 (1977).

Puff, R.D., *Phys. Rev.* **A137**, 406 (1965).

Rajagopal, A.K. and J.C. Kimball, *Phys. Rev.* **B15**, 2819 (1977).

Reinhard, P.G., *Phys. Lett.* **A169**, 281 (1992).

Rice, T.M., *Ann. Phys.* **31**, 100 (1965).

Ring, P. and P. Schuck, *The Nuclear Many-Body Problem* (Springer-Verlag, New York, 1980).

Ripka, G., *Phys. Rep.* **56**, 1 (1979).

Schick, M., *Phys. Rev.* **A3**, 1067 (1971).

Schulze, H.J., P. Schuck and N. Van Giai, *Phys. Rev.* **B61**, 8026 (2000).

Serra, Ll., R.G. Nazmitdinov and A. Puente, *Phys. Rev.* **B68**, 035341 (2003).

Shaw, R.W., *J. Phys.* **C3**, 1140.

Singwi, K.S., M.P. Tosi, R.H. Land and A. Sjolander, *Phys. Rev.* **176**, 589 (1968).

Stringari, S. and F. Dalfovo, *J. Low Temp. Phys.* **78**, 1 (1990).

Svensson, E.C., V.F. Sears, A.D.B. Woods and P. Martel, *Phys. Rev.* **21**, 3638 (1980).

Takayanagi, K. and E. Lipparini, *Phys. Rev.* **B52**, 1738 (1995).

Takayanagi, K. and E. Lipparini, *Phys. Rev.* **B54**, 8122 (1996).

Tanatar, B. and D.M. Ceperley, *Phys. Rev.* **B39**, 5005 (1989).

Taut, M. and K. Sturm, *Sol. State Commun.* **82**, 295 (1992).

Ting, C.S., T.K. Lee and J.J. Quinn, *Phys. Rev. Lett.* **34**, 870 (1975).

van Leeuwen, J.M.J., J. Groeneveld and J. De Boer, *Physica* **25**, 792 (1959).

Vignale, G., *Phys. Rev. Lett.* **74**, 3233 (1995); *Phys. Lett.* **A209**, 206 (1995).

Vignale, G. and W. Kohn, *Phys. Rev. Lett.* **77**, 2037 (1996).

Vignale, G. and W. Kohn, in *Electronic Density Functional Theory*, eds. J. Dobson, M.P. Das and G. Vignale (Plenum, New York, 1997).

vom Felde, A., J. Sprosser-Prou and J. Fink, *Phys. Rev.* **B40**, 10181 (1989).

Vosko, S.H., L. Wilk and M. Nusair, *Can. J. Phys.* **58**, 1200 (1980).

Woods, A.D.B. and R.A. Cowley, *Rep. Prog. Phys.* **36**, 1135 (1973).

Chapter 11

The Hydrodynamic and Elastic Models

11.1 The Hydrodynamic Model for Bosons

A simple way to derive the hydrodynamic equations for a system of Bosons at zero temperature is to extend the formalism of the (Hartree–Fock) density functional used in the static case, to the time-dependent problem. In the time-dependent case, one assumes that the single-particle wave function φ relative to the Bose condensate evolves in time with a phase that breaks its time reversal invariance, and generates velocity-dependent terms in the energy functional $E(\rho, \nabla\rho) = \int d\mathbf{r}\rho\epsilon(\rho, \nabla\rho)$. This single-particle wave function has the form

$$\varphi(\mathbf{r}, t) = \psi(\mathbf{r}, t)e^{is(\mathbf{r},t)}, \tag{11.1}$$

where both ψ and s are real functions. With this assumption, the system density is still given by

$$\rho = N\psi^2, \tag{11.2}$$

but the kinetic energy density $\tau = N\nabla\varphi^*\nabla\varphi$ becomes

$$\tau = \frac{1}{4}\frac{(\nabla\rho)^2}{\rho} + \rho(\nabla s)^2. \tag{11.3}$$

From Eq. (11.3) we see that the phase s determines the velocity of the fluid through the relationship

$$\mathbf{v} = \left(\frac{1}{m}\right)\nabla s. \tag{11.4}$$

In the calculation of the ground state, only zero-velocity states are considered, so that the energy is a functional of density alone, and from this we derive Hartree–Fock-like equations by the variational method. On the contrary, in the dynamic case the action of the system

$$I = \int dt \langle \Phi | H - \mu - i\frac{\partial}{\partial t} | \Phi \rangle$$

$$= \int dt \int d\mathbf{r} \left(\rho\epsilon(\rho, \nabla\rho) + \frac{1}{2m}\rho(\nabla s)^2 - \mu\rho + \rho\frac{\partial}{\partial t}s \right), \qquad (11.5)$$

where $\epsilon(\rho, \nabla\rho)$ is the same energy functional as in the static case, depends explicitly on velocity, and yields equations which describe the motion of the condensate. In Eq. (11.5), the term $\mu\rho$ ensures that the density stays normalized to the number of particles. The variation of the action integral with respect to the phase s leads to the continuity equation

$$\frac{\partial}{\partial t}\rho + \nabla(\mathbf{v}\rho) = 0, \qquad (11.6)$$

while the variation with respect to ρ leads to the following equations for s:

$$\frac{\partial}{\partial t}s + \frac{\partial}{\partial \rho}(\rho\epsilon(\rho, \nabla\rho)) - \mu + \frac{1}{2m}(\nabla s)^2 = 0. \qquad (11.7)$$

Performing the gradient of (11.7), we then obtain

$$m\frac{\partial}{\partial t}\mathbf{v} + \nabla\left(\frac{\partial}{\partial \rho}(\rho\epsilon(\rho, \nabla\rho)) + m\frac{v^2}{2}\right) = 0. \qquad (11.8)$$

Eq. (11.8) shows the irrotational nature of the superfluid motion.

For a system of N weakly interacting Bosons in a magnetic trap, Eqs. (11.8) can be derived directly from the time-dependent Gross–Pitaevskii equations by writing the order parameter Ψ of (9.122) as

$$\Psi(\mathbf{r}, t) = \sqrt{\rho(\mathbf{r}, t)}e^{is(\mathbf{r}, t)}.$$

In this case we obtain

$$m\frac{\partial}{\partial t}\mathbf{v} + \nabla\left(v_{\text{ext}} + g\rho - \frac{1}{2m\sqrt{\rho}}\nabla^2\sqrt{\rho} + m\frac{v^2}{2}\right) = 0, \qquad (11.9)$$

which is the equation corresponding to (11.8). The continuity equation is unchanged. For a harmonic confinement field, such equations hold when the condition $Na/a_{ho} \gg 1$ is satisfied (see Dalfovo et al., 1999). In this case the frequencies which are the solutions to (9.125) tend to an asymptotic value which corresponds to the collisionless hydrodynamic limit.

Equation (11.8), once linearized, takes the form

$$\frac{\partial^2}{\partial t^2}\delta\rho = \frac{1}{m}\nabla \cdot \left(\rho_0 \frac{\partial^2 \rho\epsilon}{\partial \rho^2}\bigg|_{\rho=\rho_0}\nabla\delta\rho\right), \qquad (11.10)$$

where we have written $\rho(\mathbf{r}, t) = \rho_0(\mathbf{r}) + \delta\rho(\mathbf{r}, t)$ and ρ_0 is the density of the condensate in the ground state. Eqs. (11.6) and (11.10) have the classical form of the equations of hydrodynamics.

In the case of Bosons the formulation of the equations of motion using the conjugate variables ρ and s, as described above, is fully equivalent to the zero-temperature, time-dependent density functional theory. In this theory (see also Subsection 9.9), starting from the action integral and from an energy functional $E(\rho, \nabla\rho)$, and performing variations with respect to the wave function ϕ or ϕ^*, which appears in the density, one finds a Schrödinger-like equation of the form $(H - \mu)\phi = i\frac{\partial}{\partial t}\phi$, where $H = \frac{\delta E}{\delta\phi^*}$ is an effective Hamiltonian. If we linearize the equation by writing $\phi(\mathbf{r}, t) = \phi_0(\mathbf{r}) + \delta\phi(\mathbf{r}, t)$, where $\phi_0(\mathbf{r})$ is related to the ground state, then the Hamiltonian H assumes the form $H = H_0 + \delta H$, where H_0 is the static Hamiltonian that determines the ground state, and δH (linear in $\delta\phi$) takes account of the variations of H induced by the collective motion of the system. The self-consistent solution of

$$(H_0 - \mu)\delta\phi + \delta H\phi_0 = i\frac{\partial}{\partial t}\delta\phi$$

is then completely equivalent to the solution of the hydrodynamic equations.

In homogeneous systems, remembering that the speed of sound is related to the compressibility K by Eq. (1.83) ($c^2 = 1/(Km\rho_0)$), and that the compressibility is given by the expression

$$\frac{1}{K} = \rho_0^2 \frac{\partial^2 \rho\epsilon}{\partial \rho^2}\bigg|_{\rho=\rho_0},$$

we obtain

$$\frac{\partial^2}{\partial t^2}\delta\rho = \nabla \cdot (c^2\nabla\delta\rho), \tag{11.11}$$

which immediately leads us to the phonon-like dispersion law: $\omega = cq$.

In the presence of a surface, the equations of hydrodynamics have a class of solutions different from phonons which, in the literature, are known as ripplons [see for example Pricaupenko and Treiner (1994)]. Ripplons are characterized by the dispersion relation

$$\omega^2 = \frac{\sigma}{m\rho_0}q^3, \tag{11.12}$$

where σ is the surface tension of the quantum liquid, q is the momentum parallel to the surface, and ρ_0 is the saturation density.

11.1.1 *Backflow*

In Subsection 10.8 we discussed the nonlocal effects on the dynamic properties of superfluid ^{4}He, adding to the energy functional the velocity-dependent term [see Eq. (10.66)]:

$$E_v = -\frac{m}{4} \int d\mathbf{r} d\mathbf{r}' V_J(|\mathbf{r} - \mathbf{r}'|)\rho(\mathbf{r})\rho(\mathbf{r}')(\mathbf{v}(\mathbf{r}) - \mathbf{v}(\mathbf{r}'))^2,$$

where V_J is an effective, finite range current–current interaction which plays the role of a nonlocal kinetic energy. This term transforms the equations of hydrodynamics (11.6) and (11.7) into

$$\frac{\partial}{\partial t}\rho + \nabla \cdot \rho \left(\mathbf{v} + \int d\mathbf{r}' V_J(|\mathbf{r} - \mathbf{r}'|)\rho(\mathbf{r}')(\mathbf{v}(\mathbf{r}) - \mathbf{v}(\mathbf{r}')) \right) = 0, \qquad (11.13)$$

$$\frac{\partial}{\partial t}s + \frac{\partial}{\partial \rho}(\rho\epsilon(\rho, \nabla\rho)) - \mu$$

$$+ \frac{m}{2}\left((\mathbf{v})^2 - \int d\mathbf{r}' V_J(|\mathbf{r} - \mathbf{r}'|)\rho(\mathbf{r}')(\mathbf{v}(\mathbf{r}) - \mathbf{v}(\mathbf{r}'))^2 \right) = 0. \qquad (11.14)$$

Eq. (11.13) leads to the conserved current

$$\mathbf{j}(\mathbf{r}) = \mathbf{j}_0(\mathbf{r}) + \mathbf{j}_B(\mathbf{r}), \qquad (11.15)$$

where $\mathbf{j}_0(\mathbf{r}) = \rho\mathbf{v}$, and

$$\mathbf{j}_B(\mathbf{r}) = \rho \int d\mathbf{r}' V_J(|\mathbf{r} - \mathbf{r}'|)\rho(\mathbf{r}')(\mathbf{v}(\mathbf{r}) - \mathbf{v}(\mathbf{r}')). \qquad (11.16)$$

$\mathbf{j}_B(\mathbf{r})$ is the "backflow" current, which depends on velocity and on the density near point $(\mathbf{r})$. It is evident that its contribution vanishes when the effects of the medium are not present ($\rho \to 0$).

The backflow term has been used, among others, by Dalfovo *et al.* (1995) to study the dynamics of free surfaces and thin films of ^{4}He.

11.1.2 *Compression and Surface Modes of Spherical Drops*

Eqs. (11.6) and (11.10) have been applied to study drops of ^{4}He atoms, by Casas and Stringari (1990) and Barranco and Hernandez (1994), and ^{4}He films by Pricaupenko and Treiner (1994).

In the case of ^{4}He drops, which we consider here, one has both bulk excitations (compression modes) and surface excitations. For spherical drops, both kinds of excitations are classified according to their angular momentum ℓ, which plays the role of the wave vector q, which characterizes homogeneous systems and free surfaces, and one looks for solutions to the hydrodynamic equations in the form

$$\delta\rho(\mathbf{r}, t) = \rho(\mathbf{r}, t) - \rho_0(\mathbf{r}) = \delta\rho(r, t)y_{\ell m}(\theta, \phi). \qquad (11.17)$$

Surface excitations exist only for $\ell > 1$. In fact, the monopole excitation with $\ell = 0$ exists only as a compression mode, and the surface mode with $\ell = 1$ corresponds to a rigid translation of the cluster which, due to translational invariance of the system, takes place at zero energy.

For very large drops, so that we can assume that the number N of atoms in the drop is large, the compression modes can be obtained by solving Eq. (11.11) in a sphere of homogeneous liquid ($\rho = \text{const}$) having radius $R_0 = r_0 N^{1/3}$. In this case, Eq. (11.11) has the solution

$$\delta\rho(\mathbf{r}, t) = \text{constant} \cdot j_\ell(qr) y_{\ell m}(\theta, \phi) e^{-i\omega t}, \qquad (11.18)$$

where $j_\ell(qr)$ is the spherical Bessel function and the frequency ω is given by the usual dispersion relation,

$$\omega = cq,$$

where q is determined by the boundary condition that the pressure produced by the density variation is zero at the sphere surface [see Eq. (4.88)]:

$$\delta P(r = R_0) = \frac{1}{\rho K}\delta\rho(r = R_0) = 0. \qquad (11.19)$$

Such a condition leads to the relationship

$$j_\ell(qR_0) = 0, \qquad (11.20)$$

and hence to the values $qR_0 = 3.14, 4.49, 5.76$ for the lowest energy modes with $\ell = 1, 2, 3$, respectively. In the case of $\ell = 0$, the lowest energy solution is obtained for $qR_0 = \pi$, and the compression monopole mode has an energy

$$\omega_{\ell=0} = \frac{\pi}{r_0} cN^{-1/3} = 25.6N^{-1/3}\ {}^\circ\text{K}, \qquad (11.21)$$

where for r_0 and c we have used the values of Subsection 4.9. Note that this formulation completely neglects surface and finite size effects, and as such it can be applied only to clusters with a really large number of atoms ($N \geq 500$). For smaller clusters, one needs to solve directly the differential equations of hydrodynamics.

The surface vibrations of the helium drop can be described by a set of normal coordinates $\alpha_{\ell m}$ of the system in the space of angular momentum, which are obtained by expanding the surface in spherical harmonics (Bohr and Mottelson, 1975):

$$R(\theta, \phi) = R_0 \left(1 + \sum_{\ell m} \alpha_{\ell m} y_{\ell m}^*(\theta, \phi)\right), \qquad (11.22)$$

where R_0 is the radius at equilibrium, and $R(\theta, \phi)$ is the distance between the surface and the origin. Volume conservation implies the relation

$$\alpha_0 = -\frac{1}{4\pi}\sum_{\ell m}|\alpha_{\ell m}|^2, \qquad (11.23)$$

which holds at leading order. Furthermore, from Eq. (11.22) it is easy to verify what was mentioned above, i.e. that the term with $\ell = 0$ represents a compression without change of shape, and the one with $\ell = 1$ is a translation of the whole system. Therefore, the surface vibrations of lowest order are the quadrupole ones.

In the limit of small oscillations it is possible to use a Lagrangian formulation,

$$L = \frac{1}{2} \sum_{\ell m} (D_\ell \dot{\alpha}_{\ell m}^* \dot{\alpha}_{\ell m} - C_\ell \alpha_{\ell m}^* \alpha_{\ell m}), \qquad (11.24)$$

which allows us to derive the equations of motion of the surface oscillations,

$$D_\ell \ddot{\alpha}_{\ell m} + C_\ell \alpha_{\ell m} = 0, \qquad (11.25)$$

where D_ℓ and C_ℓ are the mass parameter and the restoring force parameter of the oscillation, respectively. The equation for the oscillation frequencies is given by

$$\omega_\ell^{\text{sup}} = \sqrt{\frac{C_\ell}{D_\ell}}. \qquad (11.26)$$

The parameter C_ℓ of the restoring force of surface oscillations in a helium drop described by the mass formula (4.99) is computed starting from the variation of surface energy

$$\delta E = \frac{1}{2} \sum_{\ell m} C_\ell |\alpha_{\ell m}|^2 = \sigma \delta S, \qquad (11.27)$$

where δS is the surface variation, and σ $(a_s = 4\pi r_0^2 \sigma)$ is the surface tension of the drop. By means of standard techniques described by Bohr and Mottelson (1975) to evaluate δS as a function of the deformation coefficients $\alpha_{\ell m}$, one obtains

$$C_\ell = (\ell - 1)(\ell + 2)R_0^2 \sigma. \qquad (11.28)$$

In order to calculate the mass parameter D_ℓ, one needs to make an assumption as to the kind of flux associated with the surface oscillations. For uncompressible and irrotational fluid, for which the velocity field is the gradient of a potential $f(\mathbf{r})$, $\mathbf{v}(\mathbf{r}) = -\nabla f(\mathbf{r})$, the continuity Eq. (11.6) demands that the potential fulfill the Laplace equation $\nabla^2 f(\mathbf{r}) = 0$, with solution

$$f(\mathbf{r}) = \sum_{\ell m} a_\ell r^\ell y_{\ell m}^*(\theta, \phi). \qquad (11.29)$$

The coefficients a_ℓ and D_ℓ are connected by the surface boundary condition, which requires that on the surface the radial component of the velocity be equal to the radial velocity of surface displacement. For small displacements of the surface, such a boundary condition requires

$$-\frac{\partial}{\partial r} f(r = R_0) = \frac{dR}{dt}, \qquad (11.30)$$

from which it follows that

$$a_\ell = \frac{R_0^{2-\ell}}{\ell} \dot{\alpha}_{\ell m}. \tag{11.31}$$

Then, the kinetic energy of the fluid relative to the surface deformations becomes

$$T = \frac{1}{2} M \rho_0 \int \mathbf{v}^2(\mathbf{r}) d\mathbf{r} = \frac{1}{2} M \rho_0 \sum_{\ell m} \frac{R_0^5}{\ell} \dot{\alpha}_{\ell m}^* \dot{\alpha}_{\ell m}, \tag{11.32}$$

and, at last, comparison with (11.24) yields the result

$$D_\ell = \frac{1}{\ell} M \rho_0 R_0^5, \tag{11.33}$$

where M is the mass of a helium atom, and $\rho_0 = 3/4\pi r_0^3$.

Therefore, the frequencies of the surface modes of the drop are given by

$$\omega_\ell^{\mathrm{sup}} = \sqrt{\frac{\ell(\ell-1)(\ell+2)}{3} \frac{a_s}{Mr_0^2}} N^{-\frac{1}{2}}. \tag{11.34}$$

Using the values of Subsection 4.9 for a_s and r_0, we obtain for the energy of the quadrupole mode, which is the lowest energy surface mode, the estimate $\omega_{\ell=2}^{\mathrm{sup}} = 10.5 \, N^{-\frac{1}{2}} \, °\mathrm{K}$. Comparing such an estimate with that for the energy of the monopole compression mode of (11.21), we see that the quadrupole surface mode has a different dependence on the number of particles, and it always lies at lower energy. Therefore, it is believed that this is the lowest energy elementary excitation of spherical ^{4}He drops. Note that, apart from the drops with a small number of atoms ($N \leq 100$), there are many surface modes at lower energy than the monopole compression mode. For example, for $N = 1000$ we have $\omega_{\ell=0} = 2.56 \, °\mathrm{K}$ and $\omega_\ell^{\mathrm{sup}} = 0.33, 065, 1.00, 1.40, 1.83, 2.29$ and $2.79 \, °\mathrm{K}$, for $\ell = 2, 3, 4, 5, 6, 7$ and 8, respectively.

Assuming that in a helium drop only surface vibrations are thermally excited, and using the result (11.34), Brink and Stringari (1990) evaluated the density of states $\omega(E)$ of helium clusters starting from the relation

$$Z(\beta) = \int dE \, \omega(E) e^{-\beta E}, \tag{11.35}$$

where Z is the partition function, which is computed by assuming that the surface vibrations are thermally excited with a phonon-like population:

$$\ln Z(\beta) = -\sum_\ell \ln(1 - e^{-\beta \omega_\ell}). \tag{11.36}$$

The result for the density of states was then used to compute the evaporation rate of helium drops using the Weisskopf formula (Weisskopf, 1937).

11.1.3 Compression and Surface Modes of a Bose Gas in a Magnetic Trap

Starting from (11.9), it is also possible to derive an equation similar to (11.11) for a cold and dilute Boson gas in a magnetic trap, for which the nonsuperfluid component of the system is negligible, when the repulsive interaction among atoms is strong enough to push them outward, thus producing a very flat density. In this case, it is safe to ignore the kinetic pressure term $\nabla^2 \sqrt{\rho}/2m\sqrt{\rho}$ in (11.9), giving

$$m\frac{\partial}{\partial t}\mathbf{v} + \nabla\left(v_{\text{ext}} + g\rho + m\frac{v^2}{2}\right) = 0, \tag{11.37}$$

and thus, after linearization, an equation of the form (11.11), with $mc^2(r) = \mu - v_{\text{ext}}$. Taking for v_{ext} a harmonic potential, we have

$$m\frac{\partial^2}{\partial t^2}\delta\rho = \nabla\cdot\left(\left(\mu - \frac{m}{2}\omega_{\text{ho}}^2 r^2\right)\nabla\delta\rho\right). \tag{11.38}$$

Note that the stationary solution ($\mathbf{v} = 0$) of Eq. (11.37) coincides with the Thomas–Fermi density (4.21).

The solutions of (11.38) that have a wavelength much smaller than the system size (and a frequency much higher than the frequency ω_{ho} of the trap) propagate like sound waves. On the other hand, those with a lower frequency (of the same order as ω_{ho}) involve the motion of the whole system (Baym and Pethick, 1996), and coincide with the lowest energy solutions to the Gross–Pitaevskii Eqs. (9.125). For a spherical trap, such solutions (Stringari, 1996b) are defined in the interval $0 \leq r \leq R$, where R is the r value at which the density (4.21) vanishes: $\mu = m\omega_{\text{ho}}^2 R^2/2$. Using for μ the value (4.22), we find that $R = a_{\text{ho}}(15Na/a_{\text{ho}})^{1/5}$. They have the form

$$\delta\rho(\mathbf{r}, t) = \text{const}\, P_\ell^{(2n_r)}(r/R) r^\ell y_{\ell m}(\theta, \phi) e^{-i\omega t}, \tag{11.39}$$

where $P_\ell^{(2n_r)}$ are polynomials of degree $2n_r$, with only even powers. The frequencies of the excitation modes are given by

$$\omega(n_r, \ell) = \omega_{\text{ho}}(2n_r^2 + 2n_r\ell + 3n_r + \ell)^{1/2}, \tag{11.40}$$

and turn out to be different from the ones predicted for the system without interaction:

$$\omega(n_r, \ell) = \omega_{\text{ho}}(2n_r + \ell). \tag{11.41}$$

The surface modes are characterized by $n_r = 0$. The dipole mode ($n_r = 0, \ell = 1$) is located at the correct energy ω_{ho}, in agreement with the generalized Kohn theorem. The remaining surface modes ($n_r = 0, \ell$) are predicted at energy $\omega = \sqrt{\ell}\omega_{\text{ho}}$, which is systematically lower than the harmonic oscillator result $\ell\omega_{\text{ho}}$.

The compression mode ($n_r \neq 0$) at lowest energy is the monopole mode ($n_r = 1, \ell = 0$). Characterized by the energy $\sqrt{5}\omega_{\text{ho}}$, it is higher than the corresponding one for the noninteracting system, $2\omega_{\text{ho}}$.

For fixed N, the results of the hydrodynamic model become less and less accurate as n_r and ℓ increase because the oscillation wavelength becomes smaller and smaller, and it is no longer possible to ignore the kinetic energy term in the hydrodynamic equations.

The above results can be easily generalized to the case of nonspherical traps (Stringari, 1996b).

It is important to note that in the case of homogeneous dilute and cold gases in 3D the equations of hydrodynamics are fully equivalent to the time-dependent Gross–Pitaevskii equations, since in 3D the Bose–Einstein condensation and hydrodynamics are equivalent. This is not the case for lower-dimensional systems where the Hohenberg, Mermin and Wagner (Mermin and Wagner, 1966; Hohenberg, 1967; Mermin, 1968) theorem holds, which rules out BEC at finite temperature, and in 1D also at zero temperature (Stringari and Pitaevskii, 1991). For these systems the BEC and the hydrodynamics do not coincide, and the equations of hydrodynamics are more general than those of mean field theories. In order to solve the equations of hydrodynamics, the state equation of the system is necessary. In the case of cold dilute gases where the interaction is δ-like, such an equation, in general, is known.

11.1.4 *Compression and Surface Modes of a Superfluid Trapped Fermi Gas*

The superfluidity of trapped Fermi gases occurs at a critical temperature much smaller than the quantum degeneracy temperature fixed by the Fermi energy (see Subsection 2.7) and its observation represents a major challenge from the experimental point of view as compared to the case of a dilute Bose gas. In fact, differently from the Boson case, where the phase transition is associated with the appearance of a sharp peak in the density profile, in the Fermi case, pairing effects do not influence significantly the ground state properties of the system and one has to explore other observables sensitive to susperfluidity.

At temperatures sufficiently below T_c, the system is fully superfluid and governed, as in the case of Bosons, by the hydrodynamic equations of irrotational superfluids:

$$\frac{\partial}{\partial t}\rho + \nabla(\mathbf{v}\rho) = 0 \tag{11.42}$$

and

$$m\frac{\partial}{\partial t}\mathbf{v} + \nabla\left(v_{\text{ext}} + \mu(\rho) + m\frac{v^2}{2}\right) = 0, \tag{11.43}$$

where $\mu(\rho)$ is the chemical potential of the uniform gas evaluated at the density ρ. The above equations are expected to describe correctly excitations of the system up to energies of the order of the gap energy, corresponding to wavelengths larger than the healing length defined in Subsection 2.7. The experimental detection of the modes of excitation predicted by these equations would be a proof of the reached superfluidity of the system.

The only difference between these equations and those of the previous subsection for Bosons lies in the density dependence of the chemical potential, which for Fermions is fixed by the Fermi energy, $\mu = \hbar^2(3\pi^2\rho)^{2/3}/2m$, and for Bosons by the interaction, $\mu = g\rho$. Notice that for the Fermi gas ρ is the total density of the gas given by the sum of the two spin components. This different density dependence in μ in the two cases has important consequences on the dispersion law of the elementary excitations. By linearizing the hydrodynamic equations, as done in the Boson case of the previous subsection, one finds the result (Bruun and Clark, 1999; Baranov and Petrov, 2000)

$$\omega(n_r, \ell) = \omega_{\text{ho}} \left(\ell + \frac{4}{3}n_r(2 + \ell + n_r) \right)^{1/2}, \tag{11.44}$$

holding for an isotropic harmonic trap. The monopole oscillation occurs at the energy $2\omega_{\text{ho}}$ while in the Boson case it was $\sqrt{5}\omega_{\text{ho}}$. Conversely, the frequency of surface oscillations, being insensitive to the equation of state, is given by $\omega(\ell) = \sqrt{\ell}\omega_{\text{ho}}$, as in the Bose case. Since in the collisionless regime of the normal phase [see Eq. (11.122)] the quadrupole energy is even larger than $\simeq 2\hbar\omega_{\text{ho}}$, the emergence of a superfluid phase should be visible in the change of the quadrupole frequency, which in the superfluid phase is given by $\sqrt{2}\hbar\omega_{\text{ho}}$.

11.1.5   *The Moment of Inertia and the Scissor Mode of a Bose Gas in a Magnetic Trap*

In this subsection, starting from (11.6) and (11.37), we will calculate the moment of inertia of a cold and dilute Bose gas, in the Thomas–Fermi limit, in a trap that rotates with angular velocity Ω directed along the z axis. We will consider harmonic external fields of the type

$$V_{\text{ext}} = \sum_{i=1}^{N} \frac{m}{2}(\omega_x^2 x_i^2 + \omega_y^2 y_i^2 + \omega_z^2 z_i^2), \tag{11.45}$$

with an asymmetry $(\omega_x \neq \omega_y)$ on the rotating plane.

The moment of inertia characterizes the system response to the rotation field $-\Omega L_z$, and is defined by the ratio

$$\Theta = \frac{\langle L_z \rangle}{\Omega}, \tag{11.46}$$

between the angular momentum $\langle L_z \rangle$ induced by the rotation (along z) and the angular velocity Ω. In Eq. (11.46), the mean is over the state perturbed by the rotation.

In the rotating reference the equations of hydrodynamics become

$$\frac{\partial}{\partial t}\rho + \nabla(\rho(\mathbf{v} - \mathbf{\Omega} \times \mathbf{r})) = 0, \tag{11.47}$$

$$m\frac{\partial}{\partial t}\mathbf{v} + \nabla\left(v_{\text{ext}} + g\rho + m\frac{v^2}{2} - m\mathbf{v}\cdot(\mathbf{\Omega}\times\mathbf{r})\right) = 0, \tag{11.48}$$

and have the stationary solution (Zambelli and Stringari, 2001)

$$\mathbf{v} = \Omega\frac{\langle x^2 - y^2\rangle}{\langle x^2 + y^2\rangle}\nabla(xy). \tag{11.49}$$

Then, the angular momentum induced by the rotation along z is given by

$$\langle L_z\rangle = mN\Omega\frac{\langle x^2 - y^2\rangle^2}{\langle x^2 + y^2\rangle},$$

and the moment of inertia by the irrotational value

$$\Theta = \left(\frac{\langle x^2 - y^2\rangle}{\langle x^2 + y^2\rangle}\right)^2\Theta_{\text{rig}}, \tag{11.50}$$

where $\Theta_{\text{rig}} = mN\langle x^2 + y^2\rangle$ is the rigid value of the moment of inertia. Equation (11.50) clearly shows that in a superfluid the value of the moment of inertia is smaller than the rigid value (the irrotational and rigid values coincide only for very deformed systems with $\langle x^2\rangle \gg \langle y^2\rangle$). In the limit of small rotational velocity Ω, the moment of inertia (11.50) can be rewritten as

$$\Theta = \left(\frac{\omega_x^2 - \omega_y^2}{\omega_x^2 + \omega_y^2}\right)^2\Theta_{\text{rig}}. \tag{11.51}$$

This result is attained by putting in (11.50) the values of the mean square radii predicted by the Thomas–Fermi approximation of Subsection 4.3.2, which scale like $1/\omega^2$. We recall again that the Thomas–Fermi approximation is obtained from (11.48) by setting $\mathbf{v} = 0$ and $\Omega = 0$.

It is interesting to note that the prediction (11.51) for Θ can be obtained in a different way that makes use of linear response theory. In fact, in the limit of small Ω, the moment of inertia can be evaluated by treating the rotational field $-\Omega L_z$ as a small perturbation to the Hamiltonian

$$H = \sum_{i=1}^{N}\left[\frac{p_i^2}{2m} + \frac{m}{2}(\omega_x^2 x_i^2 + \omega_y^2 y_i^2 + \omega_z^2 z_i^2)\right] + \sum_{i<j}V(\mathbf{r}_i - \mathbf{r}_j), \tag{11.52}$$

and we obtain the result (Radomski, 1976; Stringari and Lipparini, 1980)

$$\Theta = 2\sum_n\frac{|\langle n|L_z|0\rangle|^2}{E_n - E_0}, \tag{11.53}$$

where n and E_n are the eigenstates and eigenvalues of the unperturbed Hamiltonian (11.52). Equation (11.53) shows that the moment of inertia is related to the moment $m_{-1}(L_z)$ of the excitation strength

$$S(L_z, \omega) = \sum_n |\langle n|L_z|0\rangle|^2 \delta(\omega - (E_n - E_0))$$

of the angular momentum

$$L_z = \sum_{i=1}^{N}(x_i p_i^y - y_i p_i^x)$$

(see Subsection 8.3) through the relation

$$\Theta = 2m_{-1}(L_z). \tag{11.54}$$

The angular momentum excites the system because of the asymmetric external potential, which breaks rotational invariance of Hamiltonian (11.52), and does not commute with L_z, yielding the result

$$[H, L_z] = -im(\omega_x^2 - \omega_y^2)Q, \tag{11.55}$$

where

$$Q = \sum_{i=1}^{N} x_i y_i$$

is the relevant quadrupole operator. Eq. (11.55) shows that the degrees of freedom associated with the angular-momentum and quadrupole variables are coupled by the deformation of the confinement potential, and that the moment $m_1(L_z)$ does not depend on the interaction and is given by

$$m_1(L_z) = \frac{1}{2}\langle|[L_z, [H, L_z]]|\rangle = m\frac{N}{2}(\omega_x^2 - \omega_y^2)\langle y^2 - x^2\rangle. \tag{11.56}$$

Furthermore, noting that the rigid value of the moment of inertia can be related to the $m_1(Q)$ sum rule for the quadrupole operator, and using the identity

$$\sum_{i=1}^{N} x_i^2 y_i^2 = [Q, [H, Q]], \tag{11.57}$$

it is simple to derive from Eqs. (11.54), (11.55) and (11.57) the result

$$\frac{\Theta}{\Theta_{\text{rig}}} = (\omega_x^2 - \omega_y^2)\frac{m_{-3}(Q)}{m_1(Q)}. \tag{11.58}$$

This shows that the moment of inertia of the system can be obtained by measuring the quadrupole excitations of the system itself.

The results (11.54)–(11.58) are exact results for Hamiltonian (11.52). They hold for any value of N, and for both Bose and Fermi systems.

In the case of Bose systems with repulsive interaction strong enough that the Thomas–Fermi approximation holds, and that it is possible to write the equations of hydrodynamics, the quadrupole and angular-momentum operators excite only one state at energy

$$\omega_{\text{HD}} = \sqrt{\omega_x^2 + \omega_y^2}. \tag{11.59}$$

This result generalizes for the case of an asymmetric trap the one obtained in Subsection 11.1.3 (i.e. $\sqrt{2}\omega_{\text{ho}}$) for the quadrupole mode of the system in a spherical trap. Using the result (11.59), together with Eqs. (11.56) or (11.57), it is then possible to evaluate the moments $m_{-1}(L_z)$ and $m_{-3}(Q)$, $m_1(Q)$, and derive [from either Eqs. (11.54) or (11.58)] the value of Θ, which is equal to the one we already found, (11.51), by direct solution of the equations of hydrodynamics.

The method just described for the calculation of Θ is very interesting because it evidences how important the dynamics is in determining the moment of inertia of the system. A further step toward the understanding of the problem can be made by studying the noninteracting system whose Hamiltonian is given by

$$H_0 = \sum_{i=1}^{N} \left[\frac{p_i^2}{2m} + \frac{m}{2} (\omega_x^2 x_i^2 + \omega_y^2 y_i^2 + \omega_z^2 z_i^2) \right].$$

In this model, it is easy to verify that the angular-momentum and quadrupole operators excite two modes at frequencies,

$$\omega_\pm = \omega_x \pm \omega_y, \tag{11.60}$$

whose strengths are given by the formulas

$$|\langle \omega_\pm | L_z | 0 \rangle|^2 = \frac{Nm}{4\omega_x \omega_y} \omega_\mp^2 (\omega_y \langle y^2 \rangle \pm \omega_x \langle x^2 \rangle) \tag{11.61}$$

and

$$|\langle \omega_\pm | Q | 0 \rangle|^2 = \frac{N}{4m\omega_x \omega_y} (\omega_y \langle y^2 \rangle \pm \omega_x \langle x^2 \rangle). \tag{11.62}$$

Using these results in (11.54) or (11.58), we obtain for θ the value

$$\Theta = \frac{mN}{\omega_x^2 - \omega_y^2} [(\omega_x^2 + \omega_y^2)\langle y^2 - x^2 \rangle + 2(\omega_y^2 \langle y^2 \rangle - \omega_x^2 \langle x^2 \rangle)]. \tag{11.63}$$

The results (11.60)–(11.63) are exact for the noninteracting Hamiltonian H_0. They hold for any value of N, and for both Bose and Fermi systems.

In the case of Bosons at zero temperature, when only the state $(n_x, n_y, n_z) = (0, 0, 0)$ is occupied, we have $\langle x^2 \rangle = 1/2m\omega_x$ and $\langle y^2 \rangle = 1/2m\omega_y$, and we obtain $|\langle \omega_- | L_z | 0 \rangle|^2 = 0$ and $|\langle \omega_- | Q | 0 \rangle|^2 = 0$, i.e. only the resonance at energy

$$\omega_+ = \omega_x + \omega_y \tag{11.64}$$

is excited. In this case the moment of inertia takes the value

$$\Theta = \left(\frac{\omega_x - \omega_y}{\omega_x + \omega_y}\right)^2 \Theta_{\text{rig}}. \tag{11.65}$$

The results (11.64) and (11.65) correspond to the results (11.59) and (11.51), respectively, of the hydrodynamic model. The differences are due to the interaction. However, the physics is the same, in the sense that to the unique excited state corresponds a moment of inertia which, in the spherical case, tends to zero in proportion to the square of deformation, thereby evidencing the superfluidity of the condensate.

However, the noninteracting model shows that there are two states that can be excited by quadrupole and angular-momentum operators: ω_+ and ω_-. In the Fermion case, these two resonances are always present in the system, and the state of higher energy is of the quadrupole type, in the sense that it acquires much more quadrupole strength as compared to that of the orbital angular momentum, while the lower energy state is at the orbital angular momentum type since it acquires more angular-momentum strength than quadrupole strength. The state ω_- is known in the literature as the scissor mode and has been observed (Bohle *et al.*, 1984), and theoretically predicted (Lo Iudice and Palumbo, 1978; Lipparini and Stringari, 1983, 1989; Iachello, 1984; Nojarov, Bochnacki and Faessler, 1986) for the first time in atomic nuclei. Its existence has also been predicted for metal clusters (Lipparini and Stringari, 1989b; Nesterenko *et al.*, 1999) and quantum dots (Serra, Puente and Lipparini, 1999). It has also been predicted (D. Guery-Odelin and Stringari, 1999; Zambelli and Stringari, 2001) and subsequently measured (Maragò *et al.*, 2000) in Bose condensates. The scissor mode in Fermions will be considered in forthcoming subsections. In general, it can be said that its existence strongly affects the value of the moment of inertia of the system, as we will show in this subsection for Bosons.

In the case of a Boson gas in a magnetic trap, the scissor mode exists only at temperatures higher than the critical temperature T_c. In fact, at these temperatures the situation changes drastically. The system becomes much less dense and, as a first approximation, the interaction can be neglected, and we pass from a hydrodynamic regime with only one excited state at energy (11.59) to a noninteracting regime with two excited states at energies (11.60), and with a moment of inertia given by the expression (11.63). To evaluate the quadratic radii $\langle x^2 \rangle$ and $\langle y^2 \rangle$, Stringari (1996) used the so-called macroscopic limit in which one separates the contribution of the condensate from those of the excited states. Such a limit is appropriate for temperatures KT much higher than the energies ω_x, ω_y, ω_z of the confinement potential, and leads to the result (1.69) for the depletion of the condensate. One finds that

$$N\langle x^2 \rangle = N_0(T)\frac{1}{2m\omega_x} + [N - N_0(T)]\frac{KT}{3m\omega_x^2}\frac{Q(3)}{Q(2)}, \tag{11.66}$$

where

$$Q(c) = \int_0^\infty [s^c/(e^s - 1)]ds,$$

and similarly for $\langle y^2 \rangle$. The first term of Eq. (11.66) is the contribution to $\langle x^2 \rangle$ originating from the particles in the condensate, and scales as $1/\omega_x$. The second term is the contribution from the particles out of the condensate, and scales as $1/\omega_x^2$. Inserting the result (11.66) into the expression (11.63) for the moment of inertia, we find that

$$\Theta = \epsilon_0^2 m \langle x^2 + y^2 \rangle_0 N_0(T) + m \langle x^2 + y^2 \rangle_{nc} [N - N_0(T)], \qquad (11.67)$$

where the indices at $\langle\ \rangle_0$ and $\langle\ \rangle_{nc}$ mean averaging over the condensate and non-condensate components, respectively, while ϵ_0 is the deformation parameter of the density of the condensate:

$$\epsilon_0 = \frac{\langle x^2 - y^2 \rangle_0}{\langle x^2 + y^2 \rangle_0}. \qquad (11.68)$$

ϵ_0 can also be written as $\epsilon_0 = (\omega_y - \omega_x)/(\omega_y + \omega_x)$. Eq. (11.67) shows that for temperatures higher than the critical temperature T_c, where $N_0 = 0$, the moment of inertia takes the rigid (classical) value, while at $T = 0$, when all atoms are in the condensate, it takes the irrotational value. These results have been confirmed experimentally by Maragò et al. (2000), who studied the oscillations of a Boson gas in a magnetic trap, as excited by a sudden rotation of the confinement potential. At temperatures lower than the critical temperature, the condensate oscillates with only one frequency, which is in very good agreement with the hydrodynamic prediction (11.59). Moreover, it is possible to extract from the signal (Zambelli and Stringari, 2001) a value of the moment of inertia $\Theta/\Theta_{rig} = 0.6$, which shows the superfluid nature of the system. At temperatures higher than the critical temperature, the same experiment evidences oscillations of the system at the two frequencies (11.60), and from the signal we derive a value of the moment of inertia equal to the rigid one.

11.1.6 *Vortices in the Bose Gas in a Magnetic Trap*

In the nonlinear regime, when the angular velocity of rotation Ω of the system becomes large, the term $-\Omega L_z$ may favor the creation of a vortex in the system. A vortex quantized along the z axis can be described by writing the order parameter in the form

$$\Psi(\mathbf{r}) = \Psi_v(r_\perp, z) e^{ik\psi}, \qquad (11.69)$$

where ψ is the angle around the z axis, k is an integer number, and $\Psi_v(r_\perp, z) = \sqrt{\rho(r_\perp, z)}$. The state of a vortex has a tangential velocity:

$$v = \frac{\hbar}{mr_\perp} k. \qquad (11.70)$$

The number k is the quantum circulation, and the angular momentum along z is given by $\hbar N k$. The equation for $\Psi_v(r_\perp, z)$ is derived from the Gross–Pitaevskii

Eq. (9.123), and includes a centrifugal energy term which pushes the atoms away from the rotation axis:

$$\left[-\frac{\hbar^2\nabla^2}{2m} + \frac{\hbar^2 k^2}{2mr_\perp^2} + \frac{m}{2}(\omega_\perp r_\perp^2 + \omega_z^2 z^2) + g\Psi_v^2(r_\perp, z)\right]\Psi_v(r_\perp, z)$$

$$= \mu\Psi_v(r_\perp, z). \tag{11.71}$$

Owing to such a centrifugal term, the solution to Eq. (11.71) for $k \neq 0$ must vanish on the z axis. The solution $\Psi_v(x, 0, 0)$ to this equation (Dalfovo *et al.*, 1999) for a gas of 10^4 rubidium atoms in a spherical trap with $a_{\mathrm{ho}} = 0.791$ μm and for $k = 1$, is plotted in Fig. 11.1, together with the solution without interaction. The latter is analytic and is given by

$$\Psi_v(r_\perp, z) \propto r_\perp e^{-\frac{m}{2\hbar}(\omega_\perp r_\perp^2 + \omega_z^2 z^2)}. \tag{11.72}$$

Thus, in the noninteracting case the vortex state is obtained by putting all atoms in the single-particle state with $m = 1$. Therefore, its energy is $N\hbar\omega_\perp$, plus the energy of the ground state. From the figure, it can be seen that the (repulsive) interaction significantly reduces the density as compared to the noninteracting case. The difference between the energy of the vortex state and that of the ground state allows us to calculate the critical frequency needed to create a vortex. In fact, in a system rotating at angular frequency Ω, the energy of the system is $E - \Omega L_z$, where E and L_z are as defined in the laboratory frame. For small Ω the energy is

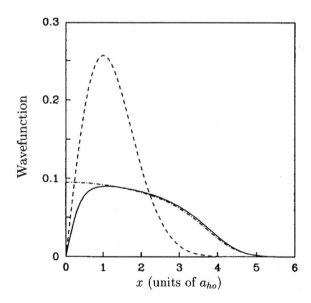

Fig. 11.1 Condensate with a vortex quantized along the z axis. The order parameter $\Psi_v(x, 0, 0)$ is plotted for a gas of 10^4 rubidium atoms in a spherical trap with $a_{\mathrm{ho}} = 0.791$ μm. The dashed–dotted line is the solution without vortex; the full line is the profile of the solution to Eq. (11.71) with a vertex; the dashed line is the noninteracting solution to Eq. (11.72).

minimum without vortex. If Ω is large enough, the vortex state becomes favored because of the term $-\Omega L_z$. The critical frequency turns out to be

$$\Omega_c = (\hbar k)^{-1} \left[\left(\frac{E}{N} \right)_k - \left(\frac{E}{N} \right)_0 \right],$$
(11.73)

where E_k is the energy of the system with a vortex having angular momentum $\hbar N k$. Lundh, Pethick and Smith (1997) found for large N the following analytic expression for Ω_c:

$$\Omega_c = \frac{5\hbar}{2mR_\perp^2} \ln \frac{0.671 R_\perp}{\xi},$$
(11.74)

where $R_\perp$ is the Thomas–Fermi radius of the density in the xy plane, which is perpendicular to the vortex line, and $\xi = (8\pi\rho a)^{-1/2}$ (with ρ the central density of the gas with no vortex) is the healing length. The critical frequency turns out to be a fraction of the oscillator frequency, and shows a decreasing behavior as a function of the number of atoms N.

The vortices in condensates have been generated and studied in several experiments (Matthews *et al.*, 1999; Madison *et al.*, 2000; Anderson *et al.*, 2000). One of the most important effects of vortices is to change the frequencies of the collective modes of the condensate, and this can be measured with high precision. This can be easily seen by studying, for example, the m_2^- sum rule of Eq. (8.39) for the exact Hamiltonian of the system

$$H = \sum_{i=1}^N \left[\frac{p_i^2}{2m} + \frac{m}{2}(\omega_\perp r_\perp^2 + \omega_z^2 z^2)_i \right] + g \sum_{i<j} \delta(\mathbf{r}_i - \mathbf{r}_j),$$

and for the quadrupole operators

$$F = \sum_j (xz - izy)_j$$

and

$$G^\dagger = F^\dagger = \sum_j (xz + izy)_j.$$

One obtains the result

$$m_2^- = \frac{\hbar^3}{m^2}\langle L_z \rangle = \frac{\hbar^4}{m^2} N k,$$
(11.75)

which shows that this sum rule is different from zero only in the presence of vortices. Zambelli and Stringari (1998) used the sum rule approach to study the splitting of the oscillation frequencies of the quadrupole mode due to the presence of vortices in the condensate. Fetter and Svidzinsky (1998) studied the same problem using the hydrodynamic model. The same authors wrote a review paper (Fetter and Svidzinsky, 2001) on the vortices in condensates in a magnetic trap.

11.2 The Fluidodynamic and Hydrodynamic Model for Fermions

Hydrodynamic equations for Fermions can be derived along the same lines as for the Bose systems by assuming that:

(i) All the single-particle orbitals evolve in time with a common complex phase as follows:

$$\varphi_i(\mathbf{r}, t) = \psi_i(\mathbf{r}, t)e^{is(\mathbf{r}, t)}, \qquad (11.76)$$

where both $\psi_i(\mathbf{r}, t)$ and $s(\mathbf{r}, t)$ are real functions;

(ii) The energy functional that describes the system is a functional $E = E(\rho)$ of the density ρ alone. This assumption implies the Thomas–Fermi approximation (4.14) for the kinetic energy density in 3D, or the equivalent in 2D:

$$\tau_{2D}(\rho) = \frac{\pi}{2m}\rho^2. \qquad (11.77)$$

Terms depending on the density derivatives can be added to the expression of $\tau(\rho)$ when necessary.

Then, the equations one derives are identical to Eqs. (11.6)–(11.8). Moreover, such equations can be easily generalized (Puente, Casas and Serra, 2000) to include cases in which the (up and down) spin densities do not oscillate in phase, for example for the spin modes, or in which it is the isospin densities which do not oscillate in phase, as in nuclei.

Recently, these equations have been solved in the cases of metal clusters (Domps, Reinhard and Suraud, 1998) and of quantum dots (Puente, Casas and Serra, 2000), in order to study the (dipole) excitations with $\ell = 1$. In Fig. 11.2 we compare the results of such calculations for quantum dots, with microscopic calculations based on the RPA. This comparison shows that in the case of dipolar excitations the hydrodynamic model works very well.

In general, the hydrodynamic model is expected to work well also in the monopole case ($\ell = 0$, compression mode), but not for all other excitation modes with $\ell > 1$. In fact, it is well known (Bertsch, 1974, 1975) that the Fermion can vibrate like an elastic medium. Such vibrations are characterized by the presence of non-diagonal terms in the stress tensor, which are absent in the hydrodynamic model. The presence of these components can be traced back microscopically to a typical quantum effect exhibited by the Fermi systems, viz. deformations of the Fermi surface characterized by being higher than multipolarity 1 in momentum space during collective motion. Such deformations affect significantly the frequencies of elementary excitations of Fermi systems with multipolarity higher than 1. Typical elastic modes of the Fermi systems are the quadrupole mode and the scissor mode. The elastic component of the quadrupole mode, for example, can also be verified in a simple way by considering a harmonic oscillator model, where the quadrupole excitation operator $Q = \sum_i r^2 Y_{2m}$ induces single-particle transitions with both

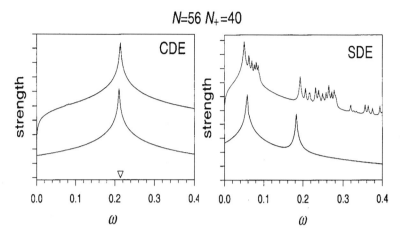

Fig. 11.2 Dipole excitation spectrum, in arbitrary logarithmic units, for density and spin-density excitations in a parabolic quantum dot with 56 electrons. The upper curve corresponds to the microscopic calculation in the RPA approximation, the lower one to the hydrodynamic calculation. In the CDE spectrum the triangle indicates the value of the frequency of the parabolic potential, ω_0. The Kohn theorem is evidently satisfied.

$\Delta N = 0$ and $\Delta N = 2$. The excitations with $\Delta N = 0$ are the microscopic counterpart of the surface vibrations of the drop model discussed in Subsection 11.1.2 while those with $\Delta N = 2$ characterize a new class of excitations with no analog in Bose systems. The restoring force related to these excitations is a pure elastic effect and is fixed by the Fermi energy, rather than by compressibility, and thus these modes are different from the compression modes discussed in Subsection 11.1.2.

The distortions of the Fermi surface with multipolarity higher than 1 are important in the bulk as well; for example, they are responsible for the difference between zero sound and first sound in ^{3}He.

In the following we will present the main characteristics of elastic vibrations of Fermi systems, using the generalized scaling model. The hydrodynamic model is obtained from the latter by just ignoring the terms related to the deformation of the Fermi surface in the derivation of the equations of motion. Moreover, the generalized scaling model was developed in nuclear physics (Bertsch, 1974, 1975; Brink and Leonardi, 1976; Stringari, 1977, 1983; Wong and McDonald, 1977; Holzwart and Eckardt, 1978, 1979; Lipparini and Stringari, 1989) and in the study of metal clusters (Lipparini, 1989; Lipparini and Stringari, 1991).

In this model, one derives fluidodynamic equations of motion, looking for solutions to the time-dependent equations in the form of the unitary transformation

$$|\Psi\rangle = e^{im \sum_i \xi(\mathbf{r}_i,t)t_i} e^{i\frac{1}{2} \sum_i (\mathbf{u}(\mathbf{r}_i,t)\cdot\mathbf{p}_i + \mathbf{p}_i\cdot\mathbf{u}(\mathbf{r}_i,t))t_i} |0\rangle, \qquad (11.78)$$

which is applied to the ground state $|0\rangle$. In Eq. (11.78), $\mathbf{p}_i$ is the momentum operator which acts on the ith particle, $\xi(\mathbf{r},t)$ is interpreted as a velocity potential, and $\mathbf{u}(\mathbf{r},t)$ is the displacement field. t_i is equal to 1 for density excitations, and equal to the third component of the spin operator σ_z in the case of the spin-density excitations (i.e. 1 for Fermions with spin up, and -1 for Fermions with spin down), or to the third component of the isospin operator τ_z in the case of isospin-density excitations (i.e. 1 for neutrons and -1 for protons). If $|0\rangle$ is a Slater determinant, then the state $|\Psi\rangle$ is a Slater determinant as well. The term in ξ introduces components which break the invariance under time reversal in $|\Psi\rangle$, without affecting the density.

The transition density associated with the transformation (11.78) is given by

$$\rho_{\mathrm{tr}}(\mathbf{r}, t) = \langle\Psi|\sum_i \delta(\mathbf{r} - \mathbf{r}_i)t_i|\Psi\rangle - \langle 0|\sum_i \delta(\mathbf{r} - \mathbf{r}_i)t_i|0\rangle$$

$$= \nabla(\mathbf{u}(\mathbf{r}, t)\rho_0(\mathbf{r})), \tag{11.79}$$

where $\rho_0(\mathbf{r})$ is the ground state density. The velocity field is expressed through ξ as

$$\mathbf{v}(\mathbf{r}, t) = \frac{1}{\rho_0(\mathbf{r})}\langle\Psi|\frac{1}{2m}\sum_i(\delta(\mathbf{r} - \mathbf{r}_i)\mathbf{p}_i + \mathrm{H.c.})t_i|\Psi\rangle = \nabla\xi(\mathbf{r}, t), \tag{11.80}$$

and is irrotational.

The fluidodynamic equations of motion can be derived using the variational principle as applied to the action integral: $\delta I = 0$, with

$$I = \int dt\langle\Psi|i\frac{\partial}{\partial t} - H|\Psi\rangle.$$

As we saw in Chapter 9, this method reproduces the TDHF or TDLDA equations if $|\Psi\rangle$ is the most general Slater determinant. As a result, using expression (11.78) for $|\Psi\rangle$, where $|0\rangle$ is a Slater determinant, we will derive approximate solutions of the TDHF (TDLDA) theory. Moreover, we will study small oscillations around the ground state so that the found solution can be considered as an approximation of the RPA theory.

Keeping only terms quadratic in ξ and $\mathbf{u}$, and neglecting the total derivatives which do not appear in the equations of motion, we find that

$$\langle\Psi|i\frac{\partial}{\partial t}|\Psi\rangle = m\int\mathbf{u}\cdot\dot{\mathbf{v}}\rho_0 d\mathbf{r} \tag{11.81}$$

and

$$\langle\Psi|H|\Psi\rangle = \langle 0|H|0\rangle + E(\mathbf{u}) + T(\xi), \tag{11.82}$$

where we have introduced the collective potential energy

$$E(\mathbf{u}) = \langle 0|e^{-i\frac{1}{2}\sum_i(\mathbf{u}\cdot\mathbf{p}_i + \mathrm{H.c.})t_i}He^{i\frac{1}{2}\sum_i(\mathbf{u}\cdot\mathbf{p}_i + \mathrm{H.c.})t_i}|0\rangle - \langle 0|H|0\rangle$$

$$= \frac{1}{2}\langle 0|\frac{1}{2}\left[\sum_i(\mathbf{u}\cdot\mathbf{p}_i + \mathrm{H.c.})t_i, \left[H, \frac{1}{2}\sum_i(\mathbf{u}\cdot\mathbf{p}_i + \mathrm{H.c.})t_i\right]\right]|0\rangle$$

$$\tag{11.83}$$

and the collective kinetic energy

$$T(\xi) = \langle 0|e^{-im\sum_i \xi t_i} H e^{im\sum_i \xi t_i}|0\rangle - \langle 0|H|0\rangle$$

$$= \frac{1}{2}m^2\langle 0|\left[\sum_i \xi t_i, \left[H, \sum_i \xi t_i\right]\right]|0\rangle. \tag{11.84}$$

Note that the terms linear in $\mathbf{u}$ and ξ vanish due to the property of the ground state $\langle 0|[H, F]|0\rangle = 0$. Assuming that the interaction term in Hamiltonian H commutes with the local and scalar operator in spin (isospin) space $\sum_i \xi_i$, and that it contributes to the commutator with the vector operator (in spin or isospin space) $\sum_i \xi t_i$, through the term [in $k_v(\rho_0)$] which depends on density, we obtain the following result for $T(\xi)$:

$$T(\xi) = \frac{1}{2m}\int v^2 \rho_0 d\mathbf{r} \quad \text{(scalar)},$$

$$T(\xi) = \frac{1}{2m}\int v^2(1 + k_v(\rho_0))\rho_0 d\mathbf{r} \quad \text{(vector)}. \tag{11.85}$$

The term $k_v(\rho_0)$ originates from possible nonlocal components of the interaction such as the terms in a and a_1 of Eq. (4.113). In the case of the functional (4.113), we obtain $k_v(\rho_0) = 2m(a - a_1)\rho_0$. Note that for the local density functionals such as those which lead to the Kohn–Sham Eqs. (4.60), we have $k_v = 0$.

Eq. (11.85) shows that the collective kinetic energy depends on the scalar potential ξ only through the velocity field $\mathbf{v} = \nabla\xi$.

The equations of motion can subsequently be derived after performing the variation with respect to the displacement field $\mathbf{u}$ and the velocity field $\mathbf{v}$:

$$\frac{\delta I}{\delta \mathbf{u}} = 0, \quad \frac{\delta I}{\delta \mathbf{v}} = 0. \tag{11.86}$$

For the scalar (s) and vector (v) modes, we find that

$$\frac{\delta E_s}{\delta \mathbf{u}} = m\dot{\mathbf{v}}\rho_0, \quad \dot{\mathbf{u}} = -\mathbf{v}, \tag{11.87}$$

and

$$\frac{\delta E_v}{\delta \mathbf{u}} = m\dot{\mathbf{v}}\rho_0, \quad \dot{\mathbf{u}} = -\mathbf{v}(1 + k_v(\rho_0)), \tag{11.88}$$

respectively. Looking for solutions of the kind $\mathbf{u}(rt) = \mathbf{u}(r)U(t)$ which oscillate at frequency ω, we arrive at the fluidodynamic equations of motion

$$\frac{\delta E_s}{\delta \mathbf{u}(\mathbf{r})} = m\omega^2\mathbf{u}(\mathbf{r})\rho_0(\mathbf{r}), \tag{11.89}$$

for the case of scalar excitations in spin (isospin) space, and

$$\frac{\delta E_v}{\delta \mathbf{u}(\mathbf{r})} = \frac{1}{1 + k_v(\rho_0)}m\omega^2\mathbf{u}(\mathbf{r})\rho_0(\mathbf{r}), \tag{11.90}$$

for vector excitations in spin (isospin) space.

Furthermore, from (11.79), (11.87) and (11.88) it is possible to derive the continuity equations for the density $\rho = \rho_\uparrow + \rho_\downarrow$ ($\rho = \rho_n + \rho_p$, $n \equiv$ neutron, and $p \equiv$ proton) in the scalar case, and for the density $\rho_v = \rho_\uparrow - \rho_\downarrow$ ($\rho_v = \rho_n - \rho_p$) in the vector case (note that in Chapter 4 we indicated ρ_v as m in the case of spin density, and as ρ_1 in the case of the isovector density of nuclear physics).

Starting from a nonpolarized system, i.e. one with $\rho_\uparrow^0 = \rho_\downarrow^0$ ($\rho_n^0 = \rho_p^0$), we obtain

$$\frac{\partial \rho}{\partial t} + \nabla(\mathbf{v}\rho_0) = 0 \tag{11.91}$$

for the scalar case, and

$$\frac{\partial \rho_v}{\partial t} + \nabla(\mathbf{v}(1 + k_v(\rho_0))\rho_0) = 0 \tag{11.92}$$

for the vector case.

In order to solve the fluidodynamic Eqs. (11.89) and (11.90), it is necessary to specify the energy functional $E = \langle \Psi | H | \Psi \rangle$ and subsequently compute the collective potential energy $E(\mathbf{u})$ of Eq. (11.83). Starting from energy functionals such as those we used in Chapter 4 to compute $E(\mathbf{u})$, one needs to know the variations $\delta\rho$, $\delta\rho_v$ and $\delta\tau$, up to terms quadratic in $\mathbf{u}$, which are induced by the transformation (11.78). The general expressions of these variations, in both the scalar and vector cases, can be found in the paper by Stringari (1983), where the formalism of generalized scaling is applied to the study of collective modes (both compression and elastic ones) of nuclear physics. In the following, for the sake of simplicity, we will limit ourselves to treating the very interesting case of zero-divergence excitations, i.e. those characterized by

$$\nabla \cdot \mathbf{u} = 0. \tag{11.93}$$

These excitation modes can be obtained by taking

$$\mathbf{u} = \frac{1}{m_v}\nabla f, \tag{11.94}$$

where f characterizes the multipolar excitation operator

$$F = \sum_{i=1}^{N} f(\mathbf{r}_i)t_i = \sum_{i=1}^{N} r_i^\ell Y_{\ell,m}(\Omega_i)t_i.$$

Note that $m_v = m$ in the scalar case, and $m_v = m/(1 + k_v)$ in the vector case and for nonlocal interactions.

The choice (11.94) automatically leads to zero-divergence excitations with

$$\rho_{\mathrm{tr}} = \frac{1}{m_v}\nabla f \cdot \nabla \rho_0,$$

because $\Delta r^\ell Y_{\ell,m} = 0$, and to the irrotational velocity fields

$$\mathbf{v} = -\dot{U}\mathbf{u} = -\dot{U}\frac{1}{m_v}\nabla f.$$

For zero-divergence excitations, the changes induced by transformation (11.78) on the densities ρ and ρ_v, and in the kinetic energy density τ, are given by

$$\delta\rho = u_k\nabla_k\rho_0 + \frac{1}{2}u_k\nabla_k u_l\nabla_l\rho_0, \quad \delta\rho_v = 0,$$

$$\delta\tau = u_k\nabla_k\tau_0 + \frac{1}{2}u_k\nabla_k u_l\nabla_l\tau_0 + \frac{1}{6}(\nabla_k u_l + \nabla_l u_k)^2\tau_0 \tag{11.95}$$

for scalar excitations ($t_i = 1$), and by

$$\delta\rho = \frac{1}{2}u_k\nabla_k u_l\nabla_l\rho_0, \quad \delta\rho_v = u_k\nabla_k\rho_0,$$

$$\delta\tau = \frac{1}{2}u_k\nabla_k u_l\nabla_l\tau_0 + \frac{1}{6}(\nabla_k u_l + \nabla_l u_k)^2\tau_0 \tag{11.96}$$

for vector excitations ($t_i = \sigma_i^z$ or τ_i^z).

From these expressions, we see that the density variations of the zero divergence excitation modes have surface terms which are proportional to the derivatives of the density ρ_0 and the kinetic energy density τ_0 of the ground state, as well as bulk terms (which appear only in the kinetic energy variation) in which the derivatives act only on the displacement fields. In homogeneous systems, the collective potential energy gets contributions only from the variation of kinetic energy, and the corresponding excitation modes are of the elastic type. In fact, in this case we find that

$$E(\mathbf{u}) = \frac{1}{10}\epsilon_F \int \left(\sum_{k,l}(\nabla_k u_l + \nabla_l u_k)^2\right)\rho_0 d\mathbf{r}, \tag{11.97}$$

which holds for both the scalar and the vector case, and is nonvanishing only for $\ell \geq 2$. The collective potential energy (11.97) has the same form as the transverse force contribution to the energy of an elastic medium with elastic Lamè constant given by $2/5\epsilon_F\rho_0$ (where ϵ_F is the Fermi energy). Therefore, the Fermi energy is the main ingredient of the restoring force of the elastic-like oscillation modes.

On the contrary, the compression modes are modes of a hydrodynamic nature, and characterized by the fact that the divergence of $\mathbf{u}$ is different from zero. Therefore, in order to derive the hydrodynamic model from the equations of generalized scaling, it is necessary to consider the most general transformations which lead to transition densities such as (11.79). For a detailed discussion on this derivation, see Stringari (1983). Here we limit ourselves to remarking that, unlike the general case, the hydrodynamic model assumes the Thomas–Fermi relation [$\tau = (3/5)(3\pi^2/2)^{2/3}\rho^{5/3}$, in 3D] as a functional relation for the kinetic energy density τ, and subsequently calculates the variations of τ starting from the variations of the density ρ. In this way, one neglects the distortions caused by the transformation (11.78) on the Fermi sphere (Jennings, 1980), which are responsible for the elastic contribution. For these compression modes, the main ingredient that appears in the restoring force $E(\mathbf{u})$ of the oscillation is not the Fermi energy. Rather, it is the compressibility for the case of scalar excitations, the symmetry energy of nuclear matter for the vector isospin excitations, and the magnetization energy for the spin-density excitations.

In nonhomogeneous systems, we have $\nabla\rho_0 \neq 0$ and $\nabla\tau_0 \neq 0$, and hence one will obtain collective potential energies which include both surface and elastic contributions. The surface terms are the only ones that remain in the dipolar case with $\ell = 1$. Therefore, the zero-divergence dipolar mode is purely a surface mode. It characterizes the giant resonances of nuclei, as well as the plasmon modes of confined and finite electron systems, such as metal clusters and quantum dots.

The choice (11.93) is particularly suited for studying very narrow resonances of finite systems, which exhaust most of the excitation strength. In fact, the transition density $\rho_{\mathrm{tr}} = \nabla f \cdot \nabla\rho_0/m_v$ and the irrotational velocity fields $\mathbf{v} = -\dot{U}\mathbf{u} = -\dot{U}(1/m_v)\nabla f$ can be obtained microscopically by assuming that a single collective state exhausts the mixed sum rules for the density operator

$$\hat{\rho} = \sum_{i=1}^{N} \delta(\mathbf{r} - \mathbf{r}_i)t_i,$$

and for the current operator $\hat{\mathbf{J}} = (1/2m)\sum_{i=1}^{N}(\mathbf{p}_i\delta(\mathbf{r} - \mathbf{r}_i) + \mathrm{H.c.})t_i$, namely

$$\sum_{n} \omega_{n0}\langle 0|F|n\rangle\langle n|\hat{\rho}|0\rangle = \frac{1}{2}\langle 0|[F,[H,\hat{\rho}]]|0\rangle = -\frac{1}{2m_v}\nabla f \cdot \nabla\rho_0 \qquad (11.98)$$

and

$$\sum_{n}(\langle 0|F|n\rangle\langle n|\hat{\mathbf{J}}|0\rangle - \langle 0|\hat{\mathbf{J}}|n\rangle\langle n|F|0\rangle) = \frac{1}{2}\langle 0|[F,\hat{\mathbf{J}}]|0\rangle = -\frac{i}{2m_v}\rho_0\nabla f, \qquad (11.99)$$

where we recall that $m_v = m$ for scalar excitations ($t_i = 1$), and $m_v = m/(1+k_v)$ for vector excitations ($t_i = \sigma_i^z$ or $t_i = \tau_i^z$) and interactions with nonlocal components.

In this case, the collective energy

$$E_{\mathrm{coll}} = \frac{1}{2}\dot{U}^2 \int u^2(\mathbf{r})\rho_0 \, d\mathbf{r} + \frac{1}{2}K(\mathbf{u})U^2, \qquad (11.100)$$

with the restoring constant K given by $K(\mathbf{u}) = 2E(\mathbf{u})$, determines the energy of the one and only collective mode through the expression

$$\omega = \sqrt{\frac{K}{M}} = \sqrt{\frac{2E(\mathbf{u})}{\int(\frac{1}{m_v}\nabla f)^2\rho_0 d\mathbf{r}}}. \qquad (11.101)$$

This energy can be interpreted as the ratio $\sqrt{m_3/m_1}$ between the sum rule cubic in energy m_3 and the f-sum rule m_1 of (8.30). In fact, for the excitation operator

$$F = \sum_{i=1}^{N} f(\mathbf{r}_i)t_i = \sum_{i=1}^{N} r_i^\ell Y_{\ell,m}(\Omega_i)t_i,$$

the following commutation rule holds,

$$[H, F] = -\frac{1}{2}\sum_{i=1}^{N}\frac{1}{m_v}(\nabla f \cdot \nabla + \nabla \cdot \nabla f)_i t_i = -\frac{1}{2}\sum_{i=1}^{N}(\mathbf{u}\cdot\nabla + \nabla\cdot\mathbf{u})_i t_i, \quad (11.102)$$

so that the transformation $e^{i\frac{1}{2}U(t)\sum_i(\mathbf{u}(\mathbf{r}_i)\cdot\mathbf{p}_i+\mathbf{p}_i\cdot\mathbf{u}(\mathbf{r}_i))t_i}|0\rangle$ of Eq. (11.78) can be rewritten as $e^{-U(t)[H,F]}|0\rangle$. The collective potential energy (11.83) then becomes

$$E(\mathbf{u}) = 1/2\langle 0|[[[F, H], H], [H, F]]|0\rangle,$$

and coincides with the sum rule cubic in energy m_3. The f-sum rule for the operator F can be easily computed and leads to the result

$$m_1 = \frac{1}{2}\langle 0|[F, [H, F]|0\rangle = \frac{1}{2}\int\left(\frac{1}{m_v}\nabla f\right)^2\rho_0 d\mathbf{r}, \quad (11.103)$$

and hence to

$$\frac{K}{M} = \frac{m_3}{m_1} \quad (11.104)$$

by comparison with (11.101).

As regards the connection with the sum rules, in general it can be said that both the elastic and the hydrodynamic models well reproduce the f-sum rule m_1; moreover, the elastic model correctly reproduces the sum rule cubic in energy, which, however, is not reproduced by the hydrodynamic model. On the contrary, the inversely energy-weighted sum rule, which is related to the static polarizability of the system, is better reproduced by the hydrodynamic model. The latter result stems from the Landau theory of Fermi liquids, which shows that the spherical shape of the Fermi sphere is preserved in the presence of a static external field, and as a consequence the linear response is correctly predicted by the hydrodynamic model in the static limit. As was discussed in Subsection 9.1, the first sound is a compression mode with a hydrodynamic nature, whose frequency is well reproduced by the ratio m_1/m_{-1}, while the zero sound at higher frequency is a mode with an elastic nature, in which the distortion of the Fermi sphere plays a crucial role, and whose frequency is well reproduced by the ratio m_3/m_1. In the following, we will analyze in detail the frequencies of some collective surface and elastic modes by using the m_3/m_1 ratio.

11.2.1 Dipolar Modes in Metal Clusters

In metal clusters described by the Hamiltonian (2.22), the zero-divergence dipolar mode with $\ell = 1$ and $m = 0$ is naturally excited by the operator $F = \sum_i z_i t_i$, where z_i is the z component of the position operator of the ith electron. Therefore, we have $f(\mathbf{r}) = z$ and $\mathbf{u} = \nabla z$, and the transformation that determines the restoring force (11.83) of the collective mode becomes

$$e^{U(t) \sum_{i=1}^{N} \nabla_z^i t_i} |0\rangle. \tag{11.105}$$

Physically this transformation corresponds in the scalar case ($t_i = 1$) to a rigid shift of the whole electron cloud, and in the vector case ($t_i = \sigma_i^z$) to a rigid shift of the cloud of electrons with spin up in the opposite direction with respect to the electrons with spin down.

In the scalar case, owing to translational invariance the only term of Hamiltonian (2.22) which contributes to the collective potential energy, and thus to m_3, is the confinement potential of the ions, which breaks such invariance through the interaction term $\int V_+(\mathbf{r})\rho(\mathbf{r})d(\mathbf{r})$ with the electrons. The second order variation in $\mathbf{u}$ of such energy is due to the second order variation of the electron density, which is given by $1/2\nabla_z^2\rho_0$ [see Eq. (11.95)]. Therefore, we obtain

$$E_s(\mathbf{u}) = m_3^s = \frac{1}{2} \int V_+(\mathbf{r})\nabla_z^2\rho_0(\mathbf{r}) \, d\mathbf{r}. \tag{11.106}$$

Performing an integration by parts and using the Poisson equation $\nabla^2 V_+ = 4\pi\rho_I$, where ρ_I is the ion charge-density distribution, we obtain (Brack, 1989)

$$E_s(\mathbf{u}) = m_3^s = \frac{2\pi}{3} \int \rho_I(\mathbf{r})\rho_0(\mathbf{r})d\mathbf{r}. \tag{11.107}$$

For a jellium sphere [see Eq. (2.21)], we then have

$$E(\mathbf{u}) = m_3^s = \frac{2\pi}{3}\rho_{\text{bulk}} \int_0^R \rho_0(\mathbf{r})dr = \frac{1}{2}\rho_{\text{bulk}} N_{in}, \tag{11.108}$$

where N_{in} is the number of electrons inside the jellium sphere of radius R. The m_1 sum rule is model-independent and, in both the scalar and vector models, it is given by [see Eq. (11.103)]

$$m_1 = \frac{N}{2}, \tag{11.109}$$

where N is the total number of electrons. Therefore, the energy of the scalar plasmon mode is given by

$$\omega_s = \sqrt{\frac{m_3^s}{m_1}} = \frac{\omega_p}{\sqrt{3}}\sqrt{\frac{N_{in}}{N}}, \tag{11.110}$$

where we have introduced the plasma frequency $\omega_p = \sqrt{4\pi\rho_{\text{bulk}}}$, which gives the plasmon energy in metals. As can be seen, the finite size effects are relevant, and

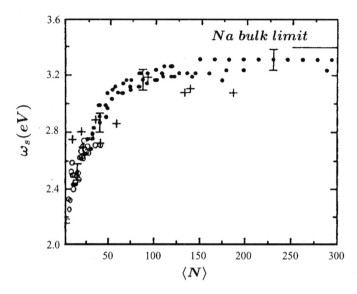

Fig. 11.3 Plasmon resonance frequency as a function of the average number of atoms, $\langle N \rangle$, in sodium clusters. The experimental data of Knight *et al.* (1985) (full circles) are compared with the result of Eq. (11.110) (open circles), and with the calculations of Beck (1984) (crosses).

strongly reduce the energy of the plasmon mode in metal clusters with respect to the bulk frequency. The expression (11.110) reproduces quite well the systematic variation with N in the experimental data of Knight *et al.* (1985) for the plasma resonance in sodium clusters, as shown in Fig. 11.3.

In the vector case, the calculation of m_3 is somewhat more complex because the interaction contributes to the sum rule. One finds that

$$E_v(\mathbf{u}) = m_3^v = \frac{1}{2}\langle 0| \left[\sum_{i=1}^{N} \nabla_z^i \sigma_i^z, \left[\sum_{i=k}^{N} \nabla_z^k \sigma_k^z, H \right] \right] |0\rangle, \tag{11.111}$$

and expanding the commutator

$$m_3^v = \frac{2\pi}{3}\langle 0| \sum_{i=1}^{N} \rho_I(\mathbf{r}_i) - \sum_{i<j=1}^{N} \delta(r_{ij})(\sigma_i^z - \sigma_j^z)^2 |0\rangle, \tag{11.112}$$

where, again, we have used the Poisson equation for the ion charge distribution, and the property $\nabla^2(1/r) = -4\pi\delta(r)$. Therefore, the final result for m_3^v is

$$m_3^v = \frac{2\pi}{3} \int (\rho_I(\mathbf{r})\rho_0(\mathbf{r}) - \rho_0(\mathbf{r})^2) d\mathbf{r}. \tag{11.113}$$

An estimate of m_3^v can be carried out using the Wood–Saxon density, $\rho_0 = \rho_b(1 + \exp((r-R)/a))^{-1}$, for ρ_0, and that of a jellium for $\rho_I(\mathbf{r})$. In this way we get

$$m_3^v \simeq 2\pi N \rho_b \frac{a}{R}, \tag{11.114}$$

and using the result (11.109) for m_1

$$\omega_v = \sqrt{\frac{m_3^v}{m_1}} = \omega_p \sqrt{\frac{a}{R}}, \qquad (11.115)$$

which shows that the vector dipole mode is a pure surface mode whose energy is always lower than that of the scalar mode. Vector surface modes of this kind have been measured in the case of quantum dots (Schuller *et al.*, 1998).

11.2.2 Spin Oscillations in Trapped Fermi Gases

It is also interesting to study the spin dipole mode excited by the operator $F = \sum_{i\uparrow} z_i - \sum_{j\downarrow} z_j$, in the case of trapped Fermi gases described by the Hamiltonian (4.41). By evaluating the m_1 and m_3 sum rules with the techniques of the previous subsection, one easily finds the result

$$\omega_{\mathrm{SD}} = \left(\omega_{\mathrm{ho}}^2 - \frac{g}{2mN} \int |\partial_z \rho|^2 \, d\mathbf{r} \right)^{1/2} \simeq \omega_{\mathrm{ho}}(1 - 0.63 k_F a) \qquad (11.116)$$

for the excitation frequency, where in the second equality we have used the ground state density (4.33) to evaluate the integral. The result (11.116) was found in the collisionless regime. However, if the collisional frequency is much larger than the frequency of the collective excitations, the effect of collisions can be important and the system goes in the hydrodynamic regime. Collisions are expected to have important consequences on the spin dipole oscillations since these excitations do not conserve the relative current between the two species and consequently give rise, in the hydrodynamic regime, to a diffusive mode. This problem was investigated by Vichi and Stringari (1999).

11.2.3 The Scalar Quadrupole Mode in Confined Systems

In what follows, we will calculate the frequency of the scalar quadrupole mode, which is excited by the operator

$$Q = \sum_{i=1}^{N} r_i^2 Y_{2m},$$

for a system whose local energy functional is of the same type as in (4.43):

$$E(\rho) = T + \int d\mathbf{r}\rho(r) \left[V_{\mathrm{ext}}(r) + \frac{1}{2}U(r) \right] + E_V(\rho), \qquad (11.117)$$

where $U(r)$ is the Hartree potential, V_{ext} is the potential of the positive charge distribution generated by the ions, and $E_V(\rho)$ is the exchange-correlation energy.

However, the functional can also describe the Boson gas in a magnetic trap by taking for V_{ext} a harmonic oscillator potential, $U = 0$, and

$$E_V(\rho) = (1/2)g \int d\mathbf{r}\rho^2(r),$$

as well as other Fermionic systems.

The energy change connected with the transformation $e^{-U(t)[H,Q]}|0\rangle$ can be easily computed, and one finds a very simple expression for $m_3(Q)$ (Bohigas, Lane and Martorell, 1979):

$$m_3(Q) = \frac{5}{\pi m^2}(E_{\text{kin}} + E_X). \tag{11.118}$$

In the above equation, E_{kin} is the expectation value of the kinetic energy operator in the ground state. E_X caters for other forms of energy. It is equal to $-E_C/5 + E_{\text{conf}}$ for charged systems [where E_C is the direct electron–electron Coulomb energy and E_{conf} is the confinement energy, given by $E_{\text{conf}} = \int d\mathbf{r}\rho(r)V_{\text{ext}}(r)$]. It is the confinement energy of the harmonic oscillator in the case of the Boson gas in a parabolic magnetic trap, and is zero for the other systems, such as nuclei.

The m_1 sum rule for the quadrupole operator is then given by

$$m_1(Q) = \frac{5}{4\pi m}N\langle r^2\rangle, \tag{11.119}$$

where

$$\langle r^2\rangle = 1/N \int r^2\rho_0 d\mathbf{r}$$

is the mean square radius of the system. Therefore, the frequency of the quadrupole scalar model in the elastic model is given by

$$\omega_Q = \sqrt{\frac{m_3(Q)}{m_1(Q)}} = \sqrt{\frac{4(E_{\text{kin}} + E_X)}{mN\langle r^2\rangle}}. \tag{11.120}$$

In the noninteracting system one has $E(X) = E_{\text{conf}}$ and $E_{\text{kin}} = E_{\text{conf}}$ from the virial theorem, and Eq. (11.120) yields the result $\omega_Q = \sqrt{8E_{\text{conf}}/mN\langle r^2\rangle}$, which, in the case of parabolic confinement, becomes the result of the harmonic oscillator $\omega_Q = 2\omega_{\text{ho}}$. For a Boson gas in a magnetic trap, and in the Thomas–Fermi limit $Na/a_{\text{ho}} \gg 1$, where the kinetic energy term is negligible, one then obtains $\omega_Q = \sqrt{2}\omega_{\text{ho}}$. For negative a, when the kinetic energy term is greater than E_{conf}, there is an increase of the quadrupole frequency with respect to that of the harmonic oscillator. Furthermore, for a charged system with a jellium confinement, neglecting the electron spillout, we have $E(X) = (9/50)e^2(N^2/R)$. In the nuclear case (where $V_{\text{ext}} = E_{\text{conf}} = 0$), we obtain $\omega_Q = \sqrt{4E_{\text{kin}}/mN\langle r^2\rangle}$. All of these results turn out to be very close to the numerical solutions to the RPA equations of the respective systems.

Finally, we note that, in the case of the trapped Fermi gas in the normal phase, described by the Hamiltonian (4.41), assuming $N_\uparrow = N_\downarrow = N$ and the same trapping frequency ω_{ho} for both species, the frequency of the quadrupole mode is still given by Eq. (11.120). Moreover, by using the virial relationship

$$2E_{kin} - 2E_{conf} + 3E_{int} = 0, \qquad (11.121)$$

holding for delta-like forces, one can write

$$\omega_Q = 2\omega_{ho}\sqrt{1 - \frac{3}{4}\frac{E_{int}}{E_{conf}}}; \qquad (11.122)$$

the interaction energy E_{int} has been calculated in Eq. (4.42) and is negative for negative scattering lengths.

11.2.4   *The Scissor Mode in Fermi Systems*

A good example of elastic behavior in Fermi systems is the scissor mode in systems with nonzero deformation. We have previously discussed such a mode for the Boson gas in an unsymmetrical magnetic trap.

A macroscopic example of such a state is provided by the following form of the displacement field relative to the electron motion in the case of metal clusters or of quantum dots, and to nucleons in the case of atomic nuclei (Lipparini and Stringari, 1989, 1991; Serra, Puente and Lipparini, 1999):

$$\mathbf{u} = \hat{\omega} \times \mathbf{r} + \delta \nabla(yz), \qquad (11.123)$$

which satisfies the condition $\nabla \cdot \mathbf{u} = 0$. In Eq. (11.123), $\hat{\omega}$ is the unit vector in the x direction (the system is assumed to be axially deformed along z), and $\delta = (3/2)(R_z^2 - R_y^2)/(R_z^2 + 2R_y^2)$ in 3D, while $\delta = (R_z^2 - R_y^2)/(R_z^2 + R_y^2)$ in 2D (where the Fermion motion is in the yz plane). This δ is the deformation of the density profile, which is assumed to have a spheroidal shape: $\rho_0 = \rho_0(x^2/R_x^2 + y^2/R_y^2 + z^2/R_z^2)$ in 3D (here R_z and $R_x = R_y$ are the radii parallel and perpendicular to the symmetry axis, respectively).

The term $\hat{\omega} \times \mathbf{r}$ of Eq. (11.123), in the scalar case and for metal clusters and quantum dots, corresponds to a rigid rotation of the electrons with respect to the distribution of positive charge which is motionless. In the vector case ($t_i = \tau_i^z$), and for nuclei, it corresponds to a rigid, counterphase rotation of protons and neutrons, as follows from transformation (11.83), which becomes

$$e^{U(t)\sum_{i=1}^{N} l_i^x t_i}|0\rangle, \qquad (11.124)$$

where l_i^x is the x component of the orbital angular momentum of the ith Fermion [scissor mode; see Fig. 11.4(a)]. If we were to include only this term in the Fermion motion, the electrons would feel a restoring force whose origin would be the Coulomb interaction with the ions, as happens in the dipole case, and the nucleons would

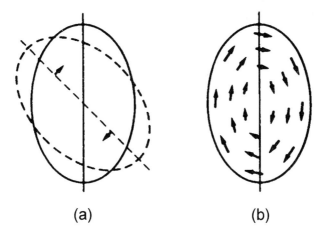

(a) **(b)**

Fig. 11.4 Displacement field for the rotational state $M1$ at low energy. Figure 11.4(a) corresponds to a rigid rotation (scissor mode) of the Fermions with respect to the confinement field. Figure 11.4(b) corresponds to a rotation within a rigid surface.

be subjected to the force resulting from the neutron–proton interaction [symmetry term, proportional to $\rho_v^2 = (\rho_n - \rho_p)^2$]. However, the Coulomb and symmetry energy cost is minimized by including the quadrupole term $\nabla(yz)$ in the displacement field. This can be seen easily by noting that the density change $\delta\rho$ and $\delta\rho_v$, given in both cases by $\nabla \cdot (\mathbf{u}\rho_0)$ [see Eq. (11.79)], becomes zero under (11.123) because $\mathbf{u} \cdot \nabla\rho_0 = 0$ in the spheroidal model. The resulting motion is shown in Fig. 11.4(b): there is a rotation within a spheroid with a rigid surface (in the nuclear case, the rotations of neutrons and protons have opposite phases) with a velocity field such that

$$\mathbf{v} \cdot \mathbf{n}|_{\text{surface}} = 0. \tag{11.125}$$

The relevant restoring force that originates during this motion is not derived from the electron–ion or neutron–proton interaction energy, but rather from the kinetic energy, and is produced by the quadrupolar component $\nabla(yz)$ of the velocity field which gives rise to a distortion of the Fermi sphere and to a collective potential energy of the type (11.97). The frequency of the resulting elastic mode is given by $\omega_{M1} = (K/M)^{1/2}$, where $K = 2E(\mathbf{u})$, with $E(\mathbf{u})$ given by Eq. (11.97), and $M = m \int \rho \mathbf{u}^2 d\mathbf{r}$, which in the limit of small deformations coincides with the rigid value of the moment of inertia θ of the rotational motion. In this limit, one finds that

$$\omega_{M1} = \delta \sqrt{\frac{4\epsilon_F}{mr_s^2}} N^{-1/3}, \quad \langle r \rangle^2 = \frac{3}{5} r_s^2 N^{2/3}, \tag{11.126}$$

in 3D, and

$$\omega_{M1} = \delta \sqrt{\frac{16\epsilon_F}{3mr_s^2}} N^{-1/2}, \quad \langle r \rangle^2 = r_s^2 N/2, \tag{11.127}$$

in 2D. The elastic mode, whose frequency is given by the above expressions, is a low energy mode with respect to the quadrupole mode we studied in the previous subsection, which goes to zero linearly with the deformation, and is the analog of the low energy mode of (11.60) for the Boson gas in a magnetic trap.

For sodium clusters ($r_s = 4$ a.u., $\epsilon_F = 3.1$ eV), $\omega_{M1} = \delta\ 4.6\ N^{-1/3}$ eV, which for N in the range 10–100 and for typical deformations $\delta = 0.2$–0.4, gives $\omega_{M1} = 0.2$–0.6 eV. This value should be compared with the dipolar plasmon frequency $\omega_s = 3.4$ eV. The scissor state lies below the threshold for particle emission, and is excited by the orbital angular momentum operator, with a strength given by

$$BM1 = \sum_k |\langle 0| \sum_{i=1}^N l_i^k |M1\rangle|^2 = \frac{4}{5}\mu_0^2\delta\sqrt{\epsilon_F r_s^2}N^{4/3}, \qquad (11.128)$$

where $\mu_0 = e\hbar/2mc$ is the Bohr magneton.

For quantum dots, the scissor mode has been studied microscopically by solving the time-dependent Kohn–Sham equations, with different initial conditions corresponding to a pure rotation, a pure quadrupole distortion, and to a combination of the two. The results relative to the time evolution of an elliptic quantum dot with 20 electrons, $r_s = 1.51$ (effective a.u.) and $\delta = 0.28$, are shown in Fig. 11.5 (Serra, Puente and Lipparini, 1999). From this figure, we see that two states are excited: the scissor mode at low energy, and the scalar quadrupole mode at high energy. The elastic macroscopic model reproduces the energies of the microscopic calculation to a very good extent.

Experimental evidence is still lacking for the existence of such excitation modes for both the metal clusters and quantum dots. On the other hand, there are many experimental results in the nuclear case, where the existence of the scissor mode has been verified in a systematic way in deformed nuclei (Bohle *et al.*, 1984). A recent experimental analysis (Enders *et al.*, 1999) showed that the energies of the low energy mode, as well as the excitation strength $M1$, are well reproduced using the moment of inertia and the gyromagnetic factors of the rotational band of the ground state. Such values of the moment of inertia are smaller than the rigid value, and so the analysis shows that only some of the nucleons participate in the rotational motion.

11.2.5 *The Moment of Inertia of Quantum Dots*

The quantum evaluation of the moment of inertia of a Fermion system, like electrons confined in an elliptic quantum dot, is not a trivial task and requires, for example in the framework of the density functional theory, the solution of the Kohn–Sham equations with an angular momentum constraint (Serra, Puente and Lipparini, 2002):

$$[h_\sigma(\rho, m) - \Omega\ell_z]\varphi_\sigma(\mathbf{r}; \Omega) = \epsilon_\sigma(\Omega)\varphi_\sigma(\mathbf{r}; \Omega), \qquad (11.129)$$

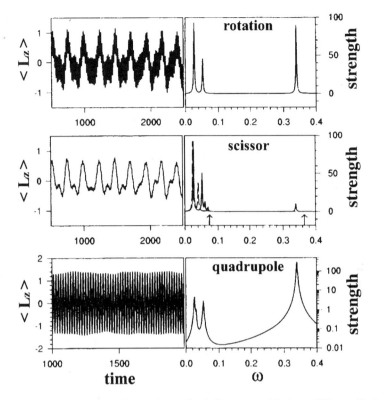

Fig. 11.5 Time evolution of an elliptic dot with 20 electrons, with three different kinds of initial conditions: pure rotation, pure quadrupolar distortion (quadrupole), and a combination of the two (scissor). The left hand panels show the $M1$ signal as a function of time, while the right hand ones show the corresponding excitation strengths in arbitrary units. The midde right hand panel also shows the strength without interaction effects (dashed line) as well as the predictions of the elastic model for the scissor and quadrupole modes (arrows).

where ℓ_z is the z component of the orbital angular momentum of the individual electron, $\sigma = \uparrow, \downarrow$ is the spin label, and the densities and magnetization are given, respectively, by $\rho = \rho_\uparrow + \rho_\downarrow$ and $m = \rho_\uparrow - \rho_\downarrow$, where

$$\rho_\sigma = \sum_i |\varphi_{i\sigma}(\mathbf{r}; \Omega)|^2.$$

Besides the kinetic energy term, the single-particle Hamiltonian h_σ in (11.129) includes the confinement potential

$$v_{\text{ext}}(\mathbf{r}) = 1/2(\omega_x^2 x^2 + \omega_y^2 y^2),$$

the Hartree potential

$$v^H(\mathbf{r}) = \int d\mathbf{r}' \rho(\mathbf{r}') / |\mathbf{r} - \mathbf{r}'|,$$

and the exchange-correlation potential

$$V_\sigma^{\mathrm{xc}}(\mathbf{r}) = \partial\mathcal{E}_{\mathrm{xc}}(\rho,m)/\partial\rho_\sigma.$$

The Kohn–Sham equations can be solved by subdividing the x–y plane into a discrete uniform grid of points, and subsequently using the methods described in detail, for example, by Puente and Serra (1999). The moment of inertia is then computed using (11.46), where the average is taken on the Slater determinant constructed by means of the solutions $\varphi_\sigma(\mathbf{r};\Omega)$ to Eq. (11.129). The results are plotted in Fig. 11.6, together with those of the noninteracting model given by (11.63), as a function of the deformation $\beta = \omega_y/\omega_x$. The average parameter of the harmonic oscillator, i.e. $\omega_0 = (\omega_x + \omega_y)/2$, is determined by $\omega_0^2 = 1/r_s^3\sqrt{N}$. In these figures, we note that there are some small oscillations at large values of the deformation, which are due to deformation shell effects; there are also some large oscillations at small values of the deformation (β close to unity) in the closed shell dots (12,20), and some divergences in the open shell dots (13,14), which are remedied by the interaction only for $N = 14$. The large oscillations at small deformation are related to the emergence, with the deformation, of spin waves in the ground state, while the divergences reflect the vanishing of the excited state energies in the denominator of (11.53) for the open shell dots, which in some cases are not remedied by the interaction. These effects might be investigated experimentally by studying the diamagnetic susceptibility of

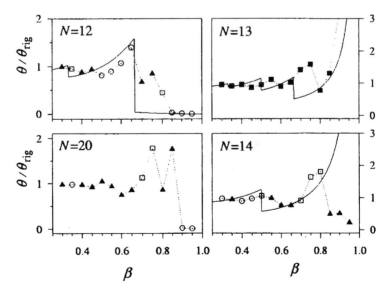

Fig. 11.6 Results for the $\theta/\theta_{\mathrm{rig}}$ ratio in elliptic quantum dots, with different numbers of electrons (N), as a function of the deformation parameter β, and for $r_s = 1.5$. Full lines are the results of the harmonic oscillator without interaction; the symbols are the LSDA results. Different symbols refer to different values of the total ground state spin: open circle ($S_z = 0$), open square (0 with spin wave), full square (1/2), full triangle (1).

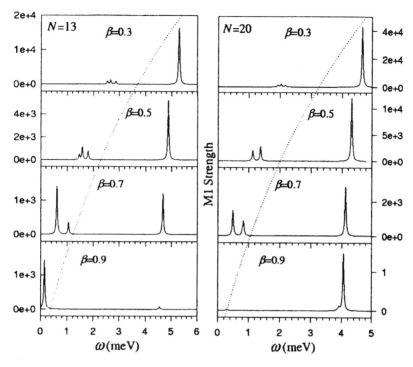

Fig. 11.7 Evolution of the strengths $M1$ with the deformation for $r_s = 1.5$, $N = 13, 20$. The dashed lines, at the intersection with the base line of each spectrum, indicate the energy ω_- in the harmonic oscillator model, while ω_+ is 6.8 meV for $N = 13$, and 6.1 meV for $N = 20$.

the dot, which is given by (see Subsection 8.7)

$$\chi_D = -\frac{e^2}{4m^2c^2}\Theta_{\rm rig}\left(1 - \frac{\Theta}{\Theta_{\rm rig}}\right). \tag{11.130}$$

Another possibility of studying the moment of inertia of Fermion systems is provided by its connection with the excitation, and in particular with the scissor mode. In Fig. 11.7 we show the evolution of the strength $M1$ with the deformation for $N = 13$ and 20. The orbital strength $M1$ is clearly separated into two regions: a high energy one, related to the quadrupole, and a low energy one, related to the scissor mode. For intermediate values of deformation ($0.5 < \beta < 0.7$), the strength is shared by the two coexisting modes, while at high and low deformations it is practically exhausted by only one mode. For large deformations the quadrupole mode survives, while for $\beta \to 1$ the dominant mode depends on θ. For systems with a divergent moment of inertia, the dominant mode is the scissor mode, while for vanishing moment of inertia the dominant mode is the quadrupole one. The figure also shows that the interaction lowers the energy of the scissor mode when compared to the prediction of the harmonic oscillator (11.60). The fractional contribution of the scissor mode to $m_{-1} = \theta/2$ and to the energy weighted sum rule (11.56) is

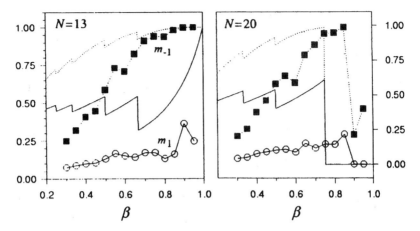

Fig. 11.8 Fractional contribution of the scissor mode to the $m_1(L_z)$ sum rule in the harmonic oscillator model (full line), and in the LSDA theory (open circles connected by a full line). Also shown is the contribution to $m_{-1}(L_z)$ in the harmonic oscillator (dotted line) and in LSDA (full squares connected by a dotted line).

shown in Fig. 11.8. This figure shows quantitatively the contribution of the scissor mode to the moment of inertia in the LSDA theory (squares), and in the harmonic oscillator (dotted line). At large deformations, we observe an appreciable decrease of such contribution due to interaction.

11.2.6   *The Vibrating Potential Model*

As discussed several times, the study of collective vibration modes, such as the surface and elastic ones which were described previously by using sum rule techniques, is carried out in a natural way within the framework of time-dependent, self-consistent theories such as the TDLDA theory of Subsection 9.9. These methods are based on the study of the variations of the average one-body potential, induced by a density oscillation. These average-potential variations produce excited states of the system, and one obtains a self-sustained collective mode, if the induced density variations are equal to the ones required to generate the oscillatory potential.

In the present subsection we will study an approximate solution to the problem, which corresponds to a first iteration of the fully self-consistent calculation. This method is known as the vibrating potential model (Rowe, 1970), and in homogeneous systems (where the translational invariance determines the pattern of density oscillations) leads to the exact solution (see Subsections 9.2). In finite systems (i.e. nuclei, quantum dots, etc.) this method provides a useful and simple approximation, and it has been widely used (Bohr and Mottelson, 1975; Lipparini and Stringari, 1991; Serra *et al.*, 1999; Nesterenko *et al.*, 1999).

For simplicity, we will limit ourselves to studying scalar excited states. The generalization to vector states is very simple.

The basic ingredients of the model are:

(i) A local functional of the scalar density ρ of the form

$$E(\rho, \tau) = \frac{1}{2m} \int \tau d\mathbf{r} + \int v(\rho) d\mathbf{r} + \frac{1}{2} \int \rho(\mathbf{r})\rho(\mathbf{r}')v(|\mathbf{r} - \mathbf{r}'|) d\mathbf{r} d\mathbf{r}', \quad (11.131)$$

where

$$\rho(\mathbf{r}) = \sum_i |\varphi_i(\mathbf{r})|^2, \quad \tau(\mathbf{r}) = \sum_i |\nabla\varphi_i(\mathbf{r})|^2,$$

and the function $v(\rho)$ includes contributions from both the external potential and the exchange-correlation one.

(ii) The density variations $\delta\rho(\mathbf{r}, t) = \alpha(t)\nabla(\rho_0(\mathbf{r})\nabla f(\mathbf{r}))$, equal to those of the generalized scaling of (11.79) with $\mathbf{u}(\mathbf{r}, t) = \alpha(t)\nabla f(\mathbf{r})$, where $\alpha(t)$ gives the oscillation amplitudes and $f(\mathbf{r})$ is the multipole operator which excites the collective mode.

The one-body Hamiltonian used to describe the collective vibrations is given by

$$H\varphi_i(\mathbf{r}) = \frac{\delta E}{\delta\varphi_i^*(\mathbf{r})} = \left(\frac{-\nabla^2}{2m} + V(\mathbf{r}, t)\right)\varphi_i(\mathbf{r}), \quad (11.132)$$

where

$$V(\mathbf{r}, t) = \frac{\partial v(\rho)}{\partial\rho} + \int \rho(\mathbf{r}')v(|\mathbf{r} - \mathbf{r}'|) d\mathbf{r}'. \quad (11.133)$$

In $H(\mathbf{r}, t)$, we can separate a static part (i.e. the one-body Hamiltonian relative to the static ground state),

$$H^0 = \frac{-\nabla^2}{2m} + \left(\frac{\partial v(\rho)}{\partial\rho}\right)_{\rho=\rho_0} + \int \rho_0(\mathbf{r}')v(|\mathbf{r} - \mathbf{r}'|) d\mathbf{r}', \quad (11.134)$$

and a dynamic time-dependent part which stems from density changes that occur during the collective motions:

$$\delta H = \delta V(\mathbf{r}, t) = \left.\frac{\partial V}{\partial\rho}\right|_{\rho=\rho_0} \delta\rho(\mathbf{r}, t)$$

$$= \left.\frac{\partial^2 v(\rho)}{\partial\rho^2}\right|_{\rho=\rho_0} \delta\rho(\mathbf{r}, t) + \int \delta\rho(\mathbf{r}', t)v(|\mathbf{r} - \mathbf{r}'|) d\mathbf{r}'. \quad (11.135)$$

If we introduce in (11.135) the density variations of the generalized scaling, we obtain

$$\delta H = \alpha(t)\left[\left.\frac{\partial^2 v(\rho)}{\partial\rho^2}\right|_{\rho=\rho_0} \nabla(\rho_0(\mathbf{r})\nabla f(\mathbf{r}))\right.$$

$$\left. + \int \nabla(\rho_0(\mathbf{r}')\nabla f(\mathbf{r}'))v(|\mathbf{r} - \mathbf{r}'|) d\mathbf{r}'\right]. \quad (11.136)$$

For excitation operators of the form $f = r^\ell Y_{\ell,m}$, for which $\nabla^2 f = 0$, we obtain

$$\delta H = \alpha(t) \left[\nabla V_0 \cdot \nabla f + \int \nabla \rho_0(\mathbf{r}') \cdot \nabla f(\mathbf{r}')) v(|\mathbf{r} - \mathbf{r}'|) d\mathbf{r}' \right], \qquad (11.137)$$

where $V_0 = \frac{\partial v(\rho)}{\partial \rho}\big|_{\rho=\rho_0}$. Defining the operator

$$\hat{Q} = \sum_{i=1}^N Q(r_i) = \sum_{i=1}^N \left(\nabla V_0 \cdot \nabla f + \int \nabla \rho_0(\mathbf{r}') \cdot \nabla f(\mathbf{r}')) v(|\mathbf{r} - \mathbf{r}'|) d\mathbf{r}' \right)_i, \qquad (11.138)$$

it is then possible to write

$$\sum_{i=1}^N \delta H(r_i) = \alpha(t)\hat{Q}. \qquad (11.139)$$

Finally, using the result

$$Q(t) = \int Q(r)\delta\rho(\mathbf{r}, t)d\mathbf{r} = \alpha(t) \int \nabla Q \cdot \nabla f \rho_0 d\mathbf{r}, \qquad (11.140)$$

it is possible to write the time-dependent dynamic term in a separable form:

$$\sum_{i=1}^N \delta H(r_i) = -kQ(t)\hat{Q}, \qquad (11.141)$$

with

$$\frac{1}{k} = \int \nabla Q \cdot \nabla f \rho_0. \qquad (11.142)$$

The solution of the linearized time-dependent equations in an oscillating external field,

$$\sum_{i=1}^N (H_0(r_i) + \delta H(r_i) + \lambda(Q(r_i)e^{i\omega t} + \text{H.c.}))|\Psi(t)\rangle = i\frac{\partial}{\partial t}|\Psi(t)\rangle, \qquad (11.143)$$

with $H_0(r_i)$ and $\delta H(r_i)$ given by Eqs. (11.134) and (11.141), respectively, can be reduced (Lipparini and Stringari, 1981) to the solution of the dispersion relation

$$2\sum_k \epsilon_k \frac{|\langle 0|\hat{Q}|k\rangle|^2}{\epsilon_k^2 - \omega^2} = \frac{1}{k}, \qquad (11.144)$$

where ϵ_k and $|k\rangle$ are the eigenvalues and eigenfunctions of the static Hamiltonian (11.134).

The oscillatory potential (11.141) can also be derived starting from a separable, effective two-body interaction (Bohr and Mottelson, 1975; Rowe, 1970) of the form

$$V = \frac{1}{2}k\sum_{i,j} Q(r_i)Q(r_j). \qquad (11.145)$$

In this case, the analysis of the present subsection is equivalent to studying the effects of RPA correlations in the many-body system, as produced by such separable interaction.

As an example of the application of the vibrating potential model (VPM), let us consider the case of quadrupole excitations in metal clusters.

For these systems, simple and significant solutions of the VPM are obtained by assuming:

(i) A complete screening between the Coulomb potential produced by the electron–electron interaction (direct term) and that of the jellium of positive charge (produced by the jellium density ρ_j), so that the static Hamiltonian becomes

$$\sum_{i=1}^{N} H^0(r_i) = \sum_{i=1}^{N} \left(\frac{-\nabla_i^2}{2m} + v_{xc}(\rho_0(r_i)) + \int \frac{(\rho_0(\mathbf{r}') - \rho_j(\mathbf{r}'))}{|\mathbf{r}_i - \mathbf{r}'|} d\mathbf{r}' \right)$$

$$\simeq \sum_{i=1}^{N} \left(\frac{-\nabla_i^2}{2m} + V_0(r_i) \right). \tag{11.146}$$

(ii) A harmonic-oscillator-like exchange-correlation one-body potential, i.e.

$$V_0(r_i) = \frac{1}{2}\omega_0^2 r_i^2, \tag{11.147}$$

where the harmonic oscillator parameter ω_0 is given by the relation

$$\omega_0 = \sqrt{\frac{2\epsilon_F}{mr_s^2}} N^{-1/3}, \tag{11.148}$$

which is obtained by equating the kinetic energy per particle as given by the harmonic oscillator model with that of the Fermi gas, and by approximating $\langle r^2 \rangle$ as $(3/5)r_s^2 N^{2/3}$.

For the quadrupole excitations we then have $f = yz$ and $Q(r) = (2\omega_0^2 - \frac{2}{5}\omega_p^2)yz$, where, in order to obtain the expression for Q, we have made use of the relations

$$\int \frac{\nabla_{z'}\rho_0(\mathbf{r}')}{|\mathbf{r} - \mathbf{r}'|} d\mathbf{r}' = \nabla_z \int \frac{\rho_0(\mathbf{r}')}{|\mathbf{r} - \mathbf{r}'|} d\mathbf{r}' \tag{11.149}$$

and

$$\int \frac{z'\rho_0(\mathbf{r}')}{|\mathbf{r} - \mathbf{r}'|} d\mathbf{r}' = z \int \frac{\rho_0(\mathbf{r}')}{|\mathbf{r} - \mathbf{r}'|} d\mathbf{r}' - \nabla_z \int \rho_0(\mathbf{r}')|\mathbf{r} - \mathbf{r}'| d\mathbf{r}', \tag{11.150}$$

as well as the results

$$\int \frac{\rho_0(\mathbf{r}')}{|\mathbf{r} - \mathbf{r}'|} d\mathbf{r}' = -\frac{1}{2}\frac{N}{R^3}r^2,$$

$$\int \rho_0(\mathbf{r}')|\mathbf{r} - \mathbf{r}'| d\mathbf{r}' = NR \left(\frac{3}{4} + \frac{1}{2}\frac{r^2}{R^2} - \frac{1}{20}\frac{r^4}{R^4} \right), \tag{11.151}$$

which hold for a uniform distribution $\rho_0(r)$. Moreover, we have

$$\frac{1}{k} = \frac{2}{3}N\left(2\omega_0^2 - \frac{2}{5}\omega_p^2\right)\langle r^2\rangle, \qquad \left|\langle 0|\sum_{i=1}^{N} y_i z_i|k\rangle\right|^2 = \frac{N\langle r^2\rangle}{6\omega_0}. \qquad (11.152)$$

Using these results, and considering that only one state at energy $2\omega_0$ contributes to the sum of (11.144), we obtain the solution for the dispersion relation

$$\omega_Q = \sqrt{2\omega_0^2 + \frac{2}{5}\omega_p^2}. \qquad (11.153)$$

The frequency (11.153) coincides with the result (11.120) provided that we use $E(X) = (9/50)e^2(N^2/R)$, $\langle r^2\rangle = 3/5NR^2$ and Eq. (11.148). Moreover, it coincides with the classical plasma frequency for the quadrupole, i.e. $\omega_Q^{cl} = \sqrt{2/5}\omega_p$, if we neglect the term in $2\omega_0$ which is produced by the quantum distortion of the Fermi sphere. Therefore, we see that, although the simplified RPA model lacks the finite-dimension effects due to electron spillout outside the jellium radius [total screening assumption (11.146)], it contains quantum effects that can be appreciable. Such effects in deformed systems give rise, for example, to the low energy scissor state which has no classical analog. The scissor state is easily obtained in the VPM using the above results for the quadrupole case, and taking into account the fact that in the deformed case two single-particle levels contribute to the sum of (11.144), with energies $\epsilon_0 = \omega_y - \omega_z = \delta\omega_0$ and $\epsilon_2 = \omega_y + \omega_z = 2\omega_0$, with the respective matrix elements

$$\left|\langle 0|\sum_{i=1}^{N} y_i z_i|\epsilon_0\rangle\right|^2 = \frac{N\langle r^2\rangle}{12\omega_0}\delta, \qquad \left|\langle 0|\sum_{i=1}^{N} y_i z_i|\epsilon_2\rangle\right|^2 = \frac{N\langle r^2\rangle}{6\omega_0}, \qquad (11.154)$$

and where we have considered only the contributions of the lowest order in the deformation δ.

Then, from the dispersion relation (11.144), we have two solutions. The higher energy one is the quadrupole plasmon excitation of (11.153), while the lower energy one has a frequency given by

$$\omega_{M1} = \sqrt{2}\omega_0\delta\left(1 + 5\frac{\omega_0^2}{\omega_p^2}\right)^{-1/2}. \qquad (11.155)$$

This solution coincides with that of the scissor mode in (11.126), as readily seen using expression (11.148) for ω_0 and neglecting the correction $5\omega_0^2/\omega_p^2$.

11.3   The Surface Vibrations of Charged Systems in 2D and 3D

In this subsection we will study the surface vibrations of a drop of charged Fermions, using the techniques developed in Subsection 11.1.2 for helium drops. The valence electrons of alkali metal clusters, and the two-dimensional electron gas confined in a quantum dot under a magnetic field, are the examples of charged quantum fluid that we will consider.

11.3.1 *Surface Vibrations of Charged Metal Clusters*

In the following, we will describe metal clusters by means of the jellium model, in which the total cluster energy is dominated by the electronic energy, and the equilibrium shape of the system corresponds to the minimum electronic energy. The positive ions have the role of screening the electrostatic potential of the valence electrons, and the shape of the positive charge background is assumed to conform to that of the electrons.

For such clusters, when the system is neutral, the total energy is well represented by formula of the liquid drop model:

$$E(N) = a_v N + a_s N^{2/3}, \tag{11.156}$$

where a_v is the mean total energy per electron in the infinite jellium, and $a_s = 4\pi r_s^2 \sigma$ is the surface energy connected to the surface tension of a semi-infinite jellium. In the case of sodium clusters, we have $r_s = 3.93$, $a_v = -2.1$ eV and $a_s = 0.57$ eV (Perdew, 1988).

Therefore, the frequencies of the surface modes of the neutral jellium drop are given by the same expression as in the case of the helium drop:

$$\omega_\ell^{\text{sup}} = \sqrt{\frac{\ell(\ell-1)(\ell+2)}{3} \frac{a_s}{Mr_s^2}} N^{-\frac{1}{2}}.$$

Since all of the cluster mass is concentrated on the positive ions, the cluster vibrations in the jellium model are due to the shape changes of positive charge background, so that the masses that appear in such formulas are the ionic masses.

If we use for a_s and r_s the above values suitable for sodium, we obtain for the energy of the quadrupole mode the estimate $\omega_{\ell=2}^{\text{sup}} = 0.01$ eV, i.e. a value that is smaller than the plasmon excitation energies by a factor of 100. The coupling between the quadrupolar surface vibrations and the plasma modes was studied by Bertsch and Tomanek (1989) and Pacheco and Broglia (1989), and accounts for the broadening of the excited states as observed in photoemission experiments.

When the cluster is charged, the Coulomb repulsion due to the extra charge contrasts the effect of the surface tension, which tends to restore the system to the original shape, and may lead to its instability. Producing a critical deformation that leads to a fission process may require a rather small amount of energy. The electrostatic energy change connected with the surface deformation of the drop has been evaluated by Lipparini and Vitturi (1990), and the calculation proceeds as follows. One starts from the usual expression

$$E_{\text{Coul}} = \frac{1}{2} \int \frac{(\rho - \rho_I)_{\mathbf{r}}(\rho - \rho_I)_{\mathbf{r}'}}{|\mathbf{r} - \mathbf{r}'|} d\mathbf{r} d\mathbf{r}', \tag{11.157}$$

for the electrostatic energy. The electronic and ionic densities, ρ and ρ_I, are written as

$$\rho(r,\theta,\phi) = \rho_0 S(R^e(\theta,\phi) - r), \quad \rho_I(r,\theta,\phi) = \rho_I S(R(\theta,\phi) - r), \tag{11.158}$$

where both $R^e(\theta,\phi)$ and $R(\theta,\phi)$ have the form (11.22), and S is the step function. Owing to the conducting nature of the metal clusters, the extra charge Z is distributed on the surface, the central densities may be assumed identical, and the ratio of the equilibrium radii R_0^e and R_0 is given by

$$\frac{R_0^e}{R_0} = \left(1 - \frac{Z}{N}\right)^{1/3}. \tag{11.159}$$

In order that this distribution of extra surface charge is conserved even when the surface is deformed, one further assumes the same form for the electron and ion densities, i.e. equal deformation parameters $\alpha_{\ell,m}$ with $\ell > 1$ for the two distributions. For small deformations $|R(\theta,\phi) - R_0| \ll R_0$, it is possible to use the expansion

$$\rho(r,\theta,\phi) = \rho_0 \Big(S(R_0 - r) + (R(\theta,\phi) - R_0)\delta(R_0 - r)$$

$$- \frac{1}{2}(R(\theta,\phi) - R_0)^2 \delta'(R_0 - r) + \cdots \Big) \tag{11.160}$$

for both of the densities (11.158). This yields $(Z/N \ll 1)$

$$E_{\text{Coul}} = \frac{Z^2}{2R_0} - \frac{1}{2}\frac{Z^2}{4\pi R_0}\sum_{\ell m}\frac{\ell^2 + 3\ell - 5}{2\ell + 1}|\alpha_{\ell m}|^2, \tag{11.161}$$

where the first term is the spherical contribution to the electrostatic energy due to the extra charge Z, and the second one gives the energy variation caused by the deformation of the surface. Therefore, the contribution of the Coulomb energy to the restoring force is $(\ell > 1)$

$$(C_\ell)_{\text{Coul}} = -\frac{Z^2}{4\pi R_0}\frac{\ell^2 + 3\ell - 5}{2\ell + 1}. \tag{11.162}$$

From Eqs. (11.28) and (11.162), we get

$$C_\ell = (C_\ell)_{\text{surf}} + (C_\ell)_{\text{Coul}}$$

$$= \frac{1}{4\pi}(\ell - 1)(\ell + 2)a_s N^{2/3} - \frac{Z^2}{4\pi r_s}\frac{\ell^2 + 3\ell - 5}{2\ell + 1}N^{-1/3} \tag{11.163}$$

for the total restoring force. As mentioned, we see that the Coulomb repulsion contrasts the effect of surface tension, and leads to instability for large values of

Z^2/N. Instability against the lowest energy quadrupole mode ($\ell = 2$) sets in for a critical value of Z^2/N given by

$$\left(\frac{Z^2}{N}\right)_{\text{crit}} = 4r_s a_s. \tag{11.164}$$

Using the values $r_s = 4$ and $a_s = 0.57$ eV, which are the relevant parameters for sodium clusters, we can predict that for the number of atoms N larger than $3.1\ Z^2$, ionized clusters with charge Z are stable against fission processes. Such a prediction, which overlooks other possible fragmentation channels, such as monomer emission which is privileged for small Z (Perdew, 1988; Baladron, 1989), is not too far from experimental results, in which fission processes are induced through reactions with photons. An exhaustive study of fission processes in metal clusters, including both theoretical and experimental aspects, is found in the review paper by Naher *et al.* (1997).

11.3.2   *Edge Vibrations of Quantum Dots*

In this subsection we will consider a drop of uncompressible liquid in two dimensions, and develop the formalism of the edge vibrations of the drop, analogous to what was done previously in three-dimensional systems (Giovanazzi, 1993). Unlike the previous subsection, where we studied the problem of instability due to excess charge in a three-dimensional drop and assumed that the shape of the jellium conforms to that of the electrons, here we consider the surface vibrations of the electron drop in a rigid confinement field. Such electronic surface vibrations correspond to the low energy excited states that were studied in Subsection 9.13.2 in a microscopic fashion and as a function of an external magnetic field acting on the quantum dot.

In analogy with what we did in Subsection 11.1.2, we may describe the edge vibrations of the two-dimensional drop by a set of normal coordinates α_{ℓ_0} of the system, in angular momentum space, by expanding the distance $R(\theta)$ of the edge from the origin in terms of the functions $e^{i\ell_0\theta}$:

$$R(\theta) = R_0\left(1 + \sum_{\ell_0} \alpha_{\ell_0} e^{i\ell_0\theta}\right), \tag{11.165}$$

where R_0 is the equilibrium radius and the sum runs over the (positive and negative) integers ℓ_0. Like in the 3D case, the term with $\ell_0 = 0$ represents a compression without change of shape, and the term with $\ell_0 = 1$ represents a translation of the whole system. The mode relative to α_2 is the first one that implies a shape change in the drop, and is the quadrupole mode.

The density variations are limited to changes in the drop profile. If $\rho(r,\theta) = \rho_0 S(R(\theta) - r)$, we have

$$\rho(r,\theta) = \rho_0 \Big(S(R_0 - r) + (R(\theta) - R_0)\delta(R_0 - r)$$

$$-\frac{1}{2}(R(\theta) - R_0)^2\delta'(R_0 - r) + \cdots \Big). \tag{11.166}$$

Surface conservation implies the relation

$$\alpha_0 = -\frac{1}{2}\sum_{\ell_0}|\alpha_{\ell_0}|^2. \tag{11.167}$$

In the limit of small oscillations, a Lagrangian formulation is possible, which allows us to derive the equations of motion of the edge oscillations in the presence of a magnetic field:

$$L = \frac{1}{2}\sum_{\ell_0}\left(D_{\ell_0}\dot{\alpha}_{\ell_0}^*\dot{\alpha}_{\ell_0} - \frac{i}{2}D_{\ell_0}\omega_c\frac{|\ell_0|}{\ell_0}(\dot{\alpha}_{\ell_0}^*\alpha_{\ell_0} - \alpha_{\ell_0}^*\dot{\alpha}_{\ell_0}) - C_{\ell_0}\alpha_{\ell_0}^*\alpha_{\ell_0}\right), \tag{11.168}$$

where the term proportional to the cyclotron frequency $\omega_c = eB/mc$ accounts for the external magnetic field applied in a direction perpendicular to the drop plane, which gives rise to a Lorentz force $-e/c\mathbf{v} \times \mathbf{B}$, with $\mathbf{v}$ the velocity of the drop edge. The equations of motion have the form

$$D_{\ell_0}\ddot{\alpha}_{\ell_0} - iD_{\ell_0}\omega_c\frac{|\ell_0|}{\ell_0}\dot{\alpha}_{\ell_0} + C_{\ell_0}\alpha_{\ell_0} = 0, \tag{11.169}$$

where D_ℓ and C_ℓ are the mass parameter and the restoring-force parameter of the oscillation, respectively. The equation for the oscillation frequency is given by

$$\omega_{\ell_0}\left(\omega_{\ell_0} + \omega_c\frac{|\ell_0|}{\ell_0}\right) = \frac{C_{\ell_0}}{D_{\ell_0}}. \tag{11.170}$$

If $\omega_{\ell_0} \ll \omega_c$, the lowest frequency is given by

$$\omega_{\ell_0} = \frac{\ell_0 C_{\ell_0}}{\omega_c|\ell_0|D_{\ell_0}}. \tag{11.171}$$

For the calculation of the mass parameter D_{ℓ_0}, we will assume (like in 3D) that the flux is uncompressible (at constant density) and irrotational, so that the velocity field is written as the gradient of some potential $\chi(\mathbf{r})$:

$$\mathbf{v}(\mathbf{r}) = -\nabla\chi(\mathbf{r}).$$

Then, the continuity Eq. (11.6) implies that the potential satisfies the Laplace equation $\nabla^2\chi(\mathbf{r}) = 0$, with the solution

$$\chi(\mathbf{r}) = \sum_{\ell_0}a_{\ell_0}r^{\ell_0}e^{i\ell_0\theta}. \tag{11.172}$$

The coefficients a_{ℓ_0} and α_{ℓ_0} are connected by the boundary condition on the edge, which demands that on the edge itself the radial component of velocity be equal to the radial velocity at which the edge moves. For small edge displacements, such a boundary condition requires

$$-\frac{\partial}{\partial r}\chi = \frac{dR}{dt},$$

as computed in $r = R_0$, from which it follows that

$$a_{\ell_0} = \frac{R_0^{2-\ell_0}}{|\ell_0|}\dot\alpha_{\ell_0}. \tag{11.173}$$

Then, the kinetic energy of the fluid due to the edge deformations becomes

$$T = \frac{1}{2}m\rho_0\int \mathbf{v}^2(r)d\mathbf{r} = \sum_{\ell_0} M\frac{R_0^2}{|\ell_0|}\dot\alpha_{\ell_0}^*\dot\alpha_{\ell_0}, \tag{11.174}$$

and comparison with (11.168) finally yields the result

$$D_{\ell_0} = \frac{2}{|\ell_0|}MR_0^2, \tag{11.175}$$

where $M = mN$ is the total mass of the electron drop.

The parameter C_{ℓ_0} of the restoring force of the edge oscillations on the drop is computed starting from the Coulomb energy variations which are connected with the electron density variation. First of all, let us consider the interaction term with the jellium,

$$\delta E_{je} = \int \delta\rho(r,\theta)V_+(r)d\mathbf{r} = -\frac{1}{2}R_0^2\sum_{\ell_0}|\alpha_{\ell_0}|^2\int(\partial_r^2\rho(r))V_+(r)d\mathbf{r}, \tag{11.176}$$

where $\partial_r\rho(r)$ is the radial derivative of the static density of the electron drop, and V_+ is the Coulomb potential generated by the jellium density $\rho_0 S(R_0 - r)$. It is immediate to show that the integral in (11.176) can be related to the electric field $\mathcal{E}(R_0)$ produced by the electrons on the jellium edge, so that

$$\delta E_{je} = \pi\rho_0 R_0^3\mathcal{E}(R_0)\sum_{\ell_0}|\alpha_{\ell_0}|^2. \tag{11.177}$$

For large values of the electron number N, one finds that $\mathcal{E}(R_0) = \rho_0\ln(\beta N)$, with $\beta = 16.4$ (Giovanazzi, Pitaevskii and Stringari, 1994). Note that $\mathcal{E}(r)$ was studied numerically in a dot in Subsection 5.6.

Let us consider now the variations of the electron–electron Coulomb interaction energy:

$$\delta E_{ee} = \int \delta\rho(r,\theta)\frac{1}{|r-r'|}\rho(r')d\mathbf{r}d\mathbf{r}' + \frac{1}{2}\int \delta\rho(r,\theta)\frac{1}{|r-r'|}\delta\rho(r',\theta')d\mathbf{r}d\mathbf{r}'. \tag{11.178}$$

This quantity becomes

$$\delta E_{ee} = \frac{1}{2}R_0^2 \sum_{\ell_0} |\alpha_{\ell_0}|^2 \int \partial_r \rho(r) \frac{\cos(\ell_0(\theta-\theta')) - \cos(\theta-\theta')}{|r-r'|} \partial_{r'} \rho(r') d\mathbf{r} d\mathbf{r}' \quad (11.179)$$

and, in the limit of large N ($N \gg \ell_0$),

$$\delta E_{ee} = -4\pi R_0^3 \rho_0^2 \sum_{\ell_0} |\alpha_{\ell_0}|^2 \sum_{k=2}^{\ell_0} \frac{1}{2k-1}, \quad (11.180)$$

where we have used

$$\int_0^{2\pi} \frac{\cos(\ell_0\theta) - \cos\theta}{\sqrt{1-\cos\theta}} = -4\sqrt{2}\sum_{k=2}^{\ell_0} \frac{1}{2k-1}, \quad (11.181)$$

which holds for $\ell_0 > 1$ (for $\ell_0 = 1$, $\delta E_{ee} = 0$). Furthermore, we have

$$\sum_{k=2}^{\ell_0} \frac{1}{2k-1} = \frac{1}{2}(\ln(\ell_0) + C + \ln 4 - 2) = \frac{1}{2}\ln(\gamma\ell_0),$$

where $C = 0.577215$ is the Euler constant, and therefore $\gamma = 0.96417$.

Finally, by combining the above results we find that

$$C_{\ell_0} = C_{\ell_0}^{je} + C_{\ell_0}^{ee} = 4\pi R_0^3 \rho_0^2 \ln(\sqrt{\beta N}) - 4\pi R_0^3 \rho_0^2 \ln(\gamma\ell_0)$$

$$= 4\pi R_0^3 \rho_0^2 \ln\left(\frac{\sqrt{\beta N}}{\gamma\ell_0}\right), \quad (11.182)$$

which holds for $\ell_0 > 1$ and large N. In the dipolar case $C_1^{ee} = 0$, thus,

$$C_1^{je} = 2\pi R_0^3 \rho_0 \mathcal{E}(R_0). \quad (11.183)$$

Using the results of Eqs. (11.170) and (11.175), we then find for the edge frequency

$$\omega_{\ell_0}\left(\omega_{\ell_0} + \omega_c \frac{\ell_0}{|\ell_0|}\right) = \frac{2\ell_0\rho_0}{mR_0}\ln\left(\frac{\sqrt{\beta N}}{\gamma\ell_0}\right), \quad (11.184)$$

for $\ell_0 > 1$ and large N, and

$$\omega_1(\omega_1 \pm \omega_c) = \frac{\mathcal{E}(R_0)}{mR_0}, \quad (11.185)$$

for $\ell_0 = 1$. Note that for parabolic confinement, the procedure that led us to (11.185) gives exactly the result $\omega_{\pm} = \sqrt{\omega_0^2 + \omega_c^2/4} \pm \omega_c/2$ of the generalized Kohn theorem of Subsection 8.9.1.

As an application of (11.184), we compute the frequency of the lowest energy solution in the case where the magnetic field is such that the filling factor is 1. In this case the electron density for large N is given by $\rho_0 = (2\pi l^2)^{-1}$ within the circle of radius $R_0 = l\sqrt{2N}$, where $l = (\hbar c/eB)^{1/2}$ is the magnetic length. For large values of ℓ_0, it is natural to bring in the parametrization $\ell_0 = qR_0 = ql\sqrt{2N}$, which follows from the condition that the oscillatory mode is stationary; this condition requires that an integer number ℓ_0 of wavelengths fit in the length of the edge. In this way we obtain the dispersion relation

$$\omega_q = \frac{q}{\pi} \ln \frac{q_0}{q}, \tag{11.186}$$

where $q_0 = \sqrt{\beta}/\sqrt{2}\gamma l = 3.0/l$. Such a relation holds in the limit where $N \gg \ell_0$, i.e. large wavelength limit, $ql \ll 1$. The $q\ln(1/q)$ dependence is typical of the long range Coulomb interaction and for jellium confinement, and has been derived by many authors (Volkov and Mikhailov, 1985; Wen 1991; Giovanazzi, Pitaevskii and Stringari, 1994).

It is interesting to note that for a neutral drop with short range interaction among its constituents, the dispersion relation is completely different from Eq. (11.186). This can be seen by calculating the parameter C_{ℓ_0} on the restoring force of the edge oscillations on the drop, starting from the variation of the edge energy

$$\delta E = \frac{1}{2} \sum_{\ell_0} C_{\ell_0} |\alpha_{\ell_0}|^2 = \lambda \delta L, \tag{11.187}$$

where δL is the edge-length variation and λ is analogous to surface tension in 3D. This restoring force is typical of short-range-interaction systems, such as the nucleon drop. However, it is also typical of electron systems under an intense magnetic field, which can lead to a fractional regime; in fact, in this case short range interactions produce eigenfunctions which coincide with the Laughlin wave functions of the fractional regime.

Using

$$\delta L = \int_0^{2\pi} d\theta (R^2(\theta) + (\partial_\theta R(\theta))^2)^{1/2} - 2\pi R_0, \tag{11.188}$$

and expanding the square root, we obtain in the limit of small oscillations (where the derivative with respect to θ is small)

$$\delta E = \lambda \int_0^{2\pi} d\theta \left(R(\theta) - R_0 + \frac{1}{2R_0} (\partial_\theta R(\theta))^2 \right) = \lambda \pi R_0 \sum_{\ell_0} (\ell_0^2 - 1)|\alpha_{\ell_0}|^2 \tag{11.189}$$

from which we derive the force parameter:

$$C_{\ell_0} = 2\lambda \pi R_0 (\ell_0^2 - 1). \tag{11.190}$$

Therefore, the frequencies of the edge modes on the drop are given by

$$\omega_{\ell_0} = \frac{1}{\omega_c} \frac{\lambda \pi}{mNR_0} \ell_0 (\ell_0^2 - 1). \tag{11.191}$$

Moreover, if we put $\ell_0 = qR_0$ we obtain

$$\omega_q = \frac{\lambda q^3}{m\omega_c \rho_0}, \qquad (11.192)$$

whose q dependence is completely different from that of (11.186).

References

Anderson, B.P. *et al.*, *Phys. Rev. Lett.* **85**, 2857 (2000).

Baladron, C. *et al.*, *Z. Phys.* **D11**, 323 (1989).

Baranov, M.A. and D.S. Petrov, *Phys. Rev.* **A62**, 041601 (2000).

Barranco, M. and E.S. Hernandez, *Phys. Rev.* **B49**, 12078 (1994).

Baym, G. and C. Pethick, *Phys. Rev. Lett.* **76**, 6 (1996).

Beck, D.E., *Phys. Rev.* **B30**, 6935 (1984).

Bertsch, G.F., *Ann. Phys.* **86**, 138 (1974); *Nucl. Phys.* **A249**, 253 (1975).

Bertsch, G.F. and D. Tomanek, *Phys. Rev.* **B40**, 2749 (1989).

Bohigas, O., A.M. Lane and J. Martorell, *Phys. Rep.* **51**, 267 (1979).

Bohle, D. *et al.*, *Phys. Lett.* **B137**, 27 (1984).

Bohr, A. and B. Mottelson, *Nuclear Structure* (Benjamin, New York, 1975), Vol. 2.

Brack, M., *Phys. Rev.* **B39**, 3533 (1989).

Brink, D.M. and R. Leonardi, *Nucl. Phys.* **A258**, 285 (1976).

Brink, D.M. and S. Stringari, *Z. Phys.* **D15**, 257 (1990).

Bruun, G. and C.W. Clark, *Phys. Rev. Lett.* **83**, 5415 (1999).

Casas, M. and S. Stringari, *J. Chem. Phys.* **87**, 5021 (1990).

Dalfovo, F., S. Giorgini, L. Pitaevskii and S. Stringari, *Rev. Mod. Phys.* **71**, 3 (1999).

Dalfovo, F., A. Lastri, L. Pricaupenko, S. Stringari and J. Treiner, *Phys. Rev.* **B52**, 1193 (1995).

Domps, A., P.G. Reinhard and E. Suraud, *Phys. Rev. Lett.* **80**, 5520 (1998).

Enders, J. *et al.*, *Phys. Rev.* **C59**, R1851 (1999).

Fetter, A.L. and A.A. Svidzinsky, *J. Phys. Cond. Matt.* **13**, R135 (2001).

Giovanazzi, S., *Eccitazioni di bordo nell'effetto Hall quantistico*, Tesi di Laurea, (1993), unpublished.

Giovanazzi, S., L. Pitaevskii and S. Stringari, *Phys. Rev. Lett.* **72**, 3230 (1994).

Guery-Odelin, D. and S. Stringari, *Phys. Rev. Lett.* **83**, 4452 (1999).

Hohenberg, P.C., *Phys. Rev.* **158**, 383 (1967).

Holzwarth, G. and G. Eckart, *Z. Phys.* **A284**, 291 (1978); *Nucl. Phys.* **A325**, 1 (1979).

Iachello, F., *Phys. Rev. Lett.* **53**, 1927 (1984).

Jennings, B.K., *Phys. Lett.* **B96**, 1 (1980).

Knight, W.D. *et al.*, *Phys. Rev.* **B31**, 2539 (1985).

Lipparini, E., lecture notes for the Workshop on Metal Clusters. (Villa Monastero, Varenna, Sept. 1989), unpublished.

Lipparini, E. and S. Stringari, *Phys. Lett.* **B130**, 139 (1983).

Lipparini, E. and S. Stringari, *Phys. Rep.* **175**, 103 (1989).

Lipparini, E. and S. Stringari, *Phys. Rev. Lett.* **63**, 570 (1989b).

Lipparini, E. and S. Stringari, *Z. Phys.* **D18**, 193 (1991).

Lipparini, E. and A. Vitturi, *Z. Phys.* **D17**, 57 (1990).

Lo Iudice, N. and F. Palumbo, *Phys. Rev. Lett.* **41**, 1532 (1978).

Lund, E., C.J. Pethick and H. Smith, *Phys. Rev.* **A55**, 2126 (1997).

Madison, K.W. *et al.*, *Phys. Rev. Lett.* **84**, 806 (2000).

Maragò, O.M. *et al.*, *Phys. Rev. Lett.* **84**, 2056 (2000).

Matthews, M.R. *et al.*, *Phys. Rev. Lett.* **83**, 2498 (1999).

Mermin, N.D., *Phys. Rev.* **176**, 250 (1968).

Mermin, N.D. and H. Wagner, *Phys. Rev. Lett.* **17**, 1133 (1966).

Naher, V. *et al.*, *Phys. Rep.* **285**, 245 (1997).

Nesterenko, V.O. *et al.*, *Phys. Rev. Lett.* **83**, 57 (1999).

Nojarov, R., Z. Bochnacki and A. Faessler, *Z. Phys.* **A324**, 289 (1986).

Pacheco, J.M. and R.A. Broglia, *Phys. Rev. Lett.* **62**, 1400 (1989).

Perdew, J.P., *Phys. Rev.* **B37**, 6175 (1988).

Pitaevskii, L. and S. Stringari, *J. Low Temp. Phys.* **85**, 377 (1991).

Pricaupenko, L. and J. Treiner, *J. Low Temp. Phys.* **96**, 19 (1994).

Puente, A., M. Casas and Ll. Serra, *Physica* **E8**, 387 (2000).

Radomski, M., *Phys. Rev.* **C14**, 1704 (1976).

Rowe, D.J., *Nuclear Collective Motion* (Methuen, London, 1970).

Schuller, C. *et al.*, *Phys. Rev. Lett.* **80**, 2673 (1998).

Serra, Ll. *et al.*, *Phys. Rev.* **B59**, 15290 (1999).

Serra, Ll., A. Puente and E. Lipparini, *Phys. Rev.* **B60**, R13966 (1999).

Stringari, S., *Nucl. Phys.* **A279**, 454 (1977); *Ann. Phys.* **151**, 35 (1983).

Stringari, S., *Phys. Rev. Lett.* **76**, 1405 (1996).

Stringari, S., *Phys. Rev. Lett.* **77**, 2360 (1996b).

Stringari, S. and E. Lipparini, *Phys. Rev.* **C22**, 884 (1980).

Svidzinsky, A.A. and A.L. Fetter, *Phys. Rev.* **A58**, 3168 (1998).

Vichi, L. and S. Stringari, *Phys. Rev.* **A60**, 4734 (1999).

Volkov, V.A. and S.A. Mikhailov, *JEPT Lett.* **42**, 556 (1985).

Weisskopf, V., *Phys. Rev.* **52**, 295 (1937).

Wen, X.G., *Phys. Rev.* **B44**, 5708 (1991).

Wong, C.Y. and J. MacDonald, *Phys. Rev.* **C16**, 1196 (1977).

Zambelli, F. and S. Stringari, *Phys. Rev. Lett.* **81**, 1754 (1998).

Zambelli, F. and S. Stringari, *Phys. Rev.* **A63**, 1 (2001).

Index